CYTOCHROME P-450, BIOCHEMISTRY, BIOPHYSICS AND ENVIRONMENTAL IMPLICATIONS

DEVELOPMENTS IN BIOCHEMISTRY

Volume 1—Mechanisms of Oxidizing Enzymes, edited by Thomas P. Singer and Raul N. Ondarza, 1978

Volume 2—Electrophoresis '78, edited by Nicholas Catsimpoolas, 1978

Volume 3—Physical Aspects of Protein Interactions, edited by Nicholas Catsimpoolas, 1978

Volume 4—Chemistry and Biology of Pteridines, edited by Roy L. Kisliuk and Gene M. Brown, 1979

Volume 5—Cytochrome Oxidase, edited by Tsoo E. King, Yutaka Orii, Britton Chance and Kazuo Okunuki, 1979

Volume 6—Drug Action and Design: Mechanism-Based Enzyme Inhibitors. edited by Thomas I. Kalman, 1979

Volume 7—Electrofocus/78, edited by Herman Haglund, John G. Westerfeld and Jack T. Ball, Jr., 1979

Volume 8—The Regulation of Coagulation, edited by Kenneth G. Mann and Fletcher B. Taylor, Jr., 1980

Volume 9—Red Blood Cell and Lens Metabolism, edited by Satish K. Srivastava, 1980

Volume 10—Frontiers in Protein Chemistry, edited by Teh-Yung Liu, Gunji Mamiya and Kerry T. Yasunobu, 1980

Volume 11—Chemical and Biochemical Aspects of Superoxide and Superoxide Dismutase, edited by J.V. Bannister and H.A.O. Hill, 1980

Volume 12—Biological and Clinical Aspects of Superoxide and Superoxide Dismutase, edited by W.H. Bannister and J.V. Bannister, 1980

Volume 13—Biochemistry, Biophysics and Regulation of Cytochrome P-450, edited by Jan-Åke Gustafsson, Jan Carlstedt-Duke, Agneta Mode and Joseph Rafter, 1980

Volume 14—Calcium-Binding Proteins: Structure and Function, edited by Frank L. Siegel, Ernesto Carafoli, Robert H. Kretsinger, David H. MacLennan and Robert H. Wasserman, 1980

Volume 15—Gene Families of Collagen and Other Proteins, edited by Darwin J. Prockop and Pamela C. Champe, 1980

Volume 16—Biochemical and Medical Aspects of Tryptophan Metabolism, edited by Osamu Hayaishi, Yuzuru Ishimura and Ryo Kido, 1980

Volume 17—Chemical Synthesis and Sequencing of Peptides and Proteins, edited by Teh-Yung Liu, Alan N. Schechter, Robert L. Heinrickson, and Peter G. Condliffe, 1981

Volume 18—Metabolism and Clinical Implications of Branched Chain Amino Acids, edited by Mackenzie Walser and John Williamson, 1981

Volume 19—Molecular Basis of Drug Action, edited by Thomas P. Singer and Raul N. Ondarza, 1981

Volume 20—Energy Coupling in Photosynthesis, edited by Bruce R. Selman and Susanne Selman-Reimer, 1981

Volume 21—Flavins and Flavoproteins, edited by Vincent Massey and Charles H. Williams, 1982

Volume 22—The Chemistry and Biology of Mineralized Connective Tissues, edited by Arthur Veis, 1982

Volume 23—Cytochrome P-450, Biochemistry, Biophysics and Environmental Implications, edited by Eino Hietanen, Matti Laitinen and Osmo Hänninen, 1982

CYTOCHROME P-450, BIOCHEMISTRY, BIOPHYSICS AND ENVIRONMENTAL IMPLICATIONS

Proceedings of the 4th International Conference on Cytochrome P-450 held in Kuopio, Finland, May 31-June 3, 1982

Editors:
EINO HIETANEN
MATTI LAITINEN
OSMO HÄNNINEN

1982

ELSEVIER BIOMEDICAL PRESS
AMSTERDAM · NEW YORK · OXFORD

ISBN 0-444-80459-5
ISSN 0165-1714

Published by:
Elsevier Biomedical Press B.V.
P.O. Box 211
100 AE Amsterdam, The Netherlands

Sole distributors for the USA and Canada:
Elsevier Science Publishing Company Inc.
52 Vanderbilt Avenue
New York, N.Y. 10017

Printed in The Netherlands

Preface

During recent years increasing evidence has accumulated on the multiplicity of cytochrome P-450 based on the purification of various cytochrome P-450'ies with affinity to a specific substrate. With the advancement of biochemistry various techniques have been rapidly adapted to the purification of cytochrome P-450 isozymes. The development of the production of monoclonal antibodies towards various cytochromes has enhanced the identification of cytochromes. Simultaneously with the success in the characterization of new cytochrome P-450 subspecies the genetic control of the cytochrome P-450 catalysed enzyme activity and its induction has made enormous progress. Hand-in-hand with the purification of various cytochromes the inducibility of this hemoprotein or its subspecies has been found to be very specific by a vast number of different compounds form environmental xenobiotics to endogenous compounds.

The significance of cytochrome P-450 catalysed polysubstrate monooxygenase system has gained increasingly interest. In former times this enzyme system was thought to metabolize only some therapeutic drugs. To day a vast number of nontherapeutic compounds are metabolized by the cytochrome P-450 system. Numerous endogenous compounds, e.g. steroid hormones, cholesterol and prostaglandins are oxidized by this hemoprotein. Whether this activity is always to the benefit of the organism is yet under debate.

The increasing environmental chemical load due the accumulation of contaminants also affects feral species which chemicals may in these species cause metabolic changes affecting possibly their capability to survive. As an indicator on the biological activity the cytochrome P-450 catalyzed reactions have often been used as model metabolic pathways to study the biological potency of the industrial compounds.

The 4th International Conference on Cytochrome P-450 held in Kuopio in Finland had emphasis on both biochemical and biophysical but also on environmental aspects and significance of cytochrome P-450 and its catalyzed reactions. Great compliments on the successful conference must be given to all participants and to voluntary personnel. The University of Kuopio had provided new excellent facilities for the conference to take place. The great gratitude is addressed to the International Scientific Advisory Board, Dr. A. Archakov, Dr. M.J. Coon, Dr. R.W. Estabrook, Dr. I.C. Gunsalus, Dr. J.-Å. Gustafsson and Dr. K. Ruckpaul for the possibility to have this conference organized in Finland. The International Union of Biochemistry, J. Vainio Foundation (Finland), Academy of Finland and the Finnish Ministry of Education are acknowledged for the positive attitude in granting funds to organize the conference and invite participants. The Finnish Societies of Physiology and Toxicology are also acknowledged for their contributions to the conference. The chairpersons of various sections, Dr. Klinger, Dr. Estabrook, Dr. Lang, Dr. Gustafsson, Dr. Gunsalus, Dr. Netter, Dr. Hildebrandt, Dr. Coon, Dr. Mannering, Dr. Omura, Dr. Ruckpaul, Dr. Sato, Dr. Kato and Dr. Orrenius all did an excellent work and kept the tight time-table as scheduled, thanks to them.

Turku, 6 June, 1982

Eino Hietanen, Matti Laitinen and Osmo Hänninen

Contents

Preface V

DIVERSIFIED FUNCTIONS OF CYTOCHROME P-450

Oxygen activation by cytochrome P-450 and other hemoproteins
D.I. Metelitza 3

A critical comparison of the oxidative activities of cytochrome P-450 and hemeproteins with respect to catalytic mechanism
R.E. White 11

The cytochrome P-450 dependent hydroxyl radical-mediated oxygenation mechanism. Implications in pharmacology and toxicology
M. Ingelman-Sundberg, A.-L. Hagbjörk, G. Ekström, Y. Terelius and I. Johansson 19

Diversified functions of cytochrome P-450
F.P. Guengerich, T.L. Macdonald, L.T. Burka, R.E. Miller, D.C. Liebler, K. Zirvi, C.B. Frederick, F.F. Kadlubar and R.A. Prough 27

Induction of P-450 isozymes which have high affinity for nitrosamines
C.S. Yang, Y.Y. Tu, R. Peng and N.A. Lorr 35

Stereoselectivity of rat liver cytochromes P-450 and P-448
S.K. Yang, M.W. Chou, P.P. Fu, P.G. Wislocki and A.Y.H. Lu 39

Role of cytochrome P-450 and NADPH cytochrome P-450 reductase in catalysis of nitrosourea denitrosation to yield nitric oxide
D.W. Potter and D.J. Reed 43

Progesterone-microsomal receptor in rat liver: properties of binding
P.S. Zacharias, R. Drangova and G. Feuer 47

The oxidation of the tryptamines and related indolylalkylamines by a reconstituted microsomal enzyme system
A. Jaccarini 53

Sharing of cytochrome P-450 isozymes during the stepwise microsomal oxidation of benzphetamine
E. Jeffery and G.J. Mannering 59

Interaction of xenobiotics and nutritional factors with drug metabolizing enzyme systems
Z. Amelizad and J.F. Narbonne 63

Inhibition of (+)-*trans*-7,8-dihydroxy-7,8-dihydrobenzo(a)pyrene metabolism by 2(3)-*tert*-butyl-4-hydroxyanisole in liver microsomes and isolated hepatocytes from mice
L. Dock, B. Jernström and P. Moldéus 67

In vitro metabolism of 6-nitrobenzo(a)pyrene: effect of the nitro substituent on the regioselectivity of the cytochrome P-450-mediated drug metabolizing enzyme system
P.P. Fu and M.W. Chou 71

Cytochrome P-450-dependent hydroxyl radical-mediated destruction of DNA
H. Luthman, A.-L. Hagbjörk and M. Ingelman-Sundberg 75

Role of radical stages in microsomal monooxygenations
V.P. Kurchenko, S.A. Usanov and D.I. Metelitza 79

Incorporation of the monooxygenase system into liposomes of dimyristoylphosphatidylcholine
G. Smettan, P.A. Kiselev, M.A. Kisel, A.A. Akhrem and K. Ruckpaul 87

Binding affinity of P_1-450 inducers for the *Ah* receptor does not correlate with the inhibition of epidermal growth factor (EGF) binding to cell-surface receptors
S.O. Kärenlampi, H.J. Eisen and D.W. Nebert 91

Cytochrome P-450 - and model systems - dependent hydroxylation of a multifunctional substrate by various oxidants
J. Leclaire, P.M. Dansette, D. Forstmeyer and D. Mansuy 95

A reconstituted hepatic cytochrome P-450 steroid 21-hydroxylase system and its stimulation by cytochrome b_5
I.R. Senciall, C.J. Embree and S. Rahal 99

Prostacyclin synthase as a cytochrome P-450 enzyme
H. Graf and V. Ullrich 103

MOLECULAR BIOLOGY AND GENETICS OF CYTOCHROME P-450

Fundamental principles of cloning
D.W. Nebert 109

Isolation and characterization of the mouse P_1-450 chromosomal gene
M. Negishi, M. Altieri, M. Nakamura, R.H. Tukey, T. Ikeda, Y.-T. Chen, T. Ohyama and D.W. Nebert 119

Evidence for a closely-related multi-gene family of phenobarbital-inducible cytochromes P-450 in rat liver microsomes
W. Levin, P.-M. Yuan, J.F. Shively, F.G. Walz Jr., G.P. Vlasuk, P.E. Thomas, C.J. Omiecinski, E. Bresnick and D.E. Ryan 127

Gene organization of phenobarbital-inducible cytochrome P-450 from rat livers
Y. Fujii-Kuriyama, Y. Mizukami, F. Kawajiri, Y. Suwa, K. Sogawa and M. Muramatsu 135

Genes for cytochrome P-450 and their regulation
M. Adesnik, E. Rivkin, A. Kumar, A. Lippman, C. Raphael and M. Atchison 143

Studies on the induction of glutathione S-transferase B and cytochrome P-448 by 3-methylcholanthrene
C.B. Pickett, C.A. Telakowski-Hopkins, A.M. Donohue and A.Y.H. Lu 149

Stimulated *de novo* synthesis of cytochromes P-450 by drugs and steroids in primary monolayer cultures of adult rat hepatocytes
P. Guzelian, S. Newman and E. Gallagher 153

Adipose tissue content modulates the response of mice to polychlorinated biphenyls, inducers of polysubstrate monooxygenases
M. Ahotupa and E. Mäntylä 157

The effect of dietary cholesterol on monooxygenation and glucuronidation reactions in control and 2,4,5,2´,4´,5´- hexachlorobiphenyl-treated C57BL and DBA/JBOnf mice
E. Mäntylä, E. Hietanen and M. Ahotupa 161

Hormonal and sex-linked factors in the induction of cytochrome P-450 by phenobarbital
R.M. Bayney, E.A. Shephard, S.F. Pike, B.R. Rabin and I.R. Phillips 165

Induction and decrease of the major phenobarbital-inducible cytochrome P-450 measured by a specific radioimmunoassay technique
E.A. Shephard, S.F. Pike, R.M. Bayney, B.R. Rabin, and I.R. Phillips 169

Molecular properties of cytosolic and nuclear forms of the *Ah* receptor
A.B. Okey and A.W. Dubé 173

Genetic regulation of mouse liver microsomal P_2-450 induction by isosafrole and 3-methylcholanthrene
T. Ohyama, R.H. Tukey, D.W. Nebert and M. Negishi 177

Induction by 3-methylcholanthrene or isosafrole of two P-450 mRNA's which share nucleotide sequence homology
T. Ikeda, M. Altieri, M. Nakamura, D.W. Nebert and M. Negishi 181

Induction of liver microsomal UDP-glucuronyltransferase by eugenol and inducers of microsomal monooxygenases in the rat
A. Yuasa and H. Yokota 185

Influence of incubation temperature on monooxygenases activities in liver of tench, suckling and adult rat after induction with β-naphthoflavon (BNF) and dichlorprop (DCP)
F. Schiller, F. Jahn and W. Klinger 189

Oxidation of benzo(a)pyrene by horseradish peroxidase-H_2O_2 intermediate
M. Yagi, S. Akagawa and H. Momoi 193

EVOLUTION OF CYTOCHROME P-450 AND ENVIRONMENTAL FACTORS IN THE CONTROL OF CYTOCHROME P-450 IN FERAL SPECIES

Higher plant cytochrome P-450: microsomal electron transport and xenobiotic oxidation
I. Benveniste, B. Gabriac, R. Fonne, D. Reichhart, J.-P. Salaün, A. Simon and F. Durst 201

Seasonal variation of cytochrome P-450 and monooxygenase activities in boreal fresh water fish
O. Hänninen, U. Koivusaari and P. Lindström-Seppä 209

Regulation of hepatic steroid and xenobiotic metabolism in fish
T. Hansson, L. Förlin, J. Rafter and J.-Å. Gustafsson 217

Characterization of the induction response of the hepatic microsomal monoxygenase (MO) system of fishes
M.J. Vodicnik and J.J. Lech 225

Variation of certain hepatic monoxygenase activities in feral winter flounder *(Pseudopleuronectes americanus)* from Maine: apparent association with induction by environmental exposure to PAH-type compounds
G.L. Foureman and J.R. Bend 233

Multiple forms of cytochrome P-450 in marine crabs
J.W. Conner and R.F. Lee 245

Cytochrome P-450 in the hepatopancreas of freshwater crayfish *Astacus astacus* L.
P. Lindström-Seppä, U. Koivusaari and O. Hänninen 251

Occurence of cytochrome P-450 in the earthworm *Lumbricus terrestris*
A. Liimatainen and O. Hänninen 255

Cytochrome P-450 and polysubstrate monooxygenase (PMSO) activities in the liver of spawning rainbow trout *(Salmo gairdneri)*
U. Koivusaari, M. Pesonen and O. Hänninen 259

Antiparasitic treatment and monooxygenases in reindeer
M. Laitinen, R. Juvonen, M. Nieminen, E. Hietanen and O. Hänninen 263

CHARACTERIZATION OF DIFFERENT CYTOCHROME P-450's

Spectral and catalytic properties of ethanol-inducible cytochrome P-450 isozyme 3a
D.R. Koop and E.T. Morgan 269

Evidence from studies on primary structure that rabbit hepatic NADPH-cytochrome P-450 reductase and isozymes 2 and 4 of cytochrome P-450 represent unique gene products
S.D. Black, G.E. Tarr and M.J. Coon 277

Rabbit microsomal cytochrome P-450 3b: structural and functional polymorphism and modulation by positive and negative effectors
E.F. Johnson, H.H. Dieter, G.E. Schwab, I. Reubi and U. Muller-Eberhard 283

Substrate specificities and stereoselectivities as probes of cytochrome P-450 isozyme composition
L.S. Kaminsky, M.J. Fasco and F.P. Guengerich 291

Turnover of different forms of microsomal cytochrome P-450 in rat liver
H. Sadano and T. Omura 299

Characterization of rat hepatic cytochrome P-450 subpopulations by anion exchange chromatography and metabolic-intermediate complex formation
M.R. Franklin, L.K. Pershing and L.M. Bornheim 307

Catalytic and structural properties of two new cytochrome P-450 isozymes from phenobarbital-induced rat liver: comparison to the major induced isozymic form
D.J. Waxman and C. Walsh 311

Separation of multiple P-450 forms by hydrophobic and anion exchange and chromatofocusing techniques
M. Pasanen and O. Pelkonen 317

Cytochromes P-450 in liver microsomes of rats pretreated with β-naphthoflavone
H.W. Strobel and P.P. Lau 321

Primary and secondary hydroxylation pattern of antipyrine in the isolated perfused rat liver preparation as affected by 3-MC
J. Böttcher, H. Bässmann and R. Schüppel 329

Studies on the purification and properties of cytochrome P-450 fractions active in cholesterol 7α-hydroxylation
H. Boström, K. Lundell and K. Wikvall 333

Rabbit microsomal cytochrome P-450 1: a cytochrome linked to variations in hepatic steroid hormone metabolism
E.F. Johnson, H.H. Dieter, G.E. Schwab, I. Reubi and U. Muller-Eberhard 337

Polychlorinated biphenyls as phenobarbitone-type inducers: structure-activity correlations
S. Safe, M.A. Campbell, I. Lambert and L. Copp 341

Use of three-phase partition as the first step in the purification of cytochrome P-450 from the yeast *Brettanomyces anomalus*
H. Nikkilä and P.H. Hynninen 345

Partial separation of two different cytochrome P-450 from rat adrenal cortex microsomes by affinity chromatography with immobilized NADPH-cytochrome c reductase as a ligand
J. Montelius and J. Rydström 349

Partial purification of a sex specific cytochrome P-450 from untreated female rats
C. MacGeoch, J. Halpert and J.-Å. Gustafsson 353

The effect of temperature on the deuterium isotope effect of the cytochrome P-450-catalyzed O- deethylation of 7-ethoxycoumarin
A.Y.H. Lu and G.T. Miwa 357

Enzyme kinetics and pharmacokinetics of dealkylation reactions after various pretreatments in mice
W. Legrum and J. Frahseck 361

In vitro and *in vivo* experiments indicating induction of mouse liver cytochrome P-450 by warfarin
L. Kling, W. Legrum, H. Reitze and K.J. Netter 365

Characterization of cytochrome P-448 from *Saccharomyces cerevisiae*
D.J. King, M.R. Azari and A. Wiseman 369

Modulation of reconstituted hydroxylase activities in biosynthesis of bile acids by protein fractions from rat and rabbit liver cytosol
I. Kalles, B. Lidström, K. Wikvall and H. Danielsson 373

Experimental approaches to determine the position of the heme chelating and substrate binding sulfhydryl groups within the sequence of $P\text{-}450_{cam}$
K.M. Dus and R.I. Murray 377

Testosterone hydroxylations catalyzed by purified rat liver cytochrome P-450 isozymes
D.J. Waxman, A. Ko and C. Walsh 381

Comparative biochemical and morphological characterization of microsomal preparations from rat, quail, trout, mussel and *Daphnia magna*
P. Ade, M.G. Banchelli Soldaini, M.G. Castelli, E. Chiesara, F. Clementi, R. Fanelli, E. Funari, G. Ignesti, A. Marabini, M. Oronesu, S. Palmero, R. Pirisino, A. Ramundo Orlando, V. Silano, A. Viarengo and L. Vittozzi 387

Purification and characterization of $P_{\tau}\text{-}450$ from mouse liver microsomes
M.A. Lang, M. Negishi, I. Stupans and D.W. Nebert 391

Characterization of the forms of cytochrome P-450 induced by *trans*-stilbene oxide and by 2-acetylaminofluorene
J. Meijer, A. Åström and J.W. Depierre 395

Isotope effect and metabolic switching in the cytochrome P-450-catalyzed O-deethylation of 7-ethoxycoumarin
A.Y.H. Lu, N. Harada, J.S.A. Walsh and G.T. Miwa 405

Steroid 11β-hydroxylase system: reconstitution and study of interactions among protein components
S.P. Martsev, I.A. Bespalov, V.L. Chashchin and A.A. Akhrem 413

Studies on sex-related difference of cytochrome P-450 in the rat: purification of a constitutive form of cytochrome P-450 from liver microsomes of untreated rats
R. Kato and T. Kamataki 421

Multiple forms of cytochrome P-450 in biosynthesis of bile acids
K. Wikvall, H. Boström, R. Hansson and K. Lundell 429

A unique form of rat liver cytochrome P-450 induced by the hypolipidaemic drug clofibrate
G.G. Gibson, P.P. Tamburini, H. Masson and C. Ioannides 437

Halogenated biphenyls as AHH inducers: effects of different halogen substituents
S. Safe, A. Parkinson, L. Robertson, S. Bandiera, T. Sawyer, L. Safe, I. Lambert, J. Andres and M.A. Campbell 441

The alkane monooxygenase system of the yeast *Lodderomyces elongisporus*: purification of the cytochrome P-450 and the NADPH cytochrome P-450 reductase and reconstitution experiments
H.-G. Müller, W.-H. Schunck, P. Riege and H. Honeck 445

Purification of human placental cytochromes P-450
O. Pelkonen and M. Pasanen 449

Comparison of the cytochrome P-450-containing monooxygenases originating from two different yeasts
M. Sauer, O. Kappeli and A. Fiechter 453

Oxidation of genetic marker drugs in human liver microsomes
C. von Bahr, C. Birgersson, M. Göransson and B. Mellström 459

Effect of oxygen concentration on formation and autoxidation of the oxyferrous intermediate of P-450 LM_4 isozyme
C. Bonfils, J.L. Saldana, C. Balny and P. Maurel 463

Influence of prenatal exposure to phenobarbital (PB) and 3-methylcholanthrene (MC) on postnatal development of monooxygenases activities and their inducibility in rats
W. Klinger, F. Jahn and E. Karge 467

Progesterone-microsomal receptor in rat liver: effect of various steroids
A. Goshal, P.S. Zacharias and G. Feuer 471

The preparation of substrate analogs for studying the mechanism of cytochrome $P\text{-}450_{cam}$ catalyzed reactions
L. Hietaniemi and P. Mälkönen 477

Multi-substrate monooxygenases - a family of similar but different catalysts
M.J. Coon 481

CONTROL AND BIOPHYSICS OF CYTOCHROME P-450 FUNCTION

Stability, conformational rigidity and life-time of microsomal redox enzymes in soluble and membrane-bound state
A.I. Archakov 487

Lipid-protein interactions as determinants of the association of cytochrome b_5 and cytochrome P-450 reductase with cytochrome P-450
B. Bösterling and J.R. Trudell 497

Possibility of activating substrates by cytochrome P-450
L.M. Weiner, Ya.Yu. Woldman and V.V. Lyakhovich 501

Regulation of the cyclic function of liver microsomal cytochrome P-450: on the role of cytochrome b_5
J. Werringloer, S. Kawano and H. Kuthan 509

Active site modification of cytochrome P-450
G.-R. Jänig, G. Smettan, J. Friedrich, R. Bernhardt, O. Ristau and K. Ruckpaul 513

Affinity modification of cytochromes P-450 and P-448
V.V. Lyakhovich, V.M. Mishin, O.A. Gromova, V.I. Popova and L.M. Weiner 517

The sixth ligand of ferric cytochrome P-450
J.H. Dawson, L.A. Andersson and M. Sono 523

Inactivation of key metabolic enzymes by P-450-linked mixed function oxidation systems
C. Oliver, L. Fucci, R. Levine, M. Wittenberger and E.R. Stadtman 531

Cholesterol side chain cleavage cytochrome P-450scc. Amino-steroids as probes of the active site structure
L.E. Vickery and J.J. Sheets 539

Resonance Raman detection of a Fe-S bond in cytochrome $P\text{-}450_{cam}$
P.M. Champion, B.R. Stallard, G.C. Wagner and I.C. Gunsalus 547

Suicide inactivation of reduced cytochrome P-450 in soluble and membrane-bound states
G.I. Bachmanova, V. Yu. Uvarov, I.P. Kanaeva and A.I. Archakov 551

Pleuromutilin derivatives as a tool for investigating the size and the shape of the active domain of Cyt P-450
I. Schuster, H. Berner and H. Egger 555

Acute ACTH stimulation of cytochrome P-450 dependent adrenal corticosteroidogenesis: discovery of a rapidly induced protein
R.J. Krueger and N.R. Orme-Johnson 559

The potential role of cytochrome b_5 as an effector in cytochrome P-450 dependent drug oxidations
P. Hlavica 563

Characterisation of *Ah* receptor mutants among benzo(a)pyrene-resistant mouse hepatoma clones
C. Legraverend, R.R. Hannah, H.J. Eisen, I.S. Owens, D.W. Nebert and O. Hankinson 567

Preparation and properties of manganese-substituted cytochrome P-450_{cam}
M.H. Gelb, W.A. Toscano, Jr. and S.G. Sligar 573

Localization of functional residues of cytochrome P-450 LM2
R. Bernhardt, G. Etzold, H. Stiel, W. Schwarze, G.-R. Jänig and K. Ruckpaul 581

Spin shift, reduction rate and N-demethylation correlation in cytochrome P-450 LM using a series of benzphetamine analogues
H. Rein, J. Blanck, M. Sommer, O. Ristau, K. Ruckpaul and W. Scheler 585

The active site structure of cytochrome P-450 as determined by extended X-ray absorption fine structure spectroscopy
J.H. Dawson, L.A. Andersson, K.O. Hodgson and J.E. Hahn 589

Microsomal electron transport. The rate of the electron transfer to oxygenated cytochrome P-450
H.H. Ruf and V. Eichinger 597

The spontaneous release of heme from hemeproteins
M.L. Smith, P.-I. Ohlsson and K.G. Paul 601

Thiol compounds as ligands of cytochrome P-450
E. Elenkova, B. Atanasov, O. Ristau, C. Jung, H. Rein and K. Ruckpaul 607

Resonance Raman study of iron coordination in the ferric low spin cytochrome P-450
P. Anzenbacher, Z. Šípal, B. Strauch and J. Twardowski 611

Comparative studies of superoxide radical generation in microsomes and reconstituted monooxygenase systems
S.K. Soodaeva, E.D. Skotzelyas, A.A. Zhukov and A.I. Archakov 615

Use of vesicles of reconstituted human cytochrome P-450 in studies of free radical production by reductive metabolism of halocarbons
B. Bösterling, J.R. Trudell and A.J. Trevor 619

Structural and mechanistic differences in quinone inhibition of microsomal drug metabolism: inhibition of NADPH-cytochrome P-450 reductase activity
E.D. Kharasch and R.F. Novak 623

Cytochrome P-$450_{11\beta}$ and P-450_{scc} systems in adrenocortical mitochondria: immunocytochemical evidence for a heterogeneous distribution among the mitochondria within a single cell
F. Mitani, Y. Ishimura, S. Izumi, N. Komatsu and K. Watanabe 627

ACTH mediated induction of cytochromes P-450 in bovine adrenal cortex
M.R. Waterman, R.E. Kramer, B. Funkenstein, J.L. McCarthy, V. Boggaram and E.R. Simpson 631

CYTOCHROME P-450, TISSUE SPECIFICITY AND TOXICITY

Cytochrome P-450 and prostaglandin synthetase catalyzed activation of paracetamol and p-phenetidine
P. Moldéus, B. Andersson, R. Larsson and B. Lindeke 637

Comparison of carcinogen metabolism in different organs and species
H. Autrup and R.C. Grafstrom 643

Purification and properties of cytochrome P-450$_{MC}$ from rat lung microsomes and its role to activation of chemical carcinogens
M. Watanabe, I. Sagami, T. Abe and T. Ohmachi 649

Regulation and role of aryl hydrocarbon hydroxylase in rat adrenal and gonads
M. Bengtsson, J. Montelius, E. Hallberg, L. Mankowitz and J. Rydström 653

Destruction of cytochrome P-450 by a rubber antioxidant
A. Zitting 657

Multiple forms of insect cytochrome P-450: the role in insecticide resistance
M. Agosin 661

Aryl hydrocarbon hydroxyase activity in rat brain mitochondria
M. Das, P.K. Seth and H. Mukhtar 671

Energy regulation and drug metabolism in hepatic and intestinal microsomes of rat and chick
Y. Zubairy, S. Govindwar, M. Soni, M. Kachole and S. Pawar 675

What are the significant toxic metabolites of styrene?
H. Vainio, F. Tursi and G. Belvedere 679

Regioselectivity of caffeine metabolism *in vivo* and peripheral blood lymphocytes in different species
G. Belvedere, F. Tursi, L. Jritano, F. Galletti and M. Bonati 689

Enzymatic reduction of liposome-embedded cytochrome P-450$_{scc}$ from bovine adrenocortical mitochondria. Effect of cholesterol concentration in membrane
T. Kimura, F. Yamakura and B. Cheng 695

Suicide inactivation of rat liver cytochrome P-450 *in vivo* and *in vitro*
J. Halpert, B. Näslund and I. Betner 701

Dietary lipids as modifiers of monooxygenase induction
E. Hietanen, M. Ahotupa, A. Heikelä and M. Laitinen 705

Effects of age on hepatic monooxygenase system in normal and toluene-induced female rats
K. Pyykkö and H. Vapaatalo 709

Interaction of imidazole with rabbit liver microsomal cytochrome P-450: induction and inhibition of activity
K.K. Hajek and R.F. Novak 713

Spectral studies of cytochrome P-450 in isolated rainbow trout liver cells
T. Andersson and L. Förlin 717

Use of a site-specific nephrotoxin for the investigation of cytochrome P-450 multiplicity and localisation in rat and mouse kidney
C.R. Wolf, J.B. Hook and E.A. Lock 721

Studies on the metabolism and toxicity of 1-naphthol in isolated hepatocytes
M.T. Smith, M. D'Arcy Doherty, J.A. Trimbrell and G.M. Cohen 725

Toxic and nontoxic pathways during metabolism of menadione (2-methyl-1,4-naphthoquinone) in isolated hepatocytes
H. Thor, M.T. Smith, P. Hartzell and S. Orrenius 729

Induction of hepatic monooxygenases by n-heptane and cyclohexane inhalation
J. Järvisalo, H. Vainio and T. Heinonen 733

The relative induction of the hepatic drug-metabolizing enzyme system by medroxyprogesterone acetate in rats
H. Saarni 739

Interaction of oleyl anilide, a component in denaturated Spanish oil, with hepatic and pulmonary cytochrome P-450
I. Roots and A.G. Hildebrandt 743

Hydration of benzo(a)pyrene 4,5-oxide by cadaver tissues
M.G. Parkki, M. Ahotupa and S. Toikkanen 749

Triacetyloleandomycin as inducer of P-450 LM_{3b} from rabbit liver microsomes
C. Bonfils, I. Dalet-Beluche, C. Dalet and P. Maurel 751

Interactions of polycyclic hydrocarbons with the estradiol receptor in rat adrenal
L. Mankowitz and J. Rydström 755

Effect of dithranol on the hepatic microsomal drug-metabolizing system in rats
A. Karppinen and N.T. Kärki 759

Unaltered metabolism of m-xylene in the presence of ethylbenzene
E. Elovaara, K. Engström and H. Vainio 765

Mitochondrial mediation of ethylmorphine-induced alterations in hepatocyte Ca^{2+} homeostasis
S.A. Jewell, G. Bellomo, H. Thor and S. Orrenius 769

The inducibility of hepatic cytochrome P-450 content and mono-oxygenases by 2,2,2-trichloro-1-(3,4-dichlorophenyl)ethyl acetate in the rat
E. Mäntylä, E. Hietanen, M. Ahotupa and H. Vainio 773

Effects of fluorenone and trinitrofluorenone on the monooxygenase, epoxide hydrolase and glutathione S-transferase activities in rabbits
E. Hietanen and H. Vainio 777

Effect of experimental diabetes on hepatic monooxygenases
E. Hietanen, R. Rauramaa and M. Laitinen 781

Metabolism of polycyclic aromatic hydrocarbons by cultured rat and human adrenal cells
E. Hallberg, A. Rane and J. Rydström 785

Metabolism of exogenous and endogenous compounds in the rat ventral prostate characterization of the enzymes involved using antibodies
T. Haaparanta, J. Halpert, H. Glaumann and J.-Å. Gustafsson 789

Redox cycling of glutathione in isolated hepatocytes
L. Eklöw, P. Moldéus and S. Orrenius 793

Response of microsomal peptides in *Dendroctonus terebrans* and rat liver to alpha-pinene and other inducers
R.A. White, Jr. and M. Agosin 797

Metabolism of chemical carcinogens in nude mouse fibroblast cultures and its effect on viral transformation
M. Laaksonen, R. Mäntyjärvi and O. Hänninen 801

Aspects on the determination of the oxidation-reduction potential of lactoperoxidase
P.-I. Ohlsson and K.G. Paul 805

CLOSING REMARKS
K. Ruckpaul 813

Author index 819

Diversified Functions of Cytochrome P-450

© 1982 Elsevier Biomedical Press B.V.
Cytochrome P-450, Biochemistry, Biophysics
and Environmental Implications, E. Hietanen,
M. Laitinen and O. Hänninen editors

OXYGEN ACTIVATION BY CYTOCHROME P-450 AND OTHER HEMOPROTEINS

DMITRIY I. METELITZA
Institute of Bioorganic Chemistry, BSSR Academy of Sciences,
Zhodinskaya, 5/2, 220600, Minsk, U.S.S.R.

INTRODUCTION

The cytochrome P-450-containing enzyme system of liver microsomes is unusual in its versatility, being capable of catalyzing the hydroxylation and other modifications of a remarkable variety of physiologically important and foreign compounds, including drugs, anestetics, pesticides, petrolium products, and carcinogens (1-4). This system includes NADPH-cytochrome P-450 reductase (EC 1.6.2.4), cytochromes P-450 (EC 1.14.14.1) and phospholipids and requares for its function NADPH and dioxygen. Cytochrome b_5 as it is shouwn, participates in the electron transport process connected with the hydroxylation. We mean the microsomal system containing all above indicated components as the full enzyme system.

During 10 years we have investigated the oxidation kinetics of various substrates in the full microsomal system - the hydroxylation of naphthalene and aniline, the oxidation of cyclohexane, the oxidative demethylation of aminopyrine and N,N-dimethylaniline (DMA), the oxidative dealkylation of anisole and p-chloranisole, as well as the microsomal oxidation of alyphatic alcohols. The data obtained were summarized and reviewed in ref. (2-5).

It is known that hydroxylating agent in the full microsomal system may be oxenoid $Fe^{3+}O \leftrightarrow Fe^{4+}O^{-}$ by analogy with Compound I for peroxidases and catalases (1-4). The oxenoid structure is not known yet. The central problem in the P-450 area is the mechanism of oxenoid interaction with various substrates: in some cases the oxenoid may react as a radical, splitting the C—H bond of substrates (1), in other cases oxenoid reacts by two electron (oxenoid) pathway with the olefins and aromatic compounds. At two electron path of reactions the oxenoid inserts the oxygen atom into the substrates without intermediate radicals liberation by analogy with the reactions of carbenes and nitrenes.

RESULTS AND DISCUSSION

We have used the inhibitoty method for the study of aniline and DMA oxidation by the rabbit liver microsomes and by a reconstituted enzyme systems containing the highly purified NADPH-cytochrome P-450 reductase, cytochrome P-450-LM2, the microsomal phospholipids, NADPH and dioxygen (6). 1-Naphthol inhibits strongly the both substrates oxidation. The inhibiting action of 1-naphthol is characterized by the complex mixed type: 1-naphthol competes with both substrates and acts in the reactions by the other mechanisms also: (see the work of V.P.Kurchenko, S.A.Usanov and D.I.Metelitza in this volume). We have concluded that 1-naphthol reacts effectively with a radicals acting in the oxidation of aniline and DMA by the full enzyme system.

In some works the additional support is obtained for the initial hydrogen abstraction of the hydrocarbons and a number of other substrates. These additional supports are discussed in (1-4):

i) an intramolecular isotope effect was invoked for the P-450-catalyzed hydroxylation of tetradeuteronorbornane, which indicates a large preference for hydrogen over deuterium abstraction (k_H/k_D= 11) s. (7); ii) mass spectral analyses of the alcohols from the P-450-catalyzed hydroxylation of tetradeuteronorbornane showed the presence of small amounts of endo-deuterium in the exo-alcohol and a deficit of similar magnitude of exo-deuterium in the endo-alcohol, i.e. the reaction did not proceed with 100% retetion of configuration, that strongly suggests the presence of a discrete carbon radical which exists long enough (7); iii) hydroxylation of methylcyclohexane by P-450-LM2 gave a mixture of isomeric methylcyclohexanols in which the ratio 1°: 2° : 3°, was 0.072 : 1 : 1.25 after statistical correction. These product distribution are expected only with the radical pathway of oxidation process (8). All these data discussed support strongly the important role of radical stages in the oxidation of some substrates by the full microsomal system.

Yet there are the data which support the two electron (or oxenoid) pathway in the full microsomal system: i) the epoxidation of unsaturated hydrocarbons and the liberation of arenoxides in aromatics hydroxylation can not be accounted for a radical scheme of oxidation process; ii) NIH-shift is observed for oxidation of substituted aromatic compounds that can not be explained by radical pathway of reaction; iii) the isotope effects for aromatics oxidation are very small (1.3-2.0) that is inconsistent with the initial hydrogen abstraction in the limiting step of oxidation process. Thus, in the full microsomal system a number of substrates (the alyphatic and alycyclic hydrocarbons, tret.amines camphor etc.) may be oxidized by the radical way, but other substrates (cyclohexene,aromatics) may be oxidized by two electron mechanism.

The important details of the mechanism of cytochrome P-450-dependent reactions may be bettrer understood at the comparative study of this biocatalyst and other hemoproteins in the same processes. There are two such processes: i) the autoxidation of oxygenated hemoproteins; ii) the oxidation of organic compounds by hydroperoxides with participation of a hemoprotein. Table 1 shows the comparative qualitative consideration of autoxidation of various hemoproteins in connection with their redox potentials (3).

TABLE 1
Redox potentials of hemoproteins and the stability of their oxygenated complexes (3)

Hemoproteins	E_o, V	$Fe^{2+}O_2$ stability	Activation energy,kcal/M
Hemoglobin	+ 0.2	very stable	17.7
Peroxidase P_7	- 0.12	stable 20 min	
HR peroxidase	- 0.25	rapidly oxidized	
S-P-450_{cam}	- 0.17	rapidly oxidized	17 - 18
P-450_{cam}	- 0.27	stable at -30°C	14 - 24
S-P-450_{scc}	- 0.30	very unstable	
P-450-LM2		very unstable	9.6

As it is seen from Table 1, the autoxidation rates of hemoproteins increase greatly with a decreasing of the redox potentials of the proteins. The redox potentials of hemoproteins are determined by the nature of the 5th axial ligand of heme iron and by apoprotein structure near protoporphyrine IX (1-4). The autoxidation process for all hemoproteins is described by three reactions

$$\text{I. } Fe^{2+} + H_2O \rightleftharpoons Fe^{3+}OH_2 + O_2^{\cdot}$$

$$\text{II. } Fe^{2+}O_2 + Fe^{2+} \rightleftharpoons Fe^{2+}O_2Fe^{2+} \longrightarrow 2\,Fe^{3+} + O_2^{2-}$$

$$\text{III. } Fe^{2+}O_2 + H \longrightarrow Fe^{3+}O_2H \longrightarrow \text{products}$$

At autoxidation of many hemoproteins superoxide anion is libera ted (3,4). Unlike all hemoproteins superoxide radicals could not be detected during autoxidation of oxygenated P-450-LM4 (10). The reaction II is unlikely for many hemoproteins, but it occurs easily in the case of syntetic iron porphyrines and cytochrome P-450_{cam}. The reaction III occurs with participation of many donors of H or e, for example, phenol, naphthol, tyrosine, nitrite etc. (3).

Many hemoproteins catalyze the hydroperoxide-dependent reactions of organic compounds oxidation (2-4). Naphthalene, aniline and cyclohexane are oxidized in the systems including the ROOH and microsomes or cytochrome P-450 (2-4). Cyclohexene is oxidized into cyclohexene oxide by tret.butyl hydroperoxide and rat liver microsomes (2). Anisole, p-Cl-anisol and DMA are demethylated effectively by ROOH and cytochrome P-450 (2-4). We have shown that cumene hydroperoxide (CHP) oxidizes aniline with participation of hemoglobin (11) or cytochrome C (12). Hemoglobin and catalse catalyze the DMA demethylation in the reaction with CHP(13)

The systematic kinetic study of 1-naphthol action inthe oxidation of DMA and aniline with participation of microsomes (6,13), cytochrome P-450-LM2 (6), hemoglobin (11), catalase (13) and cytochrome C (12) shows that antioxidant inhibits the oxidation greatly in all cases. The complex mixed type of inhibiting action of 1-naphthol in the system "CHP-microsomes (or the P-450-LM2)" supports not only the competition of inhibitor and substrates but also the participation of 1-naphthol in other processes including the radical reactions connected with the substrate oxidation. The radicals are generated from ROOH with participation of a hemoprotein. The inhibiting action of 1-naphthol is much more effective at DMA oxidation and less effective at aniline hydroxylation, e.i. the radical pathway of DMA oxidation is much more important than such a way in aniline transformation. All the hemoproteins studied give the complexes with CHP and other ROOH that is confirmed by difference spectra from which K_s were determined (Table 2).

TABLE 2.

Spectral parameters of hemoproteins complexes with ROOH (2-4).

Hemoproteins	ROOH	λ_{max},nm	λ_{min},nm	K_s, mM
Microsomes	HP of tetraline	380,440	420	
	HP of tret.butyl	380,440	418	
	CHP	380,440	418	0,043
P-450-LM2	CHP	440	415	0,120
Cytochrome C	CHP	433	413	0,083
Hemoglobin	CHP	425	405	0,100

The type of transformation of hemoproteins complexes with ROOH is determined by the nature and structure of proteins. We give the following scheme of the substrates (AH) oxidation by ROOH with participation of various hemoproteins (Fe^{3+}):

$$Fe^{3+} + ROOH \rightleftharpoons Fe^{3+}...\underset{H}{O}—OR \quad 1.$$

$$Fe^{3+}...\underset{H}{O}—OR \longrightarrow \underset{\text{Compound I}}{Fe^{3+}O} + ROH \quad 2.$$

$$\text{Compound I} + AH \longrightarrow \underset{\text{Compound II}}{Fe^{2+}O} + A^{\bullet} + H^{+} \quad 3.$$

$$\text{Compound II} + AH \longrightarrow A^{\bullet} + H_2O + Fe^{3+} \quad 4.$$

$$\text{Compound I} + AH \longrightarrow AOH + Fe^{3+} \quad 5.$$

$$Fe^{3+}...\underset{H}{O}—OR \longrightarrow Fe^{2+} + H^{+} + RO_2^{\bullet} \quad 6.$$

$$Fe^{3+}...\underset{H}{O}—OR \longrightarrow Fe^{3+}OH + RO^{\bullet} \quad 7.$$

Catalases, peroxidases and cytochromes P-450 give Comp.I (reactio n 2). Hemoglobin, myoglobin and cytochrome C do not form such Comp.I since they can not stabilize this active intermediate by aminoacide residues of their active centres. Comp.I of peroxidases and catalases oxidizes the substrates in two steps (reactions 3 and 4) by radical pathway. Comp.I of the P-450 oxidizes some substrates (cyclohexene and aromatics) by two electron way (reaction 5). Other substrates of the P-450-system (alyphatic hydrocarbons, tret.amines) are oxidized by Comp.I with liberation of radicals. Cytochrome C, myoglobin and hemoglobin generate the radicals $RO^{\bullet}$ and $RO_2^{\bullet}$ which oxidize the substrates and destruct the heme moiety and apoprotein. The pathway of the Comp.I transformation for different hemoproteins are determined by three fac tors: i) the 5th axial ligand of heme iron, ii) the structure of active centre of hemoprotein near protoporphyrin IX, iii) the nature of the substrates oxidized (the energy of C-H bond of the hydrocarbons or the ionization potentials of tret. amines).

Our data and many results achieved in other laboratories summarized and discussed in details in (1-4), allow us to divide the hemoproteins into three groups depending on the mechanism of their action in the ROOH-included reactions (3): the first group (peroxidases and catalases) oxidize the substrates by radical way with participation of Comp.I; the second group (cytochrome C, myoglobin and hemoglobin) does not form the Comp.I but generates the radicals $RO^{\bullet}$ and $RO_2^{\bullet}$ which oxidize tret.amines and aromatics; the third group includes the P-450 family which reacts not only by two electron pathway but by radical mechanism also depending on the substrate nature. Why are the cytochromes P-450 characterized by the dual reactivity depending on the substrate nature? The answer is facilitated by the following scheme which summarized many data for the hemoproteins reactions (L-redox-active ligand in inner sphere of heme iron, Fe^{3+} and Fe^{2+} -hemoproteins).

Hemoproteins and numerous model compounds may form the oxygenated complexes of two types distinguished by the coordination type and the degree of π-binding of dioxygen. In the acidic medium the complexes of both types form the Compound I (peroxidases, catalases, cytochromes P-450). Comp.I of the P-450 is strongly distinguished from other hemoproteins since Comp.I of the P-450 may be converted arbitrary into Compound II which may react only by radical way. Such transition of Comp.I into Comp.II is possible as a result of inner-sphere electron transfer from redox-active ligand to central iron ion. The absence of such ligand in the peroxidases, catalases, cytochrome C-peroxidase accounts for the nesessity of substrate participation for Comp.I -Comp II transformation in these cases. In the ROOH-dependent systems the Comp.I-Comp.II transformation is also possible without the

the substrate participation for cytochrome P-450.

$\underline{Fe^{2+} — O_2}$

π-complex of closed type   σ-complex of open type

$\bar{L}Fe^{2+}$ (O=O, π-bound) ⟷ $\bar{L}Fe^{4+}$ (O–O⁻, cyclic) $\bar{L}Fe^{2+}...O=O$ ⟷ $\bar{L}Fe^{3+}O-O^-$

$2H^+$ ↓ $-H_2O$

$\bar{L}Fe^{4+}O^- \longleftrightarrow \bar{L}Fe^{3+}O$ Comp.I

⇅

$\dot{L}Fe^{3+}O^- \longleftrightarrow \dot{L}Fe^{2+}O$ Comp.II

$\underline{Fe^{3+} — ROOH\ (or\ H_2O_2)}$

$Fe^{3+}...O(H)—OR$

heterolytic ←   → homolytic

$\bar{L}Fe^{4+}OH + RO^-$

⇅

$\dot{L}Fe^{3+}OH$

Comp.II

1. $\bar{L}Fe^{2+} + H^+ + RO_2^{\bullet}$
2. $\dot{L}Fe^{3+}OH^- + RO^{\bullet}$
3. $\dot{L}Fe^{3+}OR^- + HO^{\bullet}$

The redox-active axial ligand provides the possibility for $RO^{\bullet}$ and $RO_2^{\bullet}$ radicals liberation from ROOH without alteration of vale nce state of central ion. The homolysis of RO—OH bond is the result of inner-sphere electron transfer from thiolate ligand to iron ion of the P-450. The routes 2 and 3 of the scheme are the pseudotrimolecular reactions of inner-sphere electron transfer with participation thiolate ligand, iron ion and ROOH.

Thus, the dual reactivity of the P-450 in the full and ROOH-de pendent systems is explained by the presence of redox-active thiolate ligand which donates one electron to central ion without substrate participation. Such process of inner-sphere electron transfer makes it possible to join cytochromes P-450 the complexes of <u>partial charge transfer (PCT).</u> These complexes are characterized by a small energy difference between the ground and electron-excited states and by rapid transitions between these states The qualitative theoretic interpretation of PCT complexes is given by Dr. Yu.I.Skurlatov (14) on the base of the modern theory of outer-sphere electron transfer in polar solvents (15) and the theory of molecular donor-acceptor complexes (16). The PCT complexes are characterized by the energies of change interaction between the donor and acceptor orbitals in the range of kT $\leqslant$ V $\leqslant$ 3 kcal/mol. The PCT complexes may exist as a mixture of two mesomeric forms - i) without charge transfer and ii) with the full charge transfer. The partial charge transfer in Comp.I of the P-450 accounts for the dual reactivity of oxenoid species: with olefines and aromatics Comp.I reacts by two electron mechanism but with the hydrocarbons Comp.II reacts by radical way. Other hemoproteins are distinguished from the P-450-s since they can not form the PCT complexes as these hemoproteins have no redox-active thiolate ligand in the inner sphere of iron ion.

REFERENCES

1. White,R.E. and Coom,M.J.(1980)Ann.Rev.Biochem.,49, 315.
2. Metelitza,D.I. (1981) Uspekhi Khimii, 50,2019.
3. Metelitza,D.I. (1982) Uspekhi Khimii, 51, in press
4. Metelitza,D.I. (1982) Oxygen activation by enzyme systems. Moscow: Nauka, 254 p.
5. Metelitza,D.I. and Popova,E.M. (1979) Biokhimiya, 44,1923.
6. Kurchenko,V.P., Usanov, S.A. and Metelitza,D.I. (1981) Izv. AN BSSR, ser.khim., 6, 78.
7. Groves,J.T., McClusky,G.A., White,R.E. and Coon,M.J. (1978) Biochem.Biophys.Res.Commun., 81, 154.
8. White,R.E., Groves,J.T. and McClusky,G.A. (1979) Acta Biol.Med.Germ., 38, 475.
9. Rahimtula,A.D., O'Brien,P.J., Seifried, H.E. and Jerina,D.M. (1978) Eur.J.Biochem., 89, 133.
10. Oprian,D.D. and Coon,M.J. (1980) In: Biochemistry,Biophysics and Regulation of Cytochrome P-450. Eds.J.-A.Gustafsson et al Amsterdam: Elsivier, 323.
11. Akhrem,A.A., German,S.Yu. and Metelitza,D.I. (1978) React.Kinet. and Catal.Letters, 8, 217.
12. Metelitza,D.I., Kiselev,P.A. and Kiseleva,S.N. (1980) Kinetika i kataliz,21, 1436.
13. Akhrem,A.A., Belski,S.M. and Metelitza,D.I. (1978) Izv. AN BSSR, ser.khim.,1, 87.
14. Skurlatov, Yu.I. - Proc. Allunion Conference on the Mechanism of Catalytic Reactions. Vol.1. Moscow: Nauka,1978, p.90.
15. Dogonadze, R.R. and Kuznetzov,A.M. Kinetics of Chemical Reactions in Polar Solvents. Physical Chemistry.Kinetics. Vol.2. Moscow: VINITI, 1973.
16. Mulliken,R.S. (1952) J.Amer.Chem.Soc., 74, 811.

Cytochrome P-450, Biochemistry, Biophysics
and Environmental Implications, E. Hietanen,
M. Laitinen and O. Hänninen editors

A CRITICAL COMPARISON OF THE OXIDATIVE ACTIVITIES OF CYTOCHROME P-450 AND HEMEPROTEINS WITH RESPECT TO CATALYTIC MECHANISM

RONALD E. WHITE

Department of Pharmacology, University of Connecticut Health Center, Farmington, Connecticut 06032 (U.S.A.)

INTRODUCTION

Studies presented at this meeting by Dawson (1), as well as our own work (2), have shown that the heme iron in low spin cytochrome P-450 is ligated by cysteinate and by an oxygen ligand in the proximal and distal positions, respectively. The catalytic activity of P-450 is totally dependent on the presence of the anionic cysteinate ligand, so that preparations in which the thiolate ligand has been lost (i.e., P-420) have also lost their catalytic competence. This observation strongly suggests that the unique P-450 heme-sulfur configuration is intimately involved in the formation and/or stability of the reactive oxidizing intermediate of the P-450 catalyzed hydroxylation reaction. Therefore, mechanistic suggestions should address possible specific functions for the iron-bound thiolate.

We have presented one possible function for the thiolate in the formation of an electrophilic oxy radical during the catalysis (3). In this proposal single electron transfer from sulfur through iron to a peroxide results in homolytic cleavage of the O-O bond. Two spatially separated radical species are the consequence of this process, one an oxy radical which can abstract a hydrogen atom from the substrate, and another which may be thought of as a ferric hydroxyl radical complex and which ultimately delivers an oxygen atom to the substrate carbon radical. Sulfur might also stabilize the ferric hydroxyl complex.

$$S^{-}\ Fe^{III}\ HOOR' \longrightarrow S\cdot Fe^{III}HO^{-} + R'O\cdot$$
$$R'O\cdot + RH \longrightarrow R'OH + R\cdot$$
$$S\cdot Fe^{III}HO^{-} + R \longrightarrow S^{-}Fe^{III}\ HOR$$

Fig. 1. Proposed homolytic mechanism of hydroperoxide activation by cytochrome P-450.

Presently, the only alternative to this homolytic mechanism of oxygen activation is a heterolytic mechanism analogous to the peroxidase catalytic cycle. We have, therefore, sought to determine which of these two schemes more closely resembles the actual reaction mechanism of P-450.

$$Fe^{III}\ HOOR' \longrightarrow Fe^{V}{=}O + R'OH$$
$$Fe^{V}{=}O + RH \longrightarrow Fe^{IV}\text{-}OH + R\cdot$$
$$Fe^{IV}\text{-}OH + R\cdot \longrightarrow Fe^{III}\ HOR$$

Fig. 2. Proposed heterolytic mechanism of hydroperoxide activation by cytochrome P-450.

In initial work, reported at the Third Conference in this series, evidence was presented that P-450 did, indeed, promote the homolytic scission of peroxides (4). The experimental approach took advantage of the fact that the homolysis of a peroxyacid would generate a carboxyl radical, with a certain probability for the subsequent decarboxylation of that radical depending on the stability of the ultimate daughter radical. This phenomenon constitutes an historically applied test for homolytic processes. In the actual experiments, peroxyphenylacetic acid (PPAA) was converted to the decarboxylation product, benzyl alcohol, by exposure to P-450. When a hydroxylatable substrate was present, hydroxylation also occurred, so that two competing processes were present, substrate hydroxylation vs. peroxyacid decarboxylation. Thus, the involvement of a homolytic peroxide scission is unquestionable, at least for the fraction of reaction which leads to decarboxylation. However, the remaining fraction, leading to substrate hydroxylation, could still be proceeding through the heterolytic mechanism. The activated oxygen species may be formulated as either $[S\cdot Fe^{III}OH^- + R'O\cdot]$ or $[Fe^{V}=O]$ according to the homolytic or heterolytic proposals, respectively. This work attempts to address the remaining ambiguity about the identity of the reactive intermediate by comparing the behavior of peroxidase Compound I with that of P-450 in a set of characteristic oxidation reactions.

MATERIALS AND METHODS

Electrophoretically homogeneous rabbit liver microsomal $P\text{-}450_{LM2}$ and $P\text{-}450_{LM4}$ were prepared by the method of Coon et al. (5). NADPH-cytochrome P-450 reductase was purified from rabbit liver as described by French and Coon (6). $P\text{-}450_{LM2}$ was denatured to the P-420 form by careful heating at 60°, yielding a protein which had lost the unique P-450 ferrous carbonyl spectrum but which retained solubility. Peroxyacids were prepared by the alkaline perhydrolysis of the corresponding acid chlorides. Horseradish peroxidase (HRP) and chloroperoxidase (CPO) were obtained from Sigma Chemical Co.

The oxidation of pyrogallol to purpurogallin by hemeproteins was monitored by the absorbance increase at 420 nm, using a molar absorptivity of 3,200 M^{-1} cm^{-1}. Identification and quantitation of 1- and 2-adamantanol, 2-cyclohexen-1-ol, cyclohexene oxide, and phenol were accomplished by standard gas chromatography.

Reaction mixtures utilizing peroxide and P-450 or P-420 contained hemeprotein (1 nmol), dilauroyl glyceryl-3-phosphoryl choline (0.05 mg), one of the various substrates, potassium phosphate (0.1 mmol, pH 7.4), and either PPAA, cumene hydroperoxide (CHP), or H_2O_2 in a total volume of 1 ml. Those reactions utilizing NADPH/O_2 contained P-450_{LM2} (0.1 nmol), NADPH-cytochrome P-450 reductase, (0.3 nmol), dilauroyl glycerol-3-phosphoryl choline (0.03 mg), potassium phosphate (0.1 mmol, pH 7.4), substrate, and an NADPH generating system consisting of magnesium chloride (10 μmol), isocitric acid (5 μmol), $NADP^+$ (400 nmol), and isocitrate dehydrogenase (0.4 unit) in a total volume of 1 ml. Reaction mixtures involving peroxidases contained hemeprotein (ca. 1 nmol), potassium phosphate (0.1 mmol, pH 6.0), substrate, and hydrogen peroxide (5 mM for HRP and 20 mM for CPO) in a total volume of 1 ml. All reactions were conducted at 25°.

RESULTS

The decarboxylation reaction catalyzed by P-450_{LM2} was found to be a general phenomenon. Thus, it occurred not only with PPAA, but also with peroxy-1-methylcyclohexane carboxylic acid and with peroxy-1-adamantane carboxylic acid. In each case the decarboxylation product was the corresponding alcohol, and substrate hydroxylation occurred concurrently. We also determined that decarboxylation of PPAA was not a general property of hemeproteins. For this purpose, a common reaction was selected which each hemeprotein could catalyze. The oxidation of pyrogallol to purpurogallin by PPAA was found suitable. As shown in Table 1, each of the hemoproteins catalyzed this conversion. However, only P-450_{LM2} and P-450_{LM4} exhibited decarboxylase activity. The slight activity of P-420 could be correlated with the residual content of P-450. Neither horseradish peroxidase nor chloroperoxidase exhibited any decarboxylase activity.

TABLE 1.

ABILITY OF HEMEPROTEINS TO DECARBOXYLATE PPAA

Heme Protein	Benzyl Alcohol Produced (nmol)	Pyrogallol Consumed (nmol)	Per Cent Decarboxylation
P-450_{LM2}	7.4	55	12
P-450_{LM4}	7.9	41	16
P-420_{LM2}	0.3	31	1
CPO	0	66	0
HRP	0	34	0

The data in Table 1 demonstrate also that P-450 is a competent catalyst for a typical peroxidase-type reaction. Next we examined the ability of the peroxidases to catalyze the typical P-450 activities, namely aliphatic and aromatic hydroxylation and alkene epoxidation. Among these hemeproteins, in data not shown, only P-450 was able to hydroxylate adamantane (aliphatic hydroxylation) or benzene (aromatic hydroxylation).

The final activity determined was the oxidation of cyclohexene. P-450_{LM2} gave a mixture of cyclohexene oxide (epoxide) and 2-cyclohexen-1-ol (enol) when presented with cyclohexene, using either NADPH/O_2 or CHP (Table 2). Neither P-420_{LM2} nor HRP was able to oxidize cyclohexene to either of these products. However, CPO showed a low activity for the production of both the epoxide and the enol. Since CPO has been shown to have the same heme thiolate ligand as does P-450 (7), this activity is very interesting and will be investigated further.

TABLE 2

OXDATION OF CYCLOHEXENE

		T.O. Number (Per Min)		
Hemeprotein	Oxidant	Epoxide	Enol	Total
P-450_{LM2}	NADPH/O_2	1.6	2.6	4.2
P-450_{LM2}	CHP	10.0	11.2	21.2
P-420_{LM2}	CHP	0	0	0
HRP	H_2O_2	0	0	0
CPO	H_2O_2	1.2	2.6	3.8

DISCUSSION

In these experiments, five enzymatic activities were examined: (1) peroxidative dehydrogenation of an activated aromatic; (2) oxidative decarboxylation of peroxyacids; (3) aliphatic hydroxylation; (4) aromatic hydroxylation; and (5) olefinic epoxidation. Activity (1) is a facile reaction which is easily accomplished by all hemeproteins with an exchangeable ligand, including P-450 and P-420. The other activities represent more difficult oxidations and appear to be limited to P-450 enzymes. In particular, peroxyacid decarboxylation can now be recognized as a unique P-450 activity which must be accomodated by subsequent mechanistic proposals and which must also be exhibited by relevant synthetic model systems.

These experiments have demonstrated that the primary peroxidase reactive intermediate, Compound I, does not resemble the reactive intermediate of the P-450 catalytic cycle. Although it remains a possibility that the lack of reactivity of small molecules such as cyclohexene toward Compound I is a function of unfavorable binding by the peroxidase rather than of the intrinsic chemical behavior of Compound I, the fact that large organic peroxides as well as large organic substrates may freely enter the peroxidase active site and react there greatly attenuates the probability of this explanation. We deem it much more likely that peroxidase Compound I is simply not an adequate model of the P-450 reactive intermediate. While it is possible that the presence of a back-side thiolate ligand in a Compound I-like species could profoundly alter the behavior, the thiolate would have to act to _increase_ the reactivity of the iron-oxo species, since ordinary Compound I cannot perform the difficult oxidations. However, in the past, thiolate has been suggested to stabilize an iron-oxo intermediate (i.e., _decrease_ its reactivity), so that this notion runs counter to chemical intuition. Thus, the most likely circumstance is that the P-450 reactive intermediate is not chemically akin to Compound I. We have suggested an intermediate, resulting from peroxide homolysis, which appears to explain the extant experimental evidence. Future experiments will determine whether the homolytic mechanism is correct or whether a third type of intermediate must be invoked.

ACKNOWLEDGEMENTS

We greatfully acknowledge the expert technical assistance of Ms. Mary Beth Chudzik. This work was supported by Research Grant GM 28737 from the United States Public Health Service.

REFERENCES

1. Dawson, J.H., Andersson, L.A., and Sono, M. (1982) J. Biol. Chem. 257, 3606-3617.

2. White, R.E. and Coon, M.J. (1982) J. Biol. Chem. 257, 3073-3083.

3. White, R.E. and Coon, M.J. (1980) Ann. Rev. Biochem. 49, 315-356.

4. White, R.E., Sligar, S.G., and Coon, M.J. (1980) J. Biol. Chem. 255, 11108-11111.

5. Coon, M.J., van der Hoeven, T.A., Dahl, S.B., and Haugen, D.A. (1978) Methods Enzymol. 52, 109-117.

6. French, J.S. and Coon, M.J. (1979) Arch. Biochem. Biophys. 195, 565-577.

7. Cramer, S.P., Dawson, J.H., Hodgson, K.O., and Hager, L.P. (1978) J. Amer. Chem. Soc. 100, 728-7290.

Cytochrome P-450, Biochemistry, Biophysics and Environmental Implications, E. Hietanen, M. Laitinen and O. Hänninen editors

THE CYTOCHROME P-450-DEPENDENT HYDROXYL RADICAL-MEDIATED OXYGENATION MECHANISM. IMPLICATIONS IN PHARMACOLOGY AND TOXICOLOGY.

M. Ingelman-Sundberg, A.-L. Hagbjörk, G. Ekström, Y. Terelius and I. Johansson, Department of Physiological Chemistry, Karolinska Institute, S-104 01 Stockholm, Sweden.

INTRODUCTION

A hepatic ethanol-oxidizing system distinct from alcohol dehydrogenase was postulated in 1965 (1). Lieber and colleagues proposed that this system was of microsomal origin with cytochrome P-450 as the terminal component (2). This proposal was controversial and subsequently many years of debate followed whether the microsomal ethanol oxidation was caused by contaminating catalase in the preparations in conjunction with hydrogen peroxide formed during microsomal oxidation of NADPH (cf. (3)). However, experiments by Lu and coworkers established a role for cytochrome P-450 in this reaction (4). Extensive studies by Cederbaum and collaborators (5-7) unequivocally indicated that hydroxyl radicals produced by the liver microsomes constituted the active oxygen species mediating the microsomal oxidation of ethanol. The mechanism of the cytochrome P-450 LM2-dependent oxidation of ethanol was recently determined (8) and found to involve the action of hydroxyl radicals: Superoxide anions liberated from the oxycytochrome P-450 complex partially dismutate to hydrogen peroxide whereafter cleavage of H_2O_2 in an iron-catalyzed Haber-Weiss reaction generates hydroxyl radicals. If ethanol is present, the radicals will abstract a hydrogen atom from the alcohol; the products formed are water and another radical which in the presence of either oxygen or hydrogen peroxide is converted to acetaldehyde.

Since hydroxyl radicals are highly reactive oxygen species, it was postulated that more classes of compounds other than primary aliphatic alcohols would have the capability in being metabolized according to this oxygenation mechanism. It was found that the cytochrome P-450 LM2-dependent hydroxylation of the aromatic compounds aniline (9) and benzene (10) entirely followed this type of reaction mechanism, whereas the hydroxylation of naphtalene to a great extent was mediated by P-450 LM2-generated hydroxyl radicals (11). Compounds metabolized entirely according to ordinary P-450-dependent hydroxylation mechanisms were e.g. paranitroanisole (9), 7-ethoxycoumarin

(5), androstenedione, ethylmorphine, aminopyrine and benzo(a)pyrene (11).

In this paper we describe some properties of the cytochrome P-450-dependent hydroxyl radical-mediated oxygenation mechanism and present data indicating that benzene and ethanol induce a common cytochrome P-450 species in rabbit liver specifically effective in hydroxyl radical-mediated oxygenation of ethanol.

RESULTS AND DISCUSSION

Cytochrome P-450 LM2-dependent oxidation of aniline, benzene, deoxyribose and ethanol. Incubation of reconstituted membrane vesicles, containing cytochrome P-450 LM2, NADPH-cytochrome P-450 reductase and microsomal phospholipids at a molar ratio of about 1:0.4:1280, with aniline, benzene or deoxyribose in the presence of NADPH, revealed that the P-450-dependent oxidation of all substrates was inhibited by catalase in a similar fashion (Table I); the half-maximal inhibition was obtained when 1 U/ml of the enzyme had been introduced into the incubation mixtures. All enzymes activities were almost completely inhibited when 100 U/ml of the peroxidase had been added. Superoxide dismutase inhibited the three activities by about 50-60 %, whereas the P-450 LM2-depedent ethanol oxidation was nearly completely inhibited by SOD. The hydroxyl radical scavengers mannitol and DMSO inhibited the oxidation of benzene, deoxyribose and ethanol very effectively, whereas aniline hydroxylation was maximally 50 % inhibited by these agents. Together, the results indicate a requirement for hydrogen peroxide, superoxide anions and hydroxyl radicals in these oxygenation reactions and the mechanism for oxidation may be summarized as outlined in Fig. 1. Ferric P-450 is reduced by NADPH-cytochrome P-450 reductase and upon bindning of oxygen, autooxidation of the oxycytochrome P-450 complex results in the liberation of superoxide anions which partially dismutate to hydrogen peroxide. In an iron-catalyzed Haber-Weiss reaction

$$O_2^- + H_2O_2 \longrightarrow \cdot OH + OH^- + O_2$$

•OH is produced and interaction of this radical with the "substrate" will result in the formation of a new radical which in the presence of oxygen or hydrogen peroxide may be converted to an alcohol or an aldehyde. One may speculate that the radical intermediate in the hydroxyl radical-mediated oxidation of e.g. benzene may constitute the reactive protein binding metabolite of this aromatic compound.

Table I. Inhibition of cytochrome P-450 LM2-dependent oxygenation reactions of aniline, benzene, deoxyribose and ethanol by peroxidases and radical scavenging systems.

Substrate used a/	cat. act. b/	Inhibition caused by peroxidase c/ conc d/	max. inhib.	SOD conc d/	max. inhib.	mannitol conc. d/	max. inhib	DMSO conc. d/	max. inhib.	reference
Aniline, 0.25 mM	5.5	1 U	100%	5 μg/ml	46%	100 mM	62%	10 mM	51%	9
Benzene, 17 μM	7.1	1 U	95%	1 μg/ml	64%	10 mM	98%	2 mM	100%	10
Deoxyrib. 0.7 mM	55	1 U	96%	1 μg/ml	60%	1 mM	95%	2 mM	100%	14
Ethanol, 50 mM	6.5	0.01 U	100%	1 μg/ml	95%	10 mM	100%	e/	e/	8

a/The products measured were: from aniline, paraaminophenol; from benzene, the total amount of water-soluble products formed; from deoxyribose, the total amount of TBA-reactive material as described by Halliwell and Gutteridge (18); from ethanol, acetaldehyde as described by DeCarli and Lieber (2). b/ The oxygenation activities are expressed as nmol prod./nmol P-450, min (aniline and ethanol), pmol of products formed per nmol P-450, min (benzene) or TBA-fluorescent units obtained against a block standard containing 10-6 M Rhodamine B. c/Peroxidase used was catalase except when ethanol oxidation was determined, when horseradish peroxidase with 0.5 mM hydroquinone as reductant served as hydrogen peroxide scavenger. d/The values are expressed as the concentration giving half-maximal inhibition. e/Not determined.

Oxidation of ethanol and benzene in microsomes isolated from control and benzene-treated rabbits. Male rabbits (2.5 kg) were injected with benzene (0.9 g/kg body weight) intraperitoneally for 3 days. Liver microsomes were subsequently isolated (12) and incubated with ethanol (50 mM) or benzene (17 μM) in the presence of 0.4 mM NADPH. As seen from Table II, both substrates were oxidized at a rate 2-3-fold higher by microsomes isolated from benzene-treated than from control rabbits. The hydroxyl radical scavengers DMSO, mannitol, and catechol inhibited the oxygenations in microsomes from benzene-treated rabbits, indicating the involvement of hydroxyl radicals in the reactions. SDS-gelelectrophoretic examination of the microsomal fractions revealed the occurrence of a new protein fraction in microsomes isolated from benzene-treated rabbits, having an apparent molecular weight of 51 000 daltons (cf. Fig. 2).

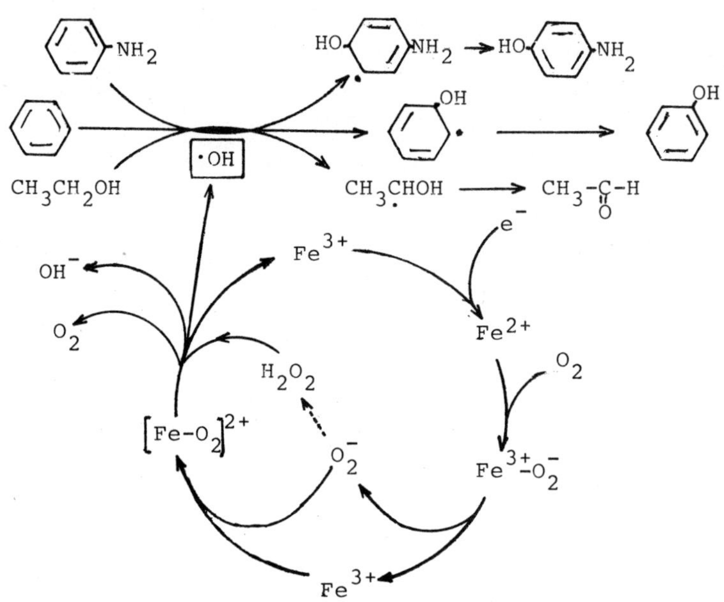

Figure 1. Proposed scheme for the cytochrome P-450-dependent hydroxyl radical-mediated oxygenation of benzene, aniline and ethanol.

Table II. Oxidation of benzene and ethanol in liver microsomes isolated from control and benzene-treated rabbits. Liver microsomes corresponding to 2 mg of microsomal protein were incubated for 20 min. at 37°C in the presence of 0.4 mM NADPH with either ethanol (50 mM) or benzene (17 µM containing 1 µCi of [14-C]benzene) in 50 mM potassium phosphate buffer, pH 7.4. The results are expressed as the mean +- S.D. of two different sets of experiments.

	benzene pmol/mg, min	ethanol nmol/mg, min
control microsomes	0.6+-0.2	2.7+-0.7
mic. from benzene-treated rabbits	2.1+-0.5	6.0+-1.5
-"-, +mannitol, 500 mM	n.d.	2.4+-0.5
-"-, + DMSO, 300 mM	0.13+-0.04	2.3+-0.2
-"-, + catechol, 1 mM	0.14+-0.09	n.d.

Purification of the benzene-inducible form of rabbit liver microsomal cytochrome P-450. The liver microsomes from benzene-treated rabbits were subjected to solubilization and polyethylene glycol fractionation essentially as described by Koop et al. (13). The fractions

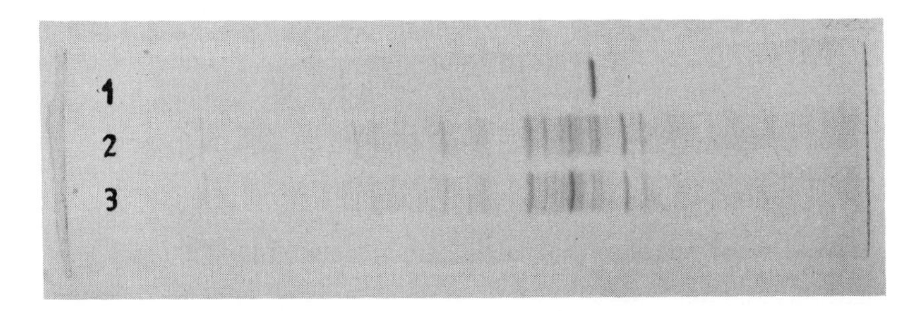

Figure 2. SDS-polyacrylamide gel electrophoresis of in lane 1, the benzene-inducible form of rabbit liver microsomal cytochrome P-450 (2 µg), in lane 2, liver microsomes isolated from benzene-treated rabbits (12 µg) and in lane 3, liver microsomes isolated from control rabbits (12 µg). Electrophoresis was performed from the left to the right in the discontinuous buffer buffer system according to Laemmli (19).

precipitating between 8 and 13 % were chromatographed in the presence of Renex 690 on DEAE-Sepharose columns according to procedures descri-bed for the DEAE-cellulose chromatography (13). The benzene-inducible protein fraction was eluted from the DEAE-columns with buffer based on 20 mM potassium sulphate (13). This fraction was treated with calcium phosphate and cholate as described for the 10 mM potassium sulphate fraction (13) and subjected to chromatography on an CM-Sepharose column (1 x 12 cm) equilibrated in 5 mM potassium phosphate buffer, pH 6.0, containing 20 % (v/v) glycerol, 0.1 mM EDTA and 2 % (v/v) Renex 690. The column was eluted with a 150 + 150 ml linear gradient of 0-0.5 M KCl in the phosphate buffer. A high spin cytochrome P-450 was eluted from the column at an ionic strength corresponding to 0.2 M KCl. This protein fraction had exactly the same electrophoretic mobility as the benzene-inducible microsomal protein fraction (Fig. 2). The purified cytochrome P-450 protein was found to be a high spin P-450 (14) with apparently identical properties to the ethanol-inducible rabbit liver microsomal cytochrome P-450 (15-18).

As is evident from Table III, the benzene-inducible form of cytochrome P-450 efficiently oxidized ethanol; the rate being about 3-fold higher than obtained with P-450 LM2 (8). However, contrary to the P-450 LM2-catalyzed reaction, the formation of acetaldehyde was not completely inhibited by the presence of 1 U/ml of Horseradish

Table III. Inhibition of ethanol and benzene oxidation catalyzed by the benzene-inducible form of cytochrome P-450 by peroxidase and radical scavengers. a/

inhibitor used	ethanol oxidation nmol/nmol, min	%inhib.	benzene oxidation pmol/nmol, min	% inhib.
none	14.5	0	5.4	0
Mannitol, 500 mM	0.7	95	0.4	92
DMSO, 100 mM	2.2	85	0.8	85
SOD, 50 µg/ml	11.0	24	2.0	63
Peroxidase b/	11.3	22	0.6	89
Thiourea, 1 mM	8.8	39	n.d.	
Thiourea, 10 mM	3.6	75	n.d.	
KMBA, 1 mM	6.2	57	n.d.	
Benzoate, 100 mM	.0.2	100	n.d.	
Catechol, 1 mM	6.4	44	n.d.	

a/ Incubations were performed with micelles, prepared by the cholate gel filtration technique from the benzene-inducible form of cytochrome P-450, NADPH-cytochrome P-450 reductase and dilauroylphosphatidylcholine in a molar ratio of about 1:0.4:133, corresponding to 0.1 nmol of P-450. The incubations were carried out for 20 min at 37°C in 50 mM potassium phosphate buffer, pH 7.4. The final concentrations were: NADPH, 0.4 mM; benzene, 17 µM; ethanol, 50 mM. b/ Catalase (100 U, benzene) and Horseradish peroxidase (1 U + 0.5 mM hydroquinone, ethanol) were used as peroxidases. n.d., not determined.

peroxidase or 50 ug/ml of superoxide dismutase in the incubation mixtures; only about 20 % inhibition was obtained. Newertheless, the hydroxyl radical scavengers mannitol, DMSO, thiourea, α-keto-γ-methiolbutyric acid (KMBA), benzoate and catechol effectively inhibited the alcohol oxidation, indicating the involvement of hydroxyl radicals in the process. Benzene oxidation by the benzene-inducible form of cytochrome P-450 was almost completely inhibited by catalase, superoxide dismutase, DMSO and mannitol, indicating the same type of oxygenation mechanism as evident in the P-450 LM2-dependent ethanol oxidation reaction (8).

In conclusion, the microsomal metabolism of ethanol and some aromatic substrates may be explained to a great extent by the capability of cytochrome P-450 in producing hydroxyl radicals in an iron-catalyzed

Haber-Weiss reaction. The physiological significance of this reaction mechanism is obscure, but the existence of a benzene and ethanol-inducible form of rabbit liver microsomal cytochrome P-450 oxidizing these compounds in a probably hydroxyl radical-mediated reaction, may indicate that some forms of cytochrome P-450 specifically utilize a similar reaction mechanism . The significance of hydroxyl radical-initiated lipid peroxidation processes, dependent on the ethanol and benzene-inducible form of rabbit liver cytochrome P-450, will be the subject of further investigations.

ACKNOWLEDGEMENTS

This work was supported by grants from Magnus Bergvalls Stiftelse and the Swedish Medical Research Council.

REFERENCES

1. Orme-Johnson, W. H. and Ziegler, D. M. (1965) Biochem. Biophys. Res. Commun. 21, 78-82
2. Lieber, C. S. and DeCarli, L. M. (1970) J. Biol. Chem. 245, 2505-2512
3. Thurman, R. G., Ley, H. G. and Scholz, R. (1972) Eur. J. Biochem. 225, 420-430
4. Miwa, G. T., Levin, W., Thomas, P. E. and Lu, A. Y. H. (1978) Arch. Biochem. Biophys. 187, 464-475
5. Cederbaum, A. I., Dicker, E., Rubin, E. and Cohen, G. (1979) Biochemistry 18, 1187-1191
6. Cederbaum, A. I., Dicker, E., Rubin, E. and Cohen, G. (1977) Biochem. Biophys. Res. Commun. 78, 1254-1262
7. Cederbaum, A. I., Miwa, G., Cohen, G., and Lu, A. Y. H. (1979) Biochem. Biophys. Res. Commun. 91, 747-754
8. Ingelman-Sundberg, M. and Johansson, I. (1981) J. Biol. Chem. 256, 6321-6326
9. Ingelman-Sundberg, M. and Ekström, G. (1982) Biochem. Biophys. Res. Commun., in press
10. Ingelman-Sundberg, M. and Johansson, I. (1982) Submitted for publication
11. Johansson, I. and Ingelman-Sundberg, M., unpublished observations
12. Ingelman-Sundberg, M., Johansson, I. and Hansson, A. (1979) Acta Biol. Med. Ger. 38, 379-388

13. Koop, D. R., Persson, A. V. and Coon, M. J. (1981) J. Biol. Chem. 256, 10704-10711

14. Ingelman-Sundberg, M. and Hagbjörk, A.-L. (1982) Xenobiotica, in press

15. Koop, D. R. and Coon, M. J. (1982) in: Kato, R. and Sato, R. (Eds.) Microsomes, Drug Oxidations and Drug Toxicities, Japan Scientific Societies Press, Tokyo, in press.

16. Ingelman-Sundberg, M., Hagbjörk, A.-L. and Johansson, I. (1982) Submitted for publication.

17. Koop, D. R., Morgan, E. and Coon, M. J. (1982) J. Biol. Chem., in press

18. Halliwell, B. and Gutteridge, J. M. C. (1981) FEBS lett. 128, 347-352

19. Laemmli, U. K. (1970) Nature 227, 680-685

Cytochrome P-450, Biochemistry, Biophysics
and Environmental Implications, E. Hietanen,
M. Laitinen and O. Hänninen editors

DIVERSIFIED FUNCTIONS OF CYTOCHROME P-450

F. PETER GUENGERICH[1], TIMOTHY L. MACDONALD[1], LEO T. BURKA[1], RANDALL E. MILLER[1], DANIEL C. LIEBLER[1], KARIMULLA ZIRVI[1], CLAY B. FREDERICK[2], FRED F. KADLUBAR[2], AND RUSSELL A. PROUGH[3]

[1]Departments of Biochemistry and Chemistry and Center in Environmental Toxicology, Vanderbilt University, Nashville, Tennessee 37232 (U.S.A.), [2]Division of Carcinogenesis, National Center for Toxicological Research, Jefferson, Arkansas 72079 (U.S.A.), and [3]Department of Biochemistry, University of Texas Health Sciences Center, Dallas, Texas 75235 (U.S.A.)

INTRODUCTION

Cytochromes P-450 constitute a family of enzymes widely distributed in nature. These enzymes carry out a variety of types of reactions and also have broad specificity with regard to substrates within a given type of reaction; e.g., C-hydroxylation. Moreover, different types of reactions are not easily associated with individual forms of P-450. A unified view of these reactions would be desirable, if possible, in order to rationalize the known activities of P-450 and to predict reactions that will occur with new compounds. While many theories have been advanced over the years as to the nature of the activated form of oxygen involved in P-450-mediated reactions (e.g., $O_2^{\dot{-}}$, $OH^{\cdot}$, singlet oxygen, hydroperoxides, etc.), the most reasonable mechanism has been proposed by Groves and others (1), where a P-450 perferryl oxygen (formally $Fe^{V}=O$) abstracts a hydrogen atom from a carbon group and then adds oxygen to that carbon. In a similar way, we view a P-450 perferryl oxygen as abstracting an electron or hydrogen atom from heteroatoms and unsaturated bonds to initiate other reactions such as heteratom oxidation, dealkylation of heteroatomic compounds, epoxidation, and oxidative group transfers.

RESULTS AND DISCUSSION

While the oxygen activated by P-450 has been thought to be electrophilic in nature, little attention has been given to the idea that primary attack in heteroatomic compounds may be at the most electron-rich points; i.e., the heteroatoms themselves. Hanzlik et al. (2) reported that compound Ia (Fig. 1) behaves as a suicide inhibitor of P-450 and suggested the upper portion of Fig. 1 as a mechanism; i.e., C-hydroxylation and imine formation. We synthesized compound Ib and found this to be even more effective than Ia as a suicide inhibitor (3). Ib does not have a position available for hydroxylation and cannot form a Schiff's base with the cyclopropyl group, nor can hydrogen be abstracted from the bridgehead carbon of the cyclopropyl moiety. Hydroxylation at the benzyl

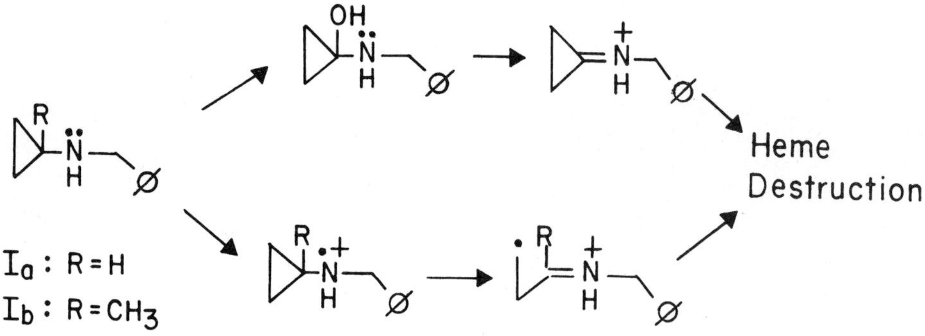

Heme Destruction

Ia : R = H

Ib : R = CH_3

Fig. 1. Possible schemes for metabolic activation and suicide destruction with cyclopropylamines.

carbon would yield benzaldehyde, which does not destroy P-450. The suicide inactivation of P-450 by Ib is consistent with the scheme shown in the lower portion of Fig. 1. In this scheme, oxygenated P-450 abstracts one electron from the nitrogen to form a cyclopropyl group α to a radical. Such systems are known to rearrange rapidly ($k > 10^8\ s^{-1}$) (4) with ring opening. Such a shift would result in the formation of a methylene radical, which would be a more potent alkylating agent for heme than would cyclopropanone or its Schiff's base (Fig. 1).

In a similar way, we examined the analogous ethers. The data presented in Table 1 indicate that metabolism is necessary for inactivation of the enzyme. Moreover, the loss of heme from a reconstituted P-450 system, using the benzyl ether (IIa), occurred with first order kinetics ($k = 0.31\ min^{-1}$ at a substrate concentration of 1 mM) and the pseudo-first order rate was saturable with increasing substrate concentration ($K_m = 0.53$ mM, extrapolated apparent inactivation rate constant = $0.56\ min^{-1}$).

Thus, the suicide inactivation that results with the methyl cyclopropyl compounds suggests that one-electron oxidation of heterocycles such as ethers and amines is the first step in P-450 oxidation. Further work will be required to isolate individual reaction products. If an oxidized heteroatom can, it probably will abstract an α-hydrogen atom to localize a radical on an adjacent carbon atom. Alternatively, a second electron oxidation may occur and a proton may be transferred. Thus carbinolamines and hemiacetals (or hemiketals) are formed and dealkylation results. However, sulfides tend not to abstract hydrogen but to accept oxygen instead to form sulfoxides.

If α-hydrogen is not available or if the oxidized heteratom is such that hydrogen transfer is unfavorable, the oxidized heteroatom may accept oxygen to form a stable oxygenated derivative. An example of such a reaction is the formation of the azoxy derivatives of azoprocarbazine (Fig. 2). As in the case of many compounds,

TABLE 1

DESTRUCTION OF CYTOCHROME P-450 DURING METABOLISM OF CYCLOPROPYL ETHERS

Incubations were carried out for 30 min at 37°C with 1 m$\underline{M}$ IIa (cyclopropyl benzyl ether) or IIb (1-methylcyclopropyl benzyl ether) and substrates were removed by dialysis prior to each analysis

Incubation system	P-450 (nmol/mg protein)	Heme (nmol/mg protein)	Aminopyrine demethylase (nmol HCHO/ min/mg protein)	Benzphetamine demethylase (nmol HCHO/ min/mg protein)
Microsomes	1.69		2.53 ± 0.03	3.06 ± 0.06
+ NADPH	1.33	2.25	2.25 ± 0.05	3.00 ± 0.04
+ IIa	1.69		2.50 ± 0.03	3.08 ± 0.04
+ IIa + NADPH	0.89	1.73	1.81 ± 0.04	2.77 ± 0.08
+ IIb	1.61		2.51 ± 0.03	3.03 ± 0.04
+ IIb + NADPH	0.79	1.54	1.57 ± 0.03	2.74 ± 0.06
Reconstituted cytochrome P-450 system (μ$\underline{M}$ P-450)				
+ NADPH	1.23			
+ II + NADPH	0.78			

azoprocarbazine is rather specifically converted to different products by different forms of P-450. Still other P-450 reactions lead to the formation of CH_4, the benzaldehyde, and the release of electrophilic structures which can bind irreversibly to protein and nucleic acids.

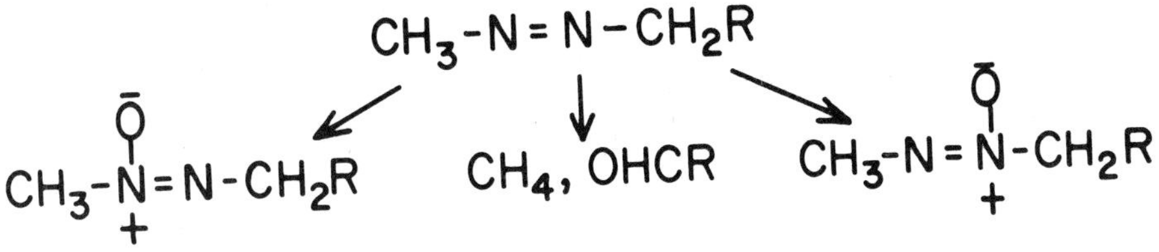

$CH_3-N=N-CH_2R$

$CH_3-\overset{\bar{O}}{\underset{+}{N}}=N-CH_2R$ $\quad$ CH_4, OHCR $\quad$ $CH_3-N=\overset{\bar{O}}{\underset{+}{N}}-CH_2R$

Fig. 2. Metabolism of azoprocarbazine ($\underline{N}$-isopropyl-α-(2-methyl azo)-$\underline{p}$-toluamide).

Another example is the conversion of 2-aminofluorene to the corresponding hydroxylamine (5). The resulting hydroxylamine is of interest because it has been shown to penetrate the nuclear membrane and bind irreversibly to chromatin DNA. The oxygenation is catalyzed both by P-450 and by microsomal flavin-containing monooxygenase. The relative contribution of the two enzymes appears to vary as a function of the animal species under consideration. However, the mechanisms of hydroxylation by the two enzymes are probably quite different. The flavin-containing monooxygenase probably utilizes a mechanism involving a 4a hydroperoxide (6). However, the P-450 mechanism is hypothesized to involve one electron oxidation prior to oxygen transfer. N-Oxygenation occurs because an α-hydrogen is not available for abstraction. The same view can be applied to N-hydroxylation of amides such as 2-acetylamino-fluorene and acetaminophen. Moreover, initial oxidation at the nitrogen of acetaminophen provides a rational explanation for the formation of an iminoquinone from acetaminophen (7).

Halogen oxides (haloso compounds) can also be formed by P-450. Alkyl haloso compounds have not been reported in the literature because of their presumed instability, but a number of theoretical (8) and experimental studies (9,10) strongly suggest their existence in oxidation of halides by chemical oxidants. Some aryl haloso compounds are reasonably stable (11,12) but are readily reduced by groups in proteins. We were able to demonstrate the formation of iodosobenzene from radiolabeled iodobenzene using purified P-450 and iodosobenzene in an exchange reaction (13). This appears to be the first example of an enzyme-mediated transfer of an oxygen to a halogen atom and may prove to be of relevance to the metabolism of the many halogenated compounds which are potential toxins and carcinogens.

Subsequent examination of the iodosobenzene system indicated that the labeled oxygen of iodosobenzene is not transferred during the formation of alcohols by P-450 because of rapid P-450-mediated exchange of the oxygen with H_2O (14). During the oxygenation of alkanes by P-450 in the presence of NADPH, NADPH-P-450 reductase, and O_2, the activated oxygen does not exchange with the solvent. We do not feel that the discrepancy in H_2O exchange is indicative of differences in the forms of activated oxygen in the two systems and have interpreted the differences as being kinetic in nature (14). We feel that perferryl oxygen does exchange rapidly with H_2O but, when formed in the vicinity of an oxidizable substrate, is committed to catalysis. These properties are consistent with the behavior of the cobalt and manganese model porphyrins of Groves (1,15).

In order to examine the mechanism of oxidation of vinyl halides, we synthesized the epoxides of vinyl chloride, vinyl bromide, trichloroethylene (TCE), and 1,1-dichloro-ethylene (DCE). The NMR and mass spectra of the first 3 epoxides were consistent with literature values. The fourth epoxide, 2,2-dichloroethylene oxide (DCE-oxide), had not been previously synthesized but the assigned structure is consistent with ^{13}C (58.3 ppm, 78.4 ppm) and ^{1}H NMR (s, 2H, δ 3.2) spectra. The breakdown products of trichloroethylene oxide (TCE-oxide) and DCE-oxide were analyzed and compared to the known microsomal products. Under acidic conditions the major products were hydrolytic products (glyoxalate and dichloroacetate from TCE-oxide, glycolate and 2-chloroacetate from DCE-oxide), while under basic conditions one-carbon products were formed (i.e., HCO_2H, CO and, in the case of DCE-oxide, HCHO). The major products observed in the P-450-catalyzed oxidations of TCE (i.e., chloral) and DCE (2-chloroacetic acid and 2,2-dichloroacetaldehyde) were observed only under extreme conditions and were not formed in the presence of P-450 (16).

We also developed a kinetic model in which TCE was converted to TCE-oxide by P-450 with a zero-order rate constant k_1, and TCE-oxide decomposed to chloral with a first-order rate constant k_2. k_2 can be calculated from the epoxide decomposition data, $k_2 = 0.693/t_{1/2}$. Measurement of chloral (and other products) in the P-450-mediated reaction yielded k_1 values, where at any time t, $[\text{chloral}] = k_1[\text{P-450}](t + [\text{P-450}](1-e^{-k_2 t}) - k_2^{-1})$. From k_1 and k_2, one can calculate the expected concentration of TCE-oxide at time t, $[\text{TCE-oxide}] = k_1 k_2^{-1}[\text{P-450}]$. The formation of TCE-oxide from $[^{14}C]$-TCE was confirmed by chromatography of an isolated p-nitrobenzylpyridine adduct. However, the experimentally-measured levels of TCE-oxide were 5-28 fold to support the obligate intermediacy of TCE-oxide in oxidation of TCE to chloral in 4 P-450 systems examined (16).

Finally, although the heme of P-450 is destroyed during the metabolism of TCE, DCE, and VC, neither 2-chloroethylene oxide (17) nor TCE-oxide (16) destroyed ferric or ferrous P-450.

Thus, TCE-oxide is neither chemically nor catalytically competent in the conversion of TCE to chloral and cannot explain the suicide inhibition. The observation with vinyl chloride and DCE support these results. A logical scheme to explain our observations would be that shown in Fig. 3. The perferryl oxygen is shown to attack the least

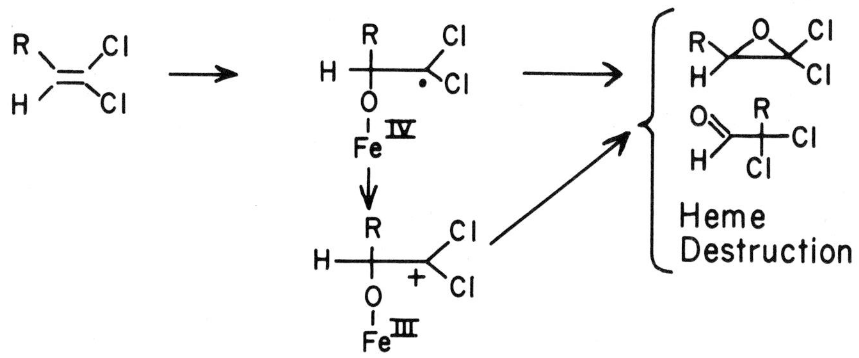

Fig. 3. Postulated scheme for metabolism of 1,1-dichloroethylene compounds.

substituted carbon of the olefin. The intermediate can be shown with either a formal Fe^{IV} or Fe^{III}, depending upon whether a radical or carbonium ion is formed at the adjacent carbon. Using simple homolytic or heterolytic electron transfer steps, the known products can easily be rationalized: i.e., epoxide formation, chlorine migration to yield chloral, and formation of heme adducts. The scheme is consistent with the structures assigned to several adducts by Ortiz de Montellano (18). The group migration and suicidal heme destruction, which do not occur with the epoxides, themselves argue that epoxidation involves a stepwise rather than a concerted mechanism. We present this scheme as a generalized mechanism for epoxidation, although variation in substituent groups may lead to changes in partitioning between epoxidation, migration, and heme destruction, the site of perferryl oxygen attack, and the electronic distribution within the

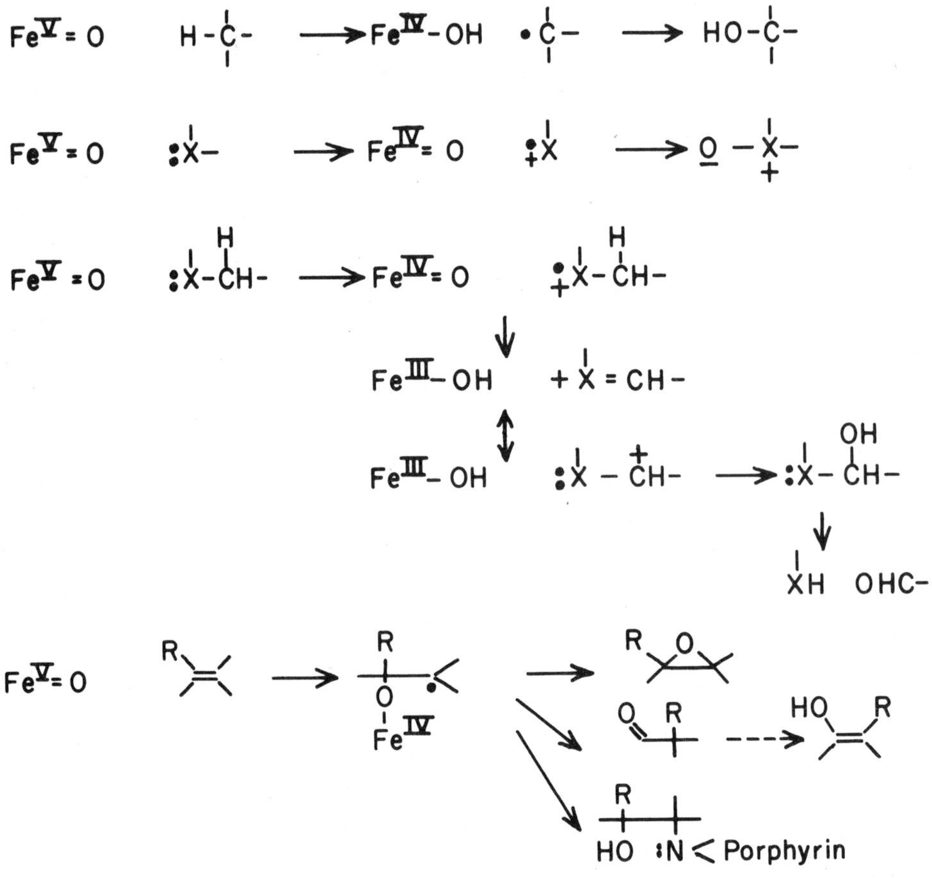

Fig. 4. Unified scheme for P-450-mediated reactions.

intermediate. The 1,2-shift has been observed before in aromatic hydroxylations and termed the "NIH Shift" (19). However, the existence of a shift can readily be explained in terms of electrophilic substitution even if epoxide formation does not occur.

Thus, in conclusion, the diverse functions of microsomal cytochrome P-450 include C-hydroxylation, N- and O-dealkylation, S- and N-hydroxylation in certain cases, epoxidation, group transfer accompanying oxidation, reduction, and suicide destruction. Many of these reactions can be explained with the view that oxygenated P-450 first abstracts an electron or hydrogen atom from a carbon, a heteroatom, or a π bond (Fig. 4). An α-hydrogen can then be abstracted from a heteroatom to create an adjacent carbon radical center which is oxygenated. In the case of heteroatoms which have no α-hydrogens or which are relatively stable after oxidation, the heteroatom can be oxygenated. All of the other known oxidative P-450 reactions can be explained by reaction of the resulting iron-oxygen complex with a carbonium ion or carbon-centered radical, and suicidal heme destruction results from the reaction of a pyrrolic nitrogen with the electrophilic carbon. Thus, the proposed mechanisms tend towards a common denominator to explain the various P-450 functions.

ACKNOWLEDGMENTS

This research was supported in part by United States Public Health Service Grants ES 00267, ES 01590, ES 02205, and ES 02702. F.P.G. is the recipient of United States Public Health Service Research Career Development Award ES 00041 and T.L.M. is an Alfred P. Sloan Fellow.

REFERENCES

1. Groves, J.T., Kruper, W.J., Jr., and Haushalter, R.C. (1980) J. Am. Chem. Soc. 102, 6375-6377.

2. Hanzlik, R.P., Kishore, V. and Tullman, R. (1979) J. Med. Chem., 22, 759.

3. Macdonald, T.L., Zirvi, K., Burka, L.T., Peyman, P. and Guengerich, F.P. (1982) J. Am. Chem. Soc., 104, 2050-2052.

4. Griller, D. and Ingold, K.V. (1980) Acct. Chem. Res., 13, 317.

5. Frederick, C., Mays, J., Kadlubar, F.F., Ziegler, D.M. and Guengerich, F.P. (1982) Cancer Res., in press.

6. Ziegler, D.M. (1980) in: Jakoby, W.B. (Ed.), Enzymatic Basis of Detoxication, Vol. 1, Academic Press, New York, pp. 201-207.

7. Hinson, J.A., Pohl, L.R., Monks, T.J., Gillette, J.R. and Guengerich, F.P. (1980) Drug Metab. Disp., 8, 289-294.

8. Wallmeier, H. and Kutzelnigg, W. (1979) J. Am. Chem. Soc., 101, 2804-2814.

9. Reich, H.J. and Peake, S.L. (1978) J. Am. Chem. Soc., 100, 4888-4889.

10. Macdonald, T.L., Narasimhan, N. and Burka, L.T. (1981) J. Am. Chem. Soc., 102, 7760-7765.

11. Banks, D.F. (1966) Chem. Rev., 66, 243-266.

12. Ngayen, T.T. and Martin, J.C. (1980) J. Am. Chem. Soc. 102, 7383-7385.

13. Burka, L.T., Thorsen, A. and Guengerich, F.P. (1980) J. Am. Chem. Soc. 102, 7615-7616.

14. Macdonald, T.L., Burka, L.T., Wright, S.T. and Guengerich, F.P. (1982) Biochem. Biophys. Res. Commun., 104, 620-625.

15. Groves, J.T., and Kruper, W.J., Jr. (1979) J. Am. Chem. Soc. 101, 7613-7615.

16. Miller, R.E. and Guengerich, F.P. (1982) Biochemistry, 21, 1090-1097.

17. Guengerich, F.P. and Strickland, T.W. (1977) Mol. Pharmacol., 13, 993-1004.

18. Ortiz de Montellano, P.R., Kunze, K.L., Beilan, H.S., and Wheeler, C. (1982) Biochemistry, 21, 1331-1339.

19. Daly, J.W., Jerina, D.M. and Witkop, B. (1972) Experientia, 28, 1129-1149.

© 1982 Elsevier Biomedical Press B.V.
Cytochrome P-450, Biochemistry, Biophysics
and Environmental Implications, E. Hietanen,
M. Laitinen and O. Hänninen editors

INDUCTION OF P-450 ISOZYMES WHICH HAVE HIGH AFFINITY FOR NITROSAMINES

CHUNG S. YANG, YITYOONG YONG TU, RENXIU PENG AND NANCY A. LORR
Department of Biochemistry, UMDNJ-New Jersey Medical School, Newark, New Jersey
U.S.A.

INTRODUCTION

The metabolism of nitrosamines has been a subject of extensive investigations because of its importance in understanding the carcinogenicity and toxicity of these compounds. Oxygenation (α-hydroxylation) has been suggested as a key step in the activation of many nitrosamines (1, 2). There is evidence indicating that the oxygenation is catalyzed by the P-450-dependent monooxygenase system (3-7) but there are also reports suggesting that other enzyme systems are involved (8-10). Some of the controversies derived from the fact that the microsomal nitrosamine-metabolizing enzymes differ from many of the commonly studied monooxygenase systems in their inducibility and susceptibility to inhibitors (8-10). These conflicts and complications are understandable if we consider the multiplicity of P-450 in microsomes. The different isozymes are known to have different structures, substrate specificities, and inducibilities. The P-450 species responsible or with high activity for nitrosamine metabolism may be different from the isozymes established in studies with classical inducers or substrates. In the present report, we describe the induction of P-450 and N-nitrosodimethylamine demethylase (NDMAd) by fasting, acetone, isopropanol, and ethanol.

MATERIALS AND METHODS

Unless otherwise stated, male Sprague-Dawley rats (body weight 45-50 g) were used. They were subjected to the following treatments: (a) fasting for 1 to 3 days, (b) acetone or isopropanol (2.5 ml/kg) administered ig as a 25% solution one day before sacrifice, and (c) 10% ethanol in drinking water for 3 days. Liver microsomes were prepared as described previously and stored at -90°C prior to use. Enzyme assays and SDS-PAGE were carried out as described previously (8, 11, 12).

RESULTS AND DISCUSSION

Fasting for 1 to 3 days caused a 2.1 to 2.4 fold increase in the microsomal NDMAd activity (Table 1). The increase was greater than those observed in the

Table 1. Induction of NDMAd and other monooxygenase activities[a]

Treatment	P-450 Content	Reductase Activity	NDMAd	Benzphetamine demethylase
1. Fast, 1 day	109	115	209	105
2. Fast, 2 days	136	119	232	107
3. Fast, 3 days	179	147	241	140
4. Acetone	160	110	300	141
5. Isopropanol	149	115	406	138
6. Ethanol	159	95	419	75

[a]Activities are expressed as % of the values of the control groups. Their P-450 content was 654-717 pmol/mg, reductase activity was 209-265 unit/mg, NDMAd activity was 1.9 to 2.2 nmol/mg/min, and benzphetamine demethylase activity was 4.2 to 4.4 nmol/mg/min.

the gross P-450 content, NADPH-cytochrome c reductase activity, and benzphetamine demethylase activity. More pronounced induction of NDMAd (3 to 4 fold) was observed by treatment with acetone, isopropanol, and ethanol. Again a large scale increase was not seen in the gross P-450 content, reductase activity, and benzphetamine demethylase activity. Lower extents of enhancement of NDMAd by these factors have been reported previously by different investigators (13-15), but the nature of the enhancement was poorly understood. In order to characterize these processes, several lines of investigation were carried out.

Kinetic analysis indicates that a low K_m form of NDMAd is induced by these treatments. With control microsomes three K_m values, 0.07, 0.38, and 38.6 mM, with corresponding V_{max} values of 1.59, 2.15, and 4.43 nmol/min/mg were observed. The V_{max} of the low K_m (= 0.07 mM) from increased 3.0 to 4.4 fold after fasting for 1 to 3 days, 4.0 to 5.5 fold after treatment with acetone or isopropanol, and 4.6 fold after treatment with ethanol. The different types of treatments appeared to have similar effects on the microsomal monooxygenase and NDMAd. Several lines of experiments suggest that the enhanced low K_m NDMAd activity is due to the induction of specific P-450 isozymes. (a) The treatments caused a "red-shift" of 0.6 nm in the CO-difference spectra of the reduced microsomal samples. (b) Protein species with M_r of 50,000 and 52,000 were induced in microsomes by the treatments (Fig. 1). The synthesis of these proteins was inhibited by $CoCl_2$ and cycloheximide. (c) The enhanced NDMAd was dependent on the microsomal NADPH-P-450 reductase. The microsomal NDMAd activity was decreased by 90% after a limited tryptic digestion and the activity was regained by preincubating purified NADPH-P-450 reductase with the trypsinized microsomes.

The results are consistent with a hypothesis that fasting, acetone, isopropanol, and ethanol induce specific P-450 isozymes which have high affinity for NDMA and possibly other nitrosamines. The mechanism of induction is not known. Since acetone and isopropanol are interconvertible in animal tissues, either compound could be the actual inducer. Isopropanol appears to be a slightly more effective inducer than acetone. During fasting and feeding with ethanol, ketone bodies are known to form and acetone is considered as a dead-end product.

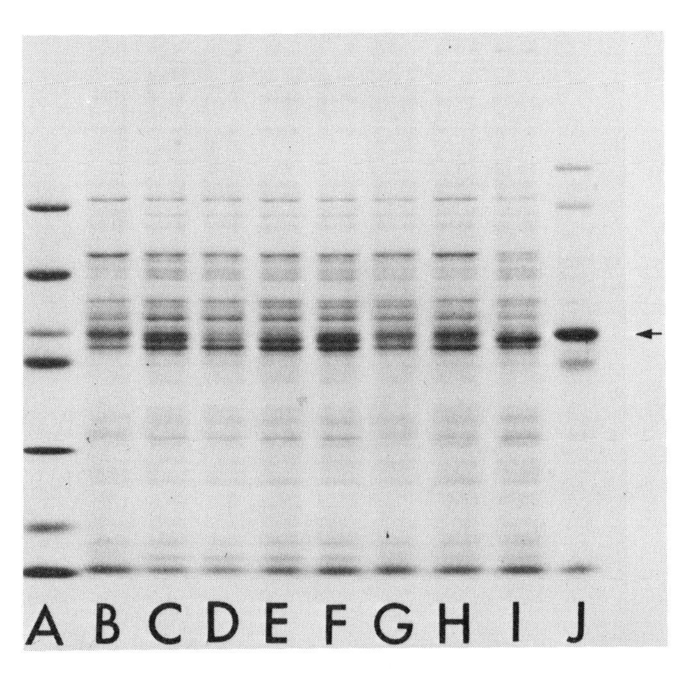

Fig. 1. Gel electrophoresis of microsomal proteins. Well A contained 7 protein standards (from bottom to top): α-lactalbumin (M_r 14,000), trypsin inhibitor (20,100), carbonic anhydrase (30,000), ovalbumin (43,000), purified P-450 (52,000), albumin (67,000), and phosphorylase b (94,000). Other wells contained microsomes: B and H, pyrazole-induced (showing an induced band with M_r of 52,000); C and F, acetone-induced (showing induced bands with M_r of 50,000 and 52,000); D and F, control; G, isopropanol-induced; I, fasting for 3 days. Well J contained partially purified P-450 from pyrazole-induced microsomes.

The similarity between the induction of NDMAd by fasting (or ethanol) and isopropanol (or acetone), suggests that isopropanol (or acetone) may be the actual inducer during fasting or ethanol consumption. Additional work is needed to substantiate this point and to characterize the induced P-450 isozymes. It appears that, in addition to xenobiotics and steroids, low molecular weight water soluble metabolites such as acetone or isopropanol may also be inducers of P-450 isozymes. Since these compounds are produced during fasting, diabetic conditions, and ethanol consumption, this type of induction may be widely occurring and of great importance in affecting carcinogenesis and cytotoxicity.

ACKNOWLEDGEMENTS

The work was supported by Grant CA-16788 from the U.S. National Cancer Institute. The authors acknowledge the excellent technical assistance of Ms. S. M. Ning.

REFERENCES

1. Magee, P.N. and Barnes, J.M. (1967) Adv. Cancer Res. 10, 163.
2. Druckrey, H., Preussman, R., and Ivankovic, S. (1969) Ann. N. Y. Acad. Sci. 163, 676.
3. Lai, D.Y. and Arcos, J.C. (1980) Life Sci. 27, 2149.
4. Lai, D.Y., Myers, S.C., Woo, Y.T., Greene, E.J., Friedman, M.A., Argus, M.F., and Arcos, J.C. (1979) Chem.-Biol. Interact. 28, 107.
5. Czygan, P., Greim, H., Garro, A.J., Hutterer, F., Schaffner, F., Popper, H., Rosenthal, O., and Cooper, D.Y. (1973) Cancer Res. 33, 2983.
6. Guengerich, F.P. (1977) J. Biol. Chem. 252, 3970.
7. Tu, Y.Y., Sonnenberg, J., Lewis, K.F., and Yang, C.S. (1981) Biochem. Biophys. Res. Commun. 103, 905.
8. Lake, B.G., Phillips, J.C., Heading, C.E., and Gangolli, S.D. (1976) Toxicol. 5, 297.
9. Rowland, I.R., Lake, B.G., Phillips, J.C., and Gangolli, S.D. (1980) Mutation Res. 72, 63.
10. Lake, B.G., Heading, C.E., Phillips, J.C., Gangolli, S.D., and Lloyd, A.G. (1974) Biochem. Soc. Trans. 2, 610.
11. Yang, C.S., Strickhart, F.S., and Kicha, L.P. (1978) Biochim. Biophys. Acta 509, 326.
12. Baskin, L.S. and Yang, C.S. (1980) Biochemistry 19, 2260.
13. Venkatesan, N., Arcos, J.C., and Argus, M.F. (1970) Cancer Res. 30, 2563.
14. Sipes, I.G., Stripp, B., Krisha, G., Maling H.M., and Gillette, J.R. (1973) 142, 237.
15. Garro, A.J., Seitz, H.K., and Lieber, C.S. (1981) Cancer Res. 41, 120.

© 1982 Elsevier Biomedical Press B.V.
Cytochrome P-450, Biochemistry, Biophysics
and Environmental Implications, E. Hietanen,
M. Laitinen and O. Hänninen editors

STEREOSELECTIVITY OF RAT LIVER CYTOCHROMES P-450 AND P-448

SHEN K. YANG[1], MING W. CHOU[1], PETER P. FU[2], PETER G. WISLOCKI[3], AND ANTHONY Y. H. LU[3]

[1]Uniformed Services University of the Health Sciences, Bethesda, Maryland 20814,
[2]National Center for Toxicological Research, Jefferson, Arkansas 72079, and
[3]Merck Sharp & Dohme Research Laboratories, Rahway, New Jersey 07065 USA

INTRODUCTION

The rat liver cytochrome P-450-containing drug-metabolizing enzyme system is known to stereoselectively metabolize polycyclic aromatic hydrocarbons (PAHs) to form optically active epoxide and dihydrodiol-epoxide intermediates (1,2). The stereoselective pathway in metabolism is an important factor in the metabolic activation of inert parent PAH to ultimate carcinogenic metabolite(s) (1,2). Interestingly, among the optically pure enantiomers of the 7,8-epoxide, trans-7,8-dihydrodiol, trans-7,8-dihydrodiol-9,10-epoxide of benzo[a]pyrene (BaP), and the 3,4-epoxide, trans-3,4-dihydrodiol and trans-3,4-dihydrodiol-1,2-epoxide of benz[a]anthracene (BA), invariably only those enantiomers that are formed predominantly in the metabolism of BaP and BA have been found to display high tumorigenic activity (2). Each of the major enantiomeric epoxide intermediates formed from BaP and BA by liver microsomes from untreated, phenobarbital (PB)-treated, and 3-methylcholanthrene (MC)-treated rats occur on the same face of the planar aromatic substrate (2). We describe in this report that the monooxygenases in liver microsomes from PB- and MC-treated immature Long-Evans rats have opposite stereoselectivity in the metabolic conversion of 8-methylbenz[a]anthracene (8-MBA) to different enantiomeric 8-MBA 8,9-epoxides. This is the first example indicating that the different forms of cytochrome P-450 may each catalyze the epoxidation reaction preferentially at a specific face of the methyl-substituted aromatic double bond of a planar monomethylated PAH molecule.

MATERIALS AND METHODS

8-MBA was incubated with liver microsomes from PB- or MC-treated immature Long-Evans rats in a 250-ml reaction mixture (pH 7.5). Each ml of reaction mixture contained 80 nmol of 8-MBA (added in 40 µl of acetone), 50 µmol of Tris-HCl, 3 µmol $MgCl_2$, 0.1 unit of glucose-6-phosphate dehydrogenase (type II, Sigma), 0.1 mg $NADP^+$, 0.65 mg glucose-6-phosphate, and 1 mg of liver microsomal protein. Incubations were carried out in the dark at 37° for 60 min.

8-MBA and its metabolites were extracted with 3 volumes of acetone/ethyl acetate (1:2,v/v). The organic layer was dehydrated with anhydrous $MgSO_4$ and subsequently evaporated to dryness under reduced pressure. The residue was redissolved in tetrahydrofuran/methanol (1:1,v/v) for reversed-phase HPLC separation of metabolites. Metabolites were separated by using a Spectra-Physics model 3500B liquid chromatograph fitted with a DuPont 0.46 cm x 25 cm Zorbax ODS column. The column was eluted at ambient temperature with a 40-min linear gradient of methanol/water (1:1,v/v) to methanol at 0.8 ml/min. Chromatographic peaks containing 8-MBA *trans*-8,9-dihydrodiol were further purified on a DuPont 0.62 cm x 25 cm Zorbax SIL column with tetrahydrofuran/hexane (2:3,v/v) as the eluting solvent.

The identification of the *trans*-8,9-dihydrodiol as a metabolite of 8-MBA by ultraviolet absorption, fluorescence, and mass spectral analyses has been reported earlier (3). The circular dichroism spectrum of 8-MBA *trans*-8,9-dihydrodiol in methanol was measured in a cell of 1 cm path length at room temperature using a Jasco model 500A spectropolarimeter equipped with a Jasco model DP-500 data processor.

RESULTS AND DISCUSSION

The circular dichroism spectra of the *trans*-8,9-dihydrodiols formed from the metabolism of 8-MBA by liver microsomes from PB- and MC-treated rats, respectively, are close to mirror images of each other (Fig. 1). The results suggest that the mechanisms of metabolic conversion of 8-MBA to the *trans*-8,9-dihydrodiol are different for the two liver microsomal preparations. There are three possible mechanisms by which different enantiomeric *trans*-8,9-dihydrodiols may be formed predominantly from the metabolism of 8-MBA by liver microsomes from PB- and MC-treated rats:

i. Epoxidation reactions catalyzed by cytochrome P-450 and P-448 enzyme systems occur on different faces of the 8,9-double bond of 8-MBA, followed by cleavage of C_9-O bonds of the 8,9-epoxide intermediates catalyzed by liver microsomal epoxide hydrolase from both MC- and PB-treated rats.

ii. Epoxidation reactions catalyzed by cytochrome P-450 and P-448 enzyme systems occur on different faces of the 8,9-double bond of 8-MBA, followed by cleavage of C_8-O bonds of the 8,9-epoxide intermediates catalyzed by liver microsomal epoxide hydrolase from both MC- and PB-treated rats.

iii. Epoxidation reactions catalyzed by cytochrome P-450 and P-448 enzyme systems occur on the same face of the 8,9-double bond of 8-MBA, followed by cleavage of C_9-O (or C_8-O) bond of the 8,9-epoxide intermediate catalyzed by liver microsomal epoxide hydrolase from MC-treated rats and by cleavage of C_8-O (or C_9-O) bond of the 8,9-epoxide intermediate catalyzed by liver microsomal epoxide hydrolase from PB-treated rats, respectively.

Mechanism iii requires that the liver microsomal epoxide hydrolases of MC- and PB-treated rats have different mechanisms of action. Current evidence (4) indicates that liver microsomal epoxide hydrolase of untreated, PB- and MC-treated rats has identical mechanism of action in the hydration of arene oxides to *trans*-dihydrodiols. Thus mechanism iii can be ruled out as a possibility.

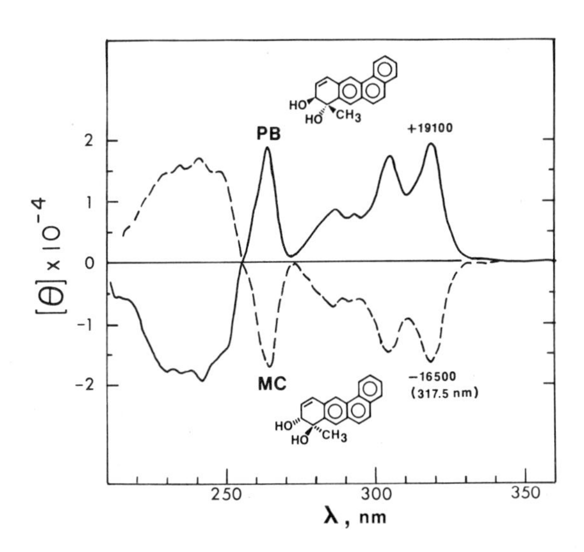

Fig. 1. Circular dichroism spectra of the (+) and (−)*trans*-8,9-dihydrodiols formed from the metabolism of 8-MBA by liver microsomes from PB- (———) and MC-treated (-----) immature Long-Evans rats. Molecular ellipticity ([Θ]) in unit of deg. $cm^2.decimole^{-1}$ was calculated based on ε = 51,200 $M^{-1}.cm^{-1}$ (methanol) at 264 nm. The absolute configuration of (−)8-MBA *trans*-8,9-dihydrodiol is arbitrarily designated as [8R,9R].

Mechanism ii requires that the water molecule activated by the catalysis of microsomal epoxide hydrolase undergoes nucleophilic attack at C_8 carbon of the 8,9-epoxide intermediate. The hydration mechanism of methyl-substituted arene oxides such as 8-MBA 8,9-epoxide by microsomal epoxide hydrolase have not been studied previously. The closest examples of methyl-substituted arene oxides that have been studied are 2,5-dimethylbenzene 1,2-oxide and 2-methylnaphthalene 1,2-oxide, but these were found not to be substrates of guinea pig liver microsomes (5). Recent studies on the hydration of monsubstituted and 1,1-disubstituted oxirane [^{18}O]oxides by rat liver microsomes indicated that epoxide hydrolase activates water molecules for nucleophilic attack at the less hindered oxirane carbon (6). Based on the available evidence (6), mechanism ii is thus less likely to be responsible for the observed results presented in this paper.

Previous results indicated that BA interacts with various forms of cytochrome P-450 in an orientation such that the epoxidation reaction occurs predominantly at only one of the two faces of the 8,9-double bond (2). This report demonstrates that cytochromes P-448 and P-450, the major induced form of cytochrome in the livers of MC- and PB-treated rats, respectively, interact with different faces of the 8,9-double bond of 8-MBA due to the presence of a methyl substituent at C_8. This is the first example indicating that a methyl substituent on a PAH molecule can drastically alter the stereoselective interaction of the substrate molecule with different forms of cytochrome P-450.

ACKNOWLEDGMENTS

This investigation was supported in part by PHS grant CA29133. We thank Miss Elizabeth Dunn for typing this manuscript.

REFERENCES

1. Yang, S. K., Roller, P. P. and Gelboin, H. V. (1978) in: Jones, P. W. and Freudenthal, R. I. (eds.) Carcinogenesis, Vol. 3: Polynuclear Aromatic Hydrocarbons, Raven Press, New York, pp. 285-301.

2. Jerina, D. M., Yagi, H., Thakker, D. R., Karle, J. M., Mah, H. D., Boyd, D. R., Gadaginamath, G., Wood, A. W., Buening, M., Chang, R. L., Levin, W. and Conney, A. H., (1978) in: Cohen, Y. (ed.), Advances in Pharmacology and Therapeutics, Vol. 9, Pergamon Press, Oxford and New York, pp. 53-62.

3. Yang, S. K., Chou, M. W., Weems, H. B. and Fu, P. P. (1979) Biochem. Biophys. Res. Commun., 90, pp. 1136-1141.

4. Thakker, D. R., Yagi, H., Levin, W., Lu, A. Y. H., Conney, A. H. and Jerina, D. M. (1977) J. Biol. Chem., 252, pp. 6328-6334.

5. Oesch, F., Kaubisch, N., Jerina, D. M. and Daly, J. W. (1971) Biochemistry, 10, pp. 4858-4866.

6. Hanzlik, R. P., Edelman, M., Michaely, W. J., Scott, G. (1976) J. Am. Chem. Soc., 98, pp. 1952-1955.

© 1982 Elsevier Biomedical Press B.V.
Cytochrome P-450, Biochemistry, Biophysics
and Environmental Implications, E. Hietanen,
M. Laitinen and O. Hänninen editors

ROLE OF CYTOCHROME P-450 AND NADPH CYTOCHROME P-450 REDUCTASE IN CATALYSIS OF NITROSOUREA DENITROSATION TO YIELD NITRIC OXIDE

D.W. POTTER AND D.J. REED
Department of Biochemistry and Biophysics, Oregon State University,
Corvallis, Oregon, 97331 (U.S.A.)

INTRODUCTION

Nitric oxide (NO) as a ligand of purified and microsomal cytochrome P-450 (R.E. Ebel et al., FEBS Lett. 55, 198, 1975) was quantitated by optical difference spectra after establishing the ferrous•NO extinction coefficient at 444-500 nm as 36 mM^{-1} cm^{-1}. Incubation of each of several nitrosoureas including CCNU, (1-(2-chloroethyl)-3-(cyclohexyl)-1-nitrosourea), and BCNU, (1,3-bis(2-chloroethyl)-1-nitrosourea), with NADPH and deoxygenated microsomes resulted in the formation of identical cytochrome P-450 ferrous•NO optical difference spectra at rates for CCNU and BCNU of 0.2 and 0.6 nmol min^{-1} mg $protein^{-1}$. Pretreatment of rats with phenobarbital (PB) increased these rates to 2.0 and 4.8 respectively. Only partial decreases in these rates were observed with SKF-525A, a naphthoflavone (ANF), metyrapone, and CO. Acrolein (0.2% v/v) treatment of microsomes decreased BCNU denitrosation rate to less than 1% of control, NADPH cytochrome P-450 reductase (FP) activity was not detectable whereas 38% of the cytochrome P-450 remained intact. Purified FP reconstituted with phospholipid, catalyzed a rapid one electron denitrosation of nitrosoureas but reconstituted purified PB cytochrome P-450 (PB P-450) lacked catalytic activity. Reconstitution of purified PB P-450 with purified FP caused stimulation of FP catalysis. The role of FMN has been studied and will be discussed in context of the role of the FP.

METHODS

PB-induced hepatic microsomes were prepared, protein and cytochrome P-450 concentrations determined, as previously described (1). PB P-450 and FP were purified according to Guengerich (2,3). FP activity was assayed at 25°C by following the rate of cytochrome c (cyt c) reduction (1 unit/μmol cyt c reduced/min with an extinction coefficient of 21 mM^{-1} cm^{-1} at 550 nm (4). The reaction mixture contained 1.5 mM potassium cyanide, 0.05 M potassium phosphate, 0.1 mM NADPH, and 50 μM cyt c, pH 7.4. The flavin composition of FP was determined by modification of the method by

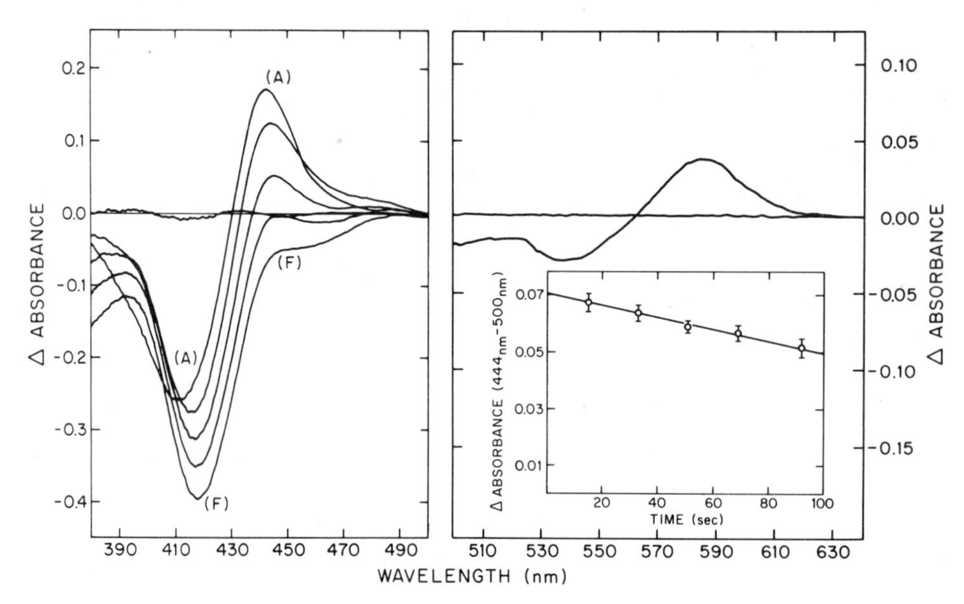

Fig. 1. P-450 ferrous•NO optical difference spectra were generated by the addition of BCNU (500 μM) into anaerobic PB-microsomes (4 mg protein, 10.0 nmol P-450, and 0.94 units FP/ml) containing 500 μM NADPH. The left panel shows scan A-F taken at 1.5 min intervals and the right panel after 14 min. P-450 ferrous•NO extinction coefficient was determined by repetitive scanning optical difference spectroscopy of anaerobic PB-microsomes (1.9 nmol P-450/ml) previously bubbled for 30s with NO and converted to ferrous•NO upon addition of 500 μM NADPH (insert). Ferrous•NO $\Delta\varepsilon$ = 36 mM^{-1} cm^{-1} at 444-500 nm was determined by linear regression analysis at time zero.

by Faeder and Siegel (5). NADPH dependent nitrosourea microsomal denitrosation was determined anaerobically at 37°C (1). [^{14}C]CCNU catalysis with purified enzyme was carried out in deoxygenated 1.0 ml reaction mixtures which contained: 20 μg (36.5 nmol) dilauroylphosphatidylcholine, 0 or 0.27 units (0 or 0.14 nmol) FP, 0 or 0.13 nmol PB P-450, 0 or 30 nmol FMN, 0.1 mmol potassium phosphate, 75 μmol glucose, 2 units catalase, 2 units glucose oxidase and 1.0 μmol NADPH, or 10 μmol sodium dithionite, pH 7.4. After 3 min preincubation, reactions were initiated with 0.5 μmol [^{14}C]CCNU in a 5 μl solution of DMSO, terminated after 10 min and analyzed by described procedure (1).

RESULTS AND DISCUSSION

Optical difference spectroscopy of anaerobic PB-induced hepatic microsomal cytochrome P-450 ferrous•NO has been previously shown to exhibit a Soret maximum at 444 nm, a single visible absorbance band maximum at 585 nm, and a

TABLE 1

RATES OF DENITROSATION OF BCNU AND CCNU UNDER ANAEROBIC CONDITIONS WITH MICROSOMAL AND PURIFIED ENZYME PREPARATIONS

Reaction Mix	Substrate and Treatment	Substrate turnover(nmol/min) Per nmol P-450	Per unit Reductase
Control Microsomes	BCNU	0.7	8.1
PB-Microsomes	BCNU	1.9	21.3
Control Microsomes	CCNU	0.2	2.6
PB-Microsomes	CCNU	0.8	9.0
PB-Microsomes	BCNU +acrolein(0.2%)	0.0	0.0
PB-Microsomes	[^{14}C]CCNU +SKF-525A(2 mM)	0.5	5.6
PB-Microsomes	[^{14}C]CCNU +Metyrapone(300 μM)	0.5	5.9
PB-Microsomes	[^{14}C]CCNU +ANF (180 μM)	0.4	4.7
PB-Microsomes	[^{14}C]CCNU +CO	0.4	4.8
FP	[^{14}C]CCNU		1.0
FP	[^{14}C]CCNU +FMN(30 μM)		3.4
FP	[^{14}C]CCNU - NADPH +$Na_2S_2O_4$(10 mM) +FMN (30 μM)		2.9
FP+PB P-450	[^{14}C]CCNU	5.4	2.7
FP+PB P-450	[^{14}C]CCNU - NADPH +$Na_2S_2O_4$(10 mM)	5.3	2.4
PB P-450	[^{14}C]CCNU-NADPH +$Na_2S_2O_4$(10 mM)	0.0	

minimum that developed at about 412 nm (1,6,7). Illustrated in Fig. 1, an identical ferrous•NO spectra was observed when nitrosourea, exemplified by BCNU, was added to NADPH and deoxygenated PB-induced hepatic microsomes. The ferrous•NO optical difference spectra was unstable at 37°C, exhibiting a complete decay of the Soret maximum at 444 nm within 14 min. The complex was stable enough for rapid quantitation of nitrosourea catalysis with microsomes. Thus, the Soret extinction coefficient was determined at 444-500 nm by extrapolating back to zero time (the lower right panel of Fig. 1).

Alternatively, the rate of $[^{14}C]$CCNU turnover was examined by the isolation of $[^{14}C]$CCU (1-(2$[^{14}C]$chloroethyl)-3-(cyclohexyl)-1-urea) from anaerobic incubations with microsomes or purified enzyme. These analyses showed stochiometric amounts of NO and $[^{14}C]$CCU being formed at a rate of 2.0 nmol/min/mg PB-microsomal protein. As shown in Table I, microsomal denitrosation was enhanced by PB pretreatment and partially decreased by cytochrome P-450 inhibitors, SKF-525A, ANF, metyrapone, and CO. However, as Table I also demonstrates, anaerobic $[^{14}C]$CCNU denitrosation was catalyzed when only FP was reconstituted with dilauroylphosphatidylcholine. Denitrosation was stimulated approximately 3-fold by 30 μM FMN or PB P-450 at concentration nearly equimolar to FP. Sodium dithionite (10 mM) was slightly less effective than NADPH (1.0 mM) as a reducing agent, which suggested that $FMNH_2$ or ferrous PB P-450 were not responsible for the stimulated denitrosation. Thus far we have attributed the observed nitrosourea denitrosation to FP alone, which interacts with FMN and P-450 in a unique and, as of yet, undefined manner.

ACKNOWLEDGEMENTS

Aided by grant CH-109 from the American Cancer Society.

REFERENCES

1. Potter, D.W., and Reed, D.J. (1982). Arch. Biochem. Biophys. 216.
2. Guengerich, F.P. (1978). Biochem. 17, 3633-3639.
3. Guengerich, F.P., and Mason, P.S. (1979). Mol. Pharmacol. 15, 154-164.
4. Williams, C.H., and Kaning, H. (1962). J. Biol. Chem. 237, 587-595.
5. Faeder, E.J. and Siegel, L.M. (1973). Analytical Biochemistry 53, 332-336.
6. Ebel, R.E., O'Keefe, D.H., and Peterson, J.A. (1975). Febs Lett. 55, 198-201.
7. O'Keefe, D.H., Ebel, R.E., and Peterson, J.A. (1978). J. Biol. Chem. 253, 3509-3516.

Cytochrome P-450, Biochemistry, Biophysics
and Environmental Implications, E. Hietanen,
M. Laitinen and O. Hänninen editors

PROGESTERONE-MICROSOMAL RECEPTOR IN RAT LIVER: PROPERTIES OF BINDING

PHILIP S. ZACHARIAS, ROSA DRANGOVA AND GEORGE FEUER
Department of Clinical Biochemistry and Pharmacology, University of Toronto, Toronto, Ontario, Canada, M5G 1L5

INTRODUCTION

Hepatocyte microsomal fractions have been well recognized as the metabolic site of many xenobiotics. The enzyme system involved in this metabolism is comprized of three major components: cytochrome P-450, reductase enzyme and phospholipid (1). While much has been established about these components, their organization and structural and functional interrelationship within the membrane still remains unsolved.

It has been found earlier that microsomes isolated from the liver of female rats contain sites that specifically bind progesterone and form a steroid-receptor complex (2). Since microsomes also represent the sites where steroids are metabolized (3), this raised the possibility that the bound progesterone may play a role in the function of these subcellular organelles. In order to test this possibility the properties of the progesterone microsomal complex was studied. In an adequate analysis of steroid protein binding one can either characterize the nature of the receptor probably a protein, or characterize the steroid constituent. The present paper reports some properties of the binding component.

METHODS

Chemicals. [1, 2-^{3}H(N)]Progesterone (sp. act. 55.7 Ci/mmol) was purchased from New England Nuclear (Montreal, Quebec), non-radioactive progesterone from Sigma Chemical Company (St. Louis, Mo), charcoal (Norit A) from Fisher Scientific (Toronto, Ont.). All other chemicals were of analytical grade preparations.

Preparation of microsomes. The liver of Wistar female rats (Hilltop Farm Laboratory Animals, Scottdale, PA) weighing 150-200g were used. Rats were sacrificed under light anaesthesia by exsanguination via the inferior vena cava. The liver was perfused with ice-cold physiological saline solution, and weighed. Lobes were cut into small pieces and homogenized with 3 vol. of medium containing sucrose, 0.25 M; Na_2EDTA, 0.001 M; tris/HCl buffer, 0.01 M, pH 7.4. The homogenate was first centrifuged at 10,000 g x 20 min. to separate postlysosomal fractions, then a clear aliquot of the supernatant centrifuged at 105.00 x 60 min to obtain microsomes. The microsomal pellet was resuspended in sucrose-EDTA-tris solution and recentrifuged. Purified microsomes were rehomogenized in the medium to produce 25% suspension/g liver. All manipulations were carried out between 0 - 4°C (4). Smooth and rough microsomes were separated by established methods.

Preparation of liposomes. Phospholipids were extracted, and liposomes prepared by established methods (5).

Progesterone binding. Microsomes (0.7 - 1 mg protein/ml) were incubated with [^{3}H] progesterone 2.5 pmol, in a mixture containing tris/HCl buffer, 0.01 M, pH 7.4; Na_2EDTA, 0.001 M; monothioglycerol, 0.01 M; and 10% glycerol in 0.5 ml total volume, in an ice-bath at 0-4°C for varying times. Bound and free steroids were separated by adding an equal volume, ice-cold dextran-coated charcoal (0.25% charcoal, 0.025% dextran), agitated for 10 min and centrifuged for 5 min at 5000 rpm at 0°C.

Measurement of radioactivity. This was done on a 100 µl aliquot of each sample in duplicates. Radioactivity was counted in a toluene-base scintillation fluid (6) using a Packard scintillation spectrometer, Model 3375. All counts were corrected to 100% efficiency by channel ratio method.

Protein content. This was measured by the modified Lowry method (7).

RESULTS

The progesterone binding molecule is heat labile and non-dialysable. Partially purified preparations reveal the protein nature of the binding component (Table 1); sulfhydryl groups play

Table 1. SODIUM CHOLATE SOLUBILIZATION AND AMMONIUM SULFATE FRACTIONATION OF MICROSOMAL PROGESTERONE RECEPTOR

FRACTION, % $(NH_4)_2SO_4$	RECOVERED PROTEIN,MG	CPM/MG PROTEIN x 10^3	ACTIVITY OF FRACTION, % OF TOTAL
Untreated microsomes	192.3	102	100.0
30-40	52.4	35	15.5
40-45	31.6	31	8.4
45-50	33.3	52	14.9
>50	51.4	140	61.3

Table 2. INTERACTION BETWEEN PHOSPHOLIPIDS AND PROGESTERONE RECEPTOR USING SOLUBILIZED AND ACETONE TREATED HEPATIC MICROSOMES

FRACTION	SPECIFIC BINDING CPM/MG PROTEIN, %
Solubilized receptor (SR)[a]	100
SR+ phosphatidylcholine (PC)	34
SR+ microsomal phospholipid (MPL)[b]	0
BSA + PC, or BSA + MPL	0
Untreated microsomes	100
Acetone powder	28
Acetone powder + PC	15
Acetone powder + MPL	0

[a] Obtained from microsomes of phenobarbital-treated rats
[b] Liposomes obtained from whole phospholipid extract of liver microsomes using octylglucoside as a detergent with subsequent dialysis.

a role in the binding (Fig. 1). Phospholipids cause an inhibitory effect (Table 2). Subfractionation of the microsomal membranes show that most of progesterone binding is in the rough fraction (Fig. 2). The formation of the progesterone microsomal complex is reduced by K^+ ions; Ca^{2+} or Mg^{2+} ions block binding in low concentrations.

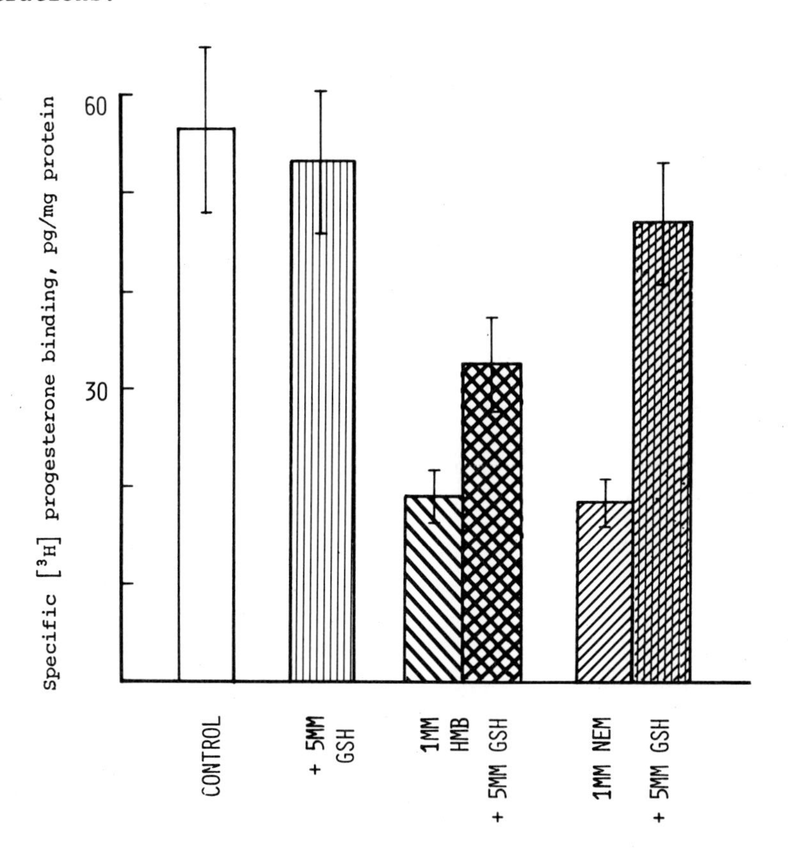

Fig. 1. Effect of sulfhydryl group blocking agents on [^{3}H] progesterone binding to hepatic microsomes. GSH, glutathione; HMB, p-hydroxymercurybenzoate; NEM, N-ethylmaleimide.

DISCUSSION

Present results with progesterone binding of rat liver microsomes show certain features similar, some dissimilar to microsomal estradiol and testosterone binding (8). The number of progesterone binding sites are 1.4-3.4 pmol/mg protein, the affinity constant $K_a = 2.2 \times 10^{-9}$ mol. The affinity between progesterone and the receptor sites is very strong. The dissociation from the high affinity binding sites K_d=30 nmol for progesterone is the same

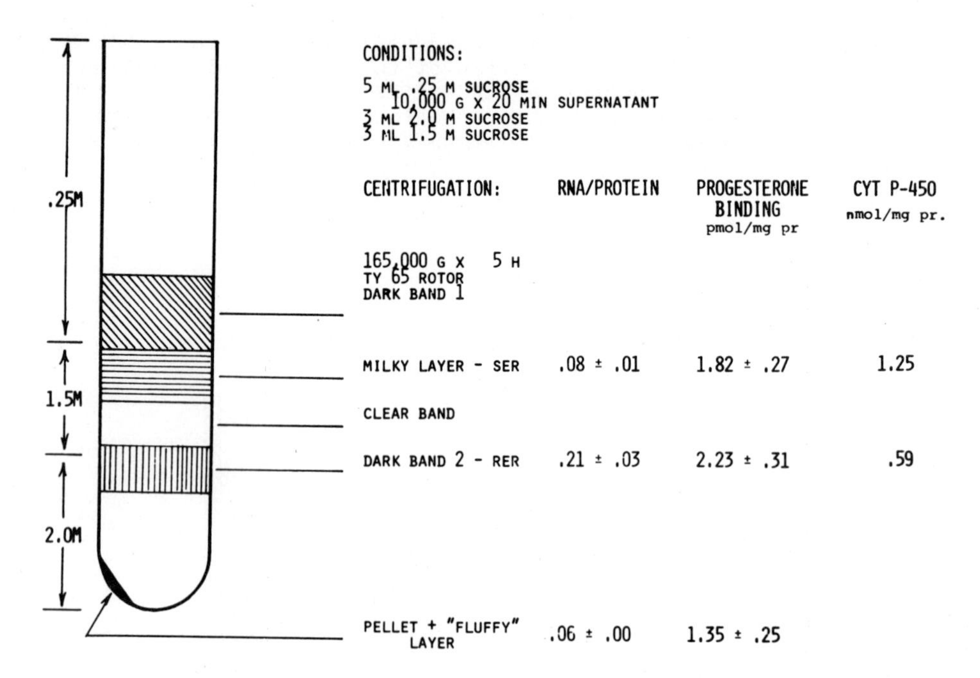

Fig. 2. Fractionation of liver microsomes by discontinuous sucrose gradient centrifugation.

magnitude as for testosterone and estradiol (9). Progesterone cannot be separated from the high affinity binding; 6-8 fold subsequent charcoal/dextrane treatment of microsomes only partially removes the bound steroid.

In contrast to the behavior of these "tight" sites, EDTA treatment of rough microsomal membranes causes a detachment of ribosomal subunits from the membrane and the degranulated membra nes show no high affinity binding for testosterone or estradiol (9). In our experiments, EDTA exerted no effect on the binding capacity of microsomes to progesterone.

Progesterone binding is associated with the protein moiety of the membrane influenced by the presence of phospholipids. In the protein part sulfhydryl groups may be responsible for the "tight" affinity. Sulfhydryl blocking agents cause a reduced binding, reversed by glutathione. The progesterone-protein association is influenced by the ion content of the medium. Addition of phospholipids to microsomes also modified progesterone binding. Removal of neutral lipids by acetone treatment causes significant reduction, chloroform/methanol treatment completely destroys progesterone binding. Addition of 5 fold excess of phosphatidylcholine to solubilized receptor brings about a decrease, addition of 3-4 fold excess of microsomal phospholipids completely blocks progesterone binding to the membrane.

These investigations give no answer as to what is the role of progesterone in the hepatic endoplasmic reticulum whether the progesterone protein binding is an essential feature of the membrane structure or has it some function significance. Further studies are directed to explore the importance of the progesterone microsomal complex binding in the liver.

ACKNOWLEDGEMENT

This research was supported by the Medical Research Council of Canada.

REFERENCES

1. Lu, A.Y.H., Junk, K.W., and Coon, M.J. (1969). Resolution of liver microsomes into three components. J. Biol. Chem. 244, 3714-3721.

2. Drangova, R. and Feuer, G. (1980). Progesterone binding by the hepatic endoplasmic reticulum of the female rat. J. Steroid Biochem. 13, 629-637.

3. Conney, A.H. and Kuntzman, R. (1971). Metabolism of normal body constituents by drug-metabolising enzymes in liver microsomes. In Handbook of Experimental Pharmacology, Ed. B.B. Brodie and J.R. Gillette. Springer Verlag, Berlin. Vol. 28/2 pp. 401-421.

4. Feuer, G., Golberg, L. and LePelley, J.R. (1965). Liver response tests. I. Exploratory studies on glucose 6-phosphatase and other liver enzymes. Fd. Cosmet. Toxicol. 3, 235-249.

5. Folch, J., Lees, M. and Sloane-Stanley G.H. (1957). A simplified method for the isolation and purification of total lipids from animal tissues. J. Biol. Chem. 226, 497-509.

6. Buehler, D.R. (1962). A simple scintillation counting technique for assaying $C^{14}O_2$ in a Warbury flask. Anal. Biochem. 4, 413-417.

7. Miller, G.L. (1959). Protein determination for large numbers of samples. Analyt. Chem. 31, 964.

8. Blyth, C.A., Freedman, R.B. and Rabin, B.R. (1971). Sex specific binding of steroid hormones to microsomal membranes of rat liver. Nature New Biol. 230, 137-139.

9. Blyth, C.A., Cooper, M.B., Roobol, A., & Rabin, B.R. (1972). The binding of steroid hormones to degranulated microsomes from rat liver endoplasmic reticulum. Eur. J. Biochem. 29, 293-300.

Cytochrome P-450, Biochemistry, Biophysics
and Environmental Implications, E. Hietanen,
M. Laitinen and O. Hänninen editors

THE OXIDATION OF THE TRYPTAMINES AND RELATED INDOLYLALKYLAMINES BY A RECONSTITUTED MICROSOMAL ENZYME SYSTEM

A. JACCARINI
Department of Pharmacology, Colleges of Medicine and Medical Sciences, King Faisal University, P.O. Box 2114, Dammam (SAUDI ARABIA)

INTRODUCTION

Tryptamines are hydroxylated in the 6-position by liver and other tissue microsomal preparations (1,2). Lysergic acid diethylamide (LSD) is metabolised to the corresponding 13 position (3). The hydroxylations of the indolylalkylamines and related compounds have assumed significance through the physiological activities of the derived phenols (4,5). We have recently demonstrated the N-hydroxylation of 2-phenyl-indole.

In the studies reported here we have investigated the 6-hydroxylation of indolylalkylamines using reconstituted fractions containing cytochrome P-450, NADPH-cytochrome c reductase and phosphatidyl choline (6,7). We have examined substrate specificity, the effect of mediators on the formation of 6-hydroxy metabolites and N-hydroxylation using 2-phenylindole as substrate (8).

MATERIALS AND METHODS

Animal pretreatment. Phenobarbital (80mg/kg) in 0.9% NaCl and 3-methylcholanthrene (20mg/kg) in olive oil were administered i.p. separately for five and three consecutive days respectively.

Isolation of microsomal fractions. Microsomal fractions from liver, lung and intestinal mucosal tissue of Wistar albino rats and New Zealand white rabbits were prepared as previously described (2). For the solubilisation of the cytochrome P-450 the microsomal pellets were resuspended in 0.1M Tris-HCl pH 7.6 containing 0.05% sodium cholate, 0.1mM dithiothreitol and 20% glycerol or in 0.15M phosphate pH 7.4 and stored at $40^{o}C$.

Purification of cytochrome P-450. Microsomes were solubilised, fractionated and partially purified by means of a procedure which involved a modification of a previously published method (9). A five-fold purification was achieved at final yield of 0.75 nmoles/mg protein. The CO-difference spectrum of dithionite reduced cytochrome P-450 showed a peak at 452 nm. A fraction containing NADPH-cytochrome C reductase was also obtained.

Analytical procedures. The 6-hydroxylation of tryptamines was determined as previously described (8). Cytochrome P-450 was measured using an extinction of $91 mM^{-1} cm^{-1}$ (10). The activity of NADPH-cytochrome c reductase was determined at $25^{o}C$ (11). Protein contents were assayed by the method of Lowry (12).

Measurement of difference spectra. Cytochrome P-450 was added to 0.15M phosphate, pH 7.4; 1.2mM NADPH; 10mM glucose-6-phosphate; 6mM $MqCl_2$; 0.7 units/ml of glucose-6-phosphate dehydrogenase. Substrate was added to the sample cuvette.

RESULTS AND DISCUSSION

Role of cytochrome P-450 in indole 6-hydroxylation. The hydroxylation of tryptamine by rabbit liver microsomal preparations is inhibited to a moderate extent in the presence of $CO-O_2$ mixtures (2). After phenobarbital pretreatment of rats and hamsters CO difference spectra of reduced liver microsomes show a slight shift in the absorption maximum of cytochrome P-450 (13).

Indole 6-hydroxylation with reconstituted systems. The effects on the activity of cytochrome P-450 after solubilisation and purification are shown in Table 1. Parallel results were obtained for lung and intestinal preparations from rabbits. The activities of rat liver, lung and intestinal mucosa correlated with values obtained for rabbit tissues.

TABLE 1

SOLUBILISATION AND PURIFICATION OF CYTOCHROME P-450

Source: Rabbit liver	Pretreatment	Total protein (mg)	P-450[a] (nmoles/ mg protein)	Reductase[b] (nmoles/min/ mg protein)
Microsomes	None	405 ± 4.6	1.31 ± 0.1	241 ± 5
Purified P-450	None	82 ± 2	0.95 ± 0.12	5.8 ± 0.3
Microsomes	3MC[c]	419 ± 4.5	1.49 ± 0.11	229 ± 3
Purified P-450	3MC	112 ± 5	1.38 ± 0.08	6.4 ± 0.1

[a]cytochrome P-450 [b]NADPH-cytochrome c reductase [c]3-methylcholanthrene. Values are means ± S.D.

Reconstitution studies (Table 2) with rats and rabbits demonstrate that cytochrome P-450 preparations can 6-hydroxylate N,N-diethyltryptamine efficiently in the presence of reductase from control or phenobarbital or 3-methylcholanthrene pretreated animals. The degree of hydroxylating activity is largely determined by the source of the cytochrome P-450 fraction. Kinetic studies show that 6-hydroxylation is linear with increasing concentration of both cytochrome P-450 and reductase, but non-linearity becomes manifest at lower cytochrome P-450 concentrations. This is in accordance with the observations previously made on the ring hydroxylation of 2-acetylaminofluorene (14). Phospholipid does not seem to be absolutely required for activity; however such a requirement has been shown to be related to the level of purification of the cytochrome P-450 fractions (15).

Substrate specificity. The reconstituted system hydroxylates numerous indolic substrates (Table 3). This is in accordance with previously reported results using 'crude' and 'washed' microsome preparations (2). However, the effect of induction seems to be lower with the purified cytochrome P-450 fraction. The hydroxylation of α-ethyltryptamine is not significantly increased with phenobarbital-induced cytochrome P-450; in contrast there is a four-fold stimulating effect with 3-methylcholanthrene. The β-carboline harman behaves similarly.

TABLE 3

SUBSTRATE SPECIFICITY OF CYTOCHROME P-450

Substrate	Source: Rat liver nmoles of hydroxy product formed/30 min Microsomes (PB)	P-450 (PB)	P-450 (3MC)
Tryptamine	43.2 ± 0.15	53.9 ± 0.16	75.3 ± 0.2
N-Acetyltryptamine	48.7 ± 0.3	61.3 ± 0.3	82.3 ± 0.15
N,N-Diethyltryptamine	42.6 ± 0.24	51.6 ± 0.27	71.5 ± 0.3
α-Ethyltryptamine	5.6 ± 0.05	5.9 ± 0.07	18.5 ± 0.08
N-Acetyltryptophan	8.3 ± 0.08	16.3 ± 0.1	17.9 ± 0.1
Harman	11.2 ± 0.13	11.3 ± 0.09	19.7 ± 0.12

TABLE 2

METABOLISM OF N,N-DIETHYLTRYPTAMINE BY MICROSOMAL FRACTIONS FROM VARIOUS SOURCES

Fractions	Pretreatment	nmoles of 6-hydroxyDET formed/30 min					
		Liver		Lung		Intestinal mucosa	
		Rabbit	Rat	Rabbit	Rat	Rabbit	Rat
Microsomes	None	46	130	6.2	3.9	0.1	4.6
P-450[a]	None	0.5	0.5	0.1	0.1	ND	0.1
Reductase[b]	None	0.5	0.5	0.1	0.1	ND	0.1
P-450 + reductase	None	10.2	28.6	2.8	1.8	ND	1.4
P-450 + reductase + lipid[c]	None	11.4	30.4	3.2	2.2	0.1	1.5
Microsomes	PB	55.8	161.2	5.8	18.6	6.8	15.6
P-450	PB	0.5	0.5	0.1	0.1	0.1	0.1
P-450 + reductase	PB	21.6	61.4	2.4	9.3	4.2	8.3
P-450 + reductase + lipid	PB	22.2	65.6	2.7	9.5	4.2	8.7
Microsomes	MC	71.4	213.2	28.6	30.8	11.8	32.2
P-450	MC	0.5	0.5	0.1	0.1	0.2	0.2
P-450 + reductase	MC	28.52	91.6	12.3	15.3	6.3	14.2
P-450 + reductase + lipid	MC	26.6	92.3	13.1	16.4	6.4	15.8
Untreated P-450 + PB reductase	-	3.4	8.1	0.8	1.2	0.5	2.4
Untreated P-450 + MC reductase	-	4.8	7.2	0.85	1.2	0.5	2.7
PB P-450 + Untreated reductase	-	23.4	84.3	12.1	18.1	5.8	17.1
MC P-450 + Untreated reductase	-	27.6	88.3	12.4	19.8	6.6	17.2
PB P-450 + MC reductase	-	26.3	91.3	12.5	18.4	5.8	18.3
MC P-450 + PB reductase	-	26.8	89.7	13.2	17.5	5.8	18.8

[a]2.1 nmoles of cytochrome P-450 [b]405 nmoles/min of reductase [c]2mg lipid. Control microsomes contained 2.1 nmoles of cytochrome P-450. All values shown are means. ND - not determined.

Effects of inhibitors (Table 4). SKF 525-A inhibited the 6-hydroxylation of tryptamine by 85% in preparations from untreated and phenobarbital-treated rabbits, but by only 23% in 3-methylcholanthrene-treated rats. A similar effect is given by CO, octylamine and p-chloromercuribenzoate. 7,8-Benzoflavone, a specific inhibitor for cytochrome P-448 (16) has no inhibitory effect when microsomes from untreated rabbits are used, a 5% effect with phenobarbital-induced preparations and 24% effect with 3-methylcholanthrene. This seems to indicate that the cytochrome P-450 induced by 3-methylcholanthrene is different from the phenobarbital-induced type. Similar effects have been reported with the 13-hydroxylation of LSD (3). The effects of KCN, metyrapone and EDTA are those normally observable for cytochrome P-450 (17).

TABLE 4

EFFECTS OF MEDIATORS ON THE CYTOCHROME P-450 DEPENDENT 6-HYDROXYLATION OF TRYPTAMINE FROM RABBIT LIVER MICROSOMAL PREPARATIONS

	A	B	C
		(percentage inhibition)	
SKF 525-A (10^{-5}M)	84.5	86	23
CO + O_2 (80; 20: v;v)	58	45	18
Metyrapone (10^{-3}M)	32	28	34
n-Octylamine (10^{-4}M)	45	42	20
7,8-benzoflavone (10^{-5}M)	0	5	24
KCN (10^{-3}M)	31	34	36
EDTA (10^{-3}M)	41	38	41.5
p- chloromercuribenzoate (10^{-4}M)	68	88	47

[A]Untreated microsomes; [B]PB P-450 + reductase + lipid; [C]MC P-450 + reductase + lipid

N-hydroxylation. Preliminary investigations with 2-phenylindole as substrate for the reconstituted system have shown that the purified preparations of cytochrome P-450 used in these studies are not effective for N-hydroxylation of this compound. A component which is absolutely required for N-hydroxylation appears to be lacking upon reconstitution of the purified fractions.

ACKNOWLEDGEMENT

This work was partially supported by a grant from the Royal University of Malta.

REFERENCES

1. Horning, E.C., Sweeley, C.C., Dalgliesh, C.E. and Kelly, W. (1959) Biochim. Biophys. Acta, 32,566.
2. Jaccarini, A. and Jepson, J.B. (1968) Biochim. Biophys. Acta, 156,347.
3. Inoue, T., Niwaguchi, T. and Murata, T. (1980) Xenobiotica 10, 913.
4. Szara, S. (1961) Experientia, 17.76.
5. Keele, C.A. (1969) Proc. Roy. Soc. B, 143,361.
6. Hiwatashi, A. and Ichikawa, Y. (1981), Biochim. Biophys. Acta, 664-33.
7. Kaminsky, L.S., Kennedy, M.W., Adams, S.M. and Guengerich, F.P. (1981) Biochemistry, 20,7379.
8. Jaccarini, A. and Felice, M.A. (1978) in: Gorrod, J.W. (Ed.), Biological Oxidation of Nitrogen, Elsevier/North-Holland Biomedical Press, Amsterdam, pp 169-175.
9. Imai, S. and Sato, A. (1974) Biochem, Biophys. Res. Commun., 60,18.
10. Omura, T. and Sato, R. (1964) J. Biol. Chem., 239,2370.
11. Baron, J. and Tephly, T.R. (1969) Molec. Pharmac., 5,10.
12. Lowry, O.H., Rosebrough, N.J., Farr, A.L., and Randall, R.S. (1951) J. Biol. Chem., 193,265.
13. Jaccarini, A. unpublished.
14. Lotlikar, P.D., Young, S.H. and Baldy, W.J. (1978) in: Gorrod, J.W. (Ed.) Biological Oxidation of Nitrogen, Elsevier/North-Holland Biomedical Press, Amsterdam, pp 185-193.
15. van der Hoeven, T.A. and Coon, M.J. (1974) J. Biol. Chem., 249, 6302.
16. Ullrich, V., Weber, P., and Wollenberg, P. (1975) Biochem. Biophys. Res. Commun., 64,808.
17. Liebman, K.C. (1969) Molec. Pharmac. 5,1.

Cytochrome P-450, Biochemistry, Biophysics and Environmental Implications, E. Hietanen, M. Laitinen and O. Hänninen editors

SHARING OF CYTOCHROME P-450 ISOZYMES DURING THE STEPWISE MICROSOMAL OXIDATION OF BENZPHETAMINE

E. JEFFERY AND G.J. MANNERING
Department of Pharmacology, University of Minnesota, Minneapolis, MN 55455 (U.S.A.)

INTRODUCTION

2-Nitroso-1-phenylpropane (NPP) and H_2O_2 are formed when 1 mM benzphetamine (Bz) is incubated with microsomes from phenobarbital-treated rats (PB-microsomes) but not from untreated rats (U-microsomes). When the concentration of Bz is decreased (e.g., to 0.025 mM), H_2O_2 and NPP formation occur in both PB- and U-microsomes. We have determined the pathway of Bz metabolism in hepatic microsomes and explained why H_2O_2 and NPP are formed in PB-microsomes but not in U-microsomes when high concentrations of Bz are used.

MATERIALS AND METHODS

Male Sprague-Dawley rats (200 - 300 g) were either untreated or administered PB (40 mg/kg/day) for 3 days prior to death. Hepatic microsomes (1 mg/ml) were incubated with Bz and norbenzphetamine (NorBz) and an NADPH generating system at 37°C in 0.1 M Tris-Cl buffer, pH 7.4. Rates of demethylation (HCHO formation,1) and H_2O_2 production (HCHO formation in the presence of catalase and methanol,2) were determined. The incubation mixtures were extracted with chloroform for the spectral quantification of methamphetamine, amphetamine, Bz, NorBz and N-benzylethyl-α-phenylnitrone (BEPnitrone) after separation by HPLC. NPP formation was estimated by the appearance of absorbance at 455 nm which results when NPP forms a complex with cytochrome P-450 (3). The partial uncoupling of the N-hydroxylation of NorBz produces H_2O_2 at approximately three times the rate of N-hydroxylation; H_2O_2 formation was used as an estimate of NorBz metabolism.

RESULTS

When U-microsomes were incubated with 1 mM Bz, the loss of Bz was stoichiometric with the accumulation of the demethylation products, NorBz and HCHO, thus

TABLE 1

APPARENT KINETIC CONSTANTS FOR BENZPHETAMINE AND NORBENZPHETAMINE METABOLISM

Five concentrations of each substrate (0.03 - 2.0 mM) were incubated as described in MATERIALS AND METHODS; inhibitor concentrations were 0.05 and 0.1 mM. Values are the mean of 3 experiments.

BENZPHETAMINE DEMETHYLATION	UNTREATED	PHENOBARBITAL
K_mBz	30 μM	34 μM
V_{max}[a]	5.7	33
K_iNorBz	120 μM	100 μM
NORBENZPHETAMINE-INDUCED H_2O_2		
K_mNorBz	200 μM	100 μM
V_{max}[a]	10	22
K_iBz	4 μM	--
BENZPHETAMINE-INDUCED H_2O_2		
K_mBz	--	180 μM
V_{max}[a]	--	19

[a]nmol/min/mg microsomal protein

showing that NorBz was not metabolized further. When PB-microsomes were incubated with 1 mM Bz, Bz disappearance was stoichiometric with HCHO formation, but less than a stoichiometric accumulation of NorBz was observed. That this was due to the further metabolism of NorBz was shown by the production of H_2O_2, BEPnitrone and NPP. When NorBz (1 mM) replaced Bz as substrate, its incubation with either U- or PB-microsomes resulted in the production of H_2O_2, BEPnitrone and NPP. These results suggested that Bz inhibits NorBz metabolism in U- but not in PB-microsomes. Kinetic studies showed this to be the case.

Although Bz is a substrate inhibitor of NorBz metabolism in U-, but not in PB-microsomes, NorBz inhibited Bz metabolism in both preparations (Table 1). We hypothesized that whereas these two substrates compete for a single enzyme in U-microsomes, an enzyme is induced by PB-treatment that preferentially metabolizes NorBz. Alternative substrate kinetic studies (4) supported this hypothesis and showed that the affinity of the constitutive enzyme was far greater for Bz than for NorBz (Table 1). This difference in affinities causes preferential metabolism of Bz in U-microsomes such that in the presence of a high concentration of Bz (1 mM), NorBz is accumulated rather than metabolized. PB-treatment did not alter the ability of NorBz to inhibit Bz metabolism (Table 1). However, no measurable inhibition of NorBz metabolism by Bz occurred. Alternative substrate kinetics showed that at low NorBz concentrations, rates of NorBz and Bz metabolism were additive and that high concentrations of NorBz prevented Bz metabolism. These results indicate a lower affinity of the induced enzyme for Bz than for NorBz.

SCHEME I

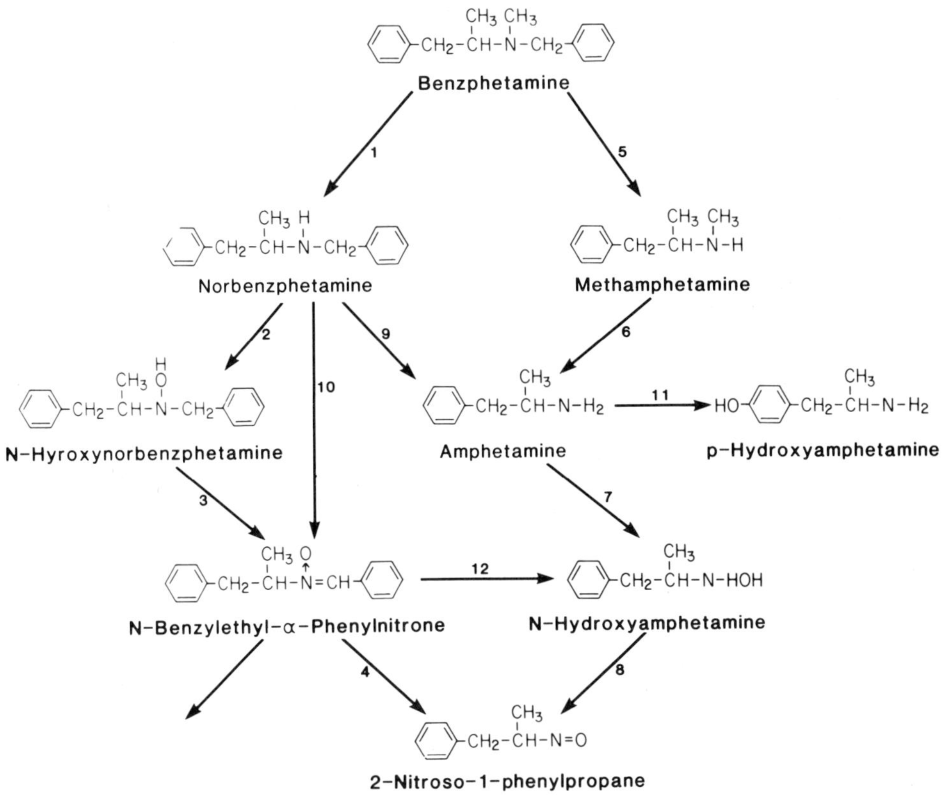

Benzphetamine

Norbenzphetamine

Methamphetamine

N-Hyroxynorbenzphetamine

Amphetamine

p-Hydroxyamphetamine

N-Benzylethyl-α-Phenylnitrone

N-Hydroxyamphetamine

2-Nitroso-1-phenylpropane

To determine the overall route of metabolism of Bz, several possible pathways were examined (scheme 1). Formaldehyde production was stoichiometric with NorBz formation in U-microsomes and linear from zero time with both preparations. These results, indicative of reaction 1 rather than reaction 5 and 6 (scheme 1), were supported by the finding that methamphetamine, amphetamine and p-hydroxyamphetamine were produced by both microsomal preparations when methamphetamine was the substrate, but none of these metabolites was formed when Bz was the substrate. Lack of amphetamine or p-hydroxyamphetamine accumulation, and the presence BEPnitrone, indicated that step 9 does not occur. The conversion of NorBz to BEPnitrone was inhibited by CO and SKF 525-A, inhibitors of cytochrome P-450, and by methimidazole and deoxycholate, inhibitors of mixed function amine oxidase (MFAO). The production of H_2O_2 in the presence of NorBz was inhibited by CO and SKF 525-A but not by methimidazole or deoxycholate; these results suggest that steps 2 and 3 are preferred over step 10, that step 2 is cytochrome

P-450-dependent and step 3 is MFAO-dependent. NPP was formed when BEPnitrone was incubated with microsomes and NADPH. BEPnitrone remained unchanged in the absence of NADPH. Both Bz and NorBz competitively inhibited the formation of NPP from BEPnitrone (step 4). In PB-microsomes, NPP formation from NorBz (steps 2,3,4) was only slightly inhibited by NorBz. These results are interpreted to mean that whereas Bz, NorBz and BEPnitrone compete as substrates for a constitutive cytochrome P-450 in U-microsomes, a PB-induced cytochrome preferentially metabolizes NorBz, thereby relieving competition and increasing the total rate of Bz metabolism in PB-microsomes. However, the rate of NPP formation by PB-microsomes depends on the amount of the constitutive enzymes involved in step 4, not on the amount of the induced enzyme involved in step 2.

In summary, of several pathways that have been considered feasible, only the pathway involving steps 1,2,3 and 4 survives the elaborated process of elimination. It is of interest that two of the feasible pathways were deficient in only a single step. When methamphetamine was incubated with microsomes, the NPP complex was formed, thus demonstrating that only step 5 is missing in the pathway involving steps 5,6,7 and 8. Step 9 is likewise the only step missing in the pathway involving steps 1,9,7 and 8.

REFERENCES

1. Sladek, N.E. and Mannering, G.J. (1969) Mol. Pharmacol., 5, 174.
2. Jeffery, E.H. and Mannering, G.J. (1979) Mol. Pharmacol., 15, 396.
3. Franklin, M.R. (1974) Mol. Pharmacol., 10, 975.
4. Cha, S. (1968) Mol. Pharmacol., 4, 621.

ACKNOWLEDGEMENTS

This research was supported by USPHS grants GM 15477 and GM 17780.

Cytochrome P-450, Biochemistry, Biophysics
and Environmental Implications, E. Hietanen,
M. Laitinen and O. Hänninen editors

INTERACTION OF XENOBIOTICS AND NUTRITIONAL FACTORS WITH DRUG METABOLIZING ENZYME SYSTEMES

Z. AMELIZAD, J.F. NARBONNE
Laboratoire de Toxicologie Alimentaire, UER de Biologie, Université de Bordeaux I, Avenue des Facultés, 33405 Talence Cedex (France)

INTRODUCTION

Studies on de novo synthesys of the cytochrome P_{450} groups of proteins are important in understanding how different stimuli such as xenobiotics and nutritional factors influence the synthesis of cytochrome P_{450} and ultimately the drug metabolizing potential of cell.

What is the influence of nutritional factors on the synthesis of the different cytochromes P_{450} forms ? Are these proteins well synthesized when amino acid dietary intake is deficient ?

With a view to answer these questions, studies have been under way to compare the two forms a and b of microsomal cytochrome P_{450} from rats fed with unbalanced diets (low protein and hight fat) by means of polyacrylamide gel electrophoresis analysis and amino acid composition.

MATERIALS AND METHODS

Treatment of animals. Weanling male rats of Sprague Dawley strain were used and treated as indicated in table 1. The inducer was Phenoclor DP6 (a french polychlorobiphenyl) incorporated in the diet at the level of 50 ppm.

TABLE 1

SCHEME OF EXPERIMENTAL PROCEDURE AND DIET COMPOSITION

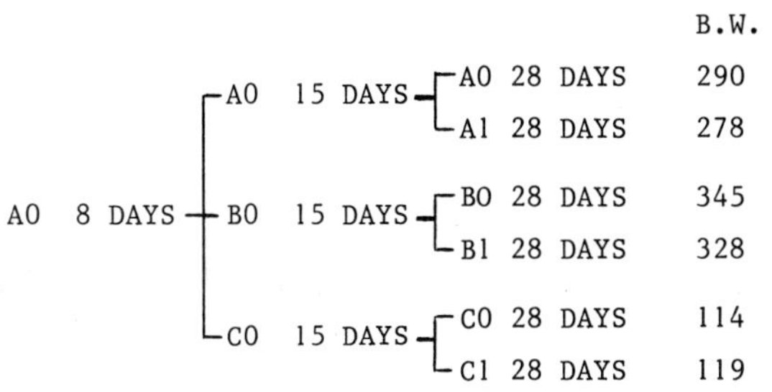

			B.W.
A0 8 DAYS	A0 15 DAYS	A0 28 DAYS	290
		A1 28 DAYS	278
	B0 15 DAYS	B0 28 DAYS	345
		B1 28 DAYS	328
	C0 15 DAYS	C0 28 DAYS	114
		C1 28 DAYS	119

DIETS	% CASEIN	% STARCH	% FAT
A	22	59	5
B	30	23	30
C	6	75	5
A1,B1,C1 = + Phenoclor DP6			

Extraction and purification of cytochrome P_{450}. Microsomes were isolated from the livers by the method of Suzuki (1) and then cleaned with 0,1 M Phosphate buffer pH 7.25, containing 20 % glycerol, 0.40 % cholate, 0.1 mM EDTA and 0.1 mM dithiothreitol (DTT) (buffer A).

Subsequentely 105 000 x g supernatant was dialysed overnight against 0.1 M phosphate buffer pH 7.25, containing 0.1 mM EDTA and 0.1 mM DTT (buffer B) and then applied to n-octylamine sepharose 4B column previously equilibrated with buffer A (2, 3). The column was washed with 2 bed volumes of buffer A and then the cytochromes were eluted with buffer A containing 0.1 mM Emulgen 913.

The cytochrome fractions were concentrated by Amicon Diaflo membrane ultrafiltration (PM 30 membrane) and dialysed 40 hours against 10 mM potassium phosphate buffer (pH 7.25) containing 20 % glycerol. Polyacrylamide gel electrophoresis was carried out in presence of 0.2 % SDS according to Ornstein (4). The relative mobilities were calculated by using molecular weight standards as described by Weber (5).

Amino acid analysis. Cytochrome samples were lyophilised and treated with 6 M HCl at 110° C for 24 hours. After removal HCl in vacuum drier, the amino acid content was determinated with a Kontron amino acid analyzer.

Cytochrome determination. Cytochromes were measured as described by Omura and Sato (6). The protein determination was carried out according with Lowry et al. (7).

RESULTS AND DISCUSSION

Cytochrome P_{450} content. Table 2 shows that prolonged protein deficiency decreases cytochrome P_{450} content in rat liver.

TABLE 2

EFFECT OF EXPERIMENTAL DIETS ON MICROSOMAL CYTOCHROME P_{450} CONTENT IN RAT LIVER

Experimental Diets	Cytochrome P_{450} nmoles/mg Mic. Prot.	Δ %
AO	0.247 ± 0.030	AO → A1 + 106***
A1	0.509 ± 0.077	
BO	0.299 ± 0.069	BO → B1 + 107***
B1	0.619 ± 0.068	AO → BO + 21*
CO	0.209 ± 0.043	CO → C1 + 168***
C1	0.561 ± 0.082	AO → CO - 15*

* $P < 0.05$ *** $P < 0.001$

This data is according with finding from Hayes and Campbell (8). In contrast the high fat diet increases the P_{450} content. This hemoprotein remains highly inducible by PCB in groups fed with unbalanced diets.

Cytochrome P_{450} analysis. In order to compare untreated and DP6 treated groups the native form of cytochrome P_{450} a is analysed (9).

In all the groups the molecular weight is unchanged and remains at 50.000 daltons (Figure 1).

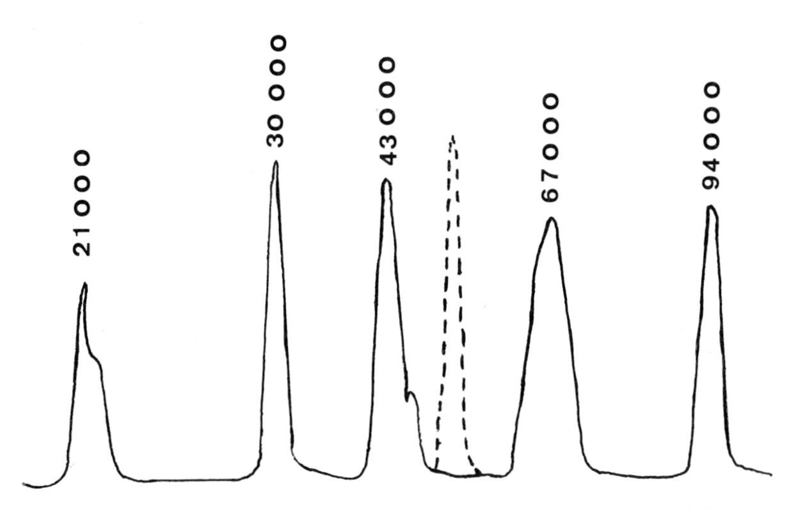

Fig 1. Densitometric tracing of SDS polyacrylamide gel of purified forms of cytochrome P_{450} a.
—— Molecular weight standards
--- P_{450} a from experimental groups A0, A1, B0, B1, C0, C1.

Cytochrome amino acid analysis (Table 3) shows that the P_{450} a induced by DP6 is like control P_{450} a. In high fat diet group the threonine content is increased.

Protein deficiency lead changes in essential amino acid composition of P_{450} a from untreated group. DP6 treatment lead the restoration of amino acid composition like control group A0.

CONCLUSION

The induction of cytochrome P_{450} by PCB is unaffected by unbalanced diets. However without treatment the nutritional factors can modify the amino acid composition of cytochrome P_{450}. These results can be correlated with changes in catalitic activities.

TABLE 3

AMINO ACID COMPOSITION AND MOLECULAR WEIGHTS OF CYTOCHROME P_{450} a FROM EXPERIMENTAL GROUPS EXPRESSE IN RESIDUES / MOLE.

	EXPERIMENTAL GROUPS					
Amino acids	A0	A1	B0	B1	C0	C1
Asp.	36	34	36	33	21	38
Thr.	51	52	73	75	52	51
Ser.	36	36	36	34	36	36
Glu.	17	18	12	16	7	19
Pro.	21	21	20	20	22	22
Gly.	39	39	37	36	37	38
Ala.	35	34	34	38	33	32
Val.	23	24	22	19	25	22
Cys.	4	4	4	4	4	4
Met.	4	4	4	4	4	3
Ileu.	20	20	18	20	49	20
Leu.	44	43	43	40	69	45
Tyr.	31	31	31	36	28	20
Phe.	23	23	23	25	8	22
Lys.	25	26	25	26	23	25
His.	11	11	11	10	10	11
Arg.	22	20	22	21	20	27
Try.	1	1	1	1	1	1
TOTAL	443	440	452	456	449	446
Mol. Weight	50 000	50 000	50 000	50 000	50 000	50 000

REFERENCES

1. Mutsuro,S. (1978) Kumamoto med. J., 31, 35-37.
2. Imai, Y. and Sato, R. (1974) J. Biochem., 75, 689-697.
3. Waener, M., La Marca, M.Y. and Neims, A.H. (1978)Drug Met. Disp., 6, 353-362.
4. Ornstein, L. (1964) Ann. N.Y. Acad. Sci., 121, 321-349.
5. Weber, K. and Osborn, M. (1969) J. Biol. Chem., 244, 4406-4412.
6. Omura, T. and Sato, R. (1964) J. Biol. Chem., 239, 2370-2378.
7. Lowry, O.H., Rosenbrough, N.J., Farr, A.L. and Randall, R.J. (1951) J. Biol. Chem., 193, 265-275.
8. Hayes, J.R. and Campbell, T.C. (1974) Biol. Pharmacol., 23, 1721-1731.
9. Botelho, L.H., Ryan, D.E. and Levin, W. (1979) J. Biol. Chem., 254, 5635-5640.
10. Narbonne, J.F. (1979) Thesis n° 631, Bordeaux France.

© 1982 Elsevier Biomedical Press B.V.
Cytochrome P-450, Biochemistry, Biophysics and Environmental Implications, E. Hietanen, M. Laitinen and O. Hänninen editors

INHIBITION OF (±)-*TRANS*-7,8-DIHYDROXY-7,8-DIHYDROBENZO(a)PYRENE METABOLISM BY 2(3)-*TERT*-BUTYL-4-HYDROXYANISOLE IN LIVER MICROSOMES AND ISOLATED HEPATOCYTES FROM MICE.

Lennart Dock, Bengt Jernström and Peter Moldéus
Department of Forensic Medicine, Karolinska Institutet, Box 60400, S-104 01 Stockholm (Sweden)

INTRODUCTION

2(3)-*tert*-butyl-4-hydroxyanisole (BHA) is a potent inhibitor of benzo(a)pyrene (BP)-induced neoplasia. This has been attributed to either an increased detoxification, due to induction of conjugation reaction pathways, or to decreased activation of BP (1). The ultimate carcinogen derived from BP is most probably 7β, 8α-dihydroxy-9α,10α-epoxy-7,8,9,10-tetrahydro-BP (anti-BPDE) which is formed through cytochrome P-450-catalyzed activation of *trans*-7,8-dihydroxy-7,8-dihydro-BP (BP-7,8-diol) (2). Since dietary BHA inhibits BP-7,8-diol-induced neoplasia (3) we have examined the effect of BHA on the microsomal activation of BP-7,8-diol to DNA-binding products. The BHA was either added directly to the incubation system or administered to the mice as a dietary constituent. We also measured the formation of water-soluble (conjugated) products of BP-7,8-diol following metabolism in isolated mouse hepatocytes in an attempt to estimate the significance of the elevated detoxification capacity in the metabolism of this carcinogen.

MATERIALS AND METHODS

Animals and chemicals. Liver microsomes and hepatocytes were prepared from female NMRI mice maintained on standard laboratory food pellets or a mixture of ground pellets and BHA (7.5 g/kg food) for two weeks as previously described (4,5). ^{14}C-, ^{3}H- and unlabelled BP-7,8-diol and ^{14}C-anti-BPDE were obtained through the Cancer Research Program of the NCI, Division of Cancer Cause and Prevention, Bethesda, MD and BHA, calf thymus DNA and NADPH were purchased from Sigma Chemical Co., St. Louis, MO.

Incubations. ^{14}C-BP-7,8-diol (10 μM, ∿5.4 mCi/mmol) was incubated with either liver microsomes (0.5 mg/ml) and NADPH (1 mM) in 50 mM Tris-HCl buffer, pH 7.5 (6) or with hepatocytes (10^6 cells/

ml) in Krebs-Hepes buffer. pH 7.5, supplemented with amino acids (5) in the absence or presence of BHA (20 μM). When DNA-binding was measured, DNA (1 mg/ml) was incubated with liver microsomes and ^{3}H-BP-7,8-diol (10 μM, ∿114 mCi/mmol) as above or with ^{14}C-anti-BPDE (20 μM, ∿5.4 mCi/mmol) in the absence of microsomes. The DNA was recovered and assayed for bound radioactivity according to published procedures (7).

Analysis of BP-7,8-diol metabolism. The reactions were terminated and products extracted into ethyl acetate as previously described for BP (4). The products remaining in the water-phase after metabolism of BP-7,8-diol in hepatocytes are predominantly glutathione-conjugates (8). The ethyl acetate extractable metabolites were separated from the unmetabolized substrate by HPLC (6).

RESULTS

Liver microsomes from both control and BHA-fed mice catalyzed the activation of BP-7,8-diol to DNA-binding products (Table).

TABLE

EFFECTS OF BHA ON THE METABOLISM OF BP-7,8-DIOL AND DNA-BINDING OF PRODUCTS FROM BP-7,8-DIOL AND ANTI-BPDE

Substrate	Microsomes	Addition	DNA-binding[a]	Metabolism[b]
BP-7,8-diol	Control	DMSO[c]	77±2	1.72±0.28
BP-7,8-diol	Control	BHA	41±4[f]	0.88±0.17[f]
BP-7,8-diol	BHA	DMSO	69±1[d]	1.27±0.34[e]
BP-7,8-diol	BHA	BHA	40±2[f]	0.66±0.15[f]
anti-BPDE	-	DMSO	1199±88	-
anti-BPDE	-	BHA	1202±35[g]	-

[a] Expressed as pmol bound/mg DNA/30 min. Means ± S.D., n=3.
[b] Expressed as nmol metabolites/0.5 mg protein/10 min. Means ± S.D., n=6-12.
[c] Dimethylsulfoxide, solvent for BHA.
[d] Significantly different from control, $p < 0.02$.
[e] $p < 0.01$
[f] $p < 0.001$
[g] Not significantly different from control.

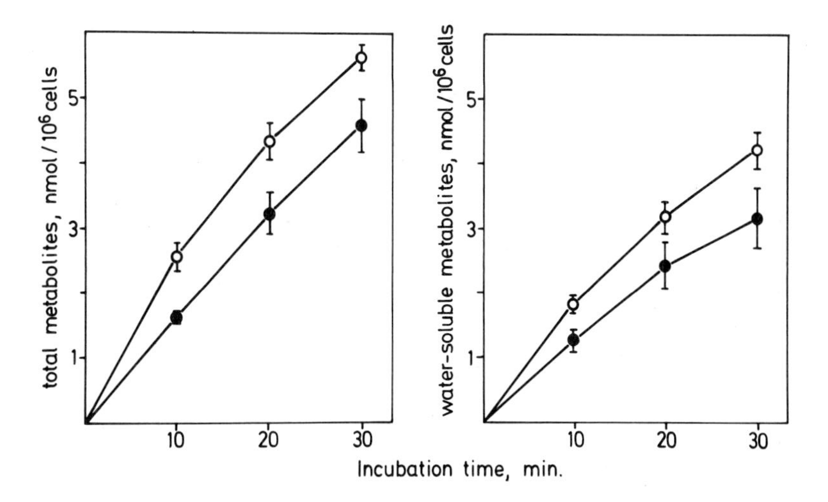

Fig. 1. Metabolism of BP-7,8-diol in isolated hepatocytes from control (o) and BHA-fed (●) mice. Mean values ± S.D. from four different cell preparations.

Fig. 2. Formation of water-soluble products of BP-7,8-diol following metabolism in isolated mouse hepatocytes. Symbols as in fig. 1.

The extent of binding could be reduced by addition of BHA to the incubation systems. Addition of BHA did not influence the binding of the pure anti-BPDE to DNA. BHA-addition had a strong inhibitory effect on the microsomal metabolism of BP-7,8-diol (Table) in accordance with previous studies using BP as the substrate (9,10).

BHA-feeding lowered the overall rate of BP-7,8-diol metabolism in isolated mouse hepatocytes (Fig. 1). Although dietary BHA strongly induced hepatic glutathione-S-transferase activities (4), a decreased formation of water-soluble metabolites of BP-7,8-diol in hepatocytes from the BHA-fed mice was observed (Fig. 2). The reason is probably the decreased rate of BP-7,8-diol activation due to BHA-feeding (c.f. Table and fig. 1) thus decreasing the amounts of anti-BPDE available for conjugation. Previous studies on the glutathione-conjugation of paracetamol support this hypothesis (5).

These data support the assumption that BHA exert its protective effect on BP carcinogenesis on the level of activation rather than on detoxification (9-11).

ACKNOWLEDGEMENTS

This investigation was supported by the Swedish Medical Research Council (B81-03X-05918-01) and NIH Grant Ca 26261.

REFERENCES

1. Wattenberg, L.W. (1980), J. Env. Pathol. Toxicol. 3, 35-52
2. Gelboin, H.V. (1980), Physiol. Rev. 60, 1107-1166
3. Wattenberg, L.W., Jerina, D.M., Lam, L.K.T. and Yagi, H. (1979), J. Natl. Cancer Inst. 62, 1103-1106
4. Dock, L., Cha, Y.-N., Jernström, B. and Moldéus, P. (1982), Carcinogenesis, 3, 15-21
5. Moldéus, P., Dock, L., Cha, Y.-N., Berggren, M. and Jernström, B. (1982), Biochem. Pharmacol. in press
6. Dock, L., Rahimtula, A., Jernström, B. and Moldéus, P. (1982), Carcinogenesis, in press
7. Jernström, B., Orrenius, S., Undeman, O., Gräslund, A. and Ehrenberg, A. (1978), Cancer Res. 38, 2600-2607
8. Jernström, B., Babson. J.R., Moldéus, P. and Reed. D.J. (1982), submitted for publication
9. Rahimtula, A.D., Zachariah, P.K. and O'Brien, P.J. (1977), Biochem. J. 164, 473-475
10. Yang, C.S., Sydor, Jr., W., Martin, M.B. and Lewis, K.F. (1981), Chem.-Biol. Interact. 37, 337-350
11. Lam, L.K.T., Fladmoe, A.V., Hochalter, J.B. and Wattenberg, L.W. (1980), Cancer Res. 40, 2824-2828

© 1982 Elsevier Biomedical Press B.V.
Cytochrome P-450, Biochemistry, Biophysics
and Environmental Implications, E. Hietanen,
M. Laitinen and O. Hänninen editors

IN VITRO METABOLISM OF 6-NITROBENZO(A)PYRENE: EFFECT OF THE NITRO SUBSTITUENT ON THE REGIOSELECTIVITY OF THE CYTOCHROME P-450-MEDIATED DRUG METABOLIZING ENZYME SYSTEM

PETER P. FU AND MING W. CHOU
NATIONAL CENTER FOR TOXICOLOGICAL RESEARCH, Jefferson, Arkansas 72079, U.S.A.

INTRODUCTION

Microsomal metabolism of polycyclic aromatic hydrocarbons by the cytochrome-P-450 monooxygenases has been studied extensively (1). However, little is known concerning the effect of nitro-aryl substitution on the metabolic regioselectivity of these enzymes. We have studied the aerobic rat liver microsomal metabolism of the mutagenic environmental pollutant, 6-nitrobenzo(a)pyrene (6-NO_2-BP), and its hydroxyl derivatives, to determine the effect of 6-nitro substitution upon the metabolic pathways in comparison to those of their respective parent hydrocarbons, benzo(a)pyrene and hydroxybenzo(a)pyrenes.

MATERIALS AND METHODS

The 500 ml incubation mixture (pH 7.5) contained 20 μmol 6-NO_2-BP, 25 mmol Tris-HCl, 15 mmol $MgCl_2$, 282 mg glucose-6-phosphate, 48 mg NADP, 50 units of glucose-6-phosphate dehydrogenase, and 500 mg rat liver microsomal protein. The incubations were performed aerobically at 37°C for 60 min, then quenched with acetone, and the metabolites were extracted with ethyl acetate. The metabolites, 1- and 3-hydroxy-6-NO_2-BP and 1- and 3-hydroxy-BP were incubated under similar conditions except on smaller scales.

[^{3}H]-6-NO_2-BP (specific activity 130 mCi/mmol, from Dr. R. Roth, Midwest Research Inst.) was incubated for 10 min in a 1.0 ml incubation volume with 0.1 mg microsomal protein which was obtained from either untreated rats, or rats pretreated with phenobarbital (PB) or 3-methylcholanthrene (MC).

Reversed-phase HPLC separations employed a DuPont Zorbax ODS column (6.2 x 250 mm) eluted with 30 min linear gradient of 75-90% methanol at 1.6 ml/min. Normal-phase HPLC separations employed a DuPont SIL column (6.2 x 250 mm) eluted with THF-hexane (1:3, v/v) at 1.5 ml/min.

Metabolites were characterized by uv-visible, mass and 500 MHz ^{1}H NMR spectral analyses.

The radioactive metabolites were separated by HPLC with added non-radioactive metabolites as uv markers. The radioactivity of each fraction (20 drops) mixed

conditions similar to those described for 6-NO_2-BP metabolism. 6-NO_2-BP-1,9-hydroquinone is the predominant metabolite of 1-hydroxy-6-NO_2-BP metabolism and 6-NO_2-BP-3,9-hydroquinone is the predominant metabolite of 3-hydroxy-6-NO_2-BP metabolism. In contrast, metabolism of 1- and 3-hydroxy-BP gave BP-1,6-quinone and BP-3,6-quinone respectively as predominant metabolites.

The specific activities and the percentages of total metabolite of each metabolite obtained from incubation of [^{3}H]-6-NO_2-BP by the microsomes from untreated rats and rats pretreated with PB, or MC microsomes are given in Table 1.

Table 1. Metabolism of [^{3}H]-6-NO_2-BP by liver microsomes obtained from untreated rats and rats pretreated with PB or MC.

Metabolite	Specific Activity[a] pmol/mg/min (% of total metabolite) Untreated	PB-treated	MC-treated
3-Hydroxy-6-NO_2-BP and 1-Hydroxy-6-NO_2-BP[b]	555 (54.5)	985 (61.2)	2086 (61.5)
6-NO_2-BP-3,9-hydroquinone	56 (5.5)	119 (7.4)	235 (6.9)
6-NO_2-BP-1,9-hydroquinone	20 (2.0)	42 (2.6)	83 (2.4)
BP-3,6-quinone	120 (11.8)	146 (9.1)	200 (5.9)
Total of Unknown	214 (21.0)	213 (13.2)	549 (16.2)
Materials in aqueous phase	53 (5.2)	105 (6.5)	242 (7.1)
% of Substrate metabolized	2.3	3.7	9.1

[a] Accuracy of these values is estimated to be ± 15% due to the loss of tritium by the hydroxylation and because hydroxylation proceeded through an NIH shift mechanism (2).

[b] The ratio of 1- and 3-hydroxy-6-NO_2-BP is approximately 1:4 based on the 500 MHz ^{1}H NMR spectral analysis of the metabolite mixture from unlabeled 6-NO_2-BP incubated with liver microsomes of rats pretreated with MC.

DISCUSSION

The results indicate that 6-NO_2-BP is aerobically metabolized by rat liver microsomes to give 3-hydroxy-6-NO_2-BP as the most abundant metabolite in all the three microsomal preparations (Table 1), and that 1- and 3-hydroxy-6-NO_2-BP are the primary metabolites which are further metabolized to give 6-NO_2-BP-1,3-hydroquinone and 6-NO_2-BP-3,9-hydroquinone. By comparison, 3-hydroxy-BP was the

with Scintisol (Isolabs) was measured by a liquid scintillation spectrometer.

RESULTS

The separation of 6-NO_2-BP metabolites by reversed-phase HPLC is shown in Figure 1A. Spectral analysis of each chromatographic peak indicated that metabolites in peaks a, b, c and e were 6-NO_2-BP-3,9-hydroquinone, 6-NO_2-BP-1,9-hydroquinone, BP-3,6-quinone, and 6-NO_2-BP, respectively. Peak d contained two metabolites which were well separated by normal-phase HPLC (Figure 1B) and were identified as 1-hydroxy-6-NO_2-BP (d-1) and 3-hydroxy-6-NO_2-BP (d-2)(3). The metabolites 1- and 3-hydroxy-6-NO_2-BP were subjected to further metabolism under

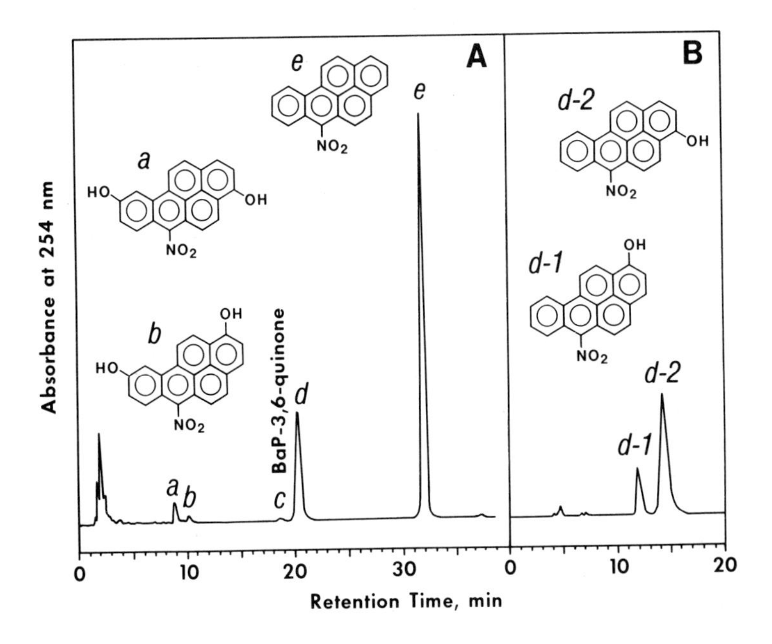

Figure 1. A. Reversed-phase HPLC separation of the organic solvent extractable metabolites. B. Normal-phase HPLC separation of meteabolites contained in peak d of Figure 1A.

predominant metabolite of BP metabolism, but substantial oxidation also occurred in the 4,5- and 7,8-regions (4). It appears, therefore, that 6-nitro substitution effectively blocked metabolism at regions *peri* (i.e., 4,5- and 7,8-) to it. This effect is presumably due to both steric and electronic constraints induced by the nitro substituent if the metabolizing enzymes approach the regions *peri* to the nitro substituent. Thus, the results indicate that the 6-nitro substituent can markedly affect the regioselectivity of the cytochrome P-450 containing drug metabolizing enzymes toward the metabolism of BP. In addition, while metabolism of 1- and 3-hydroxy-BP produced BP-1,6- and 3,6-quinone, respectively, as the predominant metabolites, metabolism of 1- and 3-hydroxy-6-NO_2-BP gave 6-NO_2-BP-1,9- and 3,9-hydroquinone.

Since 1- and 3-hydroxy-6-NO_2-BP have been shown to be more mutagenic than the substrate 6-NO_2-BP in the *Salmonella typhimurium*/microsome reversion assay, the ring oxidations are metabolic activation pathways (3). It is known that the metabolites responsible for most of the mutagenicity of BP are BP-4,5-epoxide and BP-7,8-dihydrodiol-9,10-epoxide (5). Thus, although both BP and 6-NO_2-BP are mutagenic, they have different metabolic activation pathways. For 6-NO_2-BP, it is possible that 1,9- and 3,9-hydroquinones are ultimate mutagens which may bind with cellular DNA through formation of reactive oxo radicals, species which have been suggested to be ultimate carcinogens (6).

ACKNOWLEDGEMENT

We thank L.E. Unruh for valuable technical assistance and L. Amspaugh for assistance in preparation of this manuscript.

REFERENCES

1. Gelboin, H.V. and Ts'o, P.O.P. (eds.)(1978) *Polycyclic Hydrocabons and Cancer, Vol 1, Environment, Chemistry and Metabolism*. Academic Press, New York.
2. Fu, P.P., Chou, M.W., Evans, F.E., and Yang, S.K. (1982) *Fed. Proc.*, *41*, 1730.
3. Fu, P.P., Chou, M.W., Yang, S.K., Beland, F.A., Kadlubar, F.F., Casciano, D.A., Heflich, R.H. and Evans, F.E. (1982) *Biochem. Biophys. Res. Commun. 105(3)*, 1037.
4. Holder, G., Yagi, H., Dansette, P., Jerina, D.M., Levin, W., Lu, A.Y.H., and Conney, A.H. (1974) *Proc. Natl. Acad. Sci. U.S.A.* *71*, 4356.
5. Bentley, P., Oesch, F., and Glatt, H. (1977) *Arch. Toxicol.* *39*, 65.
6. Lorentzen, R.J., Caspary, W.J., Lesko, S.A., and Ts'o, P.O.P. (1975) *Biochemistry*, *14*, 3970.

Cytochrome P-450, Biochemistry, Biophysics
and Environmental Implications, E. Hietanen,
M. Laitinen and O. Hänninen editors

CYTOCHROME P-450-DEPENDENT HYDROXYL RADICAL-MEDIATED DESTRUCTION OF DNA

HOLGER LUTHMAN+, ANN-LOUISE HAGBJÖRK AND MAGNUS INGELMAN-SUNDBERG
+Department of Biochemistry, Medical Nobel Institute and Department of Physiological Chemistry, Karolinska Institute, S-104 01 Stockholm, Sweden

INTRODUCTION

Recently, a new type of cytochrome P-450-dependent oxygenation mechanism was identified (1). In reconstituted membrane vesicles containing NADPH-cytochrome P-450 reductase and cytochrome P-450 LM2, ethanol was oxidized to acetaldehyde by hydroxyl radicals generated from the hemoprotein by an iron-catalyzed Haber-Weiss reaction between hydrogen peroxide and superoxide anions. The superoxide anions were liberated from the oxycytochrome P-450 complex and the hydrogen peroxide formed by the subsequent partial dismutation of the superoxide anions. Chronic alcohol feeding of rabbits results in the induction of cytochrome P-450-species in the liver specifically effective in ethanol oxidation (2,3). The ethanol-induced form of cytochrome P-450 oxidizes the alcohol at a rate 4-8-fold higher than the other cytochromes P-450 in a reaction that seems to be mediated by hydroxyl radicals (3).

Hydroxyl radicals are known to cause tissue damage during e.g. inflammation (4), X-irradiation (5) or after administration of cytotoxic agents (6). It was therefore of interest to investigate whether hydroxyl radical-producing species of cytochrome P-450 might mediate destruction of cellular macromolecules such as e.g. DNA.

MATERIALS AND METHODS

Lipids and enzymes were obtained from sources previously described (7). Membrane vesicles were prepared by the cholate gel filtration technique as described elsewhere (8) with the exception that 10 mM potassium phosphate buffer, pH 7.4, containing 0.2 mM EDTA and 50 mM NaCl was used instead of tris buffer (8).

Plasmid-DNA was prepared essentially as described by Birnboin and Doly (9). The plasmid pBR 322 (10) was grown in E. Coli HB 101 (11)in the presence of 3H-Thymidine. Cells were lyzed in alkali, plasmid DNA isolated and purified by neutral sucrose gradient centrifugation. The purity was checked by agarose gel electrophoresis. The DNA, consisting of a double stranded molecule 4362 base pairs in length, was linearized with Bam H 1. 0.5 µg of linear DNA (equal to about 4 000 cpm) were incubated with vesicles corresponding to 0.4 µM of cytochrome P-450 and 0.2 µM NADPH-cytochrome P-450 reductase in the phosphate buffer containing 400 µM NADPH at 37°C. The incubations (200 µl each) were terminated on ice and immediately applied on 5 ml linear alkaline sucrose gradients (5-20%) prepare in 0.7 M NaCl, 0.3 M NaOH and 5 mM EDTA. The tubes were centrifugated for 15 h at 150 000 x g at 4°C and 200 µl fractions were subsequently collected from the bottom and counted in an Intertechnique SL 32 liquid scintillation spectrometer with Aquoluma plus as scintillator liquid.

RESULTS AND DISCUSSION

Membrane vesicles containing cytochrome P-450 LM2 were mixed with 3H-labelled plasmid DNA in the presence of 0.4 mM NADPH and incubated for various time intervals at 37°C. The length of the DNA was subsequently determined by alkaline sucrose gradient centrifugation As shown in Fig 1, a time-dependent destruction of the nucleic acid occurred. After 30 min, 0.1 nmol of P-450 LM2 had degraded the DNA to a median molecular length of 400 nucleotides, as revealed by simultaneous centrifugation of radioactively labelled standard DNA molecules. This length of DNA corresponds to 20 breakages per molecule in the phosphodiester bonds of the plasmid DNA or totally 3.5 pmoles of DNA strand breakages during 30 minutes of incubation. The cytochrome P-450 LM2-dependent degradation of the DNA was effectively inhibited by the addition of catalase (10 U), superoxide dismutase (2 µg), mannitol (500mM) or cathecol (1 mM) indicating the involvement of hydrogen peroxide, superoxide anions and hydroxyl radicals in the process. In incubation mixtures where cytochrome P-450 LM2 was replaced by cytochrome b-5 in the membranes, the hydroxyl radical damage of DNA gave a median fragment length of 700 nucleotides as compared to a median length of 50 nucleotides when cytochrome P-450 LM2 was reacting under identical conditions

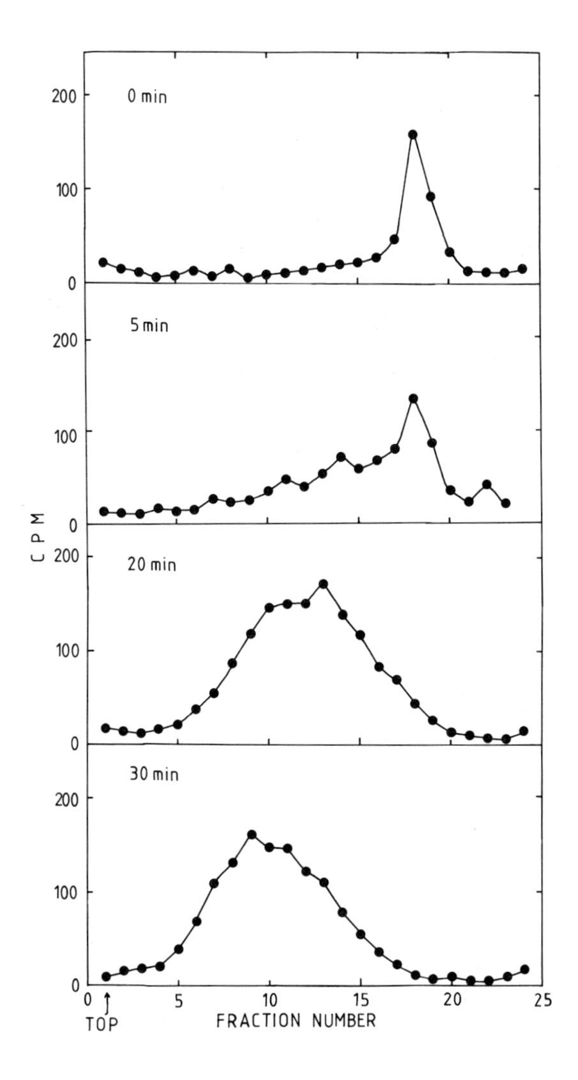

Fig. 1. Cytochrome P-450 LM2-dependent destruction of plasmid DNA as revealed by sucrose gradient centrifugation.

The finding of radical-producing forms of cytochrome P-450 utilizing the radicals for oxidation of e.g. ethanol or benzene, raises the question whether these radicals are long-lived and may contribute to cellular damage. The present investigation indicates that the radiclas are not scavenged within the active site of P-450, but may destroy macromolecules outside the membrane matrix.

ACKNOWLEDGEMENTS

This work was supported by grants from Jeanssons Stiftelser and from the Swedish Medical Research Council

REFERENCES

1. Ingelman-Sundberg, M. and Johansson, I. (1981) J. Biol. Chem. 256, 6321-6326

2. Koop, D.R. and Coon, M.J. (1982) in: Kato, R. and Sato, R. (Eds.), Microsomes, Drug Oxidations and Drug Toxicities, Japan Scientific Societies Press, Tokyo, in press

3. Ingelman-Sundberg, M., Ekström, G., Hagbjörk, A.-L., Terelius, Y. and Johansson, I. (1982) Proceedings of this conference

4. McCord, J. M. (1974) Science 185, 529-531

5. Samuni, A., Chevion, M., Halpern, Y. S., Ilan, Y. A. and Czapski, G, (1978) Radiat. Res. 75, 489-496

6. Cohen, G. and Heikkila, R. E. (1974) J. Biol. Chem. 249, 2447-2452

7. Ingelman-Sundberg, M., Haaparanta, T. and Rydström, J (1981) Biochemistry 20, 4100-4106

8. Ingelman-Sundberg, M. and Glaumann, H. (1980) Biochim. Biophys. Acta 599, 417-435

9. Birnboim, H.C. and Doly, J. (1979) Nucl. Acids Res. 7, 1513-1523

10. Bolivar, F., Rodriguez, R.L., Greene, P.J., Betlach, M.C., Heyneker, H.L., Boyer, H.W., Crosa, J.M. and Falkow, S. (1977) Gene 2, 95-113

11. Boyer, H. W. and Roulland-Dussoix, D. (1969) J. Mol. Biol. 41, 459-472

© 1982 Elsevier Biomedical Press B.V.
Cytochrome P-450, Biochemistry, Biophysics
and Environmental Implications, E. Hietanen,
M. Laitinen and O. Hänninen editors

ROLE OF RADICAL STAGES IN MICROSOMAL MONOOXYGENATIONS

VLADIMIR P.KURCHENKO, SERGEY A.USANOV AND DMITRIY I.METELITZA
Institute of Bioorganic Chemistry, BSSR Academy of Sciences,
Minsk, USSR

INTRODUCTION

Despite intensive study, there are considerable gaps in the knowlege of the mechanism of molecular oxygen activation by cytochrome P-450. The proposed mechanism of monooxygenation involves an active oxygen insertion into a substrate molecule by oxygenating species of hemeprotein, presumed to be similar to Compound I of peroxidase $Fe^{3+}O_2^{2-}$ or one of it's derivatives formed in the presence of ions $H^+ [Fe^{3+}O = Fe^{4+}O^-]$ (1,2). It was thought for a long time that the mechanisms of NADPH- and hydroperoxide-supported hydroxylations are very similar and involve the same active oxygen species $[Fe^{3+}O = Fe^{4+}O^-]$. However, recently the data have been obtained showing the differences in NADPH- and cumene hydroperoxide (CHP)-supported oxidation reactions of the same substrate by cytochrome P-450 (1,3,4). To clarify the differences in oxidation of the same substrates in NADPH- and hydroperoxide-dependent systems it is important to study the inhibition of oxidation reactions in both systems by scavengers of a radicals. The mode of inhibition and behavior of radical scavengers, undoubtedly, give a valid information about the role of radical stages during oxidation of different substrates in both systems.

In the present study, we evaluated the effect of 1-naphthol and mannitol on NADPH- and CHP-supported aniline oxidation and dimethylaniline (DMA) oxidative demethylation by liver microsomes and reconstituted system, consisting of cytochrome P-450 LM_2, NADPH-cytochrome P-450 reductase and phospholipids. Comparision of regularities of inhibition by 1-naphthol in all the systems studied enabled us to reveal similarities and differences between NADPH- and CHP-supported reactions of both substrates and also estimate the role of radical stages in these systems.

MATERIAL AND METHODS

Liver microsomes were prepared from phenobarbital-treated rabbits (5). Cytochrome P-450 LM_2 used in this study was purified from liver microsomes of PB-treated rabbits by the procedure (5) Protein concentration was determined by the method of Lowry (6) and that of cytochrome P-450 - according to Imai and Sato (7). Antibodies to highly purified cytochrome P-450 LM_2 (anti-P-450-LM_2) were prepared by immunisation of rats with antigen by the procedure described in (8). Antibodies obtained formed precipitates with cytochrome P-450 during Ouchterlony double diffusion analyses.

DMA demethylase and aniline hydroxylase activities were assayed as previously described (8) in 0.05 M Tris-HCl buffer pH 7.5 at 30°C. Reaction mixture total volume 1 ml contained 1 mM NADPH or CHP, microsomal protein (1 mg/ml) or cytochrome P-450 LM_2 and different concentrations of DMA, aniline and 1-naphthol, mannitol and anti-P-450 LM_2. In the case of inhibition by antibodies, cytochrome P-450 was preincubated 15 min in reaction mixture containing anti-P-450 LM_2. The reaction time for NADPH-supported aniline oxidation was 25 min and the reaction was characterized by the p-aminophenol produced (9). DMA oxidation was performed 6 min in NADPH-supported reaction and 1.5 min in CHP-supported reaction and was characterized by formaldehyde determination following the method (10).

RESULTS AND DISCUSSION

Mannitol added up to 0.4 mM does not inhibit the DMA (A) and aniline (B) oxidation (Fig.1). However, 1-naphthol causes effective inhibition of both substrates oxidation being used at concentration lower then 0.1 mM. As shown in Fig. 2, 1-naphthol causes effective inhibition of DMA oxidation in the system "cytochrome P-450 LM_2 and CHP", and is not so effective in three other systems: "microsomes + CHP", "microsomes + NADPH" and in reconstituted system. It is very important the mode of inhibiting effect of 1-naphthol. Despite the system used, the character of 1-naphthol inhibiting action is the same. The mixed type of inhibition is observed: an increase in inhibitor concentration results in changes of estimatited maximal rate V_{max} and apparent K_m values (Fig.3).

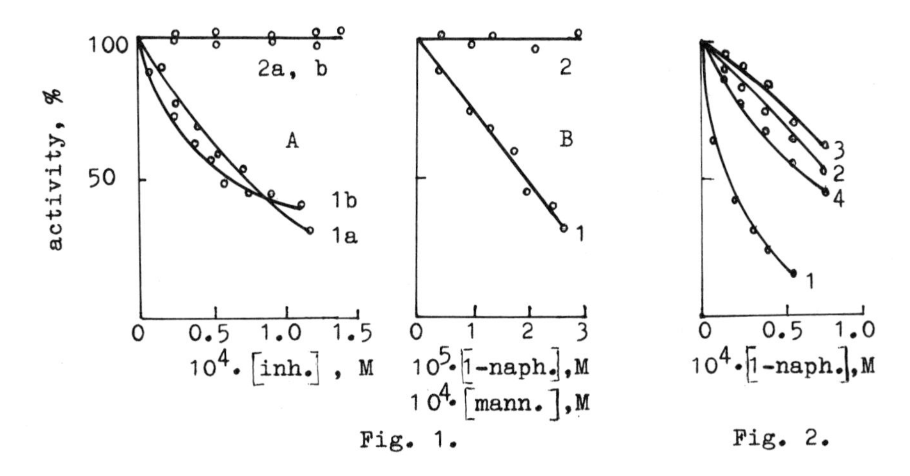

Fig. 1. Fig. 2.

Fig. 1. Inhibition of DMA (A) and aniline (B) oxidation by 1-naphthol (1) and mannitol (2): A – "microsomes + CHP" (a) and "microsomes + NADPH" (b), 3mM DMA; B – "cytochrome P-450 LM_2 + CHP", 30 mM aniline.

Fig. 2. Inhibition of DMA (3 mM) oxidation by 1-naphthol: 1 – cytochrome P-450 LM_2 (0.8 μM) + CHP (1 mM); 2 – "microsomes + CHP (1 mM); 3 – reconstituted system: cytochrome P-450 (0.8 μM), NADPH-cytochrome P-450 reductase (0.16 μM); 4 – microsomes + NADPH (1 mM).

However, the 1-naphthol concentrations which strongly inhibit DMA oxidation are much lower than the substrate concentrations used.

As shown in Fig. 4, anti-P-450 LM_2 inhibits the activity of cytochrome P-450 in both systems used, the inhibition being practically full in the case of NADPH-supported DMA oxidation by liver microsomes. However, in CHP-dependent reaction antibodies to cytochrome P-450 cause only a partial inhibition. When using both, anti-P-450 LM_2 and 1-naphthol cause more effective inhibition of CHP-supported DMA oxidation. We studied also the effect of anti-P-450 LM_2 and 1-naphthol on aniline oxidation in the sys tem "cytochrome P-450 – CHP" and DMA oxidation by liver microsomes in the NADPH- and CHP-supported oxidations. In all the systems studied a combined effect of both inhibitors is much higher than in separate inhibition experiments. However, the sum effect of both inhibitors is always lower than the additive one. Non-additivity of inhibitor effect was observed in all the systems studied for both substrates. The absence of additivity for effect of both inhibitors on DMA oxidation appears to be connected with a non-productive consumption of the inhibitor due to it's absorption on anti-cytochrome P-450 LM_2 having a relatively large molecular weight and active surface.

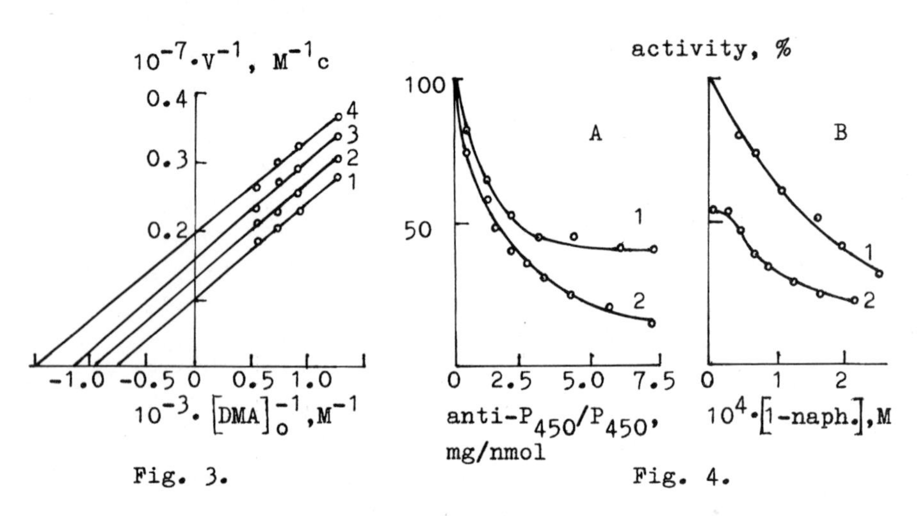

Fig. 3.

Fig. 4.

Fig. 3. Effect of 1-naphthol on DMA oxidation in the reconstituted system: cytochrome P-450 (0.8 µM), NADPH-cytochrome P-450 reductase (0.16 µM), NADPH (1 mM) and DMA (3 mM). 1 – without inhibitor, 2 – 0.4; 3 – 0.5; 4 – 0.6 uM of 1-naphthol.

Fig. 4. Inhibition of DMA oxidation by anti-P-450 LM_2 (a) and anti-P-450 LM_2 + 1-naphthol (b):a)1 – "microsomes and CHP", 2 – "microsomes + NADPH", b) 1 – "microsomes + CHP" in the presence of 1-naphthol, 2 – the same as in b, but 1-naphthol and anti-P-450 present.

Moreover, when interpreting our data it should be born in mind that 1-naphthol is oxidized by cytochrome P-450 (11) and it interacts with this hemeprotein giving rise to type I difference spectrum. A comparison of dissociation constants for the interac tion of aniline (0.14 mM) and 1-naphthol (0.6 mM) with microsomal cytochrome P-450 permits to conclude, that these compounds can compete for binding with cytochrome P-450. Concentrations of 1-naphthol used for inhibitory studies were usually much lower than those of both substrates.

A mixed type inhibition of both substrates oxidation by 1-naph thol testifies to the effect of inhibitor on different stages of oxidation process. What are these stages ? As mentioned above, this is first of all the competition of substrates and inhibitor for binding with the purified or microsomal cytochrome P-450. However, under conditions used, due to high substrate and inhibitor concentration differences it is difficult to assume that this the case, but this possibility could'nt exclude. Second, the inhibitor can compete with a substrate for active intermediate species, involved in the oxidation process. These active species can involve such active oxygenating species as $[Fe^{3+}O = Fe^{4+}O^{-}]$, transfering oxygen into the substrate molecule, or radicals formed during NADPH- or CHP-supported oxidations in microsomal or shortened system.

The possibility that 1-naphthol can compete with CHP for cytochrome P-450 binding during the CHP-supported oxidation appears to be explain different efficiency of this inhibitor in both systems, being more effective in hydroperoxide-supported oxidation reactions. Since the radical species are involved in microsomal oxidations it is difficult to exclude the possibility of 1-naphthol interaction with these radicals.

It is well known that radicals formed from CHP – $RO^{\bullet}$ and $RO_2^{\bullet}$ – can destroy cytochrome P-450 (2). If so, interacting with these radicals, 1-naphthol have to protect cytochrome P-450 from destruction, thus resulting in increase of oxidation rate. Indeed, when added to the CHP-dependent system, 1-naphthol causes a protective effect (2). However, exerting a protective effect on cytochrome P-450, 1-naphthol inhibits the DMA and aniline CHP-supported oxidations. This means, that 1-naphthol interacts with radicals, which are released in the volume and attack the substrate molecule. It should be emphasized, that the radicals formed in NADPH- and CHP-supported systems can differ significantly. There is a direct evidence for $O_2^{\bullet -}$ generation during microsomal NADPH-supported oxidation. Hydroperoxide and $O_2^{\bullet -}$ results in hydroxyl radical formation according to:

$$O_2^{\bullet -} + H_2O_2 \longrightarrow O_2 + HO^{\bullet} + HO^{-} \qquad (1)$$

Moreover, $HO^{\bullet}$ radical can be produced from H_2O_2 in the presence of reduced cytochrome P-450:

$$Fe^{2+} + H_2O_2 \longrightarrow Fe^{3+} + HO^{\bullet} + HO^{-} \qquad (2)$$

The last reaction can be catalyzed also by cytochrome P-420, practically always being present in liver microsomes. Oxidized forms of hemeproteins generate peroxy radical $HO_2^{\bullet}$:

$$Fe^{3+} + H_2O_2 \longrightarrow HO_2^{\bullet} + H^+ + Fe^{2+} \quad (3)$$

According to the reactions mentioned (1-3), relatively high stationary concentrations of $HO^{\bullet}$ radicals are present in liver microsomes. It is well known, that $HO^{\bullet}$ radical can readily oxidize many aromatic compounds including DMA and aniline. In addition, this radical effectively interacts with several scavenging agents namely, mannitol, 1-naphthol and so on. Therefore, if $HO^{\bullet}$ radical participates in DMA and aniline oxidation in NADPH- and CHP-supported systems, then the inhibitory effect of 1-naphthol on both reactions is understandable. However, the failure of mannitol to inhibit oxidation of the same substrates is difficult to explain.

In the hydroperoxide-supported systems according to reactions (2) and (3) radicals $RO^{\bullet}$ and $RO_2^{\bullet}$ are formed from the hydroperoxide present (CHP for example). One of the ways resulting in $HO^{\bullet}$ radical generation in hydroperoxide-dependent systems may be shown by the following reaction:

$$-\bar{S}-Fe^{3+}-P_{450} + ROOH \longrightarrow -\dot{S}-Fe^{3+}-P_{450} + HO^{\bullet} + RO^{-} \quad (4)$$

It should be emphasized that in this case the thiolate ligand of oxidized cytochrome P-450 plays a very important role, being an electron donor for the hydroperoxide molecule reduction. Reaction (4) appears to result in destruction cytochrome P-450 due to thiolate ligand oxidation, following of the formation by cytochrome P-420. If the radicals are formed in CHP-supported systems, then DMA and aniline can be readily oxidized by $RO^{\bullet}$ and $RO_2^{\bullet}$ radicals according to a sheme analogous to the one proposed by (4):

$$-N\begin{matrix}CH_3\\CH_3\end{matrix} \xrightarrow{RO_2^{\bullet}\ (RO^{\bullet})} -\overset{\bullet +}{N}\begin{matrix}CH_3\\CH_3\end{matrix} \xrightarrow[-H^{\bullet}]{RO^{\bullet}} -\overset{+}{N}\begin{matrix}=CH_2\\CH_3\end{matrix} \xrightarrow{H_2O} -N\begin{matrix}H\\CH_3\end{matrix} + H_2C{=}O \quad (5)$$

According to this sheme two radicals $RO^{\bullet}$ (or $RO_2^{\bullet}$) are consumed during oxidative demethylıtion of one DMA molecule. If this reaction takes place in the case of CHP-supported DMA oxidation, then the inhibitory effect of 1-naphthol on this reaction is connected with competition of 1-naphthol for $RO^{\bullet}$ and $RO_2^{\bullet}$ radicals. However, it is difficult to explain the aniline oxidation by $RO^{\bullet}$ and $RO_2^{\bullet}$ radicals, since there is no direct way leading to a p-aminophenol formation. Although, p-aminophenol is formed readily from aniline in the presence of $HO^{\bullet}$ radicals:

$$H_2N-C_6H_5 + HO^{\bullet} \longrightarrow H_2N\dot{C}_6H_5(H)(OH) \xrightarrow{HO^{\bullet}} H_2N-C_6H_4-OH + H_2O$$

The failure of antibodies to inhibit cytochrome P-450 catalyzed oxidations totally is consistent with the fact that unreacted cytochrome P-450 is active in DMA oxidation and the reaction is inhibited only in the presence of other inhibitor, 1-naphthol, according to mechanisms discribed above.

Based on the data presented one could not exclude the important role of radical stages in DMA and aniline oxidation in the NADPH- and CHP-supported reactions. This is further confirmed by the mixed type of 1-naphthol inhibition of both substrate oxidation. Along with competing with substrates for binding cytochrome P-450 1-naphthol also plays a role of radicals scavenger. However, mannitol, the other $HO^{\bullet}$ radical trapping agent, does not inhibit the oxidation reactions due to the absence of affinity to cytochrome P-450. Therefore, mannitol is far from the radical-generating center of cytochrome P-450 molecule and does not compete for radicals with substrates, which are near this center. At the same time, having high affinity for cytochrome P-450 binding, 1-naphthol compete with DMA and aniline for radicals.

The nature of radicals in NADPH- and CHP-supported reactions differs strongly. However, the same type of inhibition caused by 1-naphthol in both systems testifies that the inhibitor effectively interacts with radicals formed in NADPH- and CHP-dependent reactions.

Finally, our and literature data (1-4) allow to make a conclusion about possible role of radicals during microsomal oxidation of some substrates, that, however, does not exclude the possibili ty of these substrates oxidation through the oxenoid mechanism.

It is known that some substrates can be simultaneously oxidized by two pathways: oxenoid or two electron transfer mechanism and radical or one electron transfer mechanism. Ternary amines presum ably are dealkylated according reaction 5, i.e. through one electron oxidation. However, benzo(a)pyrene and probably aniline are oxidized simultaniously through both 1 e and 2 e transfer pathways. That the amines (DMA, aminopyrine and others) can be readily oxidized through the 1 e mechanism, means that the nonenzymatic processes can contribute to total oxidation of these compounds and mask enzymatic stages of the process. Thus, it explains the anomalously high turnover, obtained for oxidation of compounds of this type.

REFERENCES

1. White,R.E. and Coon,M.J. Ann.Rev.Biochem.,1980, 49, 315.
2. Metelitza,D.I. (1981) Uspekhi khimii, 50, 2019.
3. Capdevila,J., Estabrook,R.W. and Prough,R.A. (1980) Arch.Biochem.Biophys., 200, 186.
4. Griffin,B.W., Marth,Ch., Yasukochi,Yu. and Masters,B.S. (1980) Arch.Biochem.Biophys., 204, 397.
5. Eryomin,A.N., Usanov,S.A., Metelitza,D.I. and Akhrem,A.A. (1980) Bioorg.Chem.,Russ., 6, 757.
6. Lowry,O.H., Rosebrough,N.J., Farr,A.L. and Randall,P.J. (1951) J.Biol.Chem., 193, 265.
7. Omura,T. and Sato,R. (1964) J.Biol.Chem., 239, 2370.
8. Kurchenko,V.P., Usanov,S.A. and Metelitza,D.I. (1981) Biokhimiya, 46, 2202.
9. Fujita,T. and Mannering,G. (1973) J.Biol.Chem., 248, 8150.
10. Nash,T. (1953) Biochemical J., 55, 416.
11. Metelitza,D.I. and Popova,E.M. (1980) Biokhimiya, 45, 1379.

Cytochrome P-450, Biochemistry, Biophysics and Environmental Implications, E. Hietanen, M. Laitinen and O. Hänninen editors

INCORPORATION OF THE MONOOXYGENASE SYSTEM INTO LIPOSOMES OF DIMYRISTOYLPHOSPHATIDYLCHOLINE

G.SMETTAN, P.A.KISELEV*, M.A.KISEL*, A.A.AKHREM* and K.RUCKPAUL
Academy of Sciences,Central Institute of Molecular Biology
1115 Berlin-Buch,GDR; * Academy of Sciences,Institute of Bioorganic Chemistry, 220600 Minsk, Belorussian SSR, USSR

INTRODUCTION

The catalytic activity of the endoplasmic monooxygenatic system of mammalian liver is determined by interactions between the essential components: cytochrom P-450, NADPH dependent P-450 reductase and phospholipid. Due to its membrane bound character protein/phospholipid interactions are considered of special importance for its activity. The reconstitution of the system in liposomes of defined composition offers a useful approach to study the mutual dependences of the components. The analysis of structure/function relationships requires: (i) a defined bilayer structure of the liposomes, (ii) a complete incorporation of the protein components, (iii) distinct phase transition, (iv) and the absence of oxidation products in the membrane. Liposomes of dimyristoylphosphatidylcholine (DMPC) accomplish these requirements. The present study is aimed to analyse (1) which way the structure of the components is changed after incorporation into the membrane which is reflected in corresponding functional properties and (2) if and how the physical state of the membrane does influence the monooxygenatic system. Therefore the temperature dependence of the substrate binding to and the reduction reaction of P-450 LM2 in DMPC liposomes were studied.

METHODS AND MATERIAL

Reconstitution was achieved either by dialysis or by the gel filtration technique /1/ both in the presence of cholate. The degree of incorporation has been determined by use of DMPC labeled with ^{14}C-dimyristic acid after separating the fractions on a sepharose 4B column and analysing the fractions. In table 1 the parameters obtained from laser light scattering and electron microscopy are summarized. A molar stoichiometry phospholipid:P-450 LM2: reductase in all experiments was adjusted to 400:1:0.2. The incorporation of the proteins leads to an increase of the radius of the respective liposomes similar to published data for P-450$_{scc}$ /2/. That functionally active complexes are formed by the applied gel filtration technique is evidenced by the finding that external reductase is capable to reduce proteoliposomes with P-450 LM2 only at a rather small extent (40 %) and with a low rate constant ($< 3\cdot10^{-3}s^{-1}$). Extent (30 %) and rate are further decreased at incorporation of P-450 LM2 and reductase in separate liposomes ($< 1\cdot10^{-3}s^{-1}$). That means that obviously an exchange between different liposomes does occur only at very low rate and that an effective incorporation of P-450 and reductase can be achieved only at simultaneous incubation.

The kinetic data were evaluated using a multiexponential nonlinear estimation procedure /3/.

TABLE 1

STRUCTURAL PARAMETERS OF DMPC-LIPOSOMES AND PROTEOLIPOSOMES

Technique	Liposomes	Reconstitution %	Diameter /nm/ electron microscopy	Diameter /nm/ laser light scattering
Gelfiltration	without protein	-	15 - 40	37.4
(Na-cholate)	+ P-450 LM2	98	80 - 110	101.0
	+ reductase	95	40 - 60	53.4
	+ P-450 LM2, reductase	100	10 - 140	138.6

RESULTS AND DISCUSSION

The affinity of P-450 LM2 towards benzphetamine after incorporation into DMPC liposomes is significantly enhanced /K_s(aqueous solution) 130 µM; K_s(DMPC liposomes) 20 µM/. The increased affinity is paralleled by an increased rate of benzphetamine binding /k' (aqueous solution) $1\cdot10^5M^{-1}s^{-1}$; k' (DMPC liposomes $>5\cdot10^5M^{-1}s^{-1}$/. The activity of NADPH-P-450 reductase towards cytochrome c in the presence of DMPC was found twice that value as in aqueous solution.

In a further series of experiments the temperature dependence of the NADPH supported reduction was compared with that of dithionite reduction. By dithionite liposomal P-450 LM2 (DMPC) in the absence of substrate is reduced in a monophasic reaction without break 8-10 fold more rapidly than in aqueous solution. Dithionite reduction of the complete system (monophasic) leads to the appearance of breaks and a substrate induced rate increase.

The rate constants of the NADPH reduction in the DMPC reconstituted system are lowered about 1 order of magnitude without substrate at 20 °C as compared to the rate of the substrate bound system (Table 2). Obviously the bound substrate favours interactions between both proteins possibly via phospholipids. In contrast to microsomal P-450 (rats)/4/ which exhibits a break in the temperature dependence of the reduction only in the slow phase a break at 20 °C in the respective reaction in DMPC liposomes is observed in the fast and in the slow phase as well (Fig. 1). This break in the Arrhenius plot is in good agreement with the microcalorimetrically determined phase transition (T_c) of DMPC liposomes /5/. This result together with the temperature dependence of the dithionite reduction of P-450 LM2 evidence that phospholipid mediated interactions between reductase and their membraneous organization are reflected in the break /4/.

TABLE 2

SUBSTRATE SPECIFICITY OF THE NADPH-REDUCTION OF PROTEOLIPOSOMES

Rate constant (s^{-1})	- Benzphetamine			+ Benzphetamine		
	14.0°C	20.0°C	33.0°C	14.5°C	20.0°C	33.0°C
k_{fast}	0.0046	0.029	0.033	0.012	0.29	0.26
k_{slow}	-	0.0013	-	0.0015	0.017	0.024

Irrespective of the substrate the temperature dependence differs about one order of magnitude in dependence on the phase transition more pronounced in the fast phase. From the Arrhenius plot activation energies have been calculated which amount for the fast phase in the presence of benzphetamine to 21.8 kcal/Mol and 2.3 kcal/Mol below and above T_c, respectively. In the absence of

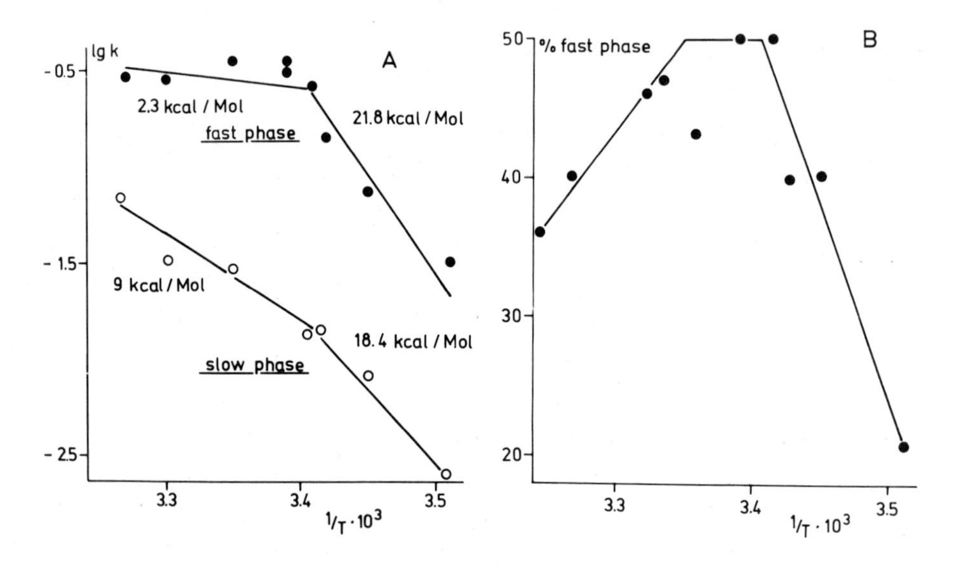

FIG. 1. Temperature dependence of the NADPH dependent reduction of P-450 LM2 reconstituted into DMPC liposomes in the presence of 1 mM benzphetamine, pH 7.4.
A: k_{fast} ●——● ; k_{slow} ○——○ ; B: phase distribution

benzphetamine the respective values are lowered to 11.5 kcal/Mol and 1.0 kcal/Mol. (i) The low activation energy of the fast phase above T_c may be explained considering phospholipid mediated interactions above the phase transition point. (ii) the decrease of the fast phase content is explicable assuming weakened interactions decisive for the formation of clusterlike structures (Fig. 1 B).

REFERENCES

1. Ingelman-Sundberg, M. and Glaumann, H. (1977) FEBS Lett. 78, 72-76

2. Yamakura, F., Kido, T. and Kimura, T. (1981) Biochim. Biophys. Acta, 649, 343-354

3. Provencher, S.W. (1976) J. Chem., Phys., 64, 2772-2777

4. Peterson, J.A., Ebel, R.E., O'Keefe, D.H., Matsubara, T. and Estabrook, R.W. (1976) J. Biol. Chem., 254, 4010-4016

5. Kiselev, P.A. et al. (1982) Biochim. Biophys. Acta (in press)

© 1982 Elsevier Biomedical Press B.V.
Cytochrome P-450, Biochemistry, Biophysics
and Environmental Implications, E. Hietanen,
M. Laitinen and O. Hänninen editors

BINDING AFFINITY OF P_1-450 INDUCERS FOR THE *Ah* RECEPTOR DOES NOT CORRELATE WITH THE INHIBITION OF EPIDERMAL GROWTH FACTOR (EGF) BINDING TO CELL-SURFACE RECEPTORS

SIRPA O. KÄRENLAMPI, HOWARD J. EISEN AND DANIEL W. NEBERT
Developmental Pharmacology Branch, National Institute of Child Health and Human Development, National Institutes of Health, Bethesda, MD 20205 (U.S.A.)

INTRODUCTION

The *Ah* locus (Fig. 1) governs the induction of numerous drug-metabolizing enzymes and other proteins by polycyclic aromatic compounds such as 3-methylcholanthrene (MeChol) and 2,3,7,8-tetrachlorodibenzo-*p*-dioxin (TCDD). The cytosolic *Ah* receptor is the gene product of the *Ah* regulatory gene (*reviewed in* Ref. 2). In addition to the multiple forms of P-450 induced by MeChol or TCDD--UDP glucuronosyltransferase, DT diaphorase, and ornithine decarboxylase induced activities have been shown to be closely associated with the Ah^b allele and therefore with the presence of detectable levels of cytosolic *Ah*

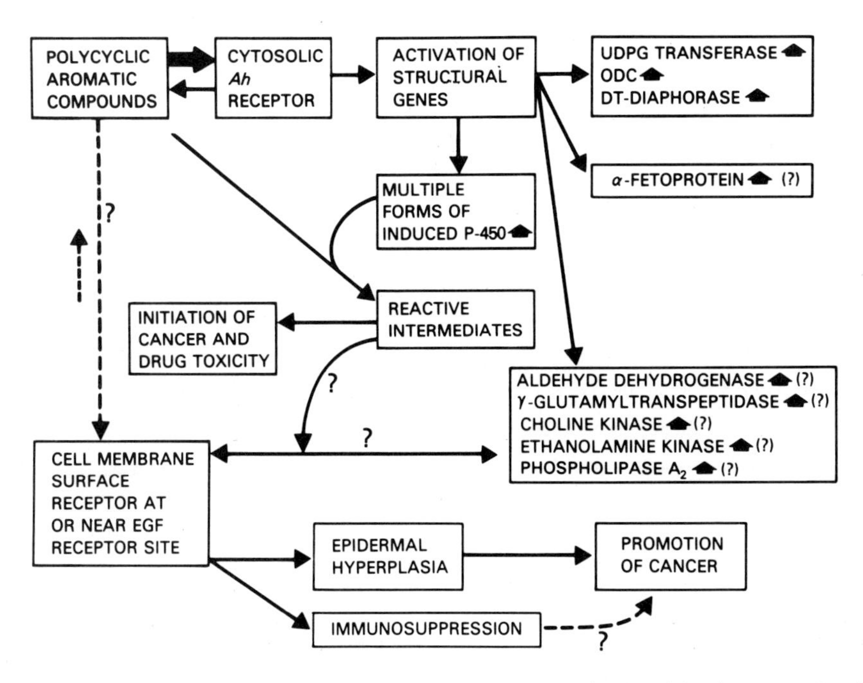

Fig. 1. Heuristic diagram showing possible interrelationship between the *Ah* receptor and the multiple responses (1). *Question marks* denote that no rigid correlation (among offspring from the $B6D2F_1$ x D2 backcross) has been determined yet.

receptor. These studies were carried out with individual MeChol-treated progeny from the (C57BL/6N)(DBA/2N)F_1 x DBA/2N (B6D2F_1 x D2) backcross. Inducers known to bind to the *Ah* receptor also enhance α-fetoprotein levels (3), aldehyde dehydrogenase (4), γ-glutamyltranspeptidase (5), choline kinase (6), ethanolamine kinase (6), and phospholipase A_2 (7) activities; most of these studies were performed in rats, however, and therefore no strict association with the murine *Ah* locus has yet been demonstrated.

It has been proposed (8, 9) that these polycyclic aromatic inducers compete with epidermal growth factor (EGF) for the EGF cell-surface receptor in the same order as that seen for compounds competing with [^{3}H]TCDD for the cytosolic *Ah* receptor. These inducers also cause epidermal keratinization (10), birth defects (11, 12), immunosuppression (*reviewed in* Ref. 2), and "promoter" effects of chemical carcinogenesis (13). The *Ah* receptor and the P_1-450 induction process occur very early in gestation--even before implantation of the mouse embryo (14, 15). All these data (Fig. 1) therefore suggest that the *Ah* system may be involved in certain growth processes such as differentiation and cancer promotion--in addition to the previously discovered phenomenon of induction of drug-metabolizing enzymes.

We chose to confirm or disprove the hypothesis of Ivanovic and Weinstein (8) by studying the five chemicals shown in Fig. 2. Large differences among these five chemicals exist in their capacity to displace [^{3}H]TCDD from the *Ah* receptor and their potency to induce P_1-450.

RESULTS AND DISCUSSION

The sources for all materials and details of all experients can be found in Refs. 17 and 18. When mouse hepatoma (Hepa-lclc7) cell cultures were exposed at room temperature to [^{125}I]EGF, maximal levels of binding were reached in about 2 h. Similar kinetics results were obtained with [^{125}I]insulin and [^{3}H]phorbol 12,13-dibutyrate binding to their respective cell-surface receptors. Beforehand we had determined for each of these three radioligands the concentrations at which saturability occurred with each of their specific cell-surface receptors (18).

We found that all five chemicals shown in Fig. 2 exerted their maximal effect on EGF binding after 24 h of incubation at 37 °C. Phorbol 12-myristate-13-acetate, on the other hand, required only 30 min of incubation at 37 °C to block EGF binding (18). Clearly the mechanisms by which polycyclic hydrocarbons and phorbol 12-myristate-13-acetate affect EGF binding to EGF cell-surface receptors are different. Following a 24-h incubation of Hepa-

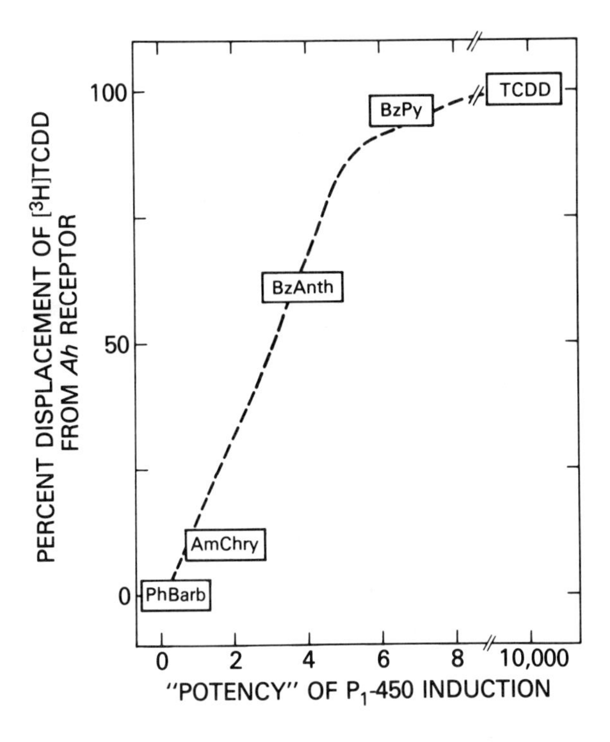

Fig. 2. Relationship between chemicals binding to the *Ah* receptor and their capacity to induce P_1-450. Values on the *ordinate* are taken from Ref. 16. Values on the *abscissa* (in arbitrary numbers) are taken from various sources in which cells in culture and liver from the intact mouse or rat were studied. PhBarb, sodium phenobarbital; AmChry, 6-aminochrysene; BzAnth, benzo[a]anthracene; BzPy, benzo[a]pyrene; TCDD, 2,3,7,8-tetrachlorodibenzo-*p*-dioxin.

1c1c7 cells with TCDD, benzo[a]pyrene, benzo[a]anthracene, 6-aminochrysene, or phenobarbital, we found no obvious rank order between their potency to displace [^{3}H]TCDD from the *Ah* receptor and their capacity to inhibit EGF binding to the cell-surface EGF receptor. These five chemicals had little effect on insulin or phorbol 12,13-dibutyrate binding in similar experiments (18).

The preceding data are consistent with toxic metabolites, rather than the parent compounds, causing the decrease in EGF binding. We chose to test this hypothesis with Hepa-1c1c7 mutant clones c2 and c4, which have decreased, and no detectable, aryl hydrocarbon hydroxylase activity, respectively (17). Benzo[a]pyrene and benzo[a]anthracene, incubated with clones c2 and c4 for 24 h at 37 °C, had little inhibitory effect on EGF binding. Incubation of the very toxic metabolite benzo[a]pyrene *trans*-7,8-diol-9,10-epoxide directly with clone c4 cells caused marked decreases in EGF binding to the EGF receptor (18).

SUMMARY

The rank order of polycyclic aromatic compounds binding to the *Ah* receptor (TCDD >> benzo[a]pyrene > benzo[a]anthracene > 6-aminochrysene > phenobarbital) is *not* correlated with the inhibitory effects of these five chemicals on EGF binding to EGF cell-surface receptors. Metabolism of these chemicals, by forms of P-450 induced during the 24-h incubation of Hepa-1c1c7 cells with these inducers, appears to play an important role in blocking EGF binding. Toxic metabolites may prevent new EGF receptors from appearing on the cell-surface. The gradual decline in EGF binding capacity during the 24-h incubation with a polycyclic aromatic compound thus may reflect EGF receptor degradation in the absence of new receptor synthesis.

REFERENCES

1. Tukey, R.H., Nakamura, M., Chen, Y.-T., Negishi, M. and Nebert, D.W. (1982) in: Sato, R. and Kato, R. (Eds.), Microsomes, Drug Oxidations, and Drug Toxicity, Japan Scientific Societies Press, Tokyo, in press.
2. Nebert, D.W., Negishi, M., Lang, M.A., Hjelmeland, L.M. and Eisen, H.J. (1982) Advanc. Genet., 21, 1.
3. Becker, F.F. and Sell, S. (1979) Cancer Res., 39, 3491.
4. Deitrich, R.A., Bludeau, P., Roper, M. and Schmuck, J. (1978) Biochem. Pharmacol., 27, 2343.
5. Gupta, B.N., McConnell, E.E., Harris, M.W. and Moore, J.A. (1981) Toxicol. Appl. Pharmacol., 57, 99.
6. Ishidate, K., Tsuruoka, M. and Nakazawa, Y. (1980) Biochim. Biophys. Acta, 620, 49.
7. Bresnick, E., Bailey, G., Bonney, R.J. and Wightman, P. (1981) Carcinogenesis, 2, 1119.
8. Ivanovic, V. and Weinstein, I.B. (1981) J. Supramol. Struct. Cell. Biochem., Suppl. 5, 232 [Abstract].
9. Ivanovic, V. and Weinstein, I.B. (1981) Nature, 293, 404.
10. Knutson, J.C. and Poland, A. (1980) Cell, 22, 27.
11. Shum, S., Jensen, N.M. and Nebert, D.W. (1979) Teratology, 20, 365.
12. Poland, A. and Glover, E. (1980) Mol. Pharmacol., 17, 86.
13. Pitot, H.C., Goldsworthy, T., Campbell, H.A. and Poland, A. (1980) Cancer Res., 40, 3616.
14. Galloway, S.M., Perry, P.E., Meneses, J., Nebert, D.W. and Pedersen, R.A. (1980) Proc. Natl. Acad. Sci. U.S.A., 77, 3524.
15. Filler, R. and Lew, K.J. (1981) Proc. Natl. Acad. Sci. U.S.A., 78, 6991.
16. Bigelow, S.W. and Nebert, D.W. (1982) Toxicol. Lett., 10, 109.
17. Legraverend, C., Hannah, R.R., Eisen, H.J., Owens, I.S., Nebert, D.W. and Hankinson, O. (1982) J. Biol. Chem., 257, 6402.
18. Kärenlampi, S.O., Eisen, H.J., Hankinson, O. and Nebert, D.W., manuscript submitted for publication.

Cytochrome P-450, Biochemistry, Biophysics
and Environmental Implications, E. Hietanen,
M. Laitinen and O. Hänninen editors

CYTOCHROME P-450 - AND MODEL SYSTEMS - DEPENDENT HYDROXYLATION OF A MULTIFUNCTIONAL SUBSTRATE BY VARIOUS OXIDANTS.

J. LECLAIRE, P.M. DANSETTE, D. FORSTMEYER and D. MANSUY
Laboratoire de Chimie, Ecole Normale Supérieure, 75231 Paris 05, France.

Oxidation of a multifunctional substrate by various forms of microsomal Cytochrome P-450 in presence of various oxidants was compared to that obtained by a ferroporphyrin based chemical oxidation system. The product pattern brings informations on the "active oxygen" species which are involved in each system and allows to test various possible mechanisms.

INTRODUCTION

The nature of the "active oxygen complexes" formed from the various forms of cytochrome P-450 with various oxidants is still incompletely known (1). When we wish to caracterize the diverse reactivities of the various cytochromes P-450 by looking at the metabolization of polyfunctional substrates, it is difficult to differenciate between the effects due to the intrinsic reactivity of the "active oxygen complex" and the proximity and positional effects due to the binding of the substrate to the active site. On the contrary, these affinity factors do not exist when the oxidation is realized with a chemical model system (2). The aim of the present study is to compare the oxidation of a multifunctional substrate by various "active oxygen complexes" obtained from various cytochromes P-450 or model systems. This approach has already been taken for some substrates (3, 4). We have decided to build a polyfunctional substrate of high hydrophobicity, with different oxidizable centers, in particular an olefinic bond, an aromatic ring and different kinds of C-H bonds. Since cytochrome P-450 has been proposed to hydroxylate C-H bond (5) and olefinic bonds (6) by the way of radical species, we wanted at the same time to test such an hypothesis with a substrate able to form a radical species of known evolution as the cyclisation of 5-hexenyl radical to cyclopentane derivatives (7). This finally directed our choice to 6-phenoxy-1-hexen (I). This molecule has a high conformational mobility and should have high affinity for cytochrome P-450 and be a good substrate with different specificities towards the different cytochrome P-450 isoenzymes.

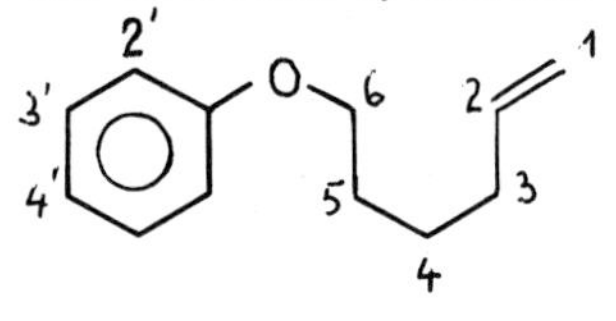

2' 3' 4' O 6 5 4 3 2 1 (I)

MATERIALS AND METHODS

Microsomes. Hepatic microsomes were prepared from livers of male Sprague-Dawley rats (220-250 g) that had been pretreated I.P. with sodium phenobarbital (PB) (80 mg/kg in saline, 3 days) and with 3-methylcholanthrene (3MC) (20 mg/kg in corn oil, 3 days).
Incubations. The incubations were performed at 37°C for 0 to 5 mn in a final volume of 1 ml phosphate buffer 0.1 M pH 7.4 containing 1 mg microsomal cytochrome P-450 and 0.5 mM substrate. The reaction was started by adding NADPH generating system (1 μmole NADP, 10 μmole G6P, 5 μmole $MgCl_2$, and 2 UI G6PDH), or oxidant (1 μmole ØIO or 1 μmole CuOOH in MeOH). The products were extracted with CH_2Cl_2 and analyzed by fused silica capillary column gas chromatography (CPSil 5,10 m x 0.32 mm ; 120°C to 250°C ; 4°C/mm) or by GC/MS with a Nermag R-10-10 system.
Oxidation models were made as described in table at a final volume of 100 μl in benzene.
6-Phenoxy-1-hexen. 6-bromo-1-hexen (Fluka) was reacted with excess of sodium phenate in DMSO (24 h reflux), extracted into hexane and distilled (Rdt 75%, Bp^{10} = 105°C). Authentics of the principal oxidation products were synthetized by classical methods and gave correct analytical data.

RESULTS

The interaction of 6-phenoxy-1-hexen with phenobarbital induced rat liver microsomal cytochrome P-450 was studied by differential spectroscopy. A type I spectrum was observed with a good affinity (K_S = 1,3 10^{-5}M). 6-Phenoxy-1-hexen was rapidly metabolized *in vitro* in the presence of oxygen and NADPH generating system with both PB or 3MC microsomes. The turnovers were similar to known good substrates (Table).

Three major metabolites were formed, representing more than 90 % of total metabolism : 6-phenoxyhexan-1,2-diol, 6-phenoxy-1-hexen-3-ol and 6-(4'-hydroxyphenoxy)-1-hexen. Adding the microsomal epoxyde hydrolase inhibitor, trichloropropene oxyde (TCPO, 0.5 mM) completely abolished the formation of the diol which was replaced by the 6-phenoxy-1,2-epoxyhexane without changing the amount of total metabolism. Trace of other alcools (∿ 1% each) and phenol and 6-hexenal (∿ 2%) were also found, showing possible metabolic attack at saturated carbon and ether carbon. However cyclopentane derivatives have not yet been identified.

In every case the epoxidation reaction was the major metabolic event (4), but was most favoured with microsomes from phenobarbital treated rats ; the allylic and aromatic hydroxylation were most favoured with microsomes from methylcholanthrene treated rats.

Similar results were obtained when microsomes were treated with iodosobenzene (ØIO) (8) or cumenehydroperoxide (CuOOH) (9) except that the reaction was linear for only one minute with ØIO. With CuOOH, the amount of epoxide reached a plateau at 2 minutes but allylic oxidation continued, showing possible change in the oxidation mechanisms.

The chemical model of cytochrome P-450 utilizing an iron porphyrin in presence of ØIO (4,10) or CuOOH (2) was used to study the oxidation of our substrate ; several porphyrins were also tested and the results obtained with chloro-mesotetraphenylporphyrinato iron ($TPPFe^{III}Cl$) are in the table.

TABLE

	Pb Pretreated [a]			3MC Pretreated [a]			$TPPFe^{III}Cl$ [b]	
	$NADPH/O_2$	ØIO	CuOOH	$NADPH/O_2$	ØIO	CuOOH	ØIO	CuOOH
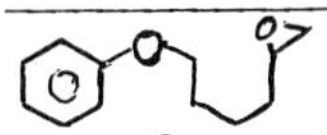	6	15,4	16	4,2	18,6	11,8	190	nd
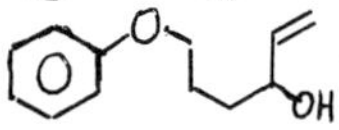 OH	1,1	5,4	4,8	2,8	9,6	6,4	148	< 5[c]
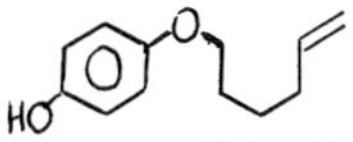 HO	0,5	nd	nd	0,8	nd	nd	nd	nd

a) Incubations conditions : 1 min at 37°C, TCPO 0.5 mM; NADPH generating system 1mM NADP, ØIO 1 mM or CuOOH 1 mM; nmole of product/min/mg protein, average of 3 experiments with 1 mg/ml of PB microsomes (2 nmole P-450/mg protein) or MC microsomes (1.4 nmole/mg protein). b) Model system : benzene containing 1 M of I, 10 mM TPPFeCl and 50 mM oxidant. Conversion are in % of porphyrin in one hour at 20°C. c) Traces of allylic ketone ; not significantly different from blank with TPPFeCl.

The products obtained when ØIO (8) is the oxidant are very similar to those obtained with cytochrome P-450 and the same oxidant, in particular the epoxide is the major product and traces of aromatic hydroxylation are also found (10). This model resembles most to oxidation by PB microsomes in the presence of ØIO.

On the contrary when CuOOH is the oxidant (2,8), no epoxide is formed and the quantity of allylic alcool obtained is not significantly different from the blank in absence of $TPPFe^{III}Cl$.

CONCLUSION

We have designed a substrate of P-450, 6-phenoxy-1-hexen, which is well suited for the study of cytochrome P-450 specificity and regioselectivity. It has a high affinity for P-450 and is efficiently metabolized in a number of hydroxylated products which can be readily analyzed and identified. The preliminary results show that the major metabolite when the epoxide hydrolase is inhibited, is the 1,2-epoxide although the proportion obtained depends of the type of induction of the microsomes and of the type of oxidant used.

The chemical oxidation system based on $TPPFe^{III}Cl$ using ØIO as the oxidant seems able to mimic most of the cytochrome P-450 basic reactions (4). Thus their "active oxygen complex" may share some similarity.

On the contrary, the model using TPPFeCl and CuOOH as the oxidant, although being able to hydroxylate saturated carbons (2,10) or ether bearing carbons (11) does not mimic cytochrome P-450. Its "active oxygen complex" is thus definitely different from that of the preceeding model (2) and of cytochrome P-450 plus NADPH O_2 or ØIO. However it is more difficult to conclude about the "active oxygen complex" formed from cytochrome P-450 and various oxidants since with the present substrate and the three oxidants tested, the qualitative product pattern was similar. The experiments are now being extented with purified forms of liver cytochrome P-450. Our finding are thus similar to those of Groves (4) with cyclohexene as substrate, although Capdevila *et al.* concluded recently that the "active oxygen complex" formed by cytochrome P-450 and CuOOH is different from the one obtained with NADPH O_2 (13).

REFERENCES

1. White, R.E. and Coon, M.J. (1980). Ann. Rev. Biochem., 49, 315-316.
2. Mansuy, D., Bartoli, J.F. and Momenteau, M. (1982). Tetrahedron Letters, in press.
3. Hewbrook, D.C. and Sligar, S.G. (1981). Biochem. Biophys. Res. Commun., 99, 530-535.
4. Groves, J.T., Krishnan, S., Avaria, G.E. and Nemo, T.E. (1980) in : Dolphin, D.T., Tabushi, I., Murkami, T. and Mc Kenna, C.E., Eds, Biomimetic Chemistry, Adv. in Chem. Series, A.C.S., Washington, D.C., vol. 191 pp. 277-289.
5. Groves, J.T., McClusky, G.A., White, R.E. and Coon, M.J. (1978) Biochem. Biophys. Res. Commun., 81, 154-160.
6. Ortiz de Montellano, P.R., Kunze, K.L., Beilan, H.S. and Wheeler, C. (1982) Biochemistry, 21, 1331-1339.
7. Julia, M. (1971). Account of Chem. Research, 4, 386-391.
8. Lichtenberger, F., Nastainczyk, W. and Ullrich, V. (1976). Biochem. Biophys. Res. Commun., 70, 939-946.
9. Rahimtula, A.D., O'Brien, P.J., Hrycay, E.G., Peterson, J.A. and Estabrook, R.W. (1974). Biochem. Biophys. Res. Commun., 62, 268-275.
10. Groves, J.T., Nemo, T.E., Myers, R.S. (1979). J. Am. Chem. Soc., 101, 1032.
11. Mansuy, D., Dansette, P.M., Pecquet, F. and Chottard, J.C. (1980). Biochem. Biophys. Res. Commun., 96, 433-439.
12. Chang, C.K. and Ebina, F. (1981). J. Chem. Soc. Commun., 778-779.
13. Capdevila, J., Estabrook, R.W., Prough, R.A. (1980). Arch. Biochem. Biophys., 200, 186-195.

Cytochrome P-450, Biochemistry, Biophysics
and Environmental Implications, E. Hietanen,
M. Laitinen and O. Hänninen editors

A RECONSTITUTED HEPATIC CYTOCHROME P-450 STEROID 21-HYDROXYLASE SYSTEM AND ITS STIMULATION BY CYTOCHROME b_5

I.R. SENCIALL, C.J. EMBREE AND S. RAHAL
Medical School, Memorial University of Newfoundland, St. John's, Newfoundland, Canada, A1B 3V6

INTRODUCTION

The adrenals are the major site of steroid C-21-hydroxylation in mammals and a cytochrome P-450 specific for this reaction has been isolated from bovine microsomal fractions (1,2). The extra-adrenal 21-hydroxylation of progesterone has recently attracted renewed interest with the observation that it occurs in human adult (3) and fetal (4) kidneys; potential target tissues for the mineralocorticoid action of the deoxycorticosterone (DOC) thereby produced

C-21-hydroxylation of progesterone is a major extra-adrenal metabolic pathway in the rabbit and occurs *in vitro* in both hepatic microsomal and mitochondrial fractions (5,6) and in kidney microsomes (7). The demonstration of cytochrome P-450 involvement with the hepatic microsomal steroid-21-hydroxylation has been equivocal in the past (8), primarily because of an apparent lack of sensitivity of the reaction to carbon monoxide inhibition. However, since submission of the present communication, Dieter *et al* (9) have elegantly identified one of six P-450 isozymes, designated form one, as a hepatic progesterone 21-hydroxylase.

The present communication also indicates that a cytochrome P-450 is involved in rabbit hepatic microsomal progesterone 21-hydroxylation, and in addition suggests that cytochrome b_5 may also participate.

MATERIALS AND METHODS

Preparation of microsomal fractions. Liver microsomes were prepared from New Zealand white rabbits stimulated with sodium phenobarbitone (0.1%) in drinking water for 5 days (8). Solubilisation and chromatography at room temperature were carried out essentially as described by Warner *et al* (10), except that both cytochrome P-450 and its reductase were eluted with 0.2M KCl in buffered detergent. Fractions were desalted on PD-10 columns and the reductase further purified and separated from cytochrome b_5 on a DEAE-Sepharose column (11). The reductase had a specific activity up to 1 mol cytochrome c reduced/min/mg protein, and cytochrome b_5 up to 23.2 nmols/mg proteins. The major P-450 fraction, designated $P\text{-}450_C$ was either used directly or after further resolution on a hydroxy apatite column into $P\text{-}450_{LM2}$ and $_{LM3}$ (11) (Specific activity up to 4.2 nmols/mg protein). All fractions were treated with Bio-beads to remove detergents before incubation.

Steroid incubations and DOC quantitation. ^{3}H-Progesterone (0.5 Ci; 101 Ci/mmol), diluted with 0.2 nmol non-labelled progesterone, was incubated for 30 min with the solubilised microsomal fractions in the presence of NADPH (100 µg); 2 ml total volume. At the end of incubation ^{14}C-DOC and non-labelled DOC

(20 μg) were added as internal standards. The incubates were extracted with ethyl acetate and the yield of DOC quantitated by TLC in chloroform-ethyl acetate (6:4 v/v), acetylation, and TLC in iso-octane-ethyl acetate (1:1 v/v) with the recovery of ^{14}C-DOC correcting for experimental losses. Spectrophotometric and protein assays were as previously reported (8).

Materials. DEAE-Sepharose CL-6B, Sephadex G-25 coarse and PD-10 columns from Pharmacia, Uppsala, Sweden. DEAE-Cellulose (DE-52) from Whatman Chemical Co.

RESULTS AND DISCUSSION

A cytochrome P-450 fraction, its reductase and cytochrome b_5 were isolated free from cross contamination.

Figure 1 shows the separation of three cytochrome P-450 fractions and the reductase fraction by DEAE-cellulose column chromatography. Only P-450_C was active in the metabolism of progesterone when combined with reductase and exhibited both 6β- and 21-hydroxylase activities.

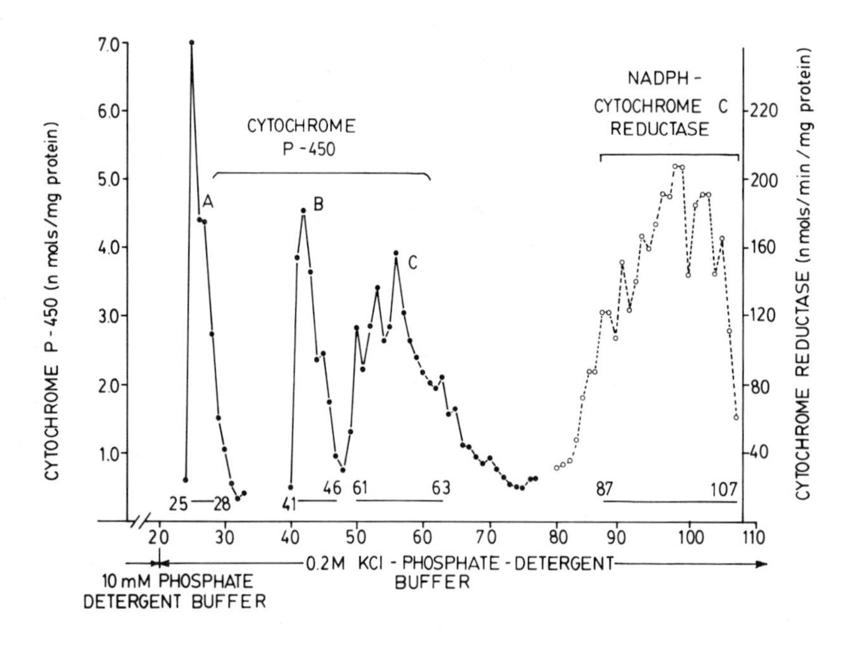

Figure 1. DEAE-Cellulose column chromatography of solubilised hepatic microsomes.

Further chromatography of P-450_C on a hydroxyapatite column gave two fractions, P-450_{LM2} and $_{LM3}$ (11), only P-450_{LM3} was active and metabolised progesterone primarily to 6 -hydroxy progesterone. Further studies were confined to P-450_C which primarily exhibited 21-hydroxylase activity; 6β-activity apparently being less active under the experimental conditions employed.

Factors affecting progesterone 21-hydroxylation. Table 1 shows that recombination of cytochrome P-450_C and reductase restored 21 hydroxylase activity not present in the individual fractions. Phospholipid did not enhance activity, whereas addition of cytochrome b_5 produced a three-fold stimulation of progesterone 21-hydroxylation.

TABLE 1
PROGESTERONE 21-HYDROXYLATION BY INDIVIDUAL AND RECOMBINED FRACTIONS.

Fraction	pmol DOC	Inhibitor	pmol DOC	%Control
P-450_C (0.248nmol)	0.09	None[1]	19.11	100
Reductase (0.1U)	0.19	CO/Air (9:1)	2.11	11.0
P-450_C + Reductase[1]	19.11	Metyrapone (10^{-4}M)	0.23	1.2
P-450_C + Reductase + Phospolipid (25 μg)	16.9	NaCN (10^{-3}M)	19.8	103.6
P-450_C + Reductase + b_5 (0.75nmol)	64.6			
b_5 + Reductase	4.9			

TABLE 2
EFFECT OF CONCENTRATION OF MICROSOMAL COMPONENTS ON PROGESTERONE 21-HYDROXYLATION

P-450_C(nmol)	pmol DOC	b_5(nmol)[a]	pmol DOC	Reductase[b] (Units)	pmol DOC
0.049	14.3	0.083	19.11	0.025	5.9
0.098	17.5	0.166	33.4	0.050	9.3
0.147	23.0	0.332	53.5	0.075	13.0
0.196	26.0	0.495	39.5	0.10	19.6

[a] Also contains P-450_C (0.412 nmols) and reductase (0.1U).
[b] Contains P-450_C (0.412 nmols).

Table 1 also shows that the reconstituted hydroxylase system was sensitive to inhibition by carbon monoxide and metyrapone, but insensitive to cyanide.

Table 2 indicates that progesterone 21-hydroxylation by reconstituted solubilised microsomal components is dependent on the concentrations of $P450_C$, NADPH cytochrome(c) P-450 reductase and cytochrome b_5 in the presence of NADPH.

CONCLUSIONS

An extra adrenal progesterone 21-hydroxylase that catalyses a major metabolic pathway in the rabbit has been reconstituted from solubilised hepatic microsmal cytochrome P-450 and NADPH cytochrome (c) P-450 reductase.

Addition of cytochrome b_5 increased hydroxylation up to three-fold under conditions where the reductase was not saturating. This suggests that cytochrome b_5, which can serve as an efficient electron donor to cytochrome P-450 (12) may also be involved in this hepatic progesterone 21-hydroxylase system.

ACKNOWLEDGEMENTS

Supported by MRC (CANADA) grant MT-5403.

REFERENCES

1. Kominami, S., Ochi, H., Kobayashi, Y. and Takemori, S. (1980) J. Biol. Chem., 255, 3386.
2. Hiwatashi, A. and Ichikawa, Y. (1981) Biochim. Biophys. Acta, 664, 33.
3. Winkel, C.A., Simpson, E.R., Milewich, L. and MacDonald, P.C. (1980) Proc. Natl. Acad. Sci. USA, 77, 7069.
4. Winkel, C.A., Casey, M.L., Simpson, E.R. and MacDonald, P.C. (1981) J. Clin. Endoc. Metab., 53 10.
5. Dey, A.C. and Senciall, I.R. (1977) Can. J. Biochem., 55, 602.
6. Bhavnani, B.R., Shah, K.N. and Solomon, S. (1972) Biochem. 11, 753.
7. Senciall, I.R. (1982) Unpublished results.
8. Senciall, I.R., Dey, A.C. and Rahal, S. (1981) J. Steroid Biochem. 14, 281.
9. Dieter, H.H., Muller-Eberhard, U. and Johnson, E.F. (1982) Biochem. Biophys. Res. Commun., 105 515.
10. Warner, M., LaMarca, M.V. and Neims, A.H. (1978) Drug Metab. Dispos. 6, 353.
11. Ingelman-Sundberg, M. and Glaumann, H. (1980) Biochim. Biophys. Acta, 599, 417.
12. Ingleman-Sundbarg, M. and Johansson, S. (1980) Biochem. Biophys. Res. Commun., 97, 582.

Cytochrome P-450, Biochemistry, Biophysics
and Environmental Implications, E. Hietanen,
M. Laitinen and O. Hänninen editors

PROSTACYCLIN SYNTHASE AS A CYTOCHROME P450 ENZYME

H. GRAF AND V. ULLRICH
Department of Physiological Chemistry, University of Saarland
6650 Homburg-Saar (Federal Republic of Germany)

INTRODUCTION

Cytochrome P450 has been identified as the oxygen activating component of many oxygenase systems. Its unusual spectral properties are due to the presence of a thiolate ligand in the fifth coordination position of the heme. We have postulated that the thiolate linkage provides the necessary activation in the active oxygen complex which can be regarded as an oxygen atom bound to the ferric cytochrome P450 (1).

In analogy to this oxenoid complex formation we have proposed that the 9,11-endoperoxide of 15-hydroxyarachidonic acid can be split and activated in a similar reaction to yield prostacyclin (2) and thromboxane (3):

$$\text{Enz-S}^{-}\cdots\text{Fe}^{III} + \overline{\underline{O}}| \longrightarrow \text{Enz-S}—\text{Fe}^{IV}—\overline{\underline{O}}\cdot$$

$$\text{Enz-S}^{-}\cdots\text{Fe}^{III} + \text{endoperoxide}(\overset{\oplus}{O}, \underset{\ominus}{O}; R_1, R_2) \longrightarrow \text{Enz-S}—\text{Fe}^{IV}—\overset{\oplus}{O}\cdot(\underset{\ominus}{O}; R_1, R_2)$$

This hypothesis was verified recently by Haurand and Ullrich (4) showing that thromboxane synthase activity in platelets co-purifies with platelet cytochrome P450. Similarly, we have identified cytochrome P450 from aortic microsomes as prostacyclin synthase (5). In the present paper we report on some properties of this enzyme.

MATERIALS AND METHODS

Pig aorta microsomes were prepared as described for liver microsomes. Solubilization of the microsomal preparation was performed by the addition of glycerol, cholic acid and 3-(1-tetradecyl-1,1-dimethylammonio)-1-propane-sulfonate (sulfobetaine) to a final concentration of 30 %, 0.55 % and 0.12 %, respectively. The supernatant was chromatographed on DEAE-Sephacel and eluted with 35 mM and 70 mM Na-phosphate buffer. The second eluate was applied to a U 46.619-conjugated Sepharose column and eluted with 0.05 % sulfobetaine. After a second affinity chromatography the cytochrome was essentially pure with a specific content of 18 nmol/mg protein.

Products of conversion of the prostaglandin endoperoxide (PGH_2) were analyzed as described (6). Optical absorption spectra were obtained by the use of an Aminco DW-2 dual wavelength photometer.

SDS-Polyacrylamide-gelelectrophoresis was performed in a Laemmli-system. 3-(1-tetradecyl-1,1-dimethylammonio)-1-propanesulfonate was synthesized according to (7).

Cholic acid and glycerol were purchased from Roth (Karlsruhe, GFR) in highest purity. 1-^{14}C-arachidonic acid was obtained from Amersham-Buchler (Braunschweig, GFR). (15S)-hydroxy-11α,9α-epoxy-methano)-prosta-5Z,13E-dienoic acid (U-46619) was a generous gift of Dr. J.Pike (Upjohn Co., Kalamazoo, USA).

RESULTS

Cytochrome P450 from pig aorta microsomes was purified by affinity chromatography about 400-fold to apparent homogeneity. The SDS-polyacrylamide gelelectrophoresis showed a main band with a calculated M_r of 49.200 (Fig. 1).

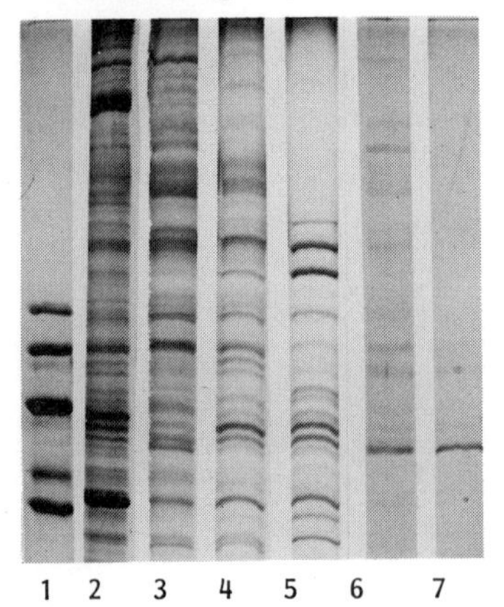

Fig. 1. SDS polyacrylamide gel-electrophoresis of aortic cytochrome P450 fractions. 1. protein standards, 2. pig aorta microsomes, 3. solubilized microsomal supernatant, 4. 1st DEAE fract., 5. 2nd DEAE fraction, 6. 1st affinity eluate, 7. 2nd affinity eluate.

Figure 2 shows the optical absorption spectra of purified cytochrome P450 pig aorta microsomes. The oxidized hemoprotein has a maximum in the γ-band at 417 nm which has not shifted after reduction but lowered in its molar extinction. After addition of CO to the reduced prostacyclin synthase a peak at 451 nm arises. The reduced enzyme is converted time dependently to a cytochrome P420-like hemoprotein with a maximal absorption at 421 nm.

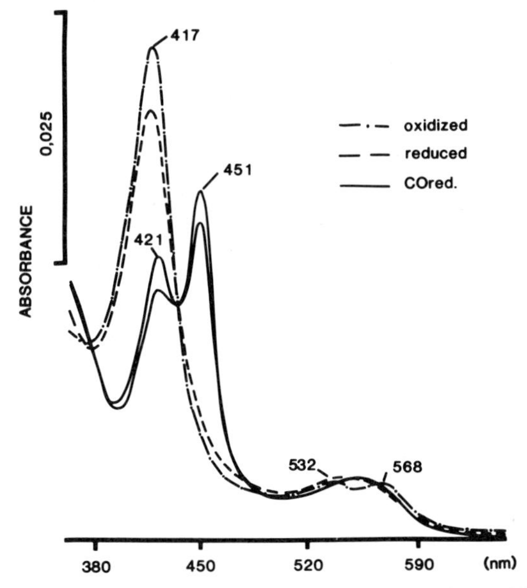

Fig. 2. Optical absorption spectra of purified cytochrome P450 from pig aortic microsomes. Prostacyclin concentration was $0.36 \cdot 10^{-6}$ mol/l, heme content of the preparation was 0.89 mol heme/mol prostacyclin synthase.

In contrast to cytochromes P450 involved in monooxygenase reactions this P450 enzyme is not reducible by NADPH in aortic microsomes (2). It also does not react spectrally with the known substrates of P450-dependent monooxygenases. It can react, however, with the prostacyclin synthase inhibitor tranylcypromine and 15-hydroperoxy arachidonic acid (results not shown).

Upon addition of 1-^{14}C-arachidonic acid aortic microsomes catalyze the formation of prostacyclin which can be monitored by TLC-determination of the stable degradation product 6-keto-prostaglandin $F_{1\alpha}$ (6). This activity co-purified with the cytochrome P450 (Table).

ENZYMIC CONVERSION OF PROSTAGLANDIN ENDOPEROXIDE (PGH_2)

Fraction	app. V_{max}	specific activity	specific content
microsomes	51.7	2.3	44.4
sol microsomes	41.1	2.9	71
2nd DEAE Frc.	61	6	144
1st Affinity	80	156	1947
2nd Affinity	109	1969	18067

Enzymic activities of prostacyclin synthase enriched fractions. Data are means of 4 determinations. V_{max}: mol prod.formed · min^{-1} ·(mol enzyme)$^{-1}$; spec. activity: mol prod.formed · min^{-1} · (mg protein)$^{-1}$; spec.content: pmol enzyme · (mg protein)$^{-1}$

From our results we conclude that aortic cytochrome P450 is identical with prostacyclin synthase.

DISCUSSION

Prostacyclin is an extremely potent vasodilator and platelet-antiaggregating factor produced from 9,11-endoperoxy-15-hydroxyarachidonic acid. The isolation of the enzyme catalyzing this isomerization has not yet been reported in literature and will have implications in drug research. Even more interesting is the impact on the mechanism of cytochrome P450 enzymes. The involvement of the heme-thiolate linkage in the isomerization supports our hypothesis on the importance of the thiolate ligand in activating the oxygen atom by the mechanism outlined in the first scheme. Since cytochrome P450 in this case only acts as an oxene transferase without undergoing reduction to the ferrous state we strongly suggest not to use the name "cytochrome" but rather "heme-thiolate protein".

REFERENCES

1. Ullrich, V. (1980) J.Molec.Catal., 7, 159-167
2. Ullrich, V., Castle, L., Weber, P. (1981) Biochem.Pharmacol.,30,2033-36
3. Ullrich, V., Castle, L., Haurand, M. (1981) Int.Symp.on Oxygenases and Oxygen Metabolism, Hakone, Academic Press, in print
4. Ullrich, V., Haurand, M. (1982) V.Int.Conf.Prostaglandins, Florence, Plenum Press, in print
5. Graf, H., Castle, L., Ullrich, V. (1982) V.Int.Conf.Prostaglandins, Abstract p. 179
6. Wlodawer, P., Hammarström, S. (1979) FEBS Letters, 97, 32-36
7. Gonenne, A., Ernst, R. (1978) Anal.Biochem., 87, 28.

Molecular Biology and Genetics of Cytochrome P-450

Cytochrome P-450, Biochemistry, Biophysics
and Environmental Implications, E. Hietanen,
M. Laitinen and O. Hänninen editors

FUNDAMENTAL PRINCIPLES OF CLONING

DANIEL W. NEBERT

Developmental Pharmacology Branch, National Institute of Child Health and Human Development, National Institutes of Health, Bethesda, MD 20205 (U.S.A.)

INTRODUCTION

To isolate any particular P-450 gene, one needs to develop an antibody and to characterize this antibody with regard to its inhibitory properties of particular monooxygenase activities. This Chapter will deal with an outline of the sequence of experiments necessary to go from a well-characterized antibody, to the correct mRNA, to the correct cDNA clone, and finally to the genomic-DNA clone. Where possible, the *Ah* locus and the murine P_1-450 gene are used as the examples.

RESULTS AND DISCUSSION

Characterizing the antibody. In 1978 in this laboratory, Dr. Negishi characterized an antibody that was extremely effective at inhibiting benzo-[a]pyrene metabolism (1). We define "cytochrome P_1-450" (Fig. 1) as that form of polycyclic-aromatic-inducible P-450 which metabolizes numerous different polycyclic hydrocarbons most effectively to ultimate carcinogenic intermediates (*reviewed in* Ref. 2). The antibody to this important catalytic activity, leading to chemical carcinogenesis, was thus called "anti-(P_1-450)."

Sizing the mRNA. Total mouse liver RNA was isolated. After two passes over an oligo(dT) column, the RNA is enriched for messenger RNA [which contains poly(A) tails]. The RNA can be separated by size (Fig. 2), either on sucrose density gradients or agarose gel electrophoresis. The mRNA from each tube of the gradient, or from each slice of the gel, can be isolated and then

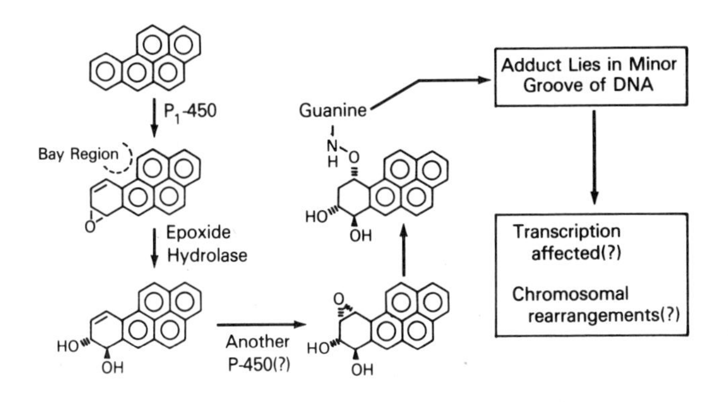

Fig. 1. Importance of P_1-450 in the metabolic steps leading to polycyclic hydrocarbon carcinogenesis.

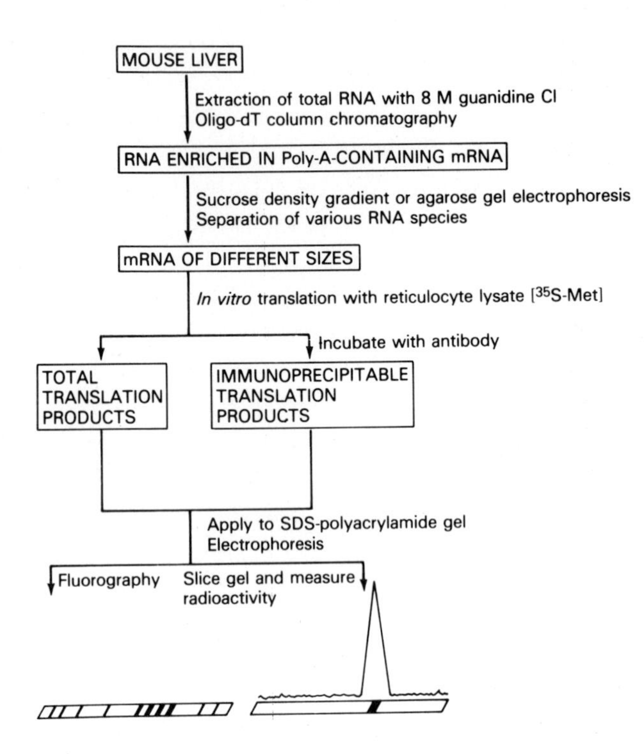

Fig. 2. Outline of methods by which mRNA for a particular protein can be sized. The method can only be as good as the antibody used for the immunoprecipitation of translation products.

translated *in vitro* with reticulocyte lysate and [^{35}S]methionine. Immunoprecipitable radioactive translation products can be quantitated by fluorography or scintillation spectrometry. From these procedures, we concluded that P_1-450 mRNA is about 23 S in size, which corresponds to 3000-3500 bp (3).

Probably the same experiment could have been done with monoclonal antibodies. However, this technique is extremely time-consuming and expensive. One also must be cautious in the interpretation of studies using monoclonal antibodies against "specific forms of P-450" (Fig. 3). In other words, one monoclonal antibody may "see" the same antigenic site on two different forms of P-450; with this single antibody, one might therefore conclude erroneously that his preparation contains only one form of P-450. On the other hand, two monoclonal antibodies may "see" two antigenic sites on a single form of P-450; with these two antibodies, one might conclude erroneously that his preparation contains two different forms of P-450. We thus believe that just a "good" antibody can provide as much information as a monoclonal antibody--without the excessive amount of time and expense.

Inserting the cDNA into a plasmid. Restriction enzymes cleave double-stranded DNA in various unique ways, recognizing "palindromes," or mirror-image areas, of DNA ranging between four and eight nucleotide bases in length. Illustrated in Fig. 4 are: *EcoRI*, which recognizes G A A T C C; *HaeIII*, which recognizes G G C C; and *PstI*, which recognizes C T G C A G. At the *bottom* is a restriction map, showing numerous restriction endonuclease

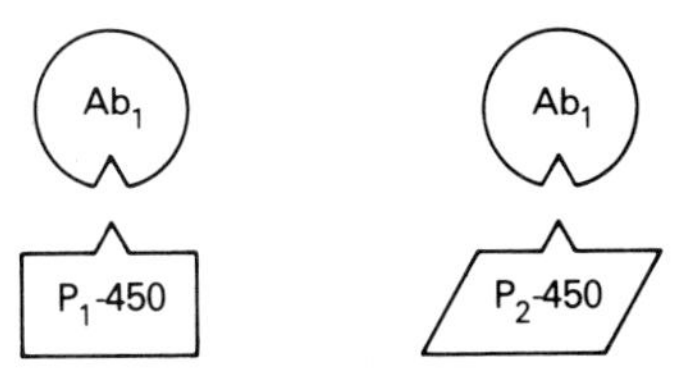

Fig. 3. Diagram to illustrate that the use of monoclonal antibodies is not a "cure-all." Data obtained from monoclonal antibodies still must be interpretated with caution.

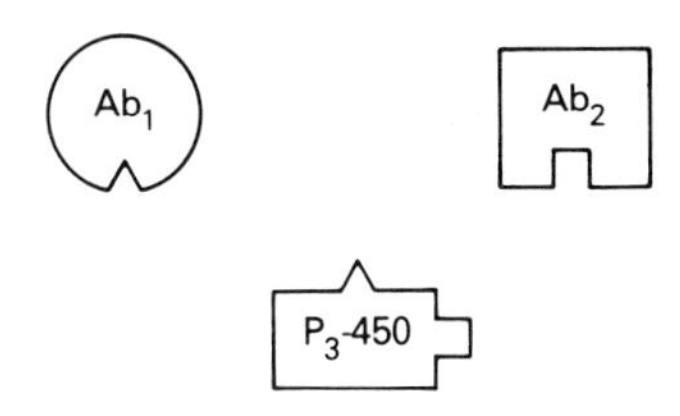

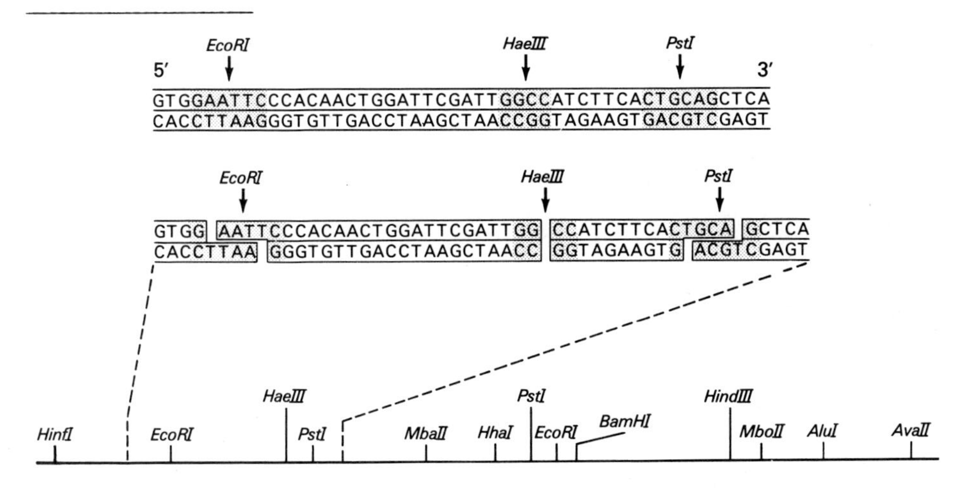

Fig. 4. Sequence of double-stranded DNA that can be cut by the restriction enzymes *EcoRI*, *HaeIII*, and *PstI*. The enlarged sequence at *top* is shown at *bottom* in the context of a more detailed restriction map.

sites. More than 300 of these unique bacterial enzymes now have been identified, and dozens are used daily in recombinant DNA research.

In order to insert a piece of mammalian DNA into the circular DNA of a plasmid, one must have mammalian DNA cleaved with a restriction enzyme (for example, *PstI*) and the plasmid DNA cleaved with the same enzyme. Control segments, or linker segments, must be added to the mammalian DNA piece, as well as to the plasmid DNA that has been cut. A poly(C) tail is usually added to the mammalian DNA fragment, and a poly(G) tail is usually added to the plasmid DNA that has been cut. These two then can be readily annealed. Next, the recombinant plasmid is inserted into *E.coli*, and the bacteria are grown in large numbers. The recombinant plasmid DNA then can be re-isolated, and the mammalian DNA insert harvested in *microgram quantities*. A single piece of mammalian DNA therefore can be "cloned," or expanded, from a single copy to billions of copies (*cf.* Refs. 4 *and* 5 *for recent reviews*).

Just such a technique was done for the P_1-450 gene (Fig. 5). Liver messenger RNA of 23 S size was isolated from 3-methylcholanthrene (MeChol)-

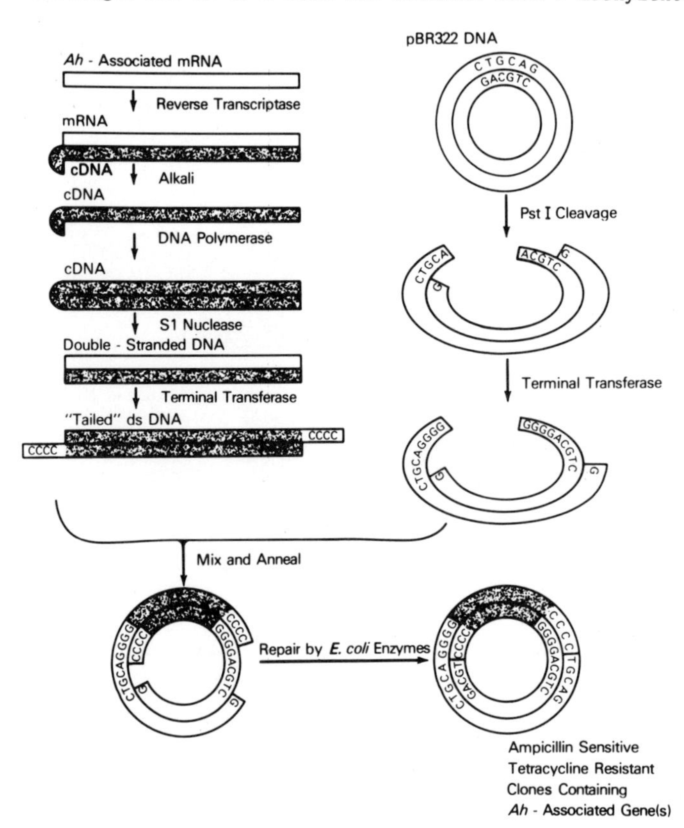

Fig. 5. Diagram of the cloning of double-stranded DNA derived from mouse liver 23 S mRNA (6).

treated C57BL/6N (B6) mice. We hoped that between 0.1% and 1% of the 23 S mRNA would be the induced P_1-450 mRNA. The 23 S mRNA was reverse-transcribed to cDNA, made double-stranded, and tailed with poly(C); this cDNA was inserted into the *PstI* site of plasmid pBR322 DNA that had been tailed with poly(G). When the double-stranded DNA is inserted into *E.coli*, bacterial enzymes fill in the nucleotide gaps. Because *PstI* breaks open the plasmid's ampicillin resistance gene but not its tetracycline resistance gene, one looks for colonies that are ampicillin-sensitive and tetracycline-resistant.

Colony hybridization. Fig. 6 summarizes the basic techniques for performing Southern or Northern blots. DNA fragments, or RNA's of different sizes, are electrophoresed on agarose gels. The migratory pattern of the nucleic acids in the gels then can be transferred directly (or "fixed" covalently) to hybridization paper. "Southern" refers to nitrocellulose paper; "Northern" refers to diazobenzyloxymethyl (or DBM) paper. The filter paper is then "probed" with radiolabeled single-stranded DNA or RNA. Following hybridization and post-hybridization washes, the paper is exposed to x-ray film to produce the blot hybridization analysis. The radiolabeled probe is then added for possible hybridization with the DNA on the filter. If hybridization occurs, one can detect this colony as a "spot" on x-ray film that has been exposed to the paper filter for several hours or days.

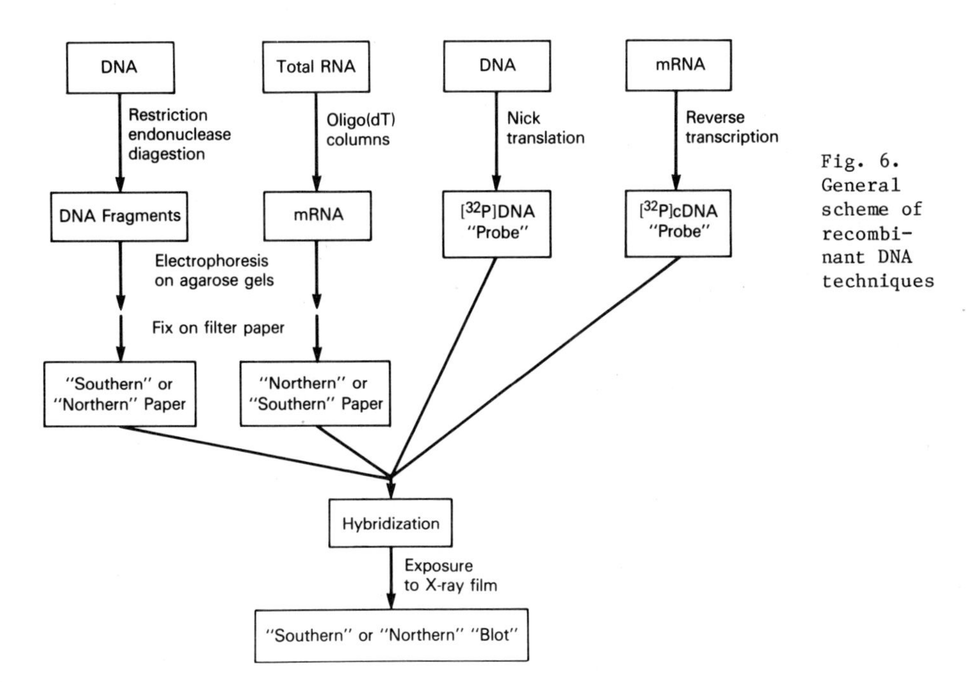

Fig. 6. General scheme of recombinant DNA techniques

Fig. 7 shows colony hybridization for the P_1-450 cDNA. Plasmid DNA from ampicillin-sensitive tetracycline-resistant *E.coli* was fixed on nitrocellulose paper. Double-stranded DNA is denatured to single-stranded DNA, usually by heat. The probe in this case was labeled cDNA freshly reverse-transcribed from 23 S mRNA from MeChol-treated B6 or DBA/2N (D2) mouse liver. Because there is at least 40 times more P_1-450 mRNA in MeChol-treated B6 than D2 mice (3), we looked for a colony that was positive in B6 and negative in D2. Two clones, numbers 46 and 68, were found in the first 72 colonies screened. Because 46 contained a larger insert than 68, we chose to study clone 46 further. Clone 30 was used in certain experiments as a positive control in both B6 and D2; clone 30 turned out to be a 1380-base pair insert representing preproalbumin cDNA. Clone 7 was used in certain experiments as a negative control in both B6 and D2; clone 7 is a 360-base pair insert of an unknown cDNA (7).

Immunologic proof for cDNA clone. "Translation arrest" is designed to test if one's antibody is associated with one's cDNA clone. Clone 46 DNA was fixed on a nitrocellulose filter, and 23 S RNA was hybridized to the filter (Fig. 8). When the pass-through mRNA was translated, there was no anti-(P_1-450)-precipitable translation product. Next, the specifically hybridized mRNA was eluted from the filter by formamide and translated; an anti-(P_1-450)-precipitable band at 55 kDa can be seen. These experiments are called

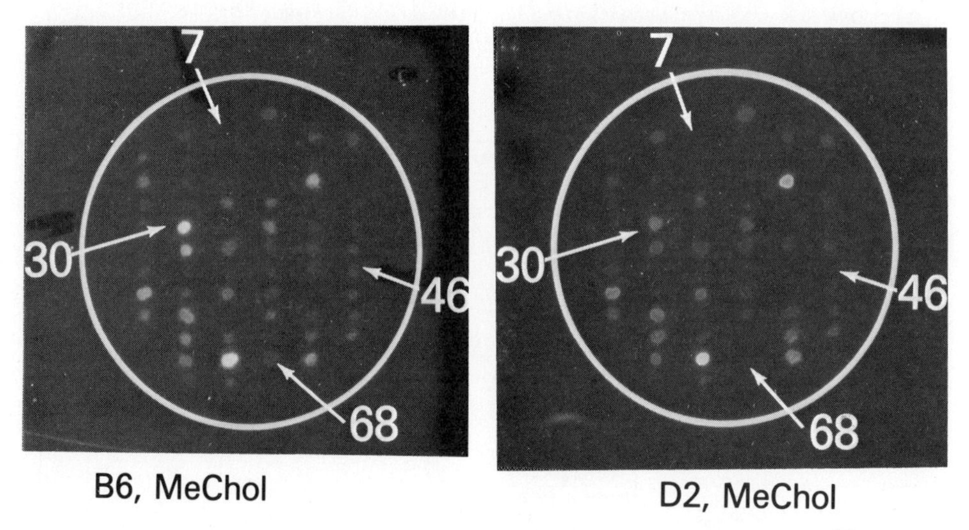

Fig. 7. Colony hybridization of cloned mouse liver DNA probed with $[^{32}P]$cDNA reverse-transcribed from MeChol-treated B6 and D2 total liver mRNA (7).

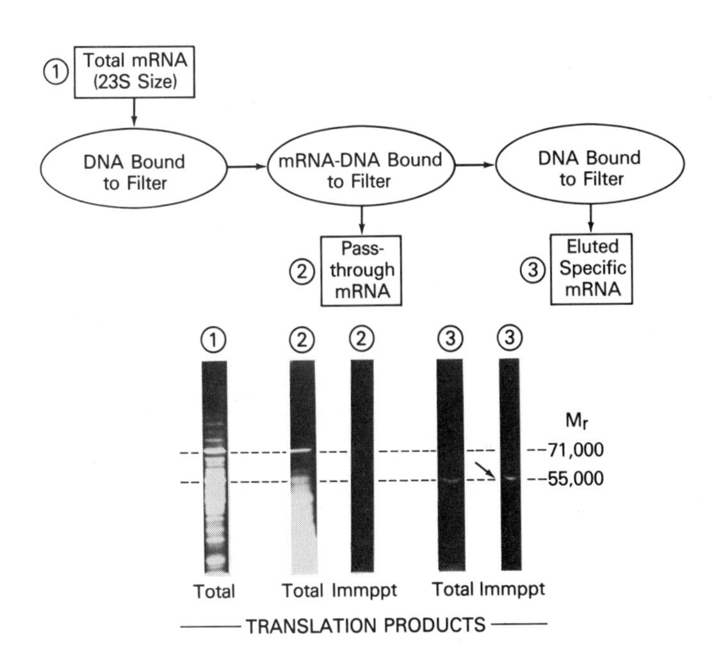

Fig. 8. Demonstration of "positive" and "negative" translation arrest, involving the mouse cDNA clone 46 and anti-(P_1-450) (7).

"positive" and "negative" translation arrest. These data provide immunologic proof that the cDNA clone isolated is associated with one's antibody. The antibody precipitates the protein that has been translated from the mRNA, and this mRNA hybridizes most specifically to the cloned cDNA.

Genetic proof for cDNA clone. Unfortunately the only proof available for rabbit or rat cDNA clones, is the immunologic criterion, because no genetic polymorphisms have been characterized in detail yet. Association of cloned cDNA only with an antibody can be uncertain, however, because most likely *all* antibodies to P-450 (including monoclonal antibodies) will cross-react with more than one form of P-450. With the mouse P_1-450 gene, we have the additional criterion of *genetic differences* that have been well characterized over the past decade. Hence, when the *Ah*-responsive F_1 heterozygote (Ah^b/Ah^d) is crossed with the *Ah*-nonresponsive D2 parent (Ah^d/Ah^d), among the progeny one finds a 50:50 distribution of *Ah*-responsive heterozygotes and *Ah*-nonresponsive homozygotes (*reviewed in* Ref. 8).

Fig. 9 shows an RNA-DNA hybridization. mRNA was isolated from *individual* mouse livers; samples 2 through 10 represent individual MeChol-treated offspring from the $B6D2F_1$ x D2 backcross. With a small piece of liver, their *Ah* phenotype was previously determined, and we know that numbers 2, 4, 5, 7, and 9 are *Ah*-responsive heterozygotes; numbers 3, 6, 8, and 10 represent *Ah*-non-

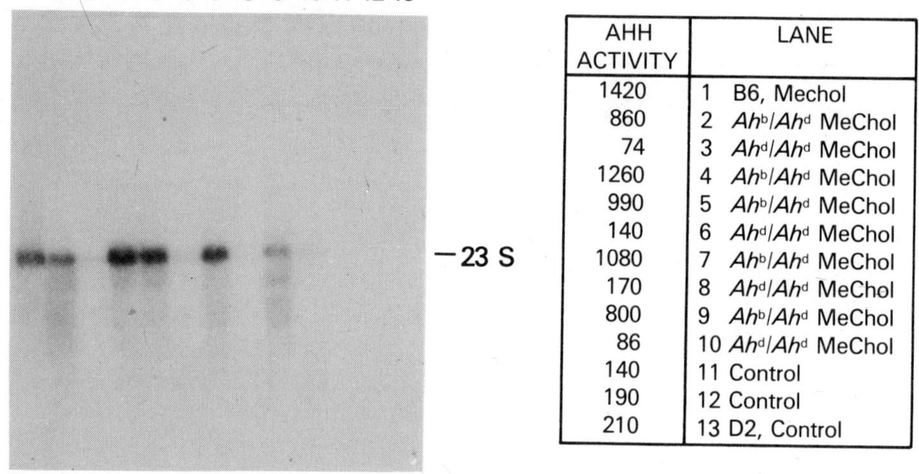

AHH ACTIVITY	LANE
1420	1 B6, Mechol
860	2 Ah^b/Ah^d MeChol
74	3 Ah^d/Ah^d MeChol
1260	4 Ah^b/Ah^d MeChol
990	5 Ah^b/Ah^d MeChol
140	6 Ah^d/Ah^d MeChol
1080	7 Ah^b/Ah^d MeChol
170	8 Ah^d/Ah^d MeChol
800	9 Ah^b/Ah^d MeChol
86	10 Ah^d/Ah^d MeChol
140	11 Control
190	12 Control
210	13 D2, Control

Fig. 9. Northern blot of mRNA from individual mice probed with clone 46 [^{32}P]DNA (9). AHH, aryl hydrocarbon (benzo[a]pyrene) hydroxylase, in units per mg of liver microsomal protein.

responsive homozygotes. The mRNA was electrophoresed and probed with labeled clone 46 DNA. There is a perfect correlation between *Ah*-responsiveness (*i.e.* induction of aryl hydrocarbon hydroxylase activity and therefore P_1-450 protein) and the presence of induced P_1-450 (23 S) mRNA that hybridizes to clone 46. This experiment therefore constitutes genetic proof that clone 46 is highly likely to be associated with the P_1-450 cDNA.

SUMMARY

A word of caution is issued: characterization of a cDNA clone by antibodies alone, without the aid of genetic differences, can lead to erroneous interpretation of data. A current illustrative example involves confusion over which cDNA clone truly represents rat P-450_b, P-450_e, and other phenobarbital-induced forms of rat P-450 (*reviewed in* Ref. 10). In this Chapter is shown the power of molecular genetics, coupled with recombinant DNA technology, in furthering our knowledge about the multiplicity of P-450's and the mechanism of P-450 induction. As demonstrated in several other Chapters in this book, this laboratory is applying the same combination of antibody-and-mouse-genetic-differences to attempt the cloning and characterization of genes for other forms of induced P-450: P_2-450 induced by isosafrole, and P_τ-450 induced by phenobarbital.

REFERENCES

1. Negishi, M. and Nebert, D.W. (1979) J. Biol. Chem., 254, 11015.

2. Pelkonen, O., Boobis, A.R. and Nebert, D.W. (1978) in: Jones, P.W. and Freudenthal, R.I. (Eds.), Carcinogenesis: Polynuclear Aromatic Hydrocarbons, Raven Press, New York, Vol. 3, pp. 383-400.

3. Negishi, M. and Nebert, D.W. (1981) J. Biol. Chem., 256, 3085.

4. Gilbert, W. and Villa-Komaroff, L. (1980) Sci. Amer., 242, 74.

5. Miller, W.L. (1981) J. Pediat., 99, 1.

6. Lang, M.A., Nebert, D.W. and Negishi, M. (1980) in: Gustafsson, J.-A., Carlstedt-Duke, J., Mode, A. and Rafter, J. (Eds.), Biochemistry, Biophysics and Regulation of Cytochrome P-450, Elsevier/North-Holland Biomedical Press, Amsterdam & New York, Vol. 13, pp. 415-422.

7. Negishi, M., Swan, D.C., Enquist, L.W. and Nebert, D.W. (1981) Proc. Natl. Acad. Sci. U.S.A., 78, 800.

8. Nebert, D.W., Negishi, M., Lang, M.A., Hjelmeland, L.M. and Eisen, H.J. (1982) Advanc. Genet., 21, 1.

9. Tukey, R.H., Nebert, D.W. and Negishi, M. (1981) J. Biol. Chem., 256, 6969.

10. Nebert, D.W. and Negishi, M. (1982) Biochem. Pharmacol., in press.

Cytochrome P-450, Biochemistry, Biophysics and Environmental Implications, E. Hietanen, M. Laitinen and O. Hänninen editors

ISOLATION AND CHARACTERIZATION OF THE MOUSE P_1-450 CHROMOSOMAL GENE

MASAHIKO NEGISHI, MARIO ALTIERI, MICHITOSHI NAKAMURA, ROBERT H. TUKEY, TOSHIHIKO IKEDA, YUAN-TSONG CHEN, TOHRU OHYAMA AND DANIEL W. NEBERT
Developmental Pharmacology Branch, National Institute of Child Health and Human Development, National Institutes of Health, Bethesda, MD 20205 (U.S.A.)

INTRODUCTION

As described in the preceding Chapter, clone 46 has been shown by both immunologic and genetic criteria to be a cloned portion of the mouse P_1-450 cDNA (1, 2). The mouse cDNA insert is 1100 bp in length (Fig. 1), with internal *XbaI* and *PstI* restriction sites. We have determined that the *PstI* site is 5'-ward of the *XbaI* site (3). Using clone 46 as the probe, we show in this Chapter (i) that much insight can be gained about the regulation of P_1-450 induction and (ii) that the mouse P_1-450 chromosomal gene has been isolated and characterized.

RESULTS AND DISCUSSION

Induction of P_1-450 mRNA quantitated by R_0t analysis. Sources for all experimental materials and animals, as well as detailed descriptions of each type of experiment, are provided in Refs. 1-4. To study mRNA levels during

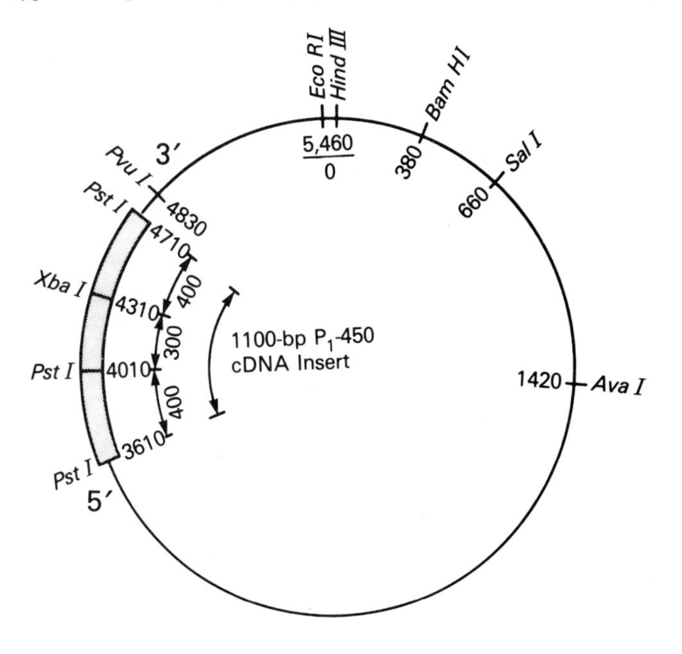

Fig. 1. Restriction map of pBR322 containing clone 46, P_1-450 cDNA (1).

polycyclic aromatic treatment, we isolated poly(A)-enriched total RNA from the livers of control B6 and D2 mice, and mice following 24 h of treatment with 3-methylcholanthrene (MeChol) or 2,3,7,8-tetrachlorodibenzo-*p*-dioxin (TCDD). The RNA was electrophoresed on agarose gels, fixed to Northern paper, and probed with clone 46. Induction of aryl hydrocarbon hydroxylase (AHH) activity, and therefore cytochrome P_1-450, occurs in *Ah*-responsive C57BL/6N (B6) mice treated with either MeChol or TCDD, whereas the induction process occurs in *Ah*-nonresponsive DBA/2N (D2) mice principally with high doses of TCDD. Large increases in P_1-450 (23 S) mRNA concentration were found in MeChol-treated and TCDD-treated B6 samples and in TCDD-treated D2 samples. Small increases in P_1-450 mRNA are measurable in MeChol-treated D2 mouse liver. These data demonstrate that P_1-450 induction is under transcriptional control (2).

More recently we have performed similar Northern hybridizations of mRNA from the livers of B6 and D2 mice treated with ten different P-450 inducers, again using clone 46 as probe (4). Fig. 2 shows these samples in which the P_1-450 mRNA has been quantitated by means of R_ot analysis, (also called excess

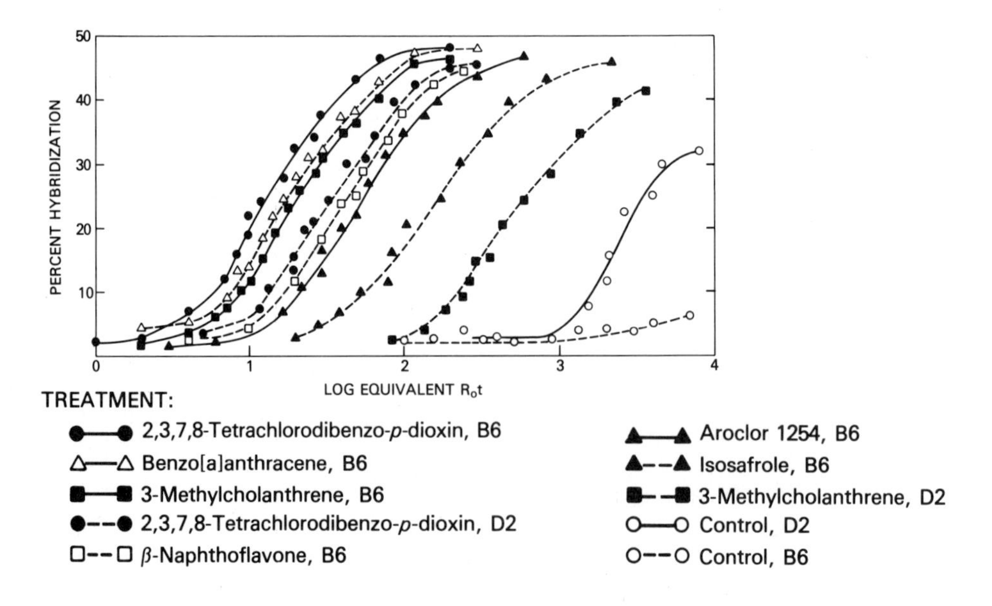

Fig. 2. Excess mRNA hybridization against nick-translated clone 46 [^{32}P]DNA. All intraperitoneally administered drugs caused maximal induction response at the dose used. Benzo[a]pyrene treatment in B6 mice (not shown) was between MeChol and β-naphthoflavone in effectiveness (4) [*Reproduced with permission from* Academic Press, Inc.].

RNA hybridization experiments). In *Ah*-responsive B6 mice, P_1-450 mRNA was induced most effectively by TCDD (more than 200-fold over controls), followed in order by benzo[a]anthracene, MeChol, benzo[a]pyrene, β-naphthoflavone, Aroclor 1254, and isosafrole. In *Ah*-nonresponsive D2 mice, TCDD was an effective inducer of P_1-450 mRNA, yet MeChol also can be seen to induce the messenger. P_1-450 mRNA was not at all induced by phenobarbital, pregnenolone 16α-carbonitrile, or *trans*-stilbene oxide. These data suggest that a number of P-450 inducers are capable of binding to the *Ah* receptors and evoking a response; the response being measured is P_1-450 mRNA induction by means of the clone 46 probe.

No effect on DNA during induction. Schimke *et al.* (5) have postulated that all forms of drug resistance might be caused by gene amplification, a process that they had found to occur for methotrexate resistance in mouse cell cultures: in response to methotrexate, the dihydrofolate reductase gene becomes amplified more than 100-fold. Because P-450 induction is a type of drug resistance, we therefore studied DNA from the livers of mice treated with the same ten P-450 inducers.

There was no evidence for increases in intensity of the blots or changes in the sizes of DNA fragments (4). We therefore conclude that, during the induction process, neither gene amplification nor some gross form of genomic re-arrangement occurs in the area of the P_1-450 gene represented by the clone 46 probe. In other hybridization experiments with B6 sperm, embryo and adult DNA (6), we also were unable to find any evidence for genomic rearrangement during development.

Isolation of recombinant λ phage. With clone 46 as the probe, a mouse plasmacytoma MOPC 41 genomic-DNA library was screened by the Benton-Davis plaque-hybridization procedure. Mouse tumor DNA had been digested by *EcoRI* and inserted into λ phage Charon 4A, and 50,000 plaques had been fixed to each filter (generous gift of Dr. Jon Seidman, National Institute of Child Health and Human Development, Bethesda). The entire library consisted of 20 filters, so that one is screening a total of about one million pieces of genomic DNA. A promising "tailed" plaque was found and purified to 100%. This positive genomic clone was named λ3NT12 and was found to be 19 kilobase pairs in length (7).

By hybridization between subclones of λ3NT12 and clone 46, we found that clone 46 hybridized to one extreme end of the mouse DNA insert and to an *EcoRI*-digested subclone pMJE12. There are two possible orientations (Fig. 3). One, shown at *top*, is that clone 46 and subclone pMJE12 are located at the 5'

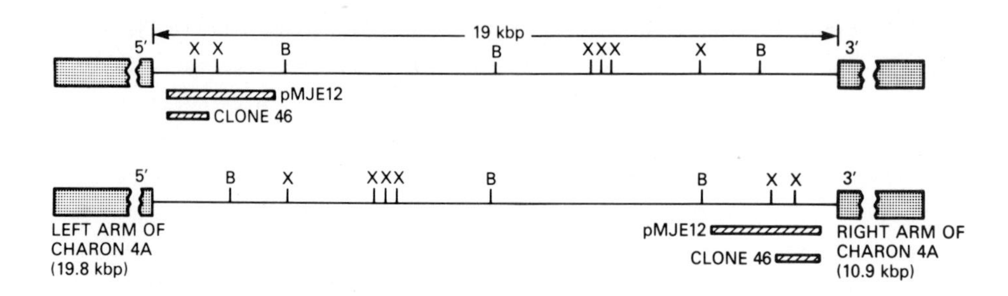

Fig. 3. Clone λ3NT12 from the mouse genomic-DNA library. Two possibilities for direction of the sense strand are shown (3). kbp, kilobase pairs [*Reproduced with permission from* Springer-Verlag].

end of λ3NT12; if this were the case, most of the P_1-450 genomic gene would be located 5'-ward of this insert and therefore not contained in λ3NT12. The other possible orientation, shown at *bottom*, is that clone 46 and subclone pMJE12 are located at the 3' end of λ3NT12; if this were the case, the entire P_1-450 chromosomal gene most likely would be contained within λ3NT12. Unfortunately, the former possibility, shown at *top*, was found to be the case (3). We therefore were forced to return to the mouse MOPC 41 genomic-DNA library to find another recombinant phage. This time we used as the ^{32}P-labeled probe subclone pMJE12, which is 3 kbp in length.

Isolation of λAhP-1. By repeating the Benton-Davis plaque hybridization technique, and characterizing three more genomic clones, we essentially have "walked up" the chromosome with clones λ3NT12, λ3NT13, λ3NT14, and finally λAhP-1. By R-loop analysis, the P_1-450 structural gene is believed to reside in the center of λAhP-1, spanning about 5 kbp (Fig. 4). The chromosomal gene of mouse P_1-450 is in the middle of λAhP-1, a clone which has a total length of 15.5 kbp. The P_1-450 genomic gene has at least 5 exons and 4 intervening sequences. The first and last exon are remarkably large, being about 1000 and 1200 bp, respectively. Total length of all exons together is about 3,000 bp.

Clone 46 probably exists in the 3'-nontranslating region of the P_1-450 gene. With the clone 46 3' probe, it should be emphasized that we are unable to hybridize this probe to any other P-450 gene in mouse genomic DNA (1). Clone 46 therefore represents a 3'-unique probe.

Various fragments of λ3NT12 and λAhP-1 have been subcloned in pBR322 and electrophoresed on nitrocellulose paper. When mRNA from MeChol-treated B6 or isosafrole-treated D2 mice was reverse-transcribed and the resulting [^{32}P]cDNA used as the probe, some of the subclones exhibit sequence homology with the

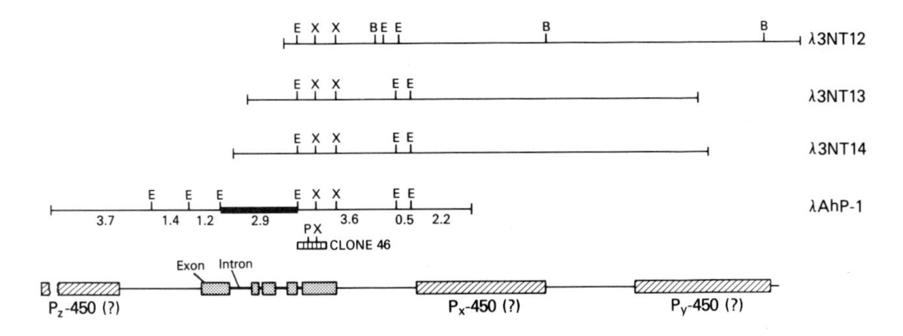

Fig. 4. Restriction maps and relative positions of four genomic-DNA clones and the cDNA clone 46. E, *EcoRI*; X, *XbaI*; B, *BamHI*; P, *PstI*. *Numbers* denote the sizes of the seven *EcoRI* fragments of λAhP-1, in kbp.

cDNA probes. A [^{32}P]cDNA probe from control mRNA does not. One possibility is that this finding represents evidence for several P-450 genes in tandem. Another possibility is that we have cross-hybridization between repetitive sequences. The 5'-nontranslating regions of mRNA occasionally have been found to have repetitive sequences (9). A third (least likely) possibility is that we have reverse-transcribed a nuclear precursor mRNA, because total liver poly(A^+)-enriched RNA was used as starting material, and again the [^{32}P]cDNA probe contains these repeating sequences.

Fig. 5 confirms the direction of the sense strand of λAhP-1. When *XbaI*-digested λAhP-1 pieces were hybridized with the 400-bp and 700-bp *PstI* fragments of clone 46 [^{32}P]DNA, the 3.8-kbp piece--but not the 0.7-kbp piece--hybridized to the 400-bp fragment. These data confirm the orientation that the *EcoRI*-digested 2.9-kbp fragment is 5'-ward of the 3.6-kbp fragment (Fig. 4).

SUMMARY

This laboratory has uncovered several important points of information with the use of the P_1-450 cDNA clone (Table 1). An association of clone 46 with the P_1-450 protein has been demonstrated by both immunologic and genetic criteria (1, 2). We know that P_1-450 induction is under transcriptional control, because increased 23 S mRNA and an intranuclear large-molecular-weight mRNA precursor occur concomitantly during the induction process (2). No evidence for gene duplication or gross form of genomic rearrangement has been found, either during induction or during development (2). P_1-450 mRNA is translated probably exclusively on membrane-bound polysomes (10). Clone 46

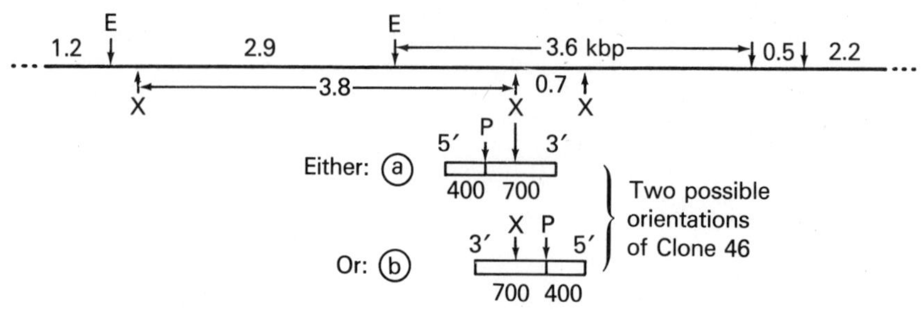

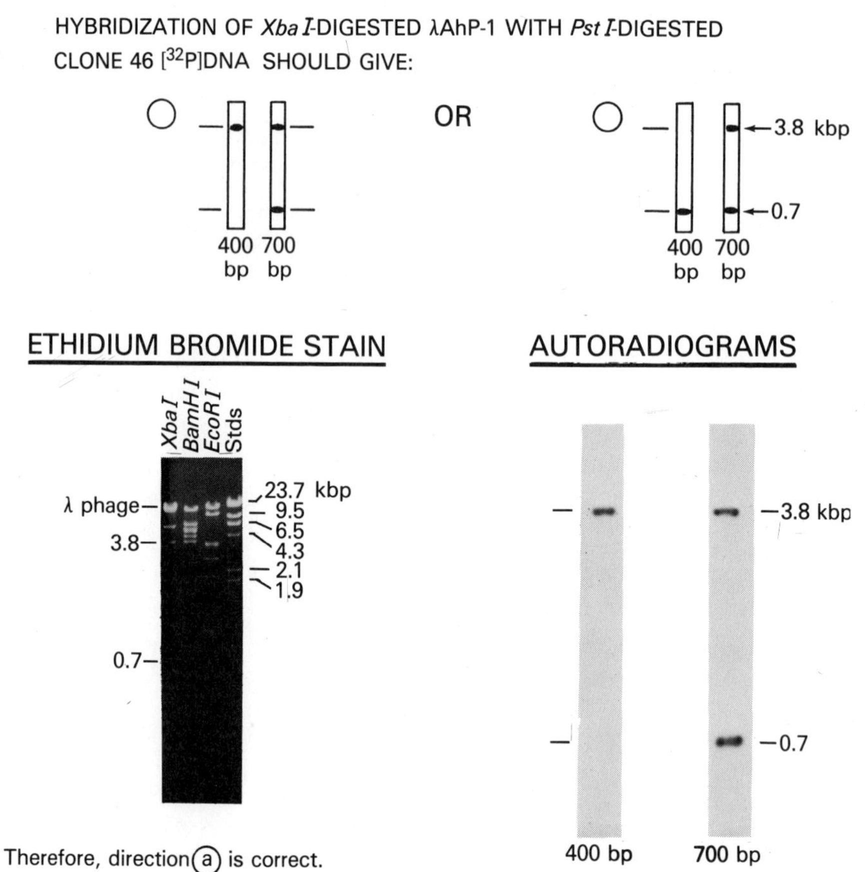

Fig. 5. Determination of direction for the sense strand of λAhP-1. E, *EcoRI*; X, *XbaI*; P, *PstI*; bp, base pairs; kbp, kilobase pairs.

TABLE 1

OUR FINDINGS WITH CLONE 46 (cDNA CLONE; 1100 bp; 3'-UNIQUE SEQUENCE)

1. Association with P_1-450 protein
    a. Negative and positive translation arrest with anti-(P_1-450)
    b. Correlation with *Ah*^{b} allele in offspring from B6D2F_1 x D2 backcross
2. P_1-450 induction under transcriptional control
    a. Association with increased mRNA (23 S) levels
    b. Correlation with intranuclear large molecular weight mRNA precursor
3. No evidence for gene amplification or gross form of genomic rearrangement
4. P_1-450 mRNA translation occurs on membrane-bound polysomes
5. Cross-hybridization with rat and rabbit DNA; rat (23 S) mRNA
6. P_1-450 gene hypomethylated in adult B6, when compared with that in adult D2, or sperm or embryonic B6
7. Other inducers (such as isosafrole) increase P_1-450 mRNA to small extent
8. P_1-450 genomic gene characterized by R-loop analysis (~5 kbp; at least 5 exons and 4 introns)
9. Localization of P_1-450 structural gene on mouse chromosome 2
10. Evidence of genes for other polycyclic-aromatic-inducible forms of P-450; perhaps P-450 genes in tandem on same chromosome
11. Excellent correlation (r=0.99) between appearance of inducer-receptor complex in nucleus and P_1-450 mRNA induction

hybridizes with rat and rabbit DNA and with rat but not rabbit mRNA; these data suggest that clone 46 hybridizes to a segment of the rabbit P_1-450 gene that is not transcribed into the messenger (6). We found that the P_1-450 gene in adult B6 is *hypo*methylated, compared with the gene in B6 sperm, B6 embryo, or adult D2 mice; this hypomethylation pattern could be related to the increased expressivity of the P_1-450 gene in B6, compared with that in D2 mice (6). Other P-450 inducers, such as benzo[a]anthracene and isosafrole, have been found to induce P_1-450 mRNA, as measured by the clone 46 probe (4). With the cDNA probe, we have isolated the chromosomal P_1-450 gene from a mouse genomic-DNA library (3). By R-loop analysis, the gene spans about 5 kilobase pairs and has at least 5 exons and 4 introns. By somatic-cell fused hybrids in culture (11), the P_1-450 structural gene has been localized to mouse chromosome number 2. We have suggestive evidence for several genes in tandem on the same chromosome; these genes appear to encode several polycyclic-aromatic-inducible forms of P-450. More work will be required to confirm this possibility, however. Lastly, we have found an excellent correlation between the intranuclear appearance of the inducer-receptor complex and the induction of P_1-450 mRNA as measured by the clone 46 probe (12). With clone λAhP-1 and the surrounding regions of this mouse chromosome, we hope to understand a great deal about the regulation of P-450 induction, the evolution of the P-450 system, and perhaps the ultimate number of P-450 forms that an individual organism is genetically *capable* of expressing. With this knowledge, we hope to gain insight into the mechanism of chemical carcinogenesis, especially since P_1-450 is directly responsible for the metabolic activation of polycyclic hydrocarbons, such as benzo[a]pyrene, to the ultimate carcinogenic intermediate, which interacts covalently with DNA. Finally, it may be possible to develop an assay, based on recombinant DNA technology, in order to assess the human *Ah* phenotype; such an assay may predict who is at increased risk for certain types of environmentally-caused cancers.

REFERENCES

1. Negishi, M., Swan, D.C., Enquist, L.W. and Nebert, D.W. (1981) Proc. Natl. Acad. Sci. U.S.A., 78, 800.
2. Tukey, R.H., Nebert, D.W. and Negishi, M. (1981) J. Biol. Chem., 256, 6969.
3. Nakamura, M., Negishi, M., Altieri, M., Chen, Y.-T. and Nebert, D.W. (1982) Eur. J. Biochem., in press.
4. Tukey, R.H., Negishi, M. and Nebert, D.W. (1982) Mol. Pharmacol., in press.
5. Schimke, R.T., Kaufman, R.J., Alt, F.W. and Kellems, R.F. (1978) Science, 202, 1051.
6. Chen, Y.-T., Negishi, M. and Nebert, D.W. (1982) DNA, in press.
7. Tukey, R.H., Nakamura, M., Chen, Y.-T., Negishi, M. and Nebert, D.W. (1982) in: Sato, R. and Kato, R. (Eds.), Microsomes, Drug Oxidations, and Drug Toxicity, Japan Scientific Societies Press, Tokyo, in press.
8. Ikeda, T., Altieri, M., Nakamura, M., Tukey, R.H., Nebert, D.W. and Negishi, M. (1982) J. Biol. Chem., in press.
9. Davidson, E.H. and Posakony, J.W. (1982) Nature, 297, 633.
10. Chen, Y.-T. and Negishi, M. (1982) Biochem. Biophys. Res. Commun., 104, 641.
11. Kärenlampi, S.O., Swan, D.C., Lalley, P.M., Negishi, M. and Nebert, D.W., manuscript in preparation.
12. Tukey, R.H., Hannah, R.R., Negishi, M., Nebert, D.W. and Eisen, H.J. (1982) Cell, in press.

© 1982 Elsevier Biomedical Press B.V.
Cytochrome P-450, Biochemistry, Biophysics
and Environmental Implications, E. Hietanen,
M. Laitinen and O. Hänninen editors

EVIDENCE FOR A CLOSELY-RELATED MULTI-GENE FAMILY OF PHENOBARBITAL-INDUCIBLE CYTOCHROMES P450 IN RAT LIVER MICROSOMES

WAYNE LEVIN[1], PAU-MIAU YUAN[2], JOHN E. SHIVELY[2], FREDERICK G. WALZ, JR.[3], GEORGE P. VLASUK[3], PAUL E. THOMAS[1], CURTIS J. OMIECINSKI[4], EDWARD BRESNICK[4] AND DENE E. RYAN[1]
[1]Hoffmann-La Roche Inc., Nutley, NJ, [2]City of Hope Research Institute, Duarte, CA, [3]Kent State University, Kent, OH, and [4]University of Vermont, Burlington, VT

INTRODUCTION

Two phenobarbital-inducible rat liver microsomal cytochromes P450 from Long Evans rats (cytochromes $P450b_{LE}$ and P450e) and a strain variant from Holtzman rats (cytochrome $P450b_{H}$) have been purified to homogeneity (1,2). Unlike other cytochromes P450 which have been purified from rat (3-5) and rabbit (6,7) liver microsomes, these 3 proteins show a remarkable degree of similarity in their biochemical, structural and immunological properties. As a result of the extensive homology among these phenobarbital-inducible cytochromes P450, only detailed molecular studies enabled us to conclude that these proteins are encoded by different structural genes and are not the result of post-translational processing of a single polypeptide.

RESULTS AND DISCUSSION

Purified cytochromes $P450b_{LE}$, P450e and a strain variant, cytochrome $P450b_{H}$, have similar but not identical minimum molecular weights (Mr ~ 52,000-53,000) in SDS-gels (Figure 1A). Cytochromes $P450b_{LE}$ and P450e differ in molecular weight by less than 1000 in the SDS gel system of Laemmli (8), and electrophoresis of a mixture of the two purified proteins yields two protein-staining bands that are barely distinguishable. However, coelectrophoresis of the strain variant cytochrome $P450b_{H}$ with P450e shows a more distinct separation of these hemoproteins compared to cytochromes $P450b_{LE}$ and P450e. This distinction indicated that cytochromes $P450b_{H}$ and $P450b_{LE}$ might also differ slightly in molecular weight. Although these strain variants cannot be resolved in this gel system, they can be separated from each other (Figure 1B) in a 6% acrylamide gel containing SDS and sodium tetradecyl sulfate (9/1). The slightly greater mobility of cytochrome $P450b_{H}$ compared to cytochrome $P450b_{LE}$ explains the greater separation of cytochromes $P450b_{H}$ and P450e compared to cytochromes $P450b_{LE}$ and P450e in the SDS gel shown in Figure 1A. Figure 2 shows a two-dimensional isoelectric focusing-SDS gel (9) of a mixture of purified cytochromes $P450b_{LE}$, $P450b_{H}$ and P450e. By this

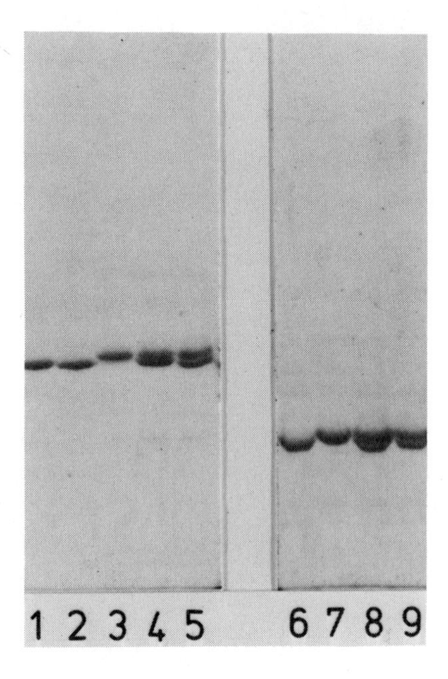

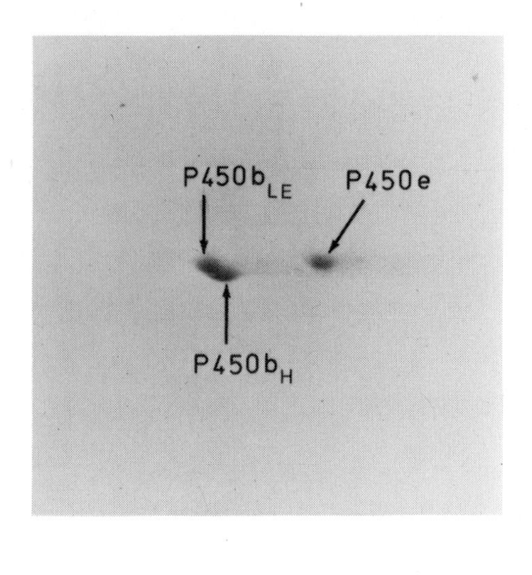

Fig. 1. Electrophoretic profile of purified cytochromes P450b and P450e in the SDS gel system of Laemmli (8). A (left), 7.5% acrylamide and SDS. Wells 1-5 contain cytochromes P450b-LE, P450b-H, P450e, a mixture of P450b-LE and P450e and a mixture of P450b-H and P450e, respectively. B (right), 6% acrylamide and a mixture of SDS and sodium tetradecyl sulfate, (9/1). Wells 6 and 7 contain cytochromes P450b-H and P450b-LE, respectively. Wells 8 and 9 contain mixtures of the two proteins, 0.5 µg (well 8) and 0.3 µg (well 9) of each protein.

Fig. 2. Two-dimensional isoelectric focusing-SDS gel of a mixture of purified cytochromes P450b-LE, P450b-H and P450e. Acidic and basic ends of the isoelectric focusing dimension are at the right and left ends of the gels, respectively. The SDS dimension is from top to bottom.

technique (9), cytochrome P450e is clearly separated from cytochromes P450b$_{LE}$ and P450b$_H$ as a result of significant differences in their isoelectric points. Cytochromes P450b$_{LE}$ and P450b$_H$ are partially resolved from each other, based on small differences in their minimum molecular weights and isoelectric points.

A comparison of the substrate specificities of purified cytochromes P450b$_{LE}$ and P450b$_H$ in the presence of saturating concentrations of NADPH-cytochrome c reductase and optimal phosphatidylcholine shows no difference in activities toward a variety of substrates (Table 1). The substrate specificity of cytochrome P450e is quite similar, although not identical, to that of cytochromes P450b but most reactions are catalyzed at only 15-25% of the rates

TABLE 1

COMPARISON OF CATALYTIC ACTIVITY OF PURIFIED CYTOCHROMES $P450b_{LE}$, $P450b_H$ and P450e

Reaction	$P450b_{LE}$	$P450b_H$ (nmol/min/nmol P450)	P450e
Benzphetamine N-demethylation	132.5	125.9	19.8
Benzo[a]pyrene hydroxylation	0.4	0.3	0.1
7-Ethoxycoumarin O-deethylation	9.6	9.3	2.0
Zoxazolamine 6-hydroxylation	3.3	4.3	1.1
Hexobarbital 3-hydroxylation	42.7	41.4	8.2
Estradiol 2-hydroxylation	0.3	0.3	0.9
Testosterone 16α-hydroxylation	2.3	2.5	0.4
7α-hydroxylation	<0.1	<0.1	<0.1
6β-hydroxylation	<0.1	<0.1	<0.1

catalyzed by cytochromes P450b. Only the 2-hydroxylation of estradiol-17β occurs at a greater rate by cytochrome P450e than cytochromes P450b. However, the poor rate of metabolism of estradiol at the 2-position by these cytochromes P450 suggests that they play a minor role in this metabolic pathway catalyzed by liver microsomes. Formation of the estradiol catechol is not increased in liver microsomes after treatment of rats with phenobarbital. Thus, unlike other rat liver cytochromes P450 which have been shown to have markedly different specificities for metabolism of the substrates shown in Table 1 (4,10), cytochromes P450b and P450e show a much more similar substrate selectivity.

The immunochemical properties of cytochrome P450 isozymes have been used extensively to demonstrate structural differences among these proteins (6,11,12). For example, rat liver cytochromes P450a, P450b and P450c appear to be antigenically distinct based on their lack of cross-reactivity with the heterologous antibodies in Ouchterlony double-diffusion analysis and the inability of the antibodies to inhibit catalytic activity of the heterologous antigens (12). Cytochromes P450c and P450d share some common immunochemical determinants but the antibodies directed against the common sites can be selectively removed by back absorption of the antibodies with the heterologous proteins (13). Thus antibodies have been prepared to each of these four isozymes

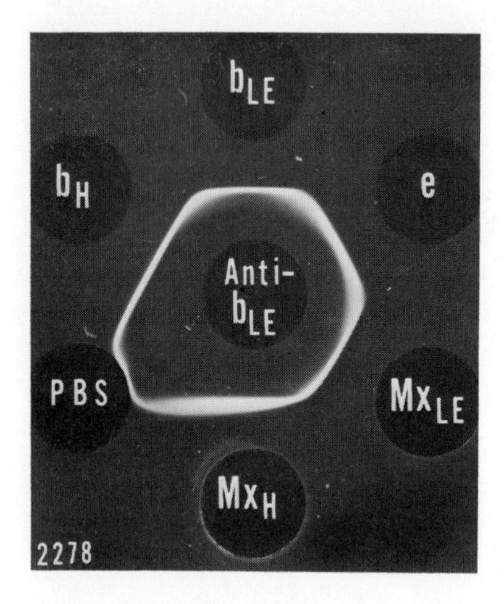

Fig. 3. Ouchterlony double-diffusion analysis of solubilized microsomes (Mx) from phenobarbital-treated Holtzman and Long Evans rats and purified cytochromes P450b and P450e. The center well contains anti-cytochrome P450b-LE. Well 5 contains phosphate buffered saline (PBS). The immunoprecipitin band in front of this well is due to fusion of the adjacent immunoprecipitin bands (P450b-H and Mx-H).

which apparently show no cross-recognition of the heterologous proteins. However, when antibody prepared against cytochrome $P450b_{LE}$ is tested against cytochromes $P450b_H$ and P450e, a single immunoprecipitin band is formed with each protein yielding a line of immunochemical identity with cytochrome $P450b_{LE}$ (Figure 3). The absence of spurs at the junction of the immunoprecipitin bands suggests that each protein shares identical immunochemical determinants. Results of immunoaffinity chromatography (14) are consistent with the apparent immunochemical identity of cytochromes $P450b_{LE}$, $P450b_H$ and P450e. A single immunoprecipitin band was also formed when anti-$P450b_{LE}$ reacted with solubilized microsomes from phenobarbital-treated Long Evans or Holtzman rats despite the fact that microsomes from Holtzman rats do not contain cytochrome $P450b_{LE}$ but only cytochromes $P450b_H$ and P450e (14).

Recent structural comparisons of cytochromes P450b and P450e using chemical or proteolytic digestion procedures have revealed small but significant differences in these proteins. When purified cytochromes $P450b_{LE}$, $P450b_H$ and P450e are excised from two-dimensional gels and submitted to ^{125}I-labeled tryptic peptide fingerprinting, a single peptide difference is observed in each protein. Peptide "s" is unique to cytochrome $P450b_{LE}$, peptide "q" is unique to cytochrome P450e, and peptide "t" is present in both cytochromes P450e and $P450b_H$ but not cytochrome $P450b_{LE}$ (14). Subsequent comparisons of peptide maps of cytochromes $P450b_{LE}$, $P450b_H$ and P450e after limited proteolytic digestion by S. aureus V8 protease, which cleaves at aspartic and glutamic acid residues, also reveal small differences in these proteins (1,2). Cyanogen bromide fragments of cytochromes $P450b_{LE}$, $P450b_H$ and P450e, generated from the cleavage of the pure proteins in SDS gels, also reveal discernible differences in these cytochromes P450 (1,2). However, using the above methods the differences observed in peptide fragmentation patterns of all three proteins are small, and these results cannot establish whether the proteins are products of separate genes or post-translational modifications (via glycosylation, phosphorylation or proteolytic processing) of a single gene product.

Two additional approaches have been taken to determine if a multi-gene family encodes cytochromes $P450b_{LE}$, $P450b_H$ and P450e. First, taking advantage of the apparent immunochemical identity of these proteins, we immunoprecipitated with anti-$P450b_{LE}$ the in vitro translation products synthesized in a rabbit reticulocyte lysate system in the presence of poly(A)$^+$-mRNA isolated from phenobarbital-treated Long Evans or Holtzman rats (15). When the immunoprecipitated ^{35}S-labeled proteins from either rat strain are subjected to two-dimensional isoelectric focusing-SDS gel electrophoresis, two radiolabeled proteins are observed (Figure 4). The two ^{35}S-labeled polypeptides encoded by liver poly(A)$^+$-mRNA from

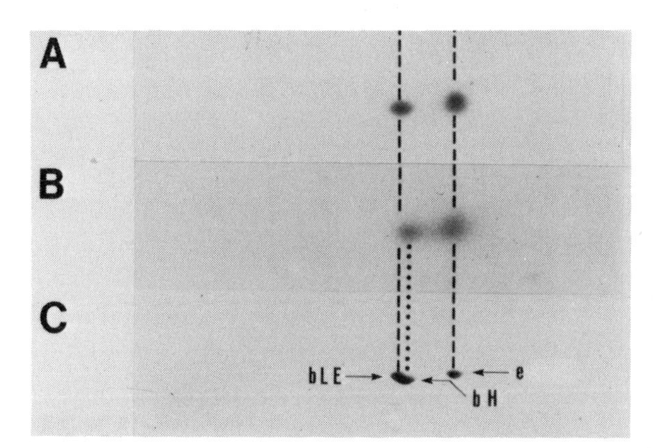

Fig. 4. Fluorographs of the same portions of two-dimensional isoelectric focusing-SDS gels containing immunoprecipiated, ^{35}S-labeled in vitro translation products using polysomal poly(A)$^+$-mRNA from livers of phenobarbital-treated Long Evans (A) or Holtzman (B) rats. C shows the same gel portion for a mixture of the purified proteins after protein staining. The dotted line indicates the isoelectric focusing coordinate of cytochrome P450b-H and the dashed lines identify the coordinates of cytochromes P450b-LE and P450e.

phenobarbital-treated Long Evans rats comigrate with purified cytochromes P450b$_{LE}$ and P450e and the two proteins encoded by mRNA from phenobarbital-treated Holtzman rats have equivalent coordinates to cytochromes P450b$_{H}$ and P450e. These results indicate that distinct mRNAs code for cytochromes P450b$_{LE}$, P450b$_{H}$ and P450e. Since the immunoprecipitated *in vitro* translation products comigrate with the purified cytochromes P450 in two-dimensional isoelectric focusing-SDS gels, it is likely these proteins do not undergo significant post-translational modification.

The second approach used to determine if these phenobarbital-inducible cytochromes P450 represent distinct gene products involves amino acid sequence determinations of the intact proteins (Edman degradation) and tryptic peptide fragments. Previous results from our laboratory have shown marked differences in the NH$_2$- and COOH-terminal sequences of cytochromes P450a, P450b, P450c and P450d from Long Evans rats (3,5). A combination of microsequence analyses of the Edman degradation products of the NH$_2$-terminal sequences, and sequence determinations of peptides isolated by HPLC, reveal that cytochromes P450b$_{LE}$ and P450e have identical NH$_2$-terminal sequences for the first 48 residues (16). However, sequence determinations of approximately 60% of the total tryptic peptide fragments of cytochromes P450b$_{LE}$ and P450e, which were resolved by HPLC, have thus far shown six amino acid differences between the two isozymes (16), each of which could have arisen from a single nucleotide change in their amino acid codons (Figure 5). Sequence determinations on other tryptic fragments of cytochromes P450b$_{LE}$ and P450e are in progress, but it is clear from the number of peptides already sequenced, that these two isozymes differ in their primary sequence by a small number of amino acids.

P450	Peptide	Sequence
b$_{LE}$	T-37-2	-Phe-Ser-Asp-Leu-Val-Pro-Ile-Gly-Val-Pro-His-Arg-
e$_{LE}$	T-32	-Phe-Ala-Asp-Leu-Ala-Pro-Ile-Gly-Leu-Pro-His-Arg-
b$_{LE}$	T-23	-Leu-Pro-Thr-Leu-Asp-Asp-Arg-
e$_{LE}$	T-20	-Glu-Ile-Asp-Glu-Val-Ile-Gly-Ser-His-Arg-Pro-Pro-Ser-Leu-Asp-Asp-Arg-
b$_{LE}$	T-34	-Ser-Lys Met-Pro-Tyr-Thr-Asp-Ala-Val-Ile-His-Glu-Ile-Gln-Arg-
e$_{LE}$	T-26-3B	-Thr-Lys Met-Pro-Tyr-Thr-Asp-Ala-Val-Ile-His-Glu-Ile-Gln-Arg-

Fig. 5. Amino acid sequences of analogous tryptic peptides from cytochromes P450b-LE and P450e. The sequence enclosed in the box is a tridecapeptide identical to that previously found in rabbit liver cytochromes P450LM2 and P450LM3b (17).

CONCLUSIONS

These studies provide the first demonstration of a closely-related gene family of phenobarbarbital-inducible cytochromes P450 in rat liver. The extensive similarities of 3 members of this family, cytochromes $P450b_{LE}$, P450e and the strain variant cytochrome $P450b_H$, were demonstrated by biochemical, immunological, peptide mapping and amino acid sequencing techniques. The identification of these hemoproteins as products of different structural genes was clearly shown by their distinct *in vitro* translation as electrophoretically unique products and by amino acid sequencing of tryptic peptides. The evidence to date suggests that cytochromes $P450b_{LE}$ and $P450b_H$ are functionally and structurally more closely related to each other than to cytochrome P450e. This evidence and preliminary genetic results (18) suggest that cytochromes $P450b_{LE}$ and $P450b_H$ are allozymes which are isozymically related to cytochrome P450e. Previously characterized rat liver microsomal cytochrome P450 isozymes (P450a, $P450b_{LE}$, P450c, P450d) were shown to differ markedly in their immunological properties, NH_2- and COOH-terminal sequences, substrate specificities and peptide maps generated proteolytically or chemically. The present results indicate that the immunochemically identical allozymes (cytochromes $P450b_{LE}$ and $P450_H$) and isozymes (cytochromes P450b and P450e), which exhibit minor structural variations and similar functionality, also contribute to the molecular complexity of these monooxygenases.

REFERENCES

1. Ryan, D.E., Thomas, P.E., Wood, A.W., Walz, F.G., Jr. and Levin, W. (1982) Fed. Proc., 41, 1403.
2. Ryan, D.E., Thomas, P.E. and Levin, W. (1982) Arch. Biochem. Biophys., in press.
3. Botelho, L.H., Ryan, D.E. and Levin, W. (1979) J. Biol. Chem., 254, 5635.
4. Ryan, D.E., Thomas, P.E. and Levin, W. (1980) J. Biol. Chem., 255, 7941.
5. Botelho, L.H., Ryan, D.E., Yuan, P-M., Kutny, R., Shively, J.E. and Levin, W. (1982) Biochemistry, 21, 1152.
6. Johnson, E.F. (1980) J. Biol. Chem., 255, 304.
7. Koop, D.E., Persson, A.V. and Coon, M.J. (1981) J. Biol. Chem., 256, 10704.
8. Laemmli, U.K. (1970) Nature, 227, 680.
9. Vlasuk, G.P. and Walz, F.G., Jr. (1980) Anal. Biochem., 105, 112.
10. Ryan, D.E., Thomas, P.E., Korzeniowski, D. and Levin, W. (1979) J. Biol. Chem., 254, 1365.
11. Thomas, P.E., Lu, A.Y.H., Ryan, D., West, S.B., Kawalek, J. and Levin, W. (1976) Mol. Pharmacol., 12, 746.
12. Thomas, P.E., Reik, L.M., Ryan, D.E. and Levin, W. (1981). J. Biol. Chem., 256, 1044.

13. Reik, L.M., Levin, W., Ryan, D.E. and Thomas, P.E. (1982) J. Biol. Chem., 257, 3950.

14. Vlasuk, G.P., Ghrayeb, J., Ryan, D.E., Reik, L., Thomas, P.E., Levin, W. and Walz, F.G., Jr. (1982) Biochemistry, 21, 789.

15. Walz, F.G., Jr., Vlasuk, G.P., Omiecinski, C.J., Bresnick, E., Thomas, P.E., Ryan, D.E. and Levin, W. (1982) J. Biol. Chem., 257, 4023.

16. Yuan, P-M., Ryan, D.E., Levin, W. and Shively, J.E. Manuscript in preparation.

17. Ozols, J., Heinemann, F.S. and Johnson, E.F. (1981) J. Biol. Chem., 256, 11405.

18. Walz, F.G., Jr., Vlasuk, G.P., Omiecinski, C.J., Bresnick, E., Levin, W., Ryan, D.E. and Thomas, P.E. (1982) Fed. Proc. 41, 1403.

Cytochrome P-450, Biochemistry, Biophysics
and Environmental Implications, E. Hietanen,
M. Laitinen and O. Hänninen editors

Gene Organization of phenobarbital-inducible cytochrome P-450 from Rat livers

Y. Fujii-Kuriyama, Y. Mizukami, K. Kawajiri[+], Y. Suwa, K. Sogawa and M. Muramatsu (Cancer Institute, Tokyo, [+]Saitama Cancer Center Research Institute, Saitama)

INTRODUCTION

Molecular multiplicity and drug induction mechanisms of cytochrome P-450 could be best understood by the investigation at the gene (DNA) level using recombinant DNA technology. We have reported the construction and identification of cDNA clones of phenobarbital-inducible cytochrome P-450 from rat livers (1, 2). From sequence analysis of these cDNA clones, we have deduced the primary amino acid sequence of the cytochrome and suggested the presence of at least two similar but distinct molecules for phenobarbital-inducible cytochrome P-450 (3). Here we report the gene dosage of the cytochrome titrated by Cot analysis, and subsequently isolation and structural analysis of its genomic clones.

RESULTS AND DISCUSSION

We titrated the gene dosage of the phenobarbital-inducible cytochrome P-450 in rat liver DNA by Cot analysis (4) using ^{32}P-labeled cDNA. Fig 1 shows the results of the hybridization kinetics. In this experiment, we used ^{3}H-labeled single-copy gene DNA as an internal standard and furthermore, we utilized cDNA probe with a very high specific activity ($1x10^8$ cpm/μg) in order to attain a true vast excess of the driving total DNA, which was necessary for an accurate determination of Cot ½ values for the single-copy DNA.

The hybridization curves obtained with the nuclease S1 assay and with the hydroxylapatite columu assay showed essentially the same pattern. Cot ½ values for the single-copy DNA and the cytochrome P-450 DNA were $2.5\text{-}2.8 \times 10^3$ and $0.5\text{-}1.0 \times 10^3$, respectively. The Cot ½ value for the cytochrome was 3 to 5 fold lower than that for the single-copy DNA, indicating that the gene number for the cytochrome P-450 is not unigue but several in rat haploid genome. The presence of several genes for the chtochrome was confirmed and more accurate quantitation was made in refereuce

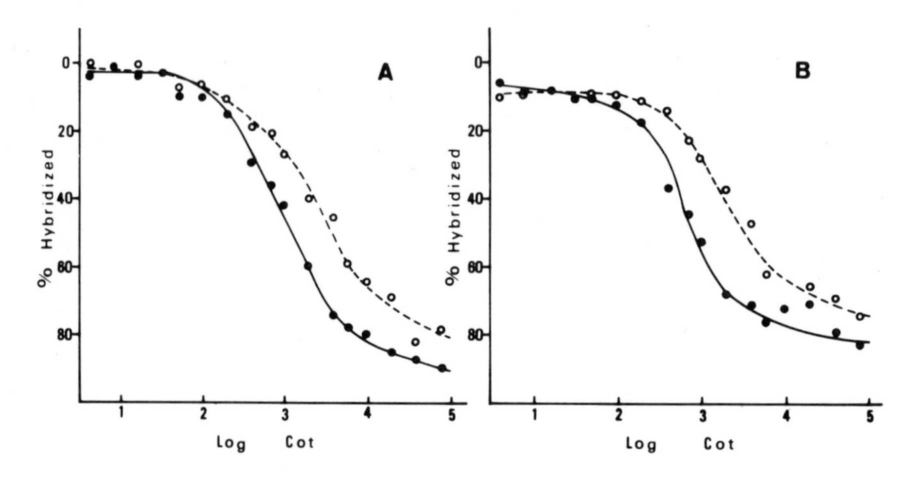

Fig. 1. Titration of cytochrome P-450 genes in total rat DNA by Cot analysis using ^{32}P-labeled P-450 cDNA as a probe. Hybrid formation was determined by hydroxylapatite method (A) and by S1 nuclease treatment.(B) •—• cytochrome P-450 o---o single copy gene

to the reassociation kinetics of standard mixtures with known gene dosage. The cloned cDNA of the cytochrome P-450 was so mixed with E, coli DNA that the content of the cDNA was 1, 3, or 6 genes per rat haploid-equivalent amount of E, coli DNA.

As shown in Fig 2, the hybridization reaction with rat DNA

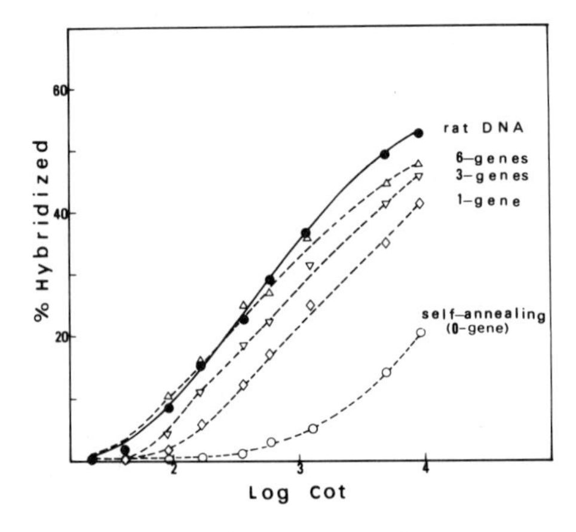

Fig. 2. Reassociation kinetics of ^{32}P-labeled P-450 cDNA with total rat DNA. Cloned cDNA of the cytochrome P-450 was so mixed with E.coli DNA that the content of the cDNA was 0, 1, 3 or 6 genes per rat haploid-equivalent amount of E.coli DNA. Hybrid formation was assayed by S1 nuclease treatment.

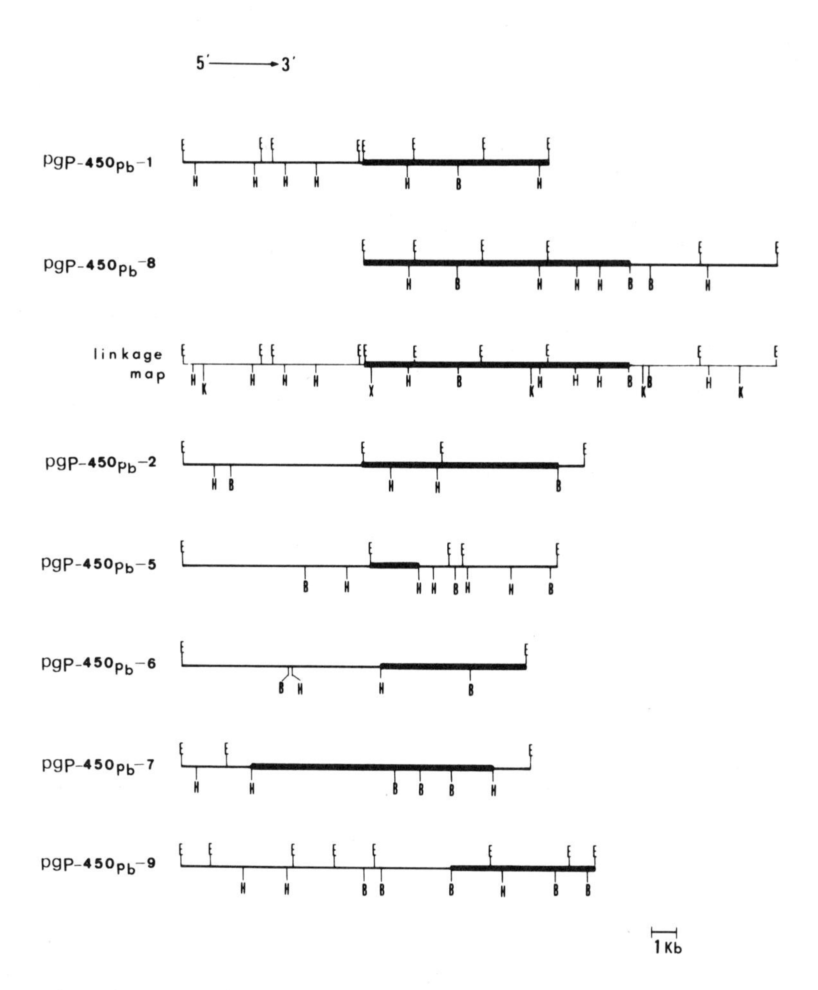

Fig. 3. Restriction cleavage maps of isolated genomic clones.
E: Eco RI, H: Hind III, B: Bom H1, K: Kpn 1, X: Xho 1

proceeded at a rate similar to that with the 6-genes standard, indicating that there exist approximately 6 genes coding for the phenobarbital inducible cytochrome P-450 in rat haploid genome.

To verify multiplicity of the cytochrome P-450 gene in the rat, we isolated genomic clones for the cytochrome by screening a rat gene library of Charon 4A with the cloned cDNA as a probe. After screening about 1 x 10^6 plaques , 9 recombinant phages hybridizing with the cDNA were isolated and designated here pg P-450 pb 1 - 9, of which 6 clones were found to be independent of one another on the basis of their restriction cleavage maps (Fig. 3) and Southern blot analysis (5) (Fig. 4).

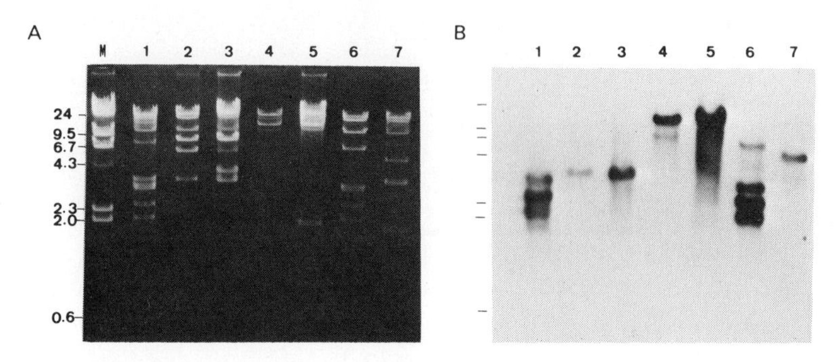

Fig. 4. Agarose gel electrophoresis (A) and subsequent Southern blot analysis of Eco RI digests of isolated P-450 genomic clones. Lane 1, pg P-450 pb 1; lane 2, pg P-450 pb 2; lane 3, pg P-450 pb 5. lane 4, pg P-450 pb 6; lane 5, pg P-450 pb 7; lane 6, pg P-450 pb 8. lane 7, pg P-450 pb 9; M, marker DNA , λDNA digested with Hind III.

It is necessary to know whether the DNA fragments in the genomic clones which hybridized with the cDNA probe are actually present in the total genomic DNA. Therefore, we performed a blot hybridization experiment with the total liver DNA. Eco R1 digested rat liver total DNA was electrophoresed on agarose in paralled with 7 cloned DNA's, transferred to the filter and hybridized with the cDNA probe. As shown in Fig. 5 (the autoradio-

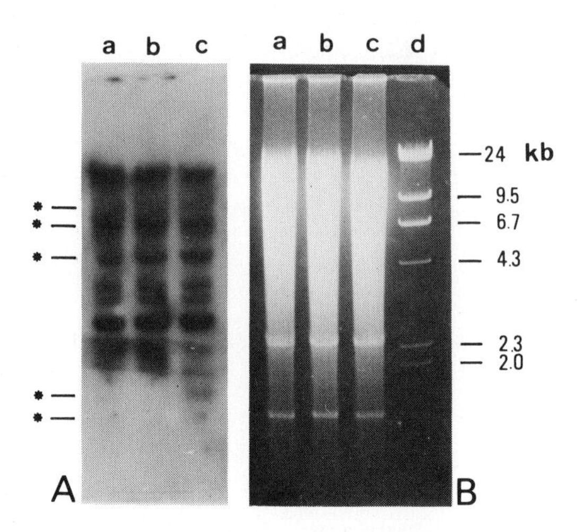

Fig.5. Agarose gel electrophoresis (B) and Southern blot analysis (A) of Eco RI digested total rat DNA. Lane a,b and c show total rat DNA digested with increased concentration of Eco RI (one,two and three fold), land d, molecular weight marker DNA (λ DNA digested with Hind III).

graphy of the cloned DNA is omitted in this figure), three different conditions of Eco R1 digestion result in the same hybridization pattern with each other, indicating the complete digestion by the enzyme. All of them showed approximately 11 visible bands. The bands equivalent to the 7 - 8 Eco R1 fragments containing cytochrome P-450 sequence in the cloned DNA's are also found in the total liver DNA. The remaining 5 bands, indicated by asterisks may show additional cytochrome P-450 sequences not represented in the isolated clones. It should be noted that the hybridization signal of 2.7 Kb band was much stronger than the others, suggesting that the nucleotide sequence of this segment is common to most of cytochrome P-450 genes. As show in Fig 3, the cleavage maps of several restriction enzymes were considerably different among the clones. How and where these differences are in the sequences of the cytochrome P-450 genes are most intriguing questions to be elucidated. It is also interesting to know whether all the gene cloned here are really expressed in livers of phenobarbital-treated rats.

We took up first pb P-450 pg-6 for further study of structural analysis because this clone was most extended toward 5' direction of the cytochrome P-450 gene. Since pg P-450 pb-6 lacked in the sequence of 3' end region of the gene, we isolated another clone (pg P-450 pb-12) whose inserted sequence covered the missing sequence in pg P-450 pb-6 by screening the rat gene library with nick-translated Eco R1-Bam H1 fragment (2.1 kb) of pg P-450 pb-6 as a probe. This clone has indeed an overlapping portion with that of pg P-450 pb-6 as revealed by restriction cleavage mapping and sequence analysis. (Fig.6)

The gene organization of the cytochrome P-450 was determined by Southern blot hybridization and sequence analysis of Maxam and Gilbert (6). As shown in Fig 6 the total length of the gene was approximately 13 Kb and was separated into 9 exons by 8 intervening sequences of various sizes.

The 3'end of the gene was determined by comparison with the cDNA sequence of pc P-450 pb-4 which had poly A sequence at the 3'terminus. Transcriptional initiation site of the gene was expected to be approximately 30 bp upstream the initiation codon as judged from the primer extension experiment (3). However the exact point has not yet been determined. S1 mapping method

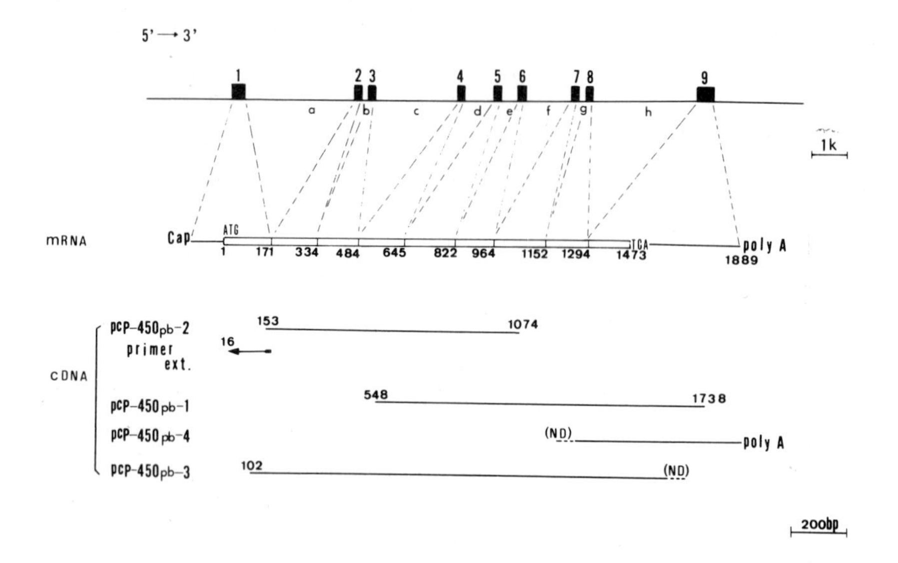

Fig. 6. Gene structure of cytochrome P-450 and its correspondence with the structive of cytochrome P-450 mRNA.

is now underway to pinpoint the transcriptional initiation site. CATAAAA sequence, probably a modified TATA box, is found 57 bp upstream the initiation codon. With aid of the gene structure, the outline of the cytochrome P-450 mRNA is depicted as shown in Fig 6. Total length of the mRNA is approximately 1920 base units plus poly A sequence, in which about 30 bases are for the leader sequence, 1473 bases for the coding sequence and 416 bases for the trailor sequence followed by the poly A stretch .

It is not yet known whether this general structure of the mRNA is applied to the mRNA for the other types of cytochrome P-450 (e.g. methylcholanthrene-inducible type et al.). Negishi et al. (7) have reported the unusually long mRNA (23S) for mouse methylcholanthrene cytochrome P_1-450. It remain to be seen whether the difference is due to the difference between species or types of cytochrome P-450.

Our preliminary results show that cDNA clone of methylcholanthrene does not appear to cross-hybridize with that of phenobarbital-inducible one under less stringent conditions of washing (1xSSC). It should be possible that genes for other types of cytochrome P-450 constitute different gene families from the one reported here.

ACKNOWLEDGEMENTS

We wish to express our gratitude to Drs T.D. Sargent, R.B. Wallace and J. Bonner for the Eco R1 rat gene library and Drs. L.L. Jagodrinsky and J. Bonner for the Hae III rat gene library. This work was supported in part by Grant-in-aids from the Ministry of Education, Science and Culture of Japan and funds obtained under Life Science Project from the Institute of Physical and Chemical Research, Japan.

REFERENCES

1. Fujii-Kuriyama, Y., Taniguchi, T., Mizukami, Y., Sakai, M., Tashiro, Y., and Muramatsu, M. (1980) Proc. Jpn. Acad. 56B 603-608
2. Fujii-Kuriyama, Y., Taniguchi, T., Mizukami, Y., Sakai, M., Tashiro, Y. and Muramatsu, M. (1981) J. Biochem. 89, 1868-1879
3. Fujii-kuriyama, Y., Mizukami, Y., Kawajiri, K., Sogawa, K., and Muramatsu, M. (1982) Proc. Natl. Acad. Sci. USA 79, 2793-2797
4. Mizukami, Y., Fujii-Kuriyama, Y. and Muramatsu, M., Submitted.
5. Southern, E. (1975) J. Mol. Biol. 98,503-517
6. Maxam, A.M. and Gilbert, W. (1977) Proc. Natl. Acad. Sci. USA 74,560-564
7. Negishi, M., Swan, D.C., Enquist, L.W. and Nebert, D.W., Proc. Natl. Acad. Sci. 78, 800-804 (1981)

Cytochrome P-450, Biochemistry, Biophysics
and Environmental Implications, E. Hietanen,
M. Laitinen and O. Hänninen editors

GENES FOR CYTOCHROME P-450 AND THEIR REGULATION

M. ADESNIK, E. RIVKIN, A. KUMAR, A. LIPPMAN, C. RAPHAEL AND M. ATCHISON
Department of Cell Biology, New York University School of Medicine, New York,
NY 10017 (U.S.A.)

We have begun to study the genes related to the one coding for a phenobarbital (PB) induced form of rat liver cytochrome P-450 and have obtained some interesting information on the size of this gene family, the relatedness of the various family members, their intron-exon organization and the extent of their transcriptional activation by PB and other inducers.

Characterization of cloned cDNA. We recently reported the isolation of cloned cDNAs bearing sequences complementary to the mRNAs coding for the phenobarbital induced forms of rat liver cytochrome P-450 (1). The insert in one of these clones, R17 is approximately 1150bp in length and is derived from the 3' half of P-450 mRNA. An open reading frame of 633 bp in the insert predicts an amino acid sequence which agrees with the partial amino acid sequence data obtained by Levin and Shively and their associates for cytochrome P-450e, the minor PB-induced form in Long-Evans rats (W. Levin and J. Shively, personal communication). In hybridization selection experiments with this cloned cDNA and mRNA from livers of PB-induced rats, followed by in vitro translation and immuneprecipitation of cell free products with either conventional or monoclonal antibodies against PB-induced cytochrome P-450, depending on the rat strain used, mRNAs coding for two (Long Evans) or three (Holtzman) electrophoretically separable polypeptides are obtained. The two mRNAs detected in Long-Evans rats presumably correspond to those coding for P-450b and P-450e in the nomenclature used by Levin and his associates (2).

Isolation of cytochrome P-450 genomic clones. Thirteen non-identical recombinant phages bearing sequences which hybridized to the cloned P-450 cDNA were isolated from a Sprague Dawley rat liver-Charon 4A genomic library. Preliminary restriction mapping using Southern blotting and hybridization to subfragments of the cloned cDNA revealed that the clones could be organized into six groups (Figure I) which are likely to represent six distinct genes.The gene represented by the clones in group I hybridizes strongly to both the 3' and 5' halves of the cDNA whereas the other genes hybridize strongly to the 5' half but only weakly to the 3' half of the cDNA. The apparent stronger homology of the cloned cDNA to the clones in group I was confirmed by the finding that when nitrocellulose filters bearing the various cloned genomic DNAs hybridized to the nick translated cDNA or <u>in vitro</u> labelled mRNA from livers of PB treated rats were washed in media of low salt concentration (0.015M NaCl), only the members of group I retained the hybridized probe. These findings strongly suggest that the gene in group I may correspond to the gene from which the R17 cDNA clone was derived. A comparison of the fine structure restriction map of clone 1, from this group,

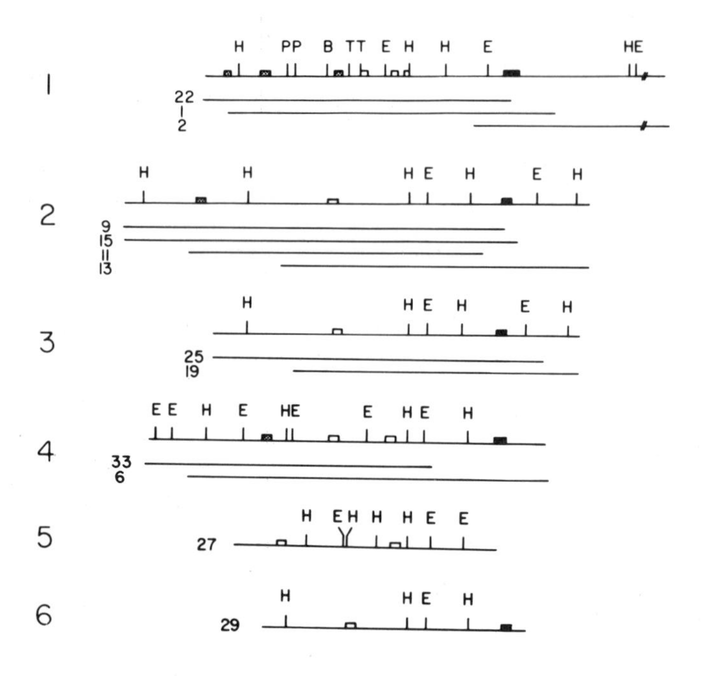

Fig. 1. Comparative restriction maps of all distinct cloned genomic segments which hybridize to the insert in cDNA clone R17. 800,000 plaques from a Sprague Dawley rat liver genomic DNA library, constructed by Linda Jagodzinski in James Bonner's laboratory were screened by hybridization with nick translated cloned cDNA. Eighteen hybridizing clones were isolated but restriction nuclease digestion data indicated that only 13 were distinct. These were mapped by single and double nuclease digestions with Eco RI (E) and Hind III (H) endonucleases using Southern blotting and hybridization to the two Hind III fragments of the R17 cDNA insert or to γ-^{32}P-labelled poly (A) + mRNA (3) to locate segments containing exonic regions and to help order the fragments. The filled in rectangles represent exons which hybridize to the 3' Hind III fragment of the cloned cDNA (see Fig. 2), the open rectangles to the exons which hybridize to the 5' Hind III fragment of the cloned cDNA and the stippled rectangles to upstream exons which hybridize to mRNA from livers of PB induced rats, but neither to the cloned cDNA nor to mRNA from livers of uninduced animals. Except for the exons in group I which hybridize to the cDNA, neither the exact size nor location of each exon relative to the flanking restriction sites are known. Some Bam H1(B), Taq 1(T) and Pst 1(P) sites in group I are shown to localize various upstream exons.

to that of the cDNA insert (Fig. 2) showed a complete coincidence of restriction sites contained in the cDNA with those in the exons of the genomic clone, taking into account the presence of introns which are also revealed by these mapping experiments. On the other hand, all the other clones which contained sequences homologous to the 3' end of the cloned cDNA, did not contain several restriction fragments which appear to be derived from the 3' terminal exon of the P-450 gene. Interestingly, all the genes contain a 70 bp HaeIII-BglII fragment found in 5' half of the cloned cDNA which in the cloned cDNA contains the sequence coding for the conserved tridecaptide found by Ozols et al (4) in two dissimilar forms of rabbit liver cytochrome P-450.

These findings further support the conclusion that clone I contains the gene coding for the PB-induced form of cytochrome P-450 whose mRNA sequence is contained in the cDNA clone R17. The restriction mapping data indicate that this gene which codes for an mRNA of 1800-2000 nucleotides, is at least 12,000 bp long and contains at least 7 exons.

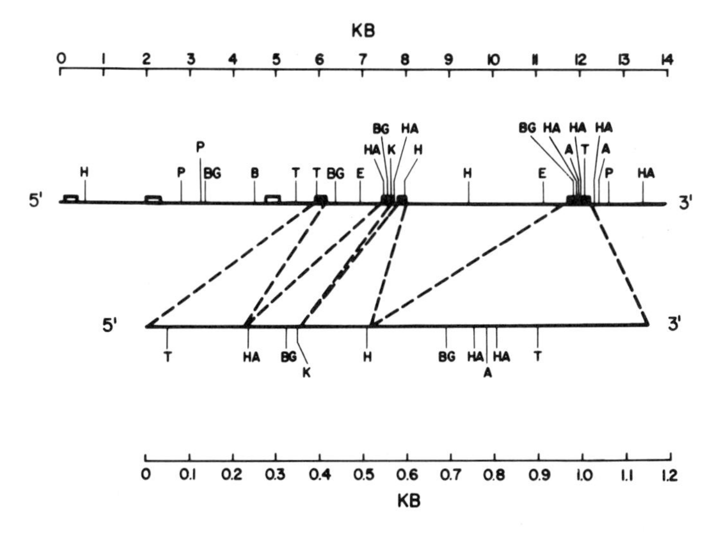

Fig. 2: Comparative fine structure restriction map of genomic clone 1 and cloned cDNA. The three Eco R1 fragments derived from the insert in clone 1 and the three Hind III subfragments of the 4.4KB central Eco R1 fragment were subcloned into pBR322. The inserts in these plasmids were then mapped by double digestion with the various endonucleases and segments containing exonic regions were identified as described in the legend to Fig. 1. The exact size and location, relative to flanking restriction sites, of exons not wholly contained within the cloned cDNA are not known. A = AvaII; B = Bam H1; BG = Bgl II; H = Hind III; HA = Hae III; K = Kpn I; P =Pst I; T = Taq I.

Transcriptional activation of P-450 genes. Hybridization of RNA labelled during *in vitro* incubation of rat liver nuclei in the presence of α-^{32}P-UTP to an excess of recombinant plasmid DNA (3) revealed a very rapid increase in the transcription of the gene or genes homologous to the cloned P-450 cDNA after a single injection of PB (Fig. 3). Increased transcription was already detectable 30 minutes after injection of PB, reached a maximum by four hours and returned to control levels by 38 hours.

This experiment which was carried out in collaboration with Dr. Eva Derman, does not indicate which of the cloned genes is activated by PB since their exons all show homology to the cloned cDNA. A semiquantitative assessment of the transcriptional level of each of the genes can be obtained, however, in a very simple experiment in which Southern blots of Eco R1 digested DNA from the various clones are probed with in vitro labelled RNA prepared from nuclei of control or PB treated rats. In this experiment transcripts of both intronic and exonic regions are measured, with the RNA complementary to intronic regions representing a great majority of specific gene transcription. The data in Fig. 4 graphically demonstrate the marked activation of the gene represented by clone 1, with the genes represented by the other clones showing a much lower rate of transcription. Quantitative hybridization experiments using subcloned fragments of clone 1, some of which contain only intronic regions of the gene, indicate that these intronic regions are transcribed at approximately the same molar rate as are the exonic regions contained in the cloned cDNA (100-200 parts per million per kb of

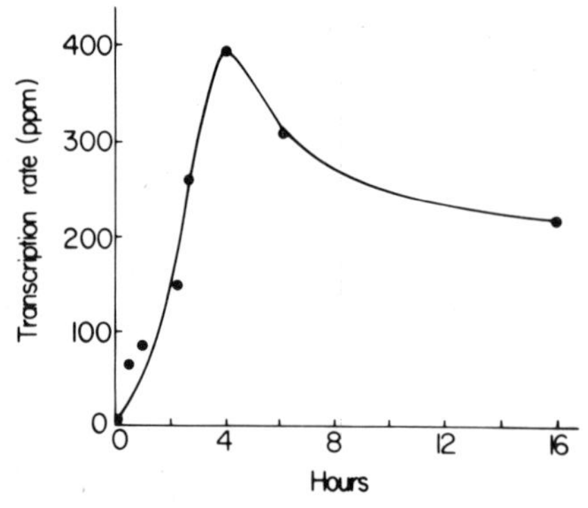

Fig. 3: Transcriptional activation of P-450 gene(s) after administration of phenobarbital. At various times after injection of phenobarbital (100mg/kg) into Long-Evans rats, nuclei were prepared, labelled *in vitro* with α-^{32}P-UTP, and the labelled RNA was extracted and hybridized to RT7 plasmid DNA (40ug) as described (5). Background hybridization to filters containing vector DNA amounted to 5-10 parts per million.

DNA). This transcription rate is approximately the same as that of the albumin gene whose mRNA product represents greater than 10% of rat liver mRNA (5). These data strongly suggest that this gene codes for P-450b, the major form of cytochrome P-450 induced by PB in livers of Long Evans rats. If this gene codes for P-450e rather than P-450b, the high rate of intronic transcription would suggest that the mRNAs for both proteins come from this same gene or that two genes coding for the two closely related proteins have highly homologous intronic sequences.

Of the genes described in this paper, one, that represented by group I, is likely to code for either P-450b or P-450e or even possibly both. We are currently sequencing various additional P-450 cDNA clones as well as the relevant regions in the group I gene to compare the mRNA and protein sequences

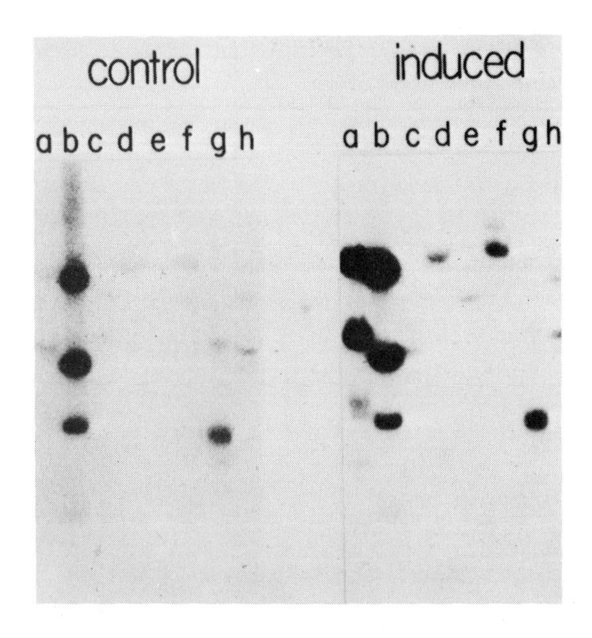

Fig. 4: Only the gene in group one is markedly activated by PB treatment. DNA from various genomic clones was cleaved with Eco R1 endonuclease, fractionated by electrophoresis on 1% agarose gels and transferred to nitrocellulose paper by the Southern blotting technique. Two replicate filters were hybridized to 10^8 cpm of fragmented (3) nuclear RNA (see legend to Fig. 3) from either a control or PB-treated rat (4 hours after injection). Lane A; clone 1; lane B, clone 2 (group 1), lanes C-F, one representative of each of the other groups. The strong hybridization of nascent nuclear RNA from uninduced animals to sequences in clone 2 (which are downstream to the 3' end of the gene in clone 1) most likely reflects hybridization to repeated sequences in the cloned DNA.

of P-450b and P-450e and to definitively identify the gene in group 1. The rat genome contains at least one and perhaps two additional genes, other than those identified here, which contain sequences complementary to P-450e cDNA as indicated by Southern blotting analysis of genomic DNA. Such a gene which by blotting experiments shows strong homology to the 3' exon and the adjacent upstream intron in clone 1, may code for P-450b or P-450e, whichever is not encoded by the group 1 gene.

Further studies are required to determine if the other genes we have isolated are pseudogenes or actually code for different forms of cytochrome P-450. None of these are likely to code for the 3-methylcholanthrene induced forms of cytochrome P-450 since in Northern blotting experiments the two electrophoretically separable mRNAs which accumulate to high levels in rat liver after 3-methylcholanthrene treatment as indicated by in vitro translation experiments, do not hybridize to the cloned P-450e cDNA even under hybridization conditions of very low stringency ($50^{o}C$, 0.9M NaCl). Furthermore, the transcription of these genes does not appear to increase in livers of animals induced with Aroclor.

ACKNOWLEDGEMENTS

We are grateful to Drs. Linda Jagodzinski and James Bonner for providing us with the rat genomic DNA library. This work was supported by Institutional grant IN-14-U from the American Cancer Society and grants GM/ES30701, GM 20277 and AG 01461 from the National Institutes of Health.

REFERENCES

1. Adesnik, M., Bar-Nun, S., Maschio,F., Zunich, M., Lippman, A. and Bard, E. (1981) J. Biol.Chem. 256:10340-10345.

2. Vlasuk, G.P., Ghrayeb, J., Ryan, D.E., Reik, L., Thomas, P.E., Levin, W. and Walz, F.J.Jr. (1982) Biochem. 21:789-798.

3. Derman, E., Krauter, K., Walling, L., Weinberger, C., Ray, M. and Darnell, J.E.Jr. (1981) Cell 23:731-739.

4. Ozols, J., Heinemann, F.S. and Johnson,E.F. (1981) J. Biol. Chem. 256: 11405-11408.

5. Taylor, J.M. and Tse, T.P.H. (1976) J. Biol. Chem. 251:7461-7467.

© 1982 Elsevier Biomedical Press B.V.
Cytochrome P-450, Biochemistry, Biophysics
and Environmental Implications, E. Hietanen,
M. Laitinen and O. Hänninen editors

STUDIES ON THE INDUCTION OF GLUTATHIONE S-TRANSFERASE B AND CYTOCHROME P-448 BY 3-METHYLCHOLANTHRENE

CECIL B. PICKETT[1], CLAUDIA A. TELAKOWSKI-HOPKINS[1], ANN MARIE DONOHUE[1] AND ANTHONY Y. H. LU[2]
From the Department of Biochemical Regulation[1] and the Department of Animal Drug Metabolism[2], Merck Sharp & Dohme Research Laboratories, Rahway, NJ 07065 (U.S.A.)

INTRODUCTION

Cytochrome P-450 and glutathione S-transferase play a major role in the metabolism of various drugs, mutagens, carcinogens and other foreign compounds. Recent studies from a number of laboratories have demonstrated that both cytochrome P-450 and glutathione S-transferase are comprised of a family of isozymes with broad overlapping substrate specificities (1-3). Consequently, the conversion of various xenobiotics to active or inactive metabolites is dependent upon the presence of different isozymes of cytochrome P-450 and glutathione S-transferase in various tissues.

In this investigation we have utilized *in vitro* translation and RNA-gel blot hybridization to determine the effect of 3-methylcholanthrene on the level of mRNA specific for glutathione S-transferase B and cytochrome P-448. Our results indicate that 3-methylcholanthrene results in a differential induction of the mRNAs specific for these two proteins.

METHODS

RNA isolation, fractionation and in vitro translation. Total rat liver RNA was isolated from 3-methylcholanthrene or phenobarbital-treated rats by the guanidine thiocyanate method of Chirgwin, et al. (4) and fractionated into poly(A^+)-RNA by oligo(dT)-cellulose column chromatography. Poly(A^+)-RNA was translated in the rabbit reticulocyte lysate cell-free system and quantitated as described previously (5).

Preparation and cloning of cDNA complementary to glutathione S-transferase B. Poly(A^+)-RNA fractions enriched in glutathione S-transferase B mRNA were reversed transcribed into single and double stranded cDNA as described by Norgard, et al. (6). The ds-cDNA was tailed with dCTP using terminal transferase. The plasmid, pBR322, was cleaved with Pst 1 and tailed with dGTP. An equimolar mixture of tailed plasmid and ds-cDNA was annealed and used to transform *E. coli* HB101 by the calcium shock procedure (7). Transformants were plated out on L-plates containing 12.5 ug/ml tetracycline.

Identification of cDNA clones complementary to glutathione S-transferase B. Recombinant colonies were screened for the presence of glutathione S-transferase B cDNA sequences by colony hybridization using a ^{32}P-cDNA probe reversed transcribed from a poly(A^{+})-RNA fraction enriched the greatest for glutathione S-transferase B mRNA. Plasmid DNA was isolated from those colonies which gave intense hybridization signals and utilized in a hybrid-selection assay to identify cDNA clones complementary to glutathione S-transferase B mRNA (8).

Filter hybridization. RNA was electrophoresed in agarose gels containing 10 mM methylmercury hydroxide and transferred to DBM paper as described by Alwine, et al. (9). After transfer, the paper containing the RNA was hybridized with a ^{32}P labeled cDNA insert complementary to glutathione S-transferase B mRNA. The cDNA clone had been labeled with ^{32}P by nick translation.

RESULTS

Induction of glutathione S-transferase B mRNA by 3-methylcholanthrene. In order to construct cDNA clones complementary to glutathione S-transferase B mRNA, we isolated total poly(A^{+})-RNA from phenobarbital-treated rats and subjected the RNA to size fractionation on a linear sucrose gradient. Each fraction of RNA was translated *in vitro* using the rabbit reticulocyte lysate cell-free translation system and immunoprecipitated with antibody against glutathione S-transferase B in order to detect the translatable mRNA. Poly(A^{+})-RNA enriched in glutathione S-transferase B mRNA were utilized for cDNA synthesis.

Double stranded cDNA was prepared using reverse transcriptase and cloned into pBR322 by G-C tailing. Recombinant clones harbouring cDNA inserts complementary to glutathione S-transferase B mRNA were identified by colony hybridization using a ^{32}P-cDNA probe reversed transcribed from poly(A^{+})-RNA enriched in glutathione S-transferase B mRNA. Approximately 30 clones were picked as candidates to contain cDNA inserts complementary to glutathione S-transferase B mRNA. These clones were analyzed further by hybrid-select translation to determine which ones were specific for glutathione S-transferase B. One clone, pGTB6, hybrid-selected glutathione S-transferase B mRNA and contained a cDNA insert of 900 bps. This clone was utilized in RNA gel-blot hybridization experiments to determine the size of glutathione S-transferase B mRNA and the degree of induction by 3-methylcholanthrene (Fig. 1). As can be seen from Figure 1, ^{32}P-pGTB6 plasmid DNA hybridizes to a single RNA species of 950 bps. Upon treatment of rats with 3-methylcholanthrene a more intense hybridization signal is seen (Figure 1, compare lane A with B). Densiometric quantitation of the autoradiographic signal indicates a 3-4 fold induction of glutathione S-transferase B mRNA. In addition, the plasmid DNA isolated from pGTB6 hybridized only to poly(A^{+})-RNA isolated from free polysomes (Fig. 1, lane C) indicating that these structures are the site of synthesis of the protein.

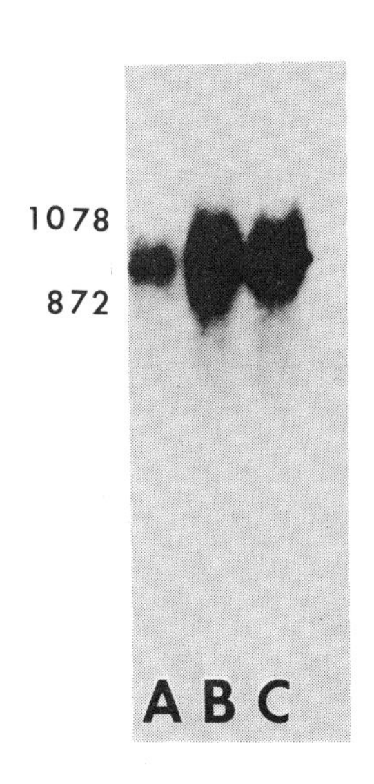

Fig. 1. Site of Synthesis and Degree of Induction of Glutathione S-Transferase B by 3-Methylcholanthrene. Five micrograms of liver poly(A^+)-RNA isolated from untreated (lane A) and 3-methylcholanthrene treated rats (lane B) or from free polysomes (lane C) were transferred from a methylmercury hydroxide/agarose gel to DBM paper and hybridized with a ^{32}P cDNA insert complementary to glutathione S-transferase B mRNA. The size markers are indicated in base pairs.

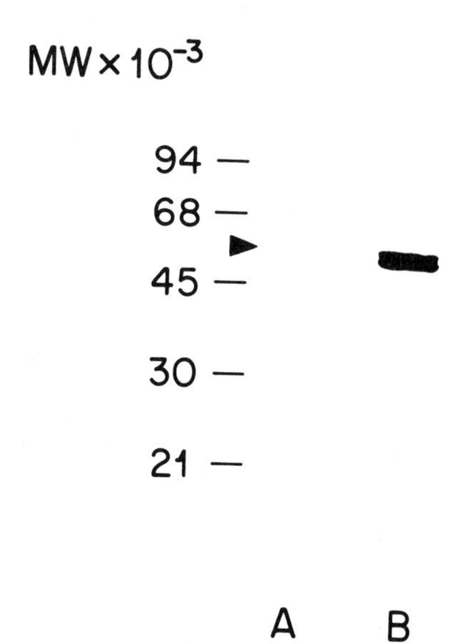

Fig. 2. Induction of Translatable Cytochrome P-448 mRNA by 3-Methylcholanthrene. Fluorogram of SDS/polyacrylamide gel electrophoresis of cytochrome P-448 immunoprecipitated from rabbit reticulocyte lysates programmed with poly(A^+)-RNA isolated from untreated rats (lane A) and 3-methylcholanthrene-treated rats (lane B). The arrowhead denotes the position of purified cytochrome P-448 in the Coomassie Blue stained gels.

Quantitation of cytochrome P-448 mRNA in untreated and 3-methylcholanthrene treated rats by in vitro translation. Total poly(A^+)-RNA from untreated rats and from rats treated with 3-methylcholanthrene was translated in the rabbit reticulocyte cell-free system. In order to detect the primary translation product of cytochrome P-448, the translation mixtures were subjected to immunoprecipitation with rabbit anti-cytochrome P-448 IgG. The immunoprecipitates obtained from these translations were analyzed by SDS/polyacrylamide gel electrophoresis and are presented in Figure 2. In untreated rats we cannot detect any functional cytochrome P-448 mRNA by this procedure (Fig. 2, lane A). Upon treatment of rats with 3-methylcholanthrene we observed that antibody to cytochrome P-448 immunoprecipitated two polypeptides (Fig. 2, lane B) with molecular weights of 55,000 and 54,000. The larger of the two comigrates with purified cytochrome P-448. The lower molecular weight polypeptide has not been identified.

DISCUSSION

In this study we have utilized a RNA gel-blot hybridization technique to determine that 3-methylcholanthrene administration to rats increases glutathione S-transferase B mRNA levels 3-4 fold and that the size of rat liver glutathione S-transferase B mRNA is 950 bps. Furthermore we have utilized in vitro translation experiments to quantitate the level of functional cytochrome P-448 mRNA in untreated and 3-methylcholanthrene treated rats. The results from these latter experiments indicate that cytochrome P-448 mRNA levels in untreated rats is very low and upon treatment by 3-methylcholanthrene becomes a major rat liver mRNA. Similar results have also been reported by Bresnick, et al. (10). Our results clearly indicate that 3-methylcholanthrene does not act in a concerted fashion but induces these two proteins very differently.

Although the molecular basis of induction of the glutathione S-transferases and cytochrome P-448 appears to be at a pretranslational level, it is unclear whether an increased transcriptional rate or a modification in post-transcriptional processing of a mRNA precursor is responsible for the accumulation of mRNA. However with the development of specific cDNA probes to the various mRNAs, these questions will be elucidated.

ACKNOWLEDGEMENTS

We would like to thank Joan Kiliyanski for assisting in the preparation of this manuscript.

REFERENCES

1. Guengerich, F.P. (1979) Pharmacol. Ther. Part A Chemother. Toxicol. Metab. Inhib., 6, 99-121.
2. Jakoby, W.B. (1978) Adv. Enzymol., 46, 383-414.
3. Lu, A.Y.H. and West, S.B. (1980) Pharmacol. Rev., 31, 277-295.
4. Chirgwin, J.M., Przybyla, A.E., MacDonald, R.J. and Rutter, W.J. (1979) Biochem., 18, 5294-5299.
5. Pickett, C.B. and Lu, A.Y.H. (1981) Proc. Natl. Acad. Sci. USA, 78, 893-891.
6. Norgard, M.V., Tocci, M.J. and Monahan, J.J. (1980) J. Biol. Chem., 255, 7665-7672.
7. Cohen, S.N., Chang, A.C.Y. and Hsu, L. (1972) Proc. Natl. Acad. Sci. USA, 69, 2110-2114.
8. Cleveland, D.W., Lopata, M.A., MacDonald, R.J., Cowan, N.J., Rutter, W.J. and Kirschner, M.W. (1980) Cell, 20, 95-105.
9. Alwine, J.C., Kemp, D.J., Parker, B.A., Reiser, J., Renart, J., Stark, G.R. and Wahl, G.M. (1979) Methods Enzymology, 68, 220-242.
10. Bresnick, E., Brosseau, M., Levin, W., Reik, L., Ryan, D.E. and Thomas, P.E. (1981) Proc. Natl. Acad. Sci. USA, 78, 4083-4987.

Cytochrome P-450, Biochemistry, Biophysics
and Environmental Implications, E. Hietanen,
M. Laitinen and O. Hänninen editors

STIMULATED de novo SYNTHESIS OF CYTOCHROMES P-450 BY DRUGS AND STEROIDS IN PRIMARY MONOLAYER CULTURES OF ADULT RAT HEPATOCYTES

PHILIP GUZELIAN, SAMMYE NEWMAN AND ERIN GALLAGHER
Medical College of Virginia, Division of Clinical Toxicology, Box 267, MCV Station, Richmond, VA 23298 (U.S.A.)

INTRODUCTION

For reasons discussed in detail elsewhere (1), it is generally believed that cultures of replicating or stationary hepatocytes derived from embryonic, malignant, or adult rodent liver seem to retain the capacity only for induction of the forms of cytochrome P-450 that predominate in fetal and extrahepatic tissues and that are especially responsive to polycyclic aromatic hydrocarbon inducers. It has been suggested that in hepatocyte cultures incubated with complex media or maintained on unusual substrata, phenobarbital (PB) "induces" cytochrome P-450 measured as CO binding hemoprotein or as catalytic oxidase activities [reviewed in (1)]. However, these methods do not provide unequivocal identification of the forms of cytochrome P-450 that accumulate in culture and are unable to distinguish between changes in synthesis versus degradation of cytochrome P-450 heme or apoprotein. By employing immunochemical techniques to measure the synthesis of individual forms of hepatic cytochrome P-450, we provide here and elsewhere (1-3) evidence that nonproliferating hepatocyte cultures maintained in serum-free medium are competent to support inducible de novo synthesis of two distinct forms of hepatic cytochrome P-450 that predominate, respectively, in rats treated with PB ($P450_{PB}$) or with the synthetic steroid, pregnenolone-16α-carbonitrile (PCN) ($P450_{PCN}$) (4).

METHODS

Hepatocyte cultures were prepared by incubating freshly isolated hepatocytes, prepared by in situ perfusion of rat liver with collagenase, in plastic 60 mm dishes precoated with rat tail collagen containing 3 ml of serum-free medium (5). Male rats (200-225 gm) were used for preparing cultures for experiments involving PB; female rats (160-180 gm) served as donors for all other cultures. All cultures were incubated in control medium for 24 hours and then transferred to medium supplemented with the agents listed in Table 1 for an additional 72-96 hours.

Synthesis of cytochromes P-450 was measured by exposing the cultures to $[^3H]$-leucine for four hours (1,2). The cells were washed, removed from the plates by scraping, and lysed by sonication. The proteins in the lysate were solubilized, and cytochrome P-450 was immunoprecipitated by addition of form-specific IgG directed against either purified $P450_{PB}$ or $P450_{PCN}$ or against non-immune IgG as control. The radioactivity in $P450_{PB}$ or $P450_{PCN}$ was determined by subjecting the immunoprecipitates to electrophoresis on polyacrylamide slab gels and counting gel slices by liquid scintillation spectrometry. Evidence that the assay is specific and quantitative has been reported elsewhere (1,2). The results are given as net radioactivity incorporated into immunoprecipitable cytochrome P-450 relative to the incorporation of isotope into total acid-precipitable protein in the cell lysate expressed as a percent.

RESULTS AND DISCUSSION

The basal rate of synthesis of the $P450_{PCN}$ in cultures maintained under standard conditions for five days was approximately 0.1% of the rate of total protein synthesis (Table 1), whereas the rate of synthesis of $P450_{PB}$ was undetectable (<0.09%), defined as less than 100 dpm's over background in the excised polyacrylamide gel band. Additions of 3-methylcholanthrene to the cultures produced no effect on either total protein synthesis or on synthesis of $P450_{PB}$ or $P450_{PCN}$. Cultures incubated for four days in the presence of PCN exhibited a dose-dependent increase in synthesis of $P450_{PCN}$ to as much as 50 times higher than that in incubated controls. Synthesis of $P450_{PB}$ remained undetectable, which suggests that PCN induces $P450_{PCN}$ specifically. Although PCN is devoid of glucocorticoid activity (6), dexamethasone or α-methylprednisolone also stimulated _de novo_ synthesis of immunoreactive $P450_{PCN}$ in these cultures. Indeed, dexamethasone was a more dramatic inducer than PCN in increasing synthesis of immunoreactive $P450_{PCN}$ to 9.5% of total protein synthesis in one experiment. This is consistent with the finding that dexamethasone is a better inducer of hepatic immunoreactive $P450_{PCN}$ in rats than is PCN (6). A non-glucocorticoid steroid, β-estradiol, was without effect on $P450_{PCN}$ synthesis in culture, and none of the steroids tested gave significant effects on the apparent rate of total protein synthesis (data not shown). It remains to be seen whether the immunoreactive $P450_{PCN}$ induced by dexamethasone in culture is identical to that induced by PCN. The cytochrome proteins induced by these steroids _in vivo_ are indistinguishable by Ouchterlony double diffusion analysis (6). Another unanswered question is whether induction of $P450_{PCN}$ by dexamethasone represents a pleiotypic response mediated by the well-known cytoplasmic glucocorticoid receptor or whether induction is mediated by a currently unrecognized receptor for which both dexamethasone and PCN are agonists.

TABLE 1

EFFECT OF DRUGS, STEROIDS, AND SELENIUM ON SYNTHESIS OF CYTOCHROMES P-450

Additions to Culture Medium	Synthesis of Cytochrome P-450 Proteins (Percent of Total Protein Synthesis)[a]	
	$P450_{PB}$	$P450_{PCN}$
None (control)	<0.09 ± 0.02 (14)	0.12 ± 0.08 (7)
3-Methylcholanthrene ($4x10^{-6}M$)	<0.05	0.06
PCN ($10^{-4}M$)	--	2.70 ± 0.84 (4)
PCN ($10^{-7}M$)	--	0.85
Dexamethasone ($10^{-5}M$)	--	7.20 ± 2.31 (3)
Dexamethasone ($10^{-6}M$)	<0.05	7.50
Dexamethasone ($10^{-7}M$)	--	1.95
Dexamethasone ($10^{-8}M$)	0.11; <0.05	0.63
α-Methylprednisolone ($10^{-5}M$)	--	2.20
β-Estradiol ($10^{-5}M$)	--	0.013
PB ($2x10^{-3}M$)	0.42 ± 0.30 (14)	--
PB ($1x10^{-4}M$)	0.19	--
Mephenytoin ($0.9x10^{-5}M$)	0.21 ± 0.05 (3)	--
Selenium ($10^{-7}M$)	<0.05	--
PB ($2x10^{-3}M$) + Selenium	0.68 ± 0.34 (13)	--
PCN ($10^{-4}M$) + Selenium	--	2.52
Single Study (Male Rat)		
PCN ($10^{-5}M$)	--	0.67
PB ($2x10^{-3}M$)	0.36	--
PCN + PB	0.21	0.36

[a]data are mean ± S.D. (n).

Additions of PB to the cultures produced a dose-dependent increase in the rates of synthesis of $P450_{PB}$ (as much as 20-fold). Whereas PB had no effect on the synthesis of $P450_{PCN}$ in culture (Table 1), PB is a modestly effective inducer of hepatic immunoreactive $P450_{PCN}$ in living rats (6). A possible explanation for this discrepancy is that PB may in some manner increase the concentration of endogenous glucocorticoids in the liver *in vivo*, and these agents, in turn, may induce $P450_{PCN}$. Steroids were without effect on induction of $P450_{PB}$ in hepatocyte cultures. In contrast, mephenytoin, a "phenobarbital-like" inducer, increased stimulated synthesis of this immunoreactive $P450_{PB}$.

Hence, for induction of both $P450_{PCN}$ and $P450_{PB}$, there is specificity, but not exclusivity, in the identified inducers.

In line with the suggestion that induction of $P450_{PB}$ may be impaired in rats maintained on a selenium-deficient diet (7), we have found that supplementation of the culture medium with selenium augmented the stimulated synthesis of $P450_{PB}$ produced by PB by approximately two-fold. Selenium was without effect on induction of $P450_{PCN}$ by PCN. Elsewhere, we have shown that the selenium content of hepatocytes incubated in control medium (no added selenium) progressively declines to less than 15% of the level in rat liver (3). Hence, selenium appears to be a permissive nutritional factor specifically needed for drug-mediated induction of at least one, but not all, forms of cytochrome P-450.

In summary, we have found that primary cultures of adult rat hepatocytes express two differentiated and specialized liver functions; namely, stimulated synthesis of $P450_{PB}$ and $P450_{PCN}$, two mature forms of hepatic cytochrome P-450. Expression of the genes controlling formation of these proteins can be evoked separately or simultaneously (Table 1) by process(es) requiring specific inducers and possibly other metabolic factors. This culture system should prove useful for defining the molecular events controlling the synthesis and degradation of cytochromes P-450.

ACKNOWLEDGEMENTS

The authors wish to thank Joyce Barwick and Nabil Elshourbagy for assistance in these experiments. We also thank Marcia Tetlak for superb secretarial assistance. This research was supported by a Program Project in Liver Metabolism grant from the National Institutes of Health (No. P01-AM18976). Dr. Guzelian is the recipient of a Research Career Development Award from the National Institutes of Health (No. K04-AM00570).

REFERENCES

1. Elshourbagy, N.A., Barwick, J.L. and Guzelian, P.S. (1981) J. Biol. Chem. 256, 6060-6068.
2. Newman, S.L., Barwick, J.L., Elshourbagy, N.A. and Guzelian, P.S. (1982) Biochem. J. 204, 281-290.
3. Newman, S.L. and Guzelian, P.S. (1982) Proc. Natl. Acad. Sci. 79, 2922-2926.
4. Elshourbagy, N.A. and Guzelian, P.S. (1980) J. Biol. Chem. 255, 1279-1285.
5. Bissell, D.M. and Guzelian, P.S. (1980) Ann. N.Y. Acad. Sci. 349, 85-98.
6. Heuman, D.M., Gallagher, E.O., Barwick, J.L., Elshourbagy, N.A. and Guzelian, P.S. (in press) Molec. Pharmacol.
7. Correia, M.A. and Burk, R.F. (1978) J. Biol. Chem. 6203-6210.

Cytochrome P-450, Biochemistry, Biophysics
and Environmental Implications, E. Hietanen,
M. Laitinen and O. Hänninen editors

ADIPOSE TISSUE CONTENT MODULATES THE RESPONSE OF MICE TO POLYCHLORINATED BIPHENYLS, INDUCERS OF POLYSUBSTRATE MONOOXYGENASES

MARKKU AHOTUPA AND EERO MÄNTYLÄ
Department of Physiology, University of Turku, Kiinamyllynkatu 10, SF-20520
Turku 52 (Finland)

INTRODUCTION

Laboratory mouse strains exhibit different responses to the exposure of foreign compounds. In "responsive" mouse strains the activities of polysubstrate monooxygenases are elevated by polycyclic aromatic hydrocarbons whereas these compounds cause only minor, if any, changes in monooxygenase activities of "nonresponsive" mice (1). "Nonresponsive" mice require a 10-fold higher dose of 2,3,7,8-tetrachlorodibenzo-p-dioxin (TCDD) than "responsive" mice for the induction of arylhydrocarbon hydroxylase (2). The susceptibility to toxicity of TCDD is shown to be inherited together with arylhydrocarbon hydroxylase responsiveness (3,4).

Recently, Decad et al. (5) reported that the adipose tissue content of a "nonresponsive" mouse strain (DBA/2J) is much greater than that of "responsive" mice (C57BL/6J). These investigators also suggested that differences in the responses of these two mouse strains to the exposure of polyhalogenated aromatic hydrocarbons (PHAs) might partly be due to the more abundant adipose tissue in the "nonresponsive" mice. The present study was designed to investigate the importance of adipose tissue content with regard to the response of laboratory mice to exposure of PHAs. For that purpose, we compared the bioaccumulation of a PCB isomer (2,4,5,2',4',5'-hexachlorobiphenyl) in "responsive" and "nonresponsive" mice, and the elevation of their hepatic polysubstrate monooxygenase activities by this isomer.

MATERIALS AND METHODS

As representatives of "responsive" and "nonresponsive" mouse strains, C57BL and DBA/JBOnf strains, respectively, were used. The animals, when fed ad libitum, weighed 22-28 g. To reduce the adipose tissue content of DBA/JBOnf mice, the food intake of animals was restricted for 10 days. The body fat content was estimated gravimetrically after diethylether extraction of homogenized whole mice. 2,4,5,2',4',5'-Hexachlorobiphenyl was purchased from RFR Corp., Hope, R.I., U.S.A. PCB, dissolved in corn oil, was administered i.p. to mice as a single dose. Control mice received an equal volume (5 ml/kg) of corn oil. The amount of PCB in liver was determined gas chromatographically after diethylether extraction (HP 5700A, ^{63}Ni EC-detector). The activities of arylhydrocarbon hydroxylase (AHH) (6) and ethoxycoumarin deethylase (ECDE) (7) were determined in freshly prepared Ca^{++}-aggregated microsomes (8).

RESULTS

The adipose tissue content of C57BL mice was much smaller than that of the DBA/JBOnf strain. When food intake of mice of the latter strain was restricted, the animals lost within 10 days about 30 % of their body weight and there was a considerable decrease in the amount of total body fat (Table 1).

TABLE 1

ADIPOSE TISSUE CONTENT, HEPATIC PCB CONCENTRATION, AND THE ACTIVITIES OF POLYSUBSTRATE MONOOXYGENASES OF C57BL AND DBA/JBOnf MICE

PCB concentration and enzyme activities were measured 5 days after a single i.p. administration of PCB in corn oil. Number of animals was 3-5.

Mice	Adipose tissue[e]	PCB[f]	AHH[g]	ECDE[h]
C57BL[a,c]	7.9 ± 1.5	59.7 ± 11.4	151 ± 13	226 ± 6
DBA/JBOnf[a,d]	14.7 ± 1.9	41.1 ± 1.9	137 ± 6	209 ± 20
DBA/JBOnf[b,d]	5.1 ± 1.3	85.9 ± 21.7	177 ± 3	378 ± 43

[a]Fed ad libitum
[b]"Fasted"
[c]100 mg/kg of PCB
[d]200 mg/kg of PCB
[g]% of the control (4.93 ± 0.73 nmol/min x g wwt)
[h]% of the control (11.6 ± 1.9 nmol/min x g wwt)
[e]% of body weight
[f]µg/g wet weight

Propably due to the greater adipose tissue content of DBA/JBOnf mice, these mice required much greater doses of PCB than C57BL mice to reach similar hepatic PCB concentrations (Fig. 1). When the adipose tissue content of DBA/JBOnf mice decreased, more PCB could be found in the liver (Table 1, Fig. 1).

There was a strong positive correlation between the hepatic PCB concentration and polysubstrate monooxygenase activities in C57BL mice treated with PCB doses $\leqq$ 100 mg/kg (Table 2), but not with DBA/JBOnf mice treated with similar doses. However, the response of "fasted" DBA/JBOnf mice with higher hepatic PCB concentrations was not different from that of the C57BL strain (Table 2).

TABLE 2

CORRELATION (r) OF THE HEPATIC PCB CONCENTRATION WITH POLYSUBSTRATE MONOOXYGENASE ACTIVITIES OF C57BL AND DBA/JBOnf MICE

Mice were treated with a single i.p. dose (10-200 mg/kg) of 2,4,5,2',4',5'-hexachlorobiphenyl. The mice were killed 5 days later. The correlations were calculated according to the least squares method. N = number of animals.

Mice	r *	N	r **	N
C57BL[a]	0.707 (2p < 0.01)	16	0.770 (2P < 0.001)	15
DBA/JBOnf[a]	0.033 (NS)	11	0.292 (NS)	11
DBA/JBOnf[b]	0.591 (2P < 0.01)	18	0.934 (2P < 0.001)	17

[a]Fed ad libitum, dose $\overline{\overline{<}}$ 100 mg/kg
[b]"Fasted", dose $\overline{\overline{<}}$ 200 mg/kg

* AHH

** ECDE

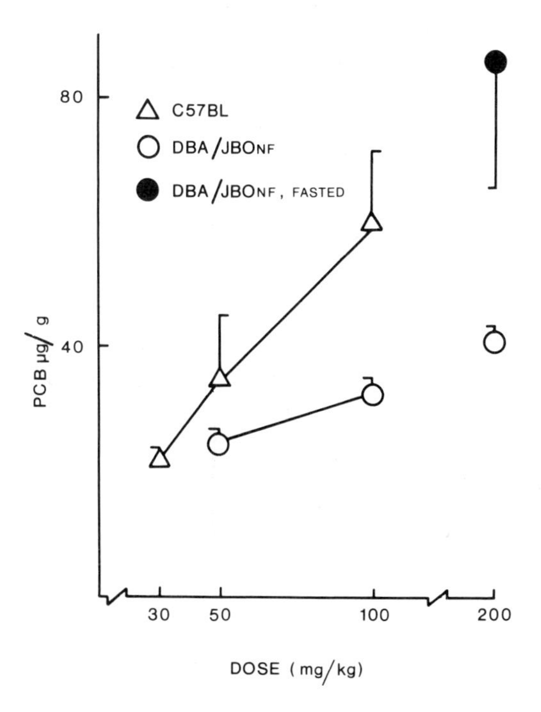

Figure 1. Hepatic PCB concentration of C57BL and DBA/JBOnf mice 5 days after a single i.p. injection of various doses. Number of animals was 3-5. PCB was analyzed gas chromatographically after diethylether extraction.

DISCUSSION

Phenobarbital treatment elevates polysubstrate monooxygenase activities in a similar way in both "responsive" and "nonresponsive" mouse strains (9). In this study we used a PCB isomer which is a representative of the phenobarbital-type PCBs (10). Therefore, differences seen in the response of C57BL and DBA/JBOnf mouse strains to exposure of PCB can hardly be explained in terms of function of the TCDD receptor. More likely, from the results of the above studies it can be concluded that the quantity of body fat strongly influences the elevation of polysubstrate monooxygenase activities by PCB, since (i) adipose tissue content of mice determined the hepatic accumulation of PCB, and (ii) the accumulation of PCB in the liver in turn determined the magnitude of elevation of hepatic polysubstrate monooxygenase activities.

ACKNOWLEDGEMENTS

This study was financially supported by grants from The Academy of Finland, and from NIH ROI ES 01684.

REFERENCES

1. Poland, A., Glover, E., Robinson, J.R. and Nebert, D.W. (1974) J. Biol. Chem. 249, 5599-5606.
2. Poland, A. and Glover, E. (1974) Mol. Pharmacol. 10, 389-398.
3. Jones, K.G. and Sweeney, G.D. (1980) Toxicol. Appl. Pharmacol. 53, 42-49.
4. Poland, A. and Glover, E. (1980) Mol. Pharmacol. 17, 86-94.
5. Decad, G.M., Birnbaum, L.S. and Matthews, H.B. (1981) Toxicol. Appl. Pharmacol. 59, 564-573.
6. DePierre, J.D., Moron, M.S., Johannesen, K.A.M. and Ernster, J. (1975) Anal. Biochem. 63, 470-484.
7. Aitio, A. (1978) Anal. Biochem. 85, 488-491.
8. Aitio, A. and Vainio, H. (1976) Acta Pharmacol. Toxicol. 39, 555-561.
9. Chhabra, R.S., Tredger, J.M., Philpot, R.M. and Fouts, J.R. (1974) 15, 123-130.
10. Goldstein, J.A. (1979) Ann.N.Y.Acad.Sci. 320, 164-178.

Cytochrome P-450, Biochemistry, Biophysics
and Environmental Implications, E. Hietanen,
M. Laitinen and O. Hänninen editors

THE EFFECT OF DIETARY CHOLESTEROL ON MONOOXYGENATION AND GLUCURONIDATION REACTIONS IN CONTROL AND 2,4,5,2´,4´,5´-HEXACHLOROBIPHENYL-TREATED C57BL AND DBA/JBOnf MICE

EERO MÄNTYLÄ, EINO HIETANEN AND MARKKU AHOTUPA
Department of Physiology, University of Turku, SF-20520 Turku 52 (Finland)

INTRODUCTION

Dietary factors influence the activities of membrane-bound drug metabolizing enzymes (1). In rats 2% cholesterol diet increases the cholesterol content of microsomal membranes and enhances polysubstrate monooxygenase and UDPglucuronosyltransferase activities (2). C57BL is a responsive and DBA/JBOnf a non-responsive mouse strain, i.e. only in responsive mouse strains polycyclic aromatic hydrocarbons cause an enhancement in several polysubstrate monooxygenase activities and UPDglucuronosyltransferase activity (3, 4). Our purpose was to study the effects of a cholesterol-free and cholesterol-rich diet on liver drug metabolizing enzymes in these two mouse strains. Moreover, since dietary cholesterol is known to modify the structure of cellular membranes we studied if these chances have an effect on the hepatic accumulation of PCB.

MATERIALS AND METHODS

C57BL and DBA/JBOnf mice (25-32 g) were used. The mice were fed on a cholesterol-free or cholesterol-rich (2% cholesterol) diet (ICN Nutritional Biochemicals, Cleveland, OH, U.S.A.) for 4 weeks. Five days before sacrifying, a group of the mice was treated i.p. (0.27 mmol/kg) with 2,4,5,2´,4´,5´-hexachlorobiphenyl (HCB) (Ultra Scientific, Hope, RI, U.S.A.) in corn oil. From the livers Ca^{2+}-aggregated microsomes were prepared (5, 6). Ethoxycoumarin deethylase (7, 8) and PPO (2,5-diphenyloxazole) hydroxylase (9) activities were determined from the microsomes. UDPglucuronosyltransferase activity with 4-methylumbelliferone as the aglycone (10, 11) was measured from digitonin activated (12) microsomes. HCB-concentrations in liver were measured with GLC (Hewlett-Packard Model 5700 A, ^{63}Ni-EC detector) after diethylether extraction. In the statistical evaluation of the results Student´s t-test was used.

RESULTS

Compared with the cholesterol-free diet, the cholesterol-rich diet increased the basal ethoxycoumarin deethylase and PPO hydroxylase activity over 2-fold in both mouse strains (Table 1). The basal UDPglucuronosyltransferase activity

TABLE 1.

HEPATIC ETHOXYCOUMARIN DEETHYLASE, PPO HYDROXYLASE AND UDPGLUCURONOSYLTRANSFERASE ACTIVITY AND PCB CONTENT IN CONTROL AND PCB-TREATED C57 AND DBA MICE FED ON A CHOLESTEROL-FREE OR 2% CHOLESTEROL DIET[a,b]

	Ethoxycoumarin deethylase	PPO hydroxylase	UDPglucuronosyl-transferase	PCB content (nmol/g)
CHOLESTEROL-FREE				
C57 (control)	3.32±0.48	0.693±0.091	230±26	-
C57 (PCB)	2.96±0.40	0.533±0.079	220±18	28.8±9.2
DBA (control)	2.94±0.28	0.750±0.031	129±18	-
DBA (PCB)	3.77±0.20	0.852±0.091	130±11	49.0±6.7
CHOLESTEROL-RICH				
C57 (control)	6.55±0.75	1.551±0.214	231±28	-
C57 (PCB)	9.72±1.13	1.937±0.181	304±33	25.6±13.3
DBA (control)[c]	6.47±0.69	2.180±0.300	205±28	-
DBA (PCB)	7.48±1.75	1.906±0.172	303±71	45.7±3.4

Significance brackets: Ethoxycoumarin deethylase — C57 (control) cholesterol-free vs. C57 (control) cholesterol-rich **; DBA (control) cholesterol-free vs. DBA (control) cholesterol-rich **; DBA (control) vs. DBA (PCB), cholesterol-free *; C57 (control) vs. C57 (PCB), cholesterol-rich *. PPO hydroxylase — C57 (control) cholesterol-free vs. C57 (control) cholesterol-rich **; DBA (control) cholesterol-free vs. DBA (control) cholesterol-rich **. UDPglucuronosyltransferase — C57 (control) vs. DBA (control), cholesterol-free *; DBA (control) cholesterol-free vs. DBA (control) cholesterol-rich *.

[a]The activities are expressed as nmol/min x g wet weight exept for PPO hydroxylase as Δfluorescence units/min x g wet weight.

[b]Means and S.E.M. ´s are shown, number of animals is 5, except in [c]n =4; * = 2p < 0.05, ** = 2p < 0.01.

was lower in DBA mice than in C57 mice with the cholesterol-free diet. The cholesterol-rich diet enhanced the basal UDPglucuronosyltransferase activity in DBA mice (1.6-fold) so that the difference between these two mouse strains disappeared (Table 1).

HCB-treatment elevated the ethoxycoumarin deethylase activity in DBA mice fed on the cholesterol-free diet (1.3-fold). With the cholesterol-rich diet the activity increased in C57 mice (1.5-fold). HCB caused a slight increase in the UDPglucuronosyltransferase activity in both mouse strains fed on the cholesterol-rich diet. The hepatic concentrations of HCB were about 28 and 47 nmol/g liver in C57 and DBA mice, respectively, with both diets (Table 1).

DISCUSSION

Cholesterol-rich diet doubled the polysubstrate monooxygenation rate both in C57 and DBA mice. This is in agreement with the results obtained with Wistar rats (2). In Wistar rats it has been also observed that cholesterol-rich diet elevates hepatic cytochrome P-450 content 1.7-2.0-fold (13, 14). It is thus possible that cholesterol exerts its effect via inducing the polysubstrate monooxygenases and a certain amount of cholesterol is needed to maintain the basal rate. On the other hand, an increase in the cholesterol content of the microsomal membranes may increase the rigidity of the membrane structure (15, 16). Further, cholesterol-rich diet elevates the microsomal phospholipid content and alters the fatty acid composition of the phospholipids (13, 15). These alterations in the properties of the membranes may facilitate substrate entrance to the monooxygenase complex and this might bring about the enhanced monooxygenation rate.

The hepatic PCB content was not affected by dietary cholesterol.

ACKNOWLEDGEMENTS

This study was supported by grants from The Foundation of University of Turku (Finland), The Academy of Finland, J. Vainio Foundation (Finland) and NIH R01 ES 01684.

REFERENCES

1. Campbell, T.C. and Hayes, J.R. (1974) Pharmac. Rev., 26, 171-197.
2. Hietanen, E., Hänninen, O., Laitinen, M. and Lang, M. (1978) Pharmacology, 17, 163-172.
3. Poland, A.P., Glover, E., Robinson, J.R. and Nebert, D.W. (1974) J. Biol. Chem., 249, 5599-5606.
4. Nebert, D.W. and Jensen, N.M. (1979) CRC Crit. Rev. Biochem., 6, 401-437.

5. Kamath, S.A. and Narayan, K.A. (1972) Anal. Biochem., 48, 59-61.
6. Aitio, A. and Vainio, H. (1976) Acta Pharmacol. Toxicol., 39, 555-561.
7. Ullrich, V. and Weber, P. (1972) Hoppe Seylers Z. Physiol. Chem., 353, 1171-1177.
8. Aitio, A. (1978) Anal. Biochem., 85, 488-491.
9. Cantrell, E.T., Abreu-Greenberg, M., Guyden, J. and Busbee, D.L. (1975) Life Sci., 17, 317-322.
10. Arias, I.M. (1962) J. Clin. Invest., 41, 2233-2245.
11. Aitio, A. (1974) Int. J. Biochem., 5, 617-621.
12. Hänninen, O. (1968) Ann. Acad. Sci. Fenn., Ser. A2H, 142, 1-96.
13. Laitinen, M. (1976) Acta Pharmacol. Toxicol., 39, 241-249.
14. Hietanen, E., Ahotupa, M., Heikelä, A. and Laitinen, M., Drug-Nutrient Interactions, in press.
15. Lang, M. (1976) Biochim. Biophys. Acta, 455, 947-960.
16. Papahadjopoulos, D., Gowden, M. and Kimelberg, H. (1973) Biochim. Biophys. Acta, 330, 8-26.

Cytochrome P-450, Biochemistry, Biophysics
and Environmental Implications, E. Hietanen,
M. Laitinen and O. Hänninen editors

HORMONAL AND SEX-LINKED FACTORS IN THE INDUCTION OF CYTOCHROME P450 BY PHENOBARBITAL

R.M. BAYNEY, E.A. SHEPHARD, S.F. PIKE, B.R. RABIN, I.R. PHILLIPS
Department of Biochemistry, University College London, Gower Street, London
WC1E 6BT, England

INTRODUCTION

Several factors have been postulated to account for sex-linked differences in the metabolism of xenobiotics by rats including the total levels and inducibility of cytochrome P450, differential substrate binding to cytochromes P450 and variations in steroid hormone levels (reviewed by Kato (1) and Colby (2)). Most reports in the literature on the sex dependence of xenobiotic induction of rat liver microsomal cytochrome P450 have measured either total cytochromes P450 by CO-difference spectra, or specific catalytic activities of the mixed function mono-oxygenase system. The latter studies are complicated by the overlapping reaction specificities of different cytochromes P450. To overcome this problem we have used a radioimmunoassay technique to quantitate specifically the major phenobarbital-inducible variant of cytochrome P450 (PB P450) in male and female rats. Although female rats had lower levels of PB P450 than males, the percentage change of this cytochrome P450 variant on induction by PB was similar in both sexes. In males, testosterone alone does not induce this cytochrome P450, but it does potentiate induction by PB.

MATERIALS AND METHODS

Animals. Rats (Crl:CD® (S.D.) B.R) (200-250 g) purchased from Charles River U.K. Limited were used in these experiments. Where indicated, animals were treated with sodium phenobarbital and testosterone propionate for four days as described (3). Treatment of castrates started seven days after the operation.

Isolation and solubilization of microsomes
Total microsomal membrane vesicles were isolated (4) and solubilized as described (5). SDS/polyacrylamide gel electrophoresis was as described (6).

Radioimmunoassay. PB P450 was quantified in solubilized microsomal membranes by an [^{125}I] radioimmunoassay technique (5,7).

RESULTS AND DISCUSSION

SDS/polyacrylamide gel electrophoresis of solubilized microsomal membrane proteins showed that a protein of the same molecular weight (52,000) as PB P450 is present and further induced by phenobarbital in both male and female rats (Fig. 1). A specific radioimmunoassay technique (5,7) using $[^{125}I]$-PB P450 as the radiolabelled antigen was used to quantitate this protein in solubilized liver microsomes of control and PB-treated male and female rats (Fig. 2; Table 1). The results showed that male rat liver contains three times the quantity of PB P450 found in female rat liver. This particular variant constitutes a greater percentage of the total P450 in males (16.5%) than in females (9.8%). Induction by phenobarbital treatment raised the levels of PB P450 10-12 fold over the basal, non-induced levels in both sexes, raising very interesting questions concerning the mechanism of the induction process. Using a similar radioimmunoassay technique no sex-specific differences were found either in the basal levels or the inducibility by PB of NADPH-dependent cyt P450 reductase (3). Due to differences in the rat breeding stock and in the treatment regime the amounts of PB P450 in control and PB-treated males found in this investigation are smaller than those reported in our other communication in these proceedings (7).

Since male and female rats were exposed to the same external environments the sex-related differences in PB P450 levels could be due to sex specific differences in the genetic expression of PB P450 and/or in the types and levels of endogenous substrates/inducers, particularly steroid hormones. In this respect, it should be noted that one of the preferred substrates of PB P450 _in vitro_ is testosterone. Castration of male rats resulted in a decrease in the basal level of PB P450 to almost the level found in females (Table 1). The inducibility of the cyt P450 variant by PB in castrates remained the same as that in males and females (i.e. ~11-fold). Treatment of castrates with testosterone did not restore the level of PB P450 to that in intact males, indicating that a testicular factor other than testosterone may be responsible for the maintenance of higher levels of PB P450 in male rats. Administration of both testosterone and PB to castrated males induced PB P450 22-fold (Table 1). Thus, although testosterone itself does not induce this cyt P450 variant, it potentiates its induction by PB. Immunological analysis of the products of translation in vitro demonstrate that the sex- and hormonal-dependent changes in the amount of this cytochrome P450 were accompanied by parallel changes in the amount of its mRNA (3). Neither castration nor testosterone treatment had any affect on the amount of NADPH-dependent cyt P450 reductase (3).

The use of a radioimmunoassay specific for a particular cyt P450 variant (PB P450) has enabled us to demonstrate a sex specific difference in the basal levels of an individual cyt P450 variant, and the ability of testosterone to potentiate its induction by a xenobiotic.

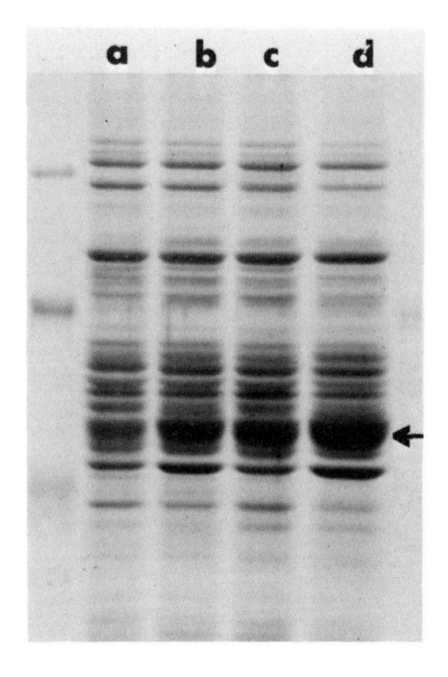

Fig. 1. SDS/polyacrylamide gel electrophoretic analysis of rat liver microsomal membrane proteins isolated from females (a), females treated with PB (b), males (c) or males treated with PB (d). The left hand track contains marker proteins of 94,000, 67,000 and 43,000 mol. wt. ← represents the position of PB P450 (52,000 mol. wt.). Only the central portion of the gel is shown.

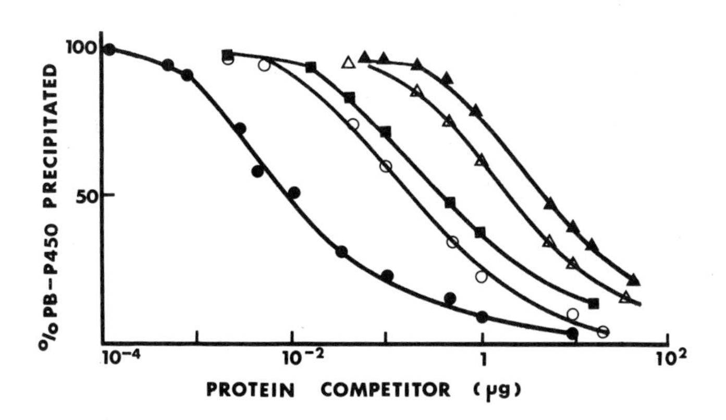

Fig. 2. Radioimmunoassay of PB P450. Radiolabelled PB P450 was competed with purified PB P450 (●), solubilized liver microsomal membrane proteins isolated from female (▲), male (△), PB-treated female (■) or PB-treated male (○) rats.

TABLE 1

SEX SPECIFICITY AND EFFECT OF TESTOSTERONE ON THE LEVELS AND INDUCIBILITY OF PB P450

Animals[a]	Treatment	PB P450[b] (nmoles mg^{-1} microsomal protein)
Female	Saline	0.05
Female	PB	0.52
Male	Saline	0.11
Male	PB	1.26
Male castrates	Oil/ethanol	0.07
Male castrates	Testosterone	0.08
Male castrates	PB	0.81
Male castrates	Testosterone/PB	1.56

[a]Pools of four animals were used in each case
[b]Results derived from radioimmunoassay

ACKNOWLEDGEMENTS

This work was supported by a grant from the Cancer Research Campaign, U.K.

REFERENCES

1. Kato, R., Drug Metab. Rev. (1974) 3, 1-32.
2. Colby, H.D. (1980) in:Advances In Sex Hormone Res. Eds. Thomas, J.A., Singhal, R.L. and Colby, H.D., pp. 27-71.
3. Bayney, R.M., Shephard, E.A., Pike, S.F., Rabin, B.R. and Phillips, I.R. Manuscript submitted.
4. Van Der Hoeven, D.A. and Coon, M.J. (1974) J. Biol. Chem. 249, 6302-6310.
5. Phillips, I.R., Shephard, E.A., Pike, S.F., Bayney, R.M. and Rabin, B.R. Manuscript submitted.
6. Phillips, I.R., Shephard, E.A., Mitani, F. and Rabin, B.R. (1981) Biochem. J. 196, 839-851.
7. Shephard, E.A., Pike, S.F., Bayney, R.M., Rabin, B.R. and Phillips, I.R., These proceedings.

Cytochrome P-450, Biochemistry, Biophysics
and Environmental Implications, E. Hietanen,
M. Laitinen and O. Hänninen editors

INDUCTION AND DECREASE OF THE MAJOR PHENOBARBITAL-INDUCIBLE CYTOCHROME P450 MEASURED BY A SPECIFIC RADIOIMMUNOASSAY TECHNIQUE

E.A. SHEPHARD, S.F. PIKE, R.M. BAYNEY, B.R. RABIN AND I.R. PHILLIPS
Department of Biochemistry, University College London, Gower Street, London,
WC1E 6BT, England

INTRODUCTION

Cytochromes P450 (cyts P450) are induced in rat liver by a variety of foreign compounds such as phenobarbital (PB) and β-naphthoflavone (NF) (1). A major requirement for studying the xenobiotic induction of cyts P450 is an assay which is specific for a single variant of cyt P450. Since cyts P450 have overlapping substrate and reaction specificities our approach to this problem has been to develop radioimmunoassays (R.I.A.) specific for individual cyt P450s. Initially we used [^{35}S]-labelled products of translation in vitro as a source of labelled antigen (2). We have now developed a more rapid, economic technique involving purified antigen labelled in vitro with ^{125}I. Both these R.I.A. systems have shown that the amount of the major PB-inducible cyt P450 variant (PB P450) is increased greatly by PB treatment but is decreased by treatment with NF.

MATERIALS AND METHODS

Animals. Male Sprague-Dawley rats (180-200g) were used in these experiments. Animals were treated with PB or NF as described (3).

Isolation and solubilization of microsomal membrane vesicles.
Total microsomal membrane vesicles were isolated (4) and solubilized by a modification of the method of Imai (5) as described (3).

Purification of PB P450 and NF P448. PB P450 and NF P448 (the major NF-inducible cyt P450) were isolated from PB-or NF-treated rats respectively, by n-octylamino sepharose 4B and DEAE cellulose column chromatography (6). Antibodies to these proteins were raised in rabbits (3).

Radioimmunoassay of PB P450 using ^{35}S-labelled PB P450.
R.I.A. was as described (2).

Radioimmunoassay of PB 450 using ^{125}I labelled PB 450.
Purified PB P450 was iodinated as described (7,3). Samples containing 2×10^4 cpm were mixed with various quantities of competitor protein and incubated overnight at 4^oC in the presence of an amount of anti-(PB P450) IgG sufficient to precipitate 50% of the labelled protein in the absence of

competitor. Antibody-antigen complexes were removed from solution by *Staphylococcus Aureus* cells (3).

RESULTS AND DISCUSSION

The [^{35}S] R.I.A. technique detects PB P450 in the range 2-200 ng but does not react with amounts of NF P448 up to 2 μg (i.e. 1000 x the minimum amount of PB P450 detectable). (Fig. 1a). Thus the cross-reactivity between NF P448 and anti-(PB P450) IgG is less than 0.1% and the IgG is highly specific for a particular P450 variant.

In the [^{125}I] R.I.A. there was somewhat higher (1.5%) cross-reactivity of NF P448 with anti-(PB P450) IgG. (Fig. 1b). To test if this small degree of cross-reactivity would interfere with the ability of the [^{125}I] R.I.A. to quantitate PB P450, we assayed PB P450 in solubilized liver microsomal membranes isolated from PB-, NF- or untreated rats. (Fig. 1c). The R.I.A. showed that PB-P450 was increased by PB from 0.003 to 0.126 mg/mg microsomal protein. This represents a 42-fold increase in the amount of PB P450 whereas total P450s (as measured by CO difference spectral analysis) were increased by only 4-fold. Although NF-treatment resulted in a 1.5-fold induction of total cyts P450 in microsomes it decreased PB-P450 by 55% to 0.0013 mg/mg microsomal protein. Essentially identical results were obtained with the [^{35}S] R.I.A. (Fig. 1d), therefore the small amount of cross-reactivity observed in the [^{125}I] R.I.A. does not affect the accuracy of this assay even when measuring a small amount of PB P450 in the presence of a relatively large amount of P448. The changes in protein levels of PB P450 observed on treatment with either PB or NF are paralleled by changes in the mRNA for this protein (3). Using a similar R.I.A. technique for the NADPH-dependent cyt P450 reductase we find that this enzyme is induced 2-fold by PB treatment but its level is not changed by NF treatment (3).

We have developed two independent R.I.A. techniques for the accurate quantitation of a specific cyt P450 variant. The results demonstrate that PB P450 is greatly increased by PB treatment but is decreased by NF. The assays can detect as little as 1 ng (~20 fmol) of PB P450, and in practise the [^{125}I] R.I.A. can quantitate cyt P450 concentrations as low as 2 pM or 0.1 ng PB cyt P450/mg protein. This is several thousand-fold more sensitive than either radial immunodiffusion (8,9) or rocket immunoelectrophoresis (10), alternative techniques used to immunochemically quantitate cyt P450 variants. Another advantage of the [^{125}I] R.I.A., is that it is experimentally very easy to operate. The competition curves are very reproducible and their precision

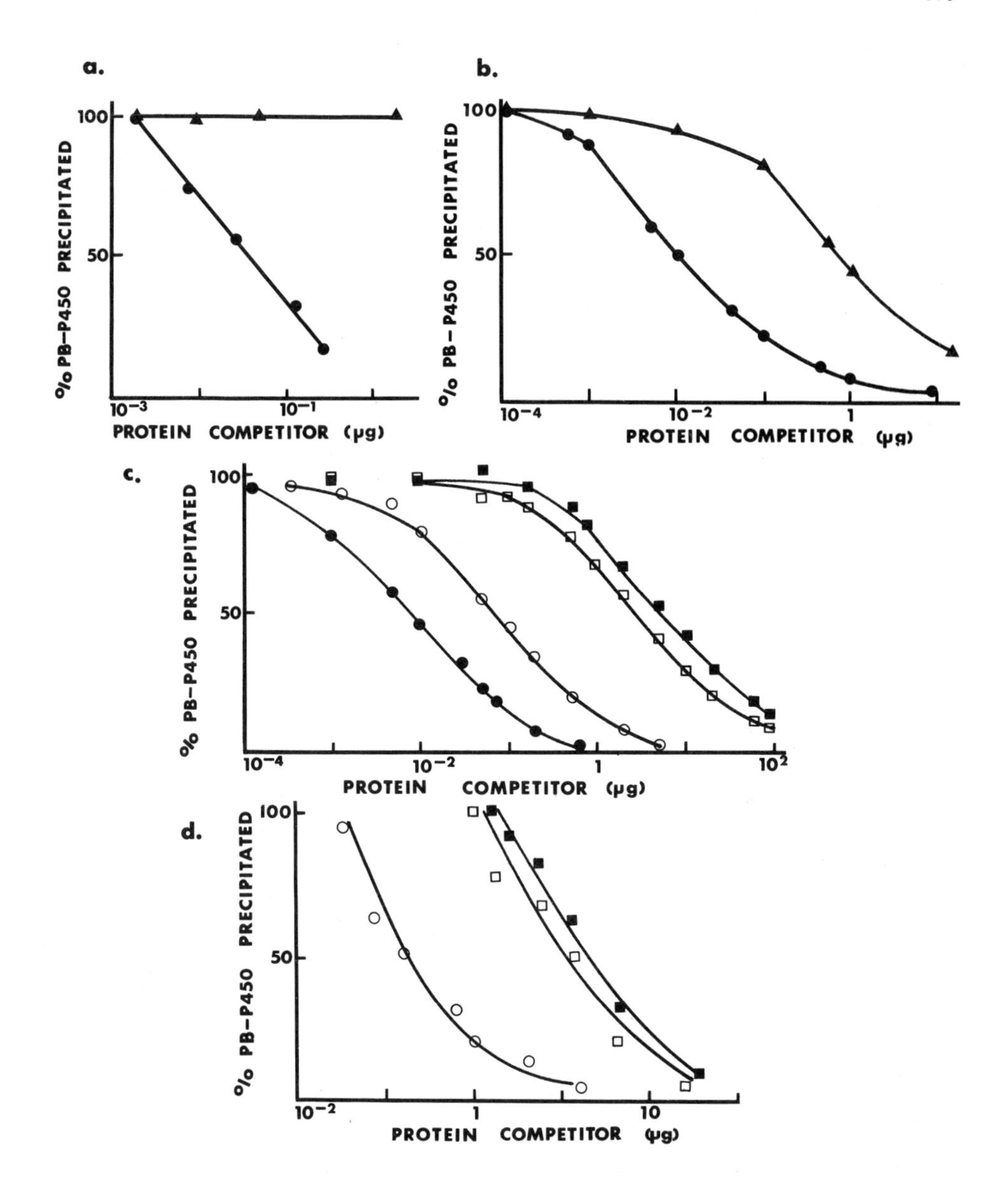

Fig. 1. Cross-reactivity of NF P448 in the [^{35}S] R.I.A. (a), and in the [^{125}I] R.I.A. (b). Radioimmunoassay of PB P450 by the [^{125}I] technique (c) or the [^{35}S] technique (d). Radiolabelled PB P450 was competed with purified PB P450 (●), purified NF P448 (▲), solubilized microsomal membrane proteins isolated from control (□), PB-treated (○), or NF-treated (■) rats.

does not decrease for the measurement of small amounts of protein. Thus the [^{125}I] R.I.A. is particularly well suited to the measurement of low quantities of specific cyt P450 variants present in control samples and in samples from animals treated with a variety of inducers. The technique should be extremely useful for the analysis of non-inducible or constitutive forms of cyt P450.

ACKNOWLEDGEMENTS

This work was supported by a grant from the Cancer Research Campaign, U.K.

REFERENCES

1. Estabrook, R.W. and Lindenlaub, E. eds. (1979) The Induction of Drug Metabolism. Schattauer Verlag, Stuttgart.
2. Phillips, I.R., Shephard, E.A., Mitani, F. and Rabin, B.R. (1981) Biochem. J. 196, 839-851.
3. Phillips, I.R., Shephard, E.A., Pike, S.F., Bayney, R.M. and Rabin, B.R. Manuscript submitted.
4. Van der Hoeven, D.A. and Coon, M.J. (1974) J. Biol. Chem. 249, 6302-6310.
5. Imai, Y. (1976) J. Biochem., 80, 267-276.
6. Guengerich, F.P. and Martin, M.V. (1980) Arch. Biochem. Biophys. 205, 365-379.
7. Bolton, A.E. and Hunter, W.M. (1973) Biochem. J., 133, 529-539.
8. Thomas, P.E., Korzeniowski, D., Ryan, D. and Levin, W. (1979) Arch. Biochem. Biophys. 192, 524-532.
9. Thomas, P.E., Reik, L.M., Ryan, D. and Levin, W. (1981) J. Biol. Chem. 256, 1044-1052.
10. Pickett, C.B., Jeter, R.L., Morin, J. and Lu, A.Y.H. (1981) J. Biol. Chem. 265, 8815-8820.

© 1982 Elsevier Biomedical Press B.V.
Cytochrome P-450, Biochemistry, Biophysics
and Environmental Implications, E. Hietanen,
M. Laitinen and O. Hänninen editors

MOLECULAR PROPERTIES OF CYTOSOLIC AND NUCLEAR FORMS OF THE *Ah* RECEPTOR

ALLAN B. OKEY and ARTHUR W. DUBÉ
Division of Clinical Pharmacology, Department of Paediatrics, Research
Institute, The Hospital for Sick Children, 555 University Avenue,
Toronto, Ontario CANADA M5G 1X8.

INTRODUCTION

The *Ah* receptor regulates induction of aryl hydrocarbon hydroxylase (AHH; cytochrome P_1-450) and other structural gene products of the *Ah* gene complex (1). Chemicals that are potent P_1-450 inducers, such as 3-methylcholanthrene (MC) and 2,3,7,8-tetrachlorodibenzo-p-dioxin (TCDD), initially bind to *Ah* receptor in the cytoplasmic compartment (2-4), after which the inducer·receptor complex translocates into the nucleus (5,6) in a step which appears to be temperature-dependent (7).

Little is known about changes in molecular characteristics of the *Ah* receptor which accompany temperature-dependent "activation" and nuclear uptake. Thus we compared properties of cytosolic and nuclear forms of the *Ah* receptor in an attempt to clarify the events which permit nuclear uptake and functional interaction of the inducer·receptor complex with chromatin.

MATERIALS AND METHODS

Preparation of Radiolabeled Cytosols and Nuclear Extracts. Cytosols were prepared from the livers of untreated C57BL/6J mice and Sprague-Dawley rats by methods previously described (4) , then incubated *in vitro* with 10 nM [^{3}H]TCDD. Radiolabeled nuclear *Ah* receptor was extracted from hepatic nuclei of animals which had been injected (IP) with [^{3}H]TCDD 18 hrs before removal of the liver (5). C57BL/6J mice each received 6 ug [^{3}H]TCDD; Sprague-Dawley rats each were injected with 18 ug [^{3}H]TCDD. Nuclear extracts were prepared from these animals as previously described (4,5). Nuclear extracts also were obtained from Hepa-1c1 cells which had been incubated in culture with 1 nM [^{3}H]TCDD for 1 hr at 37°C as previously described (7). Cytosol samples and nuclear extracts were treated with charcoal-dextran to remove unbound and loosely-bound [^{3}H]TCDD before further analysis.

Sucrose Density Gradient Analyses. Sedimentation values (*S*) for different forms of the *Ah* receptor were determined by sucrose density gradient centrifugation (4,8) using [^{14}C]-labeled bovine serum albumin as an internal sedimentation marker. Nuclear extracts were analyzed on gradients containing 0.5 M KCl. Cytosolic samples were analyzed both in the presence of 0.5 M KCl and under "low-salt" conditions.

Sephacryl S-300 Column Chromatography. Stokes radii for different forms of the *Ah* receptor were determined by gel permeation chromatography on Sephacryl S-300 columns precalibrated with standard proteins (9). Relative molecular mass (M_r) and frictional ratios (f/f_o) were calculated from data obtained by gradient centrifugation and S-300 chromatography.

RESULTS AND DISCUSSION

Molecular properties of cytosolic and nuclear forms of the *Ah* receptor from three different biological systems are summarized in Table 1. These values are estimates based on 2-4 determinations for each parameter. In both C57BL/6J mice and Sprague-Dawley rats, the cytosolic form of *Ah* receptor is larger than the form extracted from the nucleus. This difference is revealed both by velocity sedimentation on sucrose density gradients and by elution from Sephacryl S-300 gel permeation columns.

Relative molecular mass values (M_r) for cytosolic *Ah* receptor, as determined by these methods, are similar to those previously reported by Hannah et al. (9) for C57BL/6N hepatic cytosol receptor and by Carlstedt-Duke et al. (10) for rat hepatic cytosol receptor.

At low ionic strength, the cytosolic *Ah* receptor from mouse and rat liver sediments $\sim 9S$ (4,5). Sedimentation coefficients given in Table 1 were determined under high ionic strength conditions. At high ionic strength, cytosolic *Ah* receptor from rat liver dissociates to a form which sediments $\sim 7S$. Although it previously appeared (4) that *Ah* receptor from C57BL/6 mouse liver cytosol also dissociated at high ionic strength, several recent experiments do not reveal dissociation.

In Hepa-1c1 cells the sedimentation coefficient of cytosolic *Ah* receptor does not differ significantly from that of the nuclear *Ah* receptor when both are prepared from cells incubated with [^{3}H]TCDD at 37°C in culture (7). It is not known whether cytosolic receptor recovered from cells incubated at 37°C might already have undergone "activation" or molecular modification.

TABLE 1.
MOLECULAR PROPERTIES OF CYTOSOLIC AND NUCLEAR FORMS OF THE *Ah* RECEPTOR

Molecular Parameter	C57BL/6J Mice		Sprague-Dawley Rats		Hepa-1c1 Cells	
	Cytosol	Nucleus	Cytosol	Nucleus	Cytosol	Nucleus
Sedimentation Coefficient (S)[a]	9.6	5.0	7.3	5.7	5.7[b]	6.3
Stokes radius (nm)	6.0	3.9	4.4	4.1	ND	3.8
Relative Molecular Mass (M_r)	250,000	85,000	135,000	100,000	ND	90,000
Frictional Ratio (f/f_o)	1.4	1.2	1.2	1.2	ND	1.2

[a]In buffer containing 0.5 M KCl; mouse and rat cytosols were labeled in vitro with [^{3}H]TCDD at 4°C.

[b]Hepa-1c1 cytosol was obtained from cells incubated in culture with [^{3}H]TCDD at 37°C. Sedimentation values shown for Hepa-1c1 cytosol and nuclear extract are taken from reference (7).

ND = not determined.

In both C57BL/6J mice and Sprague-Dawley rats it is clear that the nuclear form of *Ah* receptor has an M_r significantly smaller than that for the cytosolic form analyzed under the same conditions. Thus it appears that the cytosolic *Ah* receptor must undergo substantial modification in the processes of "activation" and nuclear translocation. The mechanism by which this modification occurs is unknown.

ACKNOWLEDGEMENTS

Supported by a grant from the Medical Research Council of Canada. We thank Lynn M. Vella and Michelle E. Mason for expert technical assistance.

REFERENCES

1. Nebert, D.W., Eisen, H.J., Negishi, M., Lang, M.A., Hjelmeland, L.M. and Okey, A.B. (1981) Ann. Rev. Pharmacol. Toxicol. 21, 431.
2. Poland, A., Glover, E. and Kende, A. (1976) J. Biol. Chem. 251, 4936.
3. Carlstedt-Duke, J., Elfstrom, G., Snochowski, M., Hogberg, B. and Gustafsson, J.-A. (1978) Toxicol. Lett. 2, 365.
4. Okey, A.B., Bondy, G.P., Mason, M.E., Kahl, G.F., Eisen, H.J., Guenthner, T.M. and Nebert, D.W. (1979) J. Biol. Chem. 254, 11636.
5. Mason, M.E. and Okey, A.B. (1982) Eur. J. Biochem. 123, 209.
6. Poellinger, L., Kurl, R.N., Lund, J., Gillner, M., Carlstedt-Duke, J., Hogberg, B. and Gustafsson, J.-A. (1982) Biochem. Biophys. Acta. 714, 516.
7. Okey, A.B., Bondy, G.P., Mason, M.E., Nebert, D.W., Forster-Gibson, C.J., Muncan, J. and Dufresne, M.J. (1980) J. Biol. Chem. 255, 11415.
8. Tsui, H.W. and Okey, A.B. (1981) Can. J. Physiol. Pharmacol. 59, 927.
9. Hannah, R.R., Nebert, D.W. and Eisen, H.J. (1981) J. Biol. Chem. 256, 4584.
10. Carlstedt-Duke, J., Harnemo, U.-B., Hogberg, B. and Gustafsson, J.-A. (1981) Biochem. Biophys. Acta. 672, 131.

© 1982 Elsevier Biomedical Press B.V.
Cytochrome P-450, Biochemistry, Biophysics
and Environmental Implications, E. Hietanen,
M. Laitinen and O. Hänninen editors

GENETIC REGULATION OF MOUSE LIVER MICROSOMAL P_2-450 INDUCTION BY ISOSAFROLE AND 3-METHYLCHOLANTHRENE

TOHRU OHYAMA, ROBERT H. TUKEY, DANIEL W. NEBERT AND MASAHIKO NEGISHI
Developmental Pharmacology Branch, National Institute of Child Health and Human Development, National Institutes of Health, Bethesda, MD 20205 (U.S.A.)

INTRODUCTION

In DBA/2N (D2) mouse liver microsomes, isosafrole induces a form of P-450 (having a Soret maximum at ~448 nm when reduced and combined with CO) but not any detectable induced P_1-450 or aryl hydrocarbon hydroxylase (AHH) activity (1). We propose to name this form "P_2-450." In this chapter we characterize both the catalytic activities of P_2-450 and the properties of its antibody, anti-(P_2-450). With this new form of P-450, we hope to understand better the mechanism of the induction processes by both isosafrole and MeChol.

RESULTS AND DISCUSSION

Immunoprecipitable P_2-450 from mouse liver microsomes. Sources for all materials and animals, as well as details of all experimental procedures, are described in Refs. 1-4. Following $NaB[^3H]_4$ labeling of intact microsomes *in vitro*, the microsomes were solubilized with sodium cholate, immunoprecipitated, and then electrophoresed on $NaDodSO_4$-polyacrylamide gels. Anti-(P_2-450)-precipitable radioactivity (M_r~55 kDa) occurred to a small degree in control and phenobarbital-treated B6 and D2 liver microsomes. Anti-(P_2-450)-precipitable radioactivity was seven times higher in 3-methylcholanthrene (MeChol)-treated B6 than D2. This finding suggests that P_2-450 induction by MeChol operates via the *Ah* receptor (5). Isosafrole induced P_2-450 about the same in both B6 and D2 mice (4).

When anti-(P_2-450) was adsorbed on isosafrole-treated D2 liver microsomes (Fig. 1), the resulting supernatant had little capacity to immunoprecipitate P_2-450. On the first wash, only 6% of the isosafrole-induced peaks remained, whereas about 28% of the MeChol-induced B6 peak remained. With the second wash, all peaks have become negligible. These data indicate further that P_2-450, which is induced by isosafrole, is also induced by MeChol.

Isosafrole-induced microsomes from B6 or D2 mice displayed a Soret peak of the reduced hemoprotein•CO complex at about 449 nm. The highly purified P_2-450 antigen, on the other hand, exhibited its Soret peak at 448 nm, when reduced and combined with CO (4).

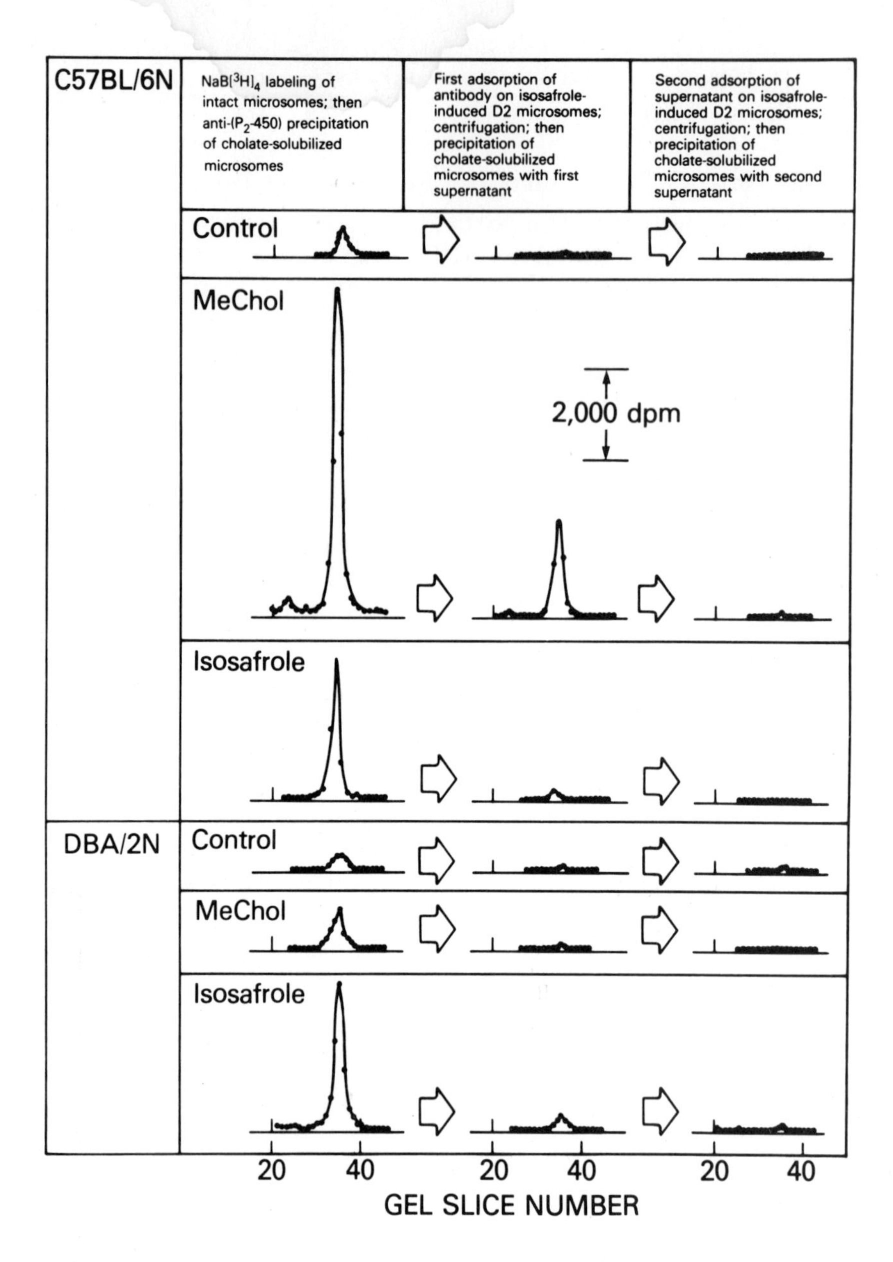

Fig. 1. Immunoprecipitation of control, MeChol-treated, or isosafrole-treated B6 or D2 microsomes by anti-(P_2-450) material that does not adsorb on isosafrole-treated D2 liver microsomes.

Catalytic activity of P_2-450. Isosafrole-induced B6 or D2 microsomes possess isosafrole metabolite(s) formed *in vivo*. Isosafrole metabolism was found *in vitro* in control, MeChol-treated, and phenobarbital-treated B6 and D2 mice, as well as isosafrole-treated B6 and D2 mice. Anti-(P_2-450) was an effective inhibitor of isosafrole metabolism *in vitro* (4). On the other hand, anti-(P_2-450) did not inhibit the metabolism of benzo[a]pyrene, acetanilide, biphenyl, or 7-ethoxycoumarin. The complete lack of inhibitory effects on AHH activity, for example, is illustrated in Fig. 2. These data indicate that P_2-450 and its antibody are more closely involved with isosafrole metabolism than other monooxygenase activities often associated with polycyclic hydrocarbon induction (*i.e.* benzo[a]pyrene, acetanilide, biphenyl, or 7-ethoxycoumarin).

Genetic expression of P_2-450 induction. B6, D2, and $B6D2F_1$ mice were tested for anti-(P_2-450)-precipitable radioactivity from liver microsomes following different doses of isosafrole. B6 levels were 2- to 3-fold greater than D2 levels, suggesting that isosafrole might use the *Ah* receptor to some degree. The F_1 levels of immunoprecipitable radioactivity were precisely intermediate between the B6 and D2 levels, indicating inheritance of P_2-450 induction is expressed as an autosomal additive trait. The results with isosafrole-treated individual offspring from the $B6D2F_1$ x D2 backcross were also consistent with additive inheritance (4). These data are in contrast to AHH induction by MeChol, which is inherited as an autosomal dominant trait (*reviewed in* Ref. 5).

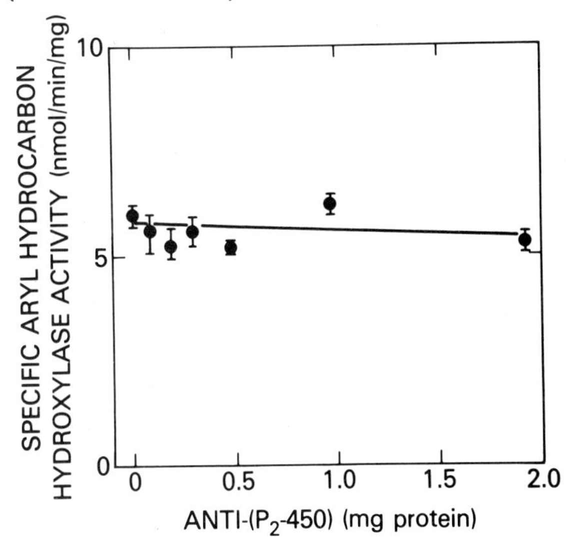

Fig. 2. Specific AHH activity in the presence of varying amounts of anti-(P_2-450). MeChol-treated B6 liver microsomes (175 μg of protein) were used.

Cloning of P_2-450 cDNA. cDNA was prepared from 2,3,7,8-tetrachlorodibenzo-*p*-dioxin-treated B6 liver 20 S mRNA and inserted into the *PstI* site of pBR322. A positive clone, p21, is 1710 bp in length; by "positive" and "negative" translation arrest, anti-(P_2-450) was shown to be associated with mRNA that hybridizes specifically to p21 DNA. Clone p21 hybridizes to pAhP-2.9, a subclone of λAhP-1. Clone p21 hybridizes to 20 S mRNA induced by isosafrole. We therefore believe that we have a cDNA clone encoding P_2-450 (6).

SUMMARY

Cytochrome P_2-450 (M_r~55 kDa) was purified from isosafrole-treated D2 mice. The CO-difference Soret peak is 448 nm for the purified protein. P_2-450 does metabolize isosafrole, but we suspect that we have yet to discover the most specific monooxygenase catalytic activity (*e.g.* an endogenous substrate) for this enzyme.

An antibody was developed against P_2-450. Immunologically it appears that P_2-450 exists in microsomes from untreated and phenobarbital-treated mice and is induced by MeChol, as well as by isosafrole. Anti-(P_2-450) does not block aryl hydrocarbon hydroxylase, acetanilide 4-hydroxylase, biphenyl 4- or 2-hydroxylase, or 7-ethoxycoumarin O-deethylase activities but does inhibit isosafrole metabolism.

P_2-450 is induced by MeChol in B6 but not D2 mice, indicating that the induction process occurs via the *Ah* receptor. P_2-450 induction by isosafrole is inherited as an autosomal additive trait and most likely uses the *Ah* receptor in evoking the induction response. A cDNA clone, p21, has been characterized and is believed to encode P_2-450.

REFERENCES

1. Tukey, R.H., Negishi, M. and Nebert, D.W. (1982) Mol. Pharmacol., in press.
2. Negishi, M. and Nebert, D.W. (1979) J. Biol. Chem., 254, 11015.
3. Negishi, M., Jensen, N.M., Garcia, G.S. and Nebert, D.W. (1981) Eur. J. Biochem., 115, 585.
4. Ohyama, T., Nebert, D.W. and Negishi, M., manuscript in preparation.
5. Nebert, D.W., Negishi, M., Lang, M.A., Hjelmeland, L.M. and Eisen, H.J. (1982) Advanc. Genet., 21, 1.
6. Tukey, R.H., Ohyama, T., Negishi, M. and Nebert, D.W. (1982) Pharmacologist, in press [Abstract].

© 1982 Elsevier Biomedical Press B.V.
Cytochrome P-450, Biochemistry, Biophysics
and Environmental Implications, E. Hietanen,
M. Laitinen and O. Hänninen editors

INDUCTION BY 3-METHYLCHOLANTHRENE OR ISOSAFROLE OF TWO P-450 mRNA's WHICH SHARE NUCLEOTIDE SEQUENCE HOMOLOGY

TOSHIHIKO IKEDA, MARIO ALTIERI, MICHITOSHI NAKAMURA,
DANIEL W. NEBERT AND MASAHIKO NEGISHI
Developmental Pharmacology Branch, National Institute of Child Health and Human Development, National Institutes of Health, Bethesda, MD 20205 (U.S.A.)

INTRODUCTION

The *Ah* locus regulates the induction of two or more forms of P-450 by polycyclic aromatic compounds such as 3-methylcholanthrene (MeChol) or 2,3,7,8-tetrachlorodibenzo-*p*-dioxin (TCDD). C57BL/6N (B6) mice are responsive to the induction process by MeChol (80 mg/kg intraperitoneally) and by TCDD (1 μg/kg intraperitoneally). The B6 allele is designated Ah^b for "aromatic hydrocarbon responsiveness." DBA/2N (D2) mice are nonresponsive even at high levels of intraperitoneal MeChol, though the induction occurs readily at TCDD concentrations of 50 or 100 μg/kg. The D2 allele is designated Ah^d for *Ah*-nonresponsiveness. The F_1 heterozygote, Ah^b/Ah^d, exhibits MeChol-induced aryl hydrocarbon hydroxylase (AHH) activity similar to the B6 parent (Ah^b/Ah^b). *Ah*-responsiveness thus is inherited as an autosomal *dominant* trait (*reviewed in* Ref. 1).

Clone 46 is a cDNA clone representing a portion of mouse cytochrome P_1-450 (that form of polycyclic-aromatic-inducible P-450 most closely associated with MeChol- or TCDD-induced AHH activity). Using clone 46 and a mouse genomic-DNA

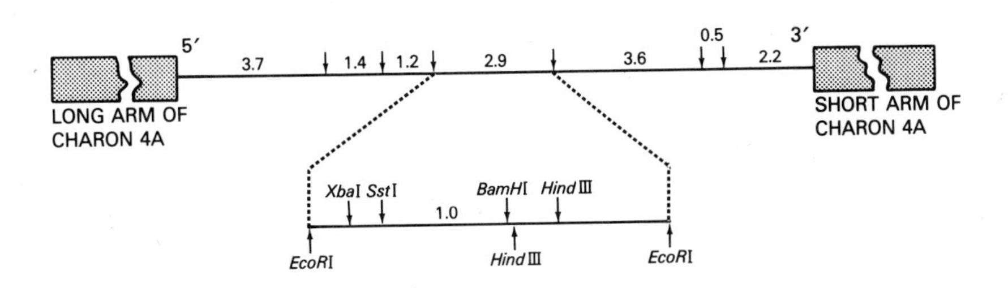

Fig. 1. Restriction map of clone λAhP-1, which spans 15.5 kbp. *Arrows* at top designate *EcoRI* cleavage sites. In the middle of clone λAhP-1, the P_1-450 genomic gene has been localized by R-loop analysis. This gene starts in the 1.2-kbp fragment and terminates in the 3.6-kbp fragment, spanning about 5 kbp and containing at least five exons and four introns. The 2.9-kbp fragment was subcloned in pBR322 and is called *pAhP-2.9*. A restriction map of pAhP-2.9 is shown at *bottom*. pAhP-2.9 was nick-translated and used, therefore, as a probe representing the 5'-region of the P_1-450 chromosomal gene.

library, we isolated λAhP-1, a 15.5-kbp clone (Fig. 1) that contains the entire P_1-450 chromosomal gene. *EcoRI* digestion yields seven fragments, ranging in size from 3.7 kbp to 0.5 kbp. The *EcoRI*-digested 2.9-kbp piece, known to be in the 5'-region of the P_1-450 genomic gene, was subcloned in pBR322 and called *pAhP-2.9*. The purpose of this study is to examine mRNA forms that hybridize to pAhP-2.9.

RESULTS AND DISCUSSION

mRNA that hybridizes to pAhP-2.9. Sources for all materials and animals, as well as details of all experimental techniques, are provided in Refs. 2-4. Hybridization of the pAhP-2.9 probe to mRNA from MeChol-treated and TCDD-treated *Ah*-responsive B6 mice (Fig. 2) showed marked increases not only in the P_1-450 (23 S) mRNA but also in a 20 S mRNA. Hybridization of the probe pAhP-2.9 to D2 mRNA showed marked increases in P_1-450 (23 S) mRNA and 20 S mRNA after TCDD pretreatment but only very slight increases after MeChol pretreatment (4). These data suggest that both the 23 S and the 20 S mRNA induction processes follow the *Ah*-responsiveness trait.

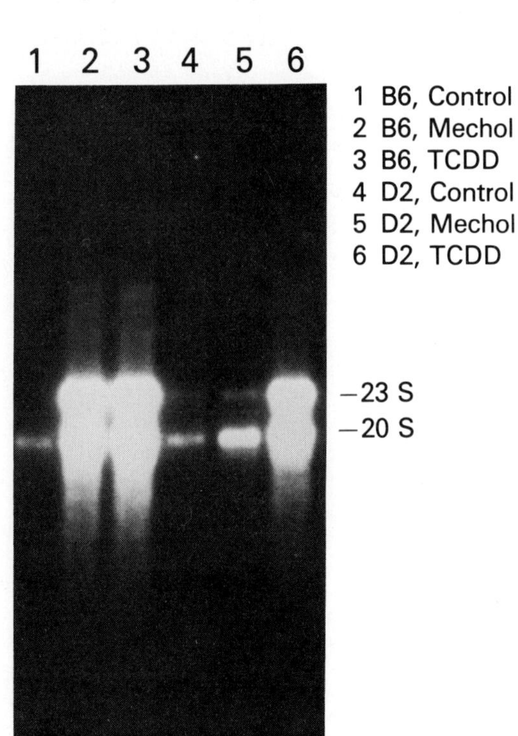

Fig. 2. Northern blot between the probe pAhP-2.9 [^{32}P]DNA and mRNA from control, MeChol-treated, or TCDD-treated B6 or D2 mice. Total liver poly(A^+)-enriched RNA was electrophoresed on diazobenzyloxymethyl (Northern) paper.

Differential sensitivity between P_1-450 mRNA and 20 S mRNA induction. MeChol was given intraperitoneally 24 h before killing at doses of 2, 20, 67, and 200 mg per kg body weight to groups of *Ah*-responsive B6 mice. We found that the 20 S mRNA is at least 10 times sensitive than the P_1-450 (23 S) mRNA to induction by MeChol in B6 mice (4). The reason for this difference in sensitivity is not known.

Association of Ah^b allele with induction of both mRNA's. The induction of both P_1-450 (23 S) mRNA and the 20 S mRNA was found to be strictly associated with the Ah^b allele. In other words, among individual MeChol-treated children from the $B6D2F_1$ x D2 backcross, AHH (P_1-450) induction was always correlated with increases in both messengers that hybridize to the pAhP-2.9 probe (4).

Isosafrole and phenobarbital treatment. When D2 and B6 mice were treated with isosafrole (Fig. 3), induction of the 20 S mRNA but not P_1-450 (23 S) mRNA occurred. No AHH induction was found in isosafrole-treated D2 mice. Phenobarbital (PhBarb) caused no rise in either the 20 S mRNA or the 23 S mRNA that hybridizes to the pAhP-2.9 probe (4).

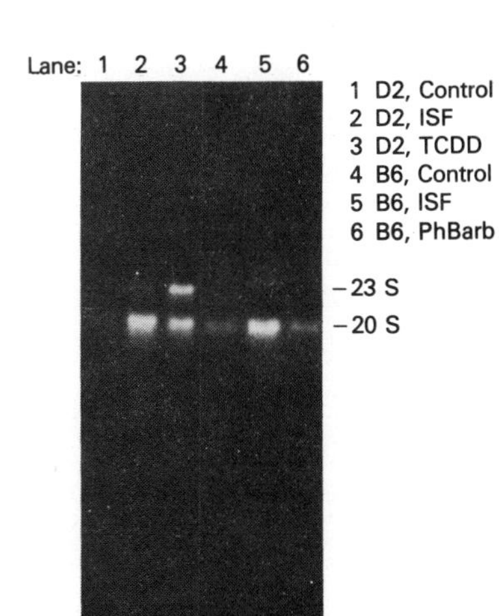

Fig. 3. Northern blot between the probe pAhP-2.9 and mRNA from control, isosafrole-treated, and PhBarb-treated D2 and B6 mice. ISF, isosafrole.

SUMMARY (Fig. 4)

MeChol (at high doses) or TCDD operate via the *Ah* receptor to induce the P_1-450 (23 S) mRNA. This mRNA is translated into P_1-450 protein, thereby causing marked increases in AHH activity that are highly specific for benzo-[a]pyrene metabolism.

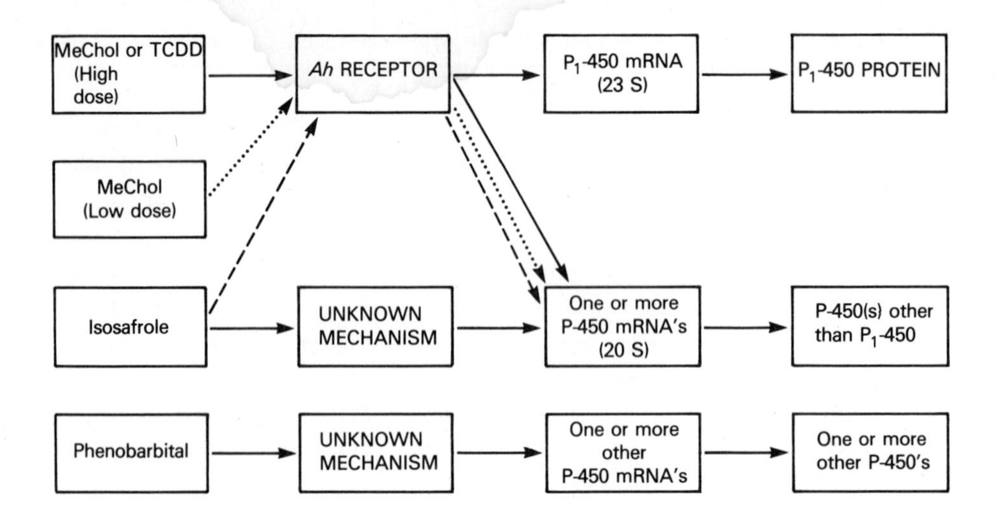

Fig. 4. Hypothetical scheme for at least three types of induction processes.

MeChol (at low doses) and isosafrole also appear to operate--at least to some extent--via the *Ah* receptor to induce one or more P-450 mRNA's 20 S in size. These mRNA's are translated into one or more forms of induced P-450 other than P_1-450. Isosafrole also operates by way of some other receptor- or nonreceptor-mediated mechanism to induce one or more forms of 20 S mRNA.

Phenobarbital operates via some other receptor- or nonreceptor-mediated process, leading to the induction of one or more P-450 mRNA's not hybridizable to our 5'-probe from the P_1-450 chromosomal gene. Translation of these mRNA's leads to the induction of one or more phenobarbital-inducible P-450 proteins. We believe that the 20 S mRNA is derived from a gene other than that represented by clone λAhP-1 but shares nucleotide sequence homology with the P_1-450 (23 S) mRNA.

REFERENCES

1. Nebert, D.W., Negishi, M., Lang, M.A., Hjelmeland, L.M. and Eisen, H.J. (1982) Advanc. Genet., 21, 1.

2. Negishi, M., Swan, D.C., Enquist, L.W. and Nebert, D.W. (1981) Proc. Natl. Acad. Sci. U.S.A., 78, 800.

3. Tukey, R.H., Nebert, D.W. and Negishi, M. (1981) J. Biol. Chem., 256, 6969.

4. Ikeda, T., Altieri, M., Nakamura, M., Tukey, R.H., Nebert, D.W. and Negishi, M. (1982) J. Biol. Chem., in press.

Cytochrome P-450, Biochemistry, Biophysics and Environmental Implications, E. Hietanen, M. Laitinen and O. Hänninen editors

INDUCTION OF LIVER MICROSOMAL UDP-GLUCURONYLTRANSFERASE BY EUGENOL AND INDUCERS OF MICROSOMAL MONOOXYGENASES IN THE RAT

AKIRA YUASA AND HIROSHI YOKOTA
Department of Veterinary Biochemistry, School of Veterinary Medicine, The College of Dairying, 582 Nishi-nopporo, Ebetsu, Hokkaido 069-01 (Japan)

INTRODUCTION

Liver microsomal UDP-glucuronyltransferase (GTase) catalyzes the glucuronidation of a variety of compounds including eugenol and bilirubin. It is known that liver microsomes contain more than one species of GTase and that administration of inducers of microsomal monooxygenases, such as phenobarbital (PB), polychlorinated biphenyls (PCBs) and 3-methylcholanthrene (3-MC), to animals leads to induction of GTase activity (1,2).

Eugenol, 4-allyl-2-methoxyphenol, is the principal constituent of clove oil and most recently in Japan, this spicery compound has become suspected of carcinogenic potential (3). Research interest of eugenol came from Finland; Hartiala *et al.* (4) reported the competitive inhibition of GTase by eugenol *in vitro*. We applied this effective glucosidurogenic compound to *in vivo* studies, which have demonstrated enhancement of GTase (4-nitrophenol as accepter substrate), UDPglucose dehydrogenase and urinary glucuronide excretion in the rat (5).

The main objective of the present series is to clarify the underlying enzyme adaptation mechanisms at the molecular level. Prior to starting to purify eugenol-induced GTase, we would like to show in this paper that eugenol serves as a substrate inducer of GTase and that its effect on the liver microsomal drug-metabolizing enzyme system *in vivo* is notably limited to a certain kind of GTase, suggesting a specific stimmulation of GTase.

MATERIALS AND METHODS

Male rats of the Wistar strain, 10-12 weeks of age, were used. They were treated by repeated oral administration at 12 h intervals with massive doses (200 mg) of eugenol. PB, PCBs and 3-MC were injected into the animals in the usual manner and doses. Bilirubin bound to bovine serum albumin was diluted in 0.1 M TES-NaOH buffer (pH 7.4) and injected intraperitoneally 1.0 mg (0.5 ml of the solution) a day for three days. All treated rats were sacrificed at the 5th day, and liver microsomes prepared in the usual manner. GTase activities towards various substrates in 0.25 % cholate- or 3.3 mM Emulgen 109P-treated

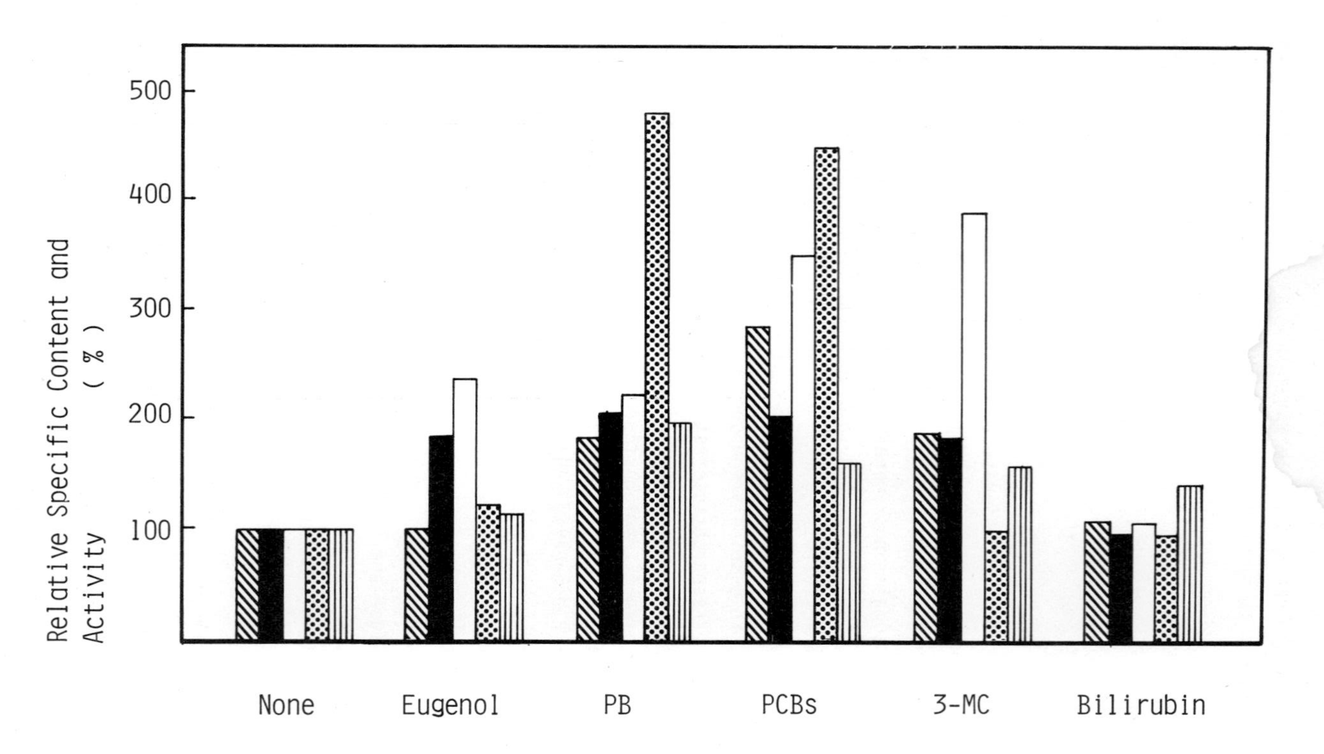

Fig. 1. Cytochrome P-450 contents and GTase activities towards four kinds of acceptor substrates in the liver microsomes of rats dosed with eugenol, bilirubin and inducers (PB, PCBs and 3-MC) of microsomal monooxygenases. Acceptor substrates for assaying GTase activities are as follows: eugenol, ■ ; 1-naphthol, □ ; 4-hydroxybiphenyl, ▩ ; bilirubin, ▥ . Cytochrome P-450 content represent by the column of ▨.

microsomes were assayed using the following methods and aglycone concentrations: 0.3 mM eugenol (6), 0.05 mM 1-naphthol (7), 0.5 mM 4-hydroxybiphenyl (8), 0.3 mM bilirubin (9). Cytochrome P-450 (P-450) was determined by the method of Omura and Sato (10).

RESULTS

Figure 1 shows induction of liver microsomal P-450 and GTase activities towards four kinds of acceptor substrates (eugenol, 1-naphthol, 4-hydroxybiphenyl and bilirubin) in the rat.

Injection of PB, PCBs, or 3-MC to the rats resulted in a significant increase in the microsomal level of P-450. Increases in *p*-nitroanisole-O-demethylation activity, cytochrome b_5 and NADPH-cytochrome P-450 reductase were also observed (data not shown). Almost all GTase activities towards the four substrates were induced remarkably by the inducers, except in the case of 4-hydroxybiphenyl conjugating activity in 3-MC induced microsomes. Thus, PB, PCBs, and 3-MC were considered as non-specific inducers for the liver microsomal drug metabolizing enzyme system including GTase. Oral administration of eugenol and injection of bilirubin, on the other hand, failed to enhance P-450 or any other microsomal electron transport system.

Treatment with eugenol, however,led to induction of eugenol and 1-naphthol glucuronidation activities without affecting GTase activities with bilirubin and 4-hydroxybiphenyl as the acceptors, and administration of bilirubin induced bilirubin glucuronidation activity alone.

DISCUSSION

Induction and heterogeneity of GTases are practically important subjects of profound interest. The present results seem to indicate that eugenol and bilirubin are capable of selectively inducing GTases possessing high activities toward them, respectively. Moreover, these observations are consistent with the findings of Bock, *et al.* (2); PB-inducible GTase activity toward 4-hydroxybiphenyl and 3-MC-inducible GTase activity toward 1-naphthol may originate from separate enzyme entities. The observed difference in substrate specificity of induced GTases suggests the possible occurrence of multiformal and multifunctional GTases, like P-450s.

Eugenol- or bilirubin-induced GTase preparation will serve as the more suitable materials in researches on mechanisms of induction and heterogeneity of GTase in that their stimulating effects on liver microsomes are notably limited and more specific than the inducers of microsomal monooxygenases such as PB, PCBs, and 3-MC.

ACKNOWLEGEMENTS

The authors are greatly indebted to Prof. R. Sato of Osaka University for his valuable advices. We would like to thank J. Hoshino, T. Fujita and N. Ohgiya of this laboratory for their helpful technical assistance.

REFERENCES

1. Dutton, G.J. (1980) Glucuronidation of Drugs and Other Compounds, CRC Press, Boca Raton, pp. 137-147.
2. Bock, K.W., Josting, D., Lilienblum, W. and Pfeil, H. (1979) Eur. J. Biochem. 98, 19-26.
3. Ishidate, M., Sofuni, T. and Yoshikawa, K. (1981) Mutagens & Toxicology, 4, 80-89.
4. Hartiala, K.J.W., Pulkkinen, M. and Ball, P. (1966) Nature, 210, 739-740.
5. Yuasa, A. In: Sato, R. and Kato, R. (Eds.), Microsomes, Drug Oxidations and Drug Toxicity, Japan Academic Societies Press, Tokyo, in press.
6. Mulder, G.J. and Van Doorn, A.B.D. (1975) Biochem. J., 151, 131-140.
7. Mackenzie, P.I. and Hänninen, O, (1980) Anal. Biochem., 109, 362-368.
8. Bock, K.W., Clausbruch, U.C.V., Kaufmann, R., Lilienblum, W., Oesch, F., Pfeil, H. and Platt, K.L. (1980) Biochem. Pharmacol., 29, 495-500.
9. Heierwegh, K.P.M., Van de Vijver, M. and Fevery, J. (1972) Biochem. J., 129, 605-618.
10. Omura, T. and Sato, R. (1964) J. Biol. Chem., 239, 2370-2378.

Cytochrome P-450, Biochemistry, Biophysics and Environmental Implications, E. Hietanen, M. Laitinen and O. Hänninen editors

INFLUENCE OF INCUBATION TEMPERATURE ON MONOOXYGENASES ACTIVITIES IN LIVER OF TENCH, SUCKLING AND ADULT RAT AFTER INDUCTION WITH ß-NAPHTHOFLAVON (BNF) AND DICHLORPROP (DCP)

FRANK SCHILLER, FLORENTINE JAHN and WOLFGANG KLINGER
Institute of Pharmacology and Toxicology, Jena, G.D.R.

INTRODUCTION

In G.D.R. water supply is dependent on the use of surface water. Contamination by industrial emission oil products, agrochemicals including herbicides etc. plays an important role. Besides chemical analysis biomonitoring is necessary, fishes seem to be especially suitable. Activity and inducibility of the cytochrome P-450 system has been used for biomonitoring successfully (1, 2). We investigated the sensitivity of a widespread fish in lakes and ponds of our region, the tench (tinca tinca) and its suitability for biomonitoring.

METHODS

Tench from a lake in the south of G.D.R. was used, weight 200 - 400 g, both sexes, age unknown, held in fresh tap water of 12 or 18°C from October - February without feeding for 2 - 4 months. For comparison 10- (poikilothermic) and 60-day-old male rats (homoiothermic) of our institute's colony breed (Uje:Wist) were used. Cytochrome P-450 content (P-450, 3),ethylmorphine N-demethylation (E-N, 4), ethoxycoumarin O-deethylation (E-O, 5) and ethoxyresorufin O-deethylation (EROD, 6) were measured in liver microsomes (P-450) or 10 000 g supernatant. BNF and DCP were added to the bassins for 96 h, rats received these compounds in 0.5 ml DMSO per 100 g body weight orally.

RESULTS

EROD activities in rat and tench are shown in Fig. 1. 60-day-old rats have a three times higher activity than 10-day-old rats. There are no significant differences in tench liver after keeping at 12 or 18°C water temperature. In tench liver about 1/6 of the activity in adult rat liver is observed. In rats the dependence on the temperature of the incubation mixture is more pronounced than in tench. The ratio EROD: E-O differs between 10-day-old rats

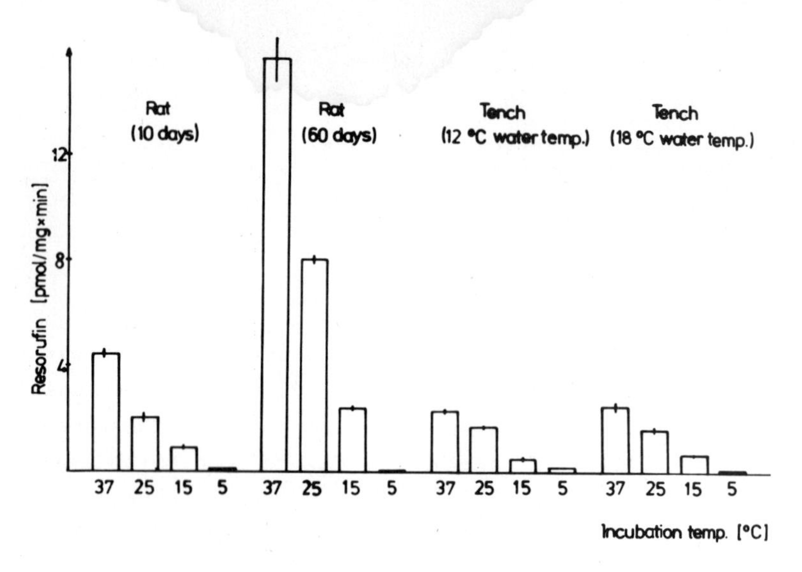

Fig. 1. EROD activities in rat and tench liver at different incubation temperatures.

and tench distinctly. Tench has much less E-O activity (1/50) than EROD activitiy (1/2) in relation to 10-day-old rats (Fig.2).

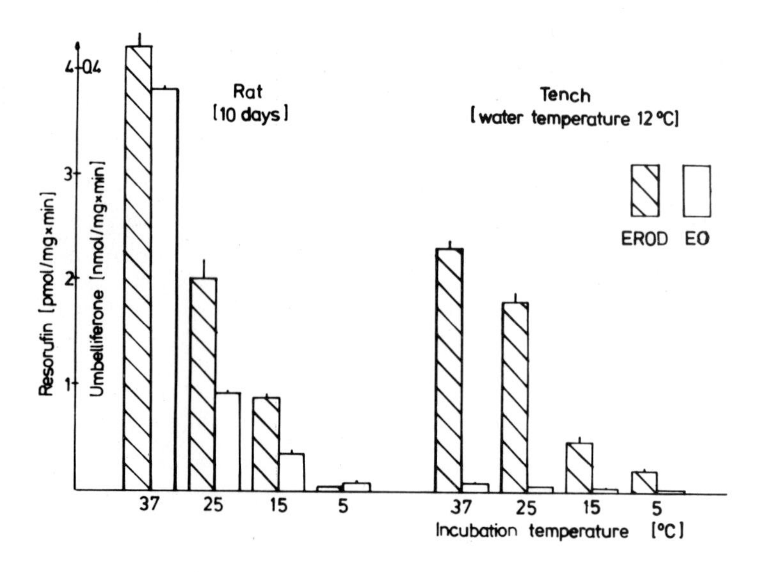

Fig. 2. Ratio EROD: E-O activities in rat (10 days old) and tench.

EROD inducibility by BNF in tench liver is dependent on concentration and temperature of incubation mixture: At low incubation temperatures (15 and 25°C) the EROD activities after 5 ppm BNF are higher than at 37°C, whereas after pretreatment with lower BNF concentrations the activities fall with falling temperature of the incubation mixture. BNF added in vitro inhibits EROD activity in tench liver depending on concentration. 10^{-6} M BNF inhibits by about 60 % at all incubation temperatures. The type of inhibition is non-competitive. DCP added in vitro up to concentrations of 10^{-4} M did not exert any effect. The influence of the herbicide DCP on EROD activities is shown in Fig. 3. DCP increases EROD (and E-O activities) as a function of concentration. Whereas after low concentrations (0.2 and 0.66 ppm) the activities decrease with decreasing incubation temperature, after the high concentration (2.0 ppm = 1/10 of $LC_{50\ (24h)}$) the stimulatory effect is independent of the incubation temperature. E-N was not detectable. In the rat DCP pretreatment a dose dependent increase in E-N, E-O and EROD was observed. The stimulation effect can be demonstrated with incubation temperatures of 37°C, at lower temperatures the effect is less pronounced or not measurable. In comparison to fish the inducing doses were much higher.
P-450 concentration is increased in rat and tench liver by DCP and BNF dependent on dose or concentration, as could be expected from the increased monooxygenase activities.

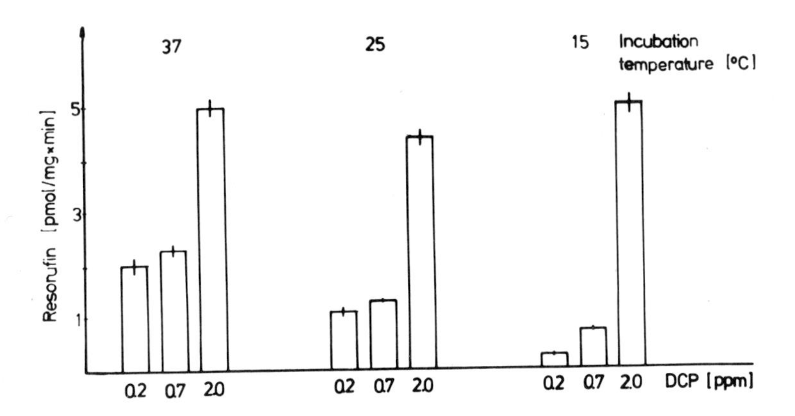

Fig. 3. Influence of DCP (various concentrations) on EROD activities in tench liver measured at different incubation temperatures.

CONCLUSIONS

1. Tench, kept at 12 and 18°C, has identical monooxygenase-activities.
2. Evidently P-450_c, as identified by EROD, contributes the major part to the mfO activities.
3. P-450_c in tench liver is readily inducible by BNF and DCP.
4. Young (poikilothermic) and adult (homoiothermic) rat liver monooxygenases show both the well known dependence on incubation temperature between 15 and 37°C.
   Tench liver monooxygenases show, especially after pretreatment with an inducer, relatively high activities at low incubation temperature.
5. Tench is a suitable biomonitor especially for inducers of the 3-methylcholanthrene type (oil products, polycyclic aromatic hydrocarbons).

REFERENCES

1. Ahokas, J.T., Kärki, N.T., Oikari, A., and Soivio, A. (1976) Bull. environ. Contam. Toxicol. 16, 270
2. Lech, J.J., and Bend, R. (1980) Environ. Health Perspect.34, 115
3. Omura, T., and Sato, R. (1964) J. biol. Chem. 239, 2370
4. Klinger, W., and Müller, D. (1977) Acta biol. med. germ. 36, 1149
5. Aitio, A. (1978) Analyt. Biochem. 85, 488
6. Pohl, R.J., and Fouts, J.R. (1980) Analyt. Biochem. 107, 150

Cytochrome P-450, Biochemistry, Biophysics and Environmental Implications, E. Hietanen, M. Laitinen and O. Hänninen editors

OXIDATION OF BENZO(a)PYRENE BY HORSERADISH PEROXIDASE-H_2O_2 INTERMEDIATE

MICHIKO YAGI[1], SHINOBU AKAGAWA[2] AND HIRONAO MOMOI[2]

[1]Department of General Education(Physics Laboratory) and [2]Dept. of Internal Medicine, Tokyo Medical and Dental University, [1]Ichikawa, Chiba 272(Japan) and [2]Bunkyo-ku, Yushima, Tokyo(Japan)

INTRODUCTION

The polycyclic hydrocarbon benzo(a)pyrene and its analog are widespread environmental polutants, and many of these compounds are potent mutagenic and carcinogenic in the presence of microsomal mixed function oxidase(1-3).

In this study, to determine a capacity of the oxidation of BP by peroxidases in several tissues, the oxidation of BP was examined kinetically by the fluorometry and spectrophotometry employing horseradish peroxidase(HRP).
The most effective and the maximum ratio of H_2O_2 to HRP for the oxidation of BP were 13.7 and 90, respectively. Km and Vmax values of H_2O_2 were 4.54uM and 12.5nM x $sec.^{-1}$, whereas those of BP were 1.82uM and 19.2nM x $sec.^{-1}$, respectively. The rate constants of the oxidation, k_5, were 6 x $10^4 M^{-1}$, $sec.^{-1}$(fluorometry) and 8 x $10^4 M^{-1}$, $sec.^{-1}$(spectrophotometry). The oxidation products of BP consist of three or more components and three of them have been identified as 1,6-, 3,6- and 6,12-quinone BP, as in the case of glutathione S-transferase B(4).

MATERIALS AND METHODS

Horseradish peroxidase(HRP) was purchased from Boehringer Company(grade 1,Lot No. 108073).

BP was obtained from Sigma Company. Stock solution of 10^{-3}M -BP was dissolved into dimethyl sulfoxide.

The study of kinetics of the oxidation of BP were conducted by the fluorometrical and the spectrophotometrical methods using emission maximum at 525nm in buffer solution(fluorometry) and at 435nm(spectrophotometry) for the formation and decomposition of compound-II of HRP-H_2O_2.

Fluorometrical and spectrophotometrical studies were carried out with Hitachi Model MPF-4 fluorometer and Hitachi Model 200-10 spectrophotometer.

The purification and identification of the oxidation products were carried out in a scale of 500 fold concentrated system than kinetic studties. The reaction mixture were incubated at 34°C, for 16hr. Then, insoluble products in the reaction medium were collected and developed for 2hr. with benzene:methanol

(9:1) on thin layer chromatography(TLC, Merck No.5641). Three components were separated under ultraviolet light(360nm). R_f of BP, emitted blue-violet fluorescence, was 0.923. R_f values of 1,6-quinone BP(yellow color) and 3,6-quinone BP (red fluorescence) were 0.91 and 0.865, respectively. These products were identified by fluorescence, absorption spectra and mass spectra.

RESULTS

Figure 1 shows a degree of oxidation of BP in several concentrations of H_2O_2 in fluorometry, to determine a capacity of H_2O_2 to the oxidation of BP.

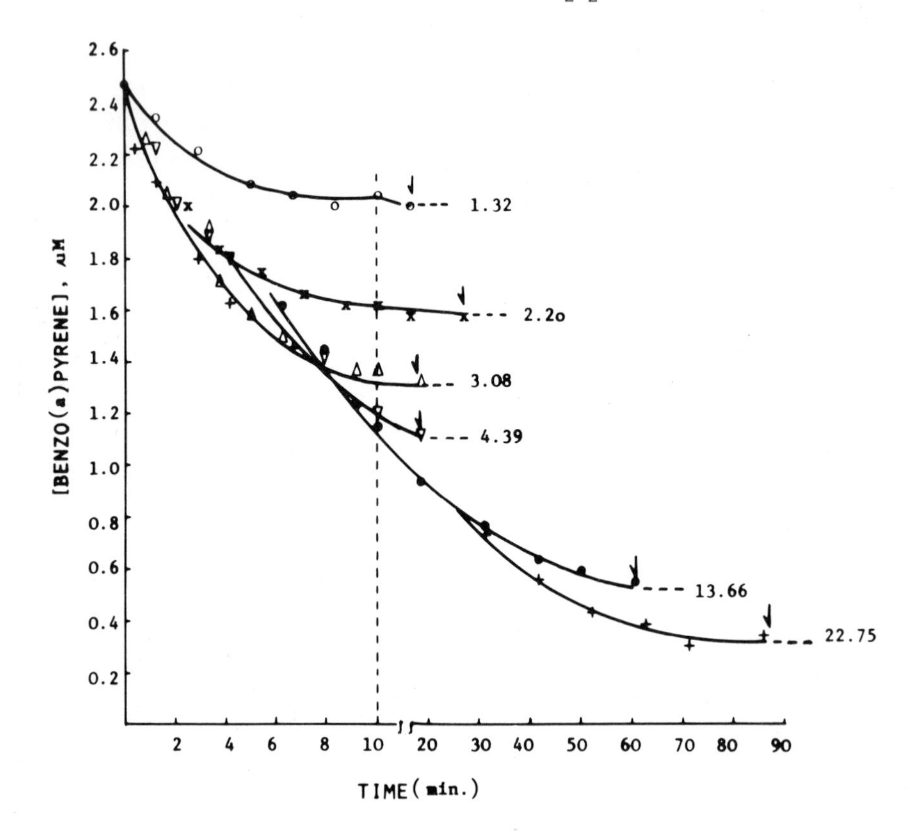

Fig. 1. The oxidation of BP in various H_2O_2 concentrations.
0.39uM-HRP, 2.48uM-BP, pH 4.7,0.1M-acetate. λexcitation 384nm and λemission 524nm. Arrows show the end point of each reaction. Numbers at side on the curves denote H_2O_2 concentrations.

Figure 2 shows the relation of the concentrations of oxidized BP versus H_2O_2 concentrations, the results of the oxidation reactions in Figure 1. Figures 3 and 4 show the kinetics of the oxidation of BP by fluorometry and spectrophotometry.

Table 1 shows the effective and the maximum ratio of H_2O_2 and HRP, and of oxidized BP and HRP to the oxidation of BP, related to Figures 1 and 2. Table 2 shows the rate constants versus BP concentrations in fluorometry and spectrophotometry. Table 3 is the properties of oxidation products.

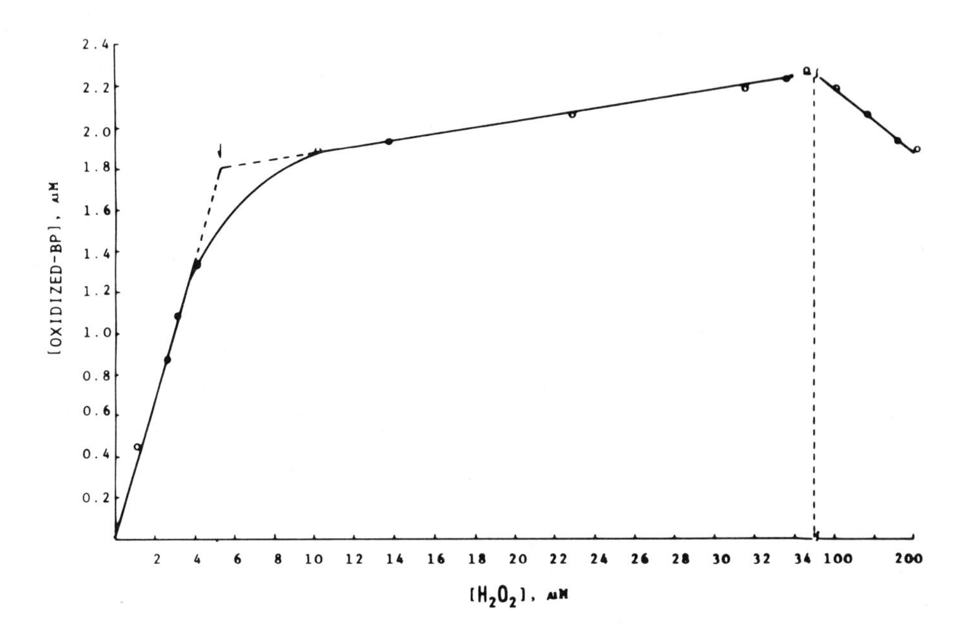

Fig. 2. Plots of oxidized BP versus H_2O_2, related to Fig.1.

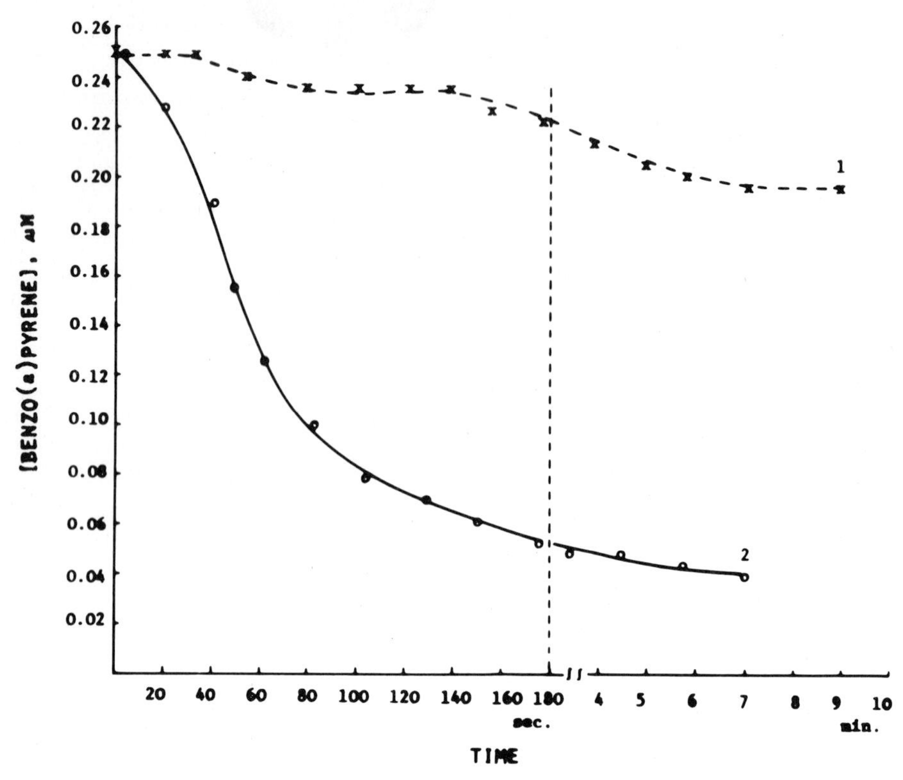

Fig. 3. Oxidation of BP by HRP-H_2O_2 intermediate, in the fluorometry.

0.39uM-HRP, 0.247uM-BP, 49uM-H_2O_2, pH 4.7,0.1M-acetate. λexcitation 384nm and λemission 524nm.
curve 1. spontaneous photoxidation of BP by the excitation beam.
curve 2. oxidation curve of BP.

TABLE 1

THE EFFECTIVE AND THE MAXIMUM RATIOS OF H_2O_2 AND HRP, AND OF OXIDIZED BP AND HRP, RELATED TO Figs. 1 AND 2.

$[H_2O_2]/[HRP]$		[oxidized-BP]/[HRP]	
effective ratio	maximum ratio	effective ratio	maximum ratio
13.74	94.15	4.58	5.7o

Effective ratio shows the ratio of HRP and the concentrations of H_2O_2 at intersections between the first and second phases of oxidized BP curve in Fig.2.

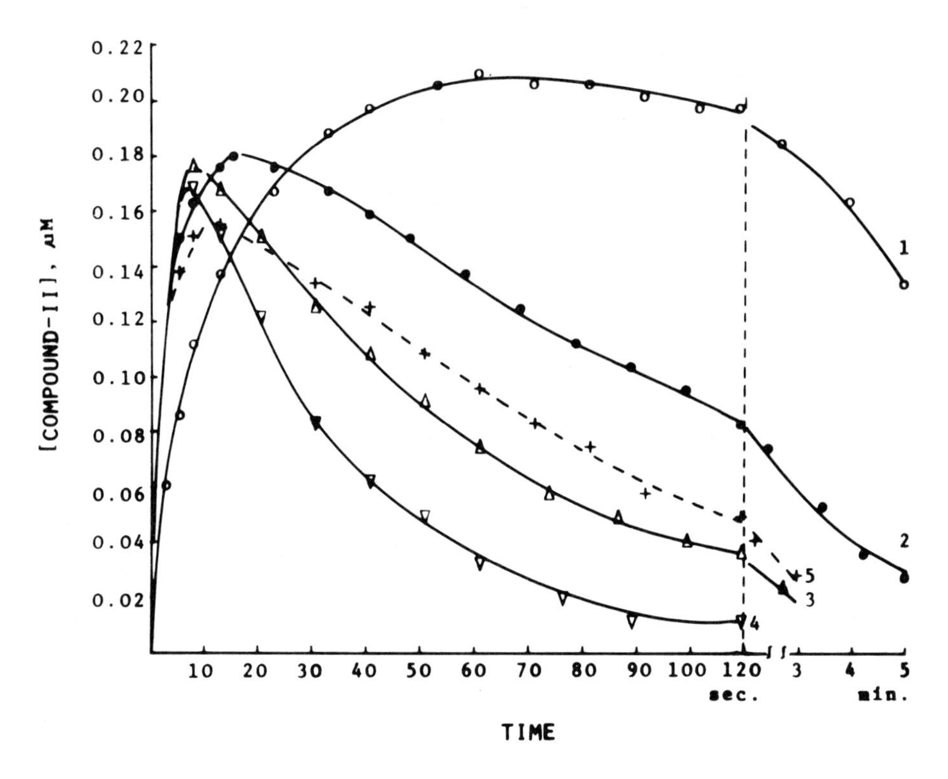

Fig. 4. Formation and decomposition curves of compound-II of HRP and H_2O_2 with and without BP.

0.87uM-HRP, 0.147uM-H_2O_2, pH 4.7,0.1M-acetate. curves: 1. BP=0, 2. 0.049uM-BP, 3. 0.248uM-BP, 4. 0.495uM-BP, 5. 0.99uM-BP. Wavelength at 435nm for compound-II.The concentrations of comp.-II were calculated taking ε_{435} = 32 mM^{-1}, cm^{-1}.

TABLE 2

THE RATE CONSTANTS OF THE OXIDATION OF BP, RELATED TO Figs. 3 AND 4.

	[BP],uM	0.149	0.249	0.348	0.497	0.992	average*
k_5, x 10^4	(a)	4.68	4.1o	1.87	1.42	0.39	6.o
M^{-1} x $sec.^{-1}$	(b)	8.5o	6.3o	5.2o	6.o	1.1o	8.o

(a) and (b) denote the values for fluorometry and spectrophotometry, respectively.

TABLE 3

PROPERTIES OF BP AND THE OXIDATION PRODUCTS

Solvents: A, benzene:methanol(9:1). B, methanol: H_2O(3:1).

Compound	TLC		Spectra			
	solvent	R_f	UV	Vis.	Emission	mass.(m/e)
BP	A	0.923	204,254, 265,284,296	345,364, 403	407.5 (524 in buff.)	
1,6-quinone	A	0.91o	219,267,272, 355^s, 390^s	435,455	non-fluo.	113,127,141,224, 225,226,253,254, 282,283
3,6-quinone	A	0.865	258,265, 318.5,342	475	590	
6,12-quinone						113,127,141,224, 225,226,253,254, <u>255</u>,282,283
Unknown material	B	0.774	295^s, 320,	420	523	

s on the numbers denotes a shoulder peak.
underline below the number indicates a characteristic peak of 6,12-quinone BP in the mass spectrum(5).

REFERENCES

1. Committee on the Biological Effect of Atmospheric Pollutants(1972) Particulate Polycyclic Organic Matter. Natl. Acad. Sci., Washington, D.C.
2. Miller, E.C. and J.A.Miller(1974) In Molecular Biology of Cancer, H.Bush(ed), Academic Press, New York, pp.377-402.
3. Borgen,A., Darvey,N., Castagnoli,N., Crocker,T.T., Rasmussen,R.E. and Wang,I.Y.(1973) J.Med. Chem.,66, 502.
4. Morgenstern,R., DePierre,J.W., Lind,C., Guthenberg,C.,Mannervik,B., and Ernster,L.(1981) Biochem. Biophys. Res. Comm.99,682.
5 McCaustland,D.J., Fischer,D.L., Kolwyck,K.C., Duncan,W.P., Wiley,J.C., Menon,C.S., Engel,J.F., Selkirk,J.K., and Roller,P.P.(1976) Carcinogenesis-A Comprehensive Survey, Vol.1. Freudenthal,R., and Jones,P.W. ed. Raven Press, New York.pp.349- 412.

---

*Average values of the rate constants were taken as values intrapolated to zero BP concentration.

Evolution of Cytochrome P-450 and Environmental Factors in the Control of Cytochrome P-450 in Feral Species

© 1982 Elsevier Biomedical Press B.V.
Cytochrome P-450, Biochemistry, Biophysics
and Environmental Implications, E. Hietanen,
M. Laitinen and O. Hänninen editors

HIGHER PLANT CYTOCHROME P-450 : MICROSOMAL ELECTRON TRANSPORT AND XENOBIOTIC OXIDATION

IRENE BENVENISTE, BRIGITTE GABRIAC, RAYMONDE FONNE, DANIELE REICHHART, JEAN-PIERRE SALAÜN, ANNICK SIMON AND FRANCIS DURST
Laboratoire de Physiologie Végétale, ERA C.N.R.S. n° 104, Université Louis Pasteur, 28 rue Goethe, 67083 Strasbourg Cedex, FRANCE.

INTRODUCTION

As a result of evolutionary pressure, plants have acquired the capacity to synthesize secondary metabolites which, by number and structural variety, exceed by far those found in animals. Many of these compounds are hydrophobic and undergo extensive oxygenation followed by conjugation and accumulation in the plant vacuole. Demonstration of cytochrome P-450 and recognition of its role in some of these reactions was slowed by a number of difficulties which still hamper the study of these enzymes in plants. Because of the excessive force required to break the rigid cell wall, plant microsomes are composed of an heterogeneous membrane population including debris from organelles other than E.R.. Chlorophyll contamination can be obviated by using white storage tissues or etiolated seedlings, but the presence of flavonoids and carotenoids which strongly interfer with spectrophotometric measurements cannot be avoided. Moreover, the concentration of cytochrome P-450 is generally low : 0.005 to 0.100 nmoles per mg protein in microsomes from non-induced plants. Nevertheless, the involvement of cytochrome P-450 is now clearly established in the oxidation of several substrates, the majority of which are physiological compounds proper to the plants.

The formation of 4-hydroxy-cinnamic acid, a precursor of flavonoids, lignin and many other phenolic compounds, is catalysed by the cinnamic acid 4-hydroxylase (CA4H) (1,2). In terms of mass of substrate processed, the CA4H is probably the most important cytochrome P-450 enzyme known today.

Several fatty acid ω- and in-chain-hydroxylases have been demonstrated by Kolattukudy and coworkers (3,4) and by our group (5,6). The physiological role of these hydroxylases remains unclear although some of them may participate in the synthesis of plant waxes.

The terpenols nerol and geraniol are converted by a cytochrome P-450 enzyme from *Catharanthus roseus* into the corresponding diols (7), which are precursors for several families of indole alkaloids.

A sequence of four oxidation steps, probably all cytochrome P-450-dependent convert kaurene to 7β-hydroxy-kaurenoic acid, a precursor of the plant hormones gibberellins (8,9). The first reaction of the catabolism of abscisic acid, an other plant hormone, is probably also catalysed by cytochrome P-450 (10). There is convincing evidence that two exogeneous compounds, 3-phenyl-1,1-dimethylurea (monuron) and 4-chloro-N-methylaniline are substrates to cytochrome P-450 enzymes in some plants (11,12).

A remarkable feature of the plant cytochrome P-450 enzymes is their high substrate specificity. This has been noted for oxidases converting kaurene into kaurenoic acid (13), for the CA4H (14,15) and for the two laurate hydroxylases that we have described (16,6). Detailed structure-activity studies showed that the two latter enzymes required both a free carboxyl group at one end and a hydrophobic rest at the other end of the molecule, and showed a high degree of chain length discrimination : fatty acids with 9, 14 and 16 atoms of carbon were not recognized, and the C10 isomer only weakly so. These observations provide a rationale for the low effectiveness of SKF 525 A, metyrapone and piperonyl butoxide against the plant monooxygenases (15), and prompted our efforts to develop suicide-substrates for the plant oxidases*. Autocatalytic destruction of cytochrome P-450 and CA4H by 1-aminobenzotriazole has been demonstrated (17) at doses that only weakly affect the laurate hydroxylases. Inactivation was also achieved with other olefins and aryloxy propargyl ethers (to be published), showing that π carbon-carbon bonds are potentially destructive for plant cytochrome P-450.

In plant storage tissues (*e.g* tubers of Jerusalem artichoke or potato), cytochrome P-450 and related activities are very low but increase markedly after slicing and on aging of the tissues (18,19). Superinduction was observed when the tissues were exposed during aging to manganese chloride, ethanol, phenobarbital and the herbicides monuron, dichlobenil (20,21) and 2,4-D (2,4-dichlorophenoxyacetic acid)(22). In etiolated pea seedlings, light, operating through the photomorphogenic photoreceptor phytochrome, increased selectively the activity of the CA4H but not that of the ω-LAH (23). Nevertheless, it is not clear at the present time wether plants possess a system equivalent to the drug-induced broad specificity hepatic monooxygenases. There is no evidence until yet that the hemoproteins implicated in the oxidation of xenobiotics in plants are distinct from the constitutive enzymes.

MATERIAL AND METHODS

Plant material. Slicing and aging of Jerusalem artichoke tuber tissues was described previously (21). *Acer pseudoplatanus* cells were a gift from Pr. J. Guern (Gif sur Yvette, France) and were cultured in liquid medium using their

* This study is done in cooperation with the Laboratory of Pr. Ortiz de Montellano, Pharmaceutical Chemistry, University of California, San Francisco, U.S.A.

conditions (24).

Preparation of microsomes. Tuber microsomes were prepared as described (21). *Acer* cells were lyophilized, pulverized with mortar and pestle at 4°C, 0.1M Na-phosphate buffer (pH 7.4) containing 15mM 2-mercaptoethanol was added (10 v/w), and the mixture was homogeneized with two strokes with an Ultra-Turrax. The 100.000g pellet was resuspended in 0.1M Na-phosphate buffer (pH 7.4) containing 1.5mM 2-mercaptoethanol and 30% (v/v) glycerol and stored at -20°C.

Enzyme assays. CA4H, ω-LAH and IC-LAH were assayed using described procedures (21,6,5). Microsomal oxidation of 2,4-D was measured as follows : microsomes from *Acer* cells (5 to 10 mg protein) were incubated at 25°C with 2,4-[2-^{14}C]D (spec. act., 56 μCi/μM) and unlabeled 2,4-D to a final concentration of 1mM, 1mM NADPH along with a glucose 6-phosphate dehydrogenase regenerating system and 0.1M Na-phosphate buffer (pH 7.4), in a final volume of 2 ml. After 60 min aerobic incubation the reaction was stopped by addition of 1 ml 4N HCl. The reaction medium was extracted two times with 10 ml benzene/diethylether (9/1,v/v). The organic phase was evaporated to dryness. The residue was dissolved in hexane/diethylether (1/1,v/v) and applied to a silica gel plate (Kieselgel F254). After developement in the solvent diethylether/petroleum ether (b.p. 35-60°C)/formic acid (70/30/1,v/v/v), the plates were scanned for radioactivity on a Berthold LB 2723 thin-layer scanner. Radioactive peaks were scraped off the plate into counting vials and 10 ml of a toluene-POPOP-PPO scintillation solution were added.

The assay for aminopyrine N-demethylation contained 5-10 mg microsomal protein, 500 μM NADPH along with a glucose 6-phosphate dehydrogenase regenerating system, 5mM aminopyrine and 0.1M Na-phosphate buffer (pH 7.4) in a final volume of 5 ml. The reaction was stopped after 30 min incubation at 30°C by the addition of 1.5 ml of a 12.5% TCA solution. The reaction mixture was clarified by centrifuging at 40.000g for 30 min, and 3 ml of the supernatant were mixed with 1 ml Nash reagent (24) and incubated for 8 min at 60°C. Spectra were recorded from 350 to 500 nm against a control assay without substrate.

Cytochromes P-450 and b_5 were measured by the method of Omura and Sato (25).

RESULTS

Microsomal electron transport during cinnamate hydroxylation. The plant cytochrome P-450 enzymes known today exhibit, as far as electron transfer is concerned, large functional similarities with the hepatic monooxygenases. Two flavoproteins, NADPH cytochrome c(P-450) reductase (that we abbreviate as Fp1) and NADH cytochrome c(b_5) reductase (abbreviated here as Fp2) are present in plant microsomes. NADPH is always the prefered electron donor, but NADPH-NADH synergism was observed in most (7,9,18) but not in all (6) cases.

However, antibodies[*] raised against NADPH-cytochrome c reductase from pig liver failed to inhibit Fpl from Jerusalem artichoke and pea microsomes as well as the NADPH-dependent hydroxylation of cinnamate and laurate in these microsomes. Moreover, no inhibition of these activities was obtained with antibodies raised against pig liver cytochromes P-450s induced by phenobarbital, naphtoflavone or ethanol. Plant microsomes fortified with pig liver cytochrome P-450 were unable to catalyse the O-dealkylation of 7-ethoxycoumarin. This activity became measurable only after further addition of purified pig liver reductase to the plant microsomes already fortified with hepatic cytochrome P-450. Although preliminary, these results indicate that the plant and animal systems differ not only by antigenic markers but also by structural features necessary to the formation of the reductase-cytochrome P-450 complex and/or transfer of reducing equivalents.

Several inhibitors of Fpl and Fp2 were used in an effort to determine the electron flow routes during hydroxylation of cinnamic acid in artichoke tuber microsomes : $NADP^+$, NAD^+, 2'AMP, 2',5'ADP, $AADP^+$ (3-aminonicotinamide adenine dinucleotide phosphate)[a], and AAD^+ (3-aminonicotinamide adenine dinucleotide)[b]. The *Ki* values obtained from kinetic experiments are listed in table 1.

TABLE 1

Ki VALUES FOR INHIBITION OF Fpl AND Fp2 BY COMPETITIVE INHIBITORS.

Inhibitor	NADPH-cytochrome c reductase	NADH-cytochrome c reductase
$NADP^+$	13[a]	25000
NAD^+	2200	1000
2'AMP	150	b
2'5'ADP	21	c
$AADP^+$	16	46000
AAD^+	45000	100

[a]Values are *Ki*s expressed in μM.
[b]Fp2 was virtually unaffected by 2'AMP.
[c]Not determined

$AADP^+$ and AAD^+ associate effectiveness and good selectivity. The effect of these compounds was tested on NADPH- and NADH-dependent CA4H activity. Figure 1

[*] Antibodies used in this study were a generous gift from Pr. V. Ullrich, Physiological Chemistry, University of Saarland, Homburg, GFR.
[a] $AADP^+$ was a generous gift from Pr. B.M. Anderson, VPI, Blacksburg, VA, U.S.A.
[b] AAD^+ was prepared by Dr. F. Schuber, Laboratoire de Chimie Enzymatique, this Institute.

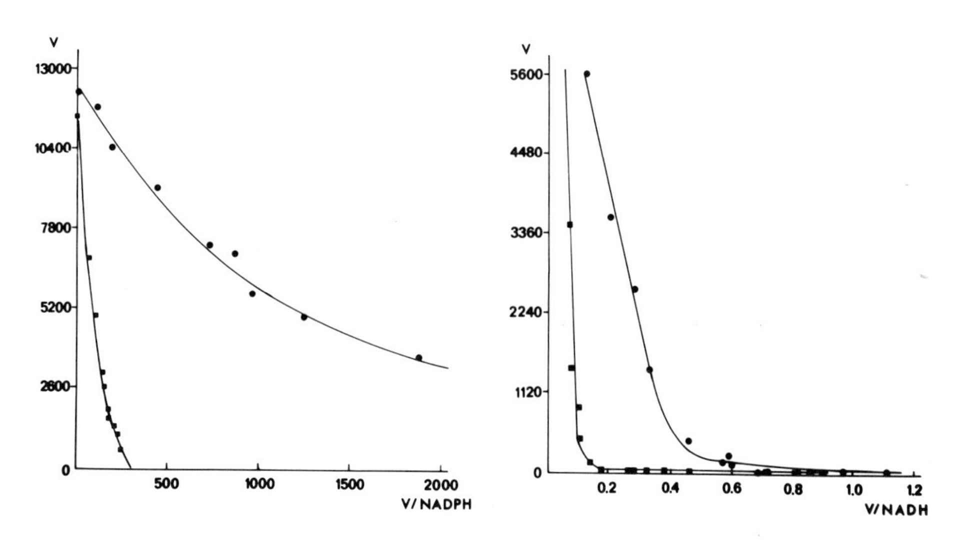

Fig. 1. Eadie-Hofstee plots of CA4H from tuber microsomes. Substrate saturation curves were determined for NADPH (A) and NADH (B), in the absence (●—●) and in the presence (■—■) of $AADP^+$. $AADP^+$ concentration was 120 µM in A and 31 µM in B.

shows that both activities were strongly inhibited by $AADP^+$. In sharp contrast, AAD^+ did not significantly affect either of the activities : values (not shown) were very near to those of the control. Similar results were obtained during the study of NADPH- and NADH-dependent *ω*-hydroxylation of lauric acid in pea microsomes (6).

The fact that NADH-dependent hydroxylation is reduced when Fp1 is inhibited but is not affected even in the case of drastic inhibition of Fp2, suggests that the rate limiting step in the transfer of electrons from NADH to cytochrome P-450 implicates Fp1. Indeed, we found that NADH is able, although with a low *Km* (28 mM), to reduce solubilized and partially purified Fp1.

Table 2 shows that AAD^+ which does not inhibit NADPH- or NADH-dependent hydroxylation reduces by 50 % the NADPH-NADH synergism on CA4H activity. Since synergism is often regarded as resulting from the more facile donation of the second electron by Fp2, one might propose that during NADH-dependent oxidation of cinnamate the first electron is transfered via Fp1, and the second via Fp2.

The Eadie-hofstee plots of CA4H activity versus NADPH or NADH concentration showed clearly biphasic kinetics (Fig. 1.). Since similar kinetics were observed for the *ω*-LAH from pea microsomes (6), one might presume that this behaviour reflects intrinsic characteristics of microsomal electron transport components in higher plants.

TABLE 2

INHIBITION OF NADPH-NADH SYNERGISM BY AAD^+

Tuber microsomes were incubated for 6 min at 25°C. NADPH and NADH were regenerated by isocitrate dehydrogenase and alcohol dehydrogenase systems, respectively.

Reducing agent and inhibitor	Concentration μM	CA4H pmoles/min.mg prot
NADPH	115	6100
NADH	11	30.5
NADPH	115	
+ NADH	11	9260
NADPH	115	
+ NADH	11	
+ AAD^+	650	7670
NADPH	115	
+ NADH	11	
+ $AADP^+$	750	3690

<u>Oxidation of xenobiotics by plant microsomes</u>. Plants are the principal ultimate recipients of pesticides and pollutants. Oxidation reactions play a major role in the bio-transformation of these compounds in plant tissues (27). However, the enzymes involved remain largely unknown, except in the few cases cited above (11,12).

Since many substituted phenylurea herbicides undergo N-demethylation during detoxication in plants, a study of N-demethylation of aminopyrine, taken as a model substrate has been undertaken. A preliminary screening showed positive responses with microsomes from ten different plant species (Salaün et al. submitted). The enzyme from tuber tissues has been characterized in some details(Fonne et al., unpublished).Demethylase activity was dependent upon reducing equivalents, NADH supporting up to 60 % of the rate achieved with NADPH. The activity was inhibited by oxidized cytochrome c, $AADP^+$ and dichlorophenolindophenol. Carbon monoxide produced a sharp reduction of N-demethylation. When microsomes were preincubated for 20 min with NADPH and 1mM 1-aminobenzotriazole, a condition producing total inactivation of CA4H (17), 30 % of demethylase activity was lost. Kinetic studies showed an apparent *Km* of 2.5mM for aminopyrine. Substrate saturation curves as a function of NADPH concentration were strongly biphasic, with apparent *Km* values ranging from 1 to 27 μM. Aminopyrine N-demethylation was increased in tuber tissues by inducers that also stimulate CA4H and IC-LAH activities (21,16). Table 3 shows that the increase in enzyme activity roughly paralleled that of cytochrome P-450. Although light reversal of CO-inhibition has not yet been determined, the evidence at hand strongly suggests that cytochrome P-450 is implicated in the N-demethylation of xenobiotics in plants.

TABLE 3

INDUCTION OF AMINOPYRINE N-DEMETHYLASE AND CYTOCHROME P-450

Artichoke tubers were sliced and aged for the time of maximal induction (48 h for water controls and phenobarbital and 72 h for $MnCl_2$) as determined previously (20,21). Preparation of microsomes and enzyme assay are described in Material and Methods.

Treatment	HCHO formed pmoles/min.mg prot.	cytochrome P-450 pmoles/mg prot.
Control	17	8
Phenobarbital (4mM)	168	154
$MnCl_2$ (25mM)	266	274

The herbicide 2,4-D is ring hydroxylated in many plants at position 4, with migration of the chloride to position 3 or 5. This NIH shift suggesting the involvement of a monooxygenase in this reaction, plant microsomes were assayed for their capacity to oxidize 2,4-D. The activities being very low in tuber and pea, *Acer* cells, known to metabolize actively 2,4-D *in vivo* (24), were used.

TABLE 4

CYTOCHROMES AND ENZYME ACTIVITIES IN MICROSOMES FROM *ACER* CELLS AND TUBER SLICES

Cells were harvested at the end of the log-phase. Tuber tissues were aged for 48 h on 25mM $MnCl_2$.

Plants	Cytochromes pmoles/mg prot.		Enzymes pmoles/min.mg prot.			
	P-450	b_5	Fpl	CA4H	LAH	2,4-DH
Acer cells	75	302	13000	182	nd.	5.6
Tuber slices	196	283	50000	825	35	nd.

Liquid cell suspension cultures retain the capacity to biosynthesize cytochromes P-450 and b_5, and to hydroxylate cinnamate (Table 4). Laurate hydroxylase was undetectable. The formation of hydroxylated 2,4-D derivatives was NADPH dependent, but remained to low to permit unequivocal determination of the enzyme involved in the non-induced cells. Preliminary work (Simon et al. unpublished) indicates that the microsomal monooxygenase system is inducible in *Acer* cells by phenobarbital.

REFERENCES

1. Benveniste, I. and Durst,F. (1974) C.R.Acad.Sc.Paris, D278,1487
2. Potts, J.R.M., Weklych, R. and Conn, E.E. (1974) J. Biol. Chem., 249, 5019
3. Croteau, R. and Kolattukudy, P.E. (1975) Arch. Biochem. Biophys., 170, 61
4. Soliday, C.L. and Kolattukudy, P.E. (1977) Plant Physiol., 59, 1116
5. Salaûn, JP., Benveniste, I., Reichhart, D. and Durst, F. (1978) Eur. J. Biochem., 90, 155
6. Benveniste, I., Salaün, JP., Simon, A., Reichhart, D., and Durst, F. (1982) Plant Physiol. in the press.
7. Madyastha, K.M., Meehan, T.D. and Coscia, C.J. (1977) Biochemistry, 15,1097
8. Coolbaugh, R.C. and Moore, T.C. (1971) Phytochemistry, 10, 2401
9. Hasson, E.P. and West, C.A. (1976) Plant Physiol., 58, 473
10. Gillard, D.F. and Walton,D.C. (1976) Plant Physiol., 58,790
11. Frear, D.S. and Swanson, H.R. (1975) Pest. Biochem. Physiol., 5, 73
12. Young, O. and Beevers, H. (1976) Phytochemistry, 15, 379
13. West, C.A. (1980) in: The Biochemistry of Plants, Vol. 2 Academic Press, NY.
14. Grisebach, H. (1977) Naturwissenschaften 64, 619
15. Benveniste, I. (1974) Thèse de Doctorat d'Etat, Strasbourg, France
16. Salaün, JP., Benveniste, I., Reichhart, D. and Durst, F. (1981) Eur. J. Biochem. 119,651
17. Reichhart, D., Simon, A., Durst, F., Mathews, J.M. and Ortiz de Montellano, P.R. (1982) Arch. Biochem. Biophys. in the press.
18. Benveniste, I., Salaün, JP. and Durst, F. (1977) Phytochemistry, 16, 69
19. Rich, P.R. and Lamb, C.J. (1977) Eur. J. Biochem. 72, 353
20. Reichhart, D., Salaün, JP., Benveniste, I. and Durst, F. (1979) Arch. Biochem. Biophys. 196, 301
21. Reichhart, D., Salaün, JP., Benveniste, I. and Durst, F. (1980) Plant Physiol. 66, 600
22. Adelé, P., Reichhart, D., Salaün, JP., Benveniste, I. and Durst, F. (1981) Plant Sci. Letters, 22, 39
23. Benveniste, I., Salaün, JP. and Durst, F. (1978) Phytochemistry, 17, 359
24. Leguay, JJ. and Guern, J. (1975) Plant Physiol. 56, 356
25. Nash, T. (1953) Biochem. J. 55, 416
26. Omura, T. and Sato, R. (1964) J. Biol. Chem. 239, 2370
27. Dohn, D.R. and Krieger, R.I. (1981) Drug Metab. Rev. 12, 119

Cytochrome P-450, Biochemistry, Biophysics
and Environmental Implications, E. Hietanen,
M. Laitinen and O. Hänninen editors

SEASONAL VARIATION OF CYTOCHROME P-450 AND MONOOXYGENASE ACTIVITIES IN BOREAL FRESH WATER FISH

OSMO HÄNNINEN, ULLA KOIVUSAARI AND PIRJO LINDSTRÖM-SEPPÄ
Deparment of Physiology, University of Kuopio, Kuopio, FINLAND

INTRODUCTION

In the boreal regions the amplitude of the temperature and light oscillations between the seasons are great. The water temperature may exceed 20°C during the summer and be around the freezing point during the winter. The fish must adapt in autumn to the cooling and during spring to the warming of waters. The annual living cycle has a period of intensive growth during summer. The reproduction affects, the metabolism at a certain season varying from species to species.

Cultivated rainbow trout provides a useful model for studies of the seasonal variation, since it is fed with controlled fodder in natural waters and its genetic background is more homogenous than that of wild species. Figure 1 shows schematically the annual life cycle of the rainbow trout (Salmo gairdnerii) in the boreal fresh waters. This species has its spawning time in spring.

Rainbow trout is capable in oxidizing xenobiotics (1, 2, 3). The present report indicates that its hepatic monooxygenase system appears to show almost an ideal temperature compensation pattern (4) during the cooling of waters in autumn. This is in contrast to UDP-glucuronosyltransferase in the same membrane structure, since this enzyme follows a non-compensation pattern (4). In spring the adaptation to the warming of waters happens together with the recovery phenomena from the spawning in mature fish. During the growth period in summer the specific monooxygenase activity diminishes while the total hepatic capacity remains uncharged due to the increase of liver weight.

MATERIALS AND METHODS

Rainbow trout of both sexes were obtained from Nilakkalohi Fish Farm Ltd, Tervo. The fish were 1^{+} or 2^{+} years old. They were fed commercial rainbow-trout fodder (Vaasan Höyrymylly Ltd, Finland).

After catching the fish were slaughtered and the livers were immediately removed, weighed and placed in ice-cold 0.25 M sucrose.

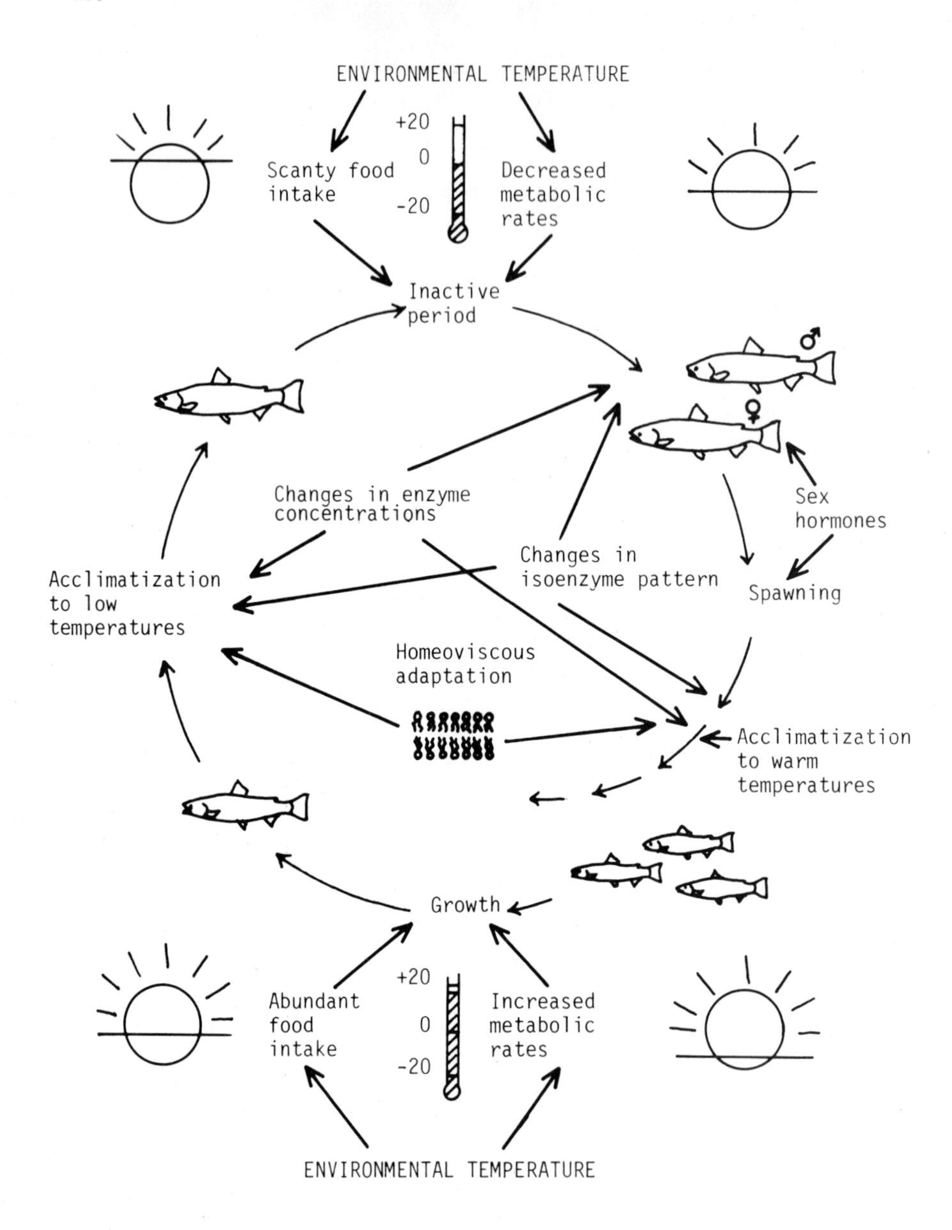

Fig. 1. A schematic presentation of the annual life cycle of rainbow trouts in boreal waters with a rapid growth period during the summer, adaptation to cooling waters in autumn, preparation to spawning during winter, spawning in spring as well as recovery from spawning and adaptation to warming of the environment in the beginning of the summer.

In the laboratory the livers were homogenized at 0-4°C by using a Potter-Elvehjem glass-Teflon homogenizer in 4 vol of 0.25 M sucrose. The homogenates were centrifuged at 10,000 g for 20 min at 4°C. From this supernatant the microsomal fraction was isolated by spinning at 105,000 g for 60 min in a Sorval OTD-2 ultracentrifuge. The pellet was resuspended in 0.25 M sucrose so that 1 ml of microsomal suspension corresponded to 1 g of liver wet weight. The content of cytochrome P-450 was measured immediately, for other measurements the microsomes were stored at -80°C.

Protein determination was carried out by the biuret method (5) as modified by Lang et al. (6) using bovine serum albumin (Sigma) as a reference protein.

The microsomal cytochrome P-450 content was measured as described by Omura & Sato (7) in a Cary 118 spectrophotometer. Aryl hydrocarbon hydroxylase activity was determined as described by Wattenberg et al. (8) and modified by Nebert & Gelboin (9). 3,4-Benzpyrene was used as a substrate and the amount of hydroxylated 3,4-benzpyrene was measured by a Perkin-Elmer spectrofluorometer using 3-hydroxy-3,4-benzpyrene as reference. UDPGlucuronosyltransferase activity was measured as described by Isselbacher (10) and Hänninen (11) using p-nitrophenol as an aglycone. The in vitro measurements were carried out at a constant temperature (18°C) and also at about the temperature where the fish are living during the particular time of the year.

RESULTS AND DISCUSSION

Cytochrome P-450 content varied during the year in the liver of 1^{+} and 2^{+} year old rainbow trout liver (Fig. 2 A and B). In autumn when the waters were cooling no significant changes in cytochrome P-450 levels were observed. In mature female fish the hepatic cytochrome P-450 content was much lower than in the immature fish or in mature males during the first months of the year in the prespawning period. After the spawning cytochrome P-450 content increased. This took place at the same time as the water temperature increased. During the growth period during summer the specific cytochrome P-450 content tended to decrease, but the total amount remained about unchanged due to the increase of the liver weight.

In autumn during the cooling of waters the monooxygenase activities like benzo(a)pyrene hydroxylase (Table 1) increased several fold, if the activity measurements were performed at constant incubation tempe-

TABLE 1.

THE BENZO(A)PYRENE HYDROXYLASE AND UDP GLUCURONOSYLTRANSFERASE ACTIVITIES IN THE LIVER OF MALE RAINBOW TROUT DURING THE COOLING OF THE WATERS DURING AUTUMN. The activities were measured either at constant incubation temperature (18°C) or at the temperature from where the fish had been caught. The benzo(a)pyrene hydroxylase activities have been calculated per unit of microsomal protein as well as unit of cytochrome P-450.

EXPERIMENTAL TEMPERATURE	AUGUST +18°C	NOVEMBER +18°C	AUGUST +20°C	NOVEMBER +5°C
BENZO(A)PYRENE HYDROXYLASE				
	***	***		
nmoles / mg micros. prot. x min.	0.017±0.001	0.057±0.006	0.021±0.002	0.017±0.002
	**	**		
nmoles / nmoles cytochr. P-450	0.16±0.01	0.64±0.11	0.19±0.01	0.17±0.02
UDP GLUCURONOSYLTRANSFERASE ACTIVITY				
			***	***
nmoles / mg prot. x min.	0.17±0.02	0.13±0.01	0.18±0.02	0.05±0.004

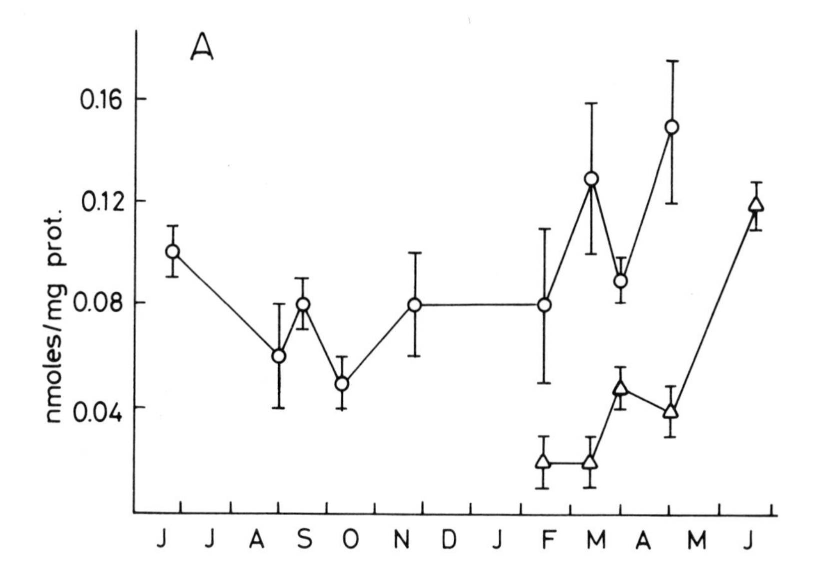

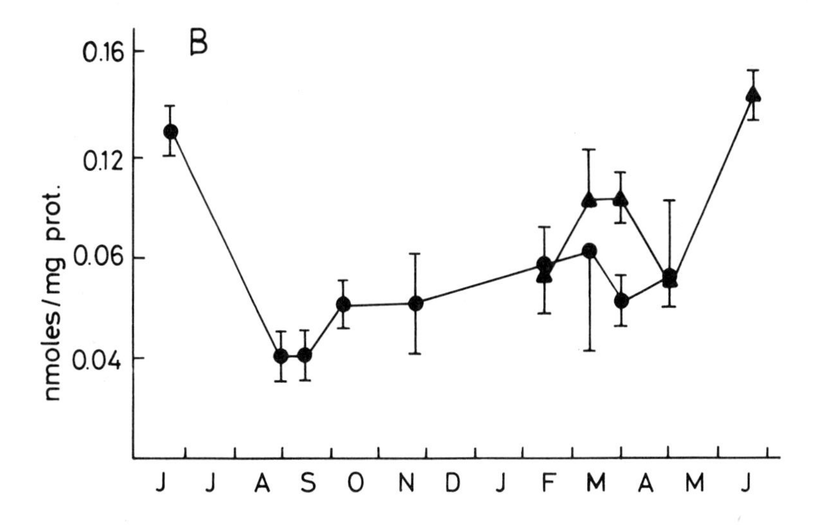

Fig. 2. The level of cytochrome P-450 in the liver of A) female rainbow trout (circles 1^+ years and triangles 2^+ years old) and B) of male rainbow trout (black balls 1^+ years and black triangels 2^+ years old). The letters indicate the names of months.

TABLE 2.

THE BENZO(A)PYRENE HYDROXYLASE ACTIVITY IN THE LIVER OF THE MALE RAINBOW TROUTS DURING THE WARMING OF WATERS IN SPRING. The activities were measured either at constant incubation temperature (18^{0}) or at the temperature at which the fish had been living. The activities have been calculated per fresh liver, microsomal protein and cytochrome P-450 units.

EXPERIMENTAL TEMPERATURE	MAY $+18^{0}C$		JUNE $+18^{0}C$	MAY $+2^{0}C$		JUNE $+13^{0}C$
nmoles / g tissue w.wt. x min.	0.20 ± 0.03	**	0.51 ± 0.09	0.04 ± 0.004	**	0.34 ± 0.07
pmoles / mg micr.prot. x min.	7.9 ± 1.3	*	19.9 ± 4.2	1.8 ± 0.35	***	13.3 ± 0.9
nmoles / nmole cyt. P-450 x min.	0.14 ± 0.04		0.14 ± 0.02	0.02 ± 0.006	***	0.09 ± 0.01

rature ($18^{0}C$). If the determinations were carried out at the temperatures where the fish had been living i.e. at $20^{0}C$ in August and at $5^{0}C$ in November, it appeared that the activities remained unchanged. Since the cytochrome P-450 content was almost constant, the catalytic ability of the monooxygenase system increased at lowered incubation temperatures. Thus the monooxygenase system appeared to follow the so called ideal temperature compensation pattern (4). The mechanism of the increased catalytic ability can be either due the changes in the membrane environment (11) or in the relative amount of hypothetical cytochrome P-450 isoenzymes or changes in other enzymes of the monooxygenase complex.

In contrast to the monooxygenase system UDPGlucuronosyltransferase activity remained constant, if measured at the constant incubation temperature ($18^{0}C$). A considerable decrease of activity was seen, if the determination was performed at the temperature in which the fish had been caught. Thus UDPGlucuronocyltransferase exhibited a non-compensation pattern (4). The two enzyme activities located in the same intracellular membrane structure appear to follow completely different response patterns.

After the adaptation to the cold water had occurred the mature trout showed a lowering of the monoxygenase activities (not illustrated) together with the cytochorome P-450 levels. During the prespawning period there was also a clear sex difference in the monooxygenase activities.

When the spawning period was over the monooxygenase activities started to increase. The increase coincided with the warming of waters. When the activities were measured at constant temperature $18^{0}C$ there was from May to June about 2.5 fold increase of specific benzo(a) pyrene hydroxylase activity but no change at all if the activity was calculated against the cytochrome P-450 content. If the benzo (a) pyrene hydroxylase activity was measured at the temperatures, where the fish had been living i.e. at $2^{0}C$ in May and at $13^{0}C$ in June there was higher than expected increase in the specific activity and there was also higher than expected increase of activity even when calculated against the cytochrome P-450 content. An increase of 4.5 fold was observed with an increase of $11^{0}C$ in the incubation temperature. The finding indicates that the catalytic properties of cytochrome P-450 were much enhanced with the increasing environmental temperature.This can be due to changes in the membrane environment as well as due to change in the cytochrome P-450 isoenzyme pattern. The last mentioned possibility is supported by the fact that monooxygenase activity before the intensive growth period helps the fish to handle the increasing amounts foreign compounds during the summer season. In wild vendace (Coregonus albula) seasonal oscillations in the xenobiotic metabolism can also be observed, although they are less clear than in the rainbow trout due to larger scatter.

The results obtained indicate that there are considerable changes in the monooxygenase activities in the liver of the rainbow trout from season to season. The molecular mechanisms of these changes without alterations of the cytochrome P-450 levels require further studies on the membrane structure as well as on the cytochrome P-450 isoenzyme pattern.

ACKNOWLEDGEMENTS

This study has been supported by a grant from the Academy of Finland (Project No 40).

REFERENCES

1. Stegeman, J.J. (1977) Fedn.Proc.Fedn.Am.Socs.exp.Biol. 36,941.
2. Stegeman, J.J. and Chevion, M. (1980) Biochem.Pharmac. 29,553.
3. Koivusaari, U., Harri, M. and Hänninen, O. (1981) Comp.Biochem. Physiol. 70C,149.
4. Precht, H. (1973) in: Precht, H., Christopherson, J., Hausel, H. and Larcher, W. (eds), Springer, Berlin, pp. 302-353.
5. Gornall, A.G., Bardawill, C.J. and David, M.M. (1949). J.biol. Chem. 177,751.
6. Lang, M., Laitinen, M., Hietanen, E. and Vainio, H. (1976) Acta pharmac.toc. 39,273.
7. Omura, T. and Sato, R. (1964) J.biol.Chem.239,2379.
8. Wattenberg, W.L., Leong, J.L. and Strand, P.J. (1962) Cancer Res. 22,1120.
9. Nebert, D.W. and Gelboin, H.V. (1968) J.biol.Chem.244,6242.
10. Isselbacher, K.J. (1956) Recent Prog.horm.Res.Commun.12,134.
11. Hänninen, O. (1968) Ann.Acad.Sci.Fenn. A 2,142, 1.
12. Koivusaari, U. (1980) Xenobiotic Biotransformation in Boreal Fresh Water Fish. Ph.Lic. Thesis, Univ.Kuopio, Finland.

Cytochrome P-450, Biochemistry, Biophysics
and Environmental Implications, E. Hietanen,
M. Laitinen and O. Hänninen editors

REGULATION OF HEPATIC STEROID AND XENOBIOTIC METABOLISM IN FISH

TIIU HANSSON[1], LARS FÖRLIN[2], JOSEPH RAFTER[1] and JAN-ÅKE GUSTAFSSON[1]
[1]Department of Medical Nutrition, Karolinska Institute, Huddinge Hospital F69, S-149 86 Huddinge and [2]Department of Zoophysiology, University of Göteborg, Box 25059, S-400 33 Göteborg, Sweden

INTRODUCTION

In recent years a great deal of information has become available with regard to cytochrome P-450-mediated metabolism in fish (1). However, little attention has been paid to the regulation of cytochrome P-450 by hormonal factors in these animals.

Sex differences in hepatic xenobiotic and steroid metabolism in rainbow trout (_Salmo gairdneri_) (2,3,4,5), brook trout (_Salvelinus fontinalis_) (2) and winter flounder (_Pseudopleuronectes americanus_) (6) have been reported. These differences only appear at certain stages of the sexual development of the fish and are manifested as higher levels of cytochrome P-450 content and several monooxygenase activities in mature male fish when compared to maturing and mature female fish and juvenile fish of both sexes.

Most species of teleost fish are subject to annual reproductive cycles with gamete development and spawning followed by periods of gonadal regression and inactivity. This is reflected in marked fluctuations in plasma steroid and gonadotropin levels where the peak levels of gonadal hormones appear during a relatively short period of the reproductive cycle. In addition, the plasma levels of gonadal steroids during the period of peak gonadal activity may be several orders of magnitude greater than those in mammals (7).

In the rat, the sexual differences in hepatic xenobiotic and steroid metabolism appear at the time of puberty and some of these differences are attributed to androgen imprinting during the neonatal period (8,9). During this period testicular androgens irreversibly imprint the hypothalamo-pituitary-liver axis thus resulting in a male type of liver metabolism. The existence of a hypophysial factor, recently identified as growth hormone or a peptide like growth hormone, responsible for the maintenance of the female type of liver metabolism has been postulated (9).

The active involvement of the hypothalamo-pituitary-gonad axis in the reproductive cycle in fish suggests that some of the reported sex differences in hepatic cytochrome P-450 system may be due to gonadal as well as to pituitary hormones.

RESULTS AND DISCUSSION

Effects of oestradiol-17β on cytochrome P-450

During the maturation period of female fish a vitellogenic protein is produced in the liver and an involvement of oestradiol-17β in the direct control of vitellogenin synthesis in fish liver has been indicated. In addition, it has been reported that the plasma levels of oestradiol-17β increase during the maturation period but decline before the onset of spawning (10).

Administration of oestradiol-17β to juvenile rainbow trout of both sexes resulted in a significant decrease in total liver microsomal cytochrome P-450 content, 6β-hydroxylation of 4-androstene-3,17-dione and metabolism of benz(a)pyrene, p-nitroanisole and ethylmorphine (Table 1) (11,12). The suppressive effect of oestradiol-17β on the cytochrome P-450-mediated reactions in trout liver was confirmed recently by Stegeman *et al.*(13) and by Vodicnik and Lech (14). The presented data are consistent with the low levels of monooxygenase activities and total cytochrome P-450 content in liver microsomes from maturing female fish indicating a role for oestradiol-17β in the regulation of hepatic cytochrome P-450 in trout. Furthermore, administration of oestradiol-17β to juvenile fish caused changes in serum constituents associated with hepatic vitellogenin synthes and increased the plasma levels of this oestrogen to levels found in maturing female fish (11,12).This suggests that the presence of low cytochrome P-450 mediated metabolism in maturing female fish may well be due to increased plasma oestradiol-17β levels.

The period of active vitellogenin synthesis in female fish is characterized by changes in hepatic endoplasmic reticulum. It is possible, therefore, that the enhanced synthesis of specific proteins in the liver of maturing female fish may interfere with the microsomal metabolism of steroids and xenobiotics. Since only 6β-hydroxylation and not the reductive metabolism of 4-androstene-3,14-dione is affected during this period and by oestradiol-17β treatment it is more likely that the decrease in the cytochrome P-450 is a result of a regulatory effect of oestradiol-17β on the cytochrome P-450 level.

TABLE 1

EFFECTS OF OESTRADIOL-17 β AMINISTRATION ON THE HEPATIC STEROID AND XENOBIOTIC METABOLISM IN JUVENILE TROUT

Cytochrome P-450 content	↓	4-Androstene-3,17-dione	
Benz(a)pyrene hydroxylase	↓	6β-Hydroxylase	↓
Ethylmorphin -N-demethylase	↓	16-Hydroxylase	—
p-Nitroanisole-O-demethylase	↓		
Ethoxyresorufin-O-deethyalse	↓	Testosterone	
Ethoxycoumarin-O-deethylase	↓	6β-Hydroxylase	↓
Benzphetamine-N-demethylase	↓	16β-Hydroxylase	—

(From Hansson and Gustafsson, 1981; Förlin and Hansson,1982; Stegeman et al., 1982; Vodicnik and Lech, 1982)
Symbols: ↓ represents a decrease in the cytochrome P-450 content or enzymeactivity, respectively, when compared to the situation in untreated juvenile trout; — represents no effect.

Further evidence for a specific effect of oestradiol-17β on cytochrome P-450 in trout was presented recently by Stegeman *et al.* who showed that a band, with the estimated molecular weight of 56 000, was nearly eliminated on the SDS-PAGE profiles of liver microsomes from oestradiol-17β-treated juvenile fish and maturing female fish (13). Finally, oestradiol-17β administration to juvenile trout increased the K_m value for p-nitroanisole-O-demethylase (13) which is consistent with the high K_m for this enzyme found in liver microsomes from maturing female fish (15).

Effects of androgens on cytochrome P-450

The androgens, 11-oxotestosterone and testosterone, seem to play an important role in the reproductive physiology of fish. High levels of these steroids have been reported in several teleost species and different ratios of testosterone and 11-oxotestosterone are present in plasma from male and female fish as well as from fish at different stages of maturity. Recently it has become evident that in some species the levels of testosterone in plasma from female fish are considerable and exceed the levels found in males (7).

Administration of 11-oxotestosterone increased the total cytochrome P-450 content but decreased the monooxygenase activities studied (16,17) (Table 2). A similar dose of testosterone had no effect on the total cytochrome P-450 content but the decrease in monooxygenase activities was as pronounced in testosterone-treated fish as that in fish given 11-oxotestosterone . A similar effect of testosterone on several liver microsomal monooxygenase activities in brook trout and rainbow trout has been reported recently by Stegeman et al.(13) and by Vodicnik and Lech (14).The response of cytochrome P-450 to testosterone in juvenile fish seem to vary depending on the dose of steroid used. Thus, low doses of this androgen seem to slightly increase the total cytochrome P-450 content whereas high doses cause a decrease or no effect on cytochrome P-450 content (14).

TABLE 2

EFFECTS OF 11-OXOTESTOSTERONE ADMINISTRATION ON THE HEPATIC STEROID AND XENOBIOTIC METABOLISM IN JUVENILE TROUT

Cytochrome P-450 content	↑	4-Androstene-3,17-dione	
Benz(a)pyrene hydroxylase	↓	6β-Hydroxylase	↓
Ethylmorphine-N-demethylase	↓	16-Hydroxylase	—
p-Nitroanisole-O-demethylase	↓		

(From Hansson 1982; Förlin and Hansson,1982)
Explanation for symbols see legend to Table 1.

The low levels of monooxygenase activities in androgen-treated juvenile fish is in contrast to the situation in mature male fish where the plasma androgen levels are high and the monooxygenase activities higher than those in female fish and juvenile fish of both sexes (2,3,4). This suggests that other factors, which may act alone or in synergism with gonadal hormones, than androgens may be involved in the regulation of cytochrome P-450 mediated metabolism in male fish. It may also be that 11-oxotestosterone induces the synthesis of a novel type of cytochrome P-450 with a low specificity for the substrates studied and/or that a specific ratio of testosterone and 11-oxotestosterone is needed to maintain the male level of monooxygenase activity in mature fish.

PITUITARY REGULATION OF CYTOCHROME P-450

Since the hypothalamo-pituitary system plays an important role in the maturational control of reproduction in fish the effects of hypophysectomy on trout liver microsomal metabolism of steroid and xenobiotic metabolism were studied. Although hypophysectomy tended to increase the total cytochrome P-450 content this treatment had no effect on the related monooxygenase activities (11,12,17). On the other hand the removal of the pituitary gland increased the activity of the liver microsomal 17-hydroxysteroid oxidoreductase active on 4-androstene-3,17-dione thus indicating that at least some of the liver microsomal enzyme activities in juvenile fish are under the influence of the pituitary gland (11).

The suppressive effect of oestradiol-17β and testosterone on the monooxygenase activities studied was also expressed in hypophysectomized trout as was the oestradiol-17β caused decrease in cytochrome P-450 content (11,12,17). This indicates that the effects of these steroids on cytochrome P-450 are direct and independent of the pituitary gland. This is in contrast to the situation in the rat where the pituitary gland has an obligatory role in gonadal steroid control of hepatic steroid and xenobiotic metabolism (9).

In the rat the development of sex differences in hepatic metabolism awaits the maturation of the hypothalamo-pituitary system. This may also be the case in fish. Since hypophysectomy of sexually mature trout is not feasible the presented experiments were performed on juvenile fish, accordingly, the role of the pituitary gland in the regulation of cytochrome P-450 in mature fish remains to be investigated.

Evidence for multiple forms of cytochrome P-450

One possible explanation for the sex differences in cytochrome P-450 dependent metabolism in the rat has centered around the multiplicity of cytochrome P-450(18).The presence of multiple forms of cytochrome P-450 in fish has not been studied to any extent.The variability in inducibility of the trout cytochrome P-450 system by common inducers in fish of different age and sex (3,19) provides evidence for the presence of different forms and/or specific ratios of multiple forms of cytochrome P-450 in male and female fish as well as in juvenile and adult fish. There is evidence that novel forms of cytochrome P-450 with high specificity for certain exogenous substrates, such as benz(a)pyrene and ethoxyresorufin, are induced by PAH-type inducers in some species of fish (1). On the other hand, the form of cytochrome P-450 induced by 3-methylcholanthrene in trout seems to be less effective in metabolizing steroid substrates than the control form (19). It is interesting to note that oestradiol-17β treatment exhibits a suppressive effect on 6β-hydroxylation but not on 16-hydroxylation of 4-androstene-3,17-dione (11) or testosterone (13) indicating that 6β- and 16-hydroxylases may represent different forms of cytochrome P-450. Finally, evidence for sex differences in cytochrome P-450 in fish includes the fact that the absorption maximum of hepatic reduced cytochrome-P-450-CO complex in spawning winter flounder males is at 448 nm while that of maturing female fish is at 450 nm (6).

Since the differences in the hepatic cytochrome P-450 in rainbow trout are confined to certain stages of the reproductive cycle and seem to be regulated by gonadal hormones one may speculate about the physiological significance of these differences. It may be suggested that the low capacity of cytochrome P-450 system in maturing female fish serves to maintain high tissue levels of oestradiol-17β during the vitellogenin synthesis in the liver. The involvement of cytochrome P-450 in oestradiol-17β metabolism in trout liver (20) and the observation by Lambert et al. (21) that the increased levels of oestradiol-17β in female fish are not parallelled by an increased 3-hydroxysteroid dehydrogenase activity in the ovaries support the contention that the increase in plasma oestradiol-17β is due to an impaired metabolism in the liver rather than to an increased biosynthesis in the ovaries.

The role of testosterone and 11-oxotestosterone in the regulation of cytochrome P-450 in fish liver is not clear. However, the oestrogen-like effect of testosterone on the monooxygenase activities and the fact that testosterone administered in high doses to female fish stimulates the vitellogenin synthesis suggest that this steroid, which occurs in considerable amounts in plasma of female fish, may play a role in the regulation of cytochrome P-450 in female fish.

Available data suggest that the mode of action of steroids on the nervous system in fish is similar to that in mammalian species and that the early presence of androgens can trigger the development of the fish pituitary (22). It may be speculated whether sex related differences in hepatic steroid and xenobiotic metabolism apparent during certain stages of development of the fish represent delayed expression of imprinting by androgens in early life of the fish.

In conclusion, gonadal steroids seem to play a significant role in the regulation of sex differences in the cytochrome P-450-mediated metabolism in trout liver whereas the role of the pituitary gland in this regulation remains to be investigated.

ACKNOWLEDGEMENTS

This work was supported by grants from the National Swedish Environment Protection Board (No. 7-334-79) and the Swedish Medical Research Council (No. 13X-2819).

REFERENCES

1. Franklin, R.B., Elcombe, C.R., Vodicnik, M.J.and Lech, J.J. (1980) Fed. Proceedings, 39, 3144.
2. Stegeman, J.J. and Chevion, M. (1980) Biochem. Pharmac. 29,553.
3. Förlin, L. (1980) Toxicol. Appl. Pharmacol. 54, 420.
4. Hansson, T. and Gustafsson, J.-Å. (1981) Gen. Comp. Endocrinol. 44, 181.
5. Koivusaari, U., Harri, N. and Hänninen, O. (1981) Comp.Biochem. Physiol. 70C, 149.
6. Stegeman, J.J. (1981) in: Gelboin, H.V. and Tsó, P.O.P. (eds), Polycyclic Hydrocarbons and Cancer , Academic Press, N.Y. Vol. 3, pp.1- 60.
7. Billard, R., Breton, B., Fostier, A., Jalabert, B. and Weil,C. (1978) in: Comparative Endocrinology , Gaillard, P.J. and Boer, H.H. (eds), Elsevier/North Holland Biomedical Press, Amsterdam, pp.37 - 48.
8. Colby, H.D. (1980) in: Advances on sex hormone research, Thomas, J.A. and Singhal, R.L. (eds), Baltimore and Munich: Urban and Scharzenberg, Vol.4, pp 27-71.
9. Gustafsson, J.-Å., Mode, A., Norstedt, G., Hökfelt, T., Sonnenschein, C., Eneroth, P. and Skett, P. (1980) in Biochemical Actions of Hormones, Litwack, G. (ed), Academic Press, New York, vol VII. pp. 47-89.
10. Whitehead, C., Bromage, N.R. and Forster, J.R.M. (1978) J. Fish. Biol. 12, 601-608.
11. Hansson, T. and Gustafsson (1981) J. Endocrinol. 90,103-112.
12. Förlin, L. and Hansson, T. (1982) J. Endocrinol. In press.
13. Stegeman, J.J., Pajor, A.M. and Thomas, P. (1982) Biochem. Pharmacol. In press.

14. Vodicnik, M.J. and Lech, J.L. (1982) J. Steroid Biochem. In press.

15. Förlin, L. and Lidman, U. (1981) Comp. Biochem. Physiol. 70C, 297 - 300.

16. Hansson, T. (1982) J. Endocrinol. 92, 409-417.

17. Förlin, L. and Hansson, T. (1982) Gen. Comp. Endocrinol.46,400.

18. Kahl, R., Buecker, M. and Netter, J. (1977) in: Microsomes and Drug oxidations , Ullrich, V. (ed), Pergamon Press, New York, pp. 177.

19. Hansson, T., Rafter, J. and Gustafsson, J.-Å. (1980) Biochem. Pharmacol. 29, 583-587.

20. Hansson, T. and Rafter, J. (1982) Gen. Comp. Endocrinol. In press.

21. Lambert, J.G.D., Bosman, G.I.C.G., van den Hurk, R. and van Oordt, P.G.W.J. (1978) Ann. Biol. Anim. Biochim. Biophys. 18, 923-927.

22. Peter, R.E. and Crim, L.W. (1979) Ann. Rev. Physiol. 41, 323-335.

Cytochrome P-450, Biochemistry, Biophysics
and Environmental Implications, E. Hietanen,
M. Laitinen and O. Hänninen editors

CHARACTERIZATION OF THE INDUCTION RESPONSE OF THE HEPATIC MICROSOMAL MONOOXYGENASE (MO) SYSTEM OF FISHES

MARY JO VODICNIK AND JOHN J. LECH
Department of Pharmacology and Toxicology and NIEHS Freshwater Biomedical Research Center, Medical College of Wisconsin, P.O. Box 26509, Milwaukee, Wisconsin 53226 (U.S.A.)

Although the biotransformation of drugs and xenobiotic chemicals in mammals has been extensively studied in the past, it is only recently that significant attention has been devoted to this phenomenon in aquatic species. Studies carried out in vitro utilizing fish tissue fractions and model substrates demonstrated both phase I and phase II biotransformation reactions, although substantial intraspecies variability exists and observed rates of metabolism are, in general, lower than those seen in mammalian tissues. There exists a good deal of evidence to indicate that the phase I reactions demonstrated in fish in vitro are catalyzed by the microsomal cytochrome(s) P-450 system. Monooxygenase activity associated with isolated hepatic microsomes from several species of fish has been demonstrated to be sensitive to inhibition by carbon monoxide and responsive to classical MO modulators such as α-naphthoflavone and metyrapone (1-3). Furthermore, it would appear that substrate specificities are similar to those observed in the mammalian system (4).

Although in vitro investigations have been extremely useful in characterizing the types of enzymes involved in the biotransformation of chemicals by fish species, an area of at least equal importance concerns the functional significance of these reactions in vivo. The information presented in Table 1 shows a partial listing of phase I reactions which have been demonstrated in a variety of fish species following in vivo exposure of animals to a number of xenobiotic chemicals.

While these data indicate that a system comparable to the cytochrome(s) P-450 system of mammals exists in fishes, it appears that some major differences exist with respect to the inducibility of microsomal MO activity between the two groups. A number of investigators have demonstrated that fish appear to be refractive to induction by phenobarbital-type agents, but that they do respond to inducers of the 3-methylcholanthrene (3-MC)-type (3-9). The latter induction was characterized by an increase in catalytic activity for cytochrome P-448-dependent substrates and an elevation of hepatic microsomal hemoprotein(s) P-450 content.

TABLE 1
PHASE I BIOTRANSFORMATION REACTIONS DEMONSTRATED IN VIVO BY SEVERAL FISH SPECIES

Biotransformation Reaction	Species	Chemical
O-Dealkylation	Fathead minnow	p-Nitrophenylethers
	Rainbow trout	Pentachloroanisole
		Fenitrothion
N-Dealkylation	Carp	Dinitramine
Oxidation	Mudsucker, sculpin	Naphthalene, benzo(a)pyrene
	Coho salmon	Naphthalene
	Rainbow trout	Methylnaphthalene
	Carp	Rotenone
	Bluegill	4-(2,4-DB)
	Mosquitofish	Aldrin, dieldrin
Hydrolysis	Catfish, bluegill	2,4-D esters
	Rainbow trout	Diethylhexylphthalate
	Pinfish	Malathion
	Mosquitofish	Parathion

The data in Table 2 illustrate the response of the hepatic microsomal MO system of rainbow trout (*Salmo gairdneri*) to a variety of inducers which were administered to the animals intraperitoneally. These compounds include classical inducing agents, as well as those which have been suggested to induce "novel" hemoprotein(s) P-450 profiles in mammals based on substrate specificity, electrophoretic mobility and biochemical and immunological characterization of the induced protein(s). In general, the data suggest that the hepatic microsomal MO system of rainbow trout is refractive to phenobarbital-type inducers, or those chemicals which resemble these agents (mirex, Kepone). 3-Methylcholanthrene-type inducers and the methylenedioxyphenyl compound, isosafrole, stimulated those catalytic activities associated with cytochrome P-448. The refractivity observed towards Kepone, mirex and 2,4,5,2',4',5'-hexachlorobiphenyl is unlikely to be due to a problem of bioavailability since their presence has been confirmed in liver extracts from pretreated rainbow trout. Furthermore, the *in vitro* addition of these inducers to the reaction mixtures for BeND and ECOD had little effect on enzyme activity, suggesting

TABLE 2

THE EFFECT OF VARIOUS INDUCING AGENTS ON HEPATIC MICROSOMAL MONOOXYGENASE ACTIVITY IN RAINBOW TROUT

Treatment[a] (Dose)	EMND[b]	BeND[c]	ECOD[d]	EROD[e]	AHH[f]	P-450
			(nmol/min/mg)			(nmol/mg)
Corn oil (1 ml/kg)	100[g]	100	100	100	100	100
Phenobarbital (65 mg/kg-4 days)	81	N.D.[h]	64	65	104	95
β-Naphthoflavone (100 mg/kg)	88	N.D.	1178*	4455*	4081*	146*
Kepone (5 mg/kg)	N.D.	80	39*	14*	N.D.	67*
Mirex (40 mg/kg)	N.D.	62*	93	150	N.D.	97
Isosafrole (175 mg/kg)	N.D.	80	496*	469*	N.D.	121
TCDD (1.2 μg/kg)	N.D.	89	1657*	1414*	N.D.	208*
Pregnenolone-16α-carbonitrile (25 mg/kg-6 days)	N.D.	269*	153*	100	N.D.	103
Aroclor 1254 (150 mg/kg)	105	49	509*	1460*	1300*	127
2,4,2',4'-tetra-chlorobiphenyl (150 mg/kg)	N.D.	95	196	340	N.D.	N.D.
3,4,3',4'-tetra-chlorobiphenyl (150 mg/kg)	N.D.	120	643*	2076*	N.D.	153*
2,4,5,2',4',5'-hexachlorobiphenyl (150 mg/kg)	N.D.	0	130	40	N.D.	63*
3,4,5,3',4'-pentachloro-biphenyl (0.6 mg/kg)	N.D.	63*	711*	2543*	N.D.	136*

[a] Sexually immature rainbow trout were pretreated i.p. with a single injection of each agent (exceptions noted above) and sacrificed 3-5 days thereafter.
[b] Ethylmorphine-N-demethylation.
[c] Benzphetamine-N-demethylation.
[d] Ethoxycoumarin-O-deethylation.
[e] Ethoxyresorufin-O-deethylation.
[f] Aryl hydrocarbon (benzo[a]pyrene) hydroxylation.
[g] Data are expressed as a percent of control activity.
[h] Not determined.
* Significantly different from corn oil-pretreated controls ($p < 0.05$).

the lack of induction to be unrelated to the inhibition of monooxygenation by these agents (4). It is of interest that pregnenolone-16α-carbonitrile (PCN) administration resulted in an induction profile which is characteristic of phenobarbital-type induction in mammals in that both BeND and ECOD were elevated. Whether the constitutive hemoprotein(s) of fish responsible for BeND activity is similar to the PCN-inducible hemoprotein isolated from the rat remains to be determined. Nonetheless, PCN may be an important agent in the further characterization of the hepatic microsomal MO system of fish species.

While it has been assumed that the induction process in fish species following treatment with 3-MC-type agents is similar to that which has been described in mammalian systems, it has only recently been demonstrated that the observed increases in catalytic activity are associated with a stimulation of amino acid incorporation into the hepatic microsomal proteins.

Figure 1 shows the fluorographic analysis of electrophoretograms of solubilized hepatic microsomes from individual rainbow trout which were pretreated with β-naphthoflavone (βNF) or corn oil followed 18 hours later by 75 μCi of ^{35}S-methionine. Microsomes were prepared 24, 48 and 72 hours following amino acid administration. Fluorography demonstrated the appearance of 3 bands with molecular weights of approximately 59,000, 57,000 and 50,000 daltons after βNF treatment which were not observed in microsomes from corn oil-pretreated animals. Following sectioning of gels into one centimeter segments, radioactivity was significantly higher in the 50,000-60,000 dalton molecular weight range in solubilized microsomes from βNF-pretreated animals when compared to controls 48 and 72 hours following ^{35}S-methionine administration (Figure 2).

It is interesting to note that βNF stimulated the incorporation of the amino acid into at least three electrophoretic protein bands whereas following densitometric scanning of Coomassie blue-stained gels, the appearance/intensification of only one band is usually reported. Since the administration of 3-MC-type inducers to rainbow trout has been shown to result in increases in a number of microsomal catalytic activities, it may be that the radiolabeled bands represent enzymes with differing substrate specificities.

While most of the studies described herein have considered the process of induction in fish from a mechanistic point of view, a question germane to aquatic toxicology is whether induction occurs among wild populations and its possible significance.

It is well established that various MO inducers have the ability to produce significant changes in the rates and routes of metabolism of a number of

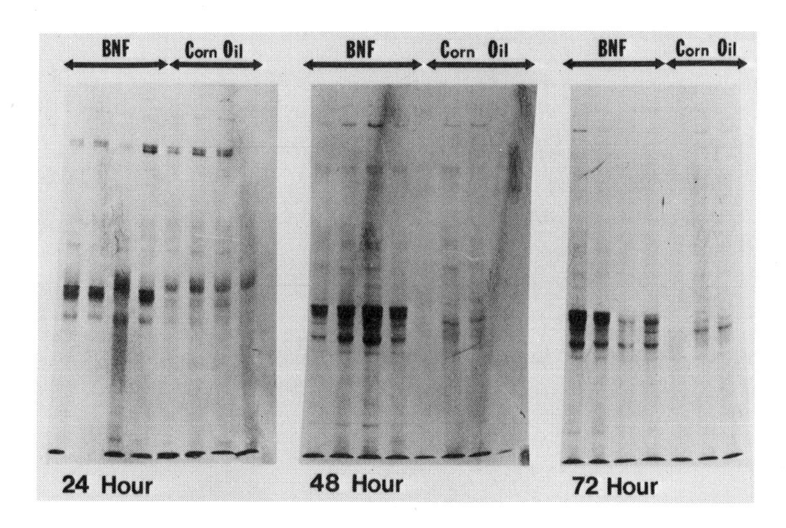

Figure 1. Fluorography of electrophoretograms of hepatic microsomes from individual rainbow trout pretreated with corn oil or BNF and ^{35}S-methionine.

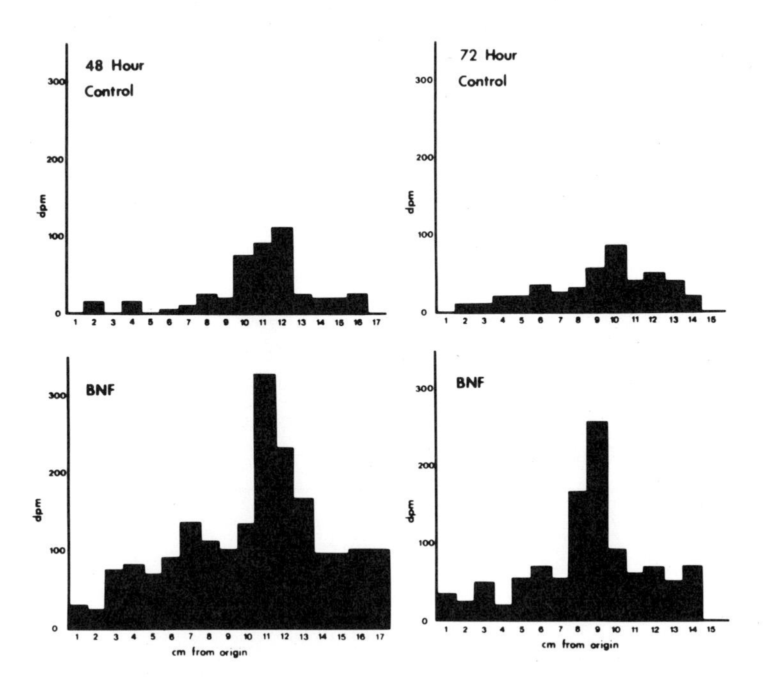

Figure 2. Radioactivity in one cm segments of electrophoretic tracks from gels of microsomes from rainbow trout pretreated with corn oil or BNF and ^{35}S-methionine.

compounds, in some cases, leading to the formation of biologically active metabolites which are more toxic than the parent compound. In vitro studies using microsomes from polycyclic aromatic hydrocarbon (PAH)-pretreated little skate and channel catfish have demonstrated alterations in the metabolism of benzo[a]pyrene (10,11). Furthermore, PAH pretreatment of rainbow trout altered rates of biliary excretion of 2-methylnaphthalene and 1,2,4-trichlorobenzene in vivo, as well as, changed the pattern of biliary metabolites of these compounds (8,12).

It has been demonstrated that covalent binding of benzo[a]pyrene metabolites to DNA is increased 10-50-fold after incubation of this compound with microsomes from 3-MC-pretreated coho salmon and lake trout (13,14). By analogy with studies in mammals, binding to DNA may be related to mutagenicity and carcinogenicity.

Finally, a number of studies in mammals and birds have suggested that exposure to environmental contaminants which are known MO inducers may alter reproductive processes by affecting sex steroid hormone metabolism (15). PCBs have been shown to cause a reduction in plasma levels of sex steroids in trout (16) and induction of MO activity by 3-MC or Clophen A50 resulted in marked changes in hepatic cytochrome(s) P-450-dependent steroid hydroxylase activities (17).

Since various investigators have demonstrated that fish from contaminated environments have elevated MO activities when compared to their counterparts from pristine waters (18-20), the effects of water-bourne chemicals known to be MO inducers in fish on the metabolism of other exogenous or endogenous compounds may be of environmental and toxicological significance.

From a mechanistic and comparative point of view, further studies on the cytochrome(s) P-450 system of fish species, particularly with respect to the refractivity of this system to phenobarbital-type inducers, may contribute to our understanding of species susceptibility to chemicals, as well as, the mechanism of action of this type of inducing agent.

REFERENCES

1. Ahokas, J.T., Pelkonen, O. and Karki, N.T. (1977) Cancer Res., 37, 3737.
2. Bend, J.R. and James, M.O. (1978) in: Malins, D. and Sargent, J.R. (Eds.), Biochem. Biophys. Perspect. Marine Biol., Academic Press, New York, Vol. 4, pp. 125-188.
3. Elcombe, C.R. and Lech, J.J. (1979) Toxicol. Appl. Pharmacol., 49, 437.
4. Vodicnik, M.J., Elcombe, C.R. and Lech, J.J. (1981) Toxicol. Appl. Pharmacol. 59, 364.

5. Buhler, D.R. and Rasmusson, M.E. (1968) Comp. Biochem. Physiol. 25, 223.
6. Bend, J.R., Pohl, R.J. and Fouts, J.R. (1973) Bull. Mt. Desert Island Biol. Lab. 13, 9.
7. Addison, R.F., Zinck, M.E. and Willis, D.E. (1977) Comp. Biochem. Physiol. 75C, 39.
8. Statham, C.N., Elcombe, C.R., Szyjka, P.A. and Lech, J.J. (1978) Xenobiotica 8,65.
9. Elcombe, C.R., Franklin, R.B. and Lech, J.J. (1979) in: Khan, M.A.Q., Lech, J.J. and Menn, J.J. (Eds.), Pesticide and Xenobiotic Metabolism in Aquatic Organisms, American Chemical Society, Washington, D.C., pp. 319-337.
10. Bend, J.R., Ball, L.M., Elmamlouk, T.H., James, M.O. and Philpot, R.M. (1979) in: Khan, M.A.Q., Lech, J.J. and Menn, J.J. (Eds.), Pesticide and Xenobiotic Metabolism in Aquatic Organisms, American Chemical Society, Washington, D.C. pp. 279-318.
11. Hinton, D.E., Klaunig, J.E., Jack, R.M., Lipsky, M.M. and Trump, B.F. (1981) in: Branson, D.R. and Dickson, K.L. (Eds.), Aquatic Toxicology and Hazard Assessment: Fourth Conference, American Society for Testing and Materials, Chicago, Illinois, pp. 226-238.
12. Melancon, M.J., Jr. and Lech, J.J. (1979) in: Marking, L.L. and Kimerle, R.A. (Eds.), Aquatic Toxicology, American Society for Testing and Materials, Philadelphia, Pennsylvania, pp. 5-22.
13. Ahokas, J.T., Saarni, H., Nebert, D.W. and Pelkonen, O. (1979) Chem.-Biol. Interactions, 25, 103.
14. Varanasi, V. and Gmur, D. (1980) Biochem. Pharmacol. 29, 753.
15. Conney, A.H., Levin, W., Jacobsen, M. and Kuntzman, R. (1973) Clin. Pharmacol. Therap. 14, 727.
16. Sivarajah, K., Franklin, C.S. and Williams, W.P. (1978) J. Fish Biol. 13, 401.
17. Hansson, T., Raftner, J. and Gustafsson, J.A. (1980) Biochem. Pharmacol. 29, 583.
18. Payne, J.F. (1976) Science 191, 945.
19. Stegeman, J.J. (1978) J. Fish. Res. Bd. Canada 36, 668.
20. Walton, D.G., Penrose, W.R. and Green, J.M. (1978) J. Fish. Res. Bd. Canada 35, 1547.

Cytochrome P-450, Biochemistry, Biophysics
and Environmental Implications, E. Hietanen,
M. Laitinen and O. Hänninen editors

VARIATION OF CERTAIN HEPATIC MONOOXYGENASE ACTIVITIES IN FERAL WINTER FLOUNDER (*Pseudopleuronectes americanus*) FROM MAINE: APPARENT ASSOCIATION WITH INDUCTION BY ENVIRONMENTAL EXPOSURE TO PAH-TYPE COMPOUNDS

G. L. FOUREMAN AND J. R. BEND
Laboratory of Pharmacology, National Institute of Environmental Health Sciences, National Institutes of Health, P.O. Box 12233, Research Triangle Park, North Carolina 27709 (U.S.A.) and Mount Desert Island Biological Laboratory, Salsbury Cove, Maine 04672 (U.S.A.)

INTRODUCTION

The cytochrome P-450-dependent monooxygenase system, essential for the oxidative metabolism and activation of many procarcinogens, has been shown to be present in the liver of every fish species thus far adequately studied (1, 2). Characteristics of hepatic microsomal cytochrome P-450-dependent monooxygenase (MO) activity in fish are similar to those in mammals although specific activities are typically lower than they are in mammalian systems (1). Reconstitution of a fish microsomal MO system (3) demonstrated requirements for lipid, a flavoprotein and cytochrome P-450 as well as NADPH and oxygen for maximal enzyme activity. Also induction of fish hepatic MO activity has been demonstrated with PAH and other PAH-type inducers; aryl hydrocarbon hydroxylase (AHH) activities in some cases were increased 30-fold over control values (4). Curiously, the phenobarbital class of inducers do not appear to induce fish MO systems and this is a major difference between fish and mammals (5, 6).

There are also several similarities in induction of the MO system by PAH in fish and mammals. In both cases there are examples where AHH and 7-ethoxyresorufin deethylase (7-ERD) activities are elevated and novel species of cytochrome P-450 apparently formed (6, 7). In some mammalian and fish species the compound 7,8-benzoflavone (ANF) has been observed to specifically inhibit AHH and 7-ERD activities *in vitro* that were induced by PAH-type compounds (8, 9).

This report describes wide variations in hepatic cytochrome P-450-dependent enzyme activities (AHH and 7-ERD) which are specifically affected by PAH-type inducers in wild winter flounder of Maine. Also, a biochemical and electrophoretic comparison of liver is made between feral flounder and flounder experimentally induced by two different PAH-type compounds.

MATERIALS AND METHODS

Winter flounder, *Pseudopleuronectes americanus*, were caught by dragnet on the SE side of Mt. Desert Island, ME, during the months of June through August, 1978-1981. After decapitation the livers were excised, gallbladders removed, and a 15% w/v liver homogenate prepared in ice-cold 1.15% KCl, 1.25 mM HEPES, pH 7.4. Tissue homogenate was used as the enzyme source for all assays. When required, microsomes were prepared by further centrifugation of this fraction (10).

In the experiments with 4-chlorobiphenyl (4-CB), flounder (200-300 g) were suspended in wire mesh baskets in 50 l aquaria equipped with free flowing seawater. Twelve-to-sixteen hours after catheterization of the urinary papillae (Sherwood Umbilical, 3.5 Fr), the fish were injected i.m. with (^{14}C)-4-chlorobiphenyl (5 mg and 9.4 μCi/kg in 0.25 ml water:ethanol:Emulphor 620, 1:1:3). Urine was collected for 24 hours. At the end of this time period, the flounder were sacrificed and the bile collected.

AHH activity was assayed according to Wattenberg *et al.* (11) at 30°C. Results of this assay are expressed in relative flourescence units (FU) where 1 FU equals the fluorescence of a 3 μg/ml solution of quinine sulfate·$2H_2O$ in 0.1 N sulfuric acid at the same wavelengths used in the assay (excitation 400 nm; emission 525 nm). Immediately prior to the addition of the substrate, 7,8-benzoflavone (ANF, 0.5 mM in 0.05 ml acetone) was added to some of the incubation mixtures. 7-ERD activity was assayed according to the fluorimetric method of Burke and Mayer (12), also at 30°C. Protein concentration was assayed according to the method of Lowry *et al.* (13) with bovine serum albumin as the standard.

PAH-type inducing agents, either 5,6-benzoflavone (BNF; 100 mg/kg i.p. in corn oil) or 1,2,3,4-dibenzanthracene (DBA; 10 mg/kg i.p. in corn oil on 3 consecutive days) were used for the treatment of feral flounder. The animals were sacrificed on the 10th day after the initial injection. The method of Laemmli and Favre (14) was used to perform SDS-polyacrylamide gel electrophoresis (SDS-PAGE) with 7.5% running and 3% stacking gels.

RESULTS

In vivo variation in the clearance of 4-chlorobiphenyl. Table 1 shows the amount of 4-chlorobiphenyl-associated radioactivity excreted in the urine and bile of several flounder 24 hr after a single i.m. dose of (^{14}C)-4-chlorobiphenyl. The amount of radioactivity present varied markedly both in urine (19-fold) and bile (nearly 30-fold). There was general agreement between the amounts of biliary and urinary excretion, fish excreting low amounts of radioactivity in urine also having low amounts of radioactivity in bile as in fish No. 1; the converse of this relationship was demonstrated by fish No. 6.

TABLE 1

URINARY AND BILIARY EXCRETION OF AN I.M. INJECTION OF ^{14}C-CHLOROBIPHENYL[a] BY WILD WINTER FLOUNDER

Fish. No.	% Dose in Urine at 24 Hr	dpm/μl Bile
1	38.2	3136
2	25.8	3710
3	15.2	1467
4	13.5	4724
5	5.8	635
6	1.7	116

[a]Five mg/kg and 9.4 μCi/kg 4-chlorobiphenyl in ethanol:water:Emulphor 620 (1:1:3)

In vitro variation of hepatic monooxygenase activities

a. Occurrence. AHH and 7-ERD activities were measured in liver homogenates of several winter flounder and are presented in Table 2. The AHH activities of these fish varied over 40-fold while 7-ERD activities varied more than 300-fold.

TABLE 2

COMPARISON OF HEPATIC BENZO(A)PYRENE HYDROXYLASE (AHH) ACTIVITY AND 7-ETHOXYRESORUFIN DEETHYLASE (7-ERD) ACTIVITY IN WHOLE HOMOGENATE FROM INDIVIDUAL WINTER FLOUNDER[a] AND THE EFFECT OF *IN VITRO* ANF ON AHH ACTIVITY

Fish No.	AHH Activity (FU/units/min/mg protein) Without ANF	With ANF[b]	7-ERD Activity (pmol/min/mg protein)
1	0.17	0.72	< 2[c]
2	0.60	0.78	26
3	1.12	0.79	67
4	1.47	1.09	75
5	1.73	0.98	130
6	2.21	0.96	67
7	2.97	1.04	116
8	7.36	2.49	617

[a]Results presented are from fish all killed and assayed on the same day
[b]7,8-benzoflavone, 0.5 mM added in 0.05 ml acetone
[c]Sensitivity of assay procedure

These activities paralleled one another; high AHH activities were associated with high 7-ERD activities. Also to be noted is the dichotomous response to *in vitro* ANF. In fish having low AHH activities (e.g., No. 1) ANF enhanced this activity whereas in fish having higher AHH activities (No. 8) ANF instead inhibited this activity. As mentioned above, ANF has been shown to specifically inhibit AHH activity induced in mammals and fish by PAH or PAH-type compounds while enhancing this activity in control (untreated) animals.

b. Characterization and statistical analysis. To investigate the recurrency of this phenomenon and to more thoroughly characterize it, the winter flounder population around Mt. Desert Island, Maine, was sampled over a four year period. The results (N = 410) were compiled and are presented in Table 3. Wide variations were noted in all years (from 76-fold to 220-fold in the case of hepatic AHH activity) and are apparent from the wide standard deviations for this enzyme activity. However, no mean AHH activity was significantly different from that of the prior or the following season. In the case of 7-ERD activity, significant differences did exist between 1979 and 1980 and between 1980 and 1981 ($p < 0.05$). Overall, 262 of the 410 flounder examined (64%) had hepatic AHH activities which were inhibited by *in vitro* ANF.

TABLE 3

SUMMATION OF SELECTED HEPATIC CYTOCHROME P-450-DEPENDENT MONOOXYGENASE ACTIVITIES[a] IN FERAL MAINE WINTER FLOUNDER OVER FOUR YEARS

Date Sampled	N	AHH activity, FU/min/mg Homogenate Protein Mean ± SD	7-ERD Activity, pmol/min/mg Homogenate Protein Mean ± SD
June-August 1978	172	1.78 ± 1.66	131.3 ± 128.0
June-August 1979	81	2.03 ± 1.88	114.2 ± 142.5[b]
June-August 1980	86	1.78 ± 1.57	236.9 ± 215.6[b]
June-August 1981	71	1.31 ± 1.23	61.6 ± 55.4

[a]Determined in liver homogenate
[b]Significantly different from the following year, Duncan Multiple Range Test ($p < 0.05$, 2 tail test)

As mentioned above, hepatic AHH and 7-ERD activities in fish are both increased by the administration of PAH-type inducers, these increases being due to concomitant increases in one or more species of cytochrome P-450. Analysis of these activities as functions of one another in a representative season (1978, Figure 1) showed them to be highly and significantly correlated with one another. Also indicated in the plot are the sexes of the fish. It should be noted that there is no apparent relationship between sex of the flounder and either of these activities. In this context it is important to note that this study was conducted after the spawning season for maine flounder.

c. Relationship to cytochrome P-450 content. Table 4 shows the cytochrome P-450 content of hepatic microsomes from several feral winter flounder, the wavelength of maximum absorption in their CO-ligated, reduced P-450 spectrum and AHH

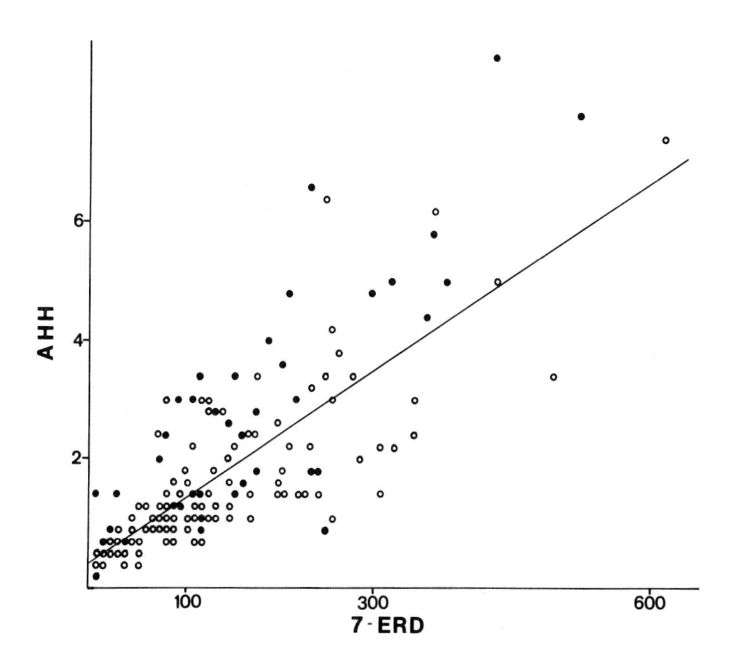

Fig. 1. Hepatic 7-ethoxyresorufin deethylase activity (7-ERD) as a function of AHH activity in male (●) and female (○) winter flounder hepatic homogenates sampled in 1978. The equation for the least squares regression line, which is included in the figure is AHH = 0.39 + 0.01056 (7-ERD); $r = 0.82$, $p < 0.001$. 172 values are represented, 35 of which are hidden (i.e. they duplicate plotted values).

activity expressed as a function of cytochrome P-450 content. With the exception of fish No. 1, there was no obvious relationship between the cytochrome P-450 content and AHH activity in the fish examined. Neither was there any consistency between the wavelength of maximum absorption of cytochrome P-450 and AHH activity although in the fish with the highest AHH activity/nmole cytochrome P-450 the wavelength was shifted hypsochromically with respect to the remainder of the fish examined.

TABLE 4

THE CYTOCHROME P-450 CONTENT, ITS WAVELENGTH OF MAXIMUM ABSORPTION AND AHH ACTIVITY OF HEPATIC MICROSOMES FROM SEVERAL WINTER FLOUNDER

Fish. No.	Cytochrome P-450 (nmol/mg protein)	λMax (nm)	AHH (FU/nmol cytochrome P-450/min)
1	0.60	448.4	6.4
2	0.25	451.0	0.7[a]
3	0.23	450.4	1.1
4	0.18	450.2	3.1
5	0.16	450.8	1.1
6	0.12	450.2	1.8
7	0.12	450.0	0.9

[a]AHH activity increased by *in vitro* addition of 0.5 mM 7,8-benzoflavone (ANF); all others were inhibited by ANF.

d. Relationship of physical characteristics. Correlation coefficients (r) were generated for several physical characteristics of the sampled fish as a function of hepatic AHH activity and are presented in Table 5. However, the highest r observed was with liver weight at -0.29. Statistically interpreted, this means less than 9% of the marked variability observed in hepatic AHH activity could be explained by this variable.

Treatment of flounder with PAH-type inducers

a. *In vitro* hepatic MO activities. Ten days after i.p. treatment of a number of wild flounder with either DBA or BNF, AHH and 7-ERD activities in all fish so treated were elevated (Table 6). It can also be seen from this table that the hepatic AHH activities in all fish so treated were inhibited by ANF. In two of these fish, the hepatic AHH activities observed were nearly twice as high as those observed in any of the over 400 feral flounder examined.

b. Electrophoretic comparison of hepatic microsomes from feral flounder with those of PAH-treated flounder. Microsomes, prepared both from the livers of fish treated with DBA and BNF (Table 6) and from several untreated flounder,

TABLE 5

CORRELATION COEFFICIENTS (r) OF SEVERAL PHYSICAL PARAMETERS OF FISH SAMPLED WITH HEPATIC AHH ACTIVITY OF WINTER FLOUNDER

Parameter	r	N
liver weight	-0.29	410
liver wt/body wt	-0.29	410
weight	-0.24	410
sex[a]	+0.19	410
month of assay	-0.19	410
length	-0.18	345
feeding status[b]	-0.15	153
gonad wt	-0.14	174
gonad wt/body wt	+0.12	174
year of assay	-0.08	410

[a]Indicates males had slightly higher hepatic AHH activities
[b]Indicates that fish with absence of particulate matter in the gut had slightly higher hepatic AHH activities.

TABLE 6

BENZO(A)PYRENE HYDROXYLASE (AHH) AND 7-ETHOXYRESORUFIN DEETHYLASE (7-ERD) ACTIVITIES IN LIVER HOMOGENATES OF DBA- AND BNF-TREATED FLOUNDER

Treatment[a]	AHH[b]		7-ERD[c]
	Without ANF	With ANF	
DBA	10.6	1.6	1147
DBA	7.7	1.5	630
DBA	4.0	1.3	1608
BNF	13.0	3.0	1787
BNF	5.8	1.3	1066
BNF	3.0	0.8	795
BNF	3.8	1.2	1259
BNF	3.9	0.9	1431

[a]Either 10 mg/kg of 1,2,3,4-dibenzanthracene (DBA) IP for 3 days or a single IP injection of 5,6-benzoflavone (BNF) at 100 mg/kg. See "Materials and Methods" for further details.
[b]Activity is expressed as FU/min/mg homogenate protein. Incubates were performed in 2 sets; 1 set with 0.5 mM 7,8-benzoflavone (ANF; in acetone) the other with just acetone.
[c]Activity is expressed as pmol resorufin formed/min/mg homogenate protein.

were compared to each other by PAGE in the presence of SDS. The patterns obtained upon electrophoresis of microsomes from two untreated flounder (Channels 1 and 2), a BNF-treated flounder (Channel 4), and a DBA-treated flounder

(Channel 5) are shown in Figure 2. In the area of the gel corresponding to approximately 57,000 MW, an intensely staining band was noted in microsomes from induced flounder. This band, at best, was faintly discernable in the microsomes from both untreated flounder. This relationship in staining intensity paralle-led the AHH activities of the respective homogenates of these fish which were 1.6, 0.7, 13.0 and 10.6 FU/min/mg for channels 1, 2, 4 and 5, respectively.

Hepatic microsomes from several untreated flounder with widely divergent hepatic AHH activities were electrophoresed in order of increasing AHH activity, left to right, as shown in Figure 3. As in the treated flounder (Figure 2), a

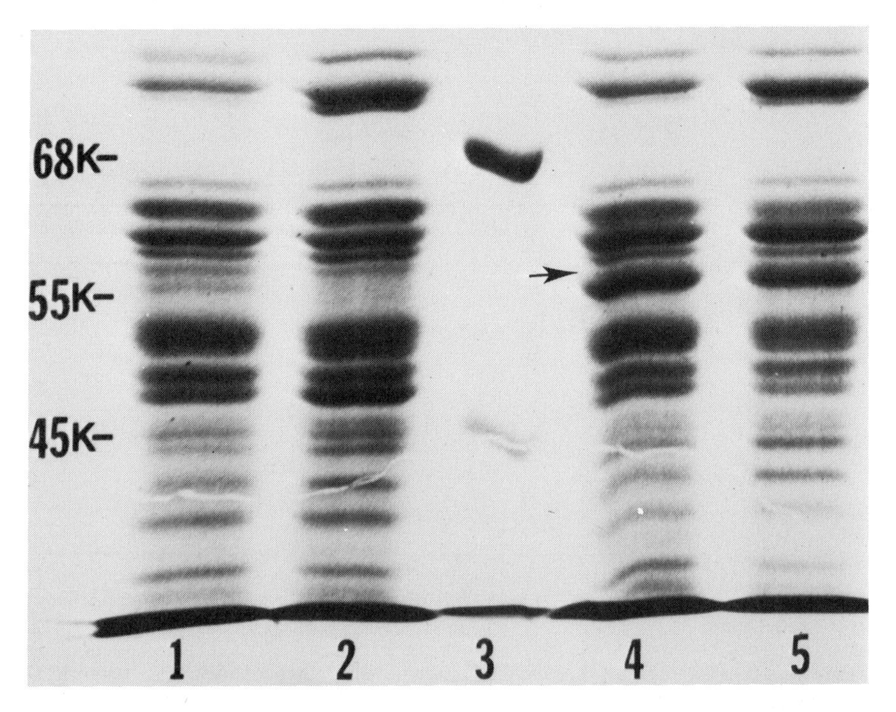

Fig. 2. SDS-PAGE of hepatic microsomes from untreated (Channels 1 and 2), a BNF-treated[a] (Channel 4) and a DBA-treated[a] (Channel 5) flounder. Standards (s) are 5 μg each of BSA (MW 68K) and ovalbumin (MW 45K). The arrow indicates the area of 57K. Microsomal protein (0.225 mg) was applied to each channel. AHH activities of the corresponding homogenate are as follows (Channel No., AHH): 1, 1.6; 2, 0.7; 4, 13.0; 5, 10.6.

[a]5,6-Benzoflavone (BNF) or 1,2,3,4-dibenzanthracene (DBA); see text for details.

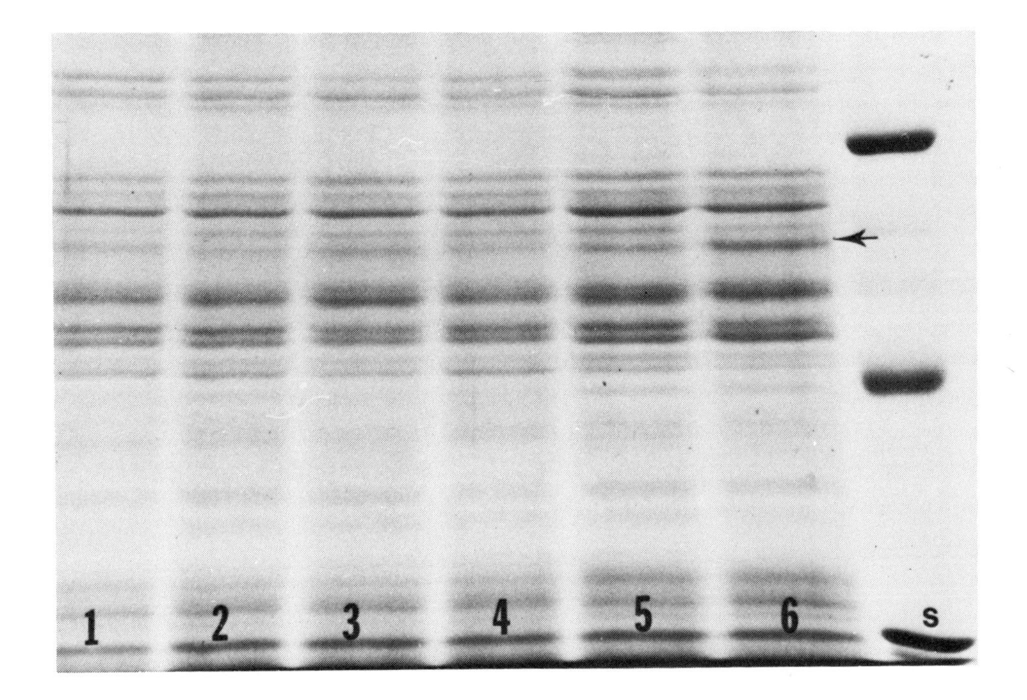

Fig. 3. SDS-PAGE of hepatic microsomes from untreated feral winter flounder. Standards (s) are, from top to bottom, 5 μg each BSA (MW 68K), ovalbumin (MW 45K) and carbonic anhydrase (MW 32K). The arrow indicates the area of 57K MW. All sample channels contain 0.05 mg of hepatic microsomal protein. AHH activity of hepatic homogenates from each flounder was as follows (Channel No., AHH): 1, 1.09; 2, 1.66; 3, 2.01; 4, 3.45; 5, 3.89; 6, 6.08.

polypeptide in the molecular weight range of 57,000 appeared to become more prominent as the AHH activity of the homogenates increased, especially in channels 5 and 6 where AHH activities were 3.9 and 6.1 FU/min/mg homogenate protein, respectively.

DISCUSSION

The wide variation observed *in vivo* in the extent of urinary and biliary excretion of 4-CB metabolites by winter flounder (Table 1) is consistent with the corresponding wide variation in the rates of cytochrome P-450-dependent AHH and 7-ERD activities (Table 2). Metabolism by the cytochrome P-450-dependent system is apparently the rate-limiting step in the clearance of 4-CB as other studies have demonstrated there are only minor individual differences in the rate of urinary excretion of polar compounds by the winter flounder (15).

Inhibition of AHH activity by ANF added *in vitro* is indicative of the presence of a form or forms of cytochrome P-450 similar to those in mammals which are induced by treatment with PAH-type compounds (8, 9). In this study, 64% (262 of 410) of the flounder tested had hepatic AHH activities that were inhibited by ANF suggesting that such forms are present in the livers of many of these wild flounder.

Further evidence for this concept is provided by the data in Table 4. In the flounder with the highest AHH activity, there was more cytochrome P-450 than in any other fish sampled. Also, the wavelength of maximum absorption of the cytochrome P-450 species in this fish was shifted towards 448 nm with respect to the other fish examined. In several fish species increases in hepatic microsomal AHH and 7-ERD activities are known to accompany both increases in cytochrome P-450 content and a spectral hypsochromic shift in a manner consistent with rats treated with PAH-type compounds (16). However, instances where elevation of cytochrome P-450-dependent activities are not accompanied by increased amounts of microsomal P-450 or by any spectral shift in the cytochrome P-450 spectrum are also common in fish treated with PAH (17) due apparently to the lack of sensitivity of these methods in detecting small amounts of catalytically active forms of cytochrome P-450. Thus, the data in Table 4 is supportive of a form (or forms) of cytochrome P-450 being present in the livers of many of these wild flounder that is(are) similar to those P-450 forms induced by the administration of PAH-type compounds. However, it should be pointed out that due to the insensitivity of these measures, neither the specific cytochrome P-450 content of microsomes nor the wavelength of maximum absorption of the cytochrome P-450 spectrum can serve as a valid parameter in fish for PAH-mediated induction of the hepatic MO system.

The ability of winter flounder to respond to PAH-type inducers is verified by the results in Table 6. Importantly, the character of the fishes' response to treatment with these compounds is qualitatively identical to that observed in many of the untreated fish from the wild, i.e. high hepatic 7-ERD and AHH activities, the latter being inhibited by ANF. Thus, many of the wild flounder examined in this study had livers with biochemical characteristics indistinguishable from flounder experimentally induced with a PAH (DBA) or a PAH-type compound (BNF).

Visualization of the end product of the cellular inductive process, i.e. the actual cytochrome P-450 polypeptide, has been achieved in mammals where multiple forms of P-450 have been isolated and purified (18). Electrophoretic patterns of hepatic microsomes from rainbow trout treated with Arochlor 1242, which had elevated AHH and 7-ERD activities, showed a novel or enriched polypeptide species near 57,000 MW. The intense band of approximately 57,000 MW present in the microsomes of PAH-treated flounder (Figure 2) is probably a form of cytochrome P-450 as it is in the molecular weight range of many other cytochrome P-450 species and it is either absent or only stained faintly in hepatic microsomes from fish with low AHH activities. A polypeptide species in this same MW range also occurred in hepatic microsomes from untreated wild flounder with high hepatic AHH activities (Channels 5 and 6, Figure 3) demonstrating that these wild flounder are electrophoretically indistinguishable from flounder treated with PAH-type inducers.

Although this major band (Figures 2 and 3) may be related to the increased AHH and 7-ERD activities, the results in Table 3 indicate that, as in mammalian species (19), there is probably at least one other PAH-inducible form of cytochrome P-450 in the livers of many of these fish. This is due to the non-parallel relationship in the fluctuations of 7-ERD and AHH activity and to the fact that 7-ERD and AHH induction is not always associated with a marked hypsochromic shift in the cytochrome P-450 spectrum.

The cause of this variation in hepatic AHH and 7-ERD activities in wild winter flounder does not appear to be due to any physical characteristic of the fish that we examined (Table 5 and Figure 1). The causative factor then could be either genetic or environmental. The likelihood that this variation may be due to an "Ah locus" similar to that demonstrated in mice (20) seems unlikely because all of the flounder treated with PAH-type compounds showed induction of both AHH and 7-ERD activities (Table 6 and unpublished data). Rather, an environmental explanation for this variation seems more plausible and in concert with the results presented here. It is known that flounder spend nearly their entire lives within bottom sediment (21) which contain varying and often appreciable amounts of PAH (22). In addition, uptake of PAH from sediment by other species of flatfish has been demonstrated (23). However, caution must be exercised in stating that the inductive response observed in some members of this species is due to exposure to PAH. As demonstrated here in flounder and elsewhere in trout (6), the biochemical and electrophoretic characteristics of teleost cytochrome P-450-dependent monooxygenase systems which are induced by flavonoids and PCBs are indistinguishable from those caused by induction with PAH. Flavones and flavonoids are widespread in nature and there is evidence for the presence of these compounds in at least 43 species of sea grasses (24). Other compounds of botanic origin which induce monooxygenase activities in mammals and which have not yet been evaluated in fish includes safroles and indoles (25). Thus, the induction observed in this feral flounder population could also be elicited by compounds biosynthesized by marine or terrestrial plants and anthropogenic compounds other than PAH.

REFERENCES

1. Bend, J.R. and James, M.O. (1978) in: Malins, D.C. and Sargent, J.R. (Eds.), Biochemical and Biophysical Perspectives in Marine Biology, 4, Academic Press, N.Y., pp. 125-188.

2. Stegeman, J. J. (1981) in: Ts'O.P. and Gelboin, H. V. (Eds.), Polycyclic Hydrocarbons and Cancer, 3, Academic Press, N.Y., pp. 1-60.

3. Bend, J.R., Ball, L.M., Elmamlouk, T.H., James, M.O. and Philpot, R.M. (1979) in: Khan, M.A.Q., Lech, J.J. and Menn, J.J. (Eds.), Pesticide and Xenobiotic Metabolism in Aquatic Organisms, ACS Press, N.Y., pp. 297-318.

4. Bend, J.R., Hall, P. and Foureman, G.L. (1976) Bull. Mt. Desert Island Biol. Lab., 16, 3.

5. Bend, J.R., Pohl, R.J. and Fouts, J.R. (1973) Bull. Mt. Desert Island Biol. Lab., 13, 9.

6. Elcombe, C.R. and Lech, J.J. (1979) Toxicol. Appl. Pharmacol., 49, 437.

7. Conney, A.H. (1967) Pharmacological Reviews, 19, 317.

8. Wiebel, F.J., Leutz, L., Diamond, L. and Gelboin, H.V. (1971) Arch. Biochem. Biophys., 144, 78.

9. Bend, J.R., James, M.O. and Dansette, P.M. (1977) Ann. N.Y. Acad. Sci. USA, 298, 505.

10. James, M.O., Bowen, E.R., Dansette, P.M. and Bend, J.R. (1979) Chem. Biol. Interact., 25, 321.

11. Wattenberg, L.W., Leong, J.L. and Strand, P.J. (1962) Cancer Res., 22, 1120.

12. Burke, M.D. and Mayer, R.T. (1974) Drug Metab. Disp., 2, 583.

13. Lowry, O.H., Rosebrough, N.J., Farr, A.L. and Randall, R.J. (1951) J. Biol. Chem., 193, 265.

14. Laemmli, U.K, and Favre, M. (1973) J. Mol. Biol., 80, 575.

15. Pritchard, J.B., Karnacky, K.J., Guarino, A.M. *et al.* (1977) Am. J. Physiol., 233, F126.

16. James, M.O. and Bend, J.R. (1980) Toxicol. Appl. Pharmacol., 54, 117.

17. Balk, L., Meijer, J., Seidegard, J., Morgenstern, R. and DePierre, J.W. (1980) Drug Metab. Disp., 8, 98.

18. Thomas, P.E., Lu, A.Y.H., Ryan, D. *et al.* (1976) Mol. Pharmacol., 12, 746.

19. Negishi, M. and Nebert, D.W. (1979) J. Biol. Chem., 254, 11015.

20. Nebert, D.W., Robinson, A., Niura, K., Kumaki, K. and Poland, A.P. (1975) J. Cell. Physiol., 85, 393.

21. Klein-MacPhee, G. (1978) NOAA Report NMFS CIRC-414, 43 p.

22. Laflamme, R.E. and Hites, R.A. (1978) Geochim. Cosmochim. Acta, 42, 289.

23. Varanasi, U. and Gmur, D.M. (1981) Aq. Toxicol., 1, 49.

24. McMillan, C.O., Zapata, O. and Escobar, L. (1980) Aquatic Botany, 8, 267.

25. Wattenberg, L.W., Loub, W.B., Lam, L.K. and Speier, J.L. (1976) Fed. Proc., 35, 1327.

Cytochrome P-450, Biochemistry, Biophysics
and Environmental Implications, E. Hietanen,
M. Laitinen and O. Hänninen editors

MULTIPLE FORMS OF CYTOCHROME P-450 IN MARINE CRABS

JOE W. CONNER AND RICHARD F. LEE
Skidaway Institute of Oceanography, P.O. Box 13687, Savannah, GA 31406 (USA)

INTRODUCTION

Crabs (Arthropoda: Crustacea:Decapoda:Brachyura) are among the commonest animals in estuaries and many species have been used in a variety of field and laboratory pollution studies (1). Cytochrome P-450 inducers, including pesticides, industrial wastes and petroleum products, enter and impact estuarine areas inhabited by various crab species.

Crabs, in common with many other vertebrate and invertebrate animals have a cytochrome P-450 mediated mixed function oxygenase (MFO) system (2,3,4,5). The crab MFO system shows similarity to the vertebrate since it is associated with the microsomal fraction and requires phospholipid, cytochrome P-450 and NADPH cytochrome P-450 reductase for activity (6). Highest MFO, i.e. benzo(a)-pyrene hydroxylase, activity in blue crabs, (Callinectes sapidus) has been found in the stomach and green gland with low activity in blood, gill, reproductive tissues, eyestalk, cardiac muscle and hepatopancreas (4). The low MFO activity in the hepatopancreas of many marine crustaceans is due to inhibition by digestive juices released during homogenization of the tissues (7). Most earlier work on crustaceans showed significant amounts of cytochrome P-420 in microsomal preparations (6,8,9). An isolation medium consisting of protease inhibitors, albumin, phenobarbital and various protecting agents allowed the isolation of cytochrome P-450 from blue crabs with no evidence of cytochrome P-420 (10).

Because of their large size and availability the two crabs we selected for our study were the spider crab, Libinia emarginata, and blue crab, Callinectes sapidus. For comparison we also partially purified the cytochrome P-450 from the lobster, Homarus americanus. Earlier studies indicated the presence of one cytochrome P-450 in the blue crab (10). Work by James and Little on spiny lobster, Panulirus argus, showed two cytochrome P-450 peaks eluted from DEAE-cellulose (11). The present work was to determine if more than one cytochrome P-450 was present in crabs and if any changes in the cytochrome P-450 occurred as a result of exposure to a mixture of polychlorinated biphenyls.

MATERIALS AND METHODS

Blue crabs and spider crabs were collected by trawling in estuaries of coastal Georgia (USA). The crabs were maintained as described in an earlier paper (6). Crabs were fed an artificial diet modified from one used by aquarists (12). For studies with polychlorinated biphenyls, Aroclor 1254 was dissolved in cod liver oil and blended into the food to give a concentration of 70 μg/g. Each crab was fed daily 2 grams of food containing Aroclor 1254 for 5 days. Controls were fed uncontaminated food.

The purification of cytochrome P-450 from microsomes was carried out using procedures slightly modified from that described by Conner and Singer (10). The microsomal pellet was solubilized with 0.4% sodium cholate and the supernatant passed down a n-octylamine Sepharose 4B column. The P-450 was eluted with 0.1% Triton N101 and the fraction containing P-450 was passed down a hydroxyapatite column. Rather than use a linear phosphate gradient as described by Conner and Singer (10) a stepwise elutution was used going from 0.012M to 0.5M potassium phosphate buffer (pH 7.3).

Cytochrome P-450 was quantified by the method of Omura and Sato (13) using an Aminco DW2A spectrophotometer. An extinction coefficient of 91 $cm^{-1}mM^{-1}$ was used although the coefficient of the crab P-450 may be different from this value. SDS-acrylamide of partially purified P-450 was performed by the method of Laemmli using a 3% stacking gel and a 9% running gel (14). Protein was determined by the method of Lowery et al. (15) with appropriate correletions for Triton (16).

RESULTS

The mean P-450 concentration in blue crabs was 0.060 ± 0.03 (n=13) nmoles P-450/mg microsomal protein, while in the spider crab the mean was 0.17 ± 0.04 (n=8) nmoles P-450/mg. An individual crab contained between 3 to 10 nmoles of P-450. The purification sequence for P-450 involved cholate digestion of the microsomes followed by passage down n-octylamine Sepharose 4B. For blue crabs, spider crabs and lobster the void volume from the affinity column contained 25%, 28% and 49%, respectively, of the P-450, while 16%, 28% and 15%, respectively, of the P-450 eluted from the column with 0.1% Triton N101. After the affinity column the fraction with the P-450 was passed down a hydroxyapatite column. From crab and lobster microsomes two cytochromes P-450 eluted from the hydroxyapatite column. Figure 1 shows the profile obtained from the blue crabs with the two peaks referred to as peaks A and B. For untreated crabs and lobster peak A was always a minor P-450 accounting for between 10 and 20% of the total P-450 eluted from the hydroxyapatite. The CO-difference spectrum of peaks A and B showed that both had absorbance maximum at 451 nm. Peaks A and B showed different electrophoretic mobility after polyacrylamide gel electrophoresis. Both peaks A and B were only partially purified and in addition to a major protein band at approximately 50,000 daltons faint bands at 24,000 daltons were observed. The highest specific content of P-450 obtained was 1.8 nmole/mg. Work is under way to further purify peaks A and B. For the blue crab only one peak was observed when the void volume fraction of the affinity column was passed down the hydroxyapatite column. This peak eluted slightly before peak B.

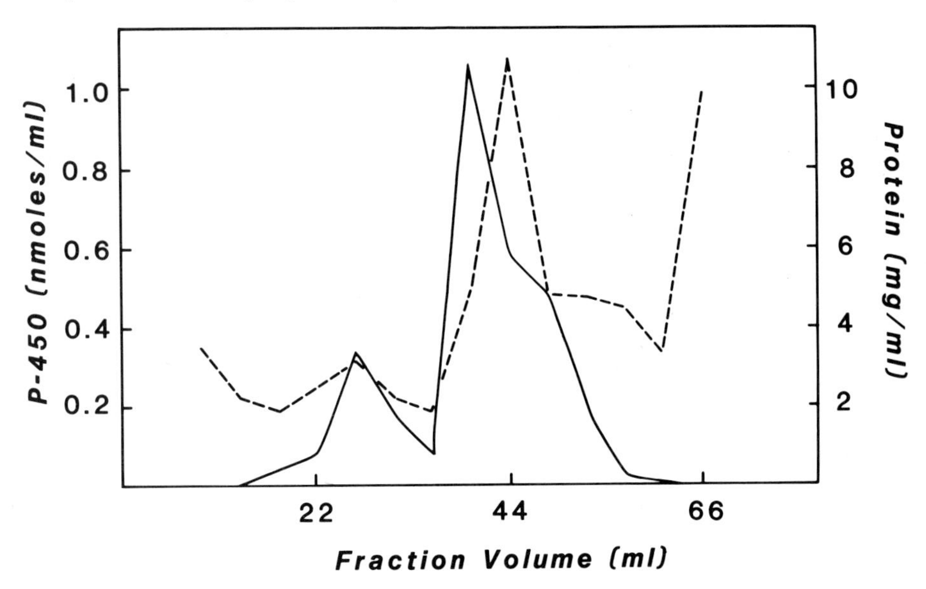

Fig. 1. Hydroxyapatite column chromatography of blue crab cytochromes P-450. The solid line represents the cytochrome P-450 content of the fraction and the dotted line the protein. Porteins were eluted by a stepwise elution going from 0.012M to 0.5M potassium phosphate buffer. The smaller peak is referred to as peak A and the larger peak as peak B.

When spider crabs were exposed to a mixture of polychlorinated biphenyls (Arochlor 1254) and the P-450s partially purified using the conditions described above, a total of four P-450s were observed (Fig. 2). The CO-difference spectrum of all peaks showed absorbance maximum at approximately 451 nm. There was no significant increase of the microsomal P-450 specific content of exposed crabs relative to controls.

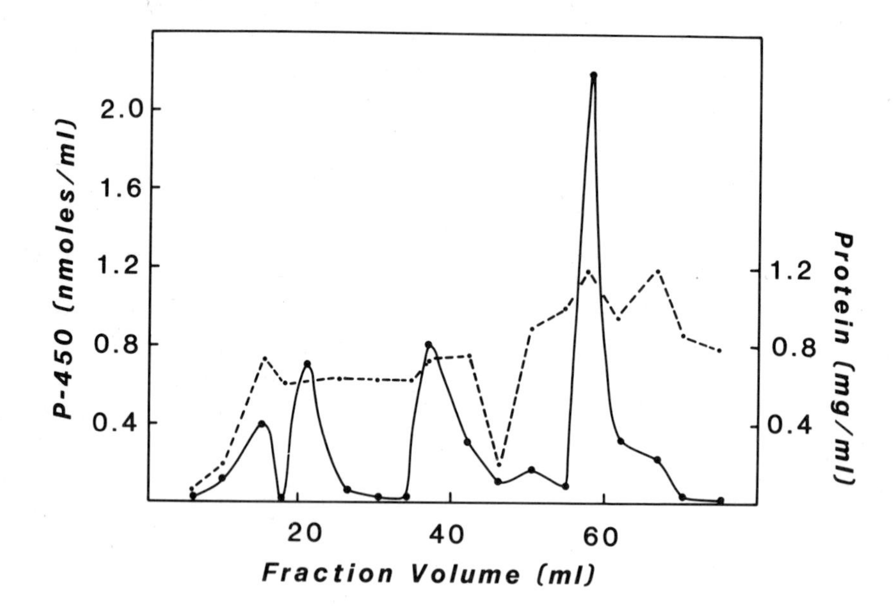

Fig. 2. Hydroxyapatite column chromatography of spider crab cytochromes P-450 after exposure to polychlorinated biphenyls. Crabs were fed food containing Aroclor 1254 daily for five days. The solid line represents the cytochrome P-450 content of the fraction and the dotted line the protein. Proteins were eluted by a stepwise elution going from 0.012M to 0.5M potassium phosphate buffer.

DISCUSSION

We obtained a partial purification of cytochromes P-450 from crabs and lobster using n-octylamine Sepharose 4B and hydroxyapatite chromatography. At least two cytochromes P-450 were observed after hydroxyapatite chromatography. James and Little using DEAE-cellulose found two cytochromes P-450 in hepatopancreas microsomes of the spiny lobster, Panulirus argus (11). The major P-450 had a molecular weight of 53,000 while the minor peak was at 48,000 daltons. We are not aware of other studies showing multiple forms of cytochrome P-450 in other marine invertebrates.

Exposure of spider crabs to polychlorinated biphenyls resulted in the production of several new forms of cytochrome P-450. Work is presently underway to characterize these new cytochrome P-450 forms and to compare them with insect and vertebrate P-450. Insect microsomes have a number of cytochromes P-450 and new forms are observed after exposure to various inducers (17, 18). Insects may have cytochromes P-450 similar to those present in crabs and lobsters, since they are all in the phyla Arthropoda.

ACKNOWLEDGEMENTS

We thank Professor Moises Agosin, Dr. Cesar Naquira and Dr. Robert White for their advice and help in the purification steps. The work was supported by NIH Research Service Award No. 1 F32ES05135-01S1 TUX (J.C.) and NSF Grant No. OCE80-17893.

REFERENCES

1. Williams, A.B. and Duke, T.W. (1979) in: Hart, C.W. and Fuller, S.L.H. (Eds.), Pollution Ecology of Estuarine Invertebrates, Academic Press, New York, pp. 171-233.
2. Bend, J.R., James, M.O., Little, P.J. and Foureman, G.L. (1981) in: Dawe, C.J. et al. (Eds.), Phyletic Approaches to Cancer, Japan Sci. Soc. Press, Tokyo, pp. 179-194.
3. James, M.O., Khan, M.A.Q. and Bend, J.R. (1979) Comp. Biochem. Physicol. 62C, 155-164.
4. Lee, R.F. (1981) Mar. Biol. Letts. 2, 87-105.
5. O'Hara, S.C.M., Corner, E.D.S., Forsberg, T.E.V. and Moore, M.N. (1982) J. Mar. Biol. Assoc. U.K. (in press).
6. Singer, S.C., March, P.E., Gonsoulin, F. and Lee, R.F. (1980) Comp. Biochem. Physiol. 65C, 129-134.
7. Pohl, R.J., Bend, J.R., Guarino, A.M. and Fouts, J.R. (1974) Drug Metab. Dispos. 2, 545-555.
8. Elmamlouk, T.H., Gessner, J. and Brownie, A.C. (1974) Comp. Biochem. Physiol. 48B, 419-425
9. Walters, J.M., Cain, R.B., Higgins, I.J. and Corner, E.D.S. (1979) J. Mar. Biol. Assoc. U.K. 59, 553-564.
10. Conner, J.W. and Singer, S.C. (1981) Aquatic Toxicol. 1, 271-278.
11. James, M.O. and Little, P.J. (1980) in: Gustafsson, J.-A., et al. (Eds.), Biochemistry, Biophysics and Regulation of Cytochrome P-450, Elsevier/North-Holland Biomedical Press, Amsterdam, pp. 113-120.
12. Lee, R.F., Singer, S.C. and Page, D.S. (1981) Aquatic Toxicol. 1, 355-365.
13. Omura , T. and Sato, R. (1964) J. Biol. Chem. 239, 2379-2385.
14. Laemmli, U.K. (1970) Nature 227, 680-685.
15. Lowry, O.H., Rosebrough, N.J., Farr, A.L. and Randall, R.J. (1951) J. Biol. Chem. 193, 265-275.
16. Mather, L.H. amd Tamplin, C.B. (1979) Anal. Biochem. 93, 139-142.
17. Hallstrom, I., Blanck, A., Grafstrom, R., Rannug, U. and Sundvall, A. (1980) in: Gustafsson, J.-A. et al. (Eds.), Biochemistry, Biophysics and Regulation of Cytochrome P-450, Elsevier/North=Holland Biomedical Press, Amsterdam, pp. 109-112.
18. Naquira, C., White, R.A. and Agosin, M. (1980) in: Gustafsson, J.-A., et al. (Eds.), Biochemistry, Biophysics and Regulation of Cytochrome P-450, Elsevier/North-Holland Biomedical Press, Amsterdam, pp. 105-108.

© 1982 Elsevier Biomedical Press B.V.
Cytochrome P-450, Biochemistry, Biophysics
and Environmental Implications, E. Hietanen,
M. Laitinen and O. Hänninen editors

CYTOCHROME P-450 IN THE HEPATOPANCREAS OF FRESHWATER CRAYFISH ASTACUS ASTACUS L.

PIRJO LINDSTRÖM-SEPPÄ, ULLA KOIVUSAARI AND OSMO HÄNNINEN
Department of Physiology, University of Kuopio, Kuopio, Finland

INTRODUCTION

The freshwater crayfish *Astacus astacus* L. has had economic significance on the Nordic Countries. The crayfish populations have, however, greatly diminished during the last two to three decades. One reason has been crayfish plague infection, but this does not fully explain the disappearance. Environmental pollutants have been suspected of contributing to this, too. *A. astacus* appears to be extremely sensitive to pollutants such as DDT (1), and it has been reported that this species practically lacks the biotransformation enzymes (2). Several other crustacean species especially those living in sea have, however, significant monooxygenase activities (3, 4, 5). Therefore one could also expect at least some biotransformation activity in *A. astacus*.

MATERIALS AND METHODS

Biological material. Freshwater crayfish of both sexes were collected from June to October from lakes near Kuopio, Finland. They were kept in water containers with air bubbling at 5°C for two to six weeks. The crayfish were 9.1 cm ($\bar{x}$) long and weighed 27.2 g. The hepatopancreas weighed 1.6 g.

Tissue preparation. The crayfish were killed by destroying the central nervous system, and tissue sampling was carried out immediately afterwards. The hepatopancreas were detached and immersed in ice-cold 0.1M potassium phosphate buffer, pH 7.4, containing 0.1M KCl, 1mM K_2EDTA, 1mM phenantroline and trypsin inhibitor (1mg/ml).

The tissues were homogenized in four volumes of the buffer. The special buffer was used to stabilize the labile enzymes and to avoid autogenous destruction by proteolytic enzymes existing in hepatopancreas. The homogenates were centrifuged at 10.000g for 20 min at 4°C, the supernatants obtained were centrifuged further at 105.000g for 60 min. The pellet (microsomal fraction) obtained was resuspended in the 0.1M potassium phosphate burrer with stabilizing agents so that 1 ml contained microsomes from 1 g tissue and 20 per cent glycerol was added for improving the storage. The microsomes were stored at -80°C.

Cytochrome P-450 levels and enzyme activities. Cytochrome P-450 content was measured immediately after the microsomes were prepared by the method of Johannesen and DePierre (1978).

Substrate binding spectra were recorded by the method of Schenkman *et al*. (1967) ethylmorphine (type I) or aniline-HCL (type II) as substrates.

Cytochrome c reductase activities (NADPH or NADH as electron donor) were determined by the method of Dallner *et al*. (1966).

The deethylation of 7-ethoxycoumarin was determined as described by Aitio (1978) and the deethylation of 7-ethoxyresorufin as described by Burke and Mayer (1974). The hydroxylation for benzo(a)pyrene was determined according to Nebert and Gelboin (1968). Demethylation of ethylmorphine was determined by using the method of Anders and Mannering (1966). All the enzyme determinations were made at 18°C.

RESULTS

TABLE 1

MONOOXYGENASE ACTIVITIES IN THE HEPATOPANCREAS OF *A. ASTACUS*

	Microsomes	Cytosol
Cytochrome P-450[a]	0.31 ± 0.07	± 0
NADPH Cytochrome c reductase[b]	2.23 ± 0.05	8.73 ± 0.19
NADH Cytochrome c reductase[b]	48.81 ± 8.14	0.43 ± 0.11
Benzo(a)pyrene hydroxylase[b]	± 0	± 0
7-Ethoxycoumarin O-deethylase[b]	0.01 ± 0.00	± 0
7-Ethoxyresorufin deethylase[b]	± 0	± 0
Ethylmorphine demethylase[b]	0.07 ± 0.05	± 0.02

[a] nmol x mg $prot.^{-1}$

[b] nmol x min^{-1} x mg $prot.^{-1}$

Freshwater crayfish was found to have considerable amounts of cytochrome P-450 in the hepatopancreas microsomes. The content (0.31 ± 0.07 nmol x mg prot.$^{-1}$, $\bar{x} \pm$ SD) was almost as high as in rat liver microsomes (0.38 ± 0.07 nmol x mg prot.$^{-1}$). The low (7-ethoxycoumarin O-deethylation and ethylmorphine demethylation) or absent (benzo(a)pyrene hydroxylation and 7-ethoxyresorufin deethylation) monooxygenase activities recorded did not, however, correlate with the high amount of cytochrome P-450 (Table 1).

The binding of type I (ethylmorphine) and type II (aniline-HCL) substrates to the enzyme cytochrome P-450, was different in crayfish hepatopancreas than in rat liver microsomes (not illustrated). There was also less NADPH than NADH cytochrome c reductase activity in crayfish hepatopancreas microsomes (Table 1). NADH supported faster demethylation of ethylmorphine and deethylation of 7-ethoxycoumarin than NADPH (Table 2). NADPH or NADH could support neither benzo(a)pyrene hydroxylation nor 7-ethoxyresorufin deethylation.

TABLE 2

THE INFLUENCE OF DIFFERENT ELECTRON DONORS ON MICROSOMAL MONOOXYGENASE ACTIVITIES IN THE HEPATOPANCREAS OF <u>A</u>. <u>ASTACUS</u>

NADPH (10mM)	3.5 ± 0.3	± 0	± 0	69.0 ± 46.7
NADH (10mM)	4.3 ± 0.6	± 0	± 0	198.0 ± 43.8

[A] 7-Ethoxycoumarin O-deethylase

[B] Benzo(a)pyrene hydroxylase

[C] 7-Ethoxyresorufin deethylase

[D] Ethylmorphine demethylase

[e] pmol x min^{-1} x mg prot.$^{-1}$

Studies with hepatopancreas subfractions added to rat liver microsomes showed that hepatopancreas cytosol contained heat labile monooxygenase inhibitor(s) (Table 3). Only traces of 7-ethoxycoumarin O-deethylase activity were found in total homogenates of extrahepatopancreatic tissues like gills, intestine and green glands (not illustrated).

TABLE 3

INFLUENCE OF THE CRAYFISH HEPATOPANCREAS SUBFRACTIONS ON THE RAT LIVER MICROSOMAL MONOOXYGENASE ACTIVITIES WITH DIFFERENT SUBSTRATES

	7-Ethoxycoumarin		Benzo(a)pyrene	
RLM[a]	100 %		100 %	
" + 25 ul M[b]	79.6	6.5 %	131.7	18.3 %
" + 25 ul S[c]	19.3	2.7 %	48.4	24.2 %
" + 25 ul HS[d]	80.6	0.0 %	147.4	7.5 %

[a]Rat liver microsomes
[b]Microsomes of crayfish hepatopancreas
[c]Crayfish hepatopancreas cytosol
[d]Heated cytosol

It can be concluded that the tissues of the crayfish (*A. astacus*) are capable of metabolizing foreign compounds at least to some extent. The monooxygenase system of this species appears to have a substrate spectrum of its own and only a few commonly used model substrates are oxidized *in vitro* despite the high cytochrome P-450 level.

AKNOWLEDGEMENTS

This study has been supported by grants from Finnish Research Council for Natural Sciences (Project No 40).

REFERENCES

1. Airaksinen, M., Valkama, E-L. and Lindqvist, O.V. (1977) in: Lindqvist, O.V. (Ed.), Freshwater crayfish 3, Kuopio, Finland, pp. 349-356.
2. Lang, M., Valkama, E-L. and Lindqvist, O.V. (1977) in: Lindqvist, O.V. (Ed.), Freshwater crayfish 3, Kuopio, Finland, pp. 343-348.
3. Brodie, B.B. and Maickel, R.P. (1962) in: Brodie, B.B. and Maickel, R.P. (Eds.), Proceedings of the I international pharmacology meeting, vol. 6, Pergamon Press, pp. 299-324.
4. Khan, M.A.Q., Coello, W., Khan, A.A. and Pinto, H. (1972) Live Sci. 11, 405-415.
5. Bend, J.R., James, M.O. and Dansette, P.M. (1977) Ann. N.Y. Adad. Sci. 298, 505-521.

Cytochrome P-450, Biochemistry, Biophysics
and Environmental Implications, E. Hietanen,
M. Laitinen and O. Hänninen editors

OCCURENCE OF CYTOCHROME P-450 IN THE EARTHWORM LUMBRICUS TERRESTRIS

ARI LIIMATAINEN AND OSMO HÄNNINEN
Department of Physiology, University of Kuopio, P.O. Box 138, SF-70101
Kuopio 10, Finland

INTRODUCTION

Earthworms of the genus Lumbricidae are an ecologically important group of soil invertebrates, which are exposed to many pesticides and pollutants. Their ecology and biology is well documented (see for example Edwards and Lofty 1977) (1), but little is known about their metabolism of xenobiotics. Nakatsugava and Nelson have shown that Lumbricus terrestris has microsomal oxidation activity in the intestine (2). In our study it is demonstrated that monooxygenase system responsible to this oxidation activity contains cytochrome P-450.

MATERIALS AND METHODS

Animals. Earthworms were collected in July from a vegetable garden in Kuopio, Finland, where no pesticides have been used. Only fully grown specimens with a clitellum were selected.

Preparation of microsomal fractions. Earthworms were rinsed with cold tap water, the head was cut off at 18 - 19 th segment. Ice cold 0.25 M sucrose was forced through the intestine to remove the soil. 3 - 4 cm of the gut was removed with a scalpel. The sample was homogenized in four times the weight volume of ice cold potassium buffer (0.1 M K-phosphate, 0.1 M KCl, 1mM K_2EDTA, 1mM dithiothreitol, 0.1 mM phenylmethylsulfonylfluoride, pH 7.4) in a Teflon-glass-homogenizer. The homogenate was centrifuged at +4 C for 10 min at 10000g. The supernatant was recentrifuged at +2 - 4 C for 60 min at 105 000g and the microsomal pellet was resuspended in potassium phosphate buffer described above so that 1 ml contained microsomes from 1 g of the gut.

In the other experiment 6.1 ml of the 10000g supernatant was applicated in a Sepharose 2B column (Pharmacia Fine Chemicals, Uppsala, Sweden; the height of the column was 195 mm, diameter 20 mm) (3). The column was eluted with a buffer (50 mM K-phosphate, 150 mM KCl, 20 % glycerol, pH 7.5, flow 19.5 ml/h). The collection of fractions was started when a clealy separated yellow zone had reached the bottom of the column. Each fraction was collected for 6 min.

Enzyme assays. Cytochrome P-450 and cytochrome b_5 were determined by the method of Omura and Sato (4), benzo(a)pyrene hydroxylase by the method of Wattenberg et al. (5) modified by Nebert and Gelboin (6), 7-ethoxyresorufin-O-deethylase by the method of Burke and Mayer (7) and 7-ethoxycoumarin-O-deethylase by the method of Ullrich and Weber (8) modified by Aitio (9). Protein content was determined by the Gornall´s biuret method (10) modified by Lang et al. (11).

RESULTS AND CONCLUSIONS

Figures 1 A and B show dithionite reduced (cytochrome b_5) and reduced carbon monoxide complex (cytochrome P-450) spectra from microsomal fractions prepared by centrifuging and by gel filtration, respectively. Microsomes prepared by centrifugation were suspected to be contaminated by hemoglobin. To confirm this the spectra of the blood of the earthworm were driven (Fig.2). The similarity between the spectra in figures 1 A and 2 indicate that hemoglobin in vivo dissolved in plasma adsorbs on microsomal fraction during the isolation process.

The amounts of cytochrome P-450 and cytochrome b_5 were about 0.86 nmol x g^{-1} fresh weight of the gut (7.7 pmol x mg^{-1} prot.) and 0.89 nmol x g^{-1} fresh weight of the gut (7.9 pmol x mg^{-1} prot.), respectively. The earthworm gut monooxygenase system was not able to hydroxylate benzo(a)pyrene or dealkylate 7-ethoxyresorufin, but it O-dealkylated 7-ethoxycoumarin about 37 pmol x min^{-1} x g^{-1} fresh weight of the gut (1.7 pmol x min^{-1} x mg^{-1} prot.).

It can be concluded from the present series of experiments that cytochrome P-450 occurs in the gut of the earthworm Lumbricus terrestris. Hemoglobin interferes the recording of cytochrome P-450 spectrum. Therefore it can be detected only in gel filtrated microsomal sample.

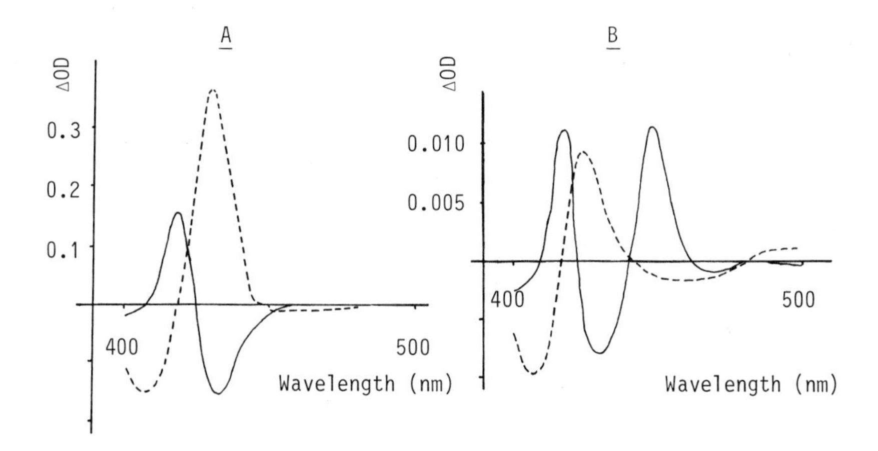

Fig. 1. The spectra of the centrifuged (A) and gel filtrated (B) microsomal fractions of the gut of Lumbricus terrestris. The centrifuged microsomes were made of a pooled sample of the guts of five earthworms, gel filtrated microsomes were made of the 10000g supernatant of the pooled guts of eleven earthworms. -------- dithionite reduced form (cytochrome b_5)
——— reduced cabon monoxide complex (cytochrome P-450)

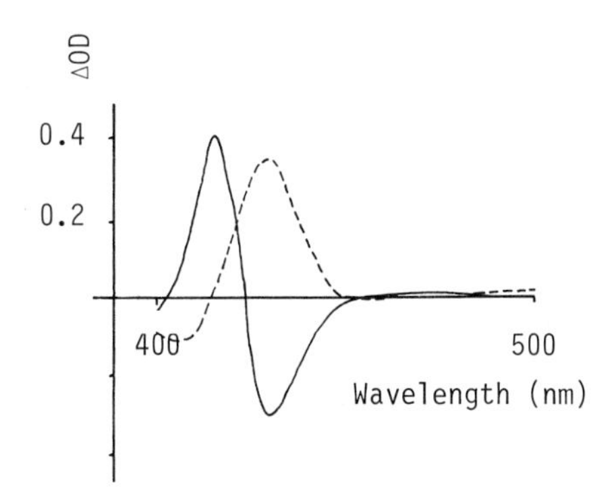

Fig. 2. The spectra of the blood of Lumbricus terrestris.
-------- dithionite reduced form
——— carbon monoxide complex

ACKNOWLEDGEMENT

This study has been supported by the Academy of Finland (project No. 40).

REFERENCES

1. Edwards, C.A. and Lofty, J.R. Biology of Earthworms. Chapman & Hall, London, 1977.
2. Nakatsugava, T. and Nelson, P.A. Environmental Toxicology of Pesticides. Matsumura, F., Brush, G.M. and Misato, T. (eds.). Academic Press, New York 1972, p. 501.
3. Tangen, O., Jonsson, J. and Orrenius, S. Isolation of Rat Liver Microsomes by Gel Filtration. Anal. Biochem. 54:597-603, 1973.
4. Omura, T. and Sato, R. The carbon monoxide binding pigment of liver microsomes. J. Biol. Chem. 239: 2379-2385, 1964.
5. Wattenberg, W.L., Leong, J.L. and Stand, P.J. Benzpyrene hydroxylase activity in the gastrointestinal tract. Cancer Res. 22:1120-1125,1962.
6. Nebert, D.W. and Gelboin, H.V. Substrate inducible microsomal arylhydrocarbon hydroxylase II. Cellular responses during enzyme induction. J. Biol. Chem. 244:6242-6249, 1968.
7. Burke, M.D. and Mayer, R.T. Ethoxyresorufin: direct fluorimetric assay of a microsomal O-dealkylation which is preferentially inducible by 3-methylcholantrene. Drug Metab. Dispos. 2 (6):583-588,1973.
8. Ullrich, B. and Weber, P. The O-dealkylation of 7-ethoxycoumarin by liver microsomes. A direct fluorometric test. Hoppe-Seyler`s Z. Physiol. Chem. 353 (7):1171-1177, 1972.
9. Aitio, A. A simple and sensitive assay of 7-ethoxycoumarin deethylation. Analyt. Biochem. 85:448-491, 1978.
10. Gornall, A.G., Bardanill, C.J. and David, M.M. Determination of serum proteins by means of the biuret reaction. J. Biol. Chem. 177:751-766,1949.
11. Lang, M., Laitinen, M., Hietanen, E. and Vainio, H. Modofication of microsomal membrane components and induction of hepatic drug biotransformation in rats on a high cholesterol diet. Acta Pharmac.Toxic. 39:273-288, 1976.

© 1982 Elsevier Biomedical Press B.V.
Cytochrome P-450, Biochemistry, Biophysics and Environmental Implications, E. Hietanen, M. Laitinen and O. Hänninen editors

CYTOCHROME P-450 AND POLYSUBSTRATE MONOOXYGENASE (PSMO) ACTIVITIES IN THE LIVER OF SPAWNING RAINBOW TROUT (*Salmo gairdneri*)

ULLA KOIVUSAARI, MAIJA PESONEN AND OSMO HÄNNINEN
Department of Physiology, University of Kuopio, P.O.B. 138, SF-70101 Kuopio 10 (Finland)

INTRODUCTION

The seasonal life cycle of the fish is accompanied by changes in environmental factors such as temperature and also by annual reproduction cycle which might affect also the xenobiotic metabolism. Seasonal variation both in xenobiotic and steroid metabolism in rainbow trout has been reported recently (1,2). Sex related differences in metabolism of xenobiotics have also been investigated in gonadally mature rainbow trout during the prespawning and spawning time (1,3). There are, however, no studies concerning the levels of cytochrome P-450 and PSMO-activities after spawning.

In present study the level of cytochrome P-450 as well as PSMO-activities in the presence of different substrates (7-ethoxycoumarin, benzo(a)pyrene, ethoxyresorufin, aminopyrine and ethylmorphine) were determined in the liver microsomes of gonadally mature rainbow trout (*Salmo gairdneri*) during and after the reproduction time from March to June.

MATERIALS AND METHODS

Animals. Gonadally mature rainbow trout of both sexes were obtained from Nilakkalohi Fish Farm Ltd in Tervo in Eastern Finland. The fish were 2+ years old and they were fed on commercial rainbow trout fodder (Vaasan Höyrymylly Ltd, Finland). The water temperature ranged from $+2^{o}C$ (in March) to $+13^{o}C$ (in June).

Assays. After isolation the microsomal fraction of fish liver was finally resuspended in 0.1 M K-phosphate buffer, pH 7.4, containing 0.15 M KCl, 1 mM K_2EDTA, 1 mM DTT and 20 % glycerol. The content of cytochrome P-450 and absorption maximum of reduced cytochrome-CO-complex were recorded immediately, for other measurements the microsomes were stored at $-80^{o}C$.

The total amount of cytochrome P-450 was measured as described by Omura and Sato (4). The microsomal polysubstrate monooxygenase

Table 1. Cytochrome P-450 content and polysubstrate monooxygenase activities in liver microsomes of prespawning and postspawning rainbow trout (*Salmo gairdneri*)

		MARCH	APRIL	MAY	JUNE
Cytochrome P-450[a]	male	0.15±0.02[c]	0.16±0.01	0.18±0.02	0.25±0.02
		***	**	*	
	female	0.04±0.01	0.11±0.01	0.12±0.01	0.22±0.02
Benzo(a)pyrene[b] hydroxylase	male	0.014±0.001	0.019±0.004	0.017±0.002	0.035±0.008
		**			
	female	0.008±0.001	0.013±0.002	0.015±0.002	0.003±0.005
7-Ethoxycoumarin[b] deethylase	male	0.040±0.005	0.086±0.017	0.062±0.015	0.067±0.017
		***	**		
	female	0.017±0.002	0.024±0.002	0.036±0.003	0.061+-0.009
Ethoxyresorufin[b] deethylase	male	0.062±0.016	0.23 ± 0.07	0.16±0.03	0.45±0.15
	female	0.072±0.025	0.16±0.03	0.24±0.07	0.33±0.09
Aminopyrine[b] demethylase	male	0.32±0.05	0.53±0:05	0.37±0.03	0.24±0.02
		**	***	**	
	female	0.09±0.01	0.21±0.04	0.22±0.03	0.31±0.11
Ethylmorphine[b] demethylase	male	0.34±0.06	0.93±0.03	0.80±0.06	0.48±0.03
		**	***		
	female	0.11±0.01	0.62±0.03	0.70±0.02	0.47±0.03

a) nmoles/mg microsomal prot. b) nmoles/mg microsomal protein x min. c) Mean values± S.E.M. from 6 individual fish. Statistical significance (Student's t-test): * $p < 0.05$, ** $p < 0.01$, *** $p < 0.001$

activity was measured by using five different substrates and methods described previously: deethylation of 7-ethoxycoumarin (5), hydroxylation of benzo(a)pyrene (6), deethylation of ethoxyresorufin (7) and demethylation of aminopyrine and ethylmorphine (8). The in vitro incubations were all carried out at +18°C, at which temperature the denaturation of enzyme protein in fish liver does not yet occur. The reactions were controlled to be linear with time and enzyme concentration.

Protein content was determined by the method of Lowry et al. (9)

RESULTS AND DISCUSSION

The rainbow trout studied had the spawning time in May. The maturation of gonads and also the increased liver-body weight ratio in female fish evidenced the preparation to spawning.

The amount of cytochrome P-450 was lowest during the prespawning period and it was significantly lower in females than in males (Table 1). The content of cytochrome P-450 increased markedly after spawning and the sex-related difference disappeared. This indicates that the sex related differences in cytochrome P-450 content are connected with certain periods on gonadal maturity and hormonal status. Sex-related differences in absorption maximum of reduced, CO-ligated cytochrome P-450, which has shown to be present in microsomes of mature rainbow trout and brook trout (3) could not be detected in present work. The absorption maximum determined in prespawning fish varied from 449 nm to 450 nm without any correlation to the sex of the fish or to the content of cytochrome P-450.

Present results indicate that the polysubstrate monooxygenase activity in prespawning and postspawning fish depends on the substrate used (Table 1). In all cases, however, the activities were at lowest level before spawning in March. Specific benzo(a)pyrene hydroxylase activity was lower in female fish but the difference between sexes occured only in March and dissappeared before the onset of spawning. Both benzo(a)pyrene hydroxylase and 7-ethoxyresorufin deethylase activities increased gradually during the spawning and postspawning time.

Although 7-ethoxyresorufin deethylase activity did not show any sex-related differences at all, metabolic rates of 7-ethoxycoumarin, aminopyrine and ethylmorphine were considerably higher in male fish when compared to female fish during prespawning time. Aminopyrine demethylase activity increased rapidly both in female and male fish from March to April and turned to decrease after spawning. The activity level measured in postspawning fish in June, however, exceeded the lowest levels measured in March. The demethylation of ethylmorphine in prespawning and spawning fish showed the greatest sex difference. In male fish the enzyme activity rapidly increased from March to April and the decreased considerably being at lowest level in June while in female fish the enzyme activity increased slowly during the spawning and postspawning time.

The disappearance of sex-related differences in postspawning fish suggests that the hormonal changes associated with reproduction may regulate the cytochrome P-450 content and PSMO-activities. The difference in changes of PSMO-activities with different substrates during the prespawning and postspawning period also indicate presence of multiple forms of cytochrome P-450 and multiple regulation systems of PSMO-activities. Simultaneously with the regression after spawning the water temperature, food intake and general metabolic rates also increase. All these factors may affect the xenobiotic metabolism in liver.

ACKNOWLEDGEMENTS

This study has been supported by grants from Finnish Research Council for Natural Sciences.

REFERENCES

1. Koivusaari U., Harri M. and Hänninen O. (1981), Comp. Biochem. Physiol. 70 C, 149-157

2. Hansson T. and Gustafsson J-Å. (1981), Gen. Comp. Endocrinol. 44, 181-188

3. Stegeman J.J. and Chevion M. (1980), Biochem. Pharmacol. 29, 553-558

4. Omura T. and Sato R. (1964), J. Biol.Chem. 239, 2379-2385

5. Aitio A. (1978), Anal. Biochem. 85, 488-491

6. Nebert D.W. and Gelboin H.V. (1968), J. Biol. Chem. 244, 6242-6249

7. Burke M.D. and Mayer R.T. (1974) Drug Metab. Disp. 2 , 583-588

8. Anders M.W. and Mannering G.J. (1966) Mol. Pharmacol. 2, 319-327

9. Lowry O.H., Rosebrough N.J., Farr A.L. and Randall R.J. (1951) J. Biol. Chem. 193, 265-275

© 1982 Elsevier Biomedical Press B.V.
Cytochrome P-450, Biochemistry, Biophysics
and Environmental Implications, E. Hietanen,
M. Laitinen and O. Hänninen editors

ANTIPARASITIC TREATMENT AND MONOOXYGENASES IN REINDEER

MATTI LAITINEN[1], RISTO JUVONEN[1], MAURI NIEMINEN[2], EINO HIETANEN[3]
AND OSMO HÄNNINEN[1]
[1]Department of Physiology, University of Kuopio, P.O.Box 138,
SF-70101 Kuopio 10, (Finland),[2]Finnish Game and Fisheries
Research Institute, Game Division, Koskikatu 33A, SF-96100 Rovaniemi 10 (Finland) and [3]Department of Physiology, University of
Turku, Kiinamyllynk. 8-10, SF-20525 Turku 52, (Finland).

INTRODUCTION

The parasites of reindeer cause great economic losses for people in northern Finland. The quantity and quality of meat is decreased as well as the well-being of the reindeer is affected. Obviously the most commonly used antiparasitic treatment is the injection of animals with an organophosphorous compound Warbex (Famphur, or o,o-dimethyl-o,p-(N,N-dimethylsulphamoyl)phenyl phosphorothioate).

In our earlier studies we have found that the metabolism of foreign compounds in the reindeer is very sensitive to different conditions (1). The seasonal variation in the amount and quality of food makes differences in the metabolic activity (2). In the present study we analyzed the effect of antiparasitic treatment on the activities of monooxygenases in the reindeer. We also tried to detect modification enzyme activities in the rat. However the treatment was not exactly the same, because reindeer had received antiparasitic medication yearly during six years and it was not possible to do so with rats.

MATERIALS AND METHODS

Warbex (American Cyanamid Company, Wayne New Jersey, USA) was injected into five reindeer (about 200 mg/animal) once a year for six years. Five untreated animals served as controls. Rats (Wistar/Af/Han/Mol/Kuo) received Warbex (35 mg/kg) s.c. three times once a week and they were killed week after the last injection. Diethylsuccinate (Oy Star AB, Tampere, Finland), the diluent of Warbex was given to one control group and the other control group had no treatment. The number of rats was five in each group.

The livers of both reindeer and rats were homogenized in four volume of 0.25 M sucrose and the homogenates were first centrifuged at 10000 x g for 15 minutes and the microsomal fraction was isolated at 105000 x g in 60 minutes. The pellets containing microsomes were resuspended to concentration 1 g liver wwt per 1 ml suspension.

The protein determination was carried out by the biuret method (3) using bovine serum albumin (Sigma Chemical Co, St.Louis, Mo, USA) as a reference. The microsomal cytochrome P-450 content was measured as described by Omura and Sato (4). The deethylation of 7-ethoxycoumarin was measured fluorometrically according to Ullrich and Weber (6) using Aitio's modification (7). The hydroxylation of 3,4- benzpyrene was measured fluorometrically according to Wattenberg (8) and Nebert and Gelboin (9).

RESULTS

The results are expressed in the next tables. In the reindeer (Table I) all measured enzyme activities were increased signifi-

TABLE 1

THE ACTIVITIES OF MONOOXYGENASES IN THE REINDEER

The activities of monooxygenases in the hepatic microsomes of the reindeer. C=control animals, W=animals treated with Warbex. Means and standard deviations are expressed. The statistical significance (Student's t-test) +=P 0.05,++=P 0.01, +++=P 0.001.

	Total activity nmol/min g liver wwt		Spesific activity nmol/min mg protein	
	C	W	C	W
Cytochrome P-450 (nmol/g or mg)	0.92+0.23	3.67+0.32	0.034+0.007	0.154+0.005
	+++		+++	
7-ethoxyresoruffin deethylase	75+32	5390+1390	3.20+0.58	224+47
	++		++	
7-ethoxycoumarin o-deethylase	0.086+.029	4.55+1.37	.0038+.0003	.188+.046
	++		++	
3,4- benzpyrene hydroxylase	72.6+1.0	304+113	2.76+0.21	12.6+4.0
	+		+	

cantly. The increase in the 3,4- benzpyrene hydroxylase was not so profound as with other enzymes. When the activities were analyzed in the rat (Tables 2 and 3) only the reaction catalyzed

TABLE 2

THE ACTIVITIES OF MONOOXYGENASES IN THE RAT

The total activities of monooxygenases in the rat are expressed. For other explanations see Table 1.

	Nontreated controls	Controls with diethylsuccinate	Treatment with Warbex
Cytochrome P-450	12.0±1.6 ——+——	14.6±1.7	12.8±1.2
7-ethoxyresorufin deethylase	2.75±0.99——++——	6.23±0.91	5.49±0.07
	└———————————	+++	———————————┘
7-ethoxycoumarin o-deethylase	16.5±3.1	13.2±4.9	14.9±3.1
3,4-benzpyrene hydroxylase	1.35±0.32	1.32±0.14	1.43±0.04

TABLE 3

THE ACTIVITIES OF MONOOXYGENASES IN THE RAT

The specific activities of monooxygenases in the rat are expressed. For other explanations see Table 1.

	Nontreated controls	Controls with diethylsuccinate	Treatment with Warbex
Cytochrome P-450	0.510±0.063	0.546±0.067	0.514±0.040
7-ethoxyresorufin deethylase	0.117±0.042—++—	0.233±0.037	0.221±0.053
	└———————————	+	———————————┘
7-ethoxycoumarin o-deethylase	0.71±0.15	0.68±0.20	0.60±0.12
3,4-benzpyrene hydroxylase	0.57±0.13	0.50±0.07	0.58±0.01

by 7-ethoxyresorufin deethylase was increased with Warbex as well as with deethylsuccinate solvent. The diethylsuccinate increased also slightly the cytochrome P-450 content in the rat (Table 2) when calculated on the liver wet weight basis.

DISCUSSION

When analyzing the activities of drug metabolizing enzymes in reindeer having antiparasitic treatment with Warbex, it looked obvious that warbex increased the activities. Also the increases in the activities were enormously great. However, when the organophosphorous compound was given to rats and the activities were analyzed both in animals having Warbex and in those having the sovent of Warbex, diethylsuccinate, it revealed that the solvent increased monooxygenase activities. Whether this is true also with reindeer, remains so far unclear. However, it is obvious that the metabolism of foreign compounds in the reindeer is very sensitive to this type of medication.

REFERENCES

1. Laitinen, M., Nieminen, M., Hietanen, E. and Hänninen, O. (1980) Acta pharmacol. toxicol. 46, 238.
2. Laitinen, M., Nieminen, M. and Hietanen, E. (1980) Proc. 2nd Reindeer/Caribou Symp., Rørus, Norway, 373.
3. Gornall, A.G., Bardawill, C.J. and David, M.M. (1949) J. biol. Chem. 177, 751.
4. Omura, T. and Sato, R. (1964) J. biol. Chem. 239, 2379.
5. Burke, M.D. and Mayer, R.T. (1974) Drug Metab. Disp. 2, 583.
6. Ullrich, B. and Weber, P. (1972) Hoppe-Seyler's Z. physiol Chem. 353, 7, 1171.
7. Aitio, A. (1978) Anal. Biochem. 85, 448.
8. Wattenberg, W.L., Leong, J.L. and Strand, P.J. (1962) Cancer Res. 22, 1120.
9. Nebert, D.W. and Gelboin, H.V. (1968) J. biol. Chem. 244, 6242.

Characterization of Different Cytochrome P-450's

Cytochrome P-450, Biochemistry, Biophysics
and Environmental Implications, E. Hietanen,
M. Laitinen and O. Hänninen editors

SPECTRAL AND CATALYTIC PROPERTIES OF ETHANOL-INDUCIBLE CYTOCHROME P-450 ISOZYME 3a

DENNIS R. KOOP AND EDWARD T. MORGAN
Department of Biological Chemistry, Medical School, The University of Michigan, Ann Arbor, Michigan 48109 (U.S.A.)

INTRODUCTION

Following the solubilization and purification of the hepatic microsomal cytochrome P-450-containing enzyme system (1), there have been numerous reports describing the purification and characterization of multiple forms of cytochrome P-450 (2). The possible role of cytochrome P-450 in the oxidation of aliphatic alcohols has been the subject of much debate. The effects of chronic ethanol administration have been well documented and include an increase in the oxidation of ethanol and the p-hydroxylation of aniline (3-6). These activities have been attributed to the induction of a unique isozyme of cytochrome P-450, but attempts to demonstrate the presence of such an isozyme have met with limited success (6-8). Ohnishi and Lieber (7) reported the solubilization and reconstitution of the ethanol-oxidizing activity and Joly and coworkers (6,8) described the partial purification of an isozyme with a high activity towards ethanol and aniline, but did not demonstrate that the activity was due to a unique isozyme. In order to demonstrate conclusively that these activities are the result of a unique form of cytochrome P-450, the enzyme must be purified to homogeneity and characterized as a distinct isozyme. This report describes the increase in a unique form of cytochrome P-450 in rabbit liver following chronic ethanol treatment and some of the unique spectral and catalytic features of the enzyme.

MATERIALS AND METHODS

Adult New Zealand male rabbits (2.0 to 2.5 kg) were given 10% (v/v)ethanol in the drinking water for 14 days with free access to Purina rabbit chow. Microsomes were isolated and the isozymes of P-450[1] were purified as previously described for isozymes 2 and 4 (9), 3b and 3c (10), and the ethanol-inducible isozyme 3a (11). In all cases the enzymes were electrophoretically

---

[1]The abbreviations used are: dilauroyl-GPC, dilauroylglyceryl-3-phosphorylcholine; SDS-PAGE, sodium dodecyl sulfate-polyacrylamide gel electrophoresis; and P-450, cytochrome P-450. The isozymes are numbered according to their electrophoretic mobilities and are referred to as isozyme or form 2, etc.

homogeneous and had specific contents ranging from 15 to 19 nmol of P-450 per mg protein. NADPH-cytochrome P-450 reductase was purified as previously described (11).

All spectral measurements were recorded with a Cary 219 spectrophotometer equipped with a Lauda K-2R water bath. The absolute spectra of isozyme 3a were obtained in the presence of 0.1 μM methylviologen to mediate the reduction of the cytochrome by dithionite (10). The binding of ethanol to isozymes 3a and 4 was measured at 20°C in the presence of 30 μg/ml dilauroyl-GPC in 150 mM potassium phosphate buffer, pH 7.4, containing 20% glycerol. The incremental absorbance change following each addition was determined at 393 nm (decrease in absorbance) and 412 nm (increase in absorbance) for isozyme 3a; the wavelength pair 392 and 418 nm was used for isozyme 4. All spectral changes were corrected for dilution, and, where applicable, the spectral binding constant was determined from double reciprocal plots of the binding data.

The activity of the various isozymes was determined in a reconstituted system in which the P-450 was the limiting component. The assay conditions were the same as have been previously described (10,11). The hydroxylation of benzo(a)pyrene was measured fluorometrically (12) in collaboration with Dr. Eric Eisenstadt of the Harvard School of Public Health. Other methods, including protein assays, P-450 determinations, electrophoretic techniques, and details of the purification are described in detail elsewhere (9-11).

RESULTS AND DISCUSSION

Since the effects of chronic ethanol administration on the hepatic P-450 system have been studied primarily in the rat, initial experiments were conducted to determine if a similar response was observed in the rabbit. When rabbits were administered 10% (v/v) ethanol as described in Materials and Methods, a 3-fold and 4-fold increase in the microsomal activity towards ethanol oxidation and aniline hydroxylation, respectively, was observed. As illustrated by the densitometric scans of the SDS-PAGE profiles of liver microsomes from untreated or ethanol-treated rabbits shown in Fig. 1, ethanol treatment results in an increase in a protein having an electrophoretic mobility slightly less than that of the phenobarbital inducible-isozyme 2. Also of interest is the change in the protein pattern in the region of isozyme 6, originally purified from livers of rabbits treated with TCDD (13). Significant quantities of this isozyme can be purified from ethanol-treated rabbits

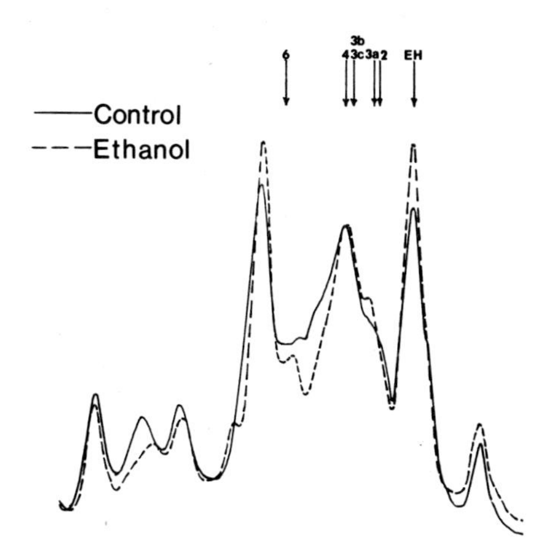

Fig. 1. Densitometric scans of SDS-polyacrylamide gels (7.5% acrylamide) of liver microsomes obtained from control or ethanol-treated animals. The molecular weight region from approximately 40,000 to 65,000 is shown. The positions of the purified isozymes of P-450 and epoxide hydratase (EH) are indicated by the arrows at the top of the figure.

and will be reported in detail elsewhere. Although isozyme 3a is increased by ethanol treatment, it still represents a very small percentage of the total cytochrome P-450 present in the microsomes and is not always readily apparent in the electrophoretic pattern of the microsomes. The purification procedure for isozyme 3a (11) permits the purification of five isozymes of P-450, epoxide hydratase, cytochrome $\underline{b}_5$, and NADPH cytochrome P-450 reductase from a single microsomal preparation. For example, in one experiment, isozymes 3a, 3b, 3c, 4 and 6 were obtained with final yields of 1, 1, 3, 17, and 1%, respectively.

Isozyme 3a is isolated in the pentacoordinate-high spin state as indicated by the absorbance maxima of the oxidized enzyme at 393 and 647 nm (Fig. 2). Isozyme 4 is also high spin when isolated, but unlike isozyme 3a does not maintain its high spin character in the presence of nonionic detergents. The reduced spectrum of 3a is similar to that of other isozymes of P-450 (10,14) with maxima at 413 and 547 nm, while the reduced-CO complex is red shifted compared to that of the other isozymes. The maximum occurs at 452 nm. As a result of the spin state of isozyme 3a, the binding of alcohols can

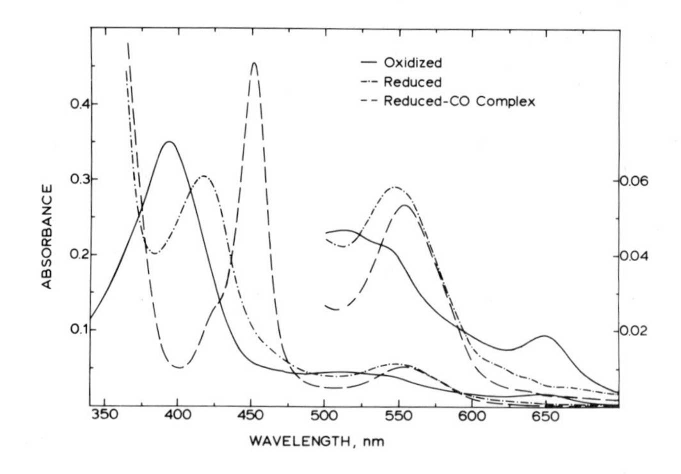

Fig. 2. Absolute spectra of purified isozyme 3a. The concentration of the cytochrome, based on the heme content, was 4.2 μM in 150 mM potassium phosphate buffer, pH 7.4, containing 20% glycerol and 0.1 mM EDTA. The spectra were recorded at 15°C.

be readily observed spectrally. Since 3a is induced by ethanol, the binding properties with ethanol were examined and compared to those of isozyme 4. The addition of increasing concentrations of ethanol to isozyme 4 results in a high to low spin transition (i.e., a reverse type I binding spectrum) with a maximum at 418 nm and an isobestic point at 406 nm. In contrast, isozyme 3a exhibits a maximum at 412 nm and an isobestic point at 403 nm. Although the low spin spectrum produced by ethanol ligation is not identical to that of isozymes 2, 3b, 3c, and 6, the similarity supports the hypothesis that an oxygen atom is coordinated as the sixth ligand in low spin forms of P-450 (14). When the absorbance change is plotted against the ethanol concentration, isozyme 3a and 4 are seen to exhibit distinct binding behavior (Fig. 3). Isozyme 4 displays a sigmoid binding curve over the concentration range tested; this behavior is not understood at the present time. As a result, an apparent binding constant of about 1 M was estimated for ethanol binding. In contrast, isozyme 3a exhibits a hyperbolic binding curve, and double reciprocal plots of the binding data yield a spectral binding constant of 250 mM, about 4 times lower than that estimated for form 4. This difference in the affinity between isozyme 3a and 4 was observed with a series of alcohols. Whether this difference reflects a unique binding site for alcohols near the heme iron remains to be determined.

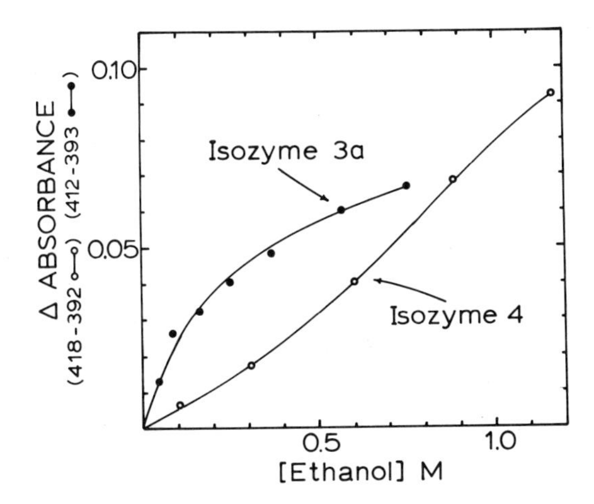

Fig. 3. Comparison of the binding of ethanol to isozymes 3a and 4. The isozymes were titrated with ethanol as described in Materials and Methods and the absorbance difference between 393 and 412 for isozyme 3a and 392 and 418 nm for isozyme 4 was determined.

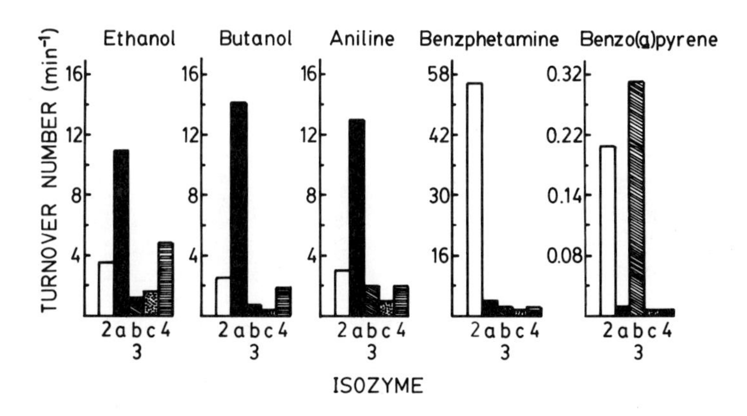

Fig. 4. Catalytic activity of purified isozymes of cytochrome P-450. Assay conditions are described in References 10 and 11.

A variety of methods have been used to demonstrate that isozyme 3a is a unique gene product. These include the determination of amino acid composition, NH_2- and COOH-terminal analysis, and peptide mapping using four different proteases (11). As previously mentioned, ethanol treatment results in an increase in the oxidation of ethanol and the hydroxylation of aniline. Thus, if isozyme 3a is responsible for the increase in the activity in microsomes, the purified isozyme should catalyze these reactions in a reconstituted system. This is indeed the case, as shown in Fig. 4. The oxidation of ethanol and butanol and the hydroxylation of aniline are catalyzed at the greatest rate by isozyme 3a; significantly lower activity is observed with the other four isozymes tested. The increase in the rate of alcohol oxidation as the carbon chain length is increased from two to four carbons is distinct from the results reported for the oxidation of alcohols by rat liver microsomes, in which a 2-fold decrease in the oxidation was reported when butanol was the substrate (15). Isozyme 3a is relatively inactive in the hydroxylation of benzo(a)pyrene and the demethylation of benzphetamine.

The mechanism by which ethanol brings about an increase in the level of isozyme 3a is not yet understood. However, the isolation of a unique isozyme of cytochrome P-450 with a high preference for alcohol oxidation from the livers of ethanol-treated rabbits provides definitive evidence that a unique isozyme is responsible for the activity in liver microsomes and gives strong support for a microsomal alcohol-oxidizing system. The physiological significance of such a system remains to be determined.

ACKNOWLEDGMENTS

This research was supported by Grant AM-10339 from the National Institutes of Health and Grant PCM8102324 from the National Science Foundation.

REFERENCES

1. Lu, A.Y.H. and Coon, M.J. (1968) J. Biol. Chem. 243, 1331-1332.

2. Lu, A.Y.H. and West, S. (1980) Pharmacol. Rev., 31, 277-295.

3. Rubin, E., Hutterer, F., and Lieber, C.S. (1968) Science 159, 1469-1471.

4. Joly, J.-G., Ishii, H., Teschke, R., Hasumura, Y., and Lieber, C.S. (1973) Biochem. Pharmacol. 22, 1532-1535.

5. Morgan, E.T., Devine, M., and Skett, P. (1981) Biochem. Pharmacol. 30, 595-600.

6. Villeneuve, J.-P., Mavier, P., and Joly, J.-G. (1976) Biochem. Biophys. Res. Commun. 70, 723-728.

7. Ohnishi, K., and Lieber, C.S. (1977) J. Biol. Chem. 252, 7124-7131.

8. Mungikar, A.M., Hetu, C., and Joly, J.-G. (1980) in: Alcohol and Aldehyde Metabolizing Systems (Thurman, R.G., Ed.) Plenum Press, New York, Vol 4., pp. 51-56.

9. Coon, M.J., van der Hoeven, T.A., Dahl, S.B., and Haugen, D.A. (1978) Methods Enzymol. 52, 109-117.

10. Koop, D.R., Persson, A.V., and Coon, M.J. (1981) J. Biol. Chem. 256, 10704-10711.

11. Koop, D.R., Morgan, E.T., Tarr, G.E., and Coon, M.J. (1982) J. Biol. Chem., in press.

12. Nebert, D.W., and Gelboin, H.V. (1968) J. Biol. Chem. 243, 6242-6249.

13. Norman, R.L., Johnson, E.F., and Muller-Eberhard, U. (1978) J. Biol. Chem. 253, 8640-8647.

14. White, R.E., and Coon, M.J. (1982) J. Biol. Chem. 257, 3073-3083.

15. Teschke, R., Hasumura, Y., and Lieber, C.S. (1975) J. Biol. Chem. 250, 7397-7404.

© 1982 Elsevier Biomedical Press B.V.
Cytochrome P-450, Biochemistry, Biophysics
and Environmental Implications, E. Hietanen,
M. Laitinen and O. Hänninen editors

EVIDENCE FROM STUDIES ON PRIMARY STRUCTURE THAT RABBIT HEPATIC NADPH-CYTOCHROME P-450 REDUCTASE AND ISOZYMES 2 AND 4 OF CYTOCHROME P-450 REPRESENT UNIQUE GENE PRODUCTS*

SHAUN D. BLACK, GEORGE E. TARR, AND MINOR J. COON
Department of Biological Chemistry, Medical School, The University of Michigan, Ann Arbor, Michigan 48109 (U.S.A.)

INTRODUCTION

Currently, many laboratories are actively engaged in the purification and characterization of multiple forms of cytochrome P-450. This family of membrane-associated *b*-type cytochromes has spectrally characteristic ferrous-carbonyl complexes, similar polypeptide molecular weights, and broad but overlapping substrate specificities toward the oxidation of a wide variety of both natural and foreign compounds. Questions of great interest in this field have focused on the total number of isozymes present in a given species, the chemical uniqueness of each of these multiple forms, and the degree to which homology might exist among similar forms from different species. Nebert *et al.* (1) have suggested that hundreds if not thousands of forms may exist, and that each isozyme, purified from animals pretreated with a given inducer, might in fact represent a distribution of different polypeptide chains in a manner much analogous to the case of the immunoglobulins. The work presented in this paper argues strongly against the lattermost proposal, and will provide evidence in favor of the distinctness of rabbit liver microsomal NADPH-cytochrome P-450 reductase and cytochrome P-450 isozymes 2 and 4. Although many possible means exist for the assessment of uniqueness, we have utilized amino acid sequencing and believe that this approach is ultimately required to provide definitive answers in this area.

---

*The isozymes of rabbit liver microsomal cytochrome P-450 are designated according to their electrophoretic mobilities, in accord with the general recommendation of the Committee on Biochemical Nomenclature of the International Union of Biochemistry. The phenobarbital- and 5,6-benzoflavone-inducible forms are referred to as P-450$_{LM_2}$ and P-450$_{LM_4}$, respectively, or simply as isozymes 2 and 4. The three isozymes characterized subsequently as having intermediate electrophoretic behavior are designated as 3a, 3b, and 3c.

MATERIALS AND METHODS

The proteins studied in this work were purified from the liver microsomes of male New Zealand white rabbits as previously described (2). Manual and automated sequencing methods are as described previously (3,4). Recent sequence studies have been carried out with peptides derived from trypsin-, proteinase A-, CNBr-, and acid-cleavage of the phenobarbital-induced isozyme 2. Purification and re-chromatography of these peptides was accomplished by means of reversed-phase high performance liquid chromatography.

RESULTS AND DISCUSSION

The amino-terminal sequences of isozymes 2 (phenobarbital-induced; M_r 48,000), 3a (ethanol-induced; M_r 51,000), 3b (constitutive; M_r 52,000), 3c (constitutive; M_r 53,000), and 6 (isosafrole-induced; M_r 58,000) are shown in Table 1. As can be readily seen, each polypeptide exhibits a unique sequence. Only isozyme 3b, at position 10, shows a possibly variable residue; however, this variability may well result from strain differences that exist within the outbred rabbit population from which the protein was isolated. The N-terminus of isozyme 4, which was previously shown to exhibit multiple residues per cycle of Edman degradation (5), has recently been shown to be the result of processing. In results to be presented elsewhere, ≥85% of the LM_4 polypeptide could be accounted for by two identical sequences which differ from one another by loss of a single amino-terminal residue (i.e., an "N" and "N-1" form). The N-terminal sequence obtained through use of *o*-phthaldialdehyde as a selective blocking agent is unique and also distinct from the other isozymes shown in Table 1.

TABLE 1

COMPARISON OF THE AMINO-TERMINAL SEQUENCES OF CYTOCHROME P-450 ISOZYMES

Isozyme	Residue Identified																								
	1				5					10					15					20					25
2	M	E	F	S	L	L	L	L	L	A	F	L	A	G	L	L	L	L	L	F	R	G	H	P	K
3a	A	V	L	G	I	T	V	A	L	L	G	W	M	V	I	L	L	F	I	S	V	W	K	Q	I
3b	M	D	L	L	I	I	L	G	I	*	L	S	E	V	V	L	L	L	L	W	G	K	A	T	
3c	M	D	L	I	F	S	L	E	T	W	V	L	L	A	A	A	L	V	T	L	Y	L	Y	G	G
6	S	D	V	L	E	L	T	D	D	N	A	A													

*Multiple residues (T or L) have been observed at this position.

At the time of this writing, the primary structure of isozyme 2 is nearly completed. Comparison of this sequence with that of the rat phenobarbital-inducible form, as determined by Fujii-Kuriyama _et al_. (6) through molecular biological techniques, shows an unexpected and striking homology; in all, the proteins are approximately 77% identical. The first 38 N-terminal residues of the rabbit and rat polypeptides are compared, in Fig. 1. Most substitutions that occur are conservative and are generally accountable as point mutations. The high degree of homology found is especially surprising in light of the previously reported lack of any strong immunochemical relationship between these two proteins. A possible explanation for this phenomenon is that the two proteins share many common antigenic sites and that these determinants are not immunogenic when one of the proteins is used to treat animals from the other species. Consequently, antibodies elicited might be specific for only the few dissimilar determinants present; comparison of the two proteins using antisera obtained in this manner would show them to be only weakly related.

```
                    1       5           10      15            20
Rabbit P-450LM2     M-E-F-S-L-L-L-L-L-A-F-L-A-G-L-L-L-L-L-F-
Rat P-450PB         M-E-P-S-I-L-L-L-L-A-L-L-V-G-F-L-L-L-L-V-

                            25          30      35
Rabbit P-450LM2     R-G-H-P-K-A-K-G-R-L-P-P-G-P-S-P-L-P
Rat P-450PB         R-G-H-P-K-S-R-G-N-F-P-P-G-P-R-P-L-P
```

Fig. 1. Comparison of the N-terminal sequences of rabbit isozyme 2 (this report) and the rat phenobarbital inducible form (6).

Another example of homology is found between the NADPH-cytochrome P-450 reductase from rabbit and rat hepatic microsomes. The amino-terminal sequences of the trypsin-solubilized reductases are aligned as shown in Fig. 2A. Within this region of 25 residues, the two proteins show approximately 80% homology, have predicted secondary structures which are nearly identical, and apparently share the same tryptic cleavage site. The homology includes a segment of at least 17 contiguous residues of exact correspondence. A portion of the N-terminal hydrophobic domains of the rabbit and rat reductases are compared in Fig. 2B. Although only Met and Leu were determined in the rat radio-sequence (8), comparison of these residues to the rabbit sequence shows that each Met and Leu of the rabbit has an exact correspondence in the rat. The remaining residues of Leu in the rat appear to correspond to different residues in the rabbit, but these can be postulated to

A

```
                      45     50      55      60      65
Rabbit Reductase     I-Q-A-P-T-S-S-S-V-K-E-S-S-F-V-E-K-M-K-K-T-G-R-N-I-
Rat Reductase        I-Q--T-T-A-P-P-V-K-E-S-S-F-V-E-K-M-K-K-T-G-R-N-I-
```

B

```
                     13
Rabbit Reductase     M-T-D-V-V-L-F-S-L-I-V-G-L-I-T-N-Y-F-L-F-
                     1
Rat Reductase        M - - - -L-L-L- -L- -L- -L-L- - - - -L- -
```

Fig. 2. Comparison of primary structural data from NADPH-cytochrome P-450 reductase of rabbit and rat. A, N-terminal sequences of trypsin-solubilized reductases. B, Sequence data from the amino-terminal hydrophobic domain. Data is taken from Refs. 3 and 8.

have arisen through conservative genetic point mutations. However, it is of interest to note that the comparison shown in Fig. 2B can be achieved only if Met_1 and Met_{13} of the rat and rabbit, respectively, are aligned.

The nearly completed sequence of isozyme 2 has also allowed the important question of microheterogeneity and uniqueness to be addressed. In results to be published elsewhere, we have shown that, thus far, no evidence for heterogeneity exists in the sequence, and that "variable residues" predicted in the sequence of the rat phenobarbital-inducible form (6) are present as unique residues in $P\text{-}450_{LM_2}$. Clearly, this rabbit hepatic microsomal protein is a unique and distinct gene product. Finally, comparison of the primary structures of rabbit isozyme 2, the aforementioned rat protein, and bacterial $P\text{-}450_{cam}$ (9) has permitted the identification of Cys_{147}, Cys_{152}, and Cys_{134} as the amino acid residues in these polypeptides, respectively, which may provide the fifth ligand to the heme prosthetic moiety.

ACKNOWLEDGMENTS

This research was supported by Grant AM-10339 from the National Institutes of Health. During a portion of these studies S.D.B. was the recipient of a Merck Predoctoral Fellowship.

REFERENCES

1. Nebert, D.W., Eisen, H.J., Negishi, M., Lang, M.A., Hjelmeland, L.M., and Okey, A.B. (1981) Ann. Rev. Pharmacol. Toxicol. 21, 431-462.

2. Haugen, D.A., and Coon, M.J. (1976) J. Biol. Chem. 251, 7929-7939.

3. Black, S.D., and Coon, M.J. (1982) J. Biol. Chem. 257, 5929-5938.

4. Tarr, G.E. (1982) in: IVth International Conference on Methods in Protein Sequencing, in press.

5. Haugen, D.A., Armes, L.G., Yasunobu, K.T., and Coon, M.J. (1977) Bio-Chem. Biophys. Res. Commun. 77, 967-973.

6. Fujii-Kuriyama, Y., Mizukami, Y., Kawajiri, K., and Muramatsu, M. (1982) Proc. Natl. Acad. Sci. U.S.A. 79, 2793-2797.

7. Thomas, P.E., Lu, A.Y.H., Ryan, D., West, S.B., Kawalek, J., and Levin, W. (1976) Mol. Pharmacol. 12, 746-758.

8. Okada, Y., Frey, A.B., Guenthner, T.M., Oesch, F., Sabatini, D.D., and Kreibich, G. (1982) Eur. J. Biochem. 122, 393-402.

9. Hanin, M., Armes, L. G., Tanaka, M., Yasunobu, K.T., Shastry, B.S., Wagner, G.C., and Gunsalus, I.C. (1982) Biochem. Biophys. Res. Commun. 105, 889-894.

Cytochrome P-450, Biochemistry, Biophysics
and Environmental Implications, E. Hietanen,
M. Laitinen and O. Hänninen editors

RABBIT MICROSOMAL CYTOCHROME P-450 3b: STRUCTURAL AND FUNCTIONAL POLYMORPHISM AND MODULATION BY POSITIVE AND NEGATIVE EFFECTORS

ERIC F. JOHNSON, HERMANN H. DIETER*, GEORGE E. SCHWAB, INGRID REUBI, and URSULA MULLER-EBERHARD
Department of Biochemistry, Scripps Clinic and Research Foundation, 10666 North Torrey Pines Road, La Jolla, CA 92037 (U.S.A.)

INTRODUCTION

Work in several laboratories using principally the rat and rabbit has resulted in the isolation and partial characterization of a number of electrophoretically homogeneous and distinct forms of microsomal cytochrome P-450 that exhibit high specific contents of cytochrome P-450 (1). However, these criteria indicate only that these preparations contain predominately cytochrome P-450 of uniform size, and they may therefore contain more than one structurally similar form of cytochrome P-450. Recent work in our laboratory indicates that most preparations of cytochrome P-450 3b prepared from outbred New Zealand White (NZW) rabbits are functionally polymorphic. In contrast, P-450 3b prepared from a genetically defined strain of rabbits, IIIVO/J, appears to lack one of these subforms. In addition, the cytochrome isolated from strain IIIVO/J exhibits allosteric regulation by a variety of steroidal compounds.

MATERIALS AND METHODS

Methods for the isolation of P-450 form 3b, its reconstitution with reductase as well as the assays for the formation of the metabolites of progesterone and biphenyl are described elsewhere, as are procedures for preparation of microsomes (2-4). [^{14}C]Progesterone was purchased from Amersham, and silca gel plates (IB2-F) were obtained from J. T. Baker Chemicals. Other steroidal compounds were purchased from Sigma. Rabbits of strain IIIVO/J were supplied by the Jackson Laboratories.

---

*Present Addresses: H. H. Dieter, Institute for Toxicology, University of Dusseldorf, W. Germany. I. Reubi, School of Medicine, University of Bern, Switzerland. U. Muller-Eberhard, Department of Pediatrics, Cornell Medical Center, New York, N.Y. U.S.A.

RESULTS

Functional Polymorphism of P-450 3b. When the metabolism of progesterone by P-450 3b was subjected to kinetic analysis, the Km for 6β-hydroxylation was found to be less than 1 μM with a Vmax of 1.0 to 2.7 μM/min/μM P-450. In contrast, Lineweaver-Burk analysis of the 16α-hydroxylation revealed a distinct biphasicity with the relative contributions of the two phases to this activity varying between preparations of 3b, as shown in figure 1. The Km of the high affinity phase is less than 5 μM whereas that of the low affinity phase is greater than 50 μM. The maximal velocities are ca. 1 and 10 μM/min/μM P-450 for the two phases respectively. This result suggests that the preparations of P-450 3b are functionally polymorphic.

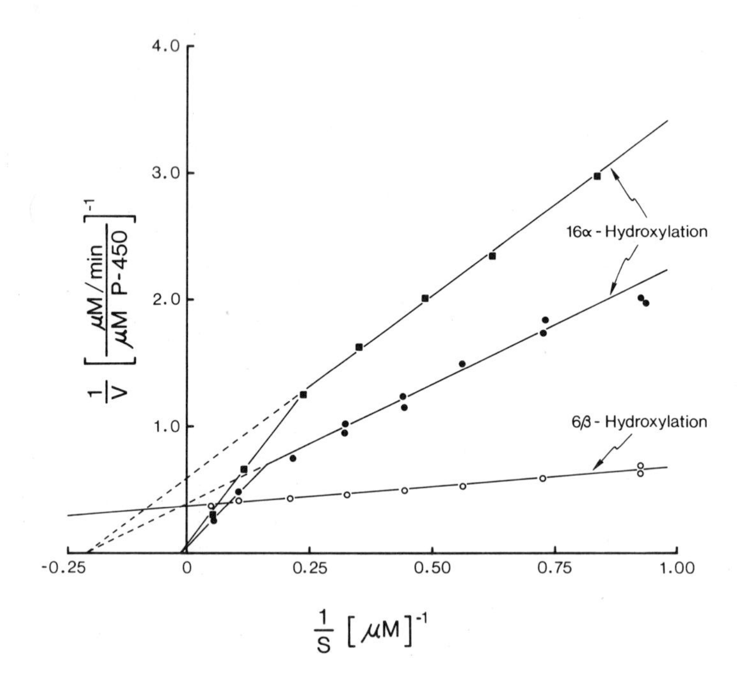

Figure 1. Lineweaver-Burk plots for 16α- and 6β-hydroxylation of progesterone. The curves for 16α-hydroxylation (solid symbols) were determined for two different preparations of P-450 3b. The values for the 6β-hydroxylase activity (open circles) were measured in concert with those shown for 16α-hydroxylation by the solid circles.

Studies with the inhibitor 16α-methylprogesterone also indicate the presence of more than one catalytic site for preparations of 3b (4). This compound is a competitive inhibitor, Ki = 1 μM, of 6β-hydroxylation catalyzed by P-450 3b. However it inhibits only the high affinity phase of 16α-hydroxylation catalyzed by this cytochrome, i.e., in the presence of 10 μM 16α-methylprogesterone the double reciprocal plot for 16α-hydroxylation is uniphasic. This suggests that 16α-methylprogesterone is a competitive inhibitor at the site which catalyzes the high affinity 16α-hydroxylation and the 6β-hydroxylation but not at the site catalyzing the low affinity 16α-hydroxylation.

This functional polymorphism was surprising since P-450 3b is highly purified as evidenced by polyacrylamide gel electrophoresis in the presence of sodium dodecyl sulfate and by the specific content of cytochrome P-450 of these preparations which indicates that they contain approximately one mole of cytochrome P-450 per mole of protein (2). However, these criteria only indicate that the preparations contain predominately cytochrome P-450 of uniform size. On the other hand, preparations of 3b routinely exhibit a single N-terminal amino acid sequence with repetitive yields which exceed 90% and initial yields which range from 70 to 90% for these determinations (6). Thus, if form 3b is polymorphic, it is likely that the putative subforms are structurally more highly related than is the case for the electrophoretically distinct cytochromes.

A Variant Form of P-450 3b. Further evidence for polymorphism is suggested by the isolation of a variant form of P-450 3b from rabbits of strain IIIVO/J (4). Whereas most microsomes prepared from outbred NZW rabbits display a Km of less than 6 μM for the 6β-hydroxylation of progesterone, all preparations from strain IIIVO/J and one preparation from an NZW rabbit displayed an apparent Km of greater than 50 μM.

Although this result suggests that those microsomes lacking the higher affinity 6β-hydroxylase activity might be deficient in P-450 3b, this is not the case since P-450 3b can be purified from these microsomes by routine procedures. The yield and purity of the cytochromes isolated from this source were similar to that of preparations obtained from our regular stock of NZW rabbits (4). More recently, we have shown using an immunoassay that the specific content of P-450 3b present in microsomes isolated from either group of animals is between 0.1 and 0.2 nmol/mg. This assay employs a monoclonal antibody directed toward P-450 3b which shows no detectable

TABLE 1

STIMULATORY EFFECT OF FOUR STEREOISOMERS OF 3-HYDROXYPREGNAN-20-ONE ON THE 16α-HYDROXYLATION OF PROGESTERONE AS CATALYZED BY P-450 3b FROM STRAIN IIIVO/J

Effector	% Increase[a]
3β-hydroxy-5β-pregnan-20-one	327 ± 1%
3β-hydroxy-5α-pregnan-20-one	136 ± 20%
3α-hydroxy-5β-pregnan-20-one	137 ± 20%
3α-hydroxy-5α-pregnan-20-one	109 ± 7%

[a]Results are expressed as a percentage of the rate catalyzed by the cytochrome in the absence of the test compound. The initial concentration of the substrate was 7 µM, and when present the concentration of the effector was 10 µM.

cross-reactivity with either P-450 2 or 4. This antibody inhibits both the 6β- and 16α-hydroxylation of progesterone as catalyzed by P-450 3b and thus appears to recognize both of the putative subforms in preparations of P-450 3b.

When the metabolism of progesterone by the P-450 3b isolated from strain IIIVO/J was examined it was observed to catalyze the 6β-hydroxylation of progesterone with a much greater Km and a lower Vmax than most preparations from NZW rabbits (4). In the case of the 16α-hydroxylation, these preparations display only the low-affinity phase when the Lineweaver-Burk plots are examined. The two types of P-450 3b will hereafter be denoted as either $6\beta^+$ or $6\beta^-$. In contrast to this difference, both the $6\beta^+$ and $6\beta^-$ forms catalyze similar rates of biphenyl 4-hydroxylation and 2-acetylaminofluorene 7-hydroxylation. These results are interpreted as indicating that the subform of P-450 3b which catalyzes 6β-hydroxylation is not expressed in the livers of the IIIVO/J rabbits.

Positive Effectors. A variety of steroidal compounds can stimulate the 16α-hydroxylation of progesterone as catalyzed by the P-450 3b prepared from strain IIIVO/J. As shown in Table 1, 3β-hydroxy-5β-pregnan-20-one is the most potent of the four stereoisomers of pregnan-3-hydroxy-20-one. Of the reduced derivatives of progesterone which we have examined, the following generalizations can be surmized. The 5β-pregnanes are more potent than the corresponding 5α-pregnanes. The 3β-hydroxy derivatives are more potent than the corresponding 3α-hydroxy compounds. The reduction of the 20-keto group to a 20α-hydroxyl moiety increases potency. The most potent

TABLE 2

VARIATIONS IN THE INHIBITION BY 16α-METHYLPROGESTERONE OF BIPHENYL HYDROXYLATION FOR PREPARATIONS OF FORM 3b EXHIBITING DIFFERENT 6β-HYDROXYLASE ACTIVITIES

6β-Hydroxylase[a] Activity	Inhibition of Biphenyl Hydroxylation[b]	Source of Preparation
0.04	0%	IIIVO/J
0.04	0%	IIIVO/J
1.0	34%	NZW
1.25	28%	NZW
2.5	54%	NZW

[a]Rates are expressed as µM/min/µM P-450, the initial concentration of progesterone was 10 µM, and assay procedures are described in references (3,4).

[b]The concentration of 16α-methylprogesterone was 20 µM, the initial concentration of biphenyl was 400 µM, and the assay procedure is described in reference (5).

stimulating agent we have investigated, 5β-pregnan-3β,20α-diol, combines all of these attributes and stimulates the 16α-hydroxylation of progesterone by approximately four-fold. Large amounts of 5β-pregnanes are produced in the liver during pregnancy, and these compounds may modulate cytochrome P-450 mediated metabolism during this period.

Variations In Occurrence of the Subforms of P-450 3b. As indicated in figure 1, the relative contributions of high and low affinity subforms to the 16α-hydroxylation of progesterone appears to vary among preparations of P-450 3b from NZW rabbits. Further evidence for this variation is provided by studies with the inhibitor 16α-methylprogesterone. This compound is only a partial inhibitor of biphenyl hydroxylation. The extent of inhibition seen is roughly correlated with the 6β-hydroxylase activity of the preparations as shown in Table 2.

Peptide Mapping. Structural comparisons of the cytochromes isolated from different strains provide information regarding the basis of functional differences such as those exhibited by the $6\beta^+$ and $6\beta^-$ P-450 3b as well as regarding structural aspects of P-450 function in general. When preparations of the $6\beta^+$ and $6\beta^-$ forms of 3b were compared by SDS polyacrylamide gel electrophoresis they exhibited a similar purity and each comprised a single major band. In addition, the $6\beta^+$ and $6\beta^-$ forms could not be distinguished when compared by one dimensional peptide mapping using SDS polyacrylamide gel electrophoresis following limited proteolytic digestion in the presence of 0.1% SDS by either S. aureus V_8 protease, chymotrypsin, or papain. This method has been shown to distinguish P-450 3b from types 2, 4, and 6 (2).

Differences were seen however when tryptic peptides derived from the two forms were compared by reverse phase chromatography. One peptide routinely observed for preparations of the $6\beta^+$ form, peptide 14 in the nomenclature of

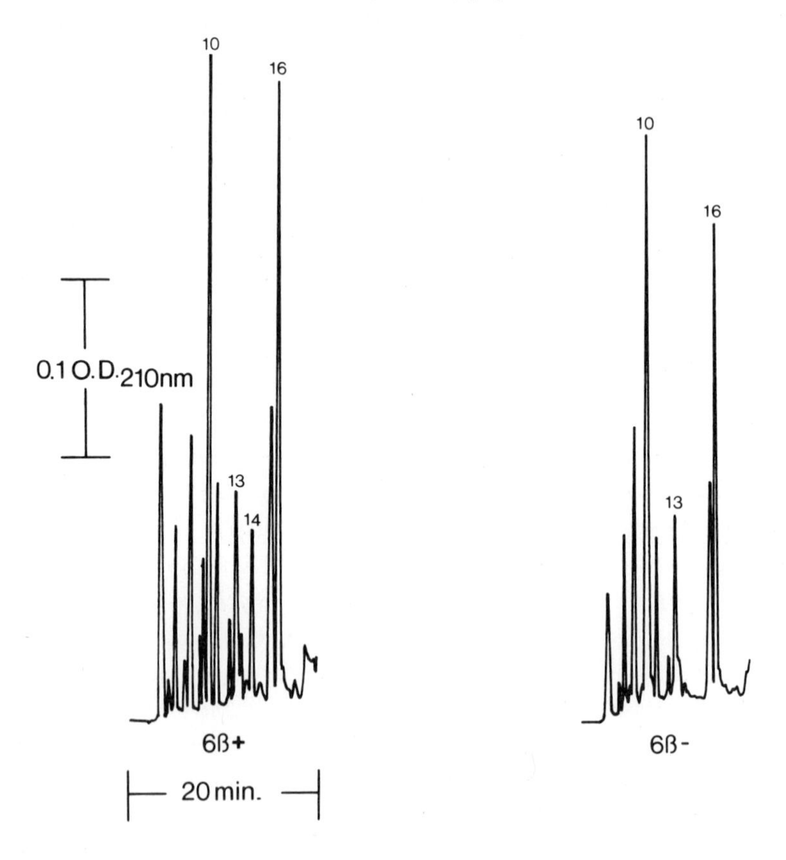

Figure 2. Portions of tryptic peptide maps of a $6\beta^+$ and a $6\beta^-$ P-450 3b. The tryptic peptides obtained from ca. 4 nmoles of each preparation were separated by reverse phase chromatography, and the elution of the peptides between 20 and 40 min was monitored by their absorbance at 210 nm (4). The peptides are numbered according to the designations used by Ozols, Heinemann, and Johnson (6). Peptide 14 which is routinely observed for P-450 3b prepared from NZW is not observed in the cytochromes prepared from strain IIIVO/J.

Ozols, Heineman, and Johnson (6), is consistently absent for each preparation of the $6\beta^-$ form we examined as illustrated by the examples shown in figure 2. In other respects the chromatograms are very similar suggesting a close structural similarity for the polymorphic forms of 3b although other less apparent differences can be seen.

DISCUSSION

The results presented here suggest that P-450 3b prepared from outbred NZW animals contains two or more structurally similar subforms. One subform of P-450 3b exhibits a high affinity for progesterone, catalyzes both the 6β- and 16α-hydroxylation of progesterone, and can be selectively inhibited by 16α-methylprogesterone. Another subform catalyzes the 16α-hydroxylation of progesterone and can be selectively stimulated by a variety of compounds which can be derived from progesterone by reductive metabolism. The extent to which the 6β-hydroxylase constitutes a fraction of the P-450 3b prepared from NZW rabbits is variable, but estimates made from the extent of inhibition of biphenyl hydroxylation by 16α-methylprogesterone suggests that the 6β-hydroxylase constitutes 20-50% of the cytochrome that catalyzes this reaction.

In contrast to most preparations of P-450 3b derived from NZW rabbits, those obtained from strain IIIVO/J appear to lack the 6β-hydroxylase subform and differ from those preparations which exhibit the 6β-hydroxylase activity by tryptic peptide mapping. It is not known whether the differences between the two types of P-450 3b arise by post-translational modification, post transcriptional processing of mRNA, or genetic diversity. If these differences reflect genetic diversity, they may indicate either genetic diversity at a single locus or multiple genes coding for P-450 3b.

In the rat, immunologic differences have been observed between preparations of a phenobarbital-inducible P-450 purified from two different strains (7). In addition, at least four immunologically similar peptides that differ by two-dimensional electrophoretic analysis and by tryptic peptide mapping are induced by phenobarbital in the rat, and interstrain differences have been noted in the expression of these proteins (8). In vitro translation experiments suggest that these proteins are encoded by different forms of mRNA (9). This has been independently demonstrated for the principal phenobarbital-inducible cytochrome in the rat by the cloning of more than one cDNA to an mRNA encoding for immunologically similar proteins. The sequences of these cDNA fragments indicate that there are multiple forms of mRNA coding for very similar but distinct forms of this protein (10). The functional significance of this polymorphism in the rat remains unknown.

The results presented here suggest that many of the techniques which have been used to distinguish the electrophoretically distinct types of P-450 may not be adequate to delineate the full extent of the diversity of P-450. The modes of regulation of the many forms of P-450 remain largely unknown, and as shown here can lead to very selective alterations in enzymatic properties and selective modulation by positive and negative effectors.

ACKNOWLEDGMENTS

This research was supported by a grant from the USPHS HD-04445. Dr. Dieter received an award from the Deutsche Forschungsgemeinschaft, and Dr. Reubi received awards from the Hartmann-Muller Stiftung of the Medical School of the University of Bern and the Janggen-Pohn Stiftung.

REFERENCES

1. Lu, A.Y.H. and West, S.B. (1980) Pharmacol. Rev. 31, 277.

2. Johnson, E.F. (1980) J. Biol. Chem. 255, 304.

3. Dieter, H.H., Muller-Eberhard, U., and Johnson, E.F. (1982) Biochem Biophys. Res. Commun. 102, 515.

4. Dieter, H.H. and Johnson, E.F. (1982) J. Biol. Chem., in press.

5. Johnson, E.F., Schwab, G.E., and Muller-Eberhard, U., (1979) Mol. Pharmacol. 15, 708.

6. Ozols, J., Heinemann, F.S., and Johnson, E.F. (1981) J. Biol. Chem., 256, 11405.

7. Guengerich, F.P., Wang, P., Mason, P.S., and Mitchell, M.B. (1981) Biochemistry 20, 2370-2378.

8. Vlasuk, G.P., Ghrayeb, J. Ryan, D.E., Reik, L., Thomas, P.E., Levin, W., and Waltz, Jr., F.G. (1982) Biochemistry 21, 789-798.

9. Waltz, Jr., F.G., Vlasuk, G.P., Omiecinski, C.J., Bresnick, E., Thomas, P.E., Ryan, D.E., and Levin, W. (1982) J. Biol. Chem. 257, 4023-4026.

10. Fujii-Kuriyama, Y., Mizukami, Y., Kawajiri, K., Sogawa, K., and Muramatsu, M. (1982) Proc. Natl. Acad. Sci. USA 79, 2793.

© 1982 Elsevier Biomedical Press B.V.
Cytochrome P-450, Biochemistry, Biophysics
and Environmental Implications, E. Hietanen,
M. Laitinen and O. Hänninen editors

SUBSTRATE SPECIFICITIES AND STEREOSELECTIVITIES AS PROBES OF CYTOCHROME P-450 ISOZYME COMPOSITION

LAURENCE S. KAMINSKY[1], MICHAEL J. FASCO[1] AND F. PETER GUENGERICH[2]
[1]New York State Department of Health, Center for Laboratories and Research, Albany, NY 12201 (U.S.A.) and [2]Vanderbilt University, Department of Biochemistry, Nashville, TN 37232 (U.S.A.)

INTRODUCTION

The ultimate assessment of the cytochrome P-450 isozyme composition of a tissue requires purification and characterization of the various isozymes (1). However, the current state of the art is such that purification is tedious and relatively inefficient with respect to the yield of isolated protein. Recently developed immunoelectrophoresis techniques more readily permit the determination of isozyme composition (2), but they require that isozymes be purified for the preparation of the antibodies, and they yield no information on isozyme functions.

An approach which yields both isozyme composition and functional information is to use substrate specificities and stereoselectivities. The anticoagulant drug warfarin (Figure 1), which yields multiple products and which is a substrate for a number of cytochrome P-450 isozymes (3-5), is an excellent candidate for such an approach. Warfarin has both aliphatic and aromatic sites which are hydroxylated by cytochromes P-450, only monohydroxylated metabolites are produced under the assay conditions used. A HPLC method has been developed which yields rapid and simultaneous analyses of all the warfarin metabolites with high sensitivity (6). Warfarin occurs in enantiomeric forms permitting the stereoselectivity of cytochrome P-450 to be probed. Antibodies to specific cytochrome P-450 isozymes can be incorporated into the assay to inhibit the formation of specific metabolites (7); and the sodium salts of warfarin and its metabolites are soluble in water, which obviates the necessity for time-consuming extractions and the use of organic vehicles for the substrate. Thus based on the regio- and stereoselectivity of metabolism and the specificity of antibody inhibition, warfarin can be used as a probe of cytochrome P-450 isozyme composition.

In this paper we describe the warfarin metabolite patterns whose formation is catalyzed by eight hepatic cytochrome P-450 isozymes from rats and two from rabbits. We use these patterns to compare variously induced rat, rabbit and mouse cytochrome P-450 isozyme compositions.

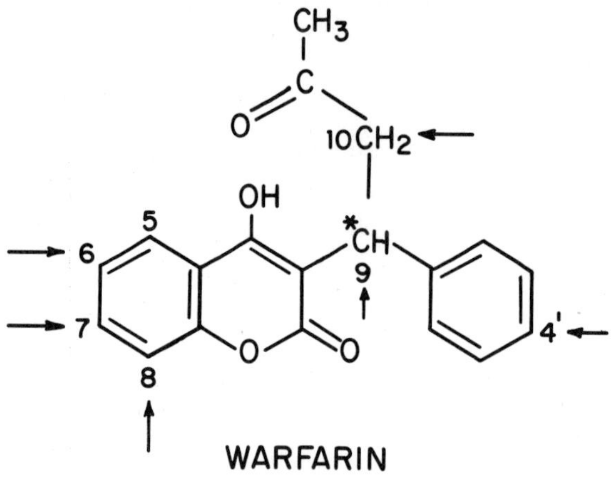

Figure 1. Warfarin. * Represents the asymmetric carbon giving rise to the R and S enantiomers. The arrows represent the cytochrome P-450-catalyzed hydroxylation sites.

MATERIALS AND METHODS

Materials. The resolution of warfarin into its enantiomers and the synthesis of the warfarin metabolites have been previously described (6, 7). Rats were male Sprague-Dawley (250 g), rabbits were male New Zealand (2 to 2.5 kg), and mice were male C57BL/6 and DBA/2 (25-30 g).

Microsomes and highly purified cytochromes P-450_{LM_2} and $_{LM_4}$ were prepared from the livers of phenobarbital (PB)- and ß-naphthoflavone (BNF)-induced rabbits respectively by Dr. M.J. Coon, as previously described (3, 8). Microsomes and highly purified hepatic cytochromes P-450 from PB-, BNF-, and pregnenolone-16α-carbonitrile (PCN)-induced rats were prepared according to the scheme in Figure 2 (F.P. Guengerich et al., in preparation). NADPH-cytochrome P-450 reductase from PB-induced rabbits and rats was prepared as previously described (9). Microsomes were prepared from untreated and BNF-induced mice as previously described (10).

Methods. The metabolism of R and S warfarin in microsomal and reconstituted systems of cytochrome P-450, NADPH-cytochrome P-450 reductase and dilauroylphosphatidylcholine and the analysis of the metabolites have been previously described in detail (7).

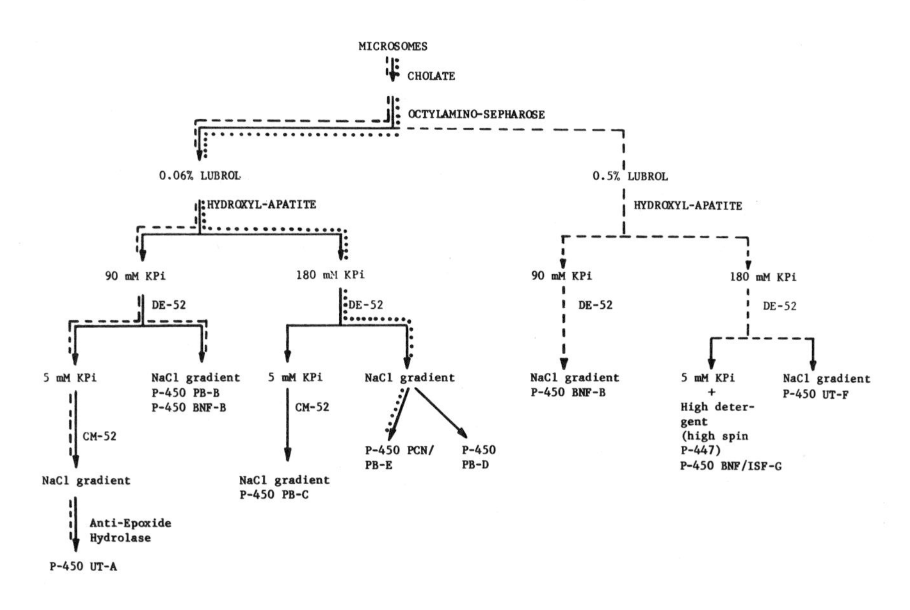

Figure 2. Purification of hepatic cytochrome P-450 isozymes from rats induced with ß-naphthoflavone (---), phenobarbital (——) or pregnenolone-16α-carbonitrile (•••••).

RESULTS

The rates of formation of the various metabolites from R and S warfarin catalyzed by highly purified rat and rabbit hepatic cytochromes P-450 in reconstituted systems are shown in Table 1. Rates of formation with rabbit liver isozymes are as previously presented (3). In PB-, BNF- and PCN-induced rats the major cytochrome P-450 isozymes, PB-B, BNF-B, and PB/PCN-E, were primarily R 4'-, R 8- and R 10-warfarin hydroxylases respectively. The R 10-hydroxylase was also a minor component in PB-induced rats. A minor low molecular weight isozyme, PB-D, induced by PB; BNF/ISF-G (high-spin), induced by BNF (or isosafrole); and UT-F from untreated rats were virtually inactive with warfarin as substrate. A constitutive isozyme which was isolated from PB- or BNF-induced rats, UT-A, was primarily a S 4'-hydroxylase. Another minor form, PB-C, from PB-induced rats was primarily a S 7-hydroxylase. Cytochrome $P\text{-}450_{LM_2}$ was primarily a S 6-hydroxylase and cytochrome $P\text{-}450_{LM_4}$ a R 6-hydroxylase.

In Table 2 the rates of formation of metabolites with microsomes from untreated (UT) and PB-, BNF-, PCN-, isosafrole (ISF)- or Aroclor 1254-induced rats; PB- or BNF-induced rabbits; or untreated and BNF-induced mice are presented.

TABLE 1

RATES OF METABOLITE FORMATION FROM R AND S WARFARIN CATALYZED BY HIGHLY PURIFIED RAT AND RABBIT CYTOCHROME P-450 ISOZYMES IN RECONSTITUTED SYSTEMS

Animal	Isozyme	Warfarin Enantiomer	Rate of Metabolite Formation[a,b]					
			Dehydro	4'-OH	6-OH	7-OH	8-OH	10-OH
Rat	UT-A[b]	R	.04	.43	.55	.07	.01	.16
		S	.01	.74	.49	.13	.02	.01
Rat	PB-B	R	.01	.13	.04	.02	.01	.02
		S	.00(4)	.03	.02	.02	.00(1)	.00(3)
Rat	BNF-B	R	.01	.03	1.37	.31	1.90	.01
		S	.02	.04	1.13	.23	.32	-
Rat	PB-C	R	.01	.06	.14	.80	.03	.01
		S	.01	.16	.20	1.17	.10	-
Rat	PB/PCN-E	R	.01	.02	.01	.01	.00(1)	.09
		S	.01	.02	.01	.02	.00(2)	.02
Rat	PB-D	R	.01	.06	.03	.01	.00(3)	.01
		S	.00(3)	.03	.02	.01	-	.01
Rat	BNF/ISF-G	R	.00(3)	.01	.01	.00(4)	.01	-
		S	.00(2)	.01	.01	.00(3)	-	-
Rat	UT-F	R	-	-	-	-	-	-
		S	.00(4)	.01	.01	.00(4)	-	-
Rabbit	LM_2	R	-	.05	.01	.00(4)	-	-
		S	-	.03	.07	.00(2)	-	-
Rabbit	LM_4	R	-	-	.05	.00(4)	.02	-
		S	-	-	.01	.01	.01	-

[a] nmol metabolite/nmol cytochrome P-450/min.
[b] Abbreviations: dehydro, 9,10-dehydrowarfarin; 4'-OH, 4'-hydroxywarfarin; 6-OH, 6-hydroxywarfarin; 7-OH, 7-hydroxywarfarin; 8-OH, 8-hydroxywarfarin; 10-OH, 10-hydroxywarfarin; UT, untreated; PB, phenobarbital; BNF, ß-naphthoflavone; PCN, pregnenolone-16α-carbonitrile; ISF, isosafrole; LM, liver microsomal.

DISCUSSION

Five of the eight rat hepatic isozymes yielded distinct metabolite profiles with warfarin, indicative of the regio- and stereoselectivity of each isozyme. While the rat and rabbit isozymes yielded many of the same

TABLE 2

RATES OF METABOLITE FORMATION FROM R AND S WARFARIN CATALYZED BY HEPATIC MICROSOMES FROM UNTREATED AND INDUCED RATS, RABBITS AND MICE

Animals	Induction	Warfarin	Rate of Metabolite Formation[a]					
		Enantiomer	Dehydro	4'-OH	6-OH	7-OH	8-OH	10-OH
Rat	UT	R	.04	.16	.10	.22	.03	.05
		S	.11	.28	.17	.13	.02	.02
Rat	PB	R	.19	.71	.56	1.87	.29	.32
		S	.34	.28	.42	.89	.14	.08
Rat	BNF	R	.02	.09	.31	.14	.58	.02
		S	.07	.12	.28	.10	.09	.01
Rat	ISF	R	.10	.19	.19	.17	.21	.12
		S	.23	.20	.16	.11	.03	.04
Rat	PCN	R	.18	.16	.18	.27	.03	.54
		S	.31	.23	.20	.19	.02	.15
Rat	Ar 1254	R	.08	.22	.88	.19	1.52	.22
		S	.15	.11	.57	.17	.16	.07
Rabbit	PB	R	-	.64	.15	.04	.01	-
		S	-	.22	.32	.04	.01	-
Rabbit	BNF	R	-	.05	.08	.07	.04	-
		S	-	.02	.06	.08	.01	-
Mouse DBA/2	UT	R	-	.21	.49	.76	.85	.17
		S	-	.12	.04	.10	-	-
Mouse DBA/2	BNF	R	-	.20	.56	.86	.98	.09
		S	-	.16	.06	.12	.07	-
Mouse C57BL/6	UT	R	.03	.21	.36	.55	.64	.16
		S	.04	.12	.05	.11	-	.02
Mouse C57BL/6	BNF	R	.04	.20	.35	.51	.55	.08
		S	.04	.09	.08	.08	-	-

[a] nmol/mg protein/min.
[b] Abbreviations and cytochrome P-450 concentrations: Rat:UT, untreated (1.10 nmol/mg protein); PB, phenobarbital (2.39 nmol/mg); BNF, ß-naphthoflavone (1.69 nmol/mg); ISF, isosafrole (1.29 nmol/mg); PCN, pregnenolone-16α- carbonitrile (1.86 nmol/mg); and Ar 1254, Aroclor 1254 (a polychlorinated biphenyl mixture) (4.28 nmol/mg). Mouse: DBA/2 UT (0.85 nmol/mg); DBA/2 BNF (1.00 nmol/mg); C57BL/6 UT (0.93 nmol/mg); C57BL/6 BNF (2.19 nmol/mg).

metabolic products, based on these metabolite profiles the two isozymes from the rabbit differed markedly from those of the rat. A major difference was that metabolism of the aliphatic side chain, to yield 10-hydroxywarfarin and dehydrowarfarin, occurred with rat but not rabbit isozymes.

Warfarin metabolism in reconstituted systems of rat or rabbit cytochromes P-450 was optimal at 1:1 molar ratios of cytochrome P-450 to NADPH-cytochrome P-450 reductase (3, 4). This contrasts sharply with the 20:1 ratio in intact microsomes. In view of this difference we have used the current data (Table 1), together with estimates of the isozyme compositions of the variously induced rat hepatic microsomes, determined by immunoelectrophoretic techniques (F.P. Guengerich et al., in preparation), to compare metabolism by microsomal (Table 2) and reconstituted systems. The sum of the isozyme concentrations determined by the immunoelectrophoretic technique exceeded the total cytochrome P-450 concentration, determined from reduced CO-bound difference spectra, by a factor of up to 2.1, possibly because of the inclusion of apoisozymes in the isozyme concentrations. Thus estimates of active isozymes by this technique could be high.

In general, for the rat hepatic system, the major warfarin metabolites whose formation was catalyzed by the major individual isozyme in a particular tissue preparation were also the major metabolites catalyzed by the corresponding intact microsomal system. However, the warfarin-metabolizing activity of the various microsomal preparations is clearly not the sum of the activities of the isozyme components of the microsomes, when determined under optimal conditions of the reconstituted system.

In microsomes from untreated rats the major isozyme is the S 4'-hydroxylase, UT-A. This isozyme is 3.5 to 7.5 fold as active in the reconstituted system as in the microsomes, depending on the specific metabolite, but the overall regio- and stereoselectivity of the purified isozymes is consistent with that of the microsomal system.

Hepatic microsomes from PB-induced rats are twice as efficient in metabolizing R warfarin, as are the component isozymes, but the components and microsomes are virtually equivalent in their metabolism of S warfarin. The major isozymes of these microsomes are PB-B, PB-E and PB-D, with PB-C being a relatively minor component. However, the high activity for warfarin metabolism exhibited by purified PB-C (Table 1) is apparently also expressed in the intact microsomes, as evidenced by the relatively high rates of formation of 7-hydroxywarfarin.

BNF-B, the major hepatic isozyme from BNF-induced rats, is between five- and sevenfold more active in reconstituted than in microsomal systems. However, both regio- and stereoselectivities of the microsomal and reconstituted systems are equivalent. This result could reflect a hindered accessibility of BNF-B for the NADPH-cytochrome P-450 reductase in intact microsomes.

The major hepatic isozyme in PCN-induced rats is PB/PCN-E, which in the microsomal system apparently has its activity amplified 3-fold relative to that of the purified reconstituted activity. This emphasizes the apparent specificity of PCN induction for PB/PCN-E, based on activity criteria.

Aroclor 1254 predominantly induces the formation of BNF-B, PB-B and PB-D but not PB-C, and this is clearly reflected in the warfarin metabolite patterns with microsomes from Aroclor 1254-induced rats.

Immunoelectrophoretic estimates of the isozyme composition of rabbit liver are not available, and thus no quantitative comparison of reconstituted and microsomal metabolism can be made. However, the regio- and stereoelectivity of warfarin metabolism by microsomes from PB-induced rabbits is very similar to that of cytochrome P-450$_{LM_2}$ (Tables 1 and 2), suggesting that this is the major isomeric component of these microsomes. The metabolite pattern with hepatic microsomes from BNF-induced rabbits shows several differences from that of cytochrome P-450$_{LM_4}$ (Tables 1 and 2), suggesting that other isozymes are also making significant contributions to the metabolic properties of these microsomes.

The warfarin metabolite patterns with Ah-responsive and -nonresponsive mice demonstrate that both strains of mice have cytochrome P-450 isozyme compositions which vary from those in the rat and the rabbit. While the two mouse strains have qualitatively similar isozyme compositions, the nonresponsive strain more actively metabolizes warfarin. The nonresponsive strain is not induced by BNF, and its isozyme composition is unaltered following BNF treatment. In the responsive strain the overall hepatic cytochrome P-450 concentration is induced 2.4-fold by BNF, but the warfarin-metabolizing capability of the hepatic microsomes decreases approximately 2-fold, implying that the newly induced isozymes do not metabolize warfarin. This latter observation sharply contrasts the effects of BNF induction on cytochrome P-450 isozyme composition in the rat and mouse.

In summary we have demonstrated that a variety of purified rat and rabbit hepatic cytochrome P-450 isozymes produce specific metabolite patterns with warfarin as substrate. The metabolite patterns are qualitatively but usually not quantitatively representative of the metabolizing properties of the isozymes in intact microsomes. Based on comparisons of warfarin metabolite patterns, it was apparent that the constitutive and induced hepatic cytochrome P-450 isozymes of rats, rabbits and two strains of mice differ.

REFERENCES

1. Guengerich, F.P. (1979) Pharmacol. Ther., Part A 6, 99.
2. Guengerich, F.P., Wang, P. and Davidson, N.K. (1982) Biochemistry 21, 1698.
3. Fasco, M.J., Vatsis, K.P., Kaminsky, L.S. and Coon, M.J. (1978) J. Biol. Chem. 253, 7813.
4. Kaminsky, L.S., Fasco, M.J. and Guengerich, F.P. (1980) J. Biol. Chem. 255, 85.
5. Pohl, L.R., Porter, W.R., Trager, W.F., Fasco, M.J. and Fenton, J.W. (1977) Biochem. Pharmacol. 26, 109.
6. Fasco, M.J., Piper, L.J. and Kaminsky, L.S. (1977) J. Chromatog. 131, 365.
7. Kaminsky, L.S., Fasco, M.J. and Guengerich, F.P. (1982) Methods Enzymol. 74, 262.
8. Coon, M.J., van der Hoeven, T.A., Dahl, S.B. and Haugen, D.A. (1978) Methods Enzymol. 52C, 109.
9. Yasukochi, Y. and Masters, B.S.S. (1976) J. Biol. Chem. 251, 5337.
10. Atlas, S.A., Boobis, A.R., Felton, J.S., Thorgeirsson, S.S. and Nebert, D.W. (1977) J. Biol. Chem. 252, 4712.

Cytochrome P-450, Biochemistry, Biophysics
and Environmental Implications, E. Hietanen,
M. Laitinen and O. Hänninen editors

TURNOVER OF DIFFERENT FORMS OF MICROSOMAL CYTOCHROME P-450 IN RAT LIVER

HIROYUKI SADANO AND TSUNEO OMURA
Department of Biology, Faculty of Science, Kyushu University,
Higashi-ku, Fukuoka, Fukuoka 812 (Japan)

INTRODUCTION

Various membrane proteins of liver microsomes are synthesized and degraded at different rates (1), and the content of an enzyme in the microsomes is maintained on a dynamic balance between its synthesis and degradation. Cytochrome P-450 in liver microsomes is inducible by various chemical inducers, and its induced increase is also effected by increased synthesis or decreased degradation, or both.

Microsomal cytochrome P-450 consists of multiple molecular species, and its induction by chemical inducers is characterized by selective increases of inducer-specific molecular species. Information about the turnover rates of various forms of cytochrome P-450 in animal liver microsomes is therefore essential for correct understanding of the mechanism of their induction.

Earlier studies (2-4) on the turnover of microsomal cytochrome P-450 in animal livers were on its heme since cytochrome P-450 was not yet purified. More recent reports describe the turnover rates of the protein portion of cytochrome P-450 in liver (5,6) or cultured cells (7,8), but the reported half lives are significantly at variance. The half-life of phenobarbital(PB)-inducible form of cytochrome P-450 in rat liver microsomes has been reported to be 40 hr (5) or 11-12 hr (6).

In this study, we examined the turnover rates of PB-inducible form and 3-methylcholanthrene(MC)-inducible form of cytochrome P-450, P-450(PB) and P-450(MC), in rat liver, and studied the effects of the inducers on their turnover. The turnover of NADPH-cytochrome P-450 reductase (fp_T) and cytochrome $\underline{b}_5$ was also examined in order to compare with our previous observations (9,10) on the stabilizing effect of PB on these microsomal proteins.

MATERIALS AND METHODS

P-450(PB) and P-450(MC) were purified from the liver microsomes of PB-treated and MC-treated rats, respectively, by the method of Harada and Omura (11). Rabbit antibodies against P-450(PB) and P-450(MC), which showed no immuno-crossreaction with each other on Ouchterlony double immunodiffusion analysis, were prepared as described previously (11). Fp_T was purified from the liver microsomes of PB-treated rats (12), and rabbit antibodies was prepared as described previously (13).

Male Sprague-Dawley rats weighing 130-200 g were used in all experiments. They were fasted for 10 hr before the injection of radioactive precursors, and 20 hr before sacrifice. PB-treatment of the animals was carried out by injecting PB (100 mg/Kg body weight) once everyday. 1-[^{14}C]-DL-Leucine (40 μCi/200 g body weight) or [^{14}C]-sodium bicarbonate (1.2 mCi/200 g body weight) was administered into rats intraperitoneally. 4-[^{14}C]-δ-Aminolevulinic acid hydrochloride (10 μCi/200 g body weight) was given to rats intravenously. At various time point after the injection of a radioactive precursor, three rats were killed and the liver microsomes were prepared. Portions of the microsomes were solubilized in a buffer containing 0.5 % sodium cholate and 0.5 % Emulgen 913, and the immunoprecipitation of P-450(PB), P-450(MC), and fp_T was performed as described previously (11). The radioactivities of the immunoprecipitates were counted by a liquid scintillation counter.

Cytochrome $\underline{b}_5$ was purified from the labeled microsomes by the procedure of Omura and Takesue (12) as modified by Ito (14). The content of cytochrome $\underline{b}_5$ was determined according to Omura and Sato (15). The purified cytochrome $\underline{b}_5$ was further subjected to polyacrylamide gel electrophoresis using 15 % polyacrylamide gel. After the electrophoresis, the band containing cytochrome $\underline{b}_5$ was cut out, dissolved in NCS solubilizer, and the radioactivity was counted. To count the radioactivities of protein and heme of the liver homogenate and microsomes, small amounts of the samples were separately precipitated with 10 % trichloroacetic acid. It was previously reported that the precipitation of microsomal protein with TCA did not release heme from cytochrome P-450 (2). Protein was determined by the method of Lowry *et al*. (16) using bovine serum albumin as a standard.

RESULTS AND DISCUSSION

Fig. 1 shows the decay of the radioactivities of fp_T, cytochrome $\underline{b}_5$, P-450(PB), and P-450(MC) of liver microsomes after a single injection of $[^{14}C]$-$NaHCO_3$ into normal rats. The radioactivities of total liver protein and total liver microsomes are also shown in the figure. Half lives of proteins were graphically calculated and are shown in Table I. In other sets of experiment, the turnover of the microsomal proteins were determined by using 1-$[^{14}C]$-

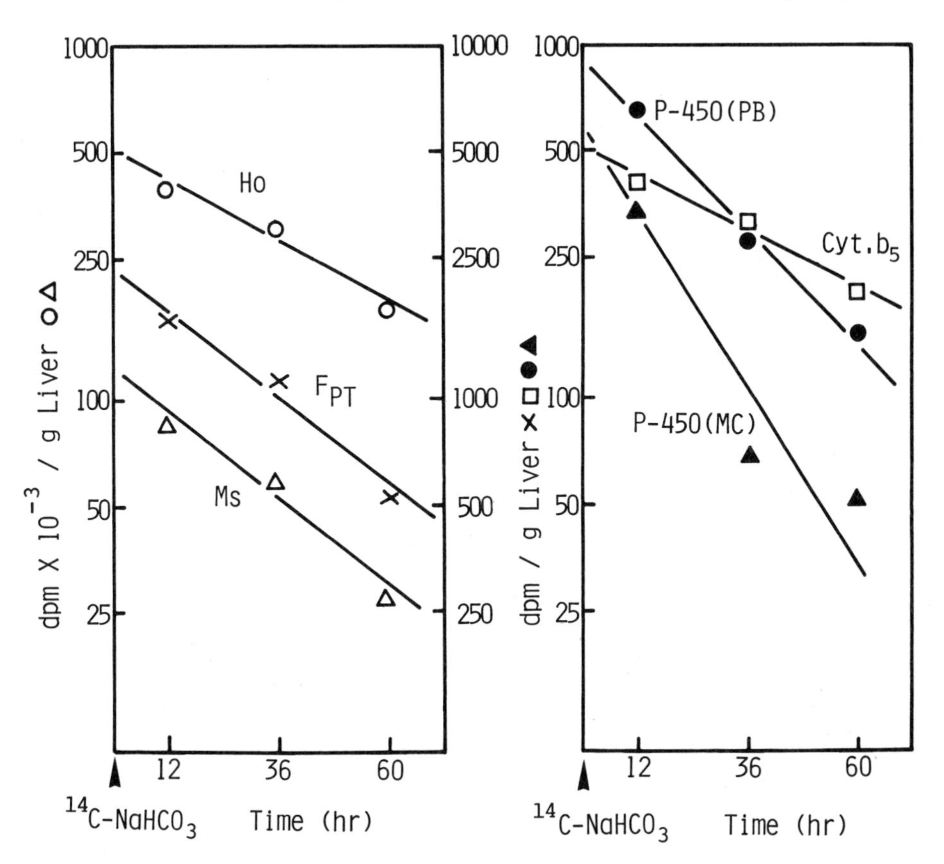

Fig. 1. Decay of radioactivities of microsomal fp_T, cytochrome $\underline{b}_5$, P-450(PB), and P-450(MC) in the liver of $[^{14}C]$-$NaHCO_3$-injected rats. The radioactive sodium bicarbonate was injected to rats intraperitoneally at time 0, and three rats were killed at each time point to determine the radioactivities of liver proteins per g of liver. The radioactivities of proteins shown in the figure at each time point represent the averages of two sets of experiment. Ho and Ms in the figure stand for liver homogenate and microsomes, respectively.

DL-leucine to label the proteins *in vivo*, and the calculated half lives, which were in good agreement with our previous determinations by the use of 1-[^{14}C]-DL-leucine (1) or guanidino-[^{14}C]-L-arginine (9), are also shown in Table 1.

It is apparent that the two forms of cytochrome P-450 are degraded *in vivo* significantly faster than the bulk of microsomal proteins, and that different forms of microsomal cytochrome P-450 are degraded at different rates. The half life of P-450(PB) was intermediate between those reported by Fujii-Kuriyama, *et al.* (5) and Kumar, *et al.* (6). It was also found that the half lives of total microsomal protein, fp_T, and cytochrome b_5 were significantly shortened from previous determinations (1,9) by the use of [^{14}C]-$HaHCO_3$, whereas those of P-450(PB) and P-450(MC) were only slightly shorter than or almost identical to those determined by the use of radioactive leucine. These observations suggested that the half life of a protein having a slow turnover rate was more significantly affected by the re-utilization of labeled leucine, and confirmed the necessity of the use of [^{14}C]-$NaHCO_3$ in determining the turnover rates of liver proteins (17).

Fig. 2 shows the decay of the radioactivities of the hemes which were bound to microsomal hemoproteins. The hemes of cytochrome b_5, P-450(PB), and P-450(MC) showed monophasic exponential decay of the radioactivities, and their half lives are summarized in

TABLE 1.

HALF LIVES OF MICROSOMAL Fp_T, CYTOCHROME b_5, P-450(PB), and P-450 (MC) IN RAT LIVER.

Rats were injected with 1-[^{14}C]-DL-leucine, [^{14}C]-$NaHCO_3$, or 4-[^{14}C]-δ-aminolevulinic acid, and the half lives of liver microsomal proteins were determined as described in Methods.

Proteins	^{14}C-Labeled precursors		
	1-Leu	$NaHCO_3$	δ-ALA
Total microsomes	50 hr	35 hr	20 hr
NADPH-P-450 reductase	50 "	35 "	— "
Cytochrome b_5	95 "	50 "	40 "
P-450(PB)	30 "	25 "	15 "
P-450(MC)	15 "	15 "	15 "

Table 1. The half life of the heme of cytochrome $\underline{b}_5$ agrees with those reported by Druyan et al. (18), Greim et al. (3), and Bock and Siekevitz (19), and was similar to that of the protein moiety determined by the use of radioactive sodium bicarbonate. The turnover rates of the heme portions of P-450(PB) and P-450(MC) were the same, indicating that the heme moiety of P-450(PB) was degraded significantly faster than its protein moiety whereas the heme and protein moieties of P-450(MC) were degraded at about the same rate. Apparently, the heme of P-450(PB) is loosely associ-

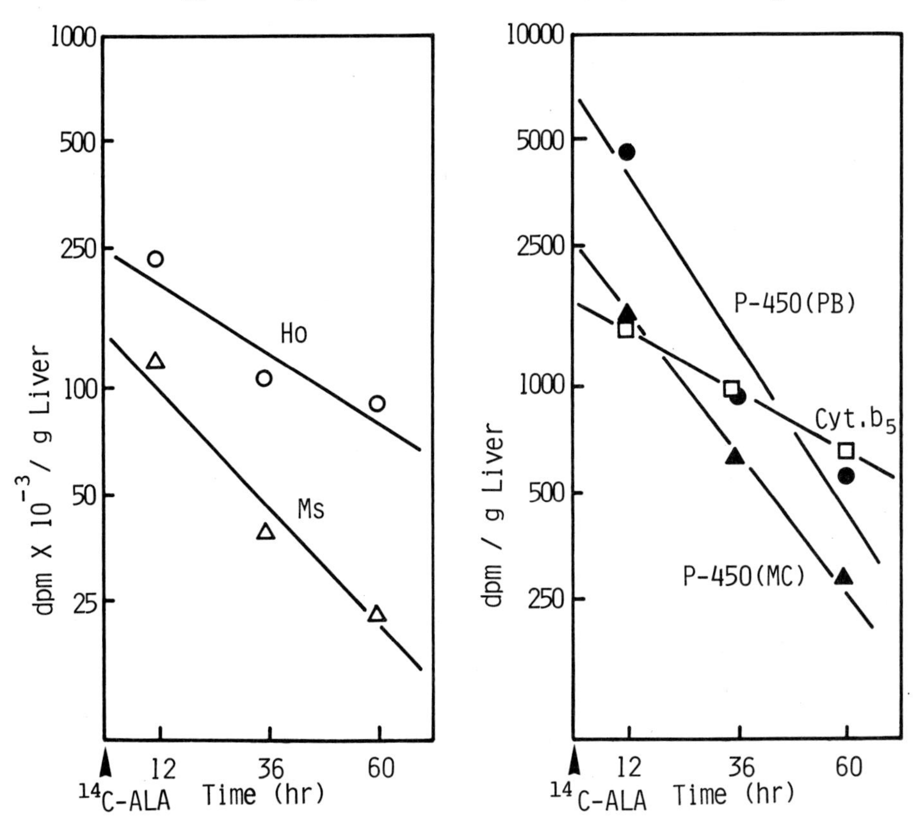

Fig. 2. Decay of radioactivities of microsomal cytochrome $\underline{b}_5$, P-450(PB), and P-450(MC) in the liver of 4-[^{14}C]-δ-aminolevulinic acid-injected rats. The radioactive δ-aminolevulinic acid was injected to rats intravenously at time 0, and three rats were killed at each time point to determine the radioactivities of liver hemoproteins per g of liver. The radioactivities of proteins shown in the figure at each time point represent the averages of two sets of experiment. Ho and Ms in the figure stand for liver homogenate and microsomes, respectively.

TABLE 2.

CONTENTS OF HEMOPROTEINS IN THE LIVER MICROSOMES OF UNTREATED AND PB-TREATED RATS.

The contents of cytochrome $\underline{b}_5$, total cytochrome P-450, P-450(PB), and P-450(MC) in the liver microsomes of untreated and PB-treated rats were calculated from data reported in a previous paper (11).

Rats	Cyt. $\underline{b}_5$	Cyt. P-450		
		total	P-450(PB)	P-450(MC)
	nmol/mg		% of total	
Normal	0.50	0.65	5-10	5-10
PB-treated (24 hr)	0.50	1.05	50-60	5

ated with the protein moiety, and undergoes reversible exchange with the heme pool in the liver cell. If the identical half life of the heme of P-450(PB) and P-450(MC) represent the turnover rate of the heme pool, the heme of P-450(MC) may also be loosely bound with the protein moiety. The dissociability of the heme of microsomal cytochrome P-450 was previously suggested by other lines of evidence (20-22).

When PB is given to rats, the content of P-450(PB) in the liver microsomes increases rapidly, and it becomes the major component of cytochrome P-450 in the microsomes (11,23) as shown in Table 2. Since we previously noticed significant stabilization of microsomal fp_T and cytochrome $\underline{b}_5$ in the liver of PB-treated rats when the turnover rates of the liver proteins were measured by labeling them with radioactive leucine or arginine (9), the effect of PB-treatment on the turnover of fp_T, cytochrome $\underline{b}_5$, P-450(PB), and P-450(MC) was examined using [^{14}C]-$NaHCO_3$ to label the proteins. The first PB dose was given to the animals 12 hr before the injection of radioactive sodium bicarbonate, when the PB-induced synthesis of P-450(PB) was optimal (24), and then the animals received PB once every day until killed. The results are summarized in Table 3 together with the half lives of the microsomal proteins in the liver of untreated rats listed for comparison.

It is clear that PB had no stabilizing action on the turnover of the microsomal proteins examined. The turnover of fp_T and cytochrome $\underline{b}_5$ was rather stimulated by PB, whereas that of P-450(PB)

TABLE 3.

HALF LIVES OF MICROSOMAL Fp_T, CYTOCHROME $\underline{b}_5$, P-450(PB), AND P-450(MC) IN THE LIVER OF PB-TREATED RATS.

The half lives of the microsomal proteins in the liver of PB-treated rats were determined by using [^{14}C]-$NaHCO_3$ to label the proteins *in vivo*. The half lives of the proteins in the liver of untreated animals are also listed in the table for comparison.

Proteins	Half lives ($NaHC^*O_3$-label)	
	Normal rats	PB-treated rats
Total microsomes	35 hr	30 hr
NADPH-P450 reductase	35 "	25 "
Cytochrome $\underline{b}_5$	50 "	30 "
P-450(PB)	25 "	20 "
P-450(MC)	15 "	20 "

and P-450(MC) was not much affected. The apparent stabilization of fp_T and cytochrome $\underline{b}_5$ by PB-treatment (9,25) was possibly caused by increased re-utilization of labeled amino acids in the liver of PB-treated animals. Stimulated synthesis is responsible for the increase of fp_T and P-450(PB) in the liver microsomes of PB-treated animals. The acceleration of the turnover of fp_T and cytochrome $\underline{b}_5$, which have slower turnover rates than cytochrome P-450, could be explained by increased autophagic activity in the liver cell, which was previously suggested by morphological observations (26). PB-stimulated autophagy will shorten the half lives of all microsomal proteins, but the effect will be more significant with proteins having slower turnover rates in the liver of untreated animals where the autophagic digestion of endoplasmic reticulum is less active. On the other hand, the rapid and different turnover rates of P-450(PB) and P-450(MC) indicate their active degradation by some specific process in the liver cell, which is to be clarified by future studies.

REFERENCES

1. Omura, T., Siekevitz, P. and Palade, G. E. (1967) J. Biol. Chem., 242, 2389-2396.
2. Levin, W. and Kuntzman, R. (1969) J. Biol. Chem., 244, 3671-3676.
3. Greim, H., Schenkman, J. B., Klotzbücher, M. and Remmer, H.

(1970) Biochim. Biophys. Acta, 201, 20-25.

4. Meyer, U. A. and Marver, H. S. (1971) Science, 121, 64-66.

5. Fujii-Kuriyama, Y., Mikawa, R., Tashiro, Y., Sakai, M. and Muramatsu, M. (1978) Seikagaku (in Japanese), 50, 870.

6. Kumar, A., Rabishankar, H. and Padmanaban, G. (1980) in: Gustafsson, J. A., Carlstedt-Duke, J., Mode, A. and Rafter, J. (Eds.), Biochemistry, Biophysics, and Regulation of Cytochrome P-450, Elsevier/North-Holland Biomedical Press, Amsterdam, pp. 423-429.

7. Guzelian, P. S. and Barwick, J. L. (1979) Biochem. J., 180, 621-630.

8. Althaus, F. R. and Meyer, U. A. (1981) J. Biol. Chem., 255, 13079-13084.

9. Kuriyama, Y., Omura, T., Siekevitz, P. and Palade, G. E. (1969) J. Biol. Chem., 244, 2017-2026.

10. Kuriyama, Y. and Omura, T. (1971) J. Biochem., 69, 659-669.

11. Harada, N. and Omura, T. (1981) J. Biochem., 89, 237-248.

12. Omura, T. and Takesue, S. (1970) J. Biochem., 67, 249-257.

13. Noshiro, M. and Omura, T. (1978) J. Biochem., 83, 61-77.

14. Ito, A. (1980) J. Biochem., 87, 67-71.

15. Omura, T. and Sato, R. (1964) J. Biol. Chem., 239, 2370-2378.

16. Lowry, O. H., Rosebrough, N. J., Farr, A. L. and Randall, R. J. (1951) J. Biol. Chem., 193, 265-275.

17. Swick, R. W. and Ip, M. M. (1974) J. Biol. Chem., 249, 6836-6841.

18. Druyan, R., Bernard, B. D. and Rabinowitz, M. (1969) J. Biol. Chem., 244, 5874-5878.

19. Bock, K. W. and Siekevitz, P. (1970) Biochem. Biophys. Res. Commun., 41, 374-380.

20. Garner, R. C. and McLean, A. E. M. (1969) Biochem. Biophys. Res. Commun., 37, 883-887.

21. Correia, M. A. and Meyer, U. A. (1975) Proc. Natl. Acad. Sci. USA., 72, 400-404.

22. Correia, M. A., Farrell, G. C., Olson, S., Wong, J. S., Schmid, R., deMontellano, P. R. O., Blian, H. S., Kunze, K. L. and Mico, B. A. (1981) J. Biol. Chem., 256, 5466-5470.

23. Thomas, P. E., Korzeniouski, D., Ryan, D. and Levin, W. (1979) Arch. Biochem. Biophys., 192, 524-532.

24. Harada, N. and Omura, T. (1982) J. Biochem., in press,

25. Jick, H. and Schuster, L. (1966) J. Biol. Chem., 241, 5366-5369

26. Bolender, R. P. and Weibel, E. R. (1973) J. Cell Biol., 56, 746-761.

© 1982 Elsevier Biomedical Press B.V.
Cytochrome P-450, Biochemistry, Biophysics
and Environmental Implications, E. Hietanen,
M. Laitinen and O. Hänninen editors

CHARACTERIZATION OF RAT HEPATIC CYTOCHROME P-450 SUBPOPULATIONS BY ANION EXCHANGE CHROMATOGRAPHY AND METABOLIC-INTERMEDIATE COMPLEX FORMATION.

MICHAEL R. FRANKLIN, LYNN K. PERSHING AND LESTER M. BORNHEIM
Departments of Pharmacology and Biochemical Pharmacology and Toxicology,
University of Utah, Salt Lake City, Utah, 84112, U.S.A.

INTRODUCTION

The purification of a single form of cytochrome P-450 from the livers of animals treated with an inducing agent is often misconstrued in implying that only a single form is induced in response to an inducing agent. By combining the separation of cytochrome P-450 subpopulations in high yield by anion exchange chromatography with their ability to sequester themselves as a metabolic intermediate (MI) complexes _in vivo_ the induction of several subpopulations by a single inducing agent can be demonstrated directly. This methodology also enables comparison of inducing agents to determine whether each has a unique induction profile or whether they fall readily into a limited classification (e.g. phenobarbital-like or polycyclic hydrocarbon-like).

METHODS

Adult male rats were treated with the inducing agents for four days at the following doses: phenobarbital, 80 mg/kg; β-naphthoflavone (BNF) 80 mg/kg; isosafrole 160 mg/kg; SKF 525-A 80 mg/kg and troleandomycin 500 mg/kg. When the cytochromes of non-induced, phenobarbital or BNF induced animals were tested for their ability to form complexes _in vivo_ (Table 2) the MI complex forming substrate was given 8 (SKF 525-A) or 12 hours (isosafrole) before death. The solubilization and separation of cytochrome P-450 subpopulations on DEAE cellulose was based on the method of Warner and Neims (1). Extinction coefficients for the Fe^{++}-CO derivative of all cytochromes was assumed to be 91 $mM^{-1}cm^{-1}$ and that for the nitrogenous (SKF 525-A and troleandomycin) MI complexes to be 65 mM^{-1} cm^{-1} (2) and 75 mM^{-1} cm^{-1} for the non-nitrogenous (isosafrole) MI complexes (3).

RESULTS

The treatment of rats for four days with one of five inducing agents increased the hepatic cytochrome P-450 concentration between 2 and 4 fold (Table 1). When the hepatic microsomal fraction was solubilized, the cytochrome P-450 subpopulations could be separated into four fractions (I to IV) by DEAE cellulose chromatography. After induction, the additional

TABLE 1

THE ELUTION OF SOLUBILIZED CYTOCHROME P-450 FROM DEAE CELLULOSE

Inducing Agent	Microsomal Cyt. P-450[a,b]	Cytochrome P-450[a] Eluting in DEAE Cellulose Fractions			
		I	II	III	IV
None	.86[11]	.10	.56	.17	.07
Isosafrole	1.50[5]	.14	1.07	.21	.08
β-Naphthoflavone	1.59[8]	.07	.91	.53	.09
Phenobarbital	1.82[12]	.27	.93	.56	.08
Troleandomycin	2.70[3]	.39	2.00	.18	.04
SKF 525-A	3.02[4]	.11	2.20	.61	.09

[a]nmoles/mg original microsomal protein.
[b]superscript = number of microsomal preparations examined.

cytochrome P-450 was found to elute in either some or all of the fractions depending upon the inducing agent. After isosafrole treatment, additional cytochrome (over that seen for uninduced microsomes) was found in only one fraction, II. Treatment with troleandomycin produced induction of cytochromes, which eluted in fractions I and II, while phenobarbital treatment increased the cytochromes eluting in fractions I, II and III. Treatment with BNF and SKF 525-A caused additional cytochrome(s) to elute in fractions II and III, although the cytochromes eluting in III after BNF differed from all others in exhibiting a Fe^{++}-CO absorbance maximum at 448 nm rather than 450nm.

Induction by the two conventional inducing agents, phenobarbital and BNF could be seen to involve more than one cytochrome P-450 subpopulation for each agent (Table 2). For each agent a cytochrome differing from that present in the uninduced animal could be seen in each of the two fractions, II and III. The induced cytochrome P-450 that elutes in fraction III after phenobarbital treatment was able to form an MI complex *in vivo* from SKF 525-A but not isosafrole. The induced cytochrome eluting in III, after BNF treatment had the reverse specificity; it formed an MI complex from isosafrole but not from SKF 525-A. The cytochrome P-450 eluting in III, in uninduced animals was unable to form MI complexes *in vivo* from either SKF 525-A or isosafrole. The additional cytochromes P-450 eluting in fraction II, after induction also differed in their ability to form MI complexes *in vivo*. After phenobarbital induction, the induced cytochrome included subpopulations which formed MI complexes from SKF 525-A but not isosafrole. After BNF

TABLE 2

THE PROPERTIES OF CYTOCHROME SUBPOPULATIONS ELUTING IN TWO DEAE CELLULOSE FRACTIONS

DEAE Cellulose Fraction	Inducing Agent	Fe^{++}-CO Absorbance Maximum (nm)	SDS-PAGE Protein Bands (MW x 10^{-3})	% Forming MI Complex In Vivo From: Isosafrole (12 hrs)	SKF 525-A (8 hrs)
III	None	450	51,54	1	2
	Phenobarbital	450	52.5	1*	13*
	β-Naphthoflavone	448	54	9*	0*
II	None	450	50	19	27
	Phenobarbital	450	49.5,51	0*	25*
	β-Naphthoflavone	450	49,50,52	70*	0*

*Induced cytochromes only (i.e. MI complex in induced minus MI complex in uninduced).

treatment the induced cytochrome included subpopulations which formed MI complexes from isosafrole but not SKF 525-A. The cytochrome P-450 eluting in this fraction from uninduced rats contained subpopulations capable of forming MI complexes from both isosafrole and SKF 525-A.

During the induction process, three of the compounds, isosafrole, troleandomycin and SKF 525-A also sequestered some of the cytochrome P-450 as an MI complex of the inducing agent (Table 3). With isosafrole, the MI complexed cytochromes were confined to those subpopulations eluting in fractions I and II, with ten times as much eluting in II as in I. With troleandomycin, the MI complexed cytochromes again eluted in fractions I and II, but while the amount eluting in II was the same as was seen with isosafrole; the amount in I was four-fold greater. When SKF 525-A was the inducing/complexing agent, the complexed cytochromes eluted in all four fractions, with the greatest amount in fractions II and III.

DISCUSSION

All the inducing agents investigated induced cytochrome subpopulations which elute from the DEAE in fraction II. Three agents (phenobarbital, BNF and SKF 525-A) also induce cytochromes which elute in fraction III and two agents, phenobarbital and troleandomycin, also induce cytochromes which elute in fraction I. Thus, all agents except isosafrole, and including the classical inducers phenobarbital and BNF, induce more than one subpopulation of cytochrome

TABLE 3

THE ELUTION OF CYTOCHROME P-450 METABOLIC INTERMEDIATE COMPLEXES FROM DEAE CELLULOSE

Inducing/MI Complex Forming Agent	Cyt. P-450[a] in MI Complexed State Eluting in DEAE Cellulose Fractions:			
	I	II	III	IV
Isosafrole	.05	.56	.02	.01
Troleandomycin	.23	.50	.00	.00
SKF 525-A	.04	1.30	.20	.03

[a]nmoles/mg original microsomal protein

P-450. Only two agents, SKF 525-A and BNF, induce identical DEAE fractions, but the presence of MI complex in fraction III after SKF 525-A induction (Table 3) and the inability of the cytochromes in this fraction to form MI complexes from SKF 525-A _in vivo_ after BNF induction (Table 2), indicate that qualitative differences exist, at least within these fraction III subpopulations.

Of the inducing agents which also form cytochrome P-450 MI complexes, only with troleandomycin does induction and MI complex formation occur in the same subpopulations. With isosafrole treatment, MI complex formation is seen in two subpopulations but induction is only seen in one of these. With SKF 525-A, MI complex formation is seen in four subpopulations, but induction is only seen in two. Thus, induction does not appear to be an attempt to replace cytochrome P-450 lost to an MI complex, or result from a stabilization of the cytochrome P-450 (in the form of an MI complex) against normal cellular turnover.

ACKNOWLEDGEMENTS

Supported by USPHS Grant No. CA 15760.

REFERENCES

1. Warner, M. and Neims, A.H. (1979). Multiple forms of ethoxyresorufin O-deethylase and benzphetamine N-demethylase in solubilized and partially resolved rat liver cytochromes P-450. Drug Metab. Dispos. 7, 188-193.
2. Franklin, M.R. (1974). The formation of a 455 nm complex during cytochrome P-450 dependent N-hydroxyamphetamine metabolism. Mol. Pharmacol. 10, 975-985.
3. Elcombe, C.R., Bridges, J.W., Nimmo-Smith, R.H. and Netter, K.J. (1975). Studies on the interaction of safrole with rat hepatic microsomes. Biochem. Pharmacol. 24, 1427-1433.

Cytochrome P-450, Biochemistry, Biophysics
and Environmental Implications, E. Hietanen,
M. Laitinen and O. Hänninen editors

CATALYTIC AND STRUCTURAL PROPERTIES OF TWO NEW CYTOCHROME P-450 ISOZYMES FROM PHENOBARBITAL-INDUCED RAT LIVER: COMPARISON TO THE MAJOR INDUCED ISOZYMIC FORM

DAVID J. WAXMAN AND CHRISTOPHER WALSH
Departments of Chemistry and Biology, Massachusetts Institute of Technology, 18-167, Cambridge, MA 02139 (U.S.A.)

INTRODUCTION

Diethylaminoethyl cellulose chromatography at room temperature has been used by numerous investigators for purification of isozymes of cytochrome P-450, including the major phenobarbital (PB)-induced isozyme present in rat liver microsomes (termed P-450 PB-4 in this study). We now report purification of three new isozymes, P-450 PB-1, P-450 PB-5 and P-450 PB-6, from PB-induced Sprague-Dawley rat liver microsomes. In addition, resolution and partial purification of isozymes PB-2/PB-3 from PB-induced rats is described. The catalytic and structural properties of PB-1 and PB-5 are presented in comparison to those of the major PB-induced isozyme, PB-4. Methods used in this study are described in greater detail in refs. 1 and 2, and in refs. therein.

RESULTS AND DISCUSSION

Isozyme resolution. Four major hemeprotein fractions, containing P-450 isozymes PB-1 to PB-6 (numbered in order of elution from Whatman DE52) were resolved using two DE52 columns [as in ref. 3 with modifications (2)]. PB-1 and PB-2/PB-3 eluted as two well-resolved peaks off the first column, with PB-4 and PB-5 (the latter containing PB-6 in the trailing edge of the peak) resolved on the second column with a salt gradient. Hemeprotein fractions were further resolved and purified by stepwise elution off BioRad HTP to give > 95% pure preparations of PB-1, PB-4, PB-5, and PB-6 (Figure 1) as well as an incompletely separated PB-2/PB-3 mixture. Experiments described below focus on three of the homogenous isozymes, PB-1, PB-4 and PB-5. P-450 isozymes PB-2 and PB-3 were both found in phenobarbital-induced as well as in uninduced rat liver and are discussed further in a report on testosterone hydroxylation included elsewhere in this volume (4). Those studies indicate that PB-3 corresponds to P-450_a, a testosterone 7α-hydroxylase (5). General properties of the six isozymes are summarized in Table 1.

Catalytic activities. Detergent-free P-450s PB-1, PB-4 and PB-5 were each reconstituted with purified P-450 reductase and cytochrome b_5 in a dilauroyl-phosphatidylcholine (DLPC) system and their enzymatic activities determined

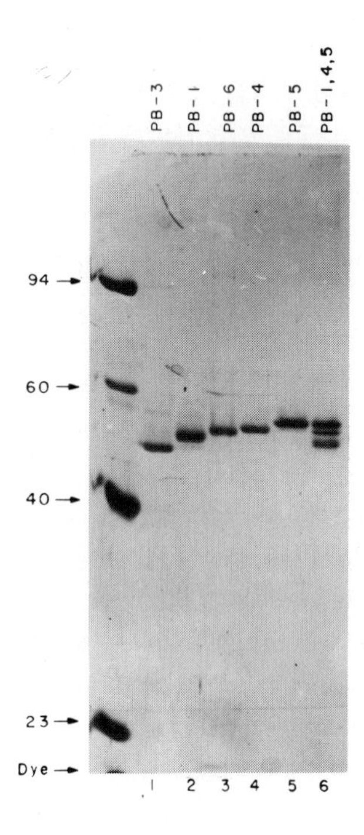

Fig. 1. SDS-PAGE using a standard Laemmli system (10% gel) of purified P-450 isozymes (as indicated), with molecular weight standards on the left. Apparent molecular weights of the isozymes are included in Table 1. PB-3 is catalytically and electrophoretically equivalent to P-450 UT-F purified by Dr. F.P. Guengerich (unpublished) and is most probably equivalent to P-450$_a$ (5) as well.

TABLE 1

P-450 ISOZYMES ISOLATED FROM PB-INDUCED RAT LIVER

	PB-1	PB-2/PB-3	PB-4	PB-5	PB-6
~ M_R (SDS-PAGE)	47,000	48,000/46,000	49,000	51,000	48,000
P-446 (ε,$mM^{-1}cm^{-1}$)[a]	0	~ 15-25	60	70	N.D.
Specific content (nmol/mg)	13.3-16.0	~ 9-11	13.0-16.8	11.0-14.2	7.4
~ Percent total[b]	10-15	10-15	40-45	15-20	~ 2-4

[a]Reduced-metyrapone *vs.* metyrapone P-446 complex. N.D. = not determined.
[b]Estimated, based on relative percent in DE52 eluents obtained from the 10-16% polyethylene glycol 6000 fraction applied to the column.

TABLE 2

CATALYTIC ACTIVITIES IN RECONSTITUTED SYSTEM[a]

Substrate (mM)	Product determined	Turnover number (nmol product/nmol P-450/min) P-450 PB-1	P-450 PB-4	P-450 PB-5
Dimethylaniline (0.5)	Formaldehyde	2.0	27	5.0
7-Ethoxycoumarin (1.0)	7-OH Coumarin	0.3	4.5	0.8
Coumarin (1.0)	7-OH Coumarin	<0.02	<0.02	< 0.02
Acetanilide (4.0)	4-OH Acetanilide	2.9	0.07	~ 0.02
Testosterone (0.1)	16αOHT + 16βOHT & Androstenedione	0.2-0.5	10.6	1.9
pTol-S-Ethyl (0.5)	pTol-SO-Ethyl	41	31	5.0
	pBzlOH-S-Ethyl	6.8	~ 0.1	<0.05
Toluene (5.0)	Benzyl alcohol	6.7	22	3.7
	_p_/_m_-Cresol	<0.5	2.4	<0.5
	_o_-Cresol	<0.5	3.4	<0.5

[a]Purified isozymes were reconstituted with purified cytochrome b_5, P-450 reductase and DLPC, and the monooxygenase activities determined essentially as described (1,2). Substrate concentrations are indicated in parenthesis.

using seven monooxygenase substrates (Table 2). It is apparent that (a) PB-4 and PB-5 display highly similar substrate specificities, with the specific catalytic activity of PB-4 approximately 5-fold greater than PB-5 and (b) PB-1 is readily distinguished from PB-4 and PB-5, exhibiting high acetanilide 4-OHase activity and high activity in the conversion of ethyl tolyl sulfide (pTol-S-Ethyl) to the corresponding benzyl alcohol (pBzlOH-S-Ethyl). These two PB-1 specific activities are both stimulated by metyrapone in microsomes (1,6). PB-1 does not form a reduced/metyrapone spectral complex (P-446, Table 1) and is not inhibited by this dipyridyl heme ligand (using 7-ethoxycoumarin as substrate). Thus the observed microsomal stimulations (1,6) might reflect increased electron flux from the limiting amounts of microsomal P-450 reductase to the microsomal P-450 isozymes (such as PB-1) which are uninhibited by metyrapone.

Effects of cytochrome b_5. Inclusion of 1.13 mol cytochrome b_5 per mol P-450 in the reconstituted system resulted in a significant increase in turnover numbers for all substrates tested (Table 3). The b_5-stimulation of the activity of PB-1 (~6 to 9-fold) was much more significant than that observed with either PB-4 or PB-5 (~1.5 to 2.0-fold higher activity in the presence of b_5). In the absence of DLPC, these b_5-stimulations were significantly diminished (to a 1.5 to 1.8-fold stimulation of PB-1) or reversed (60 to 70% inhibition of either PB-4 or PB-5 by b_5). With both PB-1 and PB-4, 1 mol b_5 per mol P-450 effected maximal stimulation, characterized by an increase in V_{max} with no significant change in K_m (for 7-ethoxycoumarin). These results are consistent with cytochrome b_5 playing a direct role in transfer of the second electron from P-450 reductase to P-450, as suggested previously by other workers.

TABLE 3

CYTOCHROME b_5-STIMULATION OF MONOOXYGENASE ACTIVITY

Substrate	Product determined	Relative Activity[a]	
		P-450 PB-1	P-450 PB-4
7-Ethoxycoumarin	7-OH Coumarin	7.7	1.8
Toluene	Benzyl alcohol		1.6
Acetanilide	4-OH Acetanilide	7.4	N.D.[b]
pTol-S-Ethyl	p-Tol-SO-Ethyl	6.6	1.5
pTol-S-Ethyl	p-BzlOH-S-Ethyl	8.5	N.D.[b]

[a]Ratio of activity in the presence of b_5 to activity in the absence of b_5. Data for P-450 PB-5 were very similar to those with PB-4 in all cases examined.
[b]Not determined. Turnover numbers with these substrates were very low for P-450 PB-4 (see Table 2).

Structural analysis. The structural relatedness of the three purified hemeproteins was assessed using three methods:

1. Sequence analysis - NH_2-terminal degradation and PTH-amino acid analysis established the following sequences for PB-4 and PB-5 (single letter amino acid code is used):

```
                                   1         1         2         2         3
            1       5              0         5         0         5         0
PB-4: H2N-M E P S I L L L L A L L V G F L L L L V R G H P K S R G N F P(P)G(P)

PB-5: H2N-M E P S I L L L L A L L V G F L L L L V R G H P K S R G N F P(P)G(P)
```

Thus the first 31 (and likely 34) residues are identical, with the sequences highly homologous to the first ~ 22 residues determined for Long Evans rat P-450_b (2 differences) as well as rabbit LM2 (6 differences) (7,8). The additional 10 residues determined in this study indicate that the hydrophobic amino terminal segment, common to several microsomal P-450s, is followed by a basic region (residues 21 to 27), suggesting a role for the NH_2-terminal region of these P-450s in membrane anchoring (also see ref. 2). Amino terminal analysis of PB-1 indicated the tentative sequence: H_2N-M D L V M L L V L ..., which is distinct from that of PB-4, PB-5 and all other published P-450 sequences including rat P-450_a, P-450_b, P-450_c, P-450_d, RLM3 and RLM5, and rabbit LM2, $LM3_b$ and $LM3_c$.(7-11)

2. Immunochemical analysis. Further evidence for a close homology between PB-4 and PB-5 was indicated by their immunocrossreactivity with antisera raised by Dr. P. Thomas, Hoffman-LaRoche, against Long Evans rat P-450_b. In this regard, P-450 PB-5 is analogous to P-450_e, described recently (12). By contrast, PB-1 did not crossreact with antisera raised against P-450s a, b, c and d. Direct comparisons by SDS-PAGE confirmed that P-450 PB-1 is distinct from each of these well-characterized hemeproteins.

3. Peptide mapping. P-450 isozymes PB-4 and PB-5 displayed highly similar but distinguishable peptide maps, whereas the peptide maps of PB-1 were apparently unrelated to those of either PB-4 or PB-5.

CONCLUSIONS

These findings establish the existence of two new P-450 isozymes in PB-induced rat liver: P-450 PB-1, which does not form a P-446 complex with metyrapone, and P-450 PB-5, a distinct isozyme closely related to P-450 PB-4, the major PB-induced isozyme studied previously. PB-1 is characterized by a marked (but not absolute) dependence on cytochrome b_5 for catalytic activity, by its catalysis of monooxygenase activities stimulated by metyrapone in liver microsomes, and by its unique NH_2-terminal sequence. P-450 PB-5, distinct from PB-4 both chromatographically and by its lower specific catalytic activity, has peptide maps which are highly similar but distinguishable from those of PB-4. This suggests that these two closely related isozymes are products of different genes. Nucleic acid sequence microheterogeneity recently observed in cDNA clones corresponding to such PB-induced polypeptides (Dr. Fujii-Kuriyama, personal communication; PNAS, in press) is consistent with this conclusion.

REFERENCES

1. Waxman, D.J., Light, D.R. and Walsh, C. (1982) Biochemistry, 21, in press, May.

2. Waxman, D.J. and Walsh, C. (1982). J. Biol. Chem., 257, in press, Aug/Sept.

3. West, S.B., Huang, M.-T., Miwa, G.T. and Lu, A.Y.H. (1979) Arch. Biochem. Biophys., 193, 42-50.

4. Waxman, D.J., Ko, A. and Walsh, C. (1982), "Testosterone hydroxylations catalyzed by purified rat liver cytochrome P-450 isozymes," This volume.

5. Ryan, D.E., Thomas, P.E., Korzeniowski, D. and Levin, W. (1979) J. Biol. Chem., 254, 1365-1374.

6. Leibman, K. (1969) Mol. Pharmacol., 5, 1-9.

7. Botelho, L.H., Ryan, D.E. and Levin, W. (1979) J. Biol. Chem., 254, 5635-5640.

8. Haugen, D.A., Armes, L.G., Yasunobu, K.T. and Coon, M.J. (1977) Biochem. Biophys. Res. Commun., 77, 969-973.

9. Botelho, L.H., Ryan, D.E., Yuan, P.-M., Kutny, R., Shively, J.E. and Levin, W. (1982) Biochemistry, 21, 1152-1155.

10. Cheng, K.C. and Schenkman, J.B. (1982) J. Biol. Chem., 257, 2378-2385.

11. Ozols, J., Heinemann, F.S. and Johnson, E.F. (1982) J. Biol. Chem., 256, 11405-11408.

12. Ryan, D.E. and Levin, W. (1981) Fed. Proc., 39, 1640.

Cytochrome P-450, Biochemistry, Biophysics
and Environmental Implications, E. Hietanen,
M. Laitinen and O. Hänninen editors

SEPARATION OF MULTIPLE P-450 FORMS BY HYDROPHOBIC AND ANION EXCHANGE AND CHROMATOFOCUSING TECHNIQUES

MARKKU PASANEN AND OLAVI PELKONEN
Department of Pharmacology, University of Oulu, Finland

INTRODUCTION

A procedure has been established previously for the partial purification of the placental cytochromes P-450 (1). This method can be used for the purification of P-450's from differently pretreated rat liver samples. In order to obtain more detailed information on the multiplicity of P-450, a new separation technique, chromatofocusing was introduced.

A report on this new method is given here, and comparisons are made between control and phenobarbital and 3-methylcholanthrene-induced rat liver microsomal P-450 material. Special interest is focused on the elution profiles during DEAE-cellulose chromatography, the molecular weights of the purified proteins, their absorption maxima over the Soret region and their elution pH during chromatofocusing.

MATERIAL AND METHODS

Young male Wistar rats were pretreated with 3-methylcholanthrene (3-MC, 40 mg/kg in corn oil i.p. once daily for three days) and with phenobarbital (PB, 0.5 g/l in drinking water for one week). The purification procedure followed that published earlier (1) with some minor modifications. Briefly, the XAD-2 step was replaced by ultrafiltration on an Amicon PM-30 membrane and Whatman DE-52 cellulose using a DEAE-Sephacel (Pharmacia Fine Chemicals). The pools from the DEAE-Sephacel were concentrated separately on the membrane as before and each was chromatofocused between pH 7.4-5.0. The chromatofocusing column (0.9 x 18 cm) was equilibrated to pH 7.4 with buffer A (0.025 M Tris-acetic acid pH 7.4, 20 % glycerol, 0.05 % Emulgen 911) and the concentrated DEAE pools were loaded onto the column, washed with the same buffer and eluted with a diluted commercial polybuffer mixture (1:10 Pharmacia Fine Chemicals Polybuffer 96 (30 %) and 74 (70 %) pH 5.0 acetic acid, 20 % glycerol and 0.05 % Emulgen 911). The elution was monitored both with absorbance determinations at 417 nm and by measuring the pH throughout the chromatogram. Those fractions in each 417 nm peak possessing the highest P-450 content were combined and concentrated as before. These concentrates were lyophilized, and electrophoresis was carried out according to Laemmli (2). Molecular weights were determined by comparing the Rf values for the unknown sample with those for a standard within the same slab gel. Protein was determined according to Lowry et al. (3).

RESULTS

The pertinent results of the purification procedure are summarized in Table 1. Briefly, hydrophobic column chromatography as a first step ensures a high recovery of P-450 for further separation. The DEAE-Sephacel produced four pools from the control rat livers and 3-MC-treated rat livers and five pools from PB-treated rat livers. The chromatofocusing technique resolved the control rat material into six pools, the PB-treated material into seven pools and the 3-MC-treated material into eight pools. The Soret maximum varies most widely within the 3-MC material, in the range 452.5-448.0 nm although this induction favours the traditional 448 nm region.

All the DEAE pools which pass the DEAE column in the void volume show similar behaviour during chromatofocusing. This also happens to those fractions eluting at low salt concentrations. Great variety is noticeable in the pH values, the molecular weights and particularly the absorbtion maxima of a unique DEAE pool, especially its "indicible forms". One significant feature is that there do not exist two identical proteins in respect of the properties investigated. Also 3-MC-DEAE 4 seems to be "highly sticky" under the precent experimental conditions i.e. it tends to aggregate during chromatofocusing, having a molecular weight of 97000 daltons. All the cytochromes having only one molecular weight figure in Table 1 exhibited one single band in electrophoresis. For those having more than one molecular weight the one underlined is the main form. Thus at least seven homogeneous proteins were purified in three chromatographic steps with good recovery and resolution.

DISCUSSION

A chromatofocusing technique for the purification of multiple forms of P-450 is introduced here. Some efforts at isoelectric focusing (4, 5) have been made earlier, but with meagre succes.

TABLE 1

MULTIPLE FORMS OF P-450 AND THEIR PROPERTIES

Fraction	NaCl (mM)	Specific Activity (nmol/mg)	Soret Maximum (nm)	Molecular Weight (1000 x)	Chromato-focusing (pH)
Control Rat					
Microsomes		0.96			
phenyl-Sepharose		2.56	451.0		
DEAE pool 1	0	2.33	451.0	53, $\underline{54.5}$	
" 2	68	6.68	450.5	53	
" 3	100	2.78	450.5	52.5	
" 4	156	4.47	449.0	55	
Chromatofocusing					
DEAE pool 1		N.D.[§]	451.0		7.40
" 2		"	450.5		"
" 3		"	450.5	40, $\underline{63}$	"
" 3		"	452.0	53	6.46
" 4		"	449.5	55	6.32
" 4		"	449.0	53	6.08
Phenobarbital Rat					
Microsomes		1.50			
phenyl-Sepharose		4.01			
DEAE pool 1	0	3.59	450.0	several	
" 2	56	10.47	450.0	"	
" 3	68	2.44	450.0	"	
" 4	80	2.38	450.0	"	
" 5	138	5.95	449.0	46, $\underline{50}$, 67	
Chromatofocusing					
DEAE pool 1		N.D.	450.0	$\underline{52}$, 55, 62	7.40
" 2		"	450.0	$\underline{52}$, $\underline{55}$, 59	"
" 3		"	450.0	$\underline{54}$, $\underline{61}$, 63	"
" 4		"	450.0	$\underline{53}$	"
" 4		"	452.0	50, $\underline{52}$, 54	6.55
" 4		"	451.5	55	6.32
" 5		"	449.0	50	6.45
3-Methylcholanthrene Rat					
Microsomes		1.13			
phenyl-Sepharose		3.55	450.0		
DEAE pool 1	0	1.55	450.5	$\underline{51}$, 53, 55	
" 2	73	4.56	450.5	$\underline{51}$, 53, 54	
" 3	128	3.36	450.5	$\underline{49}$, $\underline{52}$, 55	
" 4	183	7.91	448.0		
Chromatofocusing					
DEAE pool 1		N.D.	450.5	$\underline{51}$, 53, 55	7.40
" 2		"	449.0		"
" 2		"	452.5	49, $\underline{51}$,	6.98
" 3		"	449.0	59	7.40
" 3		"	451.0	$\underline{53}$, 59.5	6.64
" 4		"	448.0	97	6.83
" 4		"	448.0	97	6.55
" 4		"	448.0	97	6.36

[§]N.D. = not determined

Used as the final step, chromatofocusing resolved the DEAE pools, providing some homogeneous P-450 forms, as judged by gel electrophoresis. The first DEAE pools which did not focuse "properly", i.e. passed through the column in the void volume, may contain the cytochromes having high pI values detected by Vlazuk and Waltz (6). The ability to form aggregates seems to be extremely prominent after 3-MC-induction, although it happened to some extent in all groups. This caused undesirable distortion during electrophoresis by forming aggregates of 100000-250000 daltons. This phenomenon was observed earlier by Guengerich (5).

Chromatofocusing exhibited a sharp capacity for discriminating between multiple forms eluting at the same salt concentration. This can be seen when looking at the focusates from PB-DEAE pool 4, where different forms with Soret maxima of 450 nm, 452 nm and 451.5 nm can be resolved from the original pool with a Soret maximum of 450 nm. The same is apparent in 3-MC-DEAE pool 3 and its focusing products.

As far as we can judge from these preliminary experiments, chromatofocusing may become an important large-scale technique for use as a final separation tool in endeavours to separate multiple forms of P-450.

ACKNOWLEDGEMENTS

The technical assistance of Ritva Saarikoski, Irma Vartiainen and Kaisu Pulkkinen is gratefully acknowledged. This study was supported by the Finnish Foundation for Cancer Research.

REFERENCES

1. Pasanen, M. and Pelkonen, O. (1981) BBRC 103, 1310-1317.
2. Laemmli, U.K. (1970) Nature 227, 680-685.
3. Lowry, O.H., Rosebrough, N.J., Farr, A.L. and Randall, R.J. (1951) J. Biol. Chem. 265-275.
4. Warner, M., LaMarca, M.V. and Neims, A.H. (1978) Drug Metab. Dispos. 6, 353-362.
5. Guengerich, F.P. (1979) Biochem. Biophys. Acta 572, 132-141.
6. Vlasuk, G.P. and Waltz, Jr. F.G. (1982) Arch. Biochem. Biophys. 214, 248-259.

Cytochrome P-450, Biochemistry, Biophysics
and Environmental Implications, E. Hietanen,
M. Laitinen and O. Hänninen editors

CYTOCHROMES P-450 IN LIVER MICROSOMES OF RATS PRETREATED WITH β-NAPHTHOFLAVONE

HENRY W. STROBEL AND PAUL P. LAU
Department of Biochemistry and Molecular Biology, The University of Texas
Medical School, P. O. Box 20708, Houston, Texas 77025 (U.S.A.)

INTRODUCTION

The cytochrome P-450-dependent mixed function oxidase system is present and functional in many animal tissues and organs (1,2). Cytochrome P-450, the terminal oxidase of the system, catalyzes the metabolism of a broad range of substrates including steroids, fatty acids, drugs, pesticides, carcinogens and precarcinogens (1,3). Recently, several laboratories demonstrated the presence of and isolated multiple forms of cytochrome P-450 whose tissue concentrations are affected by pretreatment with inducers such as phenobarbital, 3-methylcholanthrene, β-naphthoflavone, Aroclor 1254, polychlorinated biphenyl, pregnenolone-16-carbonitrile, isosafrole and cholestyramine (4-21). Each of the cytochromes P-450 isolated has a different, but overlapping, set of substrate preferences when examined under *in vitro* reconstituted assay mixtures. Similarly, at the organ and tissue level, varying substrate preferences are observed (2). Thus, an explanation of the basis for the metabolic capabilities of the mixed function oxidase system of the various tissues would be facilitated by definition of which forms are present in a tissue under a specified set of conditions and how much of each form is present. This paper describes the isolation and purification of cytochromes P-450 from liver microsomes of rats treated with β-naphthoflavone and the quantitation of those forms in microsomes using antibodies raised to the isolated purified cytochromes.

METHODS

Purification of Cytochrome P-450

Liver microsomes from rats treated for 3 days with daily injections of β-naphthoflavone (80 mg/kg body weight in corn oil) were solubilized with 1.5% (w/v) Renex 690 and chromatographed on a DEAE-Sephadex A-25 column to separate the cytochrome P-450 containing fractions from the reductase containing fractions as previously described (22). The various forms of cytochrome P-450 were resolved from one another by chromatography on a Whatman DE-53 column (17). The forms were purified to electrophoretic homogeneity by further chromatography on CM-Sepharose or hydroxylapatite (17).

Immunochemical Quantitation

Rocket Immunoelectrophoresis. In a modification of the method of Laurel (23), a mixture containing 1% (w/v) HGT Agarose (Seakem), 0.2% (w/v) Renex 690 in barbital buffer (Sigma) was heated in a boiling water bath. About 15 ml of the melted agarose mixture was cooled to 56°C before 0.2 ml of antiserum was added to the mixture. The mixed gel was poured immediately onto the hydrophobic side of a 12 x 11 cm Marine Colloids gel bond film (FMC). After the gel solidified, 4.0 mm (i.d.) holes were punched into the gel. Samples to be analyzed were added and electrophoresed at 4°C in barbital buffer for 12-24 hrs. After electrophoresis, the gel was washed in 0.15 M NaCl for at least 1 day with 2-3 changes of wash solution. The gel was stained with Coomassie Blue and washed. The final gel film, which contained the samples and standards, was reproduced by dry copy methods. The rockets were cut out and weighed as an estimate of area under the rocket curves.

Radial Diffusion. A modification of the method used by Thomas *et al.* (24) was also used to quantitate the amount of each form of cytochrome P-450 in liver microsomes. The preparation of the gel was similar to that described for the rocket immunoelectrophoresis technique, except that 6 holes were placed at regular intervals throughout the gel rather than only at one edge. Radial immunodiffusion was conducted at 4°C for 3 days. Standards and unknowns were placed on the same gel to assure consistency. The area of each circle was measured as the square of the diameter of the outer circle made by the precipitin band minus the square of the diameter of the inner circle.

Measurement of other activities and contents were as previously described (17,22).

RESULTS AND DISCUSSION

Purification and Characterization of Cytochromes P-450

Five forms of cytochrome P-450 are resolved from solubilized liver microsomes from rats pretreated with β-naphthoflavone. As shown in Table 1, the solubilization, DEAE-Sephadex column and polyethylene glycol precipitation steps result in an overall recovery of 92% of the starting content of cytochrome P-450 free of cytochrome P-450 reductase activity. The resolving column (DE-53) separates five pools containing cytochrome P-450. The five pools (numbered in order of elution from the resolving column) account for 50.4% of the original cytochrome P-450 content of the microsomes and no other cytochrome P-450 could be detected. Each of these pools of cytochrome could be purified to a single electrophoretically pure protein by chromatography on

TABLE 1

PURIFICATION OF HEPATIC CYTOCHROMES P-450 FROM β-NAPHTHOFLAVONE-PRETREATED RATS

Step	P-450 nmol/ml	Protein mg/ml	Specific Content nmol/mg	Recovery %
Microsomes	83.5	41.3	2.02	100
Solubilization	14.28	6.15	2.32	97
DEAE-Sephadex A-25 Eluate	11.0	3.3	3.33	95
Polyethylene Glycol Concentration	34.06	7.45	4.57	92
DEAE-Cellulose				
Pool 1	2.53	0.48	5.3	9.2
" 2	2.75	0.43	6.45	10.9
" 3	0.99	0.15	6.8	4.1
" 4	0.77	0.14	5.34	4.6
" 5	3.07	0.38	8.05	21.6
P-1 CM-Sepharose	6.04	0.27	22.3	3.9
P-2 " "	2.7	0.4	6.6	7.9
P-3 " "	1.5	0.11	13	4.0
P-4 Hydroxylapatite	1.37	0.11	12.4	2.3
P-5 "	1.23	0.07	17.5	10.0

CM-Sepharose or hydroxylapatite. The pure forms together account for 28% of the original content of cytochrome P-450.

The specific contents of the purified fractions range from a high of 22.3 for Pool 1 to a low of 6.6 for Pool 2. Since each purified cytochrome P-450 pool gave rise to a single band on SDS gels and since the theoretical specific content for most cytochromes range between 17 and 22, it is likely that some of the pools contain varying amounts of apo- and holocytochrome P-450. Direct incubation of purified protein with heme, however, did not increase the specific contents of the purified cytochromes.

The subunit molecular weight of each form of cytochrome P-450 was estimated in electrophoresis in SDS gels calibrated with standards of known molecular weight. Each form was named according to its R_f in SDS gels; the form with the highest R_f (lowest subunit molecular weight) being assigned as Form 1. The minimal molecular weight and reduced carbon monoxide difference spectrum absorption maximum of each form is shown in Table 2. Each form has a unique

TABLE 2

PROPERTIES OF CYTOCHROMES P-450 OF β-NAPHTHOFLAVONE-PRETREATED RATS

Cytochrome P-450	Minimal M_r	CO-Reduced Difference λ_{max}
Form 1	47,000	452.5
" 2	50,500	449
" 3	51,500	449
" 4	53,500	447.5
" 5	56,500	447.5

peptide map when subjected to *S. aureus* V8 proteolysis and each form has a distinct set of catalytic activities (17). Antibody to each form has been raised in rabbits and the immunological cross reactivity of each form was tested with antibodies to all other forms in Ouchterlony double diffusion assays. All antibodies reacted only with the antigen to which they were prepared, but not with the other 4 antigens. Some evidence for shared antigenic determinants between forms four and five was suggested by the appearance of a slight spur of reactivity of form four with antibody to form five, cf. crossreactivity of $P\text{-}450_d$ with $P\text{-}450_c$, Ryan *et al.* (20).

Immunochemical Quantitation of Cytochromes P-450 in Liver Microsomes of β-Naphthoflavone-treated Rats

We have utilized the availability of antibodies to the five purified forms of hepatic cytochrome P-450 from β-naphthoflavone-treated rats to quantitate the amount of each form in the microsomes. Two techniques have been utilized for this purpose, rocket immunoelectrophoresis and radial immunodiffusion. These techniques have been applied to assess specific forms of cytochrome P-450 by other workers, e.g., Thomas *et al.* (24) and Pickett *et al.* (25). Using these principles, we have established both a rocket immunoelectrophoresis and radial immunodiffusion assay for the five forms of cytochrome P-450 isolated from β-naphthoflavone treated rats. The results of rocket immunoelectrophoresis of increasing amounts of purified form 5 are shown in Fig. 1, and the results of the radial immunodiffusion assay with increasing amounts of form 5 are shown in Fig. 2. Area under the rockets or size of the circles, respectively, is plotted vs. concentration of form 5 to develop standard curves as shown in Figs. 3 and 4, respectively. When these standard curves are used to quantitate the amount of form 5 in microsomes from β-naphthoflavone-treated 150 gm rats, the results agree very well, i.e., 82.7% by rocket immunoelectro-

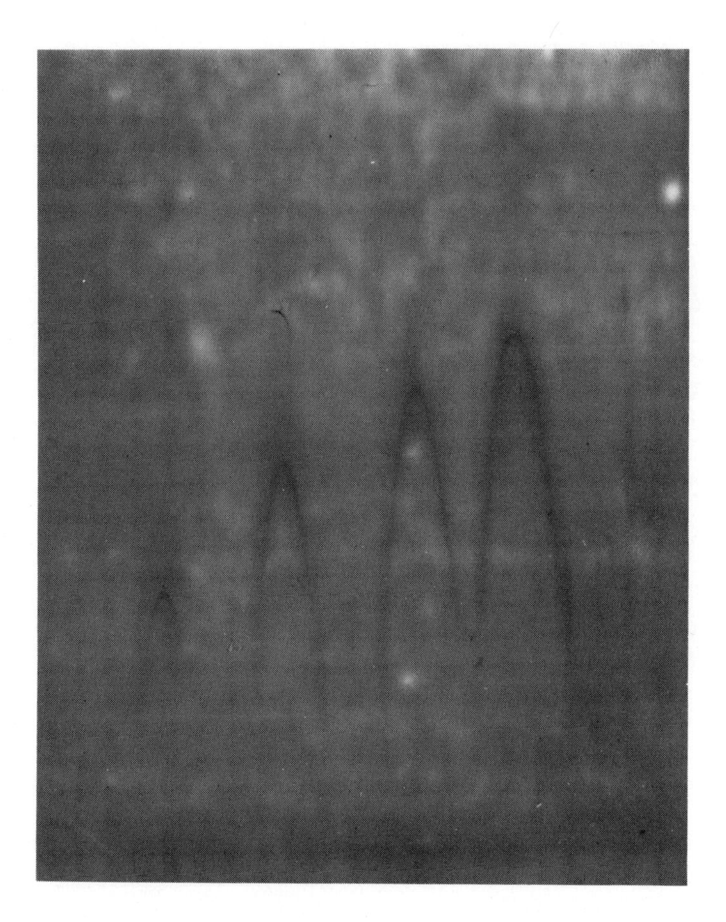

Fig. 1. Rocket immunoelectrophoretic assay using antiserum against cytochrome P-450 form 5.

phoresis and 83.3% by radial immunodiffusion. These results are also in good agreement with the determinations made by Thomas *et al.* (24) and Pickett *et al.* (25) showing that 89% and 80%, respectively, of the cytochrome P-450 in 3-methylcholanthrene-treated rats is cytochrome P-448 (P-450_c) which is identical to form 5 from β-naphthoflavone-treated rats (Lau, P.P., Pickett, C.E., Lu, A.Y.H. and Strobel, H.W., manuscript submitted).

A percent composition analysis for β-naphthoflavone-treated 25-50 gm rats was determined by developing radial immunodiffusion assays for all five forms of cytochrome P-450. The quantitation of the various forms of cytochrome P-450 in β-naphthoflavone-treated rats by radial immunodiffusion is shown in Table 3. The value obtained by radial immunodiffusion for form 5 content in 50 gm Sprague-Dawley β-naphthoflavone-treated male rats, 70%, agrees well with the

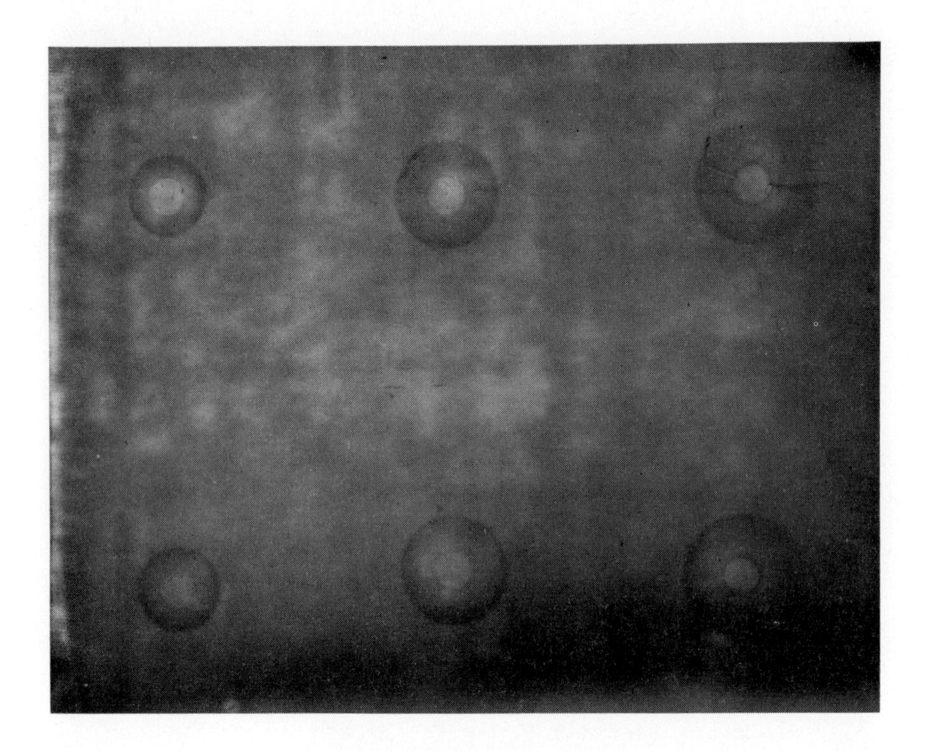

Fig. 2. Radial immunodiffusion assay using antiserum against cytochrome P-450 form 5.

value obtained by Pickett *et al.* (25) for immature Long-Evans 3-methylcholanthrene-treated male rats, 72%. When the values for each form are summed, the total composition is 158% of the specific content of cytochrome P-450 in the microsomes. Several factors may account for this overestimate. Immunological determinants shared by two or more forms of cytochrome P-450, though small, may contribute to an overestimate of each form which would, in turn, be reflected in the summation. On the other hand, it is possible that the antibody estimates include both holo- and apocytochrome P-450, whereas, the spectrophotometric quantitation measures only holoenzyme. Evidence for both of these possibilities is presented earlier in the paper. It is likely, therefore, that the overestimate by immunological methods can be attributed to several factors.

In spite of the possible overestimate of cytochrome P-450, the immunoquantitation methods can be used to provide a relative estimate of the amount of cytochromes P-450 in extrahepatic tissues. As shown in Table 4, the rocket immunoelectrophoresis assay method is used with antibody to form 5 to quanti-

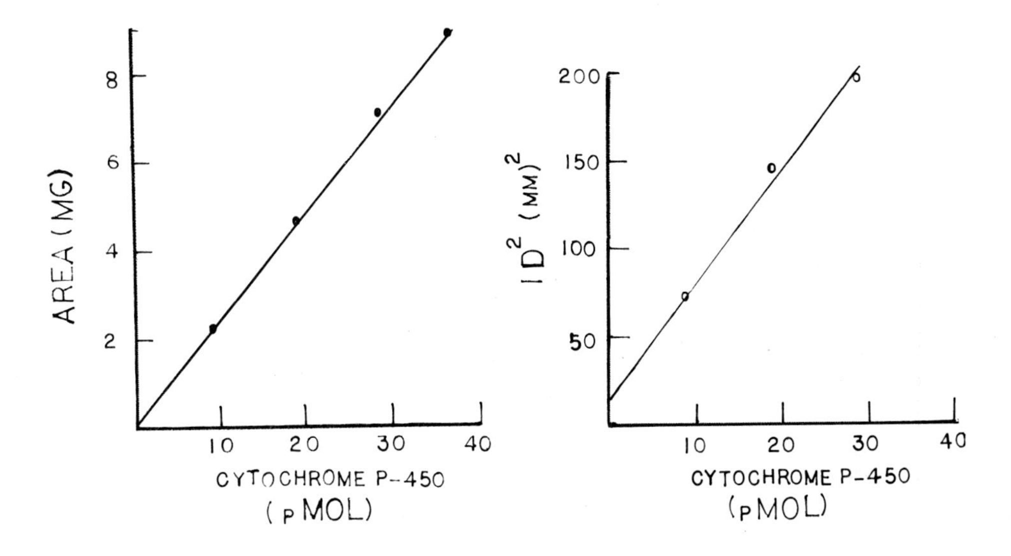

Fig. 3. Rocket immunoelectrophoresis assay standard curve.

Fig. 4. Radial immunodiffusion assay standard curve.

TABLE 3

IMMUNOQUANTITATION OF HEPATIC MICROSOMAL CYTOCHROMES P-450 FROM β-NAPHTHOFLAVONE-TREATED 50 gm RATS

Cytochrome P-450 Form:	1	2	3	4	5
Percent of Immunoprecipated P-450	21	11.5	9.6	46	70

TABLE 4

IMMUNOQUANTITATION OF CYTOCHROME P-450 FORM 5 IN EXTRAHEPATIC TISSUES OF β-NAPHTHOFLAVONE-TREATED 150 gm RATS.

Tissue	Liver	Colon	Kidney	Lung
Percent relative to liver	100	32	39	18

tate the relative amounts of form 5 in liver, colon, kidney cortex and lung microsomes. Colon and kidney microsomes from 150 gm rats have about one third the amount of form 5 found in liver, whereas, lung has only 20% as much.

The data presented suggest the utility of the immunoquantitation assays for assessing cytochrome P-450 composition in hepatic and extrahepatic tissues. In view of the possibility of overestimation, these values may be best interpreted cautiously as relative rather than absolute estimates.

ACKNOWLEDGMENT

Research supported by NCI grant CA19106, U.S. Public Health Service, DHHR.

REFERENCES

1. Gillette, J.R. (1966) Adv. Pharmacol. 4, 219-261.
2. Gram, T.E. (1980) Extrahepatic Metabolism of Drug and Other Foreign Compounds, Spectrum Press, New York.
3. Conney, A.H. (1967) Pharmacol. Rev. 19, 317-366.
4. Coon, M.J. and Vatsis, K.P. (1978) in: Gelboin, H.V. and Tso, P.D.P. (eds), Polycyclic Hydrocarbons and Cancer, Vol. I, Academic Press, New York, pp. 335-360.
5. Guengerich, F.P. (1978) J. Biol. Chem. 253, 7931-7939.
6. Guengerich, F.P. (1979) Pharmacol. Ther. A 6, 99-121.
7. Thomas, P.E., Lu, A.Y.H., Ryan, D.E., West, S.B., Kawalek, J. and Levin, W. (1976) Mol. Pharmacol. 12, 746-758.
8. Lu, A.Y.H. and West, S.B. (1980) Pharmacol. Rev. 31, 277-295.
9. Johnson, E.F. (1979) Rev. Biochem. Toxicol. 1, 1-26.
10. Daniel, R.M. and Appleby, C.A. (1972) Biochem. Biophys. Acta 275, 347-352.
11. Wang, P., Mason, P.S. and Guengerich, F.P. (1980) Arch. Biochem. Biophys. 199, 206-219.
12. Guengerich, F.P. (1977) J. Biol. Chem. 252, 3970-3979.
13. Ryan, D., Lu, A.Y.H., West, S.B. and Levin, W. (1975) J. Biol. Chem. 250, 2157-2163.
14. Ryan, D.E., Thomas, P.E., Korzeniowski, D. and Levin, W. (1979) J. Biol. Chem. 254, 1365-1374.
15. Elshourbagy, N.A. and Guzelian, P.S. (1980) J. Biol. Chem. 255, 1279-1285.
16. Saito, T. and Strobel, H.W. (1981) J. Biol. Chem. 256, 984-988.
17. Lau, P.P. and Strobel, H.W. (1982) J. Biol. Chem. 257, 5257-5262.
18. Guengerich, F.P. and Martin, M.V. (1980). Arch. Biochem. Biophys. 205, 365-379.
19. Fisher, G.J., Fukushima, H. and Gaylor, J.L. (1981). J. Biol. Chem. 256, 4388-4394.
20. Ryan, D.E., Thomas, P.E. and Levin, W. (1980) J. Biol. Chem. 255, 7941-7955.
21. Hansson, R. and Wikvall, K. (1980) J. Biol. Chem. 255, 1643-1649.
22. Dignam, J.D. and Strobel, H.W. (1977) Biochemistry 16, 1116-1123.
23. Laurel, C.-B. (1972) Scand. J. Clin. Lab. Invest. 29, 247-248.
24. Thomas, P.E., Korzeniowski, D., Ryan, D. and Levin, W. (1979) Arch. Biochem. Biophys. 192, 524-532.
25. Pickett, C.B., Jeter, R.L., Morin, J. and Lu, A.Y.H. (1981) J. Biol. Chem. 8815-8820.

Cytochrome P-450, Biochemistry, Biophysics and Environmental Implications, E. Hietanen, M. Laitinen and O. Hänninen editors

PRIMARY AND SECONDARY HYDROXYLATION PATTERN OF ANTIPYRINE IN THE ISOLATED PERFUSED RAT LIVER PREPARATION AS AFFECTED BY 3-MC

JÖRG BÖTTCHER, HEINRICH BÄSSMANN, REINER SCHÜPPEL
Inst. Pharmakol. Techn. Univ. Braunschweig, D-33-Braunschweig

INTRODUCTION

Metabolism of antipyrine comprises four different hydroxylated phase-I-metabolites, i.e. 1. 3-hydroxymethyl-antipyrine, 2. 4-hydroxy-antipyrine, 3. norantipyrine, 4. 4,4′-dihydroxy-antipyrine(1). The series of products represents four distinct types of microsomal cytochrome-P45o isoenzyme-mediated hydroxylation reactions. Antipyrine thus serves as a tool for assessing both capacity and specificity of hepatic MFO reactions and their interrelationship.

In the present study, the isolated perfused rat liver preparation has been used to examine the effects of 3-methylcholanthrene(3-MC) on the complete metabolite profile and the time course of primary vs. secondary hydroxylating processes involving the antipyrine molecule.

MATERIAL AND METHODS

Animals Male Sprague-Dawley rats (32o-37o g) were used (Lippische Versuchstieranstalt, Extertal, FRG), which were kept on Altromin^R standard chow and tap water.

Pretreatment 1 x 4o mg/kg 3-MC dissolved in olive oil was given i.p. one week before the experiment. Controls received vehicle.

Perfusion Hepatectomy and perfusion techniques have been performed according to (2). A recirculating, semisynthetic medium (3) was used containing washed bovine erythrocytes (HB = 10%). Temperature in the perfusion chamber was kept at 37° C.

Experiments were conducted for 8 h in control liver and for 6 h in 3-MC-treated liver. Initial concentration of 3-^{14}C-antipyrine was 5.o mg/1oo ml perfusate (= 6.o uCi). Samples (o.5 ml) were drawn from the blood reservoir at intervals as given in Fig. 2 and analysed for unchanged antipyrine(A) and 3-hydroxymethyl-antipyrine(3-HMA), 4-hydroxy-antipyrine(4-HA), norantipyrine(NORA), 4,4′-dihydroxy-antipyrine(4,4′-DHA) and 3-carboxy-antipyrine(3-CA)

Assay Concentration of unchanged antipyrine and of metabolites was determined, after acid hydrolysis and TLC-separation, by LSC (4, 5). Results are expressed as % of radioactivity recovered.

RESULTS AND DISCUSSION

Metabolite Profile

Metabolite profiles from perfusate of control and 3-MC-treated liver are given in Fig. 1. For comparison of metabolite pattern generated in either preparation, two sampling periods were chosen from each experiment: 2 h and 8 h in the control experiment and o.5 h and 6 h in the 3-MC-treated preparation, in order to adjust the amount of antipyrine being metabolized to an approximately equal value in both systems. Thus, different perfusion times have to be compared reflecting the difference in MFO activity in both preparations.

In both pairs of samples (o.5 h vs. 2 h and 6 h vs. 8 h), the most prominent effect of 3-MC-treatment is a sharp decrease of 3-hydroxymethyl-antipyrine being formed, the amount of which is depressed almost to zero. Correspondingly, 3-carboxy-antipyrine, originating from 3-hydroxymethyl-antipyrine, is diminished below detection limits.

In contrast, concentration of total 4-hydroxy-antipyrine and of norantipyrine is markedly augmented in both samples from 3-MC-treated liver vs. control.

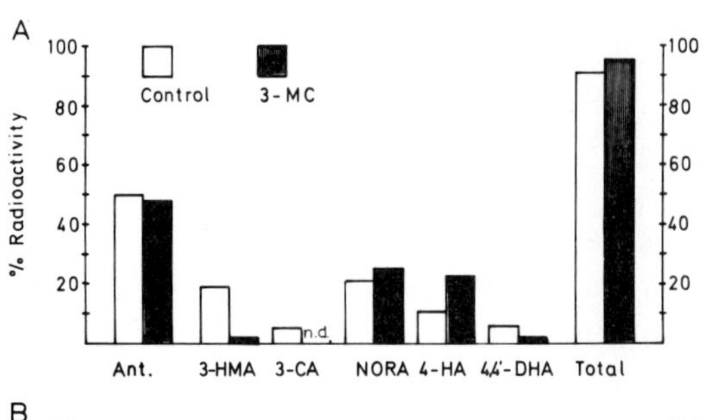

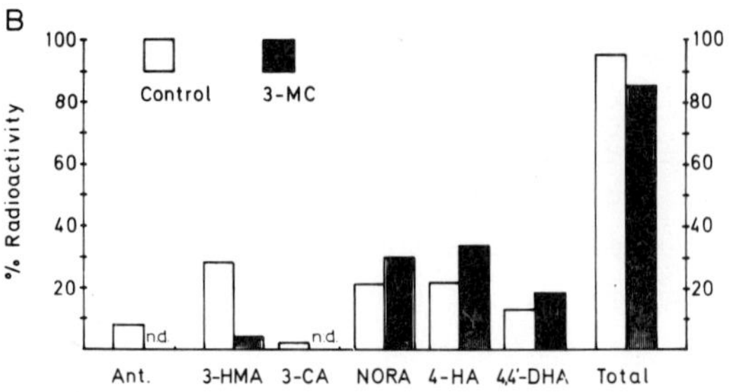

Fig. 1 Profiles of 3-^{14}C-antipyrine and metabolites in perfusate from control and 3-MC-treated liver. A: 2 h´s perfusion in control liver vs. o.5 h´s perfusion in 3-MC-treated liver. B: 8 h´s perfusion in control liver vs. 6 h´s perfusion in 3-MC-treated liver. n.d. = not detected.

These effects of 3-MC are exactly resembling those found in the urinary metabolite pattern of antipyrine in 3-MC-treated, intact rats (6, 7).

Kinetics of Primary vs. Secondary Hydroxylation

The formation of 4-hydroxy-antipyrine and of 4,4'-dihydroxy-antipyrine as main metabolites of antipyrine in the rat may be considered as a two-step process, involving two different hydroxylating reactions within the hepatic ER, attacking either the C-4 or the C-4' (p)-position of the antipyrine molecule, eventually in a fixed sequence. Denoting such a concept, 4-hydroxy-antipyrine has been tentatively adressed as a primary hydroxylated metabolite of antipyrine, whereas 4,4'-dihydroxy-antipyrine has been classified as a secondary hydroxylated metabolite (1).
To further substantiate this concept, the time course of the formation of total 4-hydroxy-antipyrine and of total 4,4'-dihydroxy-antipyrine has been followed in both preparations. Results are givenin Fig. 2. In the perfusate from control liver, there is a steady and parallel increase in the concentration of both metabolites. At the end of the experiment (8 h), there is a substantial amount of antipyrine left unmetabolized.

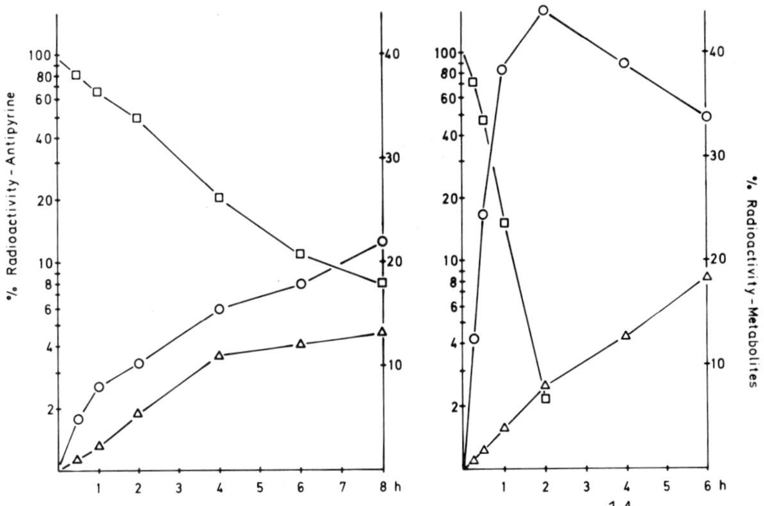

Fig. 2 Time course of concentration of 3-^{14}C-antipyrine (-□-□-) and of total 4-hydroxy-antipyrine (-o-o-) and of total 4,4'-dihydroxy-antipyrine (-△-△-) in liver perfusate. Left panel: control liver, Right panel: 3-MC-treated liver

In contrast, perfusate of 3-MC-treated liver shows a significant dissociation in the time course of either metabolites. Concentration of total 4-hydroxy-antipyrine reaches a peak at 2 h´s perfusion, and, after antipyrine being practically removed from the system, declines steadily. Only now, production of total 4,4´-dihydroxy-antipyrine becomes significant, apparently at the expense of total 4-hydroxy-antipyrine, having been formed beforehand, as can be taken from Fig. 2 (right panel). The decrease of the latter exactly mirrors the increase of the former compound. It is concluded from this experiment, that - at least in 3-MC-treated liver - recirculating conjugates of 4-hydroxy-antipyrine, in the absence of antipyrine, will readily undergo secondary hydroxylation to form conjugates of 4,4´-dihydroxy-antipyrine.
At the beginning of the experiment, high activity of cytochrome-P-448-mediated C-4-hydroxylation of antipyrine in 3-MC-treated liver apparently obviates C-4´(p)-hydroxylation, which is dependent on cytochrome-P-45o (7). Only when the substrate antipyrine is exhausted later in the experiment, C-4´-hydroxylation comes up significantly. These observations lend support to the view, that- in principal - p-hydroxylation of antipyrine is a secondary hydroxylating process in nature.

REFERENCES

1. Bässmann,H., Böttcher, J., Schüppel, R. (1979) Naunyn-Schmiedeberg´s Arch. Pharmacol. 3o9: 2o3

2. Miller, L.L., Bly, C.G., Watson, M.L., Bale, W.F. (1951) J. exp. Med. 94: 431

3. Schimassek, H. (1962) Life Sci. 11: 629

4. Böttcher, J., Bässmann, H., Schüppel, R. (1982) Drug Metab. Disp. 1o: 9o

5. Böttcher, J., Bässmann, H., Schüppel, R. (1982) J. Pharm. Pharmacol. 34: 168

6. Danhof, M., Krom, D.P., Breimer, D.D. (1979) Xenobiotica 9: 695

7. Böttcher, J., Bässmann, H., Schüppel, R. (1982) in: R.Sato(Ed.) Proc. 5. Int. Symposium on Microsomes and Drug Oxidation, Japan Scientific Societies Press, Tokyo.

Cytochrome P-450, Biochemistry, Biophysics
and Environmental Implications, E. Hietanen,
M. Laitinen and O. Hänninen editors

STUDIES ON THE PURIFICATION AND PROPERTIES OF CYTOCHROME P-450 FRACTIONS ACTIVE IN CHOLESTEROL 7α-HYDROXYLATION

HANS BOSTRÖM, KERSTIN LUNDELL AND KJELL WIKVALL
Department of Pharmaceutical Biochemistry, University of Uppsala, Uppsala, Sweden

INTRODUCTION

The first and rate-limiting step in the biosynthesis of bile acids is the 7α-hydroxylation of cholesterol. The reaction is catalyzed by a cytochrome P-450 dependent hydroxylase in liver microsomes (1). Attempts made in this laboratory to identify and isolate from rat and rabbit liver microsomes the cytochrome P-450 responsible for 7α-hydroxylation of cholesterol have shown that cholesterol 7α-hydroxylase activity is present in electrophoretically homogeneous cytochrome P-450 LM_4 from cholestyramine-treated rabbits (2). This fraction catalyzes 7α-hydroxylation of cholesterol more efficiently than rabbit liver microsomes but less efficiently than rat liver microsomes.

The present communication reports studies on the isolation and properties of cytochrome P-450 fractions active in cholesterol 7α-hydroxylation prepared from rabbit and rat liver microsomes according to different methods.

MATERIALS AND METHODS

Male New Zealand rabbits, weighing 2 to 3 kg, and rats of the Sprague-Dawley strain, weighing about 250 g were treated with cholestyramine as described previously (2).

Cytochrome P-450 LM_4 from rabbit liver microsomes was prepared as described by Coon and associates (3). The same procedure was applied to rat liver microsomes and the 300 mM potassium phosphate eluate from hydroxylapatite chromatography was used as the final cytochrome P-450 fraction. Cytochrome P-450 LM_4 from rabbit liver microsomes was further purified on a column of octylamine-Sepharose. The chromatographical conditions were those described by Imai, Sato *et al.*(4). Cytochrome P-450 LM_4I was eluted with the equilibrating buffer containing 0.46% sodium cholate. Cytochrome P-450 LM_4II was eluted with the equilibrating buffer containg 0.37% sodium cholate and 0.06% Emulgen 913. The fractions were treated with Amberlite XAD-2 and calcium phosphate (3) in order to remove Emulgen.

A cytochrome P-450 fraction was prepared from rabbit liver microsomes according to the methods described by Imai and Sato (4) for the isolation of cytochrome $P-448_1$. The same procedure was applied to rat liver microsomes with minor modifications. The fraction eluted with 300 mM phosphate in hydroxylapatite chromatography was used as the final cytochrome P-450 fraction.

NADPH-cytochrome P-450 reductase from phenobarbital-treated rabbits and rats was prepared as described by Yasukochi and Masters (5).

Optical spectra were measured with calcium phosphate-treated enzyme fractions as described by Haugen and Coon (3).

Gel electrophoresis was performed in the presence of sodium dodecyl sulfate as described by Laemmli (6) or with Pharmacia gradient slab gels, PAA 4/30, as described previously (2).

Incubation procedures and analyses of the incubation mixtures were the same as described previously (2).

RESULTS AND DISCUSSION

Cytochrome P-450 LM_4, cytochrome P-450 LM_4I, cytochrome P-450 LM_4II and cytochrome $P-448_1$ from rabbit liver all showed a single protein band with an apparent M_r= 53,000 upon gel electrophoresis. The cytochrome P-450 fraction prepared from rat liver microsomes according to the procedures of Imai and Sato (4) showed a major protein band with an apparent M_r=50,000. The cytochrome P-450 fraction prepared from rat liver microsomes according to the procedures of Coon and associates gave at least four protein bands upon gel electrophoresis. This fraction catalyzed efficient 7α-hydroxylation of cholesterol but contained only 1.5 nmol of cytchrome P-450/mg protein and was not characterized further.

Table 1 shows some properties of the electrophoretically pure cytochrome P-450 fractions. Cytochrome P-450 LM_4, cytochrome P-450 LM_4I, cytochrome P-450 LM_4II and cytochrome $P-448_1$ from rabbit liver and the cytochrome P-450 fraction from rat liver contained 15.2, 18.8, 10.8, 17.9 and 3.1 nmol of cytochrome P-450 per mg protein, respectively. The low specific cytochrome P-450 content of the fraction from rat liver is not consistent with the purity observed upon gel electrophoresis. It has been reported by Haugen and Coon that heme is lost in the preparation of cytochrome P-450 LM_4 from untreated rabbits (3). Heme losses are the most likely explanation for the lower cytochrome P-450 content in the fraction from rat liver microsomes as well as for cytochrome P-450 LM_4II.

The cytochrome P-450 fractions from rabbit liver microsomes all showed similar spectral properties and were in the high spin form. On the other hand, the cytochrome P-450 from rat liver microsomes showed absorption maxima at 418, 530 and 560 nm in the oxidized state and was thus in the low spin form.

TABLE 1

PROPERTIES OF CYTOCHROME P-450 FRACTIONS ISOLATED FROM LIVER MICROSOMES OF CHOLESTYRAMINE-TREATED RABBITS AND RATS

Fraction	Cytochrome P-450	Spectral properties	Cholesterol 7α-hydroxylation
	nmol/mg protein		pmol/nmol cyt. P-450/min
Rabbit			
Cytochrome P-450 LM_4	15.2	high spin	15
Cytochrome P-450 LM_4I	18.8	high spin	≤1
Cytochrome P-450 LM_4II	10.8	high spin	35
Cytochrome P-448_1	17.9	high spin	60
Rat			
Cytochrome P-450	3.1	low spin	90

The catalytic activities of the cytochrome P-450 fractions with respect to 7α-hydroxylation of cholesterol are summarized in Table 1. All cytochrome P-450 fractions catalyzed efficient 7α-hydroxylation except for cytochrome P-450 LM_4I. The highest activity was observed with the cytochrome P-450 fraction prepared according to Sato and associates from rat liver microsomes. Similarily, the cytochrome P-448_1 fraction prepared according to the same methods was the most active among the cytochrome P-450 fractions from rabbit liver. It should be mentioned that all cytochrome P-450 fractions also catalyzed hydroxylations of other C_{27}-steroids as well as testosterone.

The two subfractions of cytochrome P-450 LM_4, cytochrome P-450 LM_4I and cytochrome P-450 LM_4II, showed quite different cholesterol 7α-hydroxylase activity. Cytochrome P-450 LM_4I was inactive in cholesterol 7α-hydroxylation but catalyzed hydroxylations of other substrates. The substrate specificity of this fraction is similar to that of cytochrome P-450 LM_4 prepared from untreated and phenobarbital-treated rabbits as reported previously (2,7). On the other hand, cytochrome P-450 LM_4II was twice as efficient as the original cytochrome P-450 LM_4 fraction in cholesterol 7α-hydroxylation. The two subfractions had similar spectral properties and the same apparent molecular weight but differed in amino acid composition.

Taken together the results of the present study show that cytochrome P-450 fractions of high purity and with efficient cholesterol 7α-hydroxylase activity can be purified from rabbit as well as rat liver microsomes. There are apparently problems with heme losses especially in preparation of the cytochrome P-450 from rat liver microsomes. Finally, the subfractionation of cytochrome P-450 LM_4 into fractions with quite different activity towards cholesterol shows that cytochrome P-450 LM_4 from cholestyramine-treated rabbits is heterogeneous and that one component of this fraction is apparently cholesterol 7α-hydroxylase.

ACKNOWLEDGMENTS

The skilful technical assistance of Mrs Britt-Marie Johansson and Mrs Angela Lannerbro is gratefully acknowledged. This work was supported by the Swedish Medical Research Council (Project 03X-218) and by the Loo and Hans Osterman Foundation.

REFERENCES

1. Danielsson, H. and Sjövall, J. (1975) Annu. Rev. Biochem. 44, 233
2. Boström, H., Hansson, R., Jönsson, K.-H. and Wikvall, K. (1981) Eur.J. Biochem. 120, 29
3. Haugen, D.A. and Coon, M.J. (1976) j.Biol.Chem. 251, 7929
4. Imai, Y., Hashimoto-Yutsudo, C., Satake, H., Girardin, A. and Sato, R. (1980) J. Biochem. (Tokyo) 88, 489
5. Yasukochi, Y. and Masters, B.S.S. (1976) J.Biol.Chem. 251, 5337
6. Laemmli, U.K. (1970) Nature (lond.) 227, 680
7. Hansson, R. and Wikvall, K. (1980) J.Biol.Chem. 255, 1643

© 1982 Elsevier Biomedical Press B.V.
Cytochrome P-450, Biochemistry, Biophysics
and Environmental Implications, E. Hietanen,
M. Laitinen and O. Hänninen editors

RABBIT MICROSOMAL CYTOCHROME P-450 1: A CYTOCHROME LINKED TO VARIATIONS IN HEPATIC STEROID HORMONE METABOLISM

ERIC F. JOHNSON, HERMANN H. DIETER*, GEORGE E. SCHWAB, INGRID REUBI, and URSULA MULLER-EBERHARD
Department of Biochemistry, Scripps Clinic and Research Foundation, 10666 North Torrey Pines Road, La Jolla, CA 92037 (U.S.A.)

INTRODUCTION

Although tissue concentrations of three electrophoretically distinct types of rabbit microsomal cytochrome P-450, forms 2, 4 and 6, are known to be increased in a tissue and age dependent manner by administration of specific inducing agents (1-3), the modes of regulation of the occurrence of other types of rabbit microsomal cytochrome P-450 are incompletely known at this time. Recent work in our laboratory suggests that there is considerable intraspecies variation in the occurrence or properties of one of the electrophoretically distinct forms of cytochrome P-450, form 1. This cytochrome is unusual because it catalyzes the conversion of progesterone to deoxycorticosterone in the liver.

MATERIALS AND METHODS

Form 1 was isolated from the unbound material obtained from the step employing DEAE-cellulose chromatography in the procedure for the preparation of P-450 3b (4). Form 1 was purified further by chromatography on HA-agarose (LKB) using a gradient in potassium phosphate, pH 7.4, of 0.03 to 0.3 M for elution. This buffer also contained 20% glycerol, 0.1 mM EDTA and 0.3% Nonidet P-450. Procedures for removal of detergent from the final preparation were similar to those used with form 3b (4).

Methods for the reconstitution of the cytochromes and the assays for the formation of the metabolites of progesterone are described elsewhere, as are procedures for preparation of microsomes (1-5). When the metabolism of 17b-estradiol was studied, the reaction mixture was supplemented with 0.4 mM ascorbic acid to prevent the oxidation of catechol estrogens formed in

*Present Addresses: H. H. Dieter, Institute for Toxicology, University of Dusseldorf, W. Germany. I. Reubi, School of Medicine, University of Bern, Switzerland. U. Muller-Eberhard, Department of Pediatrics, Cornell Medical Center, New York, N.Y. U.S.A.

the reaction. [^{14}C]Estradiol and its metabolites were extracted from the reaction mixture into ethylacetate and were separated by thin-layer chromatography on silica gel plates previously treated with ascorbic acid as described by Gelbke and Knuppen (6).

[^{14}C]Progesterone and [^{14}C]estradiol were purchased from Amersham, and silca gel plates (IB2-F) were obtained from J. T. Baker Chemicals. Other steroidal compounds were purchased from Sigma. Rabbits of strain IIIVO/J were supplied by the Jackson Laboratories.

RESULTS

Of the various forms of cytochrome P-450 that have been purified in our laboratory from rabbit liver microsomes, form 1 is unique inasmuch as it catalyzes the conversion of estradiol and progesterone to 2-hydroxyestradiol and 21-hydroxyprogesterone respectively, each of which exhibit hormonal actions. Form 1 exhibits a rate of 5.8 μM/min/μM for the 21-hydroxylation of progesterone (10μM) which exceeded the rate exhibited by any of five other cytochromes tested, 2, 3b, 3c, 4, and 6, by more than 30-fold (5). The rate for the 2-hydroxylation of estradiol (10 μM) measured for form 1 is 3.8 μM/min/μM which also exceeded that seen for any of the other cytochromes but to a lesser degree (8-fold). Forms 4 and 6 (3) exhibited the next highest activities (ca. 0.5 μM/min/μM).

The rates for liver microsomal 21-hydroxylation of progesterone are seen to vary among outbred New Zealand White (NZW) rabbits (7). The rates of 21-hydroxylase activity displayed by 23 preparations of microsomes obtained from individual rabbits were distributed in a bimodal fashion with 8 clustered about a mean value of 3.8 μM/min/mg protein and the remainder clustered about a value of 0.36 μM/min/mg protein. This 10-fold difference exceeds those routinely seen for other activities following the induction of forms of P-450 by phenobarbital or TCDD. Male and female rabbits display both the high and the low 21-hydroxylase activity. However, the high 21-hydroxylase activity was observed more frequently among females (6 of 11) than males (2 of 12). The number of animals is not sufficient however to determine if this difference between the sexes is significant. Sex-related differences in the activities of the cytochrome P-450 monooxygenases have not been widely observed among rabbits, but have been extensively documented in the rat (8,9). In the latter species, these differences are imprinted by neonatal exposure to androgens. However, other factors are likely to be involved in the differences reported here since the differences in progesterone 21-hydroxylase are seen for both sexes in the rabbit.

The differences in progesterone 21-hydroxylase activity observed between animals may reflect differences in the occurrence or properties of P-450 1. Further support for this possibility is provided by the observation that the estradiol 2-hydroxylase activity is elevated along with progesterone 21-hydroxylase activity. In addition, P-450 1 can be isolated from animals exhibiting high 21-hydroxylase activity in greater amounts and in a more highly purified form than from animals exhibiting the low activity. However, preparations from the low 21-hydroxylase microsomes do contain P-450 1, although these preparations exhibit lower catalytic activities. We are currently characterizing monoclonal antibodies to form 1 with which to quantitate differences in the occurrence of this cytochrome and characterize potential polymorphism that may underlie the differences in activity between animals.

DISCUSSION

The product of the 21-hydroxylation of progesterone, deoxycorticosterone, can affect the retention of sodium. During the last trimester of pregnancy, the secretion of progesterone is greatly increased, and plasma levels of deoxycorticosterone rise in parallel with that of progesterone (10). However, plasma concentrations of deoxycorticosterone are not responsive to alterations in adrenal output suggesting that much of the deoxycorticosterone may be formed from circulating progesterone in extra-adrenal tissues, and thus, the production of deoxycorticosterone may not be responsive to the homeostatic mechanisms which regulate the production of aldosterone and control blood pressure.

Hypertensive conditions represent one of the major complications of pregnancy (10). The extent to which environmental, hormonal, or heritable factors which alter the extra-adrenal conversion of progesterone to deoxycorticosterone may contribute to these differences is unknown. Winkel *et al.* (11) have noted ten-fold differences in the fractional conversion of progesterone to deoxycorticosterone in both male and female human subjects. Thus, differences in 21-hydroxylase activity which have been observed among NZW rabbits may mimic a situation which occurs in the human population.

ACKNOWLEDGMENTS

This research was supported by a grant from the USPHS HD 04445. Dr. Dieter received an award from the Deutsche Forschungsgemeinschaft, and Dr. Reubi received awards from the Hartmann-Muller Stiftung of the Medical School of the University of Bern and the Janggen-Pohn Stiftung.

REFERENCES

1. Schwab, G.E., Norman, R.L., Muller-Eberhard, U., and Johnson, E.F. (1980) Mol. Pharmacol. 17, 218.

2. Norman, R.L., Johnson, E.F., and Muller-Eberhard, U. (1978) J. Biol. Chem. 253, 8640.

3. Liem, H., Muller-Eberhard, U., and Johnson, E.F. (1980) Mol. Pharmacol. 18, 565.

4. Dieter, H.H. and Johnson, E.F. (1982) J. Biol. Chem., in press.

5. Dieter, H.H., Muller-Eberhard, U., and Johnson, E.F. (1982) Biochem Biophys. Res. Commun. 102, 515.

6. Gelbke, H.P. and Knuppen, R. (1972) J. Chromat. 71, 465.

7. Dieter, H.H., Muller-Eberhard, U., and Johnson, E.F., submitted for publication.

8. Kato, R. (1974) Drug Metab. Rev. 3, 1.

9. Skett, P. and Gustafsson, J.-A (1979) Rev. Biochem. Toxicol. 1, 27 .

10. Pritchard, J.A. and MacDonald, P.C. In: Williams Obstetrics, Appleton-Century-Crofts, New York, pp. 665-700, 1980.

11. Winkel, C.A., Milewich, L., Parker, C.R., Gant, N.F., Simpson, E.R., and MacDonald, P.C. (1980) J. Clin. Invest. 66, 803.

Cytochrome P-450, Biochemistry, Biophysics
and Environmental Implications, E. Hietanen,
M. Laitinen and O. Hänninen editors

POLYCHLORINATED BIPHENYLS AS PHENOBARBITONE-TYPE INDUCERS: STRUCTURE-ACTIVITY CORRELATIONS

S. SAFE[1], M.A. CAMPBELL[2], I. LAMBERT[1] and L. COPP[2]
[1]Department of Physiology and Pharmacology, College of Veterinary Medicine, Texas A&M University, College Station, Texas 77843 (U.S.A.) and [2]Guelph-Waterloo Centre for Graduate Work in Chemistry, Department of Chemistry, Univeristy of Guelph, Guelph Ontario, Canada.

INTRODUCTION

Numerous xenobiotics including several polychlorinated biphenyl (PCB) congeners have been classified as phenobarbitone (PB)-type inducers of the hepatic drug-metabolizing enzymes (1-5). PB administered to rodents preferentially induces numerous cytochrome P-450 dependent monooxygenase activities, and produces a characteristic alteration in the relative concentrations of specific cytochrome P-450 isozymes (6). Although there are no apparent structure-activity rules for PCBs as PB-type inducers (3), the most active congeners contain a 2,3,4-trichloro-, 2,4,5-trichloro- or 2,4-dichloro- substitution pattern on at least one of the phenyl rings. This study reports the effects of structure on the activity of 2-chloro-, 4-chloro- and 2,4-dichloro- substituted PCBs as PB-type inducers, and determines the substitution patterns on the second more highly chlorinated phenyl which maximizes the inducing activity. 2,4-Disubstituted PCBs were chosen as a model since substituted halogenated biphenyls with diverse functional groups at the 2 and 4 positions can be synthesized and used as probes in order to more accurately define the effects of structure on activity.

MATERIALS AND METHODS

Chemical synthesis. The PCB congeners used in this study were prepared by the Cadogan coupling of the 2,4-dichloro-, 2-chloro- or 4-chloroanilines with excess 1,3,5-tri-, 1,2,3,5-tetra-, 1,2,3,4-tetra-, 1,2,4,5-tetra- and 1,2,3,4,5-pentachlorobenzene. The resulting PCBs were purified to > 98-99% as previously described (3-5).

Animal treatment and preparation of microsomes. Immature male Wistar rats (4-6 per group) were pretreated with the appropriate PCB congener, in corn oil, on days 1 and 3 and the animals were killed on day 6. An initial high dose of 100 $\mu mol \cdot kg^{-1}$ was used for all the PCB congeners and a lower concentration, 20 $\mu mol \cdot kg^{-1}$, was used in a second set of experiments for all PCBs exhibiting any significant induction of PB-type microsomal enzyme activity. Hepatic 100,000 x g microsomes were prepared as described (3-5).

Biochemical assays. The enzymic [eg. benzo[a]pyrene hydroxylase, NADPH cytochrome C reductase, dimethylaminoantipyrine (DMAP) N-demethylase and aldrin epoxidase], spectral and electrophoretic properties of the induced microsomes were determined as described (3,4,5,7).

RESULTS AND DISCUSSION

A summary of the chemicals used in this study is given in Figure 1. With the exception of one PCB congener, namely PCB-IV(X = Cl, Y = H), all of these compounds contained two or more ortho-chloro substituents to ensure that the AHH-inducing activity of these compounds was minimized (3). The 2,3,4,4',5-pentachlorobiphenyl has previously been reported to be a mixed-type inducer of microsomal monooxygenases (3). The results indicated that the 2-chloro- substituted PCBs, irrespective of the substitution pattern on the more highly chlorinated phenyl ring, did not elicit a PB-type induction pattern after administration of the 100 $\mu mol \cdot kg^{-1}$ dose. With the exception of 2,4,4',6-tetrachlorobiphenyl (PCB-I, X = Cl, X = H), administration of 100 $\mu mol \cdot kg^{-1}$ of the remaining PCB isomers and congeners produced an enhancement of microsomal DMAP N-demethylase and aldrin epoxidase activity consistent with that elicited by PB. In addition, when SDS slab gel electrophoresis was performed on microsomal proteins isolated from animals pretreated with PB and these PCB congeners, a similar intensification of a protein staining band at Mr 52,000 was exhibited.

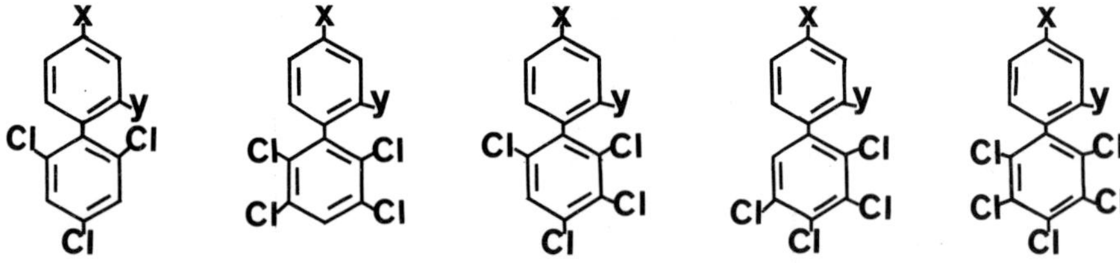

I II III IV V

i) X = Cl, Y = H

ii) X = H, Y = Cl

iii) X,Y = Cl

Figure 1.

Table 1 summarizes the effects of these compounds when administered at a lower dose (20 $\mu mol \cdot kg^{-1}$). Based on these results the following structure-activity correlations for PCBs as PB-type inducers can be formulated:

(a) the 2,4-dichloro- substituted PCBs are more active than their 2-chloro and 4-chloro analogs,

(b) the 2,4,6-trichloro substitution pattern on the more chlorinated phenyl ring gives the least active group of inducers,

(c) the activity of the PCBs as PB-type inducers increases with increasing *meta* chlorination of the more highly chlorinated phenyl ring.

TABLE 1

PCBs AS PB-TYPE INDUCERS: A SUMMARY OF ENHANCED MICROSOMAL ENZYME ACTIVITIES (20 $\mu mol \cdot kg^{-1}$)

Inducer	DMAP N-Demethylase[a]	Aldrin Epoxidase[a]
Corn oil (control)	5.65 ± 1.08	2.72 ± 0.84
PB	14.8 ± 3.2[1]	16.3 ± 6.0[1]
PCB-I (X,Y=Cl)	5.75 ± 1.22	2.79 ± 0.94
PCB-II (X=Cl,Y=H)	9.02 ± 1.30[1]	4.49 ± 0.95
PCB-II (X,Y=Cl)	9.17 ± 0.86[1]	13.4 ± 0.9[1]
PCB-III (X=Cl,Y=H)	7.82 ± 1.08	2.67 ± 0.91
PCB-III (X,Y=Cl)	7.72 ± 1.68	4.06 ± 0.46
PCB-IV (X=Cl,Y=H)	6.46 ± 0.82	2.11 ± 0.78
PCB-IV(X,Y=Cl)	10.3 ± 0.5[1]	6.64 ± 1.55[1]
PCB-V (X=Cl,Y=H)	8.91 ± 0.38[1]	12.3 ± 1.5[1]
PCB-V (X,Y=Cl)	9.62 ± 0.97[1]	11.3 ± 1.6[1]

[a]nmol product formed·mg $protein^{-1} \cdot min^{-1}$
[1]different from control at the 1% level of significance

ACKNOWLEDGEMENTS

This study was supported by the Natural Sciences and Engineering Research Council of Canada and the National Institutes of Health (1-R01-ES02798-01).

REFERENCES

1. Poland, A., Greenlee, W.F. and Kende, A.S. (1979) Ann N.Y. Acad. Sci. 320, 214-230.
2. Goldstein, J.A. (1979) Ann. N.Y. Acad. Sci. 320, 164-178.
3. Parkinson, A., Robertson, L., Safe, L. and Safe, S. (1980) Chem.-Biol. Interact., 30, 271-285.
4. Parkinson, A., Robertson, L.W., Safe, L. and Safe, S. (1980) Chem.-Biol. Interact., 35, 1-12.
5. Parkinson, A., Cockerline, R. and Safe, S. (1980) Chem.-Biol. Interact., 29, 277-289.
6. Thomas, P.E., Reick, L.M., Ryan, D.E. and Levin, W. (1981) J. Biol. Chem., 256, 1044-1052.
7. Wolff, T., Deml, E. and Wanders, H. (1979) Drug Metab. Disp., 7, 301-305.

Cytochrome P-450, Biochemistry, Biophysics
and Environmental Implications, E. Hietanen,
M. Laitinen and O. Hänninen editors

USE OF THREE-PHASE PARTITION AS THE FIRST STEP IN THE PURIFICATION OF CYTOCHROME P-450 FROM THE YEAST *BRETTANOMYCES ANOMALUS*

H. NIKKILÄ AND P.H. HYNNINEN
Department of Biochemistry, University of Kuopio, P.O. Box 138,
SF-70101 Kuopio 10 (Finland)

INTRODUCTION

Partition in nonionic polymer phase systems has been successfully used for separation of cell particles and macromolecules (1). The method is gentle and seems to be particularly suitable for the first stage fractionation of proteins on the basis of their polarity. Albertsson (2) has developed three- and four-phase polyol partition systems for the purification of hydrophobic membrane proteins which are frequently susceptible to conformational alterations during their purification. We have investigated the applicability of Albertsson's polymer phase systems to the purification of cytochrome P-450 (P-450)* from the yeast *Brettanomyces anomalus*. This method was chosen because of the low content of P-450 in this yeast and because of the known fact that polyols stabilize P-450 (3).

MATERIAL AND METHODS

Reagents. Ficoll 400 and Dextran T 500 were purchased from Pharmacia Fine Chemicals, Sweden; Polyethylene glycol 6000 (PEG), bovine serum albumin, ovalbumin and lysozyme from Sigma, U.S.A.; and Triton X-100 from BDH, England. Emulgen 911 was a gift from Kao-Atlas, Japan.

Analytical methods. Protein was measured using a modified Biuret method (4) with bovine serum albumin as a standard. The protein samples as well as the standards were first precipitated and washed with trichloroacetic acid (1). The standards were diluted in the blank PEG-phase (2). P-450 was measured according to Omura and Sato using the millimolar absorptivity of 91 $mM^{-1}cm^{-1}$ (51). The spectra were recorded with a scanning Cary model 219 spectrophotometer.

Sodium dodecyl sulphate - polyacrylamide gel electrophoresis. This was performed with minor modifications as described by Weber and Osborn (6). 7.5 % acrylamide solution was used. Bovine serum albumin (MW 66,000), ovalbumine (MW 45,000) and lysozyme (MW 14,300) were used as molecular weight markers.

---

*Abbreviations used: P-450, cytochrome P-450; PEG, polyethylene glycol; PMSF, phenylmethylsulfonylfluoride; SDS-PAGE, polyacrylamide gel electrophoresis in the presence of sodium dodecyl sulphate.

Culture conditions. *B. anomalus* (CO1) (7) was grown in the medium described by Wiseman *et al.* (8), containing 5 % glucose. Five liter Erlenmeyer flasks, four liters of medium in each, were incubated at 30°C with magnetic stirring or in a shaker (125 osc·min^{-1}). The cells were harvested at their logaritmic growth phase; *ca.* 7 grams of yeast cells were obtained per liter of medium. After washing the cells twice with cold deionized water the paste was stored at -20°C.

Breakage of the yeast cells. Ten grams (wet wt.) of yeast cells, 50 grams of glass beads (ϕ 0.45-0.5 mm) and 10 ml of buffer (0.1 M potassium phosphate buffer containing 0.1 mM dithiothreitol, 0.1 mM EDTA and 0.1 mM PMSF, pH 7.25) were shaken three times in a Braun MSK cell homogenizer (30 sec, 4000 osc·min^{-1}). After centrifugation (1250 g, 8 min) the supernatant was stored at -80°C.

Three-phase partition. Sodium chloride, 10 % Triton X-100 and 5 % Emulgen 911 were added to the 1250 g supernatant (10-14 mg protein per ml) to reach a final concentration of 1 M, 1 % and 0.2 %, respectively. The solution was gently stirred at 0°C under nitrogen for an hour. After centrifugation at 0°C (10,000 g, 15 min), the solution was mixed with a paste containing 1.1 grams of Dextran T 500, 1.2 grams of Ficoll 400, 0.8 grams of PEG and 1.9 ml of deionized water per 10 ml of the solubilizate (2). The mixture was stirred at room temperature under nitrogen to obtain a homogenous solution and then centrifugated at 20°C (10,000 g, 10 min). The "oily layer" (may be considered as a fourth phase) above the PEG-phase was collected and the proteins were precipitated with 25 % (PEG (w/v).

Microsomes. The microsomal fraction was prepared from the 1250 g supernatant by centrifugations at 10,000 g for 30 min and 105,000 g for an hour. The microsomal fraction was collected together with the fluffy layer formed over the pellet. The microsomes were diluted in the buffer (see: Breakage of the yeast cells) to obtain a protein concentration of *ca.* 10 mg·ml^{-1}. The partition was performed without adding salt to the solubilizate.

RESULTS

When partitions for 40 to 150 ml of solubilizate were performed, a purification of 25 to 60 fold with respect to the solubilizate was achieved corresponding to 0.8-2.2 nmol of P-450 per mg protein. The yield varied between 15 and 28 % (n=10). Fig. 1 presents the results of a SDS-PAGE from a preparation containing 0.8 nmol P-450 per mg protein and Table 1 shows the results of the corresponding purification. The carbon monoxide difference spectrum of a preparation containing 1.6 nmol P-450 per mg protein (Fig. 2) exhibits

TABLE 1

RESULTS FROM A THREE-PHASE PARTITION EXPERIMENT FOR THE 1250 G SUPERNATANT OF THE HOMOGENIZED YEAST CELLS.

	protein	cytochrome P-450		
	(mg)	total content (nmol)	specific content (nmol/mg)	yield (%)
Solubilizate	1608	48.0	0.030	100
The "oily layer"	13.5	10.9	0.807	22.7

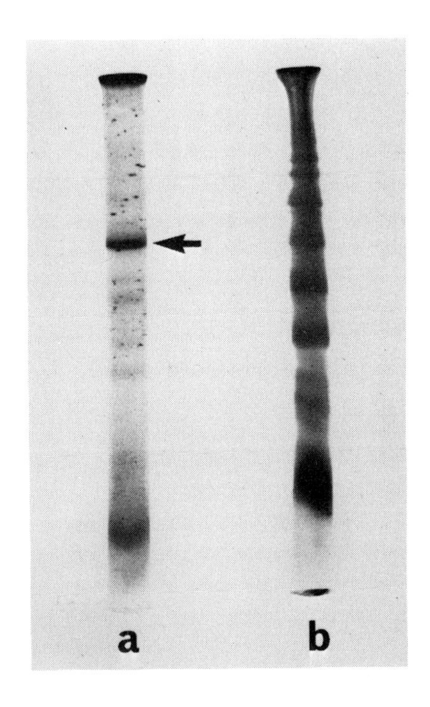

Fig. 1. SDS-PAGE of the preparations shown in Table 1. (a) The proteins precipitated from the "oily layer" (15 µg protein); (b) The solubilized 1250 g supernatant of the homogenized yeast cells (200 µg protein). The band indicated by the arrow represents an approximate molecular weight of 56,000.

ΔA
447 nm
0.06
0.04
0.02
0.00
-0.02
-0.04
380
420
460
500 nm
λ

Fig. 2. A difference spectrum of the carbon monoxide complex of reduced cytochrome P-450 partially purified from yeast cells. The precipitated preparation was diluted in 0.1 M potassium phosphate buffer (pH 7.25) containing 0.1 mM dithiothreitol, 0.1 mM EDTA and 0.1 mM PMSF (protein concentration 0.42 mg·ml^{-1}).

virtually no shoulder at 420 nm thus demonstrating that very little of the modified P-450 (P-420) was present in the preparation.

When solubilized microsomes were partitioned without adding sodium chloride, 0.5-1.0 nmol P-450 per mg protein were found in the PEG-phase with a 6 to 10 fold purification. The yield in this procedure varied between 15 and 33 % (n=3).

DISCUSSION

Variation in the purification is partly due to the heterogeneity of the starting material. Scaling up does not seem to affect either purification or yield. Losses of the enzyme during the solubilization are *ca.* 10 - 20 %. The 1 M sodium chloride is an important component in forming the phase equilibrium for the 1250 g solubilizate. It does not cause marked losses in the P-450 content when added to the solubilizate.

The aqueous phase partition with non-ionic polymers is here shown to be well applicable to the purification of the P-450. The technique is simple and rapid. It can be used with large batches of material. A drawback, however, is the relatively high cost of the polymers used.

ACKNOWLEDGEMENTS

The skilful technical assistance of Miss Raisa Malmivuori is gratefully acknowledged.

REFERENCES

1. Albertsson, P.Å. (1971) Partition of Cell Particles and Macromolecules, 2nd ed., Wiley, New York.
2. Albertsson, P.Å. (1973) Biochemistry, 12, 2525-2530.
3. Ichikawa, Y. and Yamano, T. (1967) Biochim. Biophys. Acta, 131, 490-497.
4. Stewart, P.R. (1975) Methods Cell Biol., 12, p. 114.
5. Omura, T. and Sato, R. (1964) J. Biol. Chem., 239, 2370-2378.
6. Weber, K. and Osborn, M. (1969) J. Biol. Chem., 244, 4406-4412.
7. Kärenlampi, S.O., Marin, E. and Hänninen, O.O.P. (1980) J. General Microbiol., 120, 529-533.
8. Wiseman, A., Lim, T.-K. and McCloud, C. (1975) Biochem. Soc. Trans., 3, 276-278.

Cytochrome P-450, Biochemistry, Biophysics
and Environmental Implications, E. Hietanen,
M. Laitinen and O. Hänninen editors

PARTIAL SEPARATION OF TWO DIFFERENT CYTOCHROME P-450 FROM RAT ADRENAL CORTEX MICROSOMES BY AFFINITY CHROMATOGRAPHY WITH IMMOBILIZED NADPH-CYTOCHROM C REDUCTASE AS A LIGAND

JOHAN MONTELIUS AND JAN RYDSTRÖM
Department of Biochemistry, University of Stockholm, Sweden

INTRODUCTION

Rat adrenal cortex microsomes contain, in addition to NADPH dependent and carbonmonoxide- sensitive steroid hydroxylases, a high aryl hydrocarbon hydroxylase (AHH) activity. The two polycyclic aromatic hydrocarbons benzo(a)pyrene and 7,12-dimethylbenz(a)anthracene(DMBA) are metabolized at rates three to five times greater than those of liver microsomes from untreated rats. Guenthner et al(1) have suggested that the adrenal microsomal enzyme responsible for AHH activity is similar to the 3-methylcholantrene (MC)-inducible hepatic enzyme (inhibition by α-naphtoflavone and a molecular weight around 57.000) and different from the previously characterized cytochrome P-450 (P-450) responsible for steroid hydroxylation. This suggestion is supported by the findings of Montelius et al (2) that little or no inhibition of the AHH activity was obtained with a variety of steroid hormones.

The aim of the present study is to examine the possibility of separating the AHH activity from the steroid hydroxylase activity, i.e., 21-hydroxylation of progesterone, and to clarify whether the AHH activity is associated with some other steroid hydroxylase activity.

MATERIALS AND METHODS

Preparation of the affinity gel. Purified liver NADPH-cytochrome c reductase (reductase) was dialyzed for 24h vs 2x300 volumes of 0.1M $NaHCO_3$ containing 0.5M NaCl, 8.5 ml of the dialysate containing 5.4 units of reductase (one unit= 1μmol reduced cytochrome c per min) was mixed over night at 4°C with 2.5g CNBr-activated Sepharose 4B (Pharmacia). The gel was stored in 20mM Tris-Cl buffer (pH 7.3) containing 20% glycerol and 0.1mM DTE. 10% of the added reductase activity was recovered after coupling and the gel could be stored for several months with only a slight loss of activity.

Separation procedure. Microsomes from female Sprague-Dawley rat (>180g) adrenals were solubilized with cholate and chromatographed on ω-aminooctyl-agarose (a.o.), obtained from Chem. Co. Sigma (4). Fractions containing P-450 were pooled (a.o. pool) and dialyzed over night vs 100 volumes 25mM Tris-Cl buffer (pH 7.3) containing 20% glycerol (buffer A). Emulgen 913 (final conc. 0.002%) was added to the dialysate prior to application to the reductase gel, which previouslyhad been equilibrated with buffer A containing 0.002% Emulgen 913 (buffer B). The P-450 eluted in the void volume was pooled (aff. pool I) and after washing the gel with buffer B the column was eluted with the same buffer containing 0.5 M NaCl. The salt-eluted fractions containing P-450 were pooled (aff. pool II).

Assays. Hydroxylation of progesterone and DMBA were carried out with the ^{14}C-labeled compounds in an assay mixture composed of a NADPH-generating system, 0.14 units reductase (liver), 25mM Tris-Cl buffer (pH 7.3), 5-50pmoles P-450 (added as liposomes[1] when indicated) and 100μM NADPH. After 60min the incubation was stopped and extracted with ethylacetate and the metabolites separated by TLC.

RESULTS

The a.o. pool contained, as judged by SDS-polyacrylamid gel electrophoresis, one major and three minor proteins with molecular weights of approximately 50.500, 56.500, 53.000 and 48.000, respectively (fig. 1).

Chromatography of the a.o. pool on the reductase gel appeared to bind quite selectively the 50.500 mw protein, altough a small amount of this component was also recovered in the void volume together with the three other proteins (aff. pool I, fig. 1). Salt elution of the gel resulted in the recovery of a highly purified protein with only trace amounts of the 56.500 and 53.000 proteins (aff. pool II, fig. 1).

The relatively low P-450 content in aff. pool II (Table 1), 6.44 nmoles per mg protein (theoretically 17.7 nmoles per mg P-450), indicates either that an inpurity comigrates with the

---

[1]Phospholipid vesicles were prepared by the cholate gel filtration technique (6) using a protein:lipid(dioleoylphosphatidyl-etanolamin/-cholin, 1:1) ratio of 1:10 (w/w).

P-450 and/or that inactivated P-450 is copurified.

The yield of P-450 from the hydrophobic chromatography step is usually about 60% wheras that for the affinity chromatography step varied between 50 to 80% (Table 1).

As can be seen in Table 1, the rate of hydroxylation of progesterone was drastically decreased after solubilization and a.o. chromatography. Since the yield is quite high the most plausible explanation for this decrease is that the catalytical activity is impaired, e.g., due to an improper lipid environment or unsuitable substrate. The low 21-hydroxylase activity recovered in the aff. pool II may be due to erroneous incubation conditions, or to the presence of a P-450 unrelated to 21-hydroxylase.

The specific AHH activity, i.e., hydroxylation of DMBA, decreased slightly in the a.o. pool but was greatly increased in the aff. pool I (specific activities up to 3000 pmoles per min and nmoles P-450 has been observed), in contrast the aff. pool II contained virtually no AHH activity (Table 1).

Examination of the reduced CO spectrum in the different fractions revealed that maximal absorption is changed from 449-450nm in the a.o. pool (as in the microsomes and the aff. pool II) to

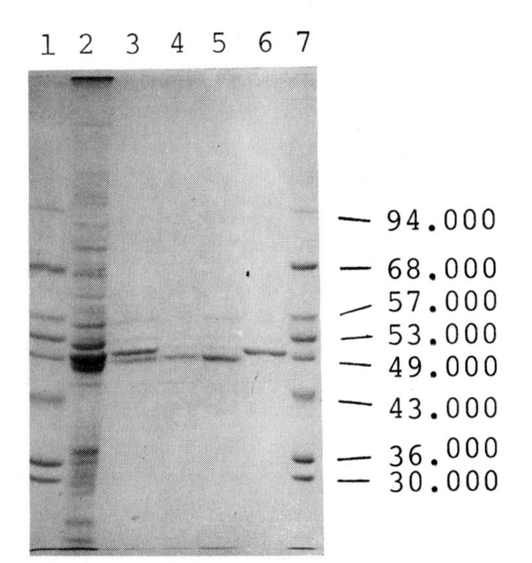

Fig. 1. 8.5% SDS-polyacrylamid gel electroforesis of the fractions indicated in Table 1: 1, reference proteins (ref.), with molecular weights indicated; 2, microsomes, 60µg; 3, a.o. pool, 6.9µg; 4, aff. pool I, 3.9µg; 5, aff. pool I, 11.6µg; 6, aff. pool II, 3.9µg; 7, ref.

TABLE 1

SEPARATION OF CYTOCHROMES P-450 FROM RAT ADRENAL

Fraction	Specific content (nmol P-450 mg^{-1})	Yield (%)	Specific activity ($pmol\ min^{-1}\ nmol\ P\text{-}450^{-1}$) Substrate Progesterone[a]	DMBA[a]
Microsomes	0.79	(100)	840	239
a.o. pool[b]	2.57	57.8	150	135
aff. pool I[b]	0.99	13.7	356	1099
aff. pool II[b]	6.44	17.6	21	9

[a]The concentration of progesterone and DMBA in the incubation mixture was 6.4μM and 7.0μM, respectively.
[b]P-450 was added as liposomes, see Materials and Methods.

447-448nm in the aff pool I. If the a.o. pool is subjected to ion exchange (not shown) this shift in absorption was accompanied by an increased relative concentration of the 56.500 protein.

These results provide additional support for the suggestion (1) that rat adrenal cortex microsomes contain a P-450-dependent AHH with a molecular weight of 56.500 and which is similar to the MC-inducible hepatic enzyme.

ACKNOWLEDGEMENT

This work was supported by the Swedish Cancer Society.

REFERENCE

1. Guenthner, T.M., Nebert, D.W. and Menard, R.H. (1979) Mol. Pharmacol., 15, 719.
2. Montelius, J., Papadopoulos, D., Bengtsson, M. and Rydström, J. (1982) Cancer Res. 42, 1479.
3. Guengerich, F.P. and Martin, M.V. (1980) Arch. Biochem. Biophys., 205, 365.
4. Ingelman-Sundberg, M., Haaparanta, T. and Rydström, J. (1981) Biochemistry, 20, 4100.

Cytochrome P-450, Biochemistry, Biophysics
and Environmental Implications, E. Hietanen,
M. Laitinen and O. Hänninen editors

PARTIAL PURIFICATION OF A SEX SPECIFIC CYTOCHROME P-450 FROM UNTREATED FEMALE RATS

CATRIONA MACGEOCH, JAMES HALPERT AND JAN-ÅKE GUSTAFSSON
Department of Medical Nutrition, Karolinska Institute, Huddinge University Hospital F69, S-141 86 Huddinge, Sweden

INTRODUCTION

A large body of evidence now indicates that cytochrome P-450 exists in multiple forms and that the variation in the metabolism of steroids and drugs probably reflects variations in the relative proportions of functionally different forms of cytochrome P-450 (1-3).One such variation is the large difference between male and female rats with regard to their hepatic steroid metabolizing enzymes. An example of such a sex-specific enzyme present in rat liver is the microsomal 15β-hydroxylase (4). This enzyme uses steroid sulphates as its substrate and has been reported to be present only in female animals. The activity of 15β-hydroxylase in adult animals is predetermined ("imprinted") in the neonatal period when testicular androgens irreversibly suppress ("masculinize") the activity of the sulphate-specific system (5,6,7). This sexual differentiation is thought to be mediated through the hypothalamo-pituitary axis. In the present study we report the partial purification of the sex-specific 15β-hydroxylase from adult female rats.

EXPERIMENTAL PROCEDURE

Solubilization of liver microsomes from male or female rats was achieved by the addition of sodium cholate to a final concentration of 0.6%. The mixture was then centrifuged at 105,000 x g for 75 min and the supernatant was applied to a 3.2 x 15 cm n-Octylamino-Sepharose 4B column, previously equilibrated with 0.1 M potassium phosphate buffer, pH 7.5, containing 20% glycerol, 1 mM EDTA (Buffer A) and 0.6% sodium cholate. The column was washed with 300 ml Buffer A containing 0.4% sodium cholate and 0.08% Lubrol. The major cytochrome P-450 fraction from the n-Octylamino-Sepharose step was dialysed overnight against 1 l 10 mM potassium phosphate buffer, pH 7.5, containing 0.1mM EDTA, 20% glycerol, 0.1% Lubrol and 0.2% sodium cholate (Buffer B). The dialysed fraction was applied to a 2 x 40 cm DEAE-Sepharose column which had been equilibrated with 400 ml Buffer B. The flow rate was 50 ml x h^{-1}. The column was washed with 250 ml Buffer B and the cytochrome P-450 fraction eluted with a 0-100 mM sodium chloride gradient in a total volume of 1000 ml Buffer B.

The major cytochrome P-450 fraction which eluted as a sharp peak at 60 mM NaCl, was dialysed for 2 h against 1 l 20% glycerol and 0.1 mM EDTA and applied to a 1 x 20 cm column of hydroxylapatite previously equilibrated with 5 mM potassium phosphate buffer, pH 7.7, containing 20% glycerol. Washing and elution of cytochrome P-450 fractions were carried out according to Philpot and Arinç (8) except that Lubrol PX was substituted for Emulgen 913. Some cytochrome P-450 was eluted with the wash buffer and was discarded. The cytochrome P-450 peak fractions from HA II and HA III eluting at 80 mM and 150 mM potassium phosphate buffer, respectively, were pooled and treated with calcium phosphate gel. All fractions were diluted 1:1 (v/v) with 20% glycerol before being added to the gel and the final dialysed enzyme preparations were then used for reconstitution studies, protein and cytochrome P-450 determinations and polyacrylamide gel electrophoresis analysis.

Steroid disulphate 15β-hydroxylase. The assay system contained the following substances: enzyme (0.2-1 nmol P-450) in Tris-HCl buffer (100 mM, pH 7.4) with EDTA to a final volume of 2 ml, cytochrome P-450 reductase (3 U/nmol P-450), dilauroylphosphatidylcholine (50 µg), sodium deoxycholate (0.24 mM) and 5α-[1,2-^{3}H]androstane-3α,17β-diol 3,17-disulphate (10^{6} dpm: 320 nmol) in 60 µl water. The reaction was started by the addition of NADPH (500 µg). Incubation was performed at 37°C for 20 min and stopped by the addition of chloroform/methanol (2:1, v/v). The extract was evaporated to dryness, subjected to solvolysis and then applied to thin-layer chromatography in the solvent system ethylacetate/cyclohexane (6:4, v/v).

RESULTS AND DISCUSSION

The results of a typical purification are shown in Table 1. We have utilized n-Octylamino-Sepharose and DEAE-Sepharose in conjunction with hydroxylapatite. This procedure reproducibly yielded cytochrome P-450 preparations with specific contents ranging from 7.4-8.9 nmol/mg protein with an overall yield of about 5%. This represents a 12-fold purification of cytochrome P-450 from hepatic microsomes. Chromatography of the cytochrome P-450 fraction eluted from DEAE-Sepharose is effective in removing most of the contaminating proteins as well as increasing the specific content of cytochrome P-450 (Fig. 1). The results indicate that the partially purified cytochrome P-450 preparation having the highest specific content, HA II, also appeared to be the most homogeneous preparation on SDS-gel electrophoresis. Only one protein (M.Wt. = 50.000) is present in the cytochrome P-450 region and only a single contaminating protein (M.Wt. = 70.000) is evident.

TABLE I

PURIFICATION OF 15β-HYDROXYLASE ACTIVITY FROM FEMALE RAT LIVER MICROSOMES

The results are from a typical experiment

Sample	Yield of P-450 (nmol)	Yield of P-450 %	Spec. content of P-450 (nmol/mg protein)[a]	Fold of P-450 purification	Activity of 15β-hydroxylase (nmol/min/nmol P-450)[b]
Microsomes	660	100	0.6-0.7	1.0	0.71-0.75
n-Octylamino-Sepharose 4B	355	55	2.5-2.8	4.0	0.94-1.06
DEAE-Sepharose	114	17	6.2-6.6	9.7	not assayed
HA II	32	5	7.4-8.9	12	3.6-4.0
HA III	47	7	5.2-5.8	8.3	0.7-0.9

[a]Represents the range of values from several preparations.
[b]Values are from duplicate determinations.

Sex differences in drug and steroid metabolism are now well established as is the involvement of cytochrome P-450 as the enzyme system responsible for the hydroxylation(s) of these compounds. However almost all previously described cytochrome P-450-catalyzed hydroxylation rections have been found to be more efficient in male than in female rat preparations. The cytochrome P-450 dependent steroid sulphate 15β-hydroxylase fraction (HA II), partially purified in this report, is therefore of a unique type in that the corresponding preparation from male rat liver microsomes was incapable of catalyzing 15β-hydroxylation of 5α-androstane-3α,17β-diol 3,17-disulphate and did not behave similarly to the female preparation when analyzed by SDS-gel electrophoresis (Fig. 2). We suggest that the female-specific SDS band of HA II, corresponding to a molecular weight of 50,000, represents the isoenzyme of cytochrome P-450 catalyzing 15β-hydroxylation of 5α-androstane-3α,17β-diol 3,17-disulphate ("cytochrome P-$450_{15\beta}$"). It is hoped that the sex-specific cytochrome P-$450_{15\beta}$ can be further purified to homogeneity to be used in studies on mechanisms of sexual differentiation of hepatic steroid metabolism.

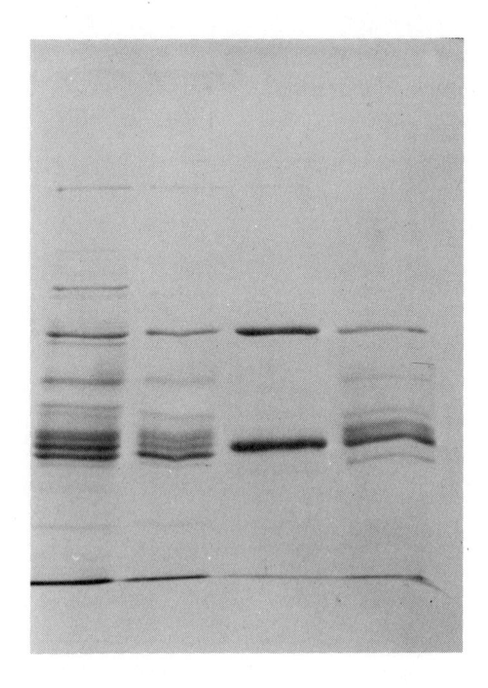

Fig. 1. Polyacrylamide gel electrophoresis of cytochrome P-450 fractions from female rat liver microsomes. The fractions were treated with sodium dodecyl sulphate and mercaptoethanol at 100° and submitted to electrophoresis in a 7.5% separation gel. The following samples were analyzed at the protein levels indicated from left to right: 1, n-octylamino-Sepharose (10 µg of protein); 2, DEAE-Sepharose (7.5 µg); 3, HA II (5 µg); 4, HA III (5 µg).

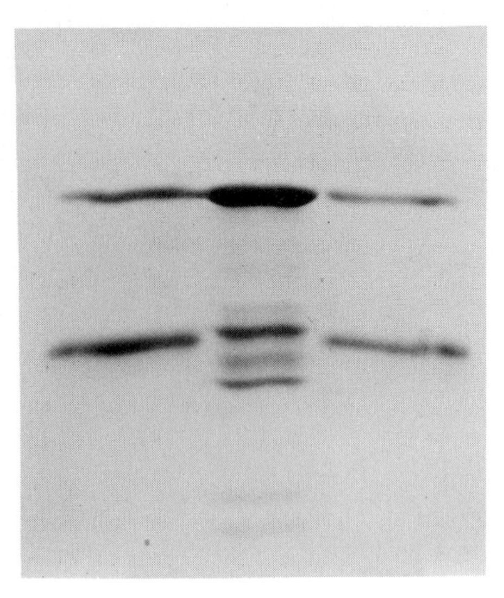

Fig. 2. Electrophoresis of partially purified male and female cytochrome P-450 as described in Methods. The different tracks represent from left to right: 1, HA II from female (10 µg of protein); 2, HA II from male (10 µg of protein); 3, HA II from female (10 µg of protein).

ACKNOWLEDGEMENTS

This work was supported by a grant from the Swedish Medical Research Council (No. 13X-2819).

REFERENCES

1. Thomas, P.E., Lu, A.Y.H., Ryan, D., West, S.B., Kawalek, J., and Levin, W. (1976) Mol. Pharmacol. 12, 746.
2. Ullrich, V., and Kremers, P. (1977) Arch. Tox. 39, 41.
3. Huang, M.T., West, S.B., and Lu, A.Y.H. (1976) J. Biol. Chem. 251, 4659.
4. Gustafsson, J.-Å., and Ingelman-Sundberg, M. (1974) J. Biol. Chem. 249, 1940.
5. Einarsson, K., Gustafsson, J.-Å., Stenberg, Å. (1973) J. Biol. Chem. 248, 4987.
6. Denef, C., and DeMoor, P. (1972) Endocrinology 91, 374.
7. Chung, L.W.K. and Chao, H. (1980) Mol. Pharmacol. 18, 543.
8. Philpot, R. and Arinç, E. (1976) Mol. Pharmacol. 12, 483.

Cytochrome P-450, Biochemistry, Biophysics
and Environmental Implications, E. Hietanen,
M. Laitinen and O. Hänninen editors

THE EFFECT OF TEMPERATURE ON THE DEUTERIUM ISOTOPE EFFECT OF THE CYTOCHROME P-450-CATALYZED O-DEETHYLATION OF 7-ETHOXYCOUMARIN

A. Y. H. LU AND G. T. MIWA
Department of Animal Drug Metabolism, Merck Sharp & Dohme Research
Laboratories, Rahway, NJ 07065 (U.S.A.)

INTRODUCTION

We have recently observed that substitution of deuterium for the α-hydrogen which undergoes bond cleavage during the O-deethylation of 7-ethoxycoumarin (7EC) results in a deuterium isotope effect and metabolic switching (1). Based on this and other results, we have proposed that the conversion of the $P450 \cdot O_2 \cdot$substrate complex to the "active oxygen species" is irreversible (2). If this is the case, then the perturbation of any steps prior to the generation of "active oxygen species" (such as electron transport, protein-protein interaction and lipid phase transition) should not affect the observed isotope effect since the cleavage of the C-H bond occurs after this irreversible step. One way to perturb the cytochrome P-450 system is to decrease the reaction rates by lowering the incubation temperature and thus decrease electron transport and protein-protein interaction. In this report, the deuterium isotope effect in cytochrome P-450-catalyzed O-deethylation of 7EC was determined as a function of temperature in order to further test our hypothesis.

MATERIALS AND METHODS

Cytochrome P-448 from 3-methylcholanthrene (3-MC)-treated rats was purified as previously described (3). The O-deethylation of non-deuterated 7-ethoxycoumarin (d_o-7EC) and deuterated 7-ethoxycoumarin (d_2-7EC) was measured fluorimetrically (4). The catalytic activity of purified cytochrome P-448 was assayed either in a non-membranous reconstituted system or in a liposomal system following the incorporation of both P-448 and the reductase into the liposomal membrane by a cholate dialysis method (5).

RESULTS AND DISCUSSION

Previous studies have established that substitution of the two hydrogens at the α-position of 7EC with deuterium results in an isotope effect in V_m but not in K_m (6). Thus, all temperature studies were carried out with a single concentration (0.25 mM) of d_o-7EC and d_2-7EC. Figure 1 shows that O-deethylation of both d_o-7EC and d_2-7EC by liver microsomes from 3-MC-treated rats and the cytochrome P-448 egg yolk liposomal system exhibited a change in activation energy at 22° to 24° in the Arrhenius plots confirming the results obtained by other investigators (7,8). Furthermore, over this temperature range the rate changed by about 10-fold. However, the deuterium isotope effects observed at

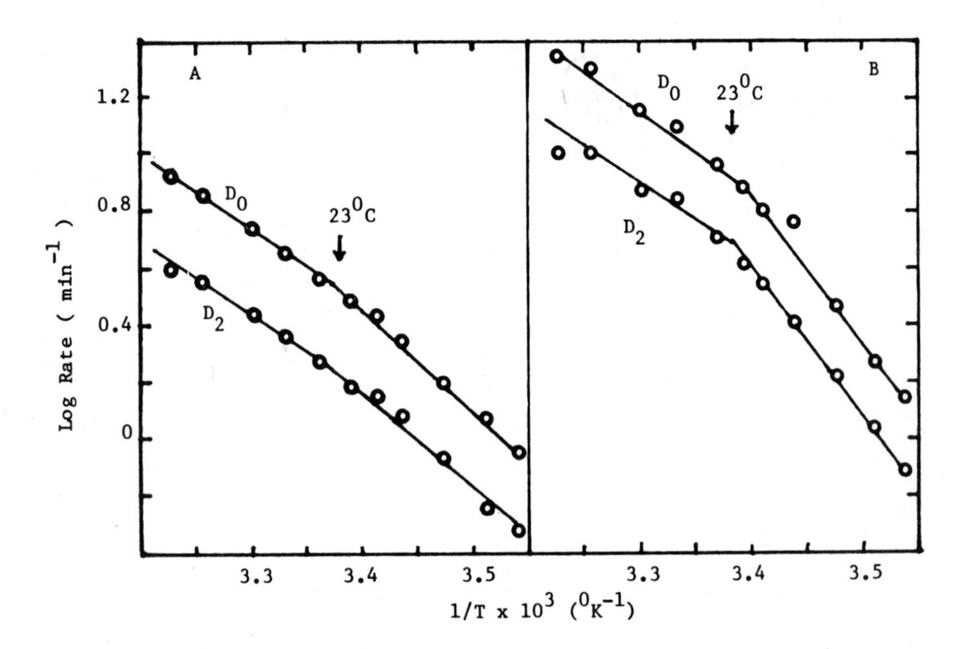

Fig. 1. Effect of temperature on the O-deethylation of d_o-7EC and d_2-7EC catalyzed by liver microsomes from 3-MC-treated rats (A) and the cytochrome P-448 egg yolk liposomal system (B).

all incubation temperatures were virtually identical resulting in two parallel plots. An examination of two other cytochrome P-448 liposomal systems (Dipalmitoyl phosphatidylcholine and dimyristoyl phosphatidylcholine) and a non-membranous reconstituted cytochrome P-448 system gave virtually the same results (Table 1).

In addition to the cytochrome P-448 system, we have also examined two other systems which gave rather large deuterium isotope effects. As shown in Table 1, isotope effects of 3.6 and 5 were observed for liver microsomes from PB-treated rats and 3-MC-treated hamster, respectively. These large isotope effect values suggest that the C-H bond cleavage step is one of the major rate-determining steps in these systems. Again, a break in the Arrhenius plots was observed at 22° to 24°, but the deuterium isotope effect in the O-deethylation of 7EC was either unchanged or only slightly decreased at temperatures below the break. The lack of a change in isotope effects even when the cytochrome P-450 system is greatly perturbed by lowering the incubation temperature is consistent

TABLE 1

EFFECT OF TEMPERATURE ON DEUTERIUM ISOTOPE EFFECT IN 7EC O-DEETHYLATION BY VARIOUS ENZYME SYSTEMS

System	k_H/k_D ($\pm$ S.D.)	
	37° to 24°	20° to 10°
A. Cytochrome P-448 system:		
Rat liver microsomes (3-MC)	2.00 $\pm$ 0.07	1.93 $\pm$ 0.07
Egg yolk PC liposomes	1.92 $\pm$ 0.07	1.78 $\pm$ 0.04
Dipalmitoyl PC liposomes	2.10 $\pm$ 0.09	1.90 $\pm$ 0.08
Dimyristoyl PC liposomes	2.23 $\pm$ 0.08	2.02 $\pm$ 0.08
Non-membranous system	2.13 $\pm$ 0.09	1.80 $\pm$ 0.09
B. Rat liver microsomes (PB)	3.66 $\pm$ 0.10	3.61 $\pm$ 0.05
C. Hamster liver microsomes (3-MC)	5.03 $\pm$ 0.55	4.42 $\pm$ 0.32

with our proposed model (2) which predicts that perturbation of any steps prior to the generation of "active oxygen species" should not affect the observed isotope effect.

ACKNOWLEDGEMENT

We wish to thank Ms. D. Sloan for her assistance in the preparation of this manuscript.

REFERENCES

1. Harada, N., Miwa, G. T. and Lu, A. Y. H. (1982) Fed. Proc. 41, 1405.

2. Lu, A. Y. H., Harada, N., Walsh, J. S. A. and Miwa, G. T., this proceeding.

3. West, S. B., Huang, M. T., Miwa, G. T. and Lu, A. Y. H. (1979) Arch. Biochem. Biophys. 193, 42-50.

4. Greenlee, W. F. and Poland, A. (1978) J. Pharmacol. Exp. Ther. 205, 596-605.

5. Kagawa, Y. and Racker, E. (1971) J. Biol. Chem. 246, 5477-5487.

6. Lu, A. Y. H., Miwa, G. T., West, S. B., Hodshon, B. J. and Garland, W. A. (1980) In: R. G. Thurman (ed.) Alcohol and Aldehyde Metabolizing Enzymes, Plenum Press, New York, Vol. 4, pp. 95-108.

7. Yang, C. S., Strickhard, F. S. and Kicha, L. P. (1977) Biochim. Biophys. Acta 465, 362-370.

8. Duppel, W. and Ullrich, V. (1976) Biochim. Biophys. Acta 426, 399-407.

Cytochrome P-450, Biochemistry, Biophysics
and Environmental Implications, E. Hietanen,
M. Laitinen and O. Hänninen editors

ENZYME KINETICS AND PHARMACOKINETICS OF DEALKYLATION REACTIONS AFTER VARIOUS PRETREATMENTS IN MICE

WOLFGANG LEGRUM AND JOCHEN FRAHSECK
Department of Pharmacology, Philipps-University, Lahnberge,
D-3550 Marburg,FRG

INTRODUCTION

In this study the exhalation of radioactive carbon dioxide derived from the ^{14}C-labelled methoxy groups of 7-methoxycoumarin and methacetin was examined. Four differently pretreated groups of mice were used: cobaltous chloride, phenobarbital, 3-methylcholanthrene and normals. Kinetic parameters of the carbon dioxide exhalation curves were compared.

MATERIALS AND METHODS

Animals. Male C57BL/6J Han mice of 20-25 g body weight were used throughout the study. They were kept on a standard laboratory pellet diet and water ad libitum at a fixed day/night cycle of 12 hours each.

Pretreatments of animals. The pretreatments consisted either of s.c. injections of 40 mg $CoCl_2$ / kg daily for two days, or of i.p. injections of 80 mg phenobarbital-Na / kg daily for three days, or of i.p. injections of 30 mg 3-methylcholanthrene in arachis oil / kg daily for two days. The group of normals did not receive any injection.

Radioactive substrates. [^{14}C]Methacetin was a gift from Dr.R. Braun, Berlin and 7-[^{14}C-methoxy]coumarin was synthesized according to the Williamson procedure. The collection of exhaled $^{14}CO_2$ from the breath was performed in microvials of 5 ml as described by Legrum and Frahseck (1).

RESULTS

The administration of the radioactively labelled methylethers 7-[^{14}C-methoxy]coumarin and [^{14}C]methacetin leads to the exhalation of $^{14}CO_2$ in mice. Both substrates used produce a biphasic slope of the exhalation curve. Figure 1, in a semilogarithmic plot, demonstrates the time course of the exhalation in normal animals. The first phase of exhalation shows a half life of 9 minutes, the second of about 2 hours for methacetin as substrate. In normal animals

61 per cent of the labelled moiety of methacetin are exhaled during the first hour. In the case of 7-methoxycoumarin as substrate (Fig.2) the resulting exhalation curve is identically shaped to that of methacetin (c.f. Fig.1). The half life for the first elimination period amounts to 13 minutes and for the second period to 220 minutes for 7-methoxycoumarin.

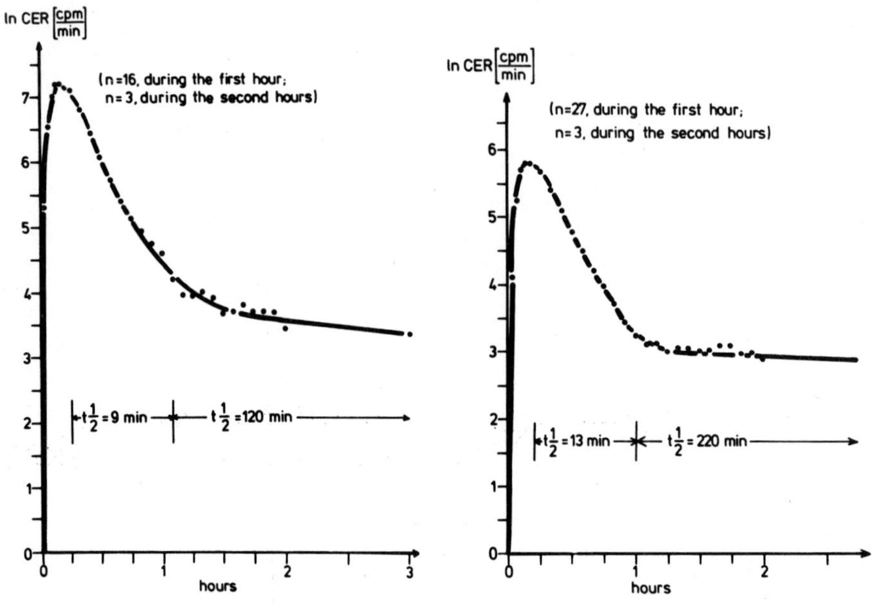

Fig. 1. Time course of the exhalation rate of $^{14}CO_2$ following a i.p. administration of [^{14}C]methacetin

Fig. 2. Time course of the exhalation rate of $^{14}CO_2$ following a i.p. administration of 7-[^{14}C-methoxy]coumarin

This fact is in so far remarkable as the radioactively labelled moiety of methacetin reveals a much higher recovery in the exhaled air during the first hour, than that of 7-methoxycoumarin, the latter yielding only 16 per cent during the first hour. The relatively small exhalation from 7-methoxycoumarin is due to the fact that other metabolic reactions than O-demethylation occur preferentially (2,3,4). These apparently diminish the relative proportion of O-demethylation.

Pretreatment of mice with cobaltous chloride doubles the exhalation of $^{14}CO_2$ derived from 7-[^{14}C-methoxy]coumarin and leads to a total exhalation of 30 per cent of the radioactivity administered. The time course of the exhalation is unchanged. This doubling of metabolism occurs in spite of a profound loss in hepatic cytochrome P-450 content (to 2/3), indicating a drastically increased activity (3-fold) towards the compound, which overcompensates the loss of the amount of cytochrome P-450 (1). Since cobalt pretreatment increases various other coumarin oxidations in vitro (5), probably the in vivo demethylation is not a result of an alteration between the metabolic pathways but rather a general elevation of activity maintaining the original metabolic pattern.

Using [^{14}C]methacetin as substrate other pretreatments for instance induction by phenobarbital or else 3-methylcholanthrene lead to enhanced maximal $^{14}CO_2$ exhalation rates ranging between 1.16 - to 1.35 -fold increases, respectively. The time course is not affected. Slight variations in the time to peak occur. However, one hour after the administration the formation of $^{14}CO_2$ ceases in all cases reaching values approximately in the range of the blanks. Injections of up to 5-fold higher doses of 7-[^{14}C-methoxy]coumarin also lead to an unchanged shape of the time curve (c.f. Fig.1 and 2). The exhalation maximum observed is proportional to the dose. We conclude that the time course of exhalation is not directly subject to the time course of the demethylation. Between the demethylation reaction generating formaldehyde and the carbon dioxide formation there is a chain of at least two enzyme catalyzed steps (6). In addition, there is the reduction of formaldehyde to methanol by the alcoholdehydrogenase. The above mentioned reaction chain causes a delay in the time course of exhalation in comparison to the O-demethylation in all cases studied. It is concluded that the original formaldehyde is accumulated in the cell (sink) for a certain period before it reaches sufficient concentrations (or the proper compartments) for the further oxidation to carbon dioxide. This latter process is assumed to occur at a more or less fixed rate independent of the primary supply by (normal or stimulated) the demethylation. For this reason it becomes understandable that modifications of the initial step by inducers show no visible effect in the time course of $^{14}CO_2$ exhalation but do so in the quantity of exhaled $^{14}CO_2$ (AUC or maximal $^{14}CO_2$ exhalation rate, CER).

REFERENCES

1. Legrum, W. and Frahseck, J. (1982) J.Pharmacol. exp. Ther.,221,
2. Kratz, F. and Staudinger, H.J. (1967) Hoppe-Seyler`s Z. Physiol. Chem., 348, 564.
3. v. Kerėkjårtó, B., Kratz, F. and Staudinger, M.J. (1964) Biochem. Z., 339, 460.
4. Mead, J.S., Smith, J.N. and Williams, R.T. (1958) Biochem. J., 68, 61.
5. Legrum, W. and Netter, K.J. (1980) Xenobiotica, 10, 271.
6. Waydhas, C., Weigl, K. and Sies, H. (1978) Eur. J. Biochem., 89, 143.

Cytochrome P-450, Biochemistry, Biophysics
and Environmental Implications, E. Hietanen,
M. Laitinen and O. Hänninen editors

IN VITRO AND IN VIVO EXPERIMENTS INDICATING INDUCTION OF MOUSE LIVER CYTOCHROME P 450 BY WARFARIN

LOTHAR KLING[1], WOLFGANG LEGRUM[1], HARALD REITZE[2], and KARL JOACHIM NETTER[1]
[1]Department of Pharmacology, Philipps-University, Lahnberge, D-3550 Marburg(FRG)
[2]Department of Zoology, Philipps-University, Lahnberge, D-3550 Marburg (FRG)

INTRODUCTION

As well as the vitamine K antagonism, warfarin shows a moderately inhibitory action on the microsomal mixed-function oxidase of rats *in vitro* (1). In this situation we investigated the effect of warfarin application on the cytochrome P 450 system in the intact animal.

The sensitivity of mice towards the rodenticide warfarin is markedly lower than that of rats (2). Therefore, *in vivo* application of warfarin was studied in mice.

MATERIAL AND METHODS

Animals. All experiments were carried out with male C57BL/6J Han mice of 20-25 g body weight. They had free access to water and to a standard laboratory pellet diet.

Pretreatment of animals. The animals were pretreated either with i.p. injections of 80 mg phenobarbital-Na/kg or oral applications of 120 mg warfarin-Na/kg, each once daily for three days.

Techniques. Liver microsomes were prepared as described by Netter (3). Cytochrome P 450 was quantitated by measuring the CO-difference spectrum of dithionite-reduced microsomal suspensions (4). Microsomal protein was determined by the method of Lowry et al. (5) with bovine serum albumin as standard. For ultrastructural studies the liver tissue was treated as described by Mothes and Seitz (6). *In vitro* metabolic studies with 7-ethoxy- and 7-methoxycoumarin as substrates were carried out according to the method published by Aitio (7). The $^{14}CO_2$-exhalation-analysis with 7-[^{14}C-methoxy]-coumarin as substrate was performed according to Legrum and Frahseck (8).

RESULTS

Pretreatment of mice with warfarin leads to an increase in microsomal cytochrome P 450 content/g liver to about 200 per cent of controls (Fig. 1.). Also the microsomal protein is raised to 130 per cent, a value not significantly different from that after a conventional phenobarbital induction (Fig. 1.).

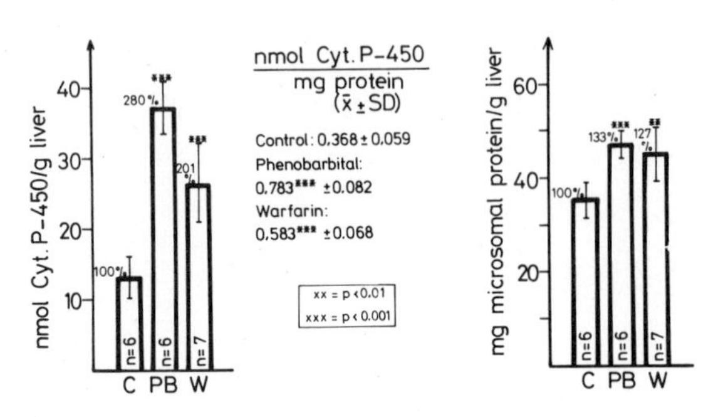

Fig. 1.

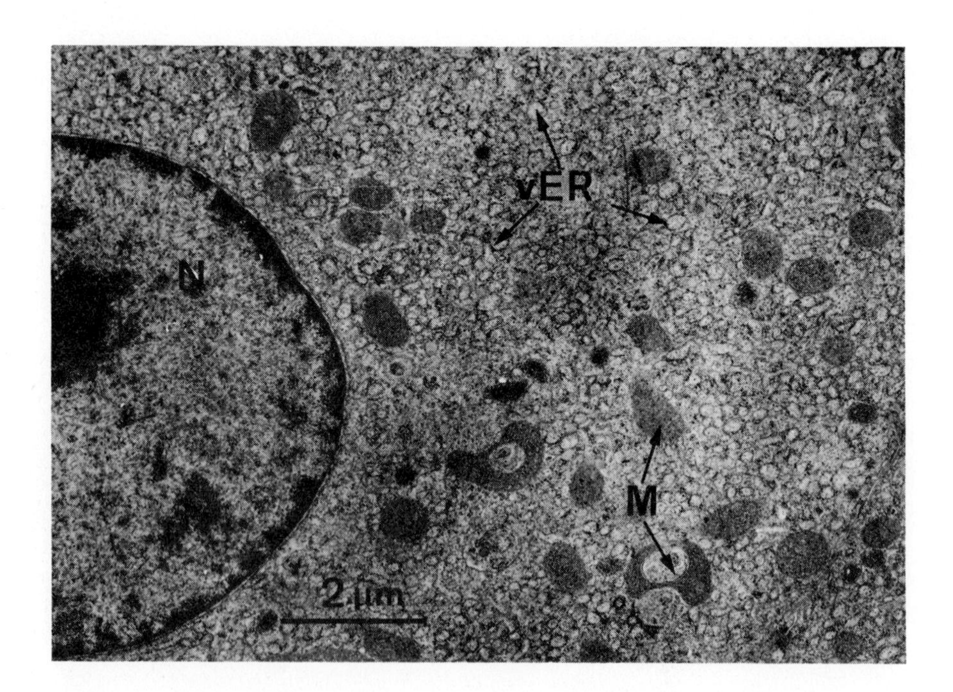

Fig. 2. Part of a hepatocyte after warfarin pretreatment: vesiculated endoplasmic reticulum (vER), mitochondrium (M), nucleus (N)

Ultrastructural studies confirm the induction phenomena showing liver cells with highly proliferated and vesiculated smooth endoplasmic reticulum (Fig. 2.).

In vitro studies with 7-ethoxy- and 7-methoxycoumarin as substrates show characteristics of warfarin-induced cytochrome P 450 in comparison to controls and phenobarbital-induced microsomes: After warfarin pretreatment the molecular activity increases in the case of 7-ethoxycoumarin as substrate to about 160 per cent (after phenobarbital to about 135 per cent) and in the case of 7-methoxycoumarin as substrate to about 130 per cent of the controls (after phenobarbital to about 140 per cent).

Inhibition experiments with metyrapone in the 7-ethoxycoumarin-deethylation-test reveal a clearly altered inhibition curve after warfarin pretreatment in comparison to controls (Fig. 3.). The warfarin-induced cytochrome P 450 is 10 times more susceptible than controls as measured by I_{50}-values.

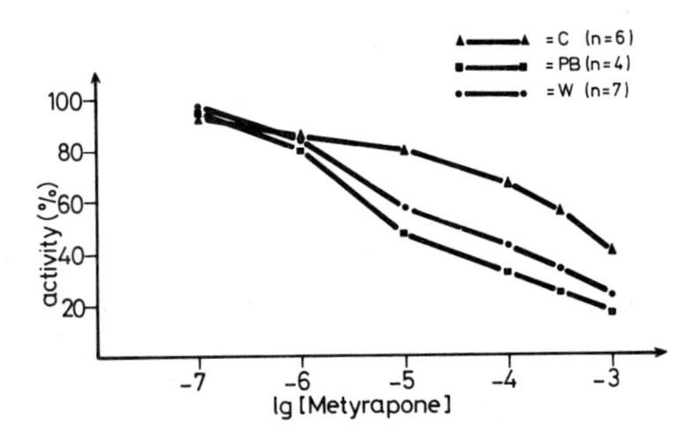

Fig. 3.

Enzyme kinetic measurements reveal K_m-values for 7-ethoxycoumarin after pretreatment with warfarin (25.3 μM) higher than in controls (17.7 μM) but much lower than after phenobarbital pretreatment (44.2 μM).

The analysis of hexobarbital binding spectra gives as a result (expressed in molecular absorption [M^{-2}]) a 2.5 fold binding affinity to warfarin-induced cytochrome P 450 (3.95×10^7) in comparison to controls (1.62×10^7), and again a 2.5 fold binding affinity to phenobarbital-induced cytochrome P 450 (9.53×10^7) in comparison to warfarin-induced microsomes.

To confirm the in vitro induction phenomena also in vivo we used 7-[^{14}C-methoxy]-coumarin in a $^{14}CO_2$-exhalation-analysis: The total $^{14}CO_2$-exhalation during 60 minutes (AUC_{0-60}) as well as the maximum $^{14}CO_2$-exhalation rate

(CER_{max}) are higher after warfarin and phenobarbital treatment (Fig. 4.). Comparison of the 1.3 fold increase in $^{14}CO_2$-exhalation with the 1.6 fold increase in cytochrome P 450 content per mouse after warfarin application suggests a decrease of the in vivo molecular activity to about 80 per cent of the control value. The discrepancy between the higher molecular activity of warfarin-induced as well as phenobarbital-induced cytochrome P 450 in vitro and the lower one in vivo is not solved yet.

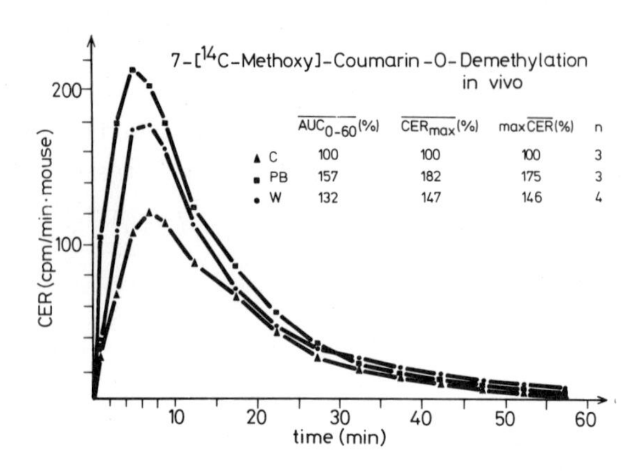

Fig. 4.

Experiments determining hexobarbital sleeping time and zoxazolamine paralysis time indicate a significant shortening of both times after warfarin pretreatment.

REFERENCES

1. Christensen, F. and Wissing, F. (1972) Biochem. Pharmacol, 21, 975
2. Lund, M. (1981) J. Hyg. Camb., 87, 101
3. Netter, K.J. (1960) Naunyn-Schmiedeberg's Arch. exp. Path. Pharmak., 238, 292
4. Omura, T. and Sato, R. (1964) J. Biol. Chem., 239, 2370
5. Lowry, O.H., Rosebrough, N.J., Farr, A.L., Randall, R.J. (1951) J. Biol. Chem., 193, 265
6. Mothes, U. and Seitz, K.A. (1981) Cell Tissue Res., 221, 339
7. Aitio, A. (1978) Anal. Biochem., 85, 488
8. Legrum, W. and Frahseck, J. (1982) J. Pharmacol. exp. Ther., 221,

Cytochrome P-450, Biochemistry, Biophysics
and Environmental Implications, E. Hietanen,
M. Laitinen and O. Hänninen editors

CHARACTERIZATION OF CYTOCHROME P-448 FROM SACCHAROMYCES CEREVISIAE.

DAVID J. KING, MAHMOOD R. AZARI AND ALAN WISEMAN
Biochemistry Division, Department of Biochemistry, University of Surrey,
Guildford, Surrey, GU2 5XH, U.K.

INTRODUCTION

The eucaryotic microorganism *Saccharomyces cerevisiae* contains a microsomally bound cytochrome P-448 which is part of a microsomal electron transport chain with many properties is common with that of mammalian liver (1, 2). This yeast cytochrome P-448 is produced under fermentative conditions (e.g. growth at high glucose concentration) which lead to the repression of mitochondrial cytochrome a + a_3 formation (3,4). Control of this process is thought to occur through cyclic AMP which exerts a repressive effect over *de novo* synthesis of cytochrome P-448, the level of cyclic AMP being determined by the glucose concentration in the growth medium in an inverse relationship (5).

This cytochrome P-448 system is capable of hydroxylating benzo(a)pyrene to a range of metabolites, principally 3-hydroxybenzo(a)pyrene, 9-hydroxybenzo(a)pyrene and 7,8-dihydro-7,8-dihydroxybenzo(a) pyrene (6). We have recently shown that this enzyme system can be induced with benzo(a)pyrene and several other compounds to produce only slightly higher levels of cytochrome P-448 but a great improvement in the efficiency of the enzyme at metabolizing benzo(a)pyrene, suggesting that more than one form of cytochrome P-450 can occur in this yeast (7). This finding has been backed up by the separation of two forms during purification studies (8). We have purified the major form of cytochrome P-448 from uninduced yeast to a high degree of purity (88-97% pure) and in this study we have undertaken to characterize this enzyme to allow a comparison with mammalian systems.

MATERIALS AND METHODS

Saccharomyces cerevisiae NCYC No. 240 was grown under glucose repression in a medium containing 1% yeast extract, 2% mycological peptone, 0.5% NaCl and 20% glucose for 44 hours at 30°C. Yeast was harvested and microsomes prepared from which cytochrome P-448 was purified as previously described (8). SDS-polyacrylamide gel electrophoresis was carried out by the method of Laemmli(9). Benzo(a)pyrene hydroxylase activity was determined by the fluorimetric method of Dehnen *et al.* (10) as modified by Woods and Wiseman (6). A reconstituted enzyme system comprising of purified cytochrome P-448, purified NADPH: cytochrome P-450 reductase and dilauroylphosphatidylcholine (1nmole cytochrome P-448 to 1 U of reductase).

RESULTS AND DISCUSSION

Cytochrome P-448 from uninduced Saccharomyces cerevisiae was purified to a specific content of 16-17.5nmole/mg protein. The homogeneity of this enzyme was shown by SDS-polyacrylamide gel electrophoresis, the molecular weight being determined as 55,500 using this method with several marker proteins. This molecular weight is similar to that of cytochrome P-448 from liver microsomes of 3-methylcholanthrene treated rats (11). Using this molecular weight the purity of our enzyme is 88-97%. Our preparation was free of NADPH:cytochrome P-450 reductase, cytochrome b_5 and cytochrome P-420. The soret peak of the reduced CO complex was at 448nm similarly to the form induced by polycyclic aromatic hydrocarbons in mammalian liver. A similarity to this enzyme is also suggested by the benzo(a)pyrene metabolite profile of the yeast enzyme (6).

The amino acid composition of yeast cytochrome P-448 was determined as shown in Table 1. This composition reveals 407 amino acid residues per molecule which leads to a molecular weight of 53,000 (a difference of 4.5% from the value determined by SDS-PAGE). The content of hydrophobic residues is 43% which is almost identical to the content of hydrophobic residues in induced and uninduced cytochromes P-450 from both rat (11) and rabbit liver (12, 13).

TABLE 1

AMINO ACID COMPOSITION OF YEAST CYTOCHROME P-448

Amino Acid	No. Residues/Molecule	Amino Acid	No. Residues/Molecule
Aspartic acid	41	Phenylalanine	20
Threonine	21	Leucine	23
Serine	24	Isoleucine	13
Glutamic acid	37	Histidine	17
Proline	23	Methionine	7
Glycine	42	Lysine	27
Alanine	27	Arginine	20
Valine	30	Tryptophan	12
Tyrosine	15	Cysteine	8
		Total	470

Benzo(a)pyrene gave rise to a type 1 binding spectrum with our purified cytochrome P-448 with an apparent spectral dissociation constant (K_s) of 50μM. Lanosterol, ethylmorphine, phenobarbital, dimethylnitrosamine and perhydrofluorene also gave rise to type 1 binding spectra. Benzo(a)pyrene hydroxylase activity of a reconstituted system comprising of purified cytochrome P-448, purified NADPH: cytochrome P-450 reductase and diluaroylphosphatidylcholine showed a K_m of 33μM and a V_{max} of 16.7 pmol 3-hydroxybenzo(a)pyrene/min/nmole P-448 when supported by NADPH. The requirement for cofactor could be replaced by cumene hydroperoxide or hydrogen peroxide generated from a glucose oxidase system, in each case the K_m and V_{max} are both increased (Table 2).

TABLE @

BENZO(a)PYRENE HYDROXYLASE ACTIVITY OF CYTOCHROME P-448 IN A RECONSTITUTED SYSTEM

	K_m (μM)	V_{max} (pmol 3-hydroxybenzo(a)pyrene/ min/nmole P-448
NADPH supported	33	16.7
Cumene hydroperoxide	125	21.9
Hydrogen peroxide (glucose oxidase)	200	33.7

A good agreement between the K_m of the NADPH supported reaction (33μM) and the benzo(a)pyrene K_s (50μM) was observed. The rate of benzo(a)pyrene hydroxylation observed with our enzyme is low compared to 3-methylcholanthrene induced (P-448) activities in rat liver, although it is comparable to forms isolated from uninduced mammalian sources (14, 15).

Yeast cytochrome P-448 is closely related to the family of enzymes from mammalian liver. The form of the enzyme which we have purified and studied in this report appears to resemble a cytochrome P-448 type of the mammalian enzyme rather than a cytochrome P-450 form, in its soret peak in the reduced CO spectrum, narrow substrate specificity and the range of benzo(a)pyrene metabolites formed. The molecular weight a amino acid composition are also closer to a P-448 form, although the activity towards benzo(a)pyrene hydroxylation is much lower than a mammalian cytochrome P-448.

ACKNOWLEDGEMENT

This work was supported in part by an S.E.R.C. CASE award to D.J.K. with May & Baker Ltd., Dagenham under the supervision of Dr. D.I. Dron and Dr. C.J. Coulson.

REFERENCES

1. Yoshida, Y., Kumaoka, H., Sato, R. (1974) J. Biochem. 75, 1201-1210.
2. Yoshida, Y., Aoyama, Y., Kumaoka, H., Kubota, S. (1977) Biochem. Biophys. Res. Comm. 78, 1005-1010.
3. Ishidate, K., Kawaguchi, K., Tagawa, T. (1969) J. Biochem. 65, 385-392
4. Ishidate, K., Kawaguchi, K., Tagawa, T., Hagihara, B. (1969) J. Biochem. 65, 375-383.
5. Wiseman, A., Lim, T.K., Woods, L.F.J. (1978) Biochim. Biophys Acta, 544, 615-623.
6. Wiseman, A., Woods, L.F.J. (1979) J. Chem. Tech. Biotechnol., 29, 320-324.
7. King, D.J., Azari, M.R., Wiseman, A. (1982) Biochem. Biophys. Res. Comm. 105, 1115-1121.
8. Azari, M.R., Wiseman, A. (1982) Anal. Biochem. in press.
9. Laemmli, U.K. (1970) Nature, 227, 680-685.
10. Dehnen, W., Tomingas, R., Ross, J. (1973) Anal. Biochem. 53, 373-381.
11. Bothelo, L.H., Ryan, D.E., Levin, W. (1979) J. Biol. Chem., 254, 5635-5640.
12. Koop, D.R., Persson, A.V., Coon, M.J. (1981) J. Biol. Chem., 256, 10704-10711.
13. Haugen, D.A., Coon, M.J. (1976) J. Biol. Chem. 251, 729-7939.
14. Ryan, D.E., Thomas, P.E., Korzeniowski, D., Levin, W. (1979) Biol. Chem. 254, 1365-1374.
15. Hashimoto, C., Imai, Y. (1976) Biochem. Biophys. Res. Comm. 68, 821-827.

Cytochrome P-450, Biochemistry, Biophysics
and Environmental Implications, E. Hietanen,
M. Laitinen and O. Hänninen editors

MODULATION OF RECONSTITUTED HYDROXYLASE ACTIVITIES IN BIOSYNTHESIS OF BILE ACIDS BY PROTEIN FRACTIONS FROM RAT AND RABBIT LIVER CYTOSOL

INGEGERD KALLES, BODIL LIDSTRÖM, KJELL WIKVALL AND HENRY DANIELSSON
Department of Pharmaceutical Biochemistry, University of Uppsala, Uppsala, Sweden

INTRODUCTION

The biosynthesis of bile acids includes a number of cytochrome P-450 dependent hydroxylations. Cholesterol 7α-hydroxylation is the first and rate-limiting step. 12α-Hydroxylation occurs at an early stage in cholic acid biosynthesis. Both hydroxylase activities are influenced by nutritional and hormonal factors (1). Recent work in this laboratory has shown that the activity of reconstituted rat liver cholesterol 7α-hydroxylase can be modulated by protein fractions from rat liver cytosol (2).

The present communication reports further studies on protein fractions from rat and rabbit liver cytosol modulating the activities of reconstituted cholesterol 7α-hydroxylase and 12α-hydroxylase systems.

MATERIALS AND METHODS

Male rabbits of the New Zealand strain, weighing about 2 kg, and male rats of the Sprague-Dawley strain, weighing about 200 g, were treated with cholestyramine, 3% (w/w) in the diet, for one week. The cytosol fraction of liver homogenate was prepared as described previously (2).

Preparation of a protein fraction from rat liver cytosol stimulating cholesterol 7α-hydroxylase was performed as follows. The cytosol was diluted with 10 mM phosphate buffer, pH 7.5, containing 0.25 M sucrose and applied to a column of hydroxylapatite in 10 mM potassium phosphate buffer, pH 7.5, containing 0.25 M sucrose. The protein, eluted with 50 mM phosphate, was further purified by DEAE-Sephacel chromatography in two steps. The columns were equilibrated, respectively, with 50 mM and 2 mM potassium phosphate buffer, pH 7.5, containing 0.25 M sucrose. Stimulatory activity was eluted with the equilibrating buffers. The protein fraction was then applied to a CM-Sephadex C-50 column equilibrated with 10 mM phosphate buffer, pH 7.5, containing 0.25 M sucrose. Fractions containing stimulatory activity were eluted with a linear KCl-gradient. The fractions were concentrated by ultrafiltration and further purified by isoelectric focusing using a pH-interval of 6 to 9.

A protein fraction from rabbit liver cytosol stimulating cholesterol 7α-hydroxylase activity was prepared by the same procedures as that from rat liver except that the cytosol was fractionated with polyethylene glycol (6 to 12%) prior to hydroxylapatite chromatography.

A fraction inhibiting 12α-hydroxylase activity was prepared from rabbit liver cytosol in the same way as described above for the 7α-hydroxylase stimulating fraction except that the inhibitory fraction was eluted from the CM-Sephadex column with 100 mM phosphate buffer, pH 7.5, containing 0.25 M sucrose.

Cytochrome P-450 (6.5 nmol/mg protein) from rat liver microsomes and cytochrome P-450 LM_4 (10.8 nmol/mg protein) from rabbit liver microsomes were prepared as described previously (3,4).

NADPH-cytochrome P-450 reductase was prepared from phenobarbital-treated rats and rabbits as described by Yasukochi and Masters (5).

Incubations and analyses of incubation mixtures were performed as described previously (4).

RESULTS AND DISCUSSION

Modulation of reconstituted cholesterol 7α-hydroxylase activity

Protein fractions which had stimulatory effect on reconstituted cholesterol 7α-hydroxylase activity were isolated from rat and rabbit liver cytosol. Table 1 shows that 0.15 mg of the fraction from rat liver cytosol stimulated 7α-hydroxylation of cholesterol catalyzed by a cytochrome P-450 fraction from rat liver microsomes two to three times. The fraction from rabbit liver cytosol (0.9 mg) similarily stimulated cholesterol 7α-hydroxylation catalyzed by cytochrome P-450 LM_4 from cholestyramine-treated rabbits two to three times.

TABLE 1

EFFECT OF PROTEIN FRACTIONS FROM RAT AND RABBIT LIVER CYTOSOL ON RECONSTITUTED CHOLESTEROL 7α-HYDROXYLASE ACTIVITY

System	7α-hydroxylated product formed
	pmol/nmol cytochrome P-450 x min
Cytochrome P-450 (rat)	133
Cytochrome P-450 + CM-Sephadex eluate (rat)	338
Cytochrome P-450 LM_4 (rabbit)	30
Cytochrome P-450 LM_4 + CM-Sephadex eluate (rabbit)	75

Further purification of the fraction from rat liver cytosol by means of isoelectric focusing resulted in the isolation of a protein fraction showing a pI of 6.7 to 6.9. This fraction gave a single protein band with an apparent M_r = 30,000 upon polyacrylamide gel electrophoresis in the presence of sodium dodecyl sulfate. Fig. 1 shows that the purified protein was dependent on reduced glutathione for stimulatory activity on cholesterol 7α-hydroxylation. In fact, the protein inhibited the reaction in the absence of reduced glutathione. In contrast to the protein fractions in the early purification steps the purified protein was not stimulatory when glutathione was replaced by dithiothreitol. The protein did not show any significant glutathione S-transferase activity and differed in molecular weight and pI from the glutathione S-transferases previously described in the litterature (6). Since it has been reported that 7α-hydroxylase activity in whole microsomes is stimulated under phosphorylating conditions (7) a series of experiments with addition of ATP, $MgCl_2$ and protein kinase was performed. Table 2 shows that the addition of these factors did not change the activity of the purified cytosol protein.

TABLE 2

EFFECT OF REDUCED GLUTATHIONE, ATP, $MgCl_2$, cAMP AND PROTEIN KINASE ON THE ACTIVITY OF PURIFIED PROTEIN FROM RAT LIVER CYTOSOL ON RECONSTITUTED CHOLESTEROL 7α-HYDROXYLASE ACTIVITY

System	7α-hydroxylated product formed			
	No addition	Glutathione	ATP, $MgCl_2$, cAMP, kinase	glutathione ATP, $MgCl_2$, cAMP, kinase
	pmol/nmol cytochrome P-450 x min			
Cytochrome P-450 (rat)	91	136	129	132
Cytochrome P-450 + protein (15 µg)	9	208	11	186

Inhibition of reconstituted 12α-hydroxylase activity by protein fraction from rabbit liver cytosol

A protein fraction which had an inhibitory effect on reconstituted 12α-hydroxylase activity was prepared from rabbit liver cytosol. Fig. 2 shows that addition of this protein fraction to reconstituted system containing cytochrome P-450 from rabbit liver microsomes markedly inhibited 12α-hydroxylation. The inhibitory effect was dependent on protein concentration. Cholesterol 7α-hydroxylation was not effected by this fraction.

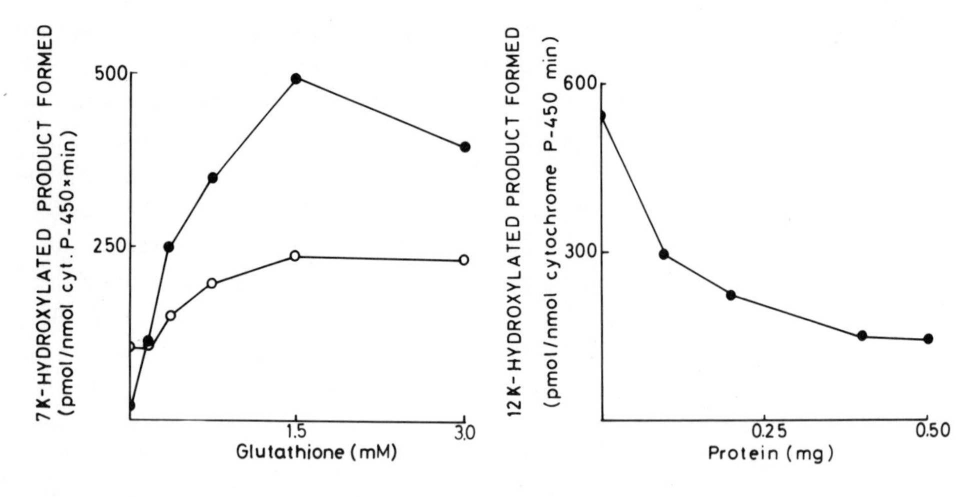

Fig. 1 Effect of reduced glutathione on the activity of purified cytosol protein on reconstituted cholesterol 7α-hydroxylase activity.

Fig. 2. Effect of protein concentration of inhibitory fraction from rabbit liver cytosol on reconstituted 12α-hydroxylase activity.

Taken together, the present results show that liver cytosol contains proteins which modulate the activity of 7α-hydroxylation of cholesterol and 12α-hydroxylation catalyzed by purified cytochrome P-450 and NADPH-cytochrome P-450 reductase.

ACKNOWLEDGMENTS

The skilful technical assistance of Mrs Anne Legnehed, Mrs Kerstin Rönnquist and Mrs Linnéa Wallsten is gratefully acknowledged. The study was supported by the Swedish Medical Research Council (Project 03X-218).

REFERENCES

1. Danielsson, H. and Sjövall, J. (1975) Annu.Rev.Biochem. 44, 233
2. Danielsson, H., Kalles, I. and Wikvall, K. (1980) Biochem.Biophys.Res.Commun. 97, 1459
3. Haugen, D.A. and Coon, M.J. (1976) J.Biol.Chem. 251, 7929
4. Kalles, I. and Wikvall, K. (1981) Biochem.Biophys.Res.Commun. 100, 1361
5. Yasukochi, Y. and Masters, B.S.S. (1976) J.Biol.Chem. 251, 5337
6. Jakoby, W.B. and Habig, W.H. (1980) in: Jakoby, W.B. (Ed.) Enzymatic Basis of Detoxication, Vol. II, Academic Press Inc., New York, pp. 63-94
7. Sanghvi, A., Grassi, E., Warty, V., Diven, W., Wight, C. and Lester, R. (1981) Biochem.Biophys.Res.Commun. 103, 886.

Cytochrome P-450, Biochemistry, Biophysics
and Environmental Implications, E. Hietanen,
M. Laitinen and O. Hänninen editors

EXPERIMENTAL APPROACHES TO DETERMINE THE POSITION OF THE HEME CHELATING AND SUBSTRATE BINDING SULFHYDRYL GROUPS WITHIN THE SEQUENCE OF $P\text{-}450_{CAM}$*

KARL M. DUS AND RALPH I. MURRAY
The Edward A. Doisy Department of Biochemistry
St. Louis University Medical School, St. Louis, MO 61304 (U.S.A.)

COMMON STRUCTURAL FEATURES OF P-450 HEMEPROTEINS

In recent years it has become of considerable interest to probe the question of whether P-450 hemeproteins show common structural features, and whether such features have biological significance. Such a comparison is of interest not only for P-450 hemeproteins isolated from a common source, but also for those located in different organelles, tissues, or species. While most structural studies of P-450s in the past have stressed differences in size, substrate specificity, N-terminal sequences, and immunological properties, we have pursued a unifying concept focusing on features which might be shared by many if not all members of the P-450 family (1). Our attempt to compare the heme environment of P-450s was based on the firm belief that the active site with the common heme prosthetic group and the structural prerequisites for a common mechanism of action should reflect this similarity more clearly than other parts of the protein. By limited BrCN cleavage of the covalently stabilized heme environment of several photoaffinity labeled P-450 hemeproteins we succeeded in isolating "core" peptides of comparable size and amino acid composition which contained most of the heme and the radioactivity imparted with the label, 1-(4'-azidophenyl)imidazole (2,3): labeled proteins include $P\text{-}450_{CAM}$, $P\text{-}450_{LM-2}$ and $P\text{-}450_{LM-4}$, both PB-induced, as well as bovine adrenocortical mitochondrial $P\text{-}450_{SCC}$ and $P\text{-}450_{11\beta}$. A radioimmunoassay based on competitive binding between ^{125}I-labeled $P\text{-}450_{CAM}$ and cold, cross-reacting antigens to anti $P\text{-}450_{CAM}$ revealed that, with the exception of $P\text{-}450_{LM-4}$, these same P-450s shared also immunochemical characteristics as determined by the extent of their cross-reactivity (4). Even the BrCN-derived hemepeptides were active in this assay, though to a lesser extent.

*Supported by NIH Grant GM-21726.

STRATEGY FOR LOCATING THE POSITIONS OF THE HEME CHELATING AND SUBSTRATE BINDING CYSTEINES

Another important feature of P-450 hemeproteins, and probably part of their common architectural design, is the presence of free sulfhydryl groups which are of great significance in heme chelation as well as substrate recognition, binding and orientation (5). This is of interest also because free SH groups are not compatible with the membrane environment of most P-450s and must be maintained at great expense. On the other hand, they provide a good chemical handle for the structural biochemist. Using $P\text{-}450_{CAM}$ as our model compound we have employed differential labeling of cysteine sulfhydryls followed by limited fragmentation using selective cleavage procedures. After isolation and analysis of the resulting fragments it was possible to identify the most important SH groups and to pinpoint their approximate position within the linear sequence of the protein.

Specifically, we have treated substrate-free $P\text{-}450_{CAM}$ with ^{14}C-labeled isobornyl bromoacetate (IBA) in a 1:1 ratio of label to protein to achieve after a 24 h incubation period ∿80% label incorporation into the substrate binding cysteine sulfhydryl (5). The labeled hemeprotein was submitted to heme extraction in the cold (6) and dialyzed. After flash evaporation the protein was cyanylated with 2-nitro-5-thiocyanobenzoic acid, dialyzed against 50% acetic acid, flash evaporated and, cleaved in 0.1 M borate, pH 9, 6 M guanidinium hydrochloride 37°C for 16 h. This digest was resolved by gel filtration on Sephadex G-50F (2x100 cm), equilibrated with 50% acetic acid.

The labeled fragment of about 80 residues was recovered in good yield and compositional analysis indicated the amino acid ratios expected for the N-terminus of the protein (7). This fragment was digested with trypsin and yielded 5 fragments which could be readily resolved by TLC. The label was associated with a peptide of 10-11 residues containing His and Trp. Since a previous NTCB digest of the unlabeled apoprotein had yielded a 50-60 residue fragment corresponding to the N-terminus it was concluded that the substrate binding CySH residue must be the first CySH in the sequence, and that it occurs between positions 50 and 60.

In search for the location of the heme chelating cysteine we applied the photoaffinity label 1-(4'-azidophenyl)imidazole (API) (2). The resulting $P\text{-}450_{CAM}$-API complex was incubated with 88% formic acid at 45°C for 40 h. The digest contained about 6 fragments which were resolved by chromatography on Sephadex G-50F (2x100 cm) equilibrated with 50% AcOH (Fig. 1). Three of the resolved bands contained significant amounts of heme (I + Ia, III, and V). Peak I contains essentially undigested hemeprotein while Ia is perhaps 100 residues shorter, probably generated by cleavage of the Asp-Pro bond 102/103 (7). Perhaps the most interesting fragment is contained in peak III. It contained ∿154 residues by amino acid analysis and had the composition of the N-terminal portion of the protein from residues 1-154. It must have resulted from selective cleavage of the Glu-Pro bond at position 154/155, although this is not normally a very acid labile sequence. We have to assume that in P-450$_{CAM}$ the folding pattern and surrounding sequences are contributing significantly to the lability of this bond. Further acid treatment of fragment III generated fragments IV and V which contained all the heme and was about 50 residues long. This is in keeping with breakage of the Asp-Pro bond at 102/103.

Since the composition of fragment V contains two cysteines it is, however, not quite clear which of these residues is the heme chelating cysteine. In further experiments we plan to cleave this fragment by NTCB at the cysteine which doesn't carry the heme to gather further evidence about this point. More direct evidence will be obtained after extensive alkylation of the free cysteine, followed by removal of heme and NTCB cleavage at what used to be the heme bearing cysteine.

Fragmentation of P-450$_{CAM}$-API Complex with 88% Formic Acid at 45°C, 40 h. Gelfiltration on Sephadex G-50 F, 2 × 100 cm

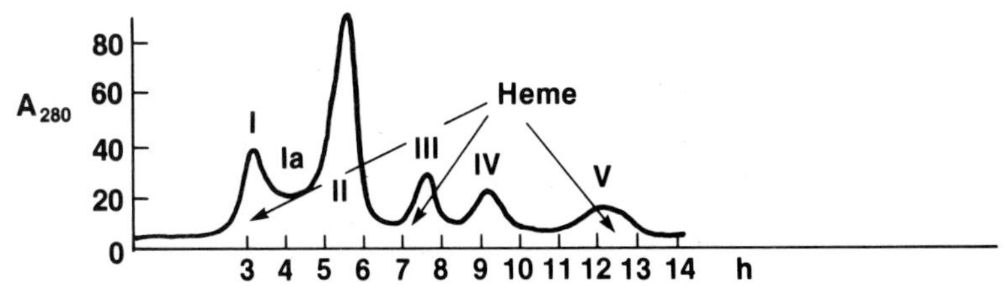

Interpretations based on Amino Acid Compositions and N-terminal Sequences:

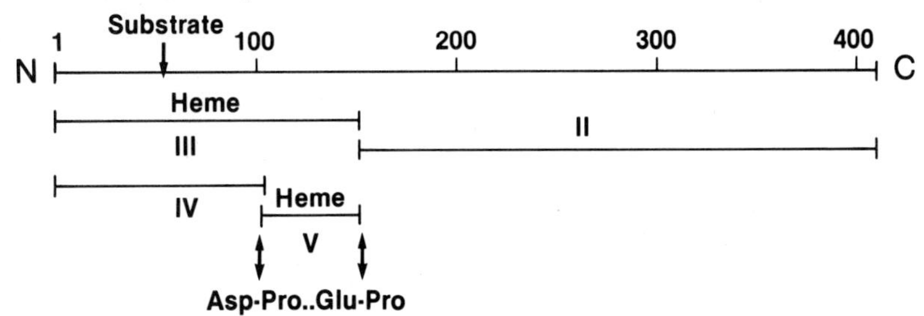

CONCLUSIONS

From these preliminary approaches we conclude that the substrate binding cysteine is located very early in the sequence, probably at position 56 from the N-terminus. This assignment is supported by correlation of the tryptic peptide carrying the IBA label with the peptide placed in this position by Haniu et al. (7). Although the precise position of the heme chelating cysteine still needs to be determined, we have narrowed the choice to residues 134 and 146 of the sequence (7) by isolating the hemepeptide comprising residues 103-154. It is gratifying to see that this is an area of extensive sequence homology between P-450_{CAM} and P-450_{LM} of PB-induced rat liver microsomes (8). Thus our predictions from various lines of structural investigations that P-450 hemeproteins show a common architectural design expressed by extensive sequence homology especially at and around their heme environments (3) have gained strong support.

REFERENCES

1. Dus, K., Litchfield, W. J., Miguel, A. G., van der Hoeven, T. A., Haugen, D. A., Dean, W. L., and Coon, M. J. (1974) Structural resemblance of cytochrome P-450 isolated from *Pseudomonas putida* and from rabbit liver microsomes, Biochim. Biophys. Res. Commun. 60, 15-21.
2. Swanson, R. A. and Dus, K. M. (1979) Specific covalent labeling of cytochrome P-450_{CAM} with 1-(4'-azidophenyl)imidazole, an inhibitor-derived photoaffinity probe for P-450 hemeproteins, J. Biol. Chem. 254, 7238-7246.
3. Dus, K. M. (1982) Toward a common structural concept for P-450 hemeproteins, in: *From Cyclotrones to Cytochromes*, (Kaplan, N. O. and Robinson, A. B., Eds.), Academic Press, Inc., pp. 231-250.
4. Dus, K. M., Litchfield, W. J., Hippenmeyer, P. J., Bumpus, J. A., Obidoa, O., Spitsberg, V., and Jefcoate, C. R. (1980) Comparative immunochemical studies of cytochrome P-450_{CAM} of *P. putida* and of cytochrome P-450_{SCC} of bovine adrenocortical mitochondria, Eur. J. Biochem. 111, 307-314.
5. Murray, R. I., Gunsalus, I. C., and Dus, K. M. (1982) Cytochrome P-450_{CAM}: model of binding and orientation of camphor derived from specific sulfhydryl interactions of isobornyl bromoacetate, J. Biol. Chem., in press.
6. Yu, C.-A., Gunsalus, I. C., Katagiri, M., Suhara, K., and Takemori, S. (1974) Cytochrome P-450_{CAM}: crystallization and properties, J. Biol. Chem. 249, 94-101.
7. Haniu, M., Armes, L. G., Tanaka, M., Yasunobu, K. T., Shastry, R. S., Wagner, G. C., and Gunsalus, I. C. (1982) The primary structure of the monoxygenase cytochrome P-450_{CAM}, Biochem. Biophys. Res. Commun. 105, 889-894.
8. Sato, R., private communication.

Cytochrome P-450, Biochemistry, Biophysics
and Environmental Implications, E. Hietanen,
M. Laitinen and O. Hänninen editors

TESTOSTERONE HYDROXYLATIONS CATALYZED BY PURIFIED RAT LIVER CYTOCHROME P-450 ISOZYMES

DAVID J. WAXMAN, ALBERT KO AND CHRISTOPHER WALSH
Departments of Chemistry and Biology, Massachusetts Institute of Technology
18-167, Cambridge, MA 02139 (U.S.A.)

INTRODUCTION

It has been suggested that steroids serve as physiological substrates for liver microsomal P-450 isozymes. Consistent with this idea is the finding that purified P-450 isozymes which exhibit broad and overlapping substrate specificities when assayed for xenobiotic metabolism display high regioselectivities during the hydroxylation of testosterone (1-3). Thus the rat isozymes P-450a, P-450b and P-450c catalyze formation of 7αOH-T (7α-hydroxytestosterone), 16αOH-T and 6βOH-T, respectively, as the principal products (3). In this study we reexamine the question of regioselectivity of testosterone hydroxylation using five P-450 isozymes isolated from phenobarbital (PB)-induced Sprague-Dawley rat liver (4). The results obtained indicate that P-450 isozymes have a somewhat broader regioselectivity for testosterone metabolism than previously reported.

EXPERIMENTAL

P-450 isozymes PB-1 to PB-5 were those described elsewhere in this volume (4) and in (5). PB-4 corresponds to the major PB-induced isozyme [e.g., P-450b (3)] and PB-5 to a closely related but structurally distinct form (4,5). PB-3 corresponds to P-450a (3) and is also equivalent to P-450 UT-F, isolated by Guengerich _et al._, (unpublished). Isozymes PB-1 (4,6) and PB-2 have apparently not been described by other investigators. Some of the experiments presented utilized a PB-2/PB-3 mixture, as indicated.

Detergent-free P-450s were reconstituted in a dilauroylphosphatidyl choline system with purified P-450 reductase and cytochrome b_5 (5,6) to give 75 nM P-450. Standard incubations with [4-^{14}C]testosterone (0.1 mM) were extracted with ethyl acetate (2 ml, twice) after 10-30' at 37°C and the hydroxylated products resolved on silica gel TLC plates developed with $CHCl_3$/ethyl acetate/ethanol, 4/1/0.7 (7). Radiolabeled products were quantitated by liquid scintillation counting. Tentative identifications were confirmed by rechromagraphy in at least two additional solvent systems and, in the case of androstenedione, by co-crystallization with an authentic sample to constant specific activity. Using these methods, the following monohydroxy testosterone derivatives were readily distinguished: 2αOH-T, 2βOH-T, 6αOH-T, 6βOH-T, 7αOH-T, 11βOH-T, 14αOH-T, 15αOH-T, 16αOH-T, 16βOH-T, 18 OH-T and androstenedione. (Standards were obtained from the MRC and from Steraloids, Inc.) Unambiguous identifications of the major enzymatic products (16αOH-T, 7αOH-T, 16βOH-T, 2αOH-T, and androstenedione) were made by GC/MS.

RESULTS AND DISCUSSION

Testosterone metabolites. Incubation of purified P-450 isozymes PB-1 to PB-5 in a reconstituted system with $[^{14}C]$testosterone resulted in an NADPH-dependent formation of six distinct enzymatic products: 5 monohydroxy testosterone derivatives and androstenedione, product of an apparent 17α-hydroxylation. Each of the five isozymes exhibited a characteristic hydroxylation pattern (Table 1). Products were identified by TLC and GC/MS (see EXPERIMENTAL).

TABLE 1

TESTOSTERONE HYDROXYLASE ACTIVITIES IN RECONSTITUTED SYSTEMS

		Turnover Number (min^{-1} $P\text{-}450^{-1}$)[a]			
Product	R_f[b]	P-450 PB-1	P-450 PB-2/PB-3	P-450 PB-4	P-450 PB-5
15αOH-T	0.23	-	-[c]	0.2 (0.07)	0.11 (0.2)
16αOH-T	0.31	0.12 (1.0)	2.6 (1.0)	3.4 (1.0)	0.56 (1.0)
7αOH-T	0.35	-	1.1 (0.4)[d]	-	0.16 (0.3)
16βOH-T	0.48	0.11 (1.0)	-	3.8 (1.1)	0.33 (0.6)
2αOH-T	0.53	-	2.1 (0.8)	-	-
Androstenedione	0.78	0.12 (1.0)	1.3 (0.5)	3.2 (0.9)	0.71 (1.3)
Total Activity		0.35	7.1	10.6	1.9

a Shown in parenthesis are activities relative to 16αOH-T formation (≡ 1.0). These values were reproducible (± 10-15%) from one enzyme preparation to the next, except where indicated. A dash indicates no activity was detected (<0.05 min^{-1} $P\text{-}450^{-1}$).

b Silica gel TLC developed with $CHCl_3$/ethyl acetate/ethanol (4/1/0.7).

c Some preparations were contaminated by T-15αOHase activity at a rate of 0.1 relative to 16αOH-T.

d T-7αOHase activity varied from ~0.19 to 1.2 relative to 16αOH-T, depending on the preparation. Further purification of PB-3 yielded T-7αOHase activity without detectable 16αOH-T, 2αOH-T and androstenedione activity. T-7αOHase activities as high as 16.5 min^{-1} $P\text{-}450^{-1}$ have been obtained with a highly purified sample of P-450 UT-F (= PB-3) obtained from Dr. F.P. Guengerich.

Inclusion of catalase (10 μg/ml), superoxide dismutase (10 μg/ml) or mannitol (0.1 M) to scavenge for reactive oxygen species (i.e., H_2O_2, O_2^- or ·OH respectively, which might arise from NADPH consumption uncoupled to hydroxylation) had minimal effect on enzymatic rates (± 5-15%) and did not alter the product ratio significantly. It is thus likely that each of the observed products (including androstenedione) arises from a specific P-450-mediated oxygenation event.

Regioselectivity of hydroxylation by P-450 PB-4. In previous studies of testosterone hydroxylation by the major PB-induced P-450 isozyme from rat liver, only 16αOH-T was identified, with no other enzymatic products detected (3,8). By contrast, highly purified P-450 PB-4 clearly catalyzed formation of significant amounts of 16βOH-T and androstenedione, in addition to 16αOH-T (Table 1). The following series of experiments confirms that the T-16βOHase and androstenedione-formation activities detected in the present study were not due to highly active, minor isozymic contaminants of PB-4:

a) effects of inhibitors - inclusion of benzphetamine, SKF-525A and metyrapone in the hydroxylase assay inhibited PB-4-mediated reactions significantly, with little change in the ratios of the three enzymatic products (Table 2).

TABLE 2

EFFECTS OF MONOOXYGENASE INHIBITORS ON TESTOSTERONE HYDROXYLASE ACTIVITY

		Relative Product Formation[a]					
Isozyme	Inhibitor[b]	Total Activity	16αOH	7αOH	16βOH	2αOH	Androstenedione
PB-1	Metyrapone	0.74	0.75	-	0.71	-	0.75
	Benzphetamine	0.57	0.60	-	0.54	-	0.67
PB-2/PB-3	Benzphetamine	0.56	0.35	0.97	-	0.39	0.32
	SKF-525A	0.58	0.08	1.38	-	0.09	0
PB-4	Metyrapone	0.17	0.13	-	0.18	-	0.24
	Benzphetamine	0.14	0.14	-	0.13	-	0.16
	SKF-525A	0.14	0.13	-	0.13	-	0.17
PB-5	Benzphetamine	0.07	0.09	0	0.09	-	0.05
	SKF-525A	0.17	0.16	0.58	0.10	-	0.16

[a] Activity in the presence of inhibitor relative to the absence of inhibitor. A dash indicates absence of the product, even in the uninhibited control.
[b] 1.0 mM final concentration for metyrapone, 0.5 mM for benzphetamine and 0.2 mM for SKF-525A.

b) suicide inactivation - the allyl-containing barbiturate secobarbital has been shown to an an isozyme-selective suicide-type inactivator, with PB-4 highly susceptible to inactivation and isozymes PB-1, PB-2/PB-3 and PB-5 resistant to inactivation (5). Incubation of PB-4 with secobarbital (10 μM) in the reconstituted system led to an NADPH- and time-dependent inactivation of testosterone hydroxylase activity. 16αOH-T, 16βOH-T and androstenedione-formation activities were each inactivated with first order kinetics characterized by $t_{1/2}$ = 40 ± 5 min (at 37°C) for all three products.

c) cytochrome b_5-stimulation - inclusion of cytochrome b_5 in the reconstituted system stimulated all three hydroxylase acivities of PB-4 to the same extent, i.e., ~2.6- to 2.8-fold.

d) effects of glycerol - in the case of PB-4 (and also PB-5) all three testosterone hydroxylase activities determined in the presence of 20% (v/v) glycerol were 45-60% lower than in the absence of glycerol. By contrast, inclusion of glycerol had little effect on the activities of PB-2 (± 10-20%) and, in fact, stimulated the T-7αOHase activity of PB-3 by about 30%.

P-450 PB-2 and P-450 PB-3. The partially purified hemeprotein fraction PB-2/PB-3 (4) was shown to consist of at least two isozymic components: PB-3, which, when purified from PB-2, catalyzed formation of a single product, 7αOH-T, and PB-2, which generated 16αOH-T, 2αOH-T and androstenedione, as well as a variable amount of 7αOH-T (Table 1). The possibility that isozymically pure PB-2 might catalyze a low level T-7αOHase activity cannot be excluded at present. The differential sensitivity of PB-2 and PB-3 to various inhibitors (Table 2) confirms the distinctive nature of the two isozymic constituents. Thus, T-7αOHase activity is benzphetamine-insensitive and is stimulated by SKF-525A in the reconstituted system. In contrast to the other isozymes used in this study, both PB-2 and PB-3 could also be isolated from uninduced rat liver microsomes. These and other data suggest that PB-2 corresponds to the neonatally imprinted T-16αOHase identified in early studies by Conney *et al.*(9).

P-450 PB-5. This isozyme, closely related to PB-4 both structurally and catalytically (4,5) yielded the same testosterone derivatives as PB-4 (Table 1). In addition, some preparations were contaminated by variable amounts of T-15αOHase and T-7αOHase activities. This latter activity is readily distinguished from the T-7αOHase activity of PB-3 in that it is sensitive to inhibition by benzphetamine as well as SKF-525A (Table 2). Thus PB-induced rat liver contains at least two distinct T-7αOHases.

P-450 PB-1. The low level testosterone hydroxylase activity catalyzed by PB-1 (Table 1) does not reflect the catalytic inactivity of this isozyme, as evidenced by the high turnover numbers obtained using several other monooxygenase substrates (4,6). Although the regioselectivity of testosterone hydroxylation by PB-1 was very similar to that of PB-4, formation of these products is not likely to reflect contamination by PB-4, as indicated by the insensitivity of the PB-1 hydroxylations to metyrapone (Table 2). This observation is consistent with the inability of dithionite reduced PB-1 to bind metyrapone to form the spectral P-446 complex detected with PB-4 and PB-5(4).

CONCLUSIONS

Testosterone hydroxylation catalyzed by five P-450 isozymes purified from PB-induced rat liver was studied in a reconstituted monooxygenase system. Six major enzymatic products were identified, with each isozyme exhibiting a characteristic hydroxylation pattern. Most notably, P-450 PB-3 was shown to be a highly specific isozyme, yielding 7αOH-T as the only enzymatic product. P-450 PB-2 which, along with PB-3, is also present in uninduced rat liver, catalyzed formation of 2αOH-T, 16αOH-T and androstenedione. P-450 PB-4, the major PB-induced isozyme, yielded 16βOH-T and androstenedione as major products, in addition to 16αOH-T, the product reported previously. Modulation of overall activity levels by inclusion of various inhibitors or suicide inactivators selective for PB-4 demonstrated that none of the three testosterone metabolites were products of contaminating isozymes. These studies indicate that purified P-450s have a somewhat broader regioselectivity for testosterone metabolism than previously recognized, establish the existence of multiple T-16αOHases and T-7αOHases, and demonstrate the usefulness of T-hydroxylation for characterization of P-450 isozymes.

REFERENCES

1. Haugen, D.A., Van der Hoeven, T.A. and Coon, M.J. (1975) J. Biol. Chem. 250, 3567-3570.

2. Huang, M.-T., West, S.B. and Lu, A.Y.H. (1976) J. Biol. Chem. 251, 4659-4665.

3. Ryan, D.E., Thomas, P.E., Korzeniowski, D. and Levin, W. (1979) J. Biol Chem. 254, 1365-1374.

4. Waxman, D.J. and Walsh, C. (1982) "Catalytic and Structural Properties of Two New Cytochrome P-450 Isozymes from Phenobarbital-induced Rat Liver: Comparison to the Major Induced Isozymic Form." This volume.

5. Waxman, D.J. and Walsh, C. (1982) J. Biol. Chem. 257, in press, Aug-Sept.

6. Waxman, D.J., Light, D.R. and Walsh, C. (1982). Biochemistry 21, in press, May.

7. Shiverick, K.T. and Neims, A.H. (1979). Drug Metab. Dispos. 7, 290-295.

8. West, S.B., Huang, M.-T., Miwa, G.T. and Lu, A.Y.H. (1979). Arch. Biochem. Biophys. 193, 42-50.

9. Conney, A.H., Levin, W., Jacobson, M., Kuntzman, R., Cooper, D.Y. and Rosenthal, O., In: Microsomes and Drug Oxidations, Gillette, et al., eds., pp. 279-302 (1969), Academic Press, N.Y.

© 1982 Elsevier Biomedical Press B.V.
Cytochrome P-450, Biochemistry, Biophysics
and Environmental Implications, E. Hietanen,
M. Laitinen and O. Hänninen editors

COMPARATIVE BIOCHEMICAL AND MORPHOLOGICAL CHARACTERIZATION OF MICROSOMAL PREPARATIONS FROM RAT, QUAIL, TROUT, MUSSEL AND *DAPHNIA MAGNA*.

P. ADE, M.G. BANCHELLI SOLDAINI, M.G. CASTELLI, E. CHIESARA, F.CLEMENTI, R. FANELLI, E. FUNARI, G. IGNESTI, A. MARABINI, M. ORONESU, S. PALMERO, R. PIRISINO, A. RAMUNDO ORLANDO, V. SILANO, A. VIARENGO, L. VITTOZZI
Applied Project "Promozione della Qualità dell'Ambiente", Operative Unit on "Comparative Metabolism", National Research Council-Istituto Superiore di Sanità, Rome, Italy.

INTRODUCTION

Rat, trout, quail, mussel and *Daphnia magna* are organisms situated on various rungs of the biological complexity ladder, recommended for toxicity and ecotoxicity testing of environmental contaminants(1,2). Knowledge of xenobiotic metabolism of these animals would not only enable a more accurate interpretation of short and long term toxicity test results, but would also facilitate extrapolation of lab scale results to external environmental conditions and/or to other types of organism. Virtually no data exist on metabolism of foreign compounds in mussel and Daphnia and further comparative metabolism data on rat, trout and quail are required.

MATERIALS AND METHODS

Animals. Livers of 200-250 g Sprague Dawley male rats (Charles River, Calco, Italy), of 200-300 g rainbow trouts (Istituto Ittiogenico, Rome, Italy), of 120-160 g male quails (*Coturnix c. Japonica*, commercial source), digestive glands of 10-11 g mussels (*Mytilus galloprovincialis*) collected from the sea near Palmaria (La Spezia, Italy) and whole bodies of female *Daphnia magna* grown at Istituto Superiore di Sanità (Rome, Italy) were used to prepare microsomes. Animals were sacrificed after at least one week acclimatization under controlled temperature, photoperiod, dietary and water conditions.
Microsome preparations. *Daphnia* and liver microsomes were prepared according to Ernster *et al.* (3). *Daphnia* homogenate contained also 1 mM EDTA and 20% Glycerol. Mussel microsomes were prepared by differential centrifugation between 30.000 x g and 150.000 x g of digestive gland homogenate in 900 mM Sucrose, 20% Glycerol, 20 mM Hepes, 1mM EDTA, 5mM DTT, 10 mM Cysteine, 0.7 mM PMSF and 120 mg/100 ml trypsin inhibitor.

TAB. III
FUNCTIONAL CHARACTERIZATION OF MICROSOMAL PREPARATIONS[a]

Parameter	Rat	Quail	Trout	Mussel	Daphnia
Cytochrome P-450 (nmoles/mg protein)	1.134 ± 0.35	0.381 ± 0.06	0.302 ± 0.04	0.047 ± 0.007	0.030 ± 0.009
Cytochrome b_5 (nmoles/mg protein)	0.742 ± 0.04	0.382 ± 0.01	0.150 ± 0.01	0.205 ± 0.040	0.310 ± 0.018
NADPH cytochrome c reductase (nmoles cytochrome reduced x mg^{-1} x min^{-1})	94.62 ± 6.5	95.57 ± 9.1	45.14 ± 2.9	11.8 ± 2.95	11.58 ± 3.05
NADH ferricyanide reductase (nmoles NADH oxidized x mg^{-1} x min^{-1})	1800 ± 166	860 ± 64	500 ± 8	140 ± 21	---
NADH cytochrome c reductase (nmoles cytochrome reduced x mg^{-1} x min^{-1})	834 ± 26	850 ± 20	202 ± 22	57,9 ± 4,3	21.86 ± 6.18
Benz (a) pyrene hydroxylase (nmoles 3-BP-OH x mg^{-1} x min^{-1})	4.005 ± 0.29	5.02 ± 0.8	1.15 ± 0.17	0.024 ± 0.002	0.014 ± 0.002
Aniline hydroxylase (nmoles p-amino-phenol x mg^{-1} x min^{-1})	1.074 ± 0.059	1.99 ± 0.13	0.034 ± 0.001[b]	not detectable	---
N,N-dimethylaniline N-demethylase (nmoles HCHO x mg^{-1} x min^{-1})	1.24 ± 0.44	3.1 ± 0.5	0.03 - 0.07	0.60 ± 0.01	---
p-dichlorobenzene hydroxylase (nmoles 2,5-dichlorophenol x mg^{-1} x min^{-1})	0.430 ± 0.44	0.056 ± 0.010	0.004	---	---

(a) Results are expressed as mean ± standard error;
(b) Tests were performed at 37° C as they showed optimal results as compared to 25° C and 15° C.

RESULTS

Mussel digestive gland, examined by electron microscopy, revealed secretory cells with numerous secretion granules, a well developed Golgi apparatus and an abundant endoplasmic reticulum. Another type of cell, similar to mammalian hepatocyte contained numerous mitochondria, free ribosomes and smooth and rough E.R. Epithelial cells of the gastrointestinal tract of Daphnia were composed of numerous membranes of smooth and rough E.R., mitochondria and lysosomes and showed well developed microvilli on the luminal membranes. Recoveries and purity of liver microsomal preparations, checked biochemically,

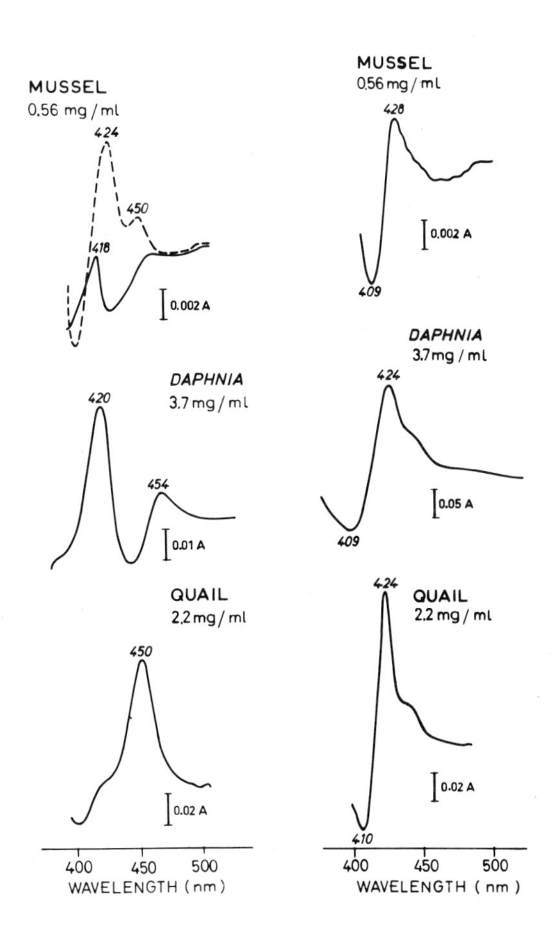

Fig. 1. Differential spectra of cytochromes P-450 (left) (4,5) and b_5 (right) (6) of microsomal preparations from mussel, Daphnia and quail.

were comparable with literature data. Mussel microsomes (3.3 mg protein/g fresh tissue) contained 10% homogenate cytochrome c oxidase; some partially disrupted mitochondria, secretory granules and plasma membrane profiles were visible contaminating smooth and rough membranes. E.M. analysis of *Daphnia* microsomes (2.2 mg protein/g fresh tissue) revealed the presence of conventional microsomal vesicles contaminated by plasma membrane fragments; no mitochondria or other cellular organelles were apparent.

Difference spectra (4-6) of cytochromes P-450 and b_5 in mussel, *Daphnia* and quail microsomes (Fig.1) were evidenced besides the well known spectra in rat and trout liver microsomes. Cytochrome P-450 in mussel microsomes was not shown by Omura and Sato's method (solid line) because of interfering cytochrome c oxidase but a clear, although weak, 450 nm-peak was apparent by the method of Ray and Estabrook (dashed line). Similarities were apparent between cytochromes P-450 of rat and quail or of trout and *Daphnia*; moreover *Daphnia* microsomes exhibited a CO-difference spectrum very like kidney cortex cytochrome P-450 (7). Difference absorbance maxima of cytochrome b_5 spectra of mussel and trout microsomes occurred at longer wavelengths as compared to the spectra of rat, quail and *Daphnia*. Cytochromes P-450 levels in mussel and *Daphnia* microsomes were 30-40 times lower than in rat, whereas cytochrome b_5 variations among species were in a narrow range. Other microsomal activities showed a general trend to be much lower in acquatic organisms (Tab. I).

REFERENCES

1. Federal Register (1979) 44 (7), January 10.
2. EEC Official Journal (1979) 22, No. L 259, pp. 10-25.
3. Ernster, L., Siekevitz, P. and Palade, G.E. (1962) J. Cell Biol., 15, 541
4. Omura, T. and Sato, R. (1964) J. Biol. Chem., 239, 2370
5. Ray, P. and Estabrook, R.W. (1970) Pharmacologist, 12, 261.
6. Garfinkel, D. (1958) Arch. Biochem. Biophys., 77, 493
7. Jacobsson, S., Thor, H. and Orrenius, S. (1970) Biochim. Biophys. Res. Commun., 39, 1073

Cytochrome P-450, Biochemistry, Biophysics
and Environmental Implications, E. Hietanen,
M. Laitinen and O. Hänninen editors

PURIFICATION AND CHARACTERIZATION OF P_{τ}-450 FROM MOUSE LIVER MICROSOMES

MATTI A. LANG[1], MASAHIKO NEGISHI[2], IEVA STUPANS[2] AND DANIEL W. NEBERT[2]
[1]EP-Research Laboratories, Pulttitie 9, Helsinki, Finland
[2]Developmental Pharmacology Branch, National Institute of Child Health and Human Development National Institutes of Health, Bethesda, MD 20205 (U.S.A.)

INTRODUCTION

The importance of having a genetic marker, in addition to a well characterized antibody toward a particular form of P-450, has been emphasized (1) as a necessary prerequisite to less ambiguous cloning studies. To this end, we have searched for and have uncovered an antibody quite specific at inhibiting coumarin 7-hydroxylase activity (2). The reason for choosing this particular catalytic activity is the genetic polymorphism described by Wood and coworkers (3-5). The DBA/2 (D2) inbred mouse strain has high constitutive coumarin 7-hydroxylase and highly inducible hydroxylase activity following phenobarbital (PhBarb) treatment. AKR (AK) and C57BL/6 (B6) inbred mouse strains have low constitutive coumarin 7-hydroxylase activities and low inducible hydroxylase activity after PhBarb treatment. Assays of individual progeny from the $AKD2F_1$ x backcross indicate coumarin 7-hydroxylase activity is inherited additively as a single autosomal trait; the *Coh* locus is located on mouse chromosome 7 linked to the albino *c* locus (5). In this Chapter we characterize an antibody that blocks quite specifically coumarin 7-hydroxylase in mice.

RESULTS AND DISCUSSION

Purification of P_{τ}-450. Sources for the materials and animals, as well as details of the experimental procedures, are provided in Refs. 2, 6, and 7. PhBarb-treated B6 liver microsomes were solubilized with sodium cholate. Octyldiamino-Sepharose 4B column chromatography, elution with Emulgen 911, hydroxylapatite column chromatography, and elution with potassium phosphate (pH 7.25) in sequence were each carried out twice. Three peaks were obtained and called ϕ, β, and τ, to relate these fractions to the inducer phenobarbital. All three peaks had specific P-450 contents between 17 and 19 nmol per mg of protein. Antibodies to each of these fractions were prepared in goats. The antibody to the third fraction was quite specific in blocking coumarin 7-hydroxylase activity. We propose to name the antigen "P_{τ}-450" and the antibody "anti-(P_{τ}-450)."

Characterization of anti-(P_τ-450). Fig. 1 shows a sharp and continuous immunoprecipitin line obtained between P_τ-450 and the undiluted antibody, as well as several dilutions of the antibody. When the antibody is exposed to cholate-solubilized liver microsomes from control or PhBarb-treated D2, AK, or B6 mice (Fig. 2), a precipitin line is formed with apparent homogeneous cross-reactivity among the three inbred strains, both control and PhBarb-treated.

Fig. 3 shows $NaDodSO_4$-polyacrylamide gel electrophoresis of liver microsomes, the purified P_τ-450, and anti-(P_τ-450)-precipitated material from.

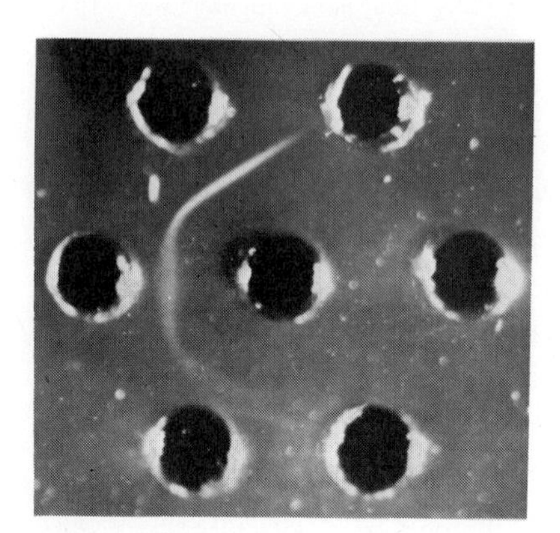

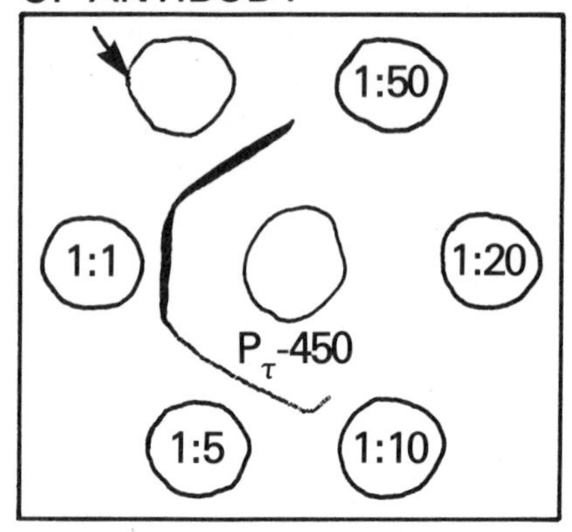

Fig. 1. Ouchterlony immunodiffusion between P_τ-450 in the center well and the undiluted and diluted anti-(P_τ-450) in the outside wells.

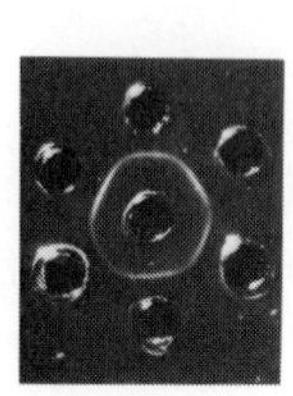

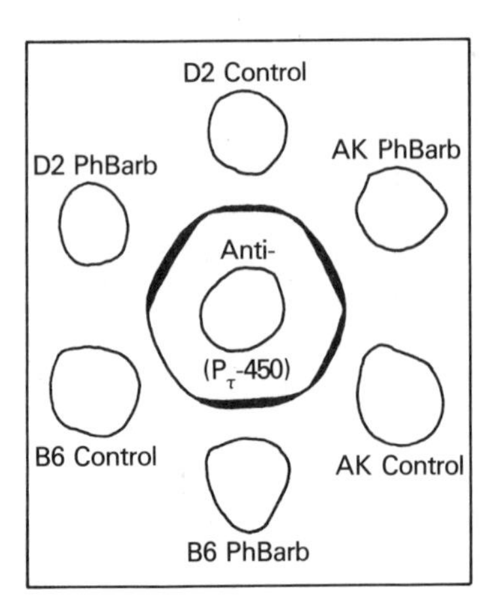

Fig. 2. Ouchterlony immunodiffusion between anti-(P_τ-450) in the center well and cholate-solubilized microsomes from control and PhBarb-treated D2, AK, and B6 mice in the outside wells.

cholate-solubilized microsomes. The purified antigen and the antibody that was developed correspond to a protein of about 51 kDa. By means of *in vitro* labeling of microsomes with $NaB[^3H]_4$ and then $NaDodSO_4$-polyacrylamide gel electrophoresis of cholate-solubilized material (2), it also was determined that the same apparent P_τ-450 exists in control D2, AK, and B6 mice and that PhBarb pretreatment enhances this (M_r~51) kDa protein between 2- and 3-fold.

Coumarin 7-hydroxylase and P_τ-450. We found that 5 mg of anti-(P_τ-450) IgG per mg of microsomal protein blocked more than 80% of coumarin 7-hydroxylase in control and PhBarb-treated mice of all three strains (2). Alternatively, 10 mg of anti-(P_τ-450) per mg of microsomal protein did not inhibit more than 15% of aminopyrine N-demethylase, ethylmorphine N-demethylase, ethoxycoumarin O-deethylase, acetanilide 4-hydroxylase, and aryl hydrocarbon hydroxylase activities (2).

Molecular biology of P_τ-450. Using anti(P_τ-450), we have precipitated a single translation product *in vitro* corresponding to P_τ-450. The size of P_τ-450 mRNA is 19 to 20 S. cDNA clones now have been prepared, as described (1), and are being screened.

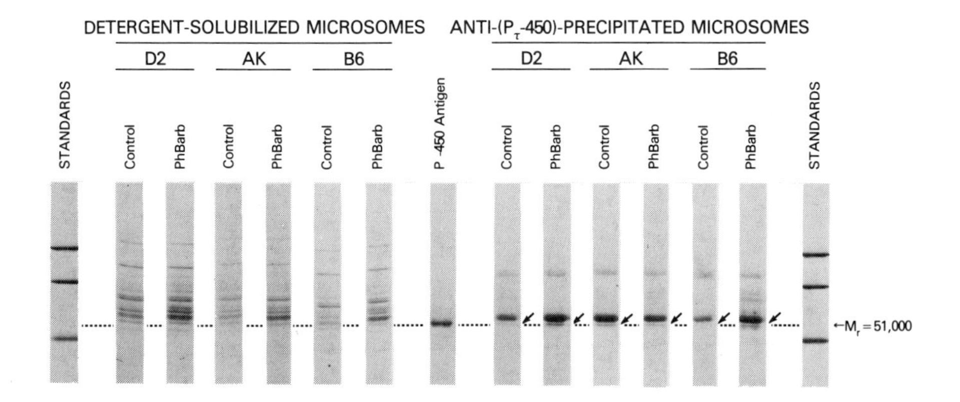

Fig. 3. $NaDodSO_4$-polyacrylamide gel electrophoresis of the intact microsomes and anti-(P_τ-450)-precipitated materials from cholate-solubilized microsomes. The major band in the latter represents goat IgG. Although another band at ~75 kDa was seen, anti-(P_τ-450)-precipitated radioactivity occurred only in the band at 51 kDa (*arrows*). Also shown is electrophoresis of the purified P_τ-450 and the standards ovalbumin (M_r=45 kDa), bovine serum albumin (M_r=66.2 kDa), and phosphorylase B (M_r=92.5 kDa). Migration is from *top* to *bottom*.

SUMMARY

Three different P-450 fractions (ϕ, β, and τ) were purified from PhBarb-treated B6 mice by a combination of hydrophobic and ion-exchange chromatography. The purified preparation from the third peak was named P_τ-450; this fraction (18.8 nmol/mg protein) exhibited a single electrophoretic band on $NaDodSO_4$-polyacrylamide gels (M_r~51 kDa). An antibody to P_τ-450 was developed. B6, D2, and AK inbred mouse strains were compared. When liver microsomes were labeled _in vitro_ with $NaB[^3H]_4$ and then detergent-solubilized and precipitated with anti-P_τ-450, a single peak (M_r~51 kDa) was found. The amount of immunoprecipitable P_τ-450 is about the same in control B6, D2 and AK, and PhBarb pretreatment increases the amount about 3-fold in all three strains. With Ouchterlony immunodiffusion, all control and PhBarb-induced forms appear identical. In studies with intact microsomes from all three control and PhBarb-treated strains _in vitro_, anti-P_τ-450 inhibits slightly (5-15%) the following activities: aminopyrine N-demethylase, ethylmorphine N-demethylase, 7-ethoxycoumarin O-deethylase, acetanilide 4-hydroxylase, and aryl hydrocarbon hydroxylase. Anti-P_τ-450 inhibits coumarin 7-hydroxylase more than 80% in both control and PhBarb-treated D2, AK, and B6 mice.

We therefore believe that anti-P_τ-450 is specific toward the PhBarb-inducible form of P-450 responsible for coumarin 7-hydroxylase in D2 but not AK mice, _i.e._ specific toward a structural gene controlled by the _Coh_ locus. From P_τ-450 mRNA, which is 19 to 20 S in size, cDNA clones have been prepared and are being screened.

REFERENCES

1. Nebert, D.W. (1982) in: Hietanen, E. (Ed.), Cytochrome P-450: Biochemistry, Biophysics and Environmental Implications, Elsevier/North-Holland Biomedical Press, Amsterdam, in press.
2. Lang, M.A., Negishi, M., Stupans, I. and Nebert, D.W., manuscript in preparation.
3. Wood, A.W. and Conney, A.H. (1974) Science, 185, 612.
4. Wood, A.W. (1979) J. Biol. Chem., 254, 5641.
5. Wood, A.W. and Taylor, B.A. (1979) J. Biol. Chem., 254, 5647.
6. Negishi, M. and Nebert, D.W. (1979) J. Biol. Chem., 254, 11015.
7. Negishi, M., Jensen, N.M., Garcia, G.S. and Nebert, D.W. (1981) Eur. J. Biochem., 115, 585.

Cytochrome P-450, Biochemistry, Biophysics
and Environmental Implications, E. Hietanen,
M. Laitinen and O. Hänninen editors

CHARACTERIZATION OF THE FORMS OF CYTOCHROME P-450 INDUCED BY trans-STILBENE OXIDE AND BY 2-ACETYLAMINOFLUORENE

JOHAN MEIJER, ANDERS ÅSTRÖM AND JOSEPH W. DEPIERRE
Department of Biochemistry, Arrhenius Laboratory, University of Stockholm,
S-106 91 Stockholm, Sweden

INTRODUCTION

The existence of several different isozymes of cytochrome P-450 differing in, among other things, substrate specificity, amino acid composition, electrophoretic mobility, terminal amino acid sequence and immunological properties is now well established. In addition, the form(s) of cytochrome P-450 induced by phenobarbital is not identical to that induced by 3-methylcholanthrene (1).

In recent years we have directed much attention towards characterizing trans-stilbene oxide (2) and 2-acetylaminofluorene (3) as inducers of drug-metabolizing enzymes. Both of these xenobiotics were found to induce hepatic levels of microsomal epoxide hydrolase and cytosolic glutathione S-transferase to a much greater extent than they induce the total amount of microsomal cytochrome P-450. However, trans-stilbene oxide and 2-acetylaminofluorene also induce cytochrome P-450 significantly and the present study was designed to determine whether these cytochrome P-450 forms are the same as those induced by phenobarbital and 3-methylcholanthrene or are previously unidentified isozymes. The criteria used to compare the different isozymes of cytochrome P-450 include the position of the absorption maximum in the difference spectrum between the reduced form of the cytochrome and the complex formed between carbon monoxide and the reduced form, substrate specificity, metabolite patterns with benzo(a)-pyrene and 2-acetylaminofluorene, sensitivity to different inhibitors, substrate binding spectra, molecular weight, immunological properties, and purification.

MATERIALS AND METHODS

Chemicals. Phenobarbital (Apoteksbolaget, Stockholm, Sweden), 3-methylcholanthrene (Eastman Kodak Company, Rochester, New York, U.S.A.), and trans-stilbene oxide (EGA-Chemie, Steinheim/Albuch, Federal Republic of Germany) were obtained from commercial sources and used without further purification. 2-Acetylaminofluorene (m.p. 192-193°) was purchased from Fluka AG (Buchs, Switzerland) and was essentially pure as judged by TLC (chloroform, chloroform/methanol (97:3) and petroleum ether (b.p. 40-60°)/acetone (7:3)). Ethoxycoumarin and ethoxyresorufin were synthesized from the hydroxy compound by ethylation with ethyl iodide (4) and ethylisocyanide was synthesized according to Jackson and McKusick (5). Benzphetamine was synthesized by Dr. Åke Pilotti of the Department of Organic Chemistry, University of Stockholm.[G-^{3}H]-Benzo(a)pyrene, [7,10-^{14}C]-benzo(a)pyrene, and 2-[9-^{14}C]-acetylaminofluorene were puchased from the Radiochemical Centre (Amersham, England) and purified before use (6,7). All other chemicals and solvents used were of reagent grade and obtained from common commercial sources.

Induction and preparation of microsomes. Male Sprague-Dawley rats (Anticimex AB, Sollentuna, Sweden) weighing 170-200 g were used throughout this study. The animals were injected in groups of 3 or 4 intraperitoneally with phenobarbital (80 mg/kg body weight in isotonic saline), 3-methylcholanthrene (20 mg/kg in corn oil), trans-stilbene oxide (400 mg/kg in corn oil), 2-acetylaminofluorene (50 mg/kg in polyethylene glycol 300), or vehicle alone once daily for 5 days. This injection schedule assures maximal induction of the activities investigated here. The total liver microsomal fraction was prepared using a standard procedure (8).

Enzyme assays. All measurements were carried out on freshly prepared microsomes. The measurement of different activities catalyzed by the cytochrome P-450 system was performed after the cytochrome P-450 content had been determined. Microsomes containing 0.05-1.0 nmol cytochrome P-450 were then used in the different measurements in order to assure linearity with time and protein. Cytochrome P-450 (9), aminopyrin N-demthylation (10), ethylmorphine N-demethylation (10), benzpethamine N-demethylation (11), benzo(a)pyrene monooxygenase (12), ethoxycoumarin O-deethylation (4), ethoxyresorufin O-deethylation (4), and the metabolite pattern obtained with benzo(a)pyrene (13) were all analyzed using published procedures. The metabolite pattern obtained with 2-acetylaminofluorene was characterized using a modification (7) of the high pressure liquid chromatographic procedure developed by Smith and Thorgeirsson (14).

Other procedures. Substrate binding spectra were performed with an Aminco DW-2 spectrophorometer at 30°. Each cuvette contained microsomes having 1 nmol cytochrome P-450 in 1 ml 50 mM Tris-Cl, pH 7.4. Aniline or hexobarbital was added to the sample cuvette to give a final concentration of 2.5 or 7.5 μM, respectively, while an equivalent volume (25 μl) of the solvent (dimethylsulfoxide) was added to the reference cuvette. The difference spectrum was recorded between 350 and 450 nm. Ethylisocyanide difference spectra were obtained by a modification of a published procedure (15). Immunological analysis was performed as described elsewhere (16). SDS-polyacrylamide gel electrophoresis was performed using slab gels as described by Laemmli (17). Protein was assayed using a slight modification of the method of Lowry and coworkers (18) with bovine serum albumin as standard.

RESULTS

Induction of microsomal cytochrome P-450. As can be seen in Table 1, both *trans*-stilbene oxide and 2-acetylaminofluorene induce the total level of cytochrome P-450 significantly, though not nearly as much as do phenobarbital and 3-methylcholanthrene. Neither *trans*-stilbene oxide nor 2-acetylaminofluorene gives rise to a blue shift in the absorption maximum of the carbon monooxide-reduced cytochrome P-450 complex, which is a first indication that these substances do not induce the same isozyme of cytochrome P-450 as does 3-methylcholanthrene.

TABLE 1

THE EFFECTS OF DIFFERENT INDUCERS ON CYTOCHROME P-450 LEVELS IN RAT LIVER MICROSOMES

Induction was performed and cytochrome P-450 assayed as described in the Materials and Methods

Treatment[a]	n[b]	nmol cytochrome P-450[c]/ mg microsomal protein	% of control	absorption[d] maximum
Control	20	0.41 ± 0.09	100	450.5 nm
PB	20	1.33 ± 0.26	324	450.0 nm
MC	20	1.36 ± 0.23	331	447.4 nm
tSBO	11	0.77 ± 0.11	187	450.0 nm
AAF	9	0.69 ± 0.09	168	450.2 nm

[a]Microsomes from untreated animals or animals receiving vehicle only (0.9% NaCl, corn oil or polyethylene glycol 300) demonstrated the same levels of cytochrome P-450 and these values were therefore pooled and designated Control. PB = phenobarbital, MC = 3-methylcholanthrene, tSBO = *trans*-stilbene oxide, AFF = 2-acetylaminofluorene.

[b]number of animals

[c]means ± standard deviations. All induced values are significantly different from the control at the level of $P < 0.001$.

[d]absorption maximum in the difference spectrum between the reduced form of the cytochrome and the complex formed between carbon monoxide and the reduced form.

TABLE 2

LEVELS OF RAT LIVER MICROSOMAL CYTOCHROME P-450-CATALYZED ACTIVITIES PREFERENTIALLY INDUCED BY PHENOBARBITAL

Rats were induced, microsomes prepared, and enzymes assayed as described in the Materials and Methods.

Treatment[a]	Activities: n[b] nmol metabolized/min-nmol cytochrome P-450[c] (% of control)					
	Aminopyrine-N-demethylase		Ethylmorphine-N-demethylase		Benzphetamine-N-demethylase	
	n		n		n	
Control	23	6.90 ± 1.07 (100)	21	8.45 ± 2.26 (100)	22	4.15 ± 1.63 (100)
PB	21	11.1 ± 1.76 (160)***	21	13.6 ± 1.16 (160)***	18	10.2 ± 1.64 (245)***
MC	23	3.22 ± 1.00 (47)***	20	3.07 ± 1.27 (36)***	18	0.54 ± 0.32 (13)***
tSBO	15	13.0 ± 2.10 (188)***	12	12.5 ± 2.36 (148)**	8	13.1 ± 2.88 (316)***
AAF	9	5.52 ± 1.21 (80)*	9	7.82 ± 1.45 (93)	9	5.93 ± 1.03 (143)**

[a] Microsomes from untreated animals or animals receiving vehicle only (0.9% NaCl, corn oil or polyetylene glycol 300) demonstrated the same activities and these values were therefore pooled and designated Control. PB = phenobarbital, MC = 3-methylcholanthrene, tSBO = *trans*-stilbene oxide, AAF = 2-acetylaminofluorene

[b] number of animals

[c] means ± standard deviations

* significantly different from Control value at the level of $P < 0.01$

** significantly different from Control value at the level of $P < 0.005$

*** significantly different from Control value at the level of $P < 0.001$

Substrate specificity. Since different isozymes of cytochrome P-450 demonstrate different substrate specificities, the activities catalyzed by the cytochrome P-450 system in liver microsomes from control and induced animals were compared. Table 2 documents the activities obtained with substrates known to be preferentially metabolized by microsomes from phenobarbital-treated rats, while in Table 3 are given three activities known to be preferentially induced by 3-methylcholanthrene. (In these and subsequent tables the values are expressed per nmol cytochrome P-450. These values can be converted to a mg microsomal protein basis by multiplying with the appropriate values from Table 1).

Investigation of the data presented in Tables 2 and 3 allows us to draw two conclusions: on the one hand, after induction with trans-stilbene oxide the substrate specificity of the liver microsomal cytochrome P-450 system closely resembles that seen after induction with phenobarbital but is strikingly different from that observed with microsomes from 3-methylcholanthrene-treated rats. On the other hand, the substrate specificity obtained after administration of 2-acetylaminofluorene differs from those seen in all of the other microsomal preparations.

Products of 2-acetylaminofluorene metabolism after induction. Table 4 presents the products of 2-acetylaminofluorene metabolism obtained with liver microsomes from rats treated with phenobarbital, 3-methylcholanthrene, 2-acetylaminofluorene, or vehicle alone. It can be seen that the pattern obtained after induction with 2-acetylaminofluorene clearly differs from that observed after phenobarbital administration and also differs in certain respects from the pattern obtained after 3-methylcholanthrene treatment. In addition Table 4 demonstrates that 2-acetylaminofluorene is a potent inducer of its own metabolism.

DISCUSSION

In addition to the data presented here we have characterized differences in the pattern of metabolites obtained with benzo(a)pyrene; the sensitivity of benzo(a)pyrene monooxygenase to the inhibitors α-naphthoflavone, metyrapone and SKF 525-A; the spectra obtained upon binding of aniline and hexobarbital; the molecular weight of cytochrome P-450 as determined by SDS-disc gel electrophoresis; and the ability of specific antibodies to inhibit cytochrome P-450 catalyzed activities using liver microsomes from rats treated with these 4 different inducers. Furthermore, we have isolated the form of cytochrome P-450 induced by trans-stilbene oxide in collaboration with Dr. F. Peter Guengerich of Vanderbilt University, Nashville, Tennessee, U.S.A. All of these findings

TABLE 3

LEVELS OF RAT LIVER MICROSOMAL CYTOCHROME P-450-CATALYZED ACTIVITIES PREFERENTIALLY INDUCED BY 3-METHYLCHOLANTHRENE

Rats were induced, microsomes prepared, and enzymes assayed as described in the Materials and Methods.

Treatment[a]	Activities: n[b] nmol metabolized/min-nmol[c] cytochrome P-450 (% of control)					
	Benzo(a)pyrene monooxygenase		Ethoxycoumarin-O-deethylase		Ethoxyresorufin-O-deethylase	
Control	24	0.44 ± 0.19 (100)	19	0.66 ± 0.15 (100)	22	0.45 ± 0.10 (100)
PB	21	1.56 ± 0.27 (355)***	19	0.86 ± 0.15 (130)	24	0.17 ± 0.03 (38)***
MC	24	4.15 ± 0.70 (943)***	16	2.13 ± 0.40 (323)***	22	32.4 ± 9.82 (7200)***
tSBO	18	1.68 ± 0.38 (382)***	9	1.42 ± 0.28 (215)***	9	0.21 ± 0.04 (47)**
AAF	9	0.64 ± 0.24 (145)	9	1.31 ± 0.27 (198)***	9	3.42 ± 0.70 (760)***

[a]Microsomes from untreated animals or animals receiving vehicle only (0.9% NaCl, corn oil or polyethylene glycol 300) demonstrated the same activities and these values were therefore pooled and designated Control. PB = phenobarbital, MC = 3-methylcholanthrene, tSBO = trans-stilbene oxide, AFF = 2-acetylaminofluorene.

[b]number of animals

[c]means ± standard deviations

* significantly different from Control value at the level of $P < 0.01$

** significantly different from Control value at the level of $P < 0.005$

*** significantly different from Control value at the level of $P < 0.001$

TABLE 4

PATTERN OF 2-ACETYLAMINOFLUORENE METABOLITES OBTAINED WITH LIVER MICROSOMES FROM INDUCED AND CONTROL RATS

Rats were induced, microsomes prepared and the metabolite pattern determined as described in the Materials and Methods. The values given represent the means ± standard deviations for six rats in each group. The values in parentheses are the percentage of the total metabolites accounted for by each individual metabolite. Control animals were injected with vehicle (0.9% NaCl, corn oil or polyethylene glycol 300) only. PB = phenobarbital, MC = 3-methylcholanthrene, AAF = 2-acetylaminofluorene. The metabolites are listed in order of their appearance (from earliest to last) on the high pressure liquid chromatogram.

	AAF metabolites formed (pmol/min-nmol cytochrome P-450)							
Treatment	Peak 1	7-hydroxy	9-hydroxy	5-hydroxy	3-hydroxy	1-hydroxy	N-hydroxy	total
Control	14.5 ± 3.3	143 ± 10	45.0 ± 3.8	2.60 ± 0.71	8.11 ± 0.92	14.2 ± 1.2	32.0 ± 4.6	259 ± 24
	(5.6)	(55.1)	(17.4)	(1.0)	(3.1)	(5.5)	(12.3)	(100)
PB	12.7 ± 3.7	177 ± 11	89.5 ± 4.9	3.12 ± 0.51	14.7 ± 2.4	4.6 ± 1.9	5.2 ± 2.3	307 ± 14
	(4.1)	(57.8)	(29.2)	(1.0)	(4.8)	(1.5)	(1.7)	(119)
MC	63.5 ± 27.3	1471 ± 280	175 ± 51	1218 ± 223	881 ± 174	22.4 ± 7.1	98.1 ± 24.3	3930 ± 815
	(1.8)	(37.3)	(4.4)	(31.0)	(22.3)	(0.6)	(2.5)	(1517)
AAF	41.7 ± 20.6	472 ± 152	131 ± 14	218 ± 104	175 ± 73	16.8 ± 7.5	92.8 ± 4.4	1162 ± 418
	(3.5)	(41.5)	(12.1)	(18.5)	(14.9)	(1.5)	(7.9)	(444)

agree with those presented above, i.e., trans-stilbene oxide induces the same isozyme of cytochrome P-450 as is induced by phenobarbital, while 2-acetylaminofluorene induces a form of cytochrome P-450 which is distinct from those induced by phenobarbital and 3-methylcholanthrene. A recent article by Levin and coworkers (19) arrived at the same conclusion with regards to trans-stilbene oxide.

In addition, the investigations presented here indicate that 2-acetylaminofluorene induces a form of cytochrome P-450 specialized for the metabolism of this substance itself, i.e., the process can be called substrate induction. The significance of this fact with regards to the carcinogenicity of 2-acetylaminofluorene cannot be assessed at the present time. Most hypotheses concerning production of the carcinogenic metabolite(s) from this xenobiotic postulate that formation of the N-hydroxy derivative is an essential step. Therefore, the fact that 2-acetylaminofluorene increases its own hydroxylation 4.5-fold may mean that this compound also increases its own toxicity and carcinogenecity.

ACKNOWLEDGEMENTS

These studies were supported by grants from the National Cancer Institute, Bethesda, Maryland, U.S.A. (grant CA 26261-03, "The metabolism of polycyclic hydrocarbons and cancer"), from the Swedish Medical Research Council, and from the Swedish Natural Science Research Council.

REFERENCES

1. Lu, A.Y.H. (1979) Drug Metab. Rev., 10, 187.
2. Seidegård, J.E., Morgenstern, R., DePierre, J.W. and Ernster, L. (1979) Biochim. Biophys. Acta, 586, 10.
3. Åström, A. and DePierre, J.W. (1981) Biochim. Biophys. Acta, 673, 225.
4. Prough, R.A., Burke, M.D. and Mayer, R.T. (1978) in Fleisher, S. and Packer, L. (Eds.), Methods in Enzymology, Academic Press, New York, Vol. L11 Part C, pp. 372-377.
5. Jackson, H.L. and McKusick, B.C. (1955) Org. Syntheses, 35, 62.
6. Seidegård, J.E., DePierre, J.W., Moron, M.S., Johannesen, K. and Ernster, L. (1977) Cancer Res., 37, 1075.
7. Åström, A. and DePierre, J.W. Carcinogenesis, in press.
8. Ernster, L., Siekevitz, P. and Palade, G.E. (1963) J. Cell Biol., 15, 541.
9. Omura, T. and Sato, R. (1964) J. Biol. Chem., 239, 2370.
10. Meijer, J., Åström, A., DePierre, J.W., Guengerich, F.P. and Ernster, L., submitted for publication.

11. Lu, A.Y.H., Strobel, H.W. and Coon, M.J. (1970) Mol. Pharmacol., 6, 213.

12. DePierre, J.W., Moron, M.S., Johannesen, K.A.M. and Ernster, L. (1975) Analyt. Biochem., 63, 470.

13. Jernström, B., Vadi, H. and Orrenius, S. (1976) Cancer Res., 36, 4107.

14. Smith, C.L. and Thorgeirsson, S.S. (1981) Analyt. Biochem., 113, 62.

15. Sladek, N.E. and Mannering, G.J. (1966) Biochem. Biophys. Res. Commun., 23, 668.

16. Guengerich, F.P., Wang, P. and Davidson, N.K. Biochemistry, in press.

17. Laemmli, U.K. (1970) Nature, 227, 680.

18. Lowry, O.H., Rosebrough, N.J., Farr, A.L. and Randall, R.J. (1951) J. Biol. Chem., 193, 265.

19. Thomas, D.E., Reik, L.M., Ryan, D.E. and Levin, W. (1981) J. Biol. Chem., 256, 1044.

Cytochrome P-450, Biochemistry, Biophysics
and Environmental Implications, E. Hietanen,
M. Laitinen and O. Hänninen editors

ISOTOPE EFFECT AND METABOLIC SWITCHING IN THE CYTOCHROME P-450-CATALYZED O-DEETHYLATION OF 7-ETHOXYCOUMARIN

A. Y. H. LU, N. HARADA, J. S. A. WALSH AND G. T. MIWA
Department of Animal Drug Metabolism, Merck Sharp & Dohme Research Laboratories, Rahway, NJ 07065 (U.S.A.)

INTRODUCTION

For many of the cytochrome P-450-catalyzed reactions, such as N- and O-dealkylations and aliphatic hydroxylation, the incorporation of an atom of molecular oxygen into the substrates requires the cleavage of a carbon-hydrogen bond. Thus, the substitution of deuterium or tritium for the hydrogen involved in the bond cleavage step and the measurement of the primary kinetic isotope effect associated with the reactions provide a valuable approach to study the mechanism of cytochrome P-450-catalyzed reactions. We have, therefore, studied both the inter- and intra-molecular isotope effects in cytochrome P-450-catalyzed N-demethylation of N,N-dimethylphentermine (1) and N,N-dimethylaniline (2) and interpreted these results in terms of how these substrates are processed by cytochrome P-450. We have now extended our studies to another cytochrome P-450-catalyzed reaction, the O-deethylation of 7-ethoxycoumarin. In this report, we describe the isotope effect for this reaction catalyzed by cytochrome P-448 and the metabolic switching to form a new metabolite when deuterium substitution slows the O-deethylation pathway.

MATERIALS AND METHODS

7-Dideuteroethoxycoumarin (d_2-7EC) was prepared by refluxing 7-hydroxycoumarin with 1,1-d_2-iodoethane and K_2CO_3 in alkaline ethanol. Highly purified cytochrome P-448 from 3-methylcholanthrene-treated rats (3) and NADPH-cytochrome P-450 reductase (4) were used in all studies unless otherwise indicated. The O-deethylation of non-deuterated 7EC (d_0-7EC) and d_2-7EC was measured fluorimetrically (5) by following the formation of the common product, 7-hydroxycoumarin (Scheme 1). When the total metabolic profile of 7EC was investigated, the reaction mixture was extracted with chloroform, and the extract was chromatographed by HPLC on a C_{18} column using a linear gradient of methanol (40% to 100%, 8 min) in water. Both substrate and metabolite were monitored by UV at A_{320}. The steady state level of oxy-cytochrome P-450 in microsomes in the presence of either d_o-7EC or d_2-7EC was determined by the method of Werringloer and Kawano (6).

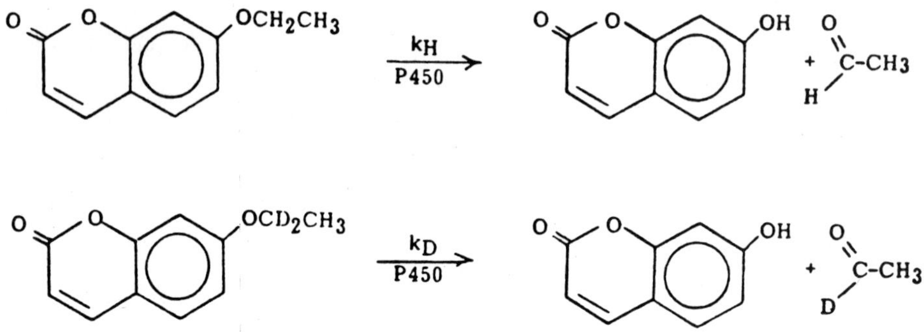

Scheme 1. Cytochrome P-450-catalyzed O-deethylation of 7-ethoxycoumarin.

RESULTS

<u>Effect of deuterium substitution on stoichiometry</u>. Substitution of deuterium for the α-hydrogen which undergoes bond cleavage during the O-deethylation of 7EC results in a deuterium isotope effect of approximately 2 in V_m, but not in K_m, for the cytochrome P-448 isozyme. Since the C-H bond cleavage step occurs late in the catalytic cycle in cytochrome P-450-catalyzed reaction, one would suspect that the isotope effect in the C-H bond cleavage following deuterium substitution could result in the accumulation of cytochrome P-450 intermediary complexes. To examine such a possibility, the steady state levels of the

oxy-cytochrome P-450-7EC complex in rat liver microsomal suspensions were determined using both d_o-7EC and d_2-7EC substrates. Table 1 shows an observed deuterium isotope effect of 2.16 for microsomes from 3-methylcholanthrene-treated rats and 3.87 for microsomes from phenobarbital-treated rats during O-deethylation of 7EC. Moreover, the steady state levels of the oxy-cytochrome P-450-substrate complex were virtually identical for both d_o-7EC and d_2-7EC, indicating the lack of accumulation of this intermediate following deuterium substitution. This is also consistent with the observation that the rate of H_2O_2 formation by the purified system remains the same despite deuterium substitution since the H_2O_2 arises from the oxy P-450 intermediates (7,8). For both d_o and d_2 substrates, the overall oxidase activity of the reconstituted cytochrome P-448 system appears to be identical judging from the same utilization of NADPH, oxygen and substrate. These results led us to believe that metabolites other than 7-hydroxycoumarin may have been formed when d_2-7EC is used as a substrate for cytochrome P-448.

TABLE 1

STEADY STATE LEVELS OF OXY-CYTOCHROME P-450-SUBSTRATE COMPLEX

Microsomal Preparation	K_H/K_D	% Cytochrome P-450 exists as oxy-P-450-substrate complex	
		d_o-7EC	d_2-7EC
3-Methylcholanthrene-treated rats	2.16	18%	18%
Phenobarbital-treated rats	3.87	30%	32%

Effect of deuterium substitution on metabolic profile. Since the fluorimetric assay (5) only measures the O-deethylation of 7EC, the total metabolic profile for 7EC was established by HPLC analysis. Fig. 1 shows that when d_o-7EC was incubated with the purified cytochrome P-448 system in the presence of NADPH, the major product was 7-hydroxycoumarin (retention time, 6.4 min), with a trace amount of another metabolite with retention time of 7.5 min. When

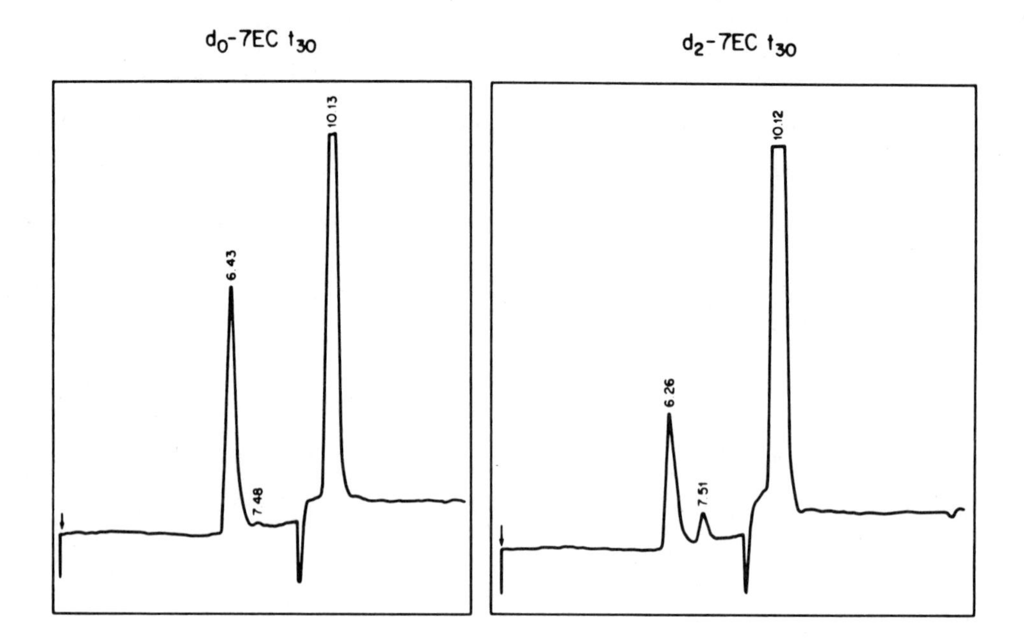

Fig. 1. Metabolism of d_0-7EC (left) and d_2-7EC (right) by the purified cytochrome P-448 system in the presence of NADPH. HPLC fractions were monitored at A_{320}. Retention times (min) are approximately 6.4 for 7-hydroxycoumarin, 7.5 for 6-hydroxy-7EC and 10.1 for 7EC.

d_2-7EC was used as the substrate, the amount of 7-hydroxycoumarin formed was decreased to about half, consistent with the observed deuterium isotope effect of 2 when 7-hydroxycoumarin was assayed fluorimetrically. In addition, considerable amount of another metabolite was formed. NMR and MS analyses have identified this metabolite as 6-hydroxy-7EC.

Metabolic switching. When the amounts of 7-hydroxycoumarin and 6-hydroxy-7EC were determined from each incubation, a pronounced metabolic switching (9) becomes quite apparent after deuterium substitution. Table 2 shows that with d_0-7EC as substrate, the ratio of the two products is 20 to 1 in favor of

TABLE 2

METABOLIC SWITCHING IN 7-ETHOXYCOUMARIN METABOLISM AS A RESULT OF DEUTERIUM SUBSTITUTION

	RELATIVE PRODUCT FORMATION	
	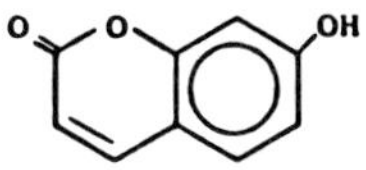	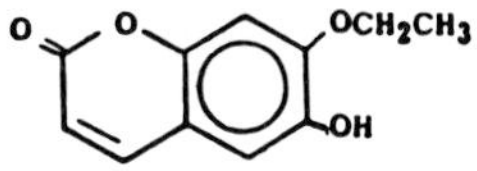
SUBSTRATE	7-HYDROXYCOUMARIN	6-HYDROXY-7-ETHOXYCOUMARIN
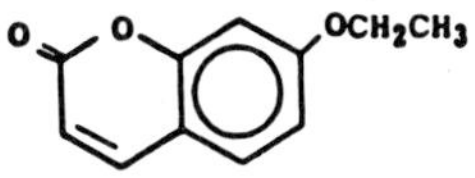 D_0 -7-ETHOXYCOUMARIN	20	1
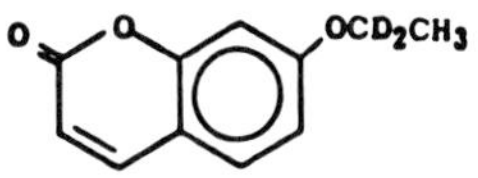 D_2 -7-ETHOXYCOUMARIN	10	5

7-hydroxycoumarin. In the case of d_2-7EC, approximately 5-fold more 6-hydroxy-7EC is formed and thus lowers the ratio of the two products to 2 to 1. Such a metabolic switching can also be demonstrated when cumene hydroperoxide was used as an oxidizing agent to replace NADPH and NADPH-cytochrome P-450 reductase. Fig. 2 again shows the production of 6-hydroxycoumarin when d_2-7EC was metabolized by cytochrome P-448 in the presence of cumene hydroperoxide.

CONCLUSION

The use of deuterium to replace hydrogen at the site of C-H bond cleavage in the O-deethylation of 7EC has yielded useful information concerning the mechanism of this cytochrome P-448-catalyzed reaction. Based on our results, several comments can be made:

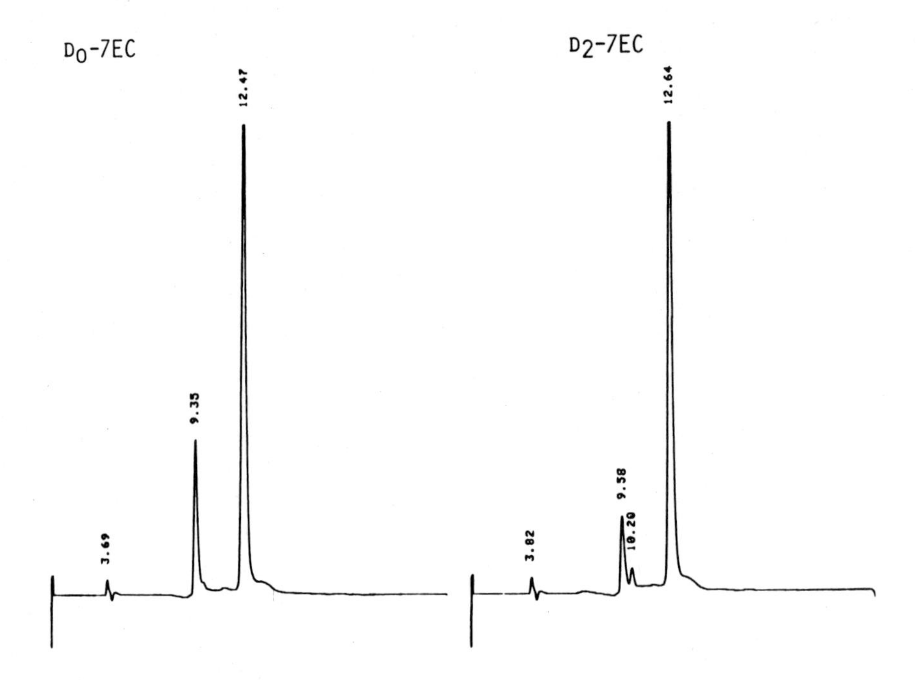

Fig. 2. Metabolism of d_o-7EC and d_2-7EC by purified cytochrome P-448 in the presence of cumene hydroperoxide. HPLC fractions were monitored at A_{320}. Retention times (min.) are approximately 9.4 for 7-hydroxycoumarin, 10.2 for 6-hydroxy-7EC and 12.5 for 7EC.

1. Although deuterium substitution results in a significant decrease in the O-deethylase activity during 7EC metabolism, it does not alter either the overall oxidase activity or the uncoupling of the 7EC-P448-O_2 complex.

2. Deuterium substitution of 7EC results in a pronounced metabolic switching away from the ethoxy group to the aromatic ring. This metabolic switching occurs in both NADPH/O_2 - and cumene hydroperoxide-supported reactions.

3. Scheme 2 shows the proposed metabolic switching mechanism. P450·S represents the P450-O_2-substrate complex following the transfer of the second electron. The formation of the active oxygen species, $P450^*$·S, is governed by k_1. In the case of d_o-7EC, the transfer of the active oxygen to the substrate

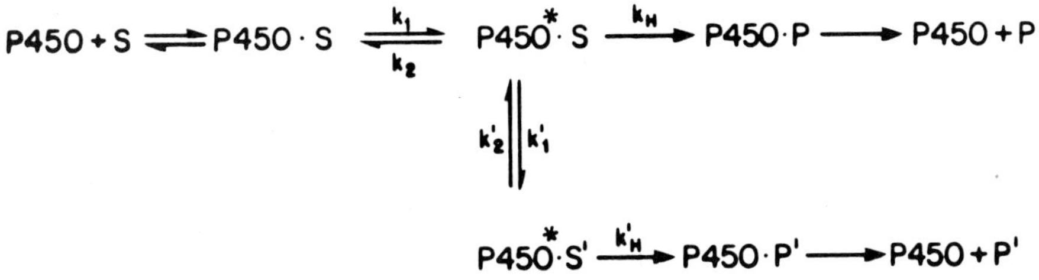

$$P450 + S \rightleftharpoons P450 \cdot S \underset{k_2}{\overset{k_1}{\rightleftharpoons}} P450^{*} \cdot S \xrightarrow{k_H} P450 \cdot P \longrightarrow P450 + P$$

$$k_2' \Big\uparrow\Big\downarrow k_1'$$

$$P450^{*} \cdot S' \xrightarrow{k_H'} P450 \cdot P' \longrightarrow P450 + P'$$

Scheme 2. Proposed metabolic switching mechanism.

results in the formation of P, 7-hydroxycoumarin. When d_2-7EC is used as substrate, the conversion of $P450^{*} \cdot S$ to $P450 \cdot P$ is decreased due to the higher energy barrier for C-D bond cleavage. In this case, the substrate can dissociate from the enzyme-active oxygen and bind to the enzyme active site again in a different orientation to form $P450^{*} \cdot S'$ which then oxidizes the substrate in a different position to form 6-hydroxy-7EC, P′. Since deuterium substitution does not cause an accumulation of oxy-cytochrome P-450 complex or an increase in uncoupling to form H_2O_2, the formation of the enzyme-active oxygen species, $P450^{*} \cdot S$, must be irreversible. That is, k_2 must be kinetically insignificant. Thus, these data suggest that the active oxygen species is irreversibly formed and, once formed, is committed to catalysis.

ACKNOWLEDGEMENT

We wish to thank Ms. D. Sloan for her assistance in the preparation of this manuscript.

REFERENCES

1. Miwa, G. T., Garland, W. A., Hodshon, B. J., Lu, A. Y. H. and Northrop, D. B. (1980) J. Biol. Chem. 255, 6049-6054.

2. Miwa, G. T., Zweig, J. S., Walsh, J. S. A. and Lu, A. Y. H. (1980) In: M. J. Coon, et al. (eds.), Microsomes, Drug Oxidations and Chemical Carcinogenesis, Academic Press, New York, Vol. 1, pp. 363-366.

3. West, S. B., Huang, M. T., Miwa, G. T. and Lu, A. Y. H. (1979) Arch. Biochem. Biophys. 193, 42-50.

4. Yasukochi, Y. and Masters, B. S. S. (1976) J. Biol. Chem. 251, 5337-5344.

5. Greenlee, W. F. and Poland, A. (1978) J. Pharmacol. Exp. Ther. 205, 596-605.

6. Werringloer, J. and Kawano, S. (1980) In: J. A. Gustafsson, et al. (eds.), Biochemistry, Biophysics and Regulation of Cytochrome P-450, Elsevier/North-Holland, Amsterdam, pp. 359-362.

7. Estabrook, R. W. and Werringloer, J. (1977) In: D. M. Jerina (ed.), Drug Metabolism Concepts, American Chemical Society, Washington, D.C. pp. 1-26.

8. Ullrich, V. and Kuthan, H. (1980) In: J. A. Gustafsson, et al. (eds.), Biochemistry, Biophysics and Regulation of Cytochrome P-450, Elsevier/North-Holland, Amsterdam, pp. 267-272.

9. Horning, M. G., Thenot, J. P., Bouwsma, O., Nowlin, J. and Lertratanangkoon, K. (1978) In: J. P. Tillement (ed.), Advances in Pharmacology and Therapeutics, Vol. 7, Biochemical Clinical Pharmacology, Pergamon Press, Oxford, pp. 245-256.

Cytochrome P-450, Biochemistry, Biophysics
and Environmental Implications, E. Hietanen,
M. Laitinen and O. Hänninen editors

STEROID 11β-HYDROXYLASE SYSTEM: RECONSTITUTION AND STUDY OF INTERACTIONS AMONG PROTEIN COMPONENTS

SERGEY P. MARTSEV, IVAN A. BESPALOV, VADIM L. CHASHCHIN AND AFANASSI A. AKHREM
Institute of Bioorganic Chemistry, Byelorussian SSR Academy of Sciences, Minsk, 220600, USSR

INTRODUCTION

Steroid 11β-hydroxylase system is localized in the inner membrane of adrenocortical mitochodria and consists of three protein components: flavoprotein adrenodoxin reductase (AR), iron sulfur protein adrenodoxin (Adx) and hemoprotein cytochrome P-450 specific for the steroid 11β-hydroxylation ($P\text{-}450_{11\beta}$) (1). So far, the properties of the 11β-hydroxylase system reconstituted using highly purified $P\text{-}450_{11\beta}$ are not systematically studied, with the only exception of substrate specificity of 11β-hydroxylase (2). However, the study of this multienzyme system is necessary for elucidation of the molecular mechanisms controling the level of a number of biologically active steroids, including glucocorticods and mineral corticoids. The present work was aimed at a general description of catalytic properties of the reconstituted 11β-hydroxylase system.

MATERIALS AND METHODS

Isolation of protein components. Beef adrenal gland were used for protein isolation. Homogeneous AR and Adx were isolated according to method (3) and had spectrophotometric indexes A_{414}/A_{280}= 0.83-0.85 for Adx and A_{272}/A_{450}= 7.2-8.2 for AR.

$P\text{-}450_{11\beta}$ was isolated by the method (4), contained 19-20 nmol of heme per 1 mg of protein and was homogeneous as judged by SDS-PAAG electrophoresis. Spectrophotometric indexes of protein were: A_{394}/A_{280}= 0.9-1.0 and $A_{394-470}/A_{418-470}$= 2.2-2.5. Monomeric molecular weight of $P\text{-}450_{11\beta}$ was determined to be 47,000±1,000 as judged by SDS-PAAG electrophoresis. N-terminal analysis revealed a single amino acid (glycine). Isoleucine (N-terminal amino acid of P-450 specific for cholesterol side chain cleavage) was not detected even in trace amounts.

Determination of 11β-hydroxylase activity. Reaction mixture

contained 0.01% Tween 20, 0.08 μM AR, 8 μM Adx, 0.1 μM P-$450_{11\beta}$, 60 μM 11-deoxycorticosterone, 1.5 μCi of 11-deoxy-[1α,2α(n) -^{3}H]-corticosterone, 120 μM NADPH, 5 mM glucose-6-phosphate, 1 unit of glucose-6-phosphate dehydrogenase in 1 ml of sodium phosphate buffer (0.01 M, pH 7.4). When concentration of one component was varied, the remaining concentrations were kept unchanged. Reaction was allowed to proceed for 4-5 min at 20°C. Reaction mixture was further processed essentially according to (2). Enzyme activity was characterized in terms of molar activity (MA) and expressed in moles of product formed per 1 s per 1 mol of P-$450_{11\beta}$.

NADPH-oxydase activity was determined by spectroscopic monitoring of a decrease in A_{340} of the solution using ε_{340}=6.2 $mM^{-1}\cdot cm^{-1}$. The reaction mixture was identical to those used in determination of 11β-hydroxylase activity with the only exception: NADPH (200 μM) was used instead of NADPH-regenerating system. Reaction was monitored during 4-5 min at 20°C.

Cytochrome c reductase activity was determined spectrophotometrically according to (5). Reaction mixture contained 0.03 μM AR, 0.9 μM Adx, 21 μM cytochrome c, 1.25 mM NADPH in 1.6 ml volume of sodium phosphate buffer (0.01 M, pH 7.4).

Catalytic activity of AR was determined by 2,6-dichlorphenol indophenol reduction test. Reaction mixture contained in 1.6 ml of sodium phosphate buffer (0.01 M, pH 7.4) 0.06 μM AR, 31 μM DCPIP and 62 μM NADPH. Reaction was followed spectrophotometrically by monitoring a decrease in A_{600} of a solution using ε_{600}= 20 $mM^{-1}\cdot cm^{-1}$. Reaction was allowed to proceed for 4-5 min at 20°C.

When studying the dependence of enzyme activities on detergent concentrations, the reaction mixture was incubated for 15 min at 20°C in a detergent solution befor initiation of the reaction.

Concentrations of P-$450_{11\beta}$ was determined by the method of Omura and Sato (6) using $\varepsilon_{450-490}$= 91 $mM^{-1}\cdot cm^{-1}$ or, alternatively, using ε_{394}= 90 $mM^{-1}\cdot cm^{-1}$ for a solution of pure oxidized protein.

RESULTS AND DISCUSSION

The reconstitution of the active 11β-hydroxylase system has been accomplished using homogeneous AR, Adx and P-$450_{11\beta}$. The ratios of the protein components in the multienzyme system necessary for its maximal catalytic activity have been determined from the experiments presented in Fig.1. Catalytic activity of 11β-hydroxylase system hyperbolically depends on concentrations of AR and Adx. The concentrations of 9 nM for AR and 280 nM for Adx provide a half-maximal reaction rate. The obtained data show that for a maximal catalytic activity to be manifested, the AR/P-$450_{11\beta}$ ratio should be closed to unity, with a 20-50-fold molar excess of Adx over the hemoprotein. Similar results were obtained earlier from the investigations of camphor hydroxylase (7) and the cholesterol side chain cleavage system (8) as well as from the study of a partially purified P-$450_{11\beta}$ (9).

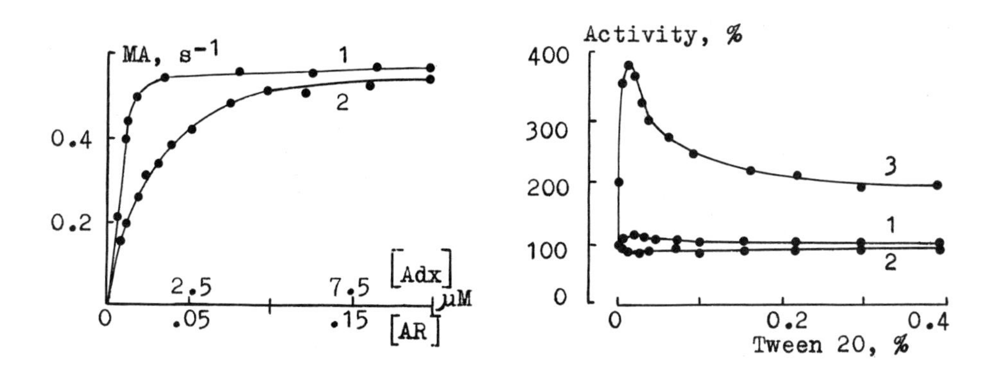

Fig. 1(left). Dependence of molar activity (MA) of 11β-hydroxylase system on Adx (1) and AR(2) concentrations.

Fig. 2(right). Effect of Tween 20 on: DCPIP reduction (1), cytochrome c reduction (2), 11β-hydroxylase activity (3).

A non-ionic detergent Tween 20 has been used for the reconstitution of the 11β-hydroxylase system, since its presence is absolutely necessary to maintain P-$450_{11\beta}$ in a soluble state. The preliminary experiments showed that of the three non-ionic detergent examined (Tween 20, Tween 80 and Tritone X-100), the first one exerts the least denaturating effect on P-$450_{11\beta}$. Since non-ionic detergents are known to alter the properties of the enzyme systems containing membrane proteins, we have investigated a relationships between catalytic parameters of the reconstituted 11β-hydroxylase system and concentrations of Tween 20. The experimental results are presented in Figs 2-4.

As follows from Fig.2, the enzyme activity of AR is practical-

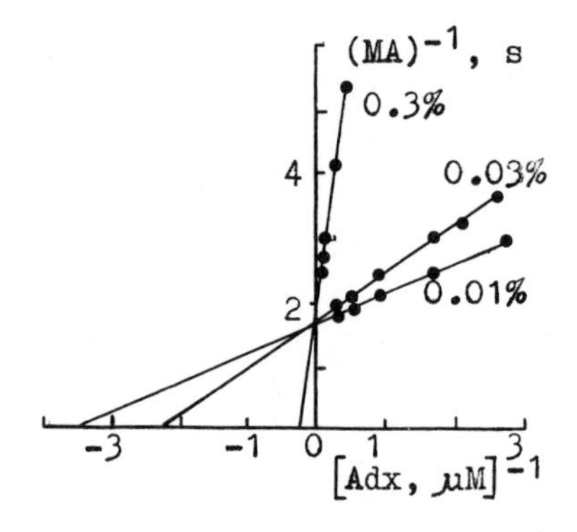

Fig. 3. Effect of Tween 20 on adrenodoxin dependence of 11β-hydroxylase activity.

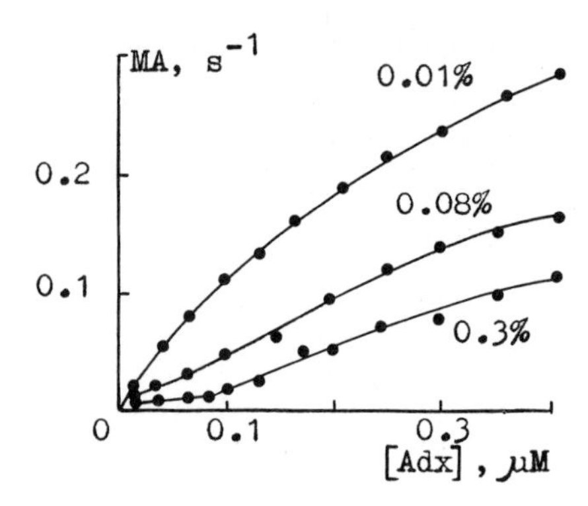

Fig. 4. Effect of Tween 20 on adrenodoxin dependence of 11β-hydroxylase activity over the range of low Adx concentrations.

ly unaffected by Tween 20 in concentrations up to 0.4%, as well as the activity of the system containing AR, Adx and cytochrome c. A substitution of the terminal component in the electron transport system, cytochrome c, by P-$450_{11\beta}$, drasticaly alters the system's response to the detergent effect, as expressed in a complex character of changes in the enzyme activity following an increase in Tween 20 concentration. The activation phase of the 11β-hydroxylase reaction has a maximum in the 0.010-0.015% region of detergent concentration. Assumingly, under these detergent concentrations P-$450_{11\beta}$ remains completely soluble over the whole catalytic cycle. The following phase of inhibition of the enzyme activity might be caused by: i) a denaturating action of Tween 20 directly on P-$450_{11\beta}$, since AR and Adx are stable against detergents; ii) a disturbance of the interaction between Adx and hemoprotein, since such interaction governs the catalytic activity of the hydroxylase system.

Fig.3 illustrates the dependence of the parameters of interaction between Adx and P-$450_{11\beta}$ on Tween 20 concentration. An increase in molar activity of the 11β-hydroxylase system following the Adx concentration growth is a process dependent on Tween 20 concentration, although the catalytic constant of the reaction ($k_{cat} = V_{max}/E_0$) remaines unchanged. The apparent K_m values of the process increases from 280 nM (0.01% Tween 20) to 5 µM (0.3% Tween 20). The picture of effector kinetics shown in Fig.3 indicates to a possible competetion between Tween 20 and Adx for binding with P-$450_{11\beta}$. We assume this interaction to be apparently competitive, since the Adx-binding site on the P-$450_{11\beta}$ molecule may be inaccessible for Adx as a result of binding of a bulky detergent micelle with a hydrophobic site of hemoprotein sterically remote from the Adx-binding site.

Of particular interest is the study of the effect of Tween 20 on the process of interaction between Adx and P-$450_{11\beta}$ over the range of Adx/P-$450_{11\beta}$ ratios close to unity (at Adx concentrations near K'_m value). Under these conditions some effects obscure at high Adx concentrations could be fully manifested. The results of these studies are presented in Fig.4. Under varying Adx concentrations in the 0.02-0.4 μM region the caracter of dependence of 11β-hydroxylase activity on Adx concentration differed from hyperbolic. To characterize the sigmoids obtained was used the parameter q similar to Hill coefficient, n, as calculated by the difference method of Kurganov (15). The q values calculated for the 0.075-0.2 μM concentration range were 0.8, 1.8 and 2.4 for curves obtained at 0.01%, 0.08% and 0.3% Tween 20 concentrations. Hence an increase in concentration of Tween 20 produces gradually pronounced sigmoids for the dependence of the activity of 11β-hydroxylase system on Adx concentration.

The above data prove that the concentration growth of Tween 20 entails a heterogenity increase of the reaction sites in the multienzyme system with respect to the interaction with Adx as revealed by higher catalytic efficiency of the enzyme system at higher Adx concentrations. The observed heterogenity of the reaction sites might be the consequence of two reasons. The first may be an inactivating detergent effect on hemoprotein and stabilizing effect of Adx. The second reason may stem from a different degree of changes in constants of interaction among AR and Adx, Adx and P-$450_{11\beta}$ upon the concentration growth of Tween 20 which, in turn, results in a change in the character of interactions among these three proteins. This assumption is based on the fact that under the effect of Tween 20 the catalytic activity of AR-Adx complex does not practically change, whereas the catalytic parameters of interaction of Adx with P-$450_{11\beta}$ change significantly (Figs 2-4).

The results presented in Figs 1 and 3-4 could be better explained based on the hypothesis that the steroid 11β-hydroxylase system functions as a dissociating binary complexes (10) (at least in the case of high concentrations of non-ionic detergents), than from the view point of other possible mechanisms. One cannot exclude the possibility that at very low concentratios of non-ionic detergents or without any, the mechanism of functioning of the 11β-hydroxylase system may be principally different from that realized at high detergent concentrations.

A simultaneous estimation of the processes of NADPH oxidation and the formation of 11β-hydroxylated product (corticosterone) enabled us to detect an uncoupling in these two processes, which suggests the formation in the reconstituted enzyme system of oxygen radicals whose formation is known for many electron transport systems, including P-450-mediated hydroxylases (11-12). It has been previously established that $O_2^{\dot{-}}$ is a primary oxygen radical which subsequently disproportionates with the formation of H_2O_2 (12).

As follows from the obtained data (Figs 5-6), in a complete enzyme system with 80-fold molar excess of Adx over hemoprotein, the substrate oxidation consumes only 25% of all reducing equivalents, whereas the formation of radical species requires the remaining 75%. In their turn, 75% of radical species formed are generated in the active site of hemoprotein and 25% - in the active site of Adx. The high spin form of hemoprotein generates radical species considerably faster, than does the low spin form. A transition from a low to high spin form induced by the addition of an excess of a substrate into the reaction medium substantially stimulates the electron transport in the system, its rate increasing 4-fold. 65% of the additional reducing equivalents are consumed by the radical process, only 35% being spent on the process of substrate hydroxylation.

A physiological significance of an uncoupling in NADPH oxidation and substrate hydroxylation in the 11β-hydroxylase system is not clearly understood. It has been recently established by Hornsby in the experiments on cultured adrenocortical cells that the substrates of 11β-hydroxylase reaction accelerate the process of destruction of $P\text{-}450_{11\beta}$, the effect being removed by antioxidants

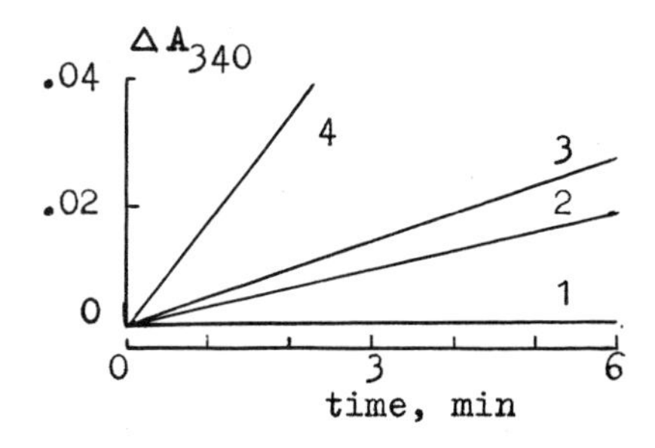

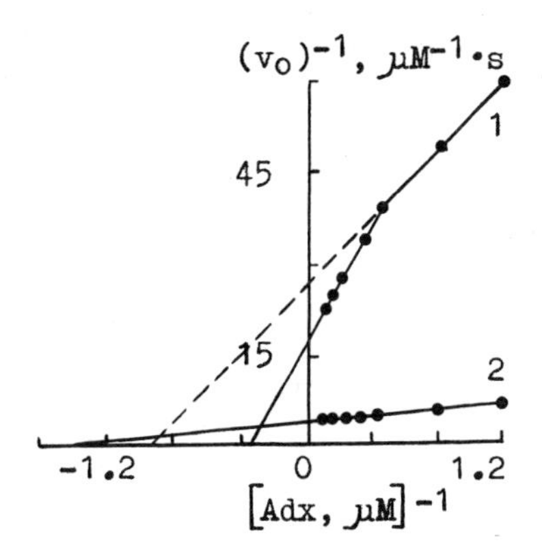

Fig. 5(left). Time course of NADPH oxydation by enzyme system containing NADPH and additionally: 1 - AR, 2 - AR+Adx, 3 - AR+Adx+P-$450_{11\beta}$, 4 - AR+Adx+P-$450_{11\beta}$+DOC. Initial rates were: 1 - 0.00, 2 - 0.040, 3 - 0.057, 4 - 0.223 ($\mu M \cdot s^{-1}$). Initial rate of corticosterone formation was 0.058 $\mu M \cdot s^{-1}$.
Fig. 6(right). Adrenodoxin dependence of NADPH oxidation in complete 11β-hydroxylase system (2) and in the system containing AR, Adx and NADPH (1).

(13). Our experimental data on the activation of the process of radical formation by the substrate form a basis for the interpretation of this phenomenon, taking into account the established fact that radical species formed as a result of uncoupling in the NADPH oxidation and substrate hydroxylation can cause a hemoprotein destruction (14). The above phenomenon may be related to the physiological regulation of the rate of substrate conversion in the 11β-hydroxylase system.

To elucidate a dependence of the radical process on the ratio of components in the 11β-hydroxylase system, we have investigated the NADPH oxidase activity of the system at concentrations of Adx varying from 0.1 to 10 μM (Fig.6). It has been established that the rate of radical formation increase with Adx concentration. The K'_m value of this process (670 nM) is close to that of 11β-hydroxylase reaction (280 nM) which provides a parallel rate increase for both processes. From these data as well as from the fact that only 25% of the total amount of radicals is formed in the active site of Adx, it follows that a change in Adx/P-$450_{11\beta}$

ratio cannot drasticaly alter the relationship between the processes of hydroxylation and formation of oxygen radicals. The intensive radical generation is primarily caused by the properties of hemoprotein rather than high molar excess of Adx over P-$450_{11\beta}$ employed in the reconstitution of the enzyme system.

REFERENCES

1. Mitani, F. (1979) Mol.Cell.Biochem., 24, 21.

2. Sato, H., Ashida, N., Suhara, K., Takemori, S., Katagiri, M. (1978) Arch.Biochem.Biophys., 190, 307.

3. Akhrem, A.A., Shkumatov, V.M., Chashchin, V.L. (1977) Bioorganic Chem., Russ.,6, 780.

4. Martsev, S.P., Chashchin, V.L., Akhrem, A.A. (1982) Biochemistry, Russ., 47 (in the press).

5. Omura, T., Sanders, E., Estabrook, R.W., Cooper, D., Rosenthal, O. (1966) Arch.Biochem.Biophys., 117, 660.

6. Omura, T., Sato, R. (1964) J.Biol.Chem., 234, 2370.

7. Tyson, C.A., Lipscomb, J.D., Gunsalus, I.C. (1972) J.Biol. Chem., 347, 5777.

8. Hanucoglu, I., Jefcoate, C.R. (1980) J.Biol.Chem., 255,3057.

9. Mitani, F., Ichiyama, A., Masuda, A., Ogata, I. (1975) J.Biol.Chem., 250, 8010.

10. Seybert, D.W., Lambeth, J.D., Kamin, H. (1978) J.Biol.Chem., 253, 8355.

11. Lipscomb, J.D., Sligar S.G., Namtvedt,M.J., Gunsalus, I.C. (1976) J.Biol.Chem., 251, 1116.

12. Kuthan, H., Tsuji, H., Graft, H., Ulrich, V., Werringloer,J., Estabrook, R.W. (1978) FEBS Letters, 91, 343.

13. Hornsby, P.J. (1980) J.Biol.Chem., 255, 4020.

14. Loosemore, M., Light, D.R., Walsh, C. (1980) J.Biol.Chem., 255, 9017.

15. Kurganov, B.I. (1978) in: Allosteric Enzymes, Nauka, Moskow, pp. 27-41.

Cytochrome P-450, Biochemistry, Biophysics and Environmental Implications, E. Hietanen, M. Laitinen and O. Hänninen editors

STUDIES ON SEX-RELATED DIFFERENCE OF CYTOCHROME P-450 IN THE RAT: PURIFICATION OF A CONSTITUTIVE FORM OF CYTOCHROME P-450 FROM LIVER MICROSOMES OF UNTREATED RATS

RYUICHI KATO AND TETSUYA KAMATAKI
Department of Pharmacology, School of Medicine, Keio University, 35 Shinanomachi, Shinjuku-ku, Tokyo 160, (Japan)

INTRODUCTION

Marked sex differences in the drug responses in the rat have been well documented. The sex differences have been clarified to be caused by differences in the activities of drug metabolizing enzymes in liver microsomes (1). Cytochrome P-450 in liver microsomes plays a central role as a drug metabolizing enzyme. Thus, sex differences in the properties of cytochrome P-450 have been extensively studied using liver microsomes. Schenkman et al. (2) demonstrated that the differences in the drug metabolizing activities could be accounted for by the difference in the affinity of drugs to cytochrome P-450. Further, Gigon et al. (3) reported that the sex difference in the N-demethylation of ethylmorphine paralleled the substrate-enhanced rate of reduction of cytochrome P-450. These studies suggested that microsomes of male and female rats contained different forms of cytochrome P-450. Recent studies, however, have provided many lines of evidence showing that there are multiple forms of cytochrome P-450 (4-6). Considering the presence of multiple forms of cytochrome P-450 in liver microsomes, we initiated this study to determine which one or more forms of cytochrome P-450 with higher drug metabolizing activities existed in larger amounts in male microsomes than in female microsomes.

We report herein that multiple forms of cytochrome P-450 are involved in the occurrence of sex difference and that one form of cytochrome P-450, namely P-450-male, purified from microsomes of male rats, exists rather specifically in male microsomes.

MATERIALS AND METHODS

Purification of cytochrome P-450 from microsomes of male and female rats. Liver microsomes from untreated male and female rats of Sprague-Dawley strain (8-11 weeks old) were solubilized with sodium cholate (7,8), and were applied to a column of ω-amino-*n*-octyl Seph. 4B. Cytochrome P-450 was eluted in two peaks (I and II fractions) by washing of the column with two buffers (9,10). Since the I-fraction was assumed to contain a form(s) of cytochrome P-450 responsible for sex difference as judged by their 7-propoxycoumarin O-depropylation activities, which will be shown below, forms of cytochrome P-450 in the I fraction were further separated by DE-52 column. This column step allowed us to separate forms of cytochrome P-450 into several peaks. The main cytochrome P-450 fraction was subjected to further purification. With the use of two hydroxylapatite columns, cytochrome P-450 was purified to a electrophoretical homogeneity. Using essentially the same purification procedure, one of each form of cytochrome P-450 could be purified from male and female microsomes. The forms of cytochrome P-450 from male and female microsomes are tentatively termed as P-450-male and P-450-female, respectively.

Preparation of a specific antibody to P-450-male. Rabbit anti-P-450-male immunoglobulin G (IgG) was prepared as described previously (8). In an attempt to remove possible antibodies raised to a common antigenic site(s) of cytochrome P-450 in female microsomes, the anti-P-450-male IgG was passed through a column on which solubilized female microsomes had been adsorbed. Most of the IgG determined as protein could be recovered in fractions eluted near a void volume.

RESULTS AND DISCUSSION

Fig. 1 shows the elution pattern of cytochrome P-450 from ω-amino-n-octyl Seph. 4B column. Cytochrome P-450 in microsomes of male and female rats was divided into two peaks (I and II fractions). The recovery of cytochrome P-450 eluted in the I fractions of male and female microsomes ranged from 23.8 to 27.5% and from 19.3 to 31.5%, respectively, and in the II fractions of male and female rats from 15.4 to 19.3% and from 14.0 to 25.2%, respectively. Cytochrome P-450 in the II fractions from male and female

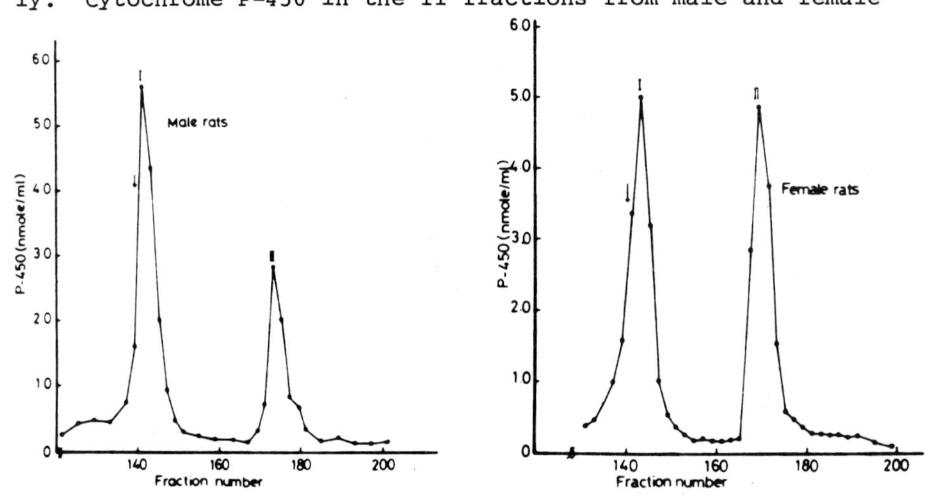

Fig. 1. Elution profiles from ω-amino-n-octyl Seph. 4B columns of cytochrome P-450 in microsomes of male (left) and female (right) rats. The I and II fractions were eluted with 10 mM potassium phosphate buffer (pH 7.4) containing 0.5% sodium cholate, 0.2% Emulgen 911, 0.1 mM EDTA, 1.0 mM dithiothreitol and 20% glycerol, and the same buffer as above except that the concentrations of phosphate and Emulgen 911 were 200 mM and 0.5%, respectively. At the points indicated by arrows in the charts, the buffer was changed (Kamataki et al., 1981).

TABLE I. 7-PROPOXYCOUMARIN O-DEPROPYLATION ACTIVITIES OF CYTOCHROME P-450 IN THE I AND II FRACTIONS OF MALE AND FEMALE RATS. The incubation mixture consisted of 0.1 nmol of cytochrome P-450, 0.5 unit of NADPH-cytochrome P-450 reductase, 15 ug of dilauroyl-L-3-phosphatidylcholine, and other necessary components as reported elsewhere (9). When necessary, 0.05 nmol of cytochrome b_5 was added.

Addition	Activity (nmol/nmol P-450/min)			
	Male I	Male II	Female I	Female II
None	0.313	0.512	0.055	0.791
Cytochrome b_5 (0.05 nmol)	0.748	0.534	0.090	0.779

rats had an absorption maximum at longer wave length than did cytochrome P-450 from corresponding I fractions: male and female I fractions showed absorption maxima at 450.5 and 449.0nm, while the II fractions from male and female rats showed at 449.0 and 448.0nm, respectively. As reported in a previous paper (11), we found that a marked sex difference was seen in the microsomal 7-propoxycoumarin O-depropylation activity in the rat. To know which one of the I and II fractions contained a form(s) of cytochrome P-450 responsible for the occurrence of sex difference, the O-depropylation activities of these partially purified preparations were compared after removing detergents with a hydroxylapatite column. As can be seen in Table I, a marked sex difference, in that male activity was higher than the female activity, was seen in the I fraction. Therefore, a form(s) of cytochrome P-450 reponsible for the sex difference was assumed to be contained in the I fraction.

Liver microsomes contain cytochrome b_5 as an electron carrier either to cytochrome P-450 (12) or cyanide sensitive factor. Accordingly, recent works from several laboratories have shown that cytochrome b_5 enhances the activity of cytochrome P-450 (13-15): the enhancement has been shown to occur depending on the forms of cytochrome P-450 and substrates employed. Addition of purified cytochrome b_5 to the incubation system containing the male and female I fractions resulted in the enhancement of the O-depropylation activities of cytochrome P-450. Therefore, most, if not all, of the forms of cytochrome P-450 in the I fractions responsible for sex difference were assumed to require cytochrome b_5 for maximal activities. This idea probably accounts for the sex difference in the NADH-synergism of NADPH-dependent drug oxidations. Correia and Mannering (16) reported that the NADH-synergism in ethylmorphine N-demethylase activity was seen in male rats to a larger extent than in female rats. In accordance with the result, we observed the NADH-synergism in 7-propoxycoumarin O-depropylase (31%) and benzphetamine N-demethylase (29%) activities in male microsomes but not in female microsomes (9). Forms of cytochrome P-450 in the male and female I fractions could be divided into several fractions by means of DE-52 column. The recoveries, specific contents, peaks in the carbon monoxide difference spectra and 7-propoxycoumarin O-depropylation activities of the isolated cytochrome P-450 fractions are shown in Table II. Major portions of cytochrome P-450 with higher specific contents were eluted in the I-b fractions in both male and female rats. However, the male I-b and other fractions did not correspond to the female fractions with respect to the peaks in the carbon monoxide difference spectra and the O-depropylation activities. Protein patterns analyzed on SDS-polyacrylamide gel electrophoresis indicated that the male and female fractions contain proteins with different molecular weights. One of each form of cytochrome P-450 in the male and female I-b fractions was purified to an apparent homogeneity on SDS-polyacrylamide gel electrophoresis by the use of hydroxylapatite columns. The forms of cytochrome P-450 purified

TABLE II. RECOVERIES, SPECIFIC CONTENTS, ABSORPTION PEAKS IN THE REDUCED-CARBON MONOXIDE DIFFERENCE SPECTRA AND 7-PROPOXYCOUMARIN O-DEPROPYLATION ACTIVITIES OF CYTOCHROME P-450 IN THE FRACTIONS ELUTED FROM DE-52 COLUMN.

Fraction		Recovery of P-450 (%)	Specific content* (nmol/mg)	Peak in CO-diff.* (nm)	O-Depropylation activity* (nmol/nmol P-450/min)
Male	I-a	7.1	0.73	451.0	0.000
	I-b	41.1	6.45	451.0	0.290
	I-c	9.9	1.87	450.5	0.463
	I-d	1.3	1.50	450.0	0.491
	I-e	4.1	1.86	449.0	0.641
	I-f	8.4	0.98	449.5	0.190
Female	I-a	5.1	1.16	450.5	0.032
	I-b	37.2	8.12	450.5	0.013
	I-c	6.6	2.94	449.5	0.050
	I-d	5.9	2.50	449.5	0.115
	I-e	2.0	ND**	449.5	ND**

* Specific content of cytochrome P-450, the absorption peak in the carbon monoxide difference spectrum and 7-propoxycoumarin O-depropylation activity were measured after concentrating the DE-52 column eluate by hydroxylapatite column.
** Not determined.

from microsomes of male and female rats were tentatively termed as P-450-male and P-450-female, respectively. The specific contents of P-450-male and P-450-female in final preparations ranged from 11.9 to 13.2 nmol/mg and from 13.6 to 16.3 nmol/mg, respectively. The peaks in carbon monoxide difference spectra were at 451.0 and 449.0 nm, and the molecular weights estimated on SDS-polyacrylamide gel electrophoresis were 52,000 and 50,000 for P-450-male and P-450-female, respectively. The catalytic activities of P-450-male and P-450-female are shown in Table III. P-450-male had higher activities than did P-450-female in all oxidation reactions except in ethylmorphine N-demethylation.

For immunochemical examinations of P-450-male and P-450-female, rabbit anti-P-450-male IgG was prepared. The anti-P-450-male IgG partially cross-reacted with P-450 female on Ouchterlony double

TABLE III. CATALYTIC ACTIVITIES OF CYTOCHROME P-450 PURIFIED FROM MALE AND FEMALE RATS.

Substrate	Activity (nmol/nmol P-450/min)	
	P-450-Male	P-450-female
7-Propoxycoumarin	0.808	0.020
Aniline	1.06	0.04
Aminopyrine	8.9	3.5
Ethylmorphine	15.9	18.2
Benzphetamine	13.9	2.4

diffusion analysis. Liver microsomes from female rats were dissolved with detergents and were adsorbed on a column of octyl Seph. 4B. To prepare a specific antibody, the IgG to P-450-male was passed through the column. The IgG thus prepared did not cross-react with P-450-female. The apparent absence of P-450-male in female microsomes could be demonstrated by Ouchterlony double diffusion and radial immunodiffusion (17) analyses, on which no detectable immunoprecipitation was formed between the purified anti-P-450-male IgG and solubilized female microsomes. The amount of P-450-male as determined by the radial immunodiffusion assay accounted for about 16% the total cytochrome P-450 in male microsomes (Fig. 2).

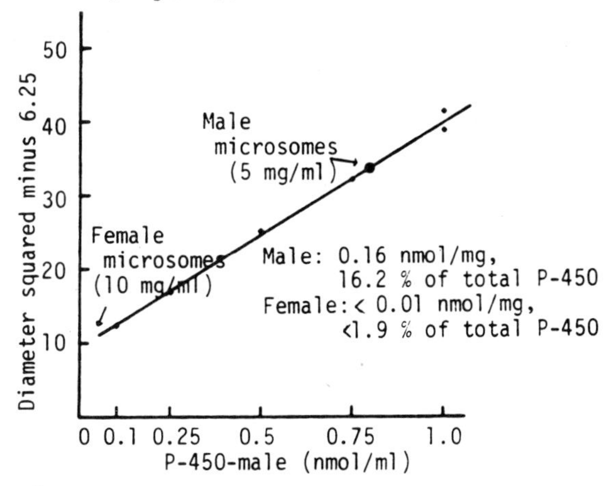

Fig. 2. Radial immunodiffusion assay for P-450-male in liver microsomes of untreated male and female rats.

SUMMARY

We showed evidence indicating that sex-related differences of drug metabolizing activities in rat liver microsomes were caused by multiple forms of cytochrome P-450. Further, we found that a particular form of cytochrome P-450, namely P-450-male, existed rather specifically in male microsomes.

ACKNOWLEDGEMENT

This work was supported by the Grant-in-Aid from the Ministry of Education, Science and Culture of Japan. The authors are indebted to Miss Kaori Maeda and Dr. Yasushi Yamazoe for their contribution in these experiments.

REFERENCES

1. Kato, R. (1974) Drug Metab. Rev., 3, 1.
2. Schenkman, J. B., Frey, I., Remmer, H. and Estabrook, R. W. (1967) Mol. Pharmac., 3, 516.
3. Gigon, P. L., Gram, T. E. and Gillette, J. R. (1968) Biochem. Biophys. Res. Commun., 31, 558.
4. Coon, M. J., Vermilion, J. L., Vatsis, K. P., French, J. S., Dean, W. L. and Haugen, D. A. (1977) in: Jerina D. M. (ED.) Drug Metabolism Concept, Amer. Chem. Soc., Washington, pp. 46-71.
5. Guengerich, F. P. (1979) Pharmac. Ther., 6, 99.
6. Lu, A. Y. H. and West, S. B. (1980) Pharmac. Rev., 31, 277.
7. Imai, Y. and Sato, R. (1974) J. Biochem., 75, 689.
8. Kamataki, T., Belcher, D. H. and Neal, R. A. (1976) Mol. Pharmac., 12, 921.
9. Kamataki, T., Maeda, K., Yamazoe, Y., Nagai, T. and Kato, R. (1981) Biochem. Biophys. Res. Commun., 103, 1.
10. Kamataki, T., Maeda, K., Yamazoe, Y., Nagai, T. and Kato, R. (1982) in R. Sato and R. Kato (Ed.) Microsomes, Drug Oxidations and Drug Toxicities, Japan Scientific Soc. Press, Tokyo, and John Wiley & Sons, London pp. 399-400.
11. Kamataki, T., Ando, M., Yamazoe, Y., Ishii, K. and R. Kato (1980) Biochem. Pharmac., 29, 1015.
12. Hildebrandt, A. and Estabrook, R. W. (1971) Archs. Biochem. Biophys. 143, 66.
13. Imai, Y. and Sato, R. (1977) Biochem. Biophys. Res. Commun., 75, 420.
14. Sugiyama, T., Miki, N. and Yamano, T. (1979) Biochem. Biophys. Res. Commun., 90, 715.
15. Kuwahara, S. and Omura, T. (1980) Biochem. Biophys. Res. Commun., 96, 1562.
16. Correia, M. A. and Mannering, G. J. (1973) Mol. Pharmac., 9, 470.
17. Thomas, P. E., Korzniowski, D., Ryan, D. and Levin, W. (1979) Archs. Biochem. Biophys., 192, 524.

Cytochrome P-450, Biochemistry, Biophysics and Environmental Implications, E. Hietanen, M. Laitinen and O. Hänninen editors

MULTIPLE FORMS OF CYTOCHROME P-450 IN BIOSYNTHESIS OF BILE ACIDS

KJELL WIKVALL, HANS BOSTRÖM, RONNIE HANSSON AND KERSTIN LUNDELL
Department of Pharmaceutical Biochemistry, University of Uppsala, Uppsala, Sweden

INTRODUCTION

Cholesterol and other C_{27}-steroids involved in the biosynthesis of bile acids are major physiological substrates for the microsomal cytochrome P-450 system of the liver. Recent work in this laboratory with electrophoretically homogeneous cytochrome P-450 fractions from rabbit liver microsomes has shown that cytochrome P-450 LM_4 prepared as described by Coon and associates (1) catalyzes hydroxylations of C_{27}-steroids in bile acid biosynthesis (2,3) as well as of a great number of other substrates (4).

The cytochrome P-450 LM_4 fraction is considered to be homogeneous according to a number of criteria (1,5). However, the properties of the hydroxylations in the biosynthesis of bile acids are such that it appears difficult to reconcile these with a single species of cytochrome P-450 LM_4. Work in this laboratory has demonstrated that the catalytic activity of the cytochrome P-450 LM_4 fraction is influenced by different pretreatments of the animals. Thus, cholesterol 7α-hydroxylation has only been observed with preparations from cholestyramine-treated rabbits whereas 12α- and 25-hydroxylations are catalyzed also by preparations from untreated, phenobarbital-treated animals (2,3). A difference in sulfhydryl group dependency for the cholesterol 7α-hydroxylase activity compared to the other C_{27}-steroid hydroxylase activities present in the cytochrome P-450 LM_4 fraction has also been observed (6).

The present communication presents evidence that multiple forms of cytochrome P-450 are involved in the biosynthesis of bile acids.

MATERIALS AND METHODS

Sources and preparations of the various materials and labeled substrates have been described previously (2,3).

Male rabbits of the New Zealand strain, weighing 2 to 3 kg, were used. The animals were used untreated or treated with phenobarbital (2), ß-naphthoflavone (1), cholestyramine (3) or were starved for 72 h prior to killing.

Cytochrome P-450 LM_4 from liver microsomes of differently treated rabbits was prepared as described by Coon and associates (1). Two hundred nmol of cytochrome P-450 LM_4 from cholestyramine-treated rabbits were dissolved in

100 mM potassium phosphate buffer, pH 7.25, containing 20% (v/v) glycerol, 1 mM EDTA and 0.7% (w/v) sodium cholate, and subjected to octylamine-Sepharose chromatography (column, 1 x 10 cm) as described by Imai and Sato (7). A minor fraction of cytochrome P-450 was eluted with the equilibrating buffer containing 0.46% sodium cholate. The main part of cytochrome P-450 was eluted with the equilibrating buffer containing 0.37% sodium cholate and 0.06% (W/v) Emulgen 913. Additional cytochrome P-450 was eluted by increasing the concentration of Emulgen to 0.2%. The cytochrome P-450 containing fractions eluted with the equilibrating buffer containing 0.46% sodium cholate were pooled to give cytochrome P-450 LM_4I and the fractions eluted with the buffer containing 0.2% Emulgen were pooled to give cytochrome P-450 LM_4II. All cytochrome P-450 fractions were treated with Amberlite XAD-2 and calcium phosphate (1) in order to remove the Emulgen. The fractions were dialyzed against 150 mM potassium phosphate buffer, pH 7.4, containing 20% glycerol and 0.1 mM EDTA.

NADPH-cytochrome P-450 reductase was prepared from phenobarbital-treated rabbits as described by Yasukochi and Masters (8) and had a specific activity of 55 units/mg protein (2).

Optical spectra were measured with calcium phosphate-treated enzyme fractions as described by Haugen and Coon (1) using a Cary 219 spectrophotometer.

Gel electrophoresis was performed in the presence of sodium dodecyl sulfate as described by Laemmli (9) or with Pharmacia gradient slab gels, PAA 4/30, as described previously (2).

One-dimensional peptide mapping was performed after limited proteolytic digestion with papin, chymotrypsin or trypsin according to Cleveland et al. (10). Electrophoresis was performed in the presence of sodium dodecyl sulfate using gradient slab gels.

Amino acid analyses were carried out on enzyme samples, 30-40 µg, that had been extensively dialyzed against 50 mM phosphate buffer, pH 7.4. The protein samples were hydrolyzed in 6 M HCl for 24, 48 and 72 h and the amino acid composition was determined with a Durrum D-500 amino acid analysator. Half-cystin was determined by performate oxidation and tryptophan according to Penke et al. (11).

Incubation procedures and analyses of the incubation mixtures were the same as described previously (3).

RESULTS AND DISCUSSION

Cytochrome P-450 LM_4 and 12α- and 25-hydroxylations of 5β-cholestane-3α,7α-diol.

Table 1 summarizes the 12α- and 25-hydroxylase activities in cytochrome P-450 LM_4 fractions from phenobarbital-treated, β-naphthoflavone-treated, untreated and starved rabbits and in the corresponding microsomal fractions.

TABLE 1

12α- AND 25-HYDROXYLASE ACTIVITIES IN PREPARATIONS OF LIVER MICROSOMAL CYTOCHROME P-450 LM_4 FRACTIONS FROM PHENOBARBITAL-TREATED, β-NAPHTHOFLAVONE-TREATED, UNTREATED AND STARVED RABBITS

Fraction	Cytochrome P-450 (specific content)	5β-cholestane-3α,7α-diol	
		12α-OH	25-OH
	nmol/mg protein	pmol/nmol cytochrome P-450/min	
Phenobarbital-treated			
Microsomes	3.5	32	11
Cytochrome P-450 LM_4	19.0	112	20
β-Naphthoflavone-treated			
Microsomes	3.1	54	15
Cytochrome P-450 LM_4	15.0	130	22
Untreated			
Microsomes	1.9	48	13
Cytochrome P-450 LM_4	15.9	105	22
Starved			
Microsomes	2.0	124	12
Cytocrhome P-450 LM_4	14.0	481	24

As expected, the rate of 12α-hydroxylation with microsomes from starved rabbits was three to four times faster than with microsomes from phenobarbital-treated, β-naphthoflavone-treated and untrated rabbits. The rate of 12α-hydroxylation with the cytochrome P-450 LM_4 fraction from starved animals was three to four times faster than with the other cytochrome P-450 LM_4 fractions. There were no significant differences in the rates of 12α-hydroxylation catalyzed by preparations from untreated, phenobarbital-treated and β-naphthoflavone-treated rabbits. The rate of 25-hydroxylation was not effected by starvation nor by any of the other treatments. The cytochrome P-450 LM_4 fractions contained 14 to 19 nmol of cytochrome P-450 per mg of protein and were all apparently homogeneous upon polyacrylamide gel electrphoresis having the same apparent molecular weight (M_r= 53,000).

The properties of cytochrome P-450 LM_4 from starved rabbits, which was the most active fraction in 12α-hydroxylation, were studied in more detail in a series of experiments and compared with the properties of a preparation from phenobarbital-treated rabbits. Table 2 shows that the absolute absorption spectra of the two cytochrome P-450 LM_4 preparations were similar in the reduced state and for the reduced carbon monoxide complex but differed in the oxidized state.

TABLE 2

OPTICAL CHARACTERISTICS OF ISOLATED CYTOCHROME P-450 FRACTIONS

Fraction	Oxidized	Reduced	Reduced-CO complex
	λ max	λ max	λ max
Cytochrome P-450 LM_4 (phenobarbital-treated)	396 645	415 545	448 549
Cytochrome P-450 LM_4 (starved)	416 533 569 641	417 547	448 550

Cytochrome P-450 LM_4 from phenobarbital-treated rabbits showed the same spectral properties as described previously (1,2) and was thus in the high spin form. Cytochrome P-450 LM_4 from starved rabbits was a mixture of a high spin and a low spin form. It should be mentioned that both preparations were extensively treated with calcium phosphate in order to remove the non-ionic detergent Renex 690 (1).

Cytochrome P-450 LM_4 from starved rabbits had similar amino acid composition as cytochrome P-450 LM_4 from phenobarbital-treated animals but differed in the content of lysine, arginine, proline. Also peptide mapping experiments revealed minor differences between the two cytochrome P-450 LM_4 fractions. The fraction from starved rabbits showed additional peptides after proteolytic digestion which were not observed with the fraction from phenobarbital-treated animals. These findings indicate the presence of additional protein(s) in the cytochrome P-450 LM_4 fraction from starved rabbits.

There are several results in the present study which show that the properties of 12α-hydroxylase differ from those of the bulk of cytochrome P-450 LM_4. The most obvious explanation for this is the cytochrome P-450 LM_4 fraction, prepared as described by Coon and associates (1) is heterogeneous and contains another cytochrome P-450 with the same apparent molecular weight which is responsible for 12α-hydroxylation and which is stimulated by starvation. Results indicating a heterogeneity of the cytochrome P-450 LM_4 fraction are the stimulating effect of starvation on the 12α-hydroxylation but not on the 25-hydroxylation as well as the lack of stimulation of 12α-hydroxylation by β-naphthoflavone. The differences in physical properties between cytochrome P-450 LM_4 from starved rabbits and cytochrome P-450 LM_4 from phenobarbital-treated rabbits lend further support to the contention that 12α-hydroxylation is catalyzed by a specific species of cytochrome P-450 present in the cytochrome P-450 LM_4 fraction.

Isolation of subfractions with different substrate specificity from cytochrome P-450 LM_4.

Cytochrome P-450 LM_4 from cholestyramine-treated rabbits catalyzes in addition to the hydroxylations catalyzed by preparations from untreated, phenobarbital-treated and β-naphthoflavone-treated animals also an efficient 7α-hydroxylation of cholesterol. Chromatography of cytochrome P-450 LM_4 from cholestyramine-treated rabbits on octylamine-Sepharose resulted in the isolation of two chromatographically different subfractions, cytochrome P-450 LM_4I and cytochrome P-450 LM_4II. Table 3 shows that the yield of cytochrome P-450 compared to the original cytochrome P-450 LM_4 was 15 and 16% in cytochrome P-450 LM_4I and cytochrome P-450 LM_4II, respectively. The major part - 54% - of the cytochrome P-450 applied to the octylamine-Sepharose column appeared in a fraction intermediate to cytochromes P-450 LM_4I and II.

TABLE 3

HYDROXYLASE ACTIVITIES IN PREPARATION OF SUBFRACTIONS FROM CYTOCHROME P-450 LM_4

Fraction	Cytochrome P-450	Cytochrome P-450 (yield)	Hydroxylation 7α	12α	25	(Testosterone) 6β
	nmol/mg protein	%	pmol/nmol cytochrome P-450/min			
Cytochrome P-450 LM_4	15.2	100	23	120	109	250
Cytochrome P-450 LM_4I	18.8	15	≤ 1	160	120	327
Intermediate fraction	15.5	54	12	206	107	358
Cytochrome P-450 LM_4II	10.8	16	35	40	50	164

The specific cytochrome P-450 content was 15.2 nmol/mg protein in the original cytochrome P-450 LM_4 and 18.8, 15.5 and 10.8 nmol/mg protein, respectively in cytochrome P-450 LM_4I, intermediate fraction and cytochrome P-450 LM_4II. The fractions were apparently homogeneous upongel electrophoresis in the presence of sodium dodecyl sulfate in a polyacrylamide gradient gel system (2) as well as in the system of Laemmli (9). The three fractions all showed the same apparent molecular weight as the original cytochrome P-450 LM_4 fraction (M_r=53,000).

The catalytical activities of the original cytochrome P-450 LM_4 fraction and of the subfractions are summarized in Table 3. The hydroxylations studied were the 7α-hydroxylation of cholesterol, the 12α-hydroxylation of 5β-cholestane-3α, 7α-diol, the 25-hydroxylation of 5β-cholestane-3α,7α,12α-triol and the 6β-hydroxylation of testosterone. The original cytochrome P-450 LM_4 fraction catalyzed all these hydroxylations. Cytochrome P-450 LM_4I was inactive in 7α-hydroxylation of cholesterol but catalyzed the other hydroxylations as efficiently as cytochrome P-450 LM_4. On the other hand, cytochrome P-450 LM_4II catalyzed 7α-hydroxylation of cholesterol more efficiently than the original cytochrome P-450 LM_4. This fraction also showed 12α, 25 and 6β-hydroxylase activities but these activities were lower than with cytochrome P-450 LM_4I as well as with the original cytochrome P-450 LM_4. The intermediate fraction catalyzed the same hydroxylations and to similar rates as cytochrome P-450 LM_4. To ascertain that the different substrate specificity of cytochromes P-450 LM_4I and II was not due to the presence of phospholipid or detergent in the fractions a series of experiments was performed with addition of ether extracts or boiled extracts from one fraction to the other. No changes in catalytic activity of cytochromes P-450 LM_4I and II were observed upon the addition experiments. The catalytic activity of the two subfractions were effected differently by the presence of the non-ionic detergent Emulgen 913. Table 4 shows the effect of addition of Emulgen to a final concentration of 0.2% (w/v) in the incubations mixtures.

TABLE 4
EFFECT OF EMULGEN ON THE CATALYTIC ACTIVITY OF CYTOCHROMES P-450 LM_4I AND II.

Fraction	Hydroxylation			
	7α	12α	25	(Testosterone) 6β
	pmol/nmol cytochrome P-450/min			
Cytochrome P-450 LM_4I				
+ Emulgen 913	≤1	≤1	≤1	≤1
Cytochrome P-450 LM_4II				
+ Emulgen 913	30	≤1	≤1	≤1

Cytochrome P-450 LM_4I was completely inactive towards all substrates studied in the presence of Emulgen. Similarily, the 12α, 25 and 6β-hydroxylase activities in cytochrome P-450 LM_4II were completely inhibited by Emulgen. However, the cholesterol 7α-hydroxylase activity of cytochrome P-450 LM_4II was not effected.

Cytochrome P-450 LM_4I and cytochrome P-450 LM_4II showed the same spectral properties and were in the high spin form. Thus, both fractions showed absorbance maxima in the oxidized state at 393 nm and 640 nm. The abosrbance maxima for the reduced carbon monoxide complex was at 447 nm for both fractions.

Table 5 shows the results of amino acid analyses of cytochrome P-450 LM_4I and cytochrome P-450 LM_4II.

TABLE 5

AMINO ACID ANALYSIS OF CYTOCHROME P-450 LM_4I AND CYTOCHROME P-450 LM_4II

Amino acid	Number of residues per molecule			
	Cytochrome P-450 LM_4I		Cytochrome P-450 LM_4II	
	Variation	Integer	Vatiation	Integer
Asx	38.38 ± 0.47	38	38.33 ± 1.24	38
Threonine	22.91 ± 0.53	23	24.26 ± 0.44	24
Serine	36.60 ± 0.53	37	35.69 ± 0.46	36
Glx	45.50 ± 0.47	46	41.49 ± 1.19	42
Proline	31.20 ± 1.07	31	27.03 ± 1.01	27
Glycine	31.82 ± 1.40	32	32.66 ± 0.49	33
Alanine	31.69 ± 0.61	32	31.44 ± 0.41	31
$Cys(O_3H)$	6.73 ± 1.36	7	7.93 ± 0.72	8
Valine	33.73 ± 0.43	34	35.76 ± 0.49	36
Methionine	9.66 ± 0.60	10	12.79 ± 0.68	13
Isoleucine	24.66 ± 0.49	25	24.49 ± 0.71	25
Leucine	52.18 ± 1.11	52	52.60 ± 0.88	53
Tyrosine	10.66 ± 0.67	11	14.47 ± 0.16	15
Phenylalanine	29.54 ± 1.03	29	28.73 ± 0.40	29
Histidine	11.67 ± 0.40	12	12.24 ± 0.21	12
Lysine	24.04 ± 0.36	24	29.05 ± 0.87	29
Tryptophan	8.02 ± 1.11	8	7.18 ± 0.57	7
Arginine	32.68 ± 0.70	33	25.87 ± 0.32	26
Total		484		484

The two fractions differed with respect to content of arginine, lysine, tyrosine, proline and glutamic acid. Limited proteolysis with papain, chymotrypsin or trypsin of the two subfractions resulted in similar but not identical peptide patterns after gel electrophoresis.

The results of the present investigation show that the cytochrome P-450 LM_4 fractions from cholestyramine-treated rabbits contains at least two species of cytochrome P-450 with different amino acid composition and with different substrate specificity towards C_{27}-steroids in bile acid biosynthesis. The experiments with cytochrome P-450 LM_4 fractions from differently pretreated rabbits lend further support for the concept that $7\alpha,12\alpha$ and 25-hydroxylation of C_{27}-steroids in the biosynthesis of bile acids are catalyzed by different species of cytochrome P-450.

ACKNOWLEDGEMENTS

The skilful technical assistance of Mrs Britt-Marie Johansson, Mrs Angela Lannerbro and Mrs Anne Legnehed is gratefully acknowleged. We also wish to express our gratitude to Dr David Eaker and Mr Ragnar Thorselius for performing the amino acid analyses and to Dr Ingemar Björk for placing his Cary 219 spectrophotometer to our disposal. This work was supported by the Swedish Medical Research Council (project 03X-218), the Loo and Hans Osterman Foundation, the Fortia Foundation and the I.F. Foundation for Pharmaceutical Research.

REFERENCES

1. Haugen, D.A. and Coon, M.J. (1976) J.Biol.Chem. 251, 7929
2. Hansson, R. and Wikvall, K. (1980) J.Biol.Chem. 255, 1643
3. Boström, H., Hansson, R., Jönsson, K-H. and Wikvall, K. (1981) Eur.J. Biochem. 120, 29
4. Koop, D.R., Persson, A.V. and Coon, M.J. (1981) J.Biol.Chem. 256, 10704
5. Dean, W.L. and Coon, M.J. (1977) J.Biol.Chem. 252, 3255
6. Kalles, I. and Wikvall, K. (1981) Biochem.Biophys.Res.Commun. 100, 1361
7. Imai, Y. and Sato, R. (1974)
8. Yasukochi, Y. and Masters, B.S.S. (1976) J.Biol.Chem. 251, 5337
9. Laemmli, U.K. (1970) Nature (Lond.) 227, 680
10. Cleveland, D.W., Fischer, S.G., Irschner, M.W. and Laemmli, U.K. (1977) J.Biol.Chem. 252, 1102
11. Penke, B., Ferenczi, R. and Kovacs, K. (1974) Anal.Biochem. 60, 45

© 1982 Elsevier Biomedical Press B.V.
Cytochrome P-450, Biochemistry, Biophysics
and Environmental Implications, E. Hietanen,
M. Laitinen and O. Hänninen editors

A UNIQUE FORM OF RAT LIVER CYTOCHROME P450 INDUCED BY THE HYPOLIPIDAEMIC DRUG CLOFIBRATE

G. GORDON GIBSON, PAUL P. TAMBURINI, HILARY MASSON AND COSTAS IOANNIDES
Department of Biochemistry, Division of Pharmacology and Toxicology, University of Surrey, Guildford, Surrey GU2 5XH, England, U.K.

INTRODUCTION

Many chemicals are known to induce liver microsomal cytochrome P450 (1), and in general, two major forms of the hemoprotein have been isolated and characterised from mammalian systems (2), one form induced by the phenobarbital class of drugs (termed PB-P450, $P450_{LM2}$ or P450 b) and the other form induced by xenobiotics such as 3-methylcholanthrene or β-naphthoflavone (termed BNF-P450, 3MC-P450, $P450_{LM4}$ or P450 c). These two multiple forms of cytochrome P450 have been distinguished from each other based on observed differences in their monomeric molecular weights, spectral properties, substrate specificities, immunological properties, catalytic activities and limited amino acid sequence data (2).

Recently, we have reported the induction and purification of liver microsomal cytochrome P450 in rats pretreated with the hypolipidaemic drug clofibrate (3) and it is the purpose of this communication to compare and contrast some properties of this highly-purified, clofibrate-induced hemoprotein with PB-P450 and BNF-P450. Based on several criteria, we postulate that clofibrate pretreatment induces a unique form of cytochrome P450 which is readily distinguishable from both PB-P450 and BNF-P450.

MATERIALS AND METHODS

NADPH-cytochrome P450 reductase, PB-P450 and BNF-P450 were purified from male Wistar rat liver as previously described (4) and clofibrate-induced cytochrome P450 was isolated by the method of Gibson, et al. (3). All hemoproteins were purified to a specific content of 12 to 18 nmol cytochrome P450/mg protein. SDS-PAGE analysis of the purified cytochrome P450 preparations was carried out by the method of Laemmli (5) on calibrated 10% polyacrylamide Slab gels and limited proteolysis of the purified hemoproteins was performed as described by Cleveland, et al. (6) in the presence of either S. aureus V8 protease, chymotrypsin or papain. Reconstitution of cytochrome P450-dependent enzyme activities was carried out as previously described (7) and drug and fatty acid metabolic oxidation products were determined by the method of Gibson, et al. (3).

RESULTS AND DISCUSSION

As can be seen from SDS-PAGE analysis of BNF-P450, PB-P450 and the highly-purified clofibrate induced cytochrome P450 (FIG.1), one major band is present in all the hemoprotein preparations. Furthermore, using a Ferguson plot analysis of calibrated polyacrylamide slab gels, differences were observed in the monomeric molecular weights of the three cytochromes corresponding to 56,000, 53,300 and 51,500 for the BNF-P450, PB-P450 and clofibrate-induced cytochromes respectively. The three cytochrome preparations also differed from each other in their polypeptide fragmentation patterns after limited digestion with different proteases (Fig.2), again indicating significant structural differences between the three cytochromes.

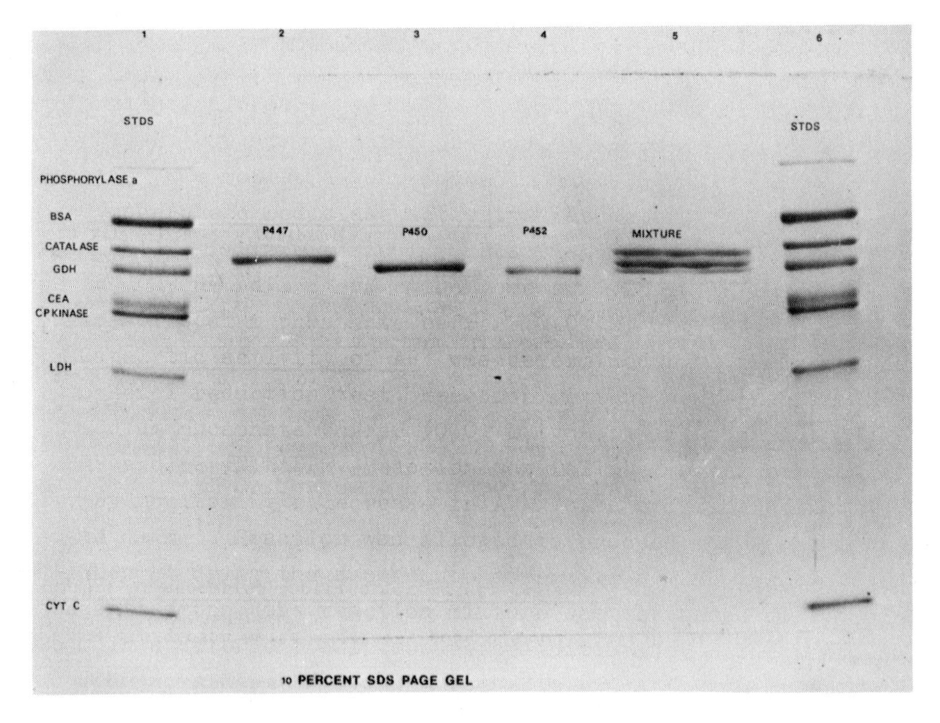

Fig.1: SDS-Page Analysis of Purified Cytochromes P447, P450 and P452

Functional differences (substrate specificities) were also noted between the three cytochrome preparations (Table 1) in that the clofibrate-induced cytochrome P450 readily catalysed the hydroxylation of lauric acid in a reconstituted system compared to BNF-P450 and PB-P450 which had poor activity towards this fatty acid. The clofibrate-induced cytochrome was inactive towards the other substrates tested

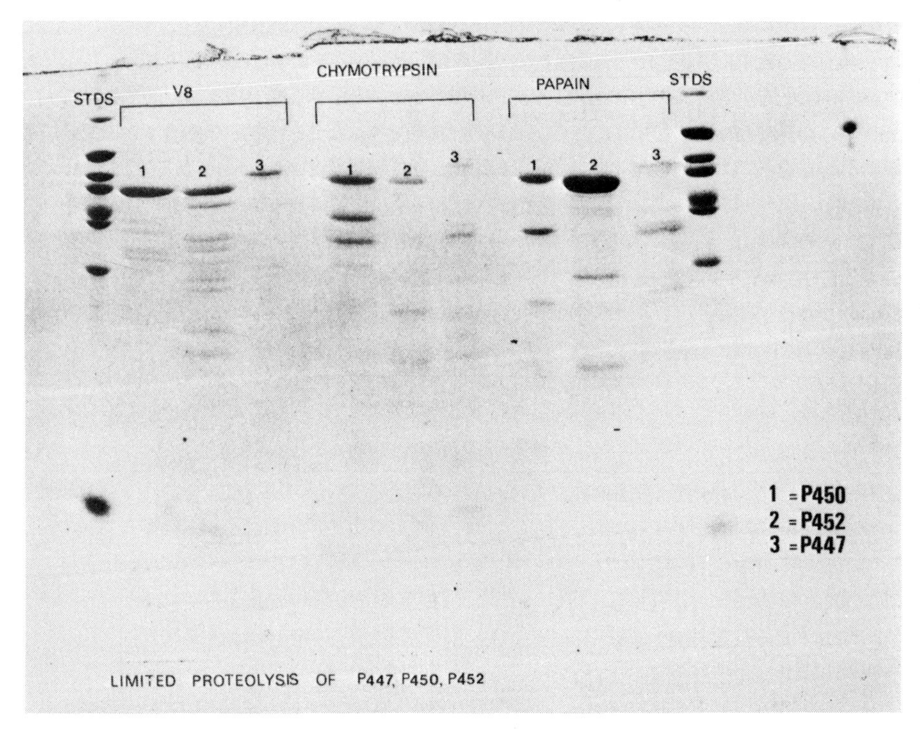

Fig.2: Cleveland Proteolysis Analysis of Purified Cytochromes P447, P450 and P452

(including benzphetamine which represents activity at the confident minimum level of detection) and therefore appeared specific for lauric acid oxidation. Furthermore, we have previously shown that the clofibrate-induced cytochrome P450 specifically catalyses the 12-hydroxylation of lauric acid with smaller amounts of the 11-hydroxy metabolite being formed (3), and therefore exhibits a regioselectivity towards lauric acid oxidation.

In conclusion, we have presented structural and functional data which gives credence to our hypothesis that clofibrate pretreatment induces a form of liver microsomal cytochrome P450 that is distinct from the PB-P450 and BNF-P450 forms of the hemoprotein, a conclusion further substantiated by our previous work on this protein (3). The biological significance of the induction of this unique form of cytochrome P450 with respect to the pharmacological and toxicological properties of clofibrate remains to be established.

TABLE 1

RECONSTITUTION OF DRUG AND FATTY ACID OXIDATIVE ACTIVITIES IN THE PRESENCE OF HIGHLY-PURIFIED FORMS OF RAT LIVER CYTOCHROME P450[a]

Enzyme Activity		Form of Cytochrome P450[b]		
		BNF-P450	PB-P450	CLOF-P450
Lauric acid hydroxylase (11 + 12)		2.2	6.7	43.0
Benzphetamine N-demethylase		39.0	281.0	27.0
Testosterone hydroxylase	7α	NDA	NDA	NDA
	16α	NDA	5.2	NDA
	6β	1.2	4.6	NDA
	2β	NDA	3.0	NDA
Ethoxyresorufin de-ethylase		16.4	TRACE	NDA

[a] All activities expressed as nmol product formed/nmol cytochrome/min

[b] BNF-P450, PB-P450 and CLOF-P450 are those major forms of the hemoproteins induced by β-naphthoflavone, phenobarbital and clofibrate respectively.

NDA: No detectable activity

ACKNOWLEDGEMENTS

This work was funded by the SRC and MRC of the United Kingdom.

REFERENCES

1. Parke, D.V. (1975) in: Parke, D.V. (ed.), Enzyme Induction, Plenum Press, New York, pp.207-271.
2. Lu, A.Y.H. and West, S.B. (1980) Pharmacol. Rev., 31, 277-295.
3. Gibson, G.G., Orton, T.C. and Tamburini, P.P. (1982) Biochem. J., 203, 161-168.
4. Guengerich, F.P. (1978) J. Biol. Chem., 253, 7931-7939.
5. Laemmli, U.K. (1970) Nature, 227, 680-685.
6. Cleveland, D.W., Fischer, S.G., Kirschner, M.W. and Laemmli, U.K. (1977) J. Biol. Chem., 252, 1102-1106.
7. Haugen, D.A., Van der Hoeven, T.A. and Coon, M.J. (1975) J. Biol. Chem., 250, 3567-3570.

© 1982 Elsevier Biomedical Press B.V.
Cytochrome P-450, Biochemistry, Biophysics
and Environmental Implications, E. Hietanen,
M. Laitinen and O. Hänninen editors

HALOGENATED BIPHENYLS AS AHH INDUCERS: EFFECTS OF DIFFERENT HALOGEN SUBSTITUENTS

S. SAFE[1], A. PARKINSON[2], L. ROBERTSON[1], S. BANDIERA[2], T. SAWYER[1], L. SAFE[1], I. LAMBERT[1], J. ANDRES[1] and M.A. CAMPBELL[2]

[1]Department of Physiology and Pharmacology, College of Veterinary Medicine, Texas A&M University, College Station, Texas 77843 (U.S.A.) and [2]Guelph-Waterloo Centre for Graduate Work in Chemistry, Department of Chemistry, University of Guelph, Guelph Ontario, Canada.

INTRODUCTION

Numerous halogenated aromatic compounds including specific polychlorinated dibenzo-*p*-dioxins (PCDDs), polychlorinated biphenyls (PCBs) and polybrominated biphenyls (PBBs) elicit a number of common toxic and biologic properties. Moreover, for the most toxic members within each class, eg. 2,3,7,8-tetrachlorodibenzo-*p*-dioxin (TCDD), 3,3',4,4',5-pentachlorobiphenyl and 3,3',4,4',5,5'-hexabromobiphenyl, there is an excellent correlation between their toxicity, their potency as inducers of cytochrome P-448-dependent monooxygenases (aryl hydrocarbon hydroxylase, AHH) and their avidity to bind to a hepatic cytosolic receptor protein (1). The effects of structure on the activity of PCB and PBB congeners are similar: PCBs which induce microsomal AHH must be substituted in both *para* and at least two *meta* positions whereas PBBs require bromine substitution at both *para* and at only one *meta* position; the addition of one or two *ortho* substituents to those congeners defined above tends to diminish but not necessarily eliminate their AHH-inducing activity (2,3). This paper reports the relative activities of the different halogen substituents using 4'-halo-2,3,4,5-tetrachlorobiphenyls and 3,3',4,4'-tetrahalobiphenyl isostereomers as structural probes. The chlorinated analogs of both halogenated biphenyls are inducers of microsomal AHH (2,3).

MATERIALS AND METHODS

Chemical synthesis and purification. The 4'-halo-2,3,4,5-tetrachlorobiphenyls were synthesized by the diazo coupling of the 4-haloanilines with 1,2,3,4-tetrachlorobenzene. The 3,3',4,4'-tetrahalobiphenyls were synthesized by the diazo coupling of the appropriate 3,4-dihaloaniline and *o*-dihalobenzene or by diazotization of 3,3'-dichlorobenzidine. The compounds were purified (> 98-99%) as described (2,3).

Animal treatment and preparation of microsomes. For the biochemical assays, immature male Wistar rats (3-5 per group) were pretreated with the appropriate chemicals on days 1 and 3 and the animals killed on day 6. Hepatic 100,000 x g microsomes were prepared as described (2,3). Thymic involution was determined 14 days after administration of a single dose of 150 μmole/kg of the chemical.

Biochemical assays. The enzymic (eg. benzo[a]pyrene hydroxylase, 4-chlorobiphenyl hydroxylase, dimethylaminoantipyrine N-demethylase, ethoxyresorufin O-deethylase and aldrin epoxidase), spectral and electrophoretic properties of the induced microsomes were determined as described (2,3). The induction of AHH activity in rat hepatoma H-4-II E cells (4) and the binding affinities of the halogenated biphenyls for the cytosolic receptor protein (5) were also measured as described.

RESULTS AND DISCUSSION

Table 1 summarizes the activities of five isosteric 3,3',4,4'-tetrahalobiphenyls containing from 0-4 chlorine or bromine substituents and provides a direct measurement of the effects of Cl versus Br. The results indicate that the toxicity, as measured by thymic atrophy, and AHH-inducing activity of these isostereomers increases with increasing bromine content. All the 3,3',4,4'-tetrahalobiphenyls maximally induced microsomal AHH activity after administration of 50 μmol/kg. However, at a lower dose level (10 μmol/kg) the differences in activity are apparent. Similarly, these isostereomers all cause thymic involution in the rat, but the biphenyls containing more bromines produced a proportionally greater decrease in thymus weights. Preliminary results also indicate that the more highly brominated compounds are more active as inducers of cytochrome P-448 associated enzyme activity in rat hepatoma H-4-II E cells and bind more avidly to the receptor protein.

4'-Halo-2,3,4,5-tetrachlorobiphenyls are less potent and less toxic than the 3,3',4,4'-tetrahalobiphenyls. However, this substitution pattern, which contains a single, laterally-substituted, halogen atom on one phenyl ring, is ideally suited as a probe for the effects of structure on activity. The EC_{50} values for ethoxyresorufin O-deethylase induction in cell cultures and receptor protein binding are summarized in Table 2. The relative potencies of the halogen substituents follow the order I > Br > Cl > F for both assays, although the differences in activity were more pronounced using the enzyme induction assay. The order I > Br > Cl > F holds for the relative polarizability (6), molecular volume, lipophilicity and inverse electronegativity of these functional groups and it is not clear which of these parameters is critical for interaction with the receptor. Current research in our laboratory is focused on evaluating the relative importance of the physico-chemical parameters of the lateral substituent groups in determining the biologic and toxic effects of halogenated biphenyls.

TABLE I.

HALOGENATED BIPHENYLS AS AHH INDUCERS

Halogenated Biphenyl	Relative Activities	
	AHH Induction[a]	Toxicity[b]
Br, Br (left ring) – Br, Br (right ring)	100	100
Cl, Br (left ring) – Br, Br (right ring)	67	70
Cl, Br (left ring) – Cl, Br (right ring)	87	60
Cl, Br (left ring) – Cl, Cl (right ring)	32	52
Cl, Cl (left ring) – Cl, Cl (right ring)	10	10

[a]male Wistar rat (10 μmol·kg^{-1} dose); [b]male Wistar rat (150 μmol·kg^{-1})

TABLE 2.

Cl, Cl, Cl, Cl (left ring) – X (right ring)	AHH Induction[c]	Receptor Binding[d]
X = I	100	100
X = Br	13	60
X = Cl	1.5	36
X = F	0.03	6

[c,d]from EC_{50} values for ethoxyresorufin O-deethylase activity in rat hepatoma H-4-II E cells, and competitive receptor binding assays, respectively.

ACKNOWLEDGEMENTS

This study was supported by the Natural Sciences and Engineering Research Council of Canada and the National Institutes of Health (1-RO1 ES02798-01).

REFERENCES

1. Poland, A., Greenlee, W.F. and Kende, A.S. (1979) Ann. N.Y. Acad. Sci. 320, 214-230.
2. Parkinson, A., Robertson, L., Safe, L. and Safe, S. (1980) Chem.-Biol. Interact., 30, 271-285.
3. Robertson, L.W., Parkinson, A., Campbell, M.A. and Safe, S. (1982) Chem. -Biol. Interact., in press.
4. Sawyer, T. and Safe, S. PCB Isomers and Congeners (1982) Toxicol. Letters, in press.
5. Bandiera, S., Safe, S. and Okey, A. (1982) Chem.-Biol. Interact., in press.
6. Albro, P.W. and McKinney, J.D. (1981) Chem.-Biol. Interact., 34, 373-378.

Cytochrome P-450, Biochemistry, Biophysics
and Environmental Implications, E. Hietanen,
M. Laitinen and O. Hänninen editors

THE ALKANE MONOOXYGENASE SYSTEM OF THE YEAST LODDEROMYCES ELONGISPORUS: PURIFICATION OF THE CYTOCHROME P-450 AND THE NADPH CYTOCHROME P-450 REDUCTASE AND RECONSTITUTION EXPERIMENTS

H.-G. MÜLLER, W.-H. SCHUNCK, P. RIEGE AND H. HONECK
Dept. of Applied Enzymology, Central Institute of Molecular Biology, Berlin-Buch, German Democratic Republic

INTRODUCTION

For the first time in 1968 Cardini and Jurtshuk (1) proposed for an alkane-utilizing Corynebacterium that cytochrome P-450 could be the alkane-hydroxylating enzyme. Some years later Lebeault et al. (2) suggested the same for a eucaryotic microorganism, the yeast Candida tropicalis. However, a convincing proof resisting the severe criteria according to Cooper et al. (3) and Burke (4) was not furnished for the cytochrome P-450 of alkane-utilizing microorganisms up to the present time. Within the last two years we have been successful in the purification of the alkane-monooxygenase system of the yeast Lodderomyces elongisporus consisting of cytochrome P-450 and NADPH-cytochrome P-450-reductase and in the reconstitution of the active alkane-monooxygenase system. This enzyme system is induced during the growth on n-alkanes (5).

MATERIALS AND METHODS

Yeast cells were grown on n-alkanes as the only carbon and energy source. After the disruption of the cells in a Dyno-Mill and a low speed centrifugation to separate the cell debris the membrane fraction containing the enzymes was obtained by a Ca^{2+}-mediated sedimentation. Detailed instructions have already been published elsewhere (6-8).

RESULTS AND DISCUSSION

Cytochrome P-450 and NADPH-cytochrome P-450 are constituents of the microsomal fraction (0.3 - 0.4 nmoles/mg protein and 0.6 - 0.8 µmoles reduced cytochrome c/min/mg protein, respectively). For the purification of cytochrome P-450 the solubilization was carried out with Na-cholate (final concentration 0.8 %). Subsequently, the cytochrome P-450 was obtained by affinity chromatography on -aminooctyl-Sepharose 4B. In the most cases already one step-procedure results in an electrophoretically homogeneous preparation. Sometimes a rechromatography for removing small impurities is necessary. Finally the purification was completed by a calcium phosphate gel step for removing the detergents (7).

Decisive for the high effectivity of the procedure, which represents a modification of the method of Imai and Sato (9) is the combination of specific binding of the cytochrome P-450 in presence of 1.2 % Na-cholate at low ionic strength (10 mM K-phosphate, pH 7.25) and of specific elution with the detergent Tween 20 (0.5 % in presence of 0.6 % Na-cholate) (7).

Some characteristics of the protein are listed in Table 1. The spectral parameters indicate that the purified yeast cyt P-450 is mainly in the low spin state. By addition of substrate (hexandecane) and increase of the ionic strength (200 mM phosphate) a partial transformation to the high-spin state is reached.

In contrast to cytochrome P-450 the solubilization of the reductase is carried out with Na-deoxycholate, which destroys the former. In a subsequent two-step-procedure (8) with DEAE-Sephacel and hydroxylapatite we obtain an electrophoretically homogeneous protein with a specific activity of nearly 70 µmoles reduced cytochrome c x min^{-1} x mg^{-1} protein. Likewise, to deplete detergents, the reductase was adsorbed to calcium phosphate gel and after washing eluted with 0.4 M potassium phosphate buffer.

Decisive for the success in purifying the reductase was the chosen combination of ionic and non-ionic detergents (0.1 % Na-cholate, 0.3 % Präwozell WON-100) in connection with an addition of 1 µM FAD and FMN to all buffers used after the solubilization.

Some characteristics of the protein are listed in Table 1. In oxidized form the reductase shows a typical flavoprotein spectrum with peaks at 272, 380 and 453 nm. In addition to the cytochrome c the purified reductase also catalyzes the reduction of other artificial electron acceptors, e.g. dichlorophenolindophenol and potassium ferricyanide. The ability to reduce cytochrome P-450 can be proved both by the formation of the CO-complex and by the detection of hexadecane hydroxylation in the reconstituted

TABLE 1

Some of the most important properties of cytochrome P-450 and NADPH-cytochrome P-450 reductase of alkane-grown Lodderomyces elongisporus

Enzyme	Molecular weight	Prosthetic group	Specific[a] content	Specific activity
NADPH-cytochrome P-450 reductase	79 000	1 FAD 1 FMN	-	60-70[b]
Cytochrome P-450	53 500	1 heme	12 - 17	= 6[c] (= 0.8)[d]

a) nmoles cytochrome P-450/mg protein
b) μmoles reduced cytochrome c/mg protein/ min.
c) nmoles hydroxylated hexadecane/nmole cytochrome P-450/min in the reconstituted enzyme system consisting of 0.18 nmoles cytochrom P-450 cyt P_L450, 0.67 units reductase, 250 nmoles NADPH, 100 nmoles 1-^{14}C-hexadecane, 160 nmoles Präwocell WON 100
d) () without Präwocell WON 100

alkane-monooxygenase system. Even for the former experiment the presence of substrate and the detergent Präwocell WON 100 is necessary.

The omission of one protein component leads to a complete loss of activity. Low concentrations of the non-ionic detergent Präwocell WON 100 are stimulating (Table 1). This result suggest a similar important role of lipid in the cytochrome P-450-dependent alkane monooxygenase of yeast (see also (10)) as it was established for the liver microsomal system.

The only detectable product of hexadecane conversion by the reconstituted system is the primary alcohol. With lesser purified fractions, e.g. microsomes, likewise only intermediates (primary alcohol, fatty acids) are detectable, which are compatible with the terminal hydroxylation of the alkane. Therefore, a subterminal hydroxylation should be very improbable.

Using the reconstituted system no conversion of benzphetamine, aminophenazone, and ethyl morphine was detected. Furthermore, microsomes of L. elongisporus are unable to hydroxylate several steroids. Likewise, benzo(a)pyrene was not transformed.

All the results demonstrate both a high substrate specificity and a high regioselectivity of the cytochrome P-450 containing alkane monooxygenase system of the yeast L. elongisporus. The reconstitution experiments represent the first convincing proof that cytochrome P-450 and NADPH cyt P-450 reductase actually form the alkane monooxygenase of an alkane-utilizing microorganism as it was already proposed earlier (1,2,6,10,11).

REFERENCES

1. Cardini, G. and Jurtshuk, P. (1968) J. Biol. Chem. 243, 6070-6072.
2. Lebeault, J.M., Lode, E.T. and Coon, M.J. (1971) Biochem. Biophys. Res. Commun. 42, 413-419.
3. Cooper, D.Y., Schleyer, H., Levin, S.S., Eisenhardt, R.H., Novack, B.G. and Rosenthal, O. (1979) Drug Metabol. Rev., 10, 153-185.
4. Burke, M.D. (1981) Biochem. Pharmacol. 30, 181-187.
5. Mauersberger, S., Schunck, W.-H., Müller, H.-G. (1981) Z. allgem. Mikrobiol. 21, 313-321.
6. Müller, H.-G., Schunck, W.-H., Riege, P. and Honeck, H. (1979) Acta biol. med. germ. 38, 345-349.
7. Riege, P., Schunck, W.-H., Honeck, H. and Müller, H.-G. (1981) Biochem. Biophys. Res. Commun. 94, 527-534.
8. Honeck, H., Schunck, W.-H., Riege, P. and Müller, H.-G. (1982) Biochem. Biophys. Res. Commun., submitted.
9. Imai, Y. and Sato, R. (1974) J. Biochem. 75, 689-697.
10. Duppel,W., Lebeault, J.-M. and Coon, M.J. (1973) Eur. J. Biochem. 36, 583-592.
11. Schunck, W.-H., Riege, P. and Kuhl, R. (1978) Pharmazie 33, 412-414.

© 1982 Elsevier Biomedical Press B.V.
Cytochrome P-450, Biochemistry, Biophysics
and Environmental Implications, E. Hietanen,
M. Laitinen and O. Hänninen editors

PURIFICATION OF HUMAN PLACENTAL CYTOCHROMES P-450

OLAVI PELKONEN AND MARKKU PASANEN
Department of Pharmacology, University of Oulu, SF-90220 Oulu 22, (Finland)

INTRODUCTION

Human placental cytochrome P-450 (P-450) participates in the biotransformation of steroids and exogenous compounds. On the basis of indirect evidence, several different P-450s have been postulated in both placental endoplasmic reticulum (microsomes) and mitochondria (1). It seems likely that the aromatization of androgens, cholesterol side chain cleavage and estrogen 2-hydroxylation as well as several xenobiotic oxidations are catalyzed by different P-450s or that different regulatory signals are involved in the expression of these biotransformations. The matter could be unequivocally settled by the purification of placental P-450, but it has proved difficult (1). Our primary purpose was to try to purify the human placental P-450 associated with the induction of aryl hydrocarbon hydroxylase (AHH) activity by maternal cigarette smoking. AHH induction in placenta exhibits several unusual features. For example, the associated P-450 has never been detected with certainty, either spectrally or in other ways (1). Also the extent of induction may be very large, which makes this system interesting from the regulation point of view (2). In the course of these studies we developed methods to purify placental P-450, but apparently the proteins purified thus far are primarily involved in steroid biotransformations. So the final goal, purification of AHH-associated P-450 from human placenta, still remains elusive.

MATERIALS AND METHODS

The procedure for the purification of placental P-450 follows that published previously (3), but we have introduced some minor modifications which improve the recovery. The mitochondria and microsomes were isolated by the standard centrifugation schedule. The fractions were solubilized with 0.5 % cholate and 0.1 % Emulgen 911 for 1 h and diluted with one volume 10 mM potassium phosphate buffer (pH 6.8) containing 20 % glycerol and 0.1 % cholate, and then centrifuged at 100 000 x g for 30 min. The supernatant was made up to 0.5 M with NaCl, and then applied to the phenyl-Sepharose column equilibrated with 10 mM potassium phosphate buffer (pH 6.8) containing 20 % glycerol, 0.05 % Emulgen 911, 0.1 % cholate and 0.5 M NaCl. The column was washed with the equilibrating buffer until A_{417} was below 0.01. The elution was started with 10 mM potassium phosphate buffer (pH 7.4) containing 20 % (mitochondria) or 60 % (microsomes) glycerol, 1 % Emulgen 911 and 0.1 % cholate. P-450 eluted in a single peak. Pooled fractions were concentrated by ultrafiltration on an Amicon PM-30 membrane. The concentrate was applied to an anion exhange column (1.5 x 30 cm, Whatman DE-52 or DEAE-Sephacel), washed with the equilibrating buffer (10 mM potassium phosphate pH 7.4, 20 % glycerol, 0.05 % Emulgen 911 and 0.1 % cholate). The elution was started with a continuous salt gradient (0-300 mM NaCl) and the elution was monitored with a spectrophotometer at 417 nm. The most active fractions were combined and concentrated by ultrafiltration. The cytochrome content was determined according to the method of Greim et al. (4).

RESULTS AND DISCUSSION

Table 1 demonstrates a compilation of the purification data from more than 10 different placentas. The concentration of P-450 in microsomal or mitochondrial fractions is rather low when compared with even control rat liver microsomes. Consequently, the success in the purification is necessarily dependent on the solubilization of original subcellular fraction and on the efficiency of the first chromatography step. Solubilization has earlier been the step which often resulted in a drastic loss of P-450. Now losses during this step have been about 20 to 50 per cent for mitochondria and 45 to 60 per cent for microsomes, so that from an average placenta about 60 and 160 nmoles of microsomal and mitochondrial P-450, respectively, are available for further purification. Microsomal and mitochondrial P-450 behave differently in the hydrophobic chromatography on phenyl-Sepharose. The mitochondrial P-450 can be eluted with a lower Emulgen concentration than the microsomal P-450. Also the requirement of glycerol for the optimal elution behaviour is different with microsomal and

TABLE 1

PURIFICATION OF HUMAN PLACENTAL CYTOCHROMES P-450

Variation of values when starting from a single placenta.

Fraction	Total Content (nmoles)	Specific Activity (nmol/mg)	Purification	Yield %
Mitochondrial fraction	110-210	0.10-0.14	1.0	100
Solubilized supernatant	60-150	0.10-0.15	1.0	50-80
phenyl-Sepharose eluate	35-100	0.7-1.1	6-10	30-50
DE-52 eluate	20-50	4-11	50-110	15-25
Microsomal fraction	40-90	0.06-0.10	1.0	100
Solubilized supernatant	20-50	0.06-0.10	1.0	40-55
phenyl-Sepharose	15-25	0.4-0.8	6-10	20-30
DE-52 eluate	8-15	3.0-5.7	40-70	10-15

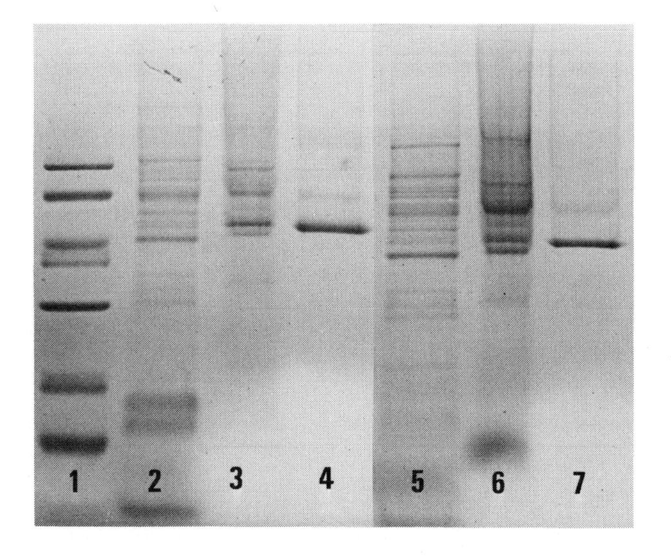

Fig. 1. SDS/Polyacrylamide-gel electrophoresis showing the purification of both placental mitochondrial and microsomal P-450s. Electrophoresis was carried out according to Laemmli (5) and 12.6 % gels were used. Samples include: lane 1: standards (Mw from top: 94000,67000,43000 (2 bands),30000,20100 and 14400); lanes 2-4: mitochondrial fraction (70 ug), phenyl-Sepharose eluate (60 ug) and DEAE-eluate (30 ug); lanes 5-7: microsomal fraction (70 ug), phenyl-Sepharose eluate (60 ug) and DEAE-eluate (30 ug).

mitochondrial P-450. The recovery of P-450 from the material applied to the hydrophobic column is good, usually more than 50 per cent. The subsequent concentrating step on an Amincon membrane results in a small loss of P-450. The DEAE-cellulose column chromatography also exhibits an about fifty-per-cent recovery with both mitochondrial and microsomal P-450. The concentrates are electrophoretically almost homogeneous, although the specific content is far below the theoretical maximum (fig. 1). A possible reason could be the loss of heme during the purification. On the other hand, the addition of hemin to partially purified preparations could not increase the spectrally detectable P-450 or catalytic activity. The characterization of the purified preparations has not advanced very far yet, because the amount of the purified P-450s has been too small for extensive investigations (table 1). It is apparent that these purified P-450s do not participate in xenobiotic biotransformations, because 7-ethoxycoumarin O-deethylase activity present in native subcellular fractions is lost during subsequent steps. Spectral interaction studies give some indication that both microsomal and mitochondrial P-450s are involved in steroid biotransformations.

ACKNOWLEDGEMENTS

The technical assistance of Irma Vartiainen and Ritva Saarikoski and the secretarial assistance of Leena Pyykkö is gratefully acknowledged. This study was supported by the Finnish Foundation for Cancer Research and The Academy of Finland.

REFERENCES

1. Juchau, M.R. (1980) Pharmac. Ther. 8, 501.
2. Pelkonen, O., Vähäkangas, K. & Kärki, N.T. (1980) in: Biochemistry, Biophysics and Regulation of Cytochrome P-450, Gustafsson, J.-Å, Carlstedt-Duke, J., Mode, A. & Rafter, J. (eds.), Elsevier/North-Holland, Amsterdam, pp. 163-170.
3. Pasanen, M. and Pelkonen, O. (1981) Biochem. Biophys. Res. Commun. 103, 1310.
4. Greim, H., Schenkman, M., Klotzbucher, M. and Remmer, H. (1970) Biochem. Biophys. Acta 201, 21.
5. Laemmli, U.K. (1970) Nature 227, 680.

Cytochrome P-450, Biochemistry, Biophysics
and Environmental Implications, E. Hietanen,
M. Laitinen and O. Hänninen editors

COMPARISON OF THE CYTOCHROME P450-CONTAINING MONOOXYGENASES ORIGINATING FROM TWO DIFFERENT YEASTS

MARIJA SAUER, OTHMAR KAPPELI AND A. FIECHTER
Department of Biotechnology, Swiss Federal Institute of Technology, Hönggerberg, 8093 Zürich, Switzerland

INTRODUCTION

Cytochrome P450 is the hemoprotein part of a mixed function oxygenase system that has been found in a great variety of organisms. It is involved in different hydroxylation and dealkylation reactions on drugs, steroids, fatty acids or carcinogens.

The occurence of cytochrome P450 in microorganisms is related to more specific metabolic functions. In Candida tropicalis, it is part of the n-alkane hydroxylating system (1, 2). Little is known about its functions in Saccharomyces type yeasts where it is induced at semi-anaerobic growth conditions (3). However, by isolating and characterizing the components of the complete enzyme system, it proved to be analogous to the electron transport system of liver microsomes. These similarities open the possibility of using the cytochrome P450 system from yeasts as a model for metabolic and toxicological studies.

The objective of our work is the isolation of an active monooxygenase system from Candida tropicalis and Saccharomyces uvarum. The main question is whether the systems from the two yeasts are identical or whether there are organism-specific differences.

MATERIALS AND METHODS

Organisms. The yeasts used in this study were Candida tropicalis (ATCC 32113), which is a hydrocarbon-assimilating yeast, and Saccharomyces uvarum (Department of Biotechnology, Swiss Federal Institute of Technology).

Cultivation. The cells were cultivated in continuous culture as described previously (4).

Isolation of microsomal fraction. From the cells of C. tropicalis and S. uvarum respectively, spheroplasts were prepared by enzymatic digestion of the cell wall in isotonic buffer. The spheroplasts were lysed by a short sonic treatment, and the homogenate was subsequently centrifuged for 2000 x g for 10 min. and 12000 x g for 15 min. An additional centrifugation step was carried out with the homogenate from C. tropicalis (25000 x g for 15 min.) since hydrocarbon-grown cells are rich in microbodies (5). From the 12000 x g and 25000 x g supernatant, the microbodies were precipitated by the addition of 16 mM $CaCl_2$ and collected by centrifugation at 15000 x g for 15 min.

Hydroxylation and demethylation assay. Hydroxylation activity was measured by incubating the microsomal fraction with 1 - ^{14}C - hexadecane and determining the amount of oxidized products by thin layer chromatography. The main product formed was always palmitic acid.

Demethylation was assessed by determing the formaldehyde resulting from the demethylation reaction. The substrates tested were Aminopyrine, p-Nitroanisole and caffeine.

RESULTS AND DISCUSSION

Enrichment of microsomal fraction by $CaCl_2$ precipitation. By the $CaCl_2$ precipitation method, approximately 90% and over 80% of the cytochrome P450 from cell free extracts of C. tropicalis and S. uvarum respectively were recovered. The specific cytochrome P450 content of the enriched microsomal fraction from C. tropicalis and S. uvarum was 270 pmol mg^{-1} and 320 pmol mg^{-1} respectively (Table 1), which is at least as high as that obtained by density gradient centrifugation (6, 7, 8).

TABLE 1
Enrichment of cytochrome P450 by $CaCl_2$ precipitation of the microsomal fraction of C. tropicalis and S. uvarum.

Fraction	Specific cytochrome P450 content (pmol mg^{-1})	
	C. tropicalis	S. uvarum
12000 x g sup.	--	36.6
25000 x g sup.	27.6	--
$CaCl_2$ microsomes	221.3	323.6

Reduced CO-difference spectra. The reduced CO-difference spectra of the two microsomal fractions are shown in Fig. 1. It is noteworthy that there is virtually no cytochrome P420 present in either preparation and also no influence of cytochrome oxidase is indicated (no shift in absorption maximum to higher wavelengths or absorption at 425 to 430 nm). However, the cytochrome P450 of S. uvarum has its absorption maximum at 448 nm, whereas that of C. tropicalis is exactly at 450 nm, which indicates that the two yeasts yield two different cytochrome P450 classes.

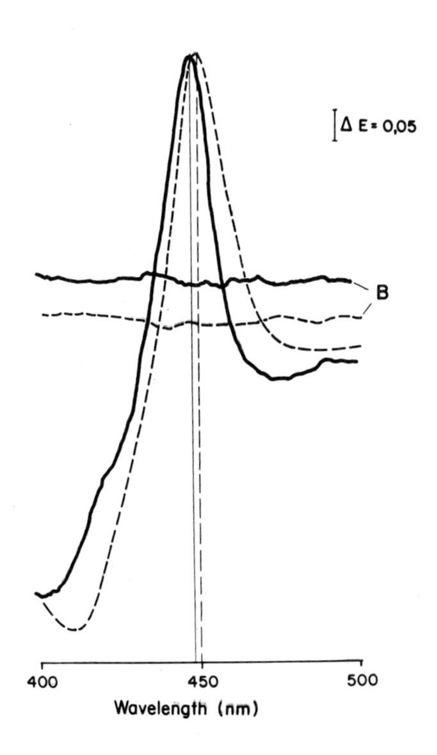

Fig. 1. Reduced CO-difference spectrum of the microsomal fraction C. tropicalis (---) and S. uvarum (——). Absorption maximum at 450 and 448 respectively.

Content of constituents of the microsomal electron transport system. Important constituents of the microsomal electron transport system, besides cytochrome P450, are the NADPH-cytochrome c reductase and cytochrome b_5. The microsomal fractions of both yeasts contained these components, but at different concentrations (Table 2).

TABLE 2

Content of constituents of the microsomal electron transport system in microsomes isolated from *C. tropicalis* and *S. uvarum*.

Constituent	*C. tropicalis*	*S. uvarum*
Cytochrome b_5 (pmol mg^{-1})	518.6	183.2
NADPH-cytochrome c reductase (nmol min^{-1} mg^{-1})	826.2	210.0

In vitro hydroxylation and demethylation activity. Hydroxylation activity was determined by measuring the conversion of hexadecane to palmitic acid. Demethylation activity was assessed by the liberation of formaldehyde when aminopyrine, p-nitroanisole or caffeine was incubated with the microsomal fraction.

TABLE 3

Hydroxylation and demethylation activity of the microsomal fractions from *C. tropicalis* and *S. uvarum*. The hydroxylation assay was carried out with NADPH as cofactor, while in the demethylation assay a cofactor regeneration system was used (glucose oxidase). Time of incubation: 1 h.

Activity	*C. tropicalis*	*S. uvarum*
	(pmol min^{-1} mg^{-1}	protein)
Hexadecane hrdroxylation	130	0
Demethylation of:		
Aminopyrine	0	241
p-Nitroanisole	0	177
Caffeine	0	61

Table 3 indicates that the two systems differ in their catalytic activity. Hydroxylation was observed with microsomes from C. tropicalis only, which has to be expected since the system is induced by long chain n-alkanes. With the microsomal fraction of S. uvarum, no hydroxylation activity was detected. However, good demethylation activity was obtained with aminopyrine, p-nitroanisole and caffeine. With these substrates, no formation of formaldehyde was measured. In conclusion, the presented data shows that the two yeast species yield two classes of monooxygenase systems.

REFERENCES

1. Lebeault, J.M., Lode, E.T. and Coon, M.J. (1971) Biochem. Biophys. Res. Commun., 42, 413.
2. Gmünder, F.K., Käppeli, O. and Fiechter, A. (1981) Eur. J. Appl. Microbiol. Biotechnol., 12, 135.
3. Ishidate, K., Kawagudchi, K., Tagawa, K. and Hagihara, B. (1969) J. Biochem. (Tokyo), 65, 375.
4. Gmünder, F.K., Käppeli, O. and Fiechter, A. (1981) Eur. J. Appl. Microbiol. Biotechnol., 12, 129.
5. Osumi, M., Miwa, N., Teranishi, Y., Tanaka, A. and Fukui, S. (1974) Arch. Microbiol., 99, 181.
6. Gallo, M., Roche, B. and Azoulay, E. (1976) Biochem. Biophys. Acta, 419, 425.
7. Schunk, W.-H., Riege, P. Blasig, R., Honeck, H. and Müller, H.-G. (1978) Acta Biol. Med. Germ., 37, K3.
8. Yoshida, A., Kumaoka, H. and Sato, R. (1974) J. Biochem. (Tokyo), 75, 1201.

Cytochrome P-450, Biochemistry, Biophysics
and Environmental Implications, E. Hietanen,
M. Laitinen and O. Hänninen editors

OXIDATION OF GENETIC MARKER DRUGS IN HUMAN LIVER MICROSOMES

CHRISTER VON BAHR, CAROL BIRGERSSON, MONICA GÖRANSSON AND BRITT MELLSTRÖM
Department of Clinical Pharmacology, Karolinska Institute, Huddinge Hospital,
S-141 86 Huddinge, Sweden

INTRODUCTION

A few % of Caucasian populations have decreased capacity to oxidize debrisoquine (D), sparteine (S) and nortriptyline (NT) (1-3) and some other drugs. This defect in oxidation of D (1) and S (2) seems to be inherited as an autosomal recessive trait and it may depend on a lacking or abnormal cytochrome P-450 -isozyme. This paper describes our initial efforts to characterize this in human livers by studying the interaction of D, NT and propranolol (P) (Fig 1) with microsomes and by enzyme purification.

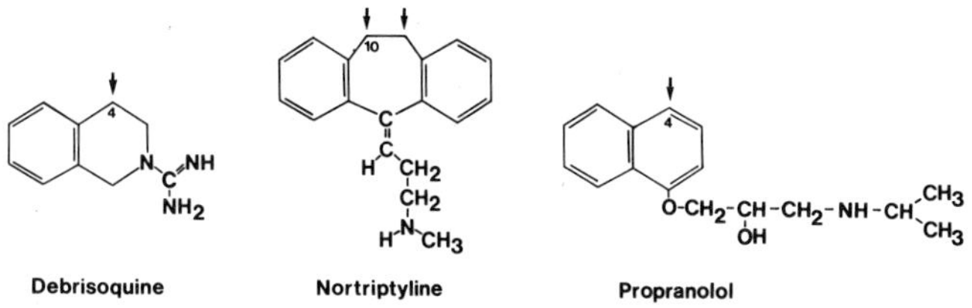

Debrisoquine Nortriptyline Propranolol

Fig 1

MATERIALS AND METHODS

Livers were obtained shortly after circulatory arrest from renal transplant donors and microsomes were prepared (4). The metabolites of D (5), NT (6) and P (7) were quantitated by GC-MC and HPLC.

RESULTS AND DISCUSSION

D, NT and P gave a type I spectral change on addition to microsomes indicating interaction with cytochrome P-450. We have earlier found covariation of certain in vitro drug oxidation rates among human livers suggesting a common regulation of the enzyme(s) involved (4). Since we thought that NT could serve as a convenient genetic marker drug we developed a method for quantitation of its main metabolite 10-hydroxynortriptyline (10-OH-NT) (6). The formation of 10-OH-NT was linear for 60 min and up to a microsomal protein concentration of 1 mg/ml. On incubation of D, NT and P with microsomes the formation of 4-OH-D was 19-95 pmol/mg · min (n = 6), of 10-OH-NT 12-57 (n = 15), of 4 OH-P 80-490 and of N-deisopropyl-P (nor-P) 80-460 (n = 6). Formation of 10-OH-NT correlated with that of 4-OH-D ($r = 0.95$; $p < 0.05$, $n = 6$) and 4-OH-P ($r = 0.83$; $p < 0.05$, $n = 6$) (Fig 2) but not that of nor-P ($r = 0.18$; NS). This indicates that the former reactions, but not the N-dealkylation of P, are similarily regulated.

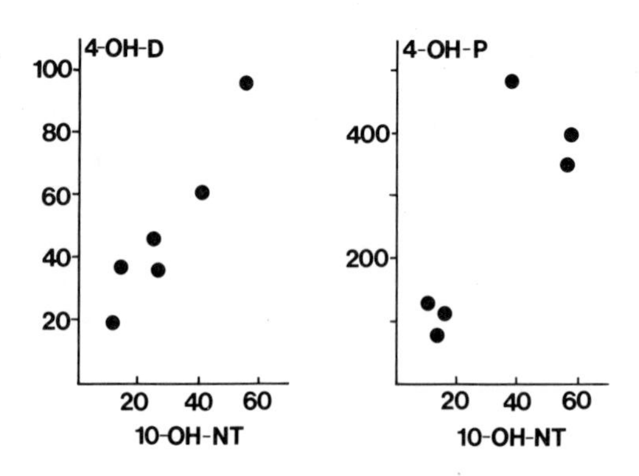

Fig 2
Relationship between 4-hydroxylation of debrisoquine (4-OH-D) and propranolol (4-OH-P) and 10-hydroxylation of nortriptyline (10-OH-NT). Unit: pmol/mg protein·min.

To get a hint whether a common cytochrome P-450-enzyme could catalyze the above described oxidations we studied the influence of D, P and other compounds on the 10-hydroxylation of NT. Table I shows that D and P inhibited 10-OH-NT formation. P was an efficient inhibitor compatible with a fairly low K_m for its own oxidation (∿ 10 μM) (7) whereas rather high concentrations of D were needed, probably due to its moderate affinity for the enzyme (K_m ∿ 130 μM) (8).

Table 1

EFFECT OF DRUGS ON THE 10-HYDROXYLATION OF NORTRIPTYLINE

Inhibitor	IC_{50} (μM)
Debrisoquine	2000
Propranolol	40
Sparteine	400
Encainide	500

[a]/Concentration giving 50 % inhibition

Also sparteine (S) and encainide inhibited NT hydroxylation. Our data are similar to recent findings that D competetively inhibit the metabolism of S (9), and altogether they indicate that D, NT and P may be, at least partly, hydroxylated by a common cytochrome P-450. Encainide was used since its O-dealkylation (10), like that of phenacetine (11), is similarily regulated as D-hydroxylation. p-nitroanisol is also oxidized by O-dealkylation. Interestingly, high concentrations of this compound increased the formation of 10-OH-NT, especially with microsomes from a liver induced by pentobarbital. The mechanism for this is not known.

A few human livers exposed to "inducers" (barbiturates, phenytoin) did not form 10-OH-NT and 4-OH-D faster than nonexposed livers. This agrees with findings that phenobarbital treatment of rats can cause a slightly decrease of the 10-hydroxylation of NT (12) and 4-hydroxylation of D (not shown). Treatment of rats with polycyclic hydrocarbons can increase both reactions (8, 12).

In preliminary experiments we have tried to purify the hypothetical cytochrome P-450-isozyme selectively by using analogues of potential genetic marker drugs as affinity chromatography ligands. Promising results were obtained when demethyl-NT (DNT) was coupled to activated CH-Sepharose 4 B. When solubilized (0.6 % cholate, 0.2 % Emulgen 911 and 20 % glycerol) microsomes were applied to the column cytochrome P-450 was retained and could be eluted with 0.4 % Emulgen 911. The peak fraction (417 nm) elicited a reduced CO binding spectrum (peak 450 nm). The cytochrome P-450 concentration was 6 nmol/mg protein. By using Sepharose 6 B a more selective retention of cytochrome P-450 could be obtained (17 % of total) which could be eluted with 1 mM NT (Fig 3).

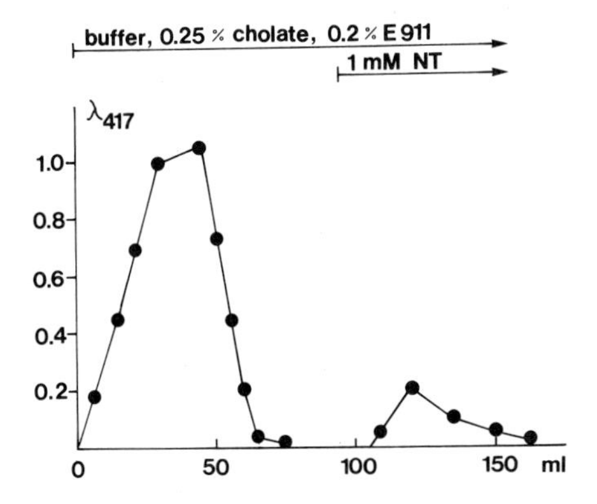

Fig 3. Chromatography of solubilized human liver microsomes on DNT-Sepharose 6 B. 111 nmoles of cytochrome P-450 was added. 71 nmoles was eluted in first peak and 19 in second.

This study shows that 4-, 10- and 4-hydroxylation of D, NT and P, respectively, covariate among different human livers, indicating a common regulation. Inhibition experiments indicate that the reactions can be catalyzed by the same cytochrome P-450 species, which may not be markedly induced by barbiturate or phenytoin treatment. Some O-dealkylations are regulated in the same way as D-hydroxylation, but no N-dealkylations have been found to behave similarily. Preliminary experiments indicate possibilities for purification of the genetically governed hypothetical human cytochrome P-450-isozyme.

ACKNOWLEDGMENTS

Supported by the Swedish Medical Research Council (14x-05577) and Nordiska Samfundets Stiftelse för vetenskaplig forskning utan djurförsök.

REFERENCES

1. Mahgoub,A., Idle, J.R., Dring, L.G., Lancaster, K.and Smith, R.L. (1977) Lancet 2, 584.
2. Eichelbaum, M., Spannbrucker, N., Steincke, B. and Dengler, H.J. (1979) Eur.J.Clin. Pharmacol. 16, 183.
3. Mellström, B., Bertilsson, L., Säwe, J., Schulz, H.-U. and Sjöqvist,F. (1981) Clin.Pharmacol.Ther. 30,189.
4. von Bahr, C., Groth, L-G, Jansson,H., Lundgren,G., Lind, M. and Glaumann, H. (1980) Clin. Pharmacol.Ther.27, 711.
5. Kahn, G.C., Boobis, A.R., Murray, S., Brodie, M.J. and Davies, D.S. (1982) Br. J. Clin. Pharmacol. submitted.
6. Mellström, B., Bertilsson, L., Göransson, M. and von Bahr, C. (1982) Drug Metabol. Dispos., submitted.
7. von Bahr, C., Hermansson, J. and Lind, M. (1982) J.Pharmacol. Exp. Ther., in press.
8. Kahn, C.G., Boobis, A.R., Murray, S., Plummer, S., Brodie, M.J. and Davies, D.S. (1981) Br. J. Clin. Pharmacol. 13, 594 P.
9. Otton, S.V., Inaba, T., Mahon, W.A. and Kalow, W. (1982). Can. J. Physiol. Pharmacol. 60, 102.
10. Woosley, R.L., Roden, D.M., Duff, H.J., Carey, D.L., Wood, A.J.J. and Wilkinson, G.R. (1981) Clin. Res. 29, 501 A.
11. Sloan, J.P., Mahgoub, A., Lancaster, R., Idle, J.R. and Smith, R.L. (1978) Br. Med. J. 2, 655.
12. von Bahr, C., Schenkman, J.B. and Orrenius, S. (1972) Xenobiotica 2, 89.

Cytochrome P-450, Biochemistry, Biophysics
and Environmental Implications, E. Hietanen,
M. Laitinen and O. Hänninen editors

EFFECT OF OXYGEN CONCENTRATION ON FORMATION AND AUTOXIDATION OF THE OXY-FERROUS INTERMEDIATE OF P-450 LM_4 ISOZYME

CLAUDE BONFILS, JEAN LOUIS SALDANA, CLAUDE BALNY AND PATRICK MAUREL
INSERM U 128, B.P. 5051, 34033 Montpellier Cedex (France)

INTRODUCTION

In 1972, two groups Peterson et al (1) and Gunsalus et al (2), demonstrated with bacterial cytochrome P-450 (P-450_{cam}) the occurence of a stable oxyferrous intermediate ($Fe^{2+}O_2$). This intermediate, formed after mixing reduced P-450_{cam} with molecular oxygen, was characterized by various parameters such as spectral maxima (418 and 555 nm), second order rate constant of formation ($k = 0.77\ 10^6$ $M^{-1}\ sec^{-1}$), dissociation constant ($K_d = 1.4\ 10^{-6}$ M) and first order rate constant of autoxidation ($k = 10^{-4} sec^{-1}$ pH 7.4 and 4 °C). In the meantime comparative studies were undertaken with P-450 from liver microsomes (P-450 LM) (3). Successful solubilization, separation and purification of distinct P-450 isozymes renewed the interest in the study of the $Fe^{2+}O_2$ intermediate of P-450 LM.

It is now clear from previous reports (4)-(7), that the spectrum of the $Fe^{2+}O_2$ intermediate of P-450 LM resembles that observed with the bacterial enzyme, (λ_{max} = 420 and 558 nm) (5). However, with P-450 LM the intermediate is unstable and autoxidizes in a complex multiphasic process. Because of the apparent lack of data on this aspect, we have undertaken a study of the effect of oxygen concentration on both formation and autoxidation of the $Fe^{2+}O_2$ intermediate of P-450 LM. In this communication, some of our recent data concerning the P-450 LM_4 isozyme are presented.

MATERIALS AND METHODS

Electrophoretically homogeneous P-450 LM_4 (16-18 nmol/mg) was prepared according to ref. (8). The anaerobic stopped flow apparatus used in this work was described elsewhere (9). LM_4, 6 to 10 µM in 0.1 M K-phosphate pH 7.4 was photoreduced in one seringe (10 mM EDTA, 2 µM deazalumiflavin), whereas the other seringe contained the same solution, with variable oxygen concentration, LM_4 and deazalumiflavin being omitted. After rapid mixing (dead time 3 msec) of an equal amount (150 µl) of both solutions, the optical data were collected, stored and analyzed as described previously (9). Oxygen concentration was varied by adding, within the second seringe (whose content had been previously deoxygenated under argon) known amounts of an air saturated solution.

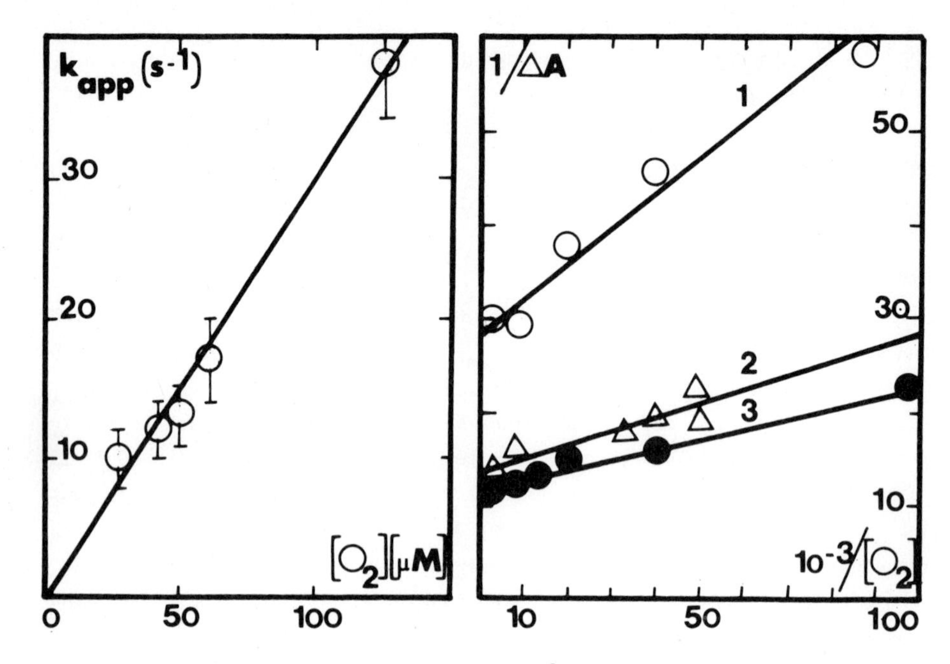

Fig. 1. (left). Apparent rate constant of $Fe^{2+}O_2$ formation as a function of O_2 concentration. LM_4 concentration 3.4 μM.

Fig. 2. (right). Reciprocal plot of absorbance change 500 msec after mixing (plateau) against O_2 concentration. (1) = 435/700 nm, pH 6.2, LM_4 : 3.75 μM, K_d = 14 μM ; (2) = 430/390 nm, pH 7.4, LM_4 : 3.4 μM, K_d = 10 μM ; (3) 430/390 nm pH 7.4, 80 μM benzo(a)pyrene, LM_4 : 3.4 μM, K_d = 8.3 μM.

RESULTS AND DISCUSSION

As in the case of P-450 LM_2, a low temperature study was carried out on the oxyferrous intermediate of isozyme LM_4 (detailled report will be published elsewhere). Similar results were observed in particular λ_{max} $(Fe^{2+}O_2)$ = 420 and 555 nm ; ε $(Fe^{2+}O_2)$ is 80 $\pm$ 2 mM^{-1} cm^{-1} at 420 nm and 14 $\pm$ 2 mM^{-1} cm^{-1} at 555 nm ; autoxidation proceeds through a biphasic process with clearly well defined isosbestic points at 412 nm and 538 nm. These data allowed us to select convenient wavelengths for further investigation, and were confirmed by stopped flow experiments carried out in aqueous buffers at 2 °C.

When reduced LM_4 is rapidly mixed with oxygenated buffer and the optical change at 430 nm analyzed, it first appears a very fast increase, a plateau is reached within 200 msec and a slower decrease follows. According to our low temperature spectra, the increase represents the formation of the oxyferrous

TABLE I

AUTOXIDATION DATA FOR P-450 LM_4

Assay conditions : 0.1 M K-phosphate pH 7.4, 10 mM EDTA ; 2 °C. LM_4 concentration after mixing : 3.4 μM. ΔA absorbance change at 430/390 between initial (Fe^{2+}) and equilibrium ($Fe^{2+}O_2$ + Fe^{2+}) level, where Fe^{3+} is neglegible. Ψ_f and Ψ_s = relative contribution of fast and slow process. k_f and k_s = rate constant for fast and slow process (in sec^{-1}).

O_2 (μM)	30	50	125	500
ΔA	0.059	0.060	0.063	0.071
Ψ_f (%)	40	37	43	46
Ψ_s (%)	60	63	57	54
k_f	0.20	0.20	0.19	0.20
$k_s \times 10^3$	17	15	11	8

intermediate and the subsequent decrease, the return to the oxidized form. Formation of the $Fe^{2+}O_2$ species is a quasi first-order process with respect to P-450 (oxygen concentration being always in excess). In Fig. 1., we have plotted the apparent rate constant of $Fe^{2+}O_2$ formation as a function of oxygen concentration (after mixing). The results show that k_{app} is proportional to O_2 concentration as expected from a formation reaction first-order with oxygen. From this graph we calculated the second-order reaction rate constant for $Fe^{2+}O_2$ formation : $k_1 = 0.33\ 10^6\ M^{-1}\ sec^{-1}$. This result is in reasonable agreement with the value ($0.77\ 10^6\ M^{-1}\ sec^{-1}$) reported for P-$450_{cam}$ at 4 °C (1). As oxygen concentration increases from 10 μM to 500 μM, the absorbance change between initial level (Fe^{2+}) and the plateau ($Fe^{2+}O_2$ + Fe^{2+} at equilibrium) increases (Table I). Double reciprocal plots from such experiments are reported in Fig. 2. These plots allowed us to determine the dissociation constant K_d of the equilibrium $Fe^{2+} + O_2 \rightleftarrows Fe^{2+}O_2$. Change in pH, or substrate (here benzo(a)pyrene) do not significantly modify the K_d which remains in the range of 10 μM, i.e. an order of magnitude higher than that observed with P-450_{cam}. From k_1 and K_d we obtain a first-order dissociation rate constant of 3.3 sec^{-1}.

Kinetics of $Fe^{2+}O_2$ autoxidation were also investigated as a function of oxygen concentration (from 30 μM to 500 μM). Results are reported Table I. As previously found with LM_2 (5) and LM_4 by Oprian and Coon (7), autoxidation is a biphasic first-order process. The rate constants observed here appear to be lower than those reported in (7). This desagreement could originate from the large difference in LM_4 concentration used in the present (3-5 μM) and previous work (80 μM) (7). As oxygen concentration increases, the relative amplitude of each phase remains reasonably constant, as well as the rate constant of the fast phase. However, it was repetitively found that the rate constant of the slow phase decreased as oxygen concentration increased, (similar finding was made in the presence of substrates). The reason for this finding is not yet understood, but it must be recalled that such observation was also made with hemoglobin (10). According to this result and the observation (not shown here) that the relative amplitude of both phases is substrate and pH dependent (Ψ_s 40 %, Ψ_f 60 % at pH 6.2), it is suggested that the biphasic process originates from the existence of two populations of cytochrome, in equilibrium, having different mechanism of autoxidation.

REFERENCES

1. Peterson, J.A., Ishimura, Y. and Walker Griffin, B. (1972) Arch. Biochem. Biophys. 149, 197-208.
2. Tyson, C.A., Lipscomb, J.D. and Gunsalus, I.C. (1972) J. Biol. Chem. 247, 5777-5784.
3. Estabrook, R.W., Hildebrandt, A.G., Baron, J., Netter, K.J. and Leibman, K. (1971) Biochem. Biophys. Res. Commun. 42, 132-139.
4. Guengerich, F.P., Ballou, D.P. and Coon, M.J. (1976) Biochem. Biophys. Res. Commun. 70, 951-956.
5. Bonfils, C., Debey, P. and Maurel, P. (1979) Biochem. Biophys. Res. Commun. 88, 1301-1307.
6. Bonfils, C., Balny, C., Douzou, P. and Maurel, P. (1980) In Biochemistry, Biophysics and Regulation of Cytochrome P-450, pp. 559-564. (Gustafsson, J.A. et al, eds). Elsevier/North Holland Biomedical Press.
7. Oprian, D.D. and Coon, M.J. (1980) In Biochemistry, Biophysics and Regulation of Cytochrome P-450, pp. 323-330. (Gustafsson, J.A. et al, eds). Elsevier/North Holland Biomedical Press.
8. Haugen, D.A. and Coon, M.J. (1976) J. Biol. Chem. 251, 7929-7939.
9. Bonfils, C., Balny, C. and Maurel, P. (1981) J. Biol. Chem. 256, 9457-9465.
10. Wallace, W.J. and Caughey, W.S. (1979) In Biochemical and Clinical aspects of Oxygen, pp. 69-86. (Caughey, W.S., ed). Academic Press.

© 1982 Elsevier Biomedical Press B.V.
Cytochrome P-450, Biochemistry, Biophysics
and Environmental Implications, E. Hietanen,
M. Laitinen and O. Hänninen editors

INFLUENCE OF PRENATAL EXPOSURE TO PHENOBARBITAL (PB) AND 3-METHYLCHOLANTHRENE (MC) ON POSTNATAL DEVELOPMENT OF MONOOXYGENASES ACTIVITIES AND THEIR INDUCIBILITY IN RATS

WOLFGANG KLINGER, FLORENTINE JAHN, techn. Ass. ELKE KARGE
Institute of Pharmacology and Toxicology, Löbderstr. 1,
DDR-6900 Jena, G.D.R.

INTRODUCTION

Age is a critical variable in enzyme induction (1, 2). Moreover pre- and perinatal exposure to inducing agents might influence postnatal development and inducibility during a long or even the whole lifespan (2) by a kind of imprinting on the transscriptional or translational level, which is not due to longlasting induction phenomena in the postnatal period after pretreatment of the mother with potent inducers with an extremely long biological half life time, e.g. TCDD (3). Preceding experiments had shown, that there might be a kind of booster effect when rats, after a first exposure to inducing agents in the suckling period, were treated with the same inducer after a long period in adulthood.
Soyka (2) could not detect an influence of MC injected late in pregnancy to mice on monooxygenase basal activity and inducibility by MC and PB up to an age of 2 years. The question remained whether the clear ontogenetic changes of monooxygenase reactions in the rat are influenced by treatment in early and possibly more sensitive stages of pregnancy.

METHODS

In the 1st, 2nd or 3rd week of pregnancy rats received 3 x 60 mg/kg PB i.p., ethylmorphine N-demethylation (E-N, 4) and ethoxycoumarin O-deethylation (E-O, 5) activities were determined in the offspring after birth and PB pretreatment (1 x 60 mg/kg i.p.) and at the 10th, 30th and 60th day of life after PB pretreatment (3 x 60 mg/kg i.p.).

RESULTS

Minor alterations have been observed for E-N in all age groups and its inducibility by PB, Figs. 1 - 4. The basal activities of

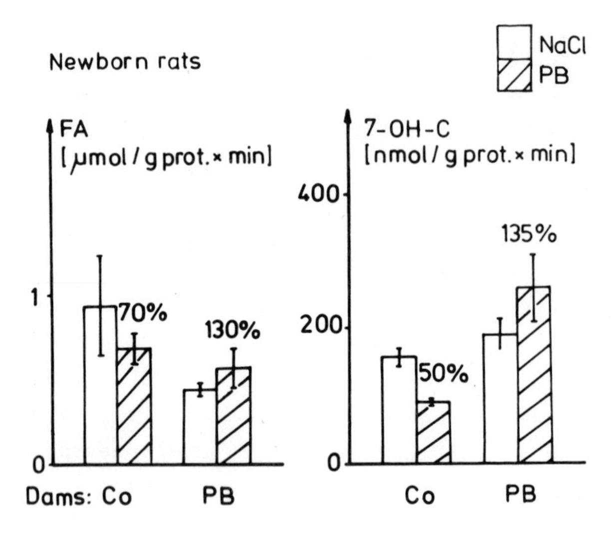

Fig. 1. Influence of PB pretreatment in pregnancy on E-N and E-O in newborn rats after pretreatment with PB.
FA = formaldehyde, 7-OH- C = 7-OH-coumarin.

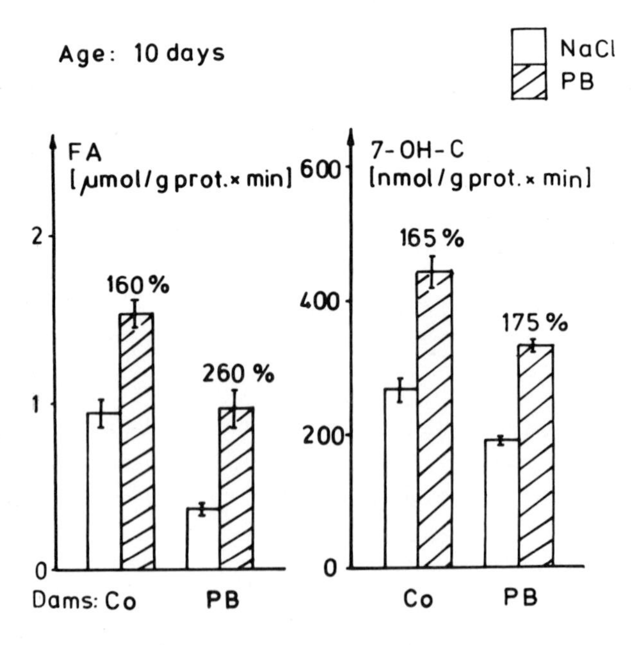

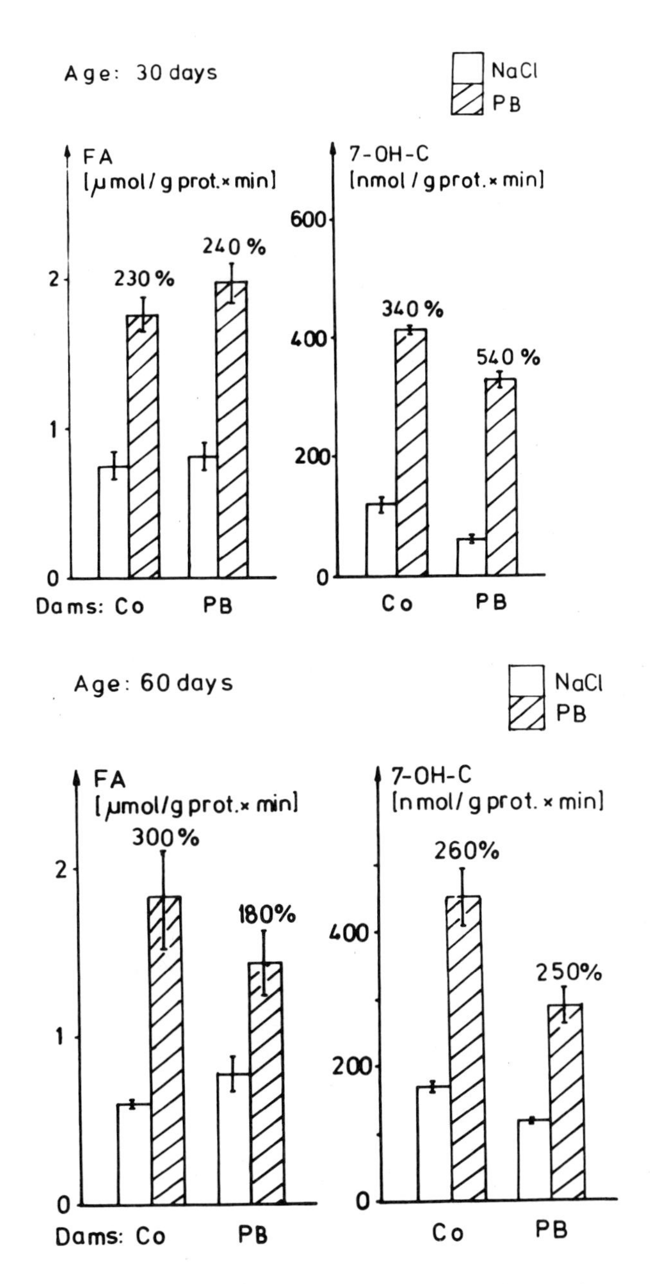

Fig. 2 - 4. Influence of PB pretreatment in pregnancy on E-N and E-O in 10-, 30 and 60-day-old rats after pretreatment with PB.

E-O were reduced in all age groups after PB pretreatment of the dams, and the inducibility by PB was increased. As E-N mainly reflects the PB inducible P-450 subfraction b and E-O both the PB and the MC inducible subfractions b and c we can conclude, that pretreatment of the dams with PB did not influence the ontogenic development and inducibility of $P\text{-}450_b$, but unexpectedly depressed ontogenic development of $P\text{-}450_c$. Induction with PB in the offspring of PB pretreated dams enhanced the $P\text{-}450_b$-mediated part of E-O. The values were absolutely in the same range as in controls, but in relation to the lower basal levels the inducibility is higher. So the effects of prenatal influence of MC on monooxygenases and the P-450 subfractions and their interrelations are of special interest. The first, preliminary results indicate that the basal E-O values in the offspring of MC treated dams 20 days after birth are higher than the control values (offspring of dams, treated during pregnancy with the solvent only). The inducibility of E-O with MC was enhanced significantly, whereas E-N values were not influenced. Thus the former findings on increased inducibility of P-450 concentration and monooxygenase activities after pretreatment of suckling rats with PB or MC can be confirmed and it can be added that after pretreatment during gestation not only inducibility but also basal activities are influenced in the offspring.

REFERENCES

1. Klinger, W., Müller, D., and Kleeberg, U. (1979) in: Estabrook, R.W., and Lindenlaub, E. (Ed.), The Induction of Drug Metabolism, F.K. Schattauer Verlag, Stuttgart - New York, p. 517
2. Soyka, L.F. (1981) in: Soyka, L.F., and Redmond, G.P. (Eds.), Drug Metabolism in the Immature Human, Raven Press New York, p. 101
3. Lucier, G.W. (1976) Environ. Health Perspect. 18, 25
4. Klinger, W., and Müller, D. (1977) Acta biol. med. germ. 36, 1149
5. Aitio, A. (1978) Analyt. Biochem. 85, 488

© 1982 Elsevier Biomedical Press B.V.
Cytochrome P-450, Biochemistry, Biophysics
and Environmental Implications, E. Hietanen,
M. Laitinen and O. Hänninen editors

PROGESTERONE-MICROSOMAL RECEPTOR IN RAT LIVER: EFFECT OF VARIOUS STEROIDS

AMIT GOSHAL, PHILIP S. ZACARIAS AND GEORGE FEUER
Department of Clinical Biochemistry and Pharmacology, University of Toronto, Toronto, Ontario, Canada, M5G 1L5.

INTRODUCTION

Contrasting effects have been demonstrated between reduced and hydroxy derivatives of progesterones on various microsomal hydroxylase and demethylase activities of the rat liver. The contrasting actions indicated a possible regulatory role of this steroid hormone on the activity of enzymes bound to the hepatic endoplasmic reticulum (1). Our recent finding that endoplasmic reticulum membranes contain receptor sites that specifically bind progesterone (2) raised the possibility that in the molecular action of steroid hormones the formation of the highly specific progesterone-microsomal complex may represent a key stage. In order to characterize this binding in this paper the contribution of the steroid component was analyzed. The affinity of various progesterones and other steroid hormones for receptor sites of the hepatic endoplasmic reticulum in the female rat was examined, aiming at mainly to reveal certain structural features of the steroid required for the specific interaction with microsomal protein.

METHODS

Chemicals. [1,2-^{3}H(N)] Progesterone (sp. act. 55.7 Ci/mmol) was obtained from New England Nuclear (Montreal, Quebec), non-radioactive steroids from Sigma Chemical Company (St. Louis, Mo), contraceptive steroids courtesy of Ortho Pharmaceuticals, (Toronto, Ont.) and Upjohn Company (Toronto, Ont.). 16α-Hydroxyprogesterone was synthesized in our laboratory (3). All other chemicals were of analytical grade preparations.

Preparation of microsomes. This and the determination of progesterone binding have been carried out as described in the previous paper (4).

Measurement of relative binding affinity. This was determined by applying unlabelled competitor steroids in concentrations ranging from 0 to 20000 nmol/l; radioactive progesterone was kept constant at 20 nmol/l in each sample. [^{3}H]Progesterone binding to microsomes was considered as 100% in the absence of unlabelled progesterone or other steroid competitors, and the molar concentrations of the unlabelled steroid competitor causing 50% reduction of binding was established on a semilog plot.

Radioactivity was measured as described in a previous paper (2).

Protein content was determined by the modified Lowry method (5).

RESULTS

Various steroid hormones exerted little competition for progesterone binding sites (Table 1). Estradiol, cortisol, testosterone and dihydrotestosterone showed low action on progesterone

TABLE 1. RELATIVE BINDING AFFINITY OF VARIOUS STEROIDS TO MICROSOMAL RECEPTORS IN THE LIVER OF THE FEMALE RAT

STEROID	RELATIVE BINDING, %
Progesterone	100.0
β-Estradiol	1.8
Mestranol (17α-ethynyl-3-methoxy-$\Delta_{1,3,5}$-estratrien-17β-ol)	1.0
17α-Ethynyl estradiol	4.7
Testosterone	9.1
Dihydrotestosterone	3.5
5α-Androstan-17β-ol-3-one	0.0
Cortisol	6.4
Cholesterol	0.0

TABLE 2. RELATIVE BINDING AFFINITY OF VARIOUS PROGESTERONES TO MICROSOMAL RECEPTORS IN THE LIVER OF THE FEMALE RAT

STEROID	RELATIVE BINDING, %
Progesterone (Δ_4-pregnen-3,20-dione)	100.0
5α-Pregnan-3,20-dione	47.6
5β-Pregnan-3,20-dione	20.7
Promegestone (Δ_4,6-Pregnen-17α-methyl-3, 20-dione)	27.0
Δ_4-Pregnen-20β-ol-3-one	2.8
Δ_4-Pregnen-3β,20β-diol	2.1
Pregnenolone (Δ_5-pregnen-3β-ol-20-one)	14.3
5α-Pregnen-3β-ol-20-one	5.2
5β-Pregnan-3α-ol-20-one	0[a]
5α-Pregnan-3β,20β-diol	0
5β-Pregnen-3α,20β-diol	0
5β-Pregnane-3α,17α-diol-20-one	0[a]
11α-Hydroxyprogesterone	21.0
16α-Hydroxyprogesterone	7.9
17α-Hydroxyprogesterone	26.4
16α,17α-Epoxyprogesterone	2.7
17α-Hydroxypregnenolone	8.0
16-Methylprogesterone	3.8
6-Methyl-17α-hydroxyprogesterone	6.7
6-Methylpregnenolone	1.2

[a] Increased progesterone binding

binding, cholesterol had no effect at all. Testosterone was the strongest competitor, but it only replaced 9% of progesterone binding sites. Similarly the ability of various progesterone derivatives to inhibit [^{3}H] progesterone binding was very low with the exception of 5α-pregnan-3, 20 dione which replaced 48% of the progesterone binding sites; 11α-hydroxy-progesterone resulted 21%,

17α-hydroxyprogesterone 26% and promegestone 27% reduction (Table 2). Some progesterone derivatives caused an increased binding at high concentrations. This observation could be explained that probably compounds with structures somewhat similar to progesterone if added to microsomes in excess amounts suppressed the dissociation of the progesterone-microsomal complex resulting in an enhanced progesterone binding (Fig. 1).

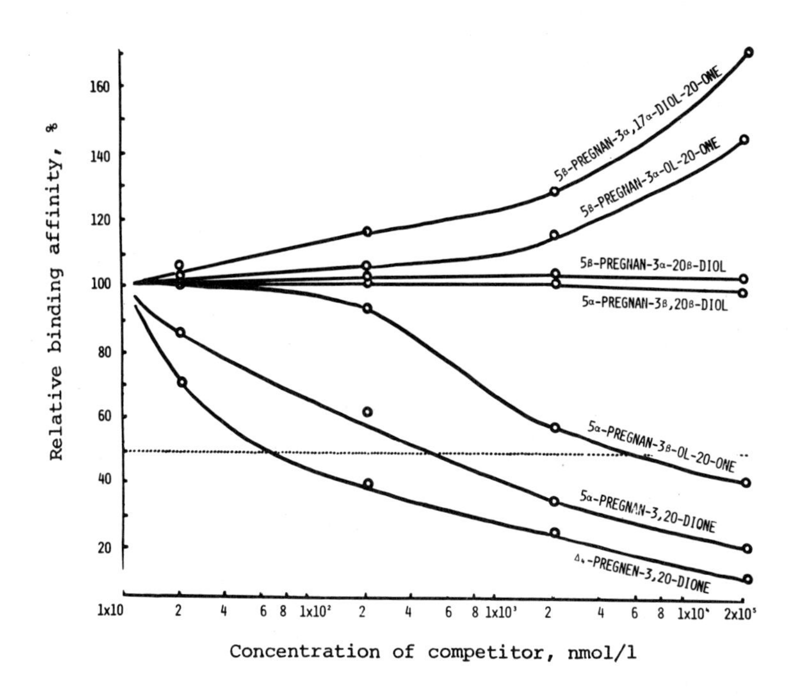

Fig. 1. Effect of various progesterone derivatives on progesterone binding to hepatic microsomes.

DISCUSSION

Determination of the relative binding affinity of various progesterones and other steroids in liver microsomes of the female rat as compared to progesterone binding has shown that the introduction of any substituent results in a decrease in binding affinity. This indicates that the binding site is highly specific for progesterone. 5α-Pregnan-3,20-dione gave the highest relative binding affinity of about half of the progesterone binding. It has also been shown that this compound exerts a high binding affinity to progesterone-binding globulin. The effect was attributed to the flatness of this steroid molecule (6). In this study it has been found that the binding of 5α-pregnan-3,20-dione was actually greater than progesterone. This is not the case in our experiments suggesting that binding of progesterone to the hepatic endoplasmic may be more specific than to progesterone-binding globulin.

Reduction of the oxo group on carbon position 3 results in a significant decrease in binding affinity. This is in agreement with the data that the 3-oxo group of progesterone is essential in the binding to progesterone-binding globulin (6). This is attributed to the presence of a hydrogen donor group in the binding site which is located close to the 3 oxo group of the ligand (7).

Binding of the C-5 region is extremely tight since the addition of a hydrogen at this position significantly decreases binding, although not to the same extent as reducing the 3 keto group.

Binding at the C-6 region is also extremely tight since the addition of methyl groups in this region reduces the binding affinity of both pregnenolone and 17α-hydroxyprogesterone. These results are in contrast to previous studies (6), that addition of a methyl group in this region increased the binding affinity of the steroid to progesterone-binding globulin. They attributed this to the increase in hydrophobic binding which the methyl groups provided. Obviously, the binding must have been looser because no steric hindrance was evident, as was the case in this study.

Binding at the C_{16}-C_{17} region is essentially hydrophobic since introduction of hydroxy groups reduces the binding affinity to a greater extent than with the addition of methyl groups. Addition of an epoxy group further decreases binding affinity. Binding at the C-20 region may be hydrophilic since reduction of this group lowers the binding affinity of 5α-pregnan-3β-ol-20-one from 5.2% to zero.

ACKNOWLEDGEMENT

This research was supported by the Medical Research Council of Canada.

REFERENCES

1. Feuer, G., Kardish, R., and Farkas, R. (1977). Differential action of progesterones on hepatic microsomal activities in the rat. Biochem. Pharmacol. 26, 1495-1499.

2. Drangova, R., and Feuer, G. (1980). Progesterone binding by the hepatic endoplasmic reticulum of the female rat. J. Steroid Biochem. 13, 629-637.

3. Cole, W. and Julian, P.L. (1954). Steroids XIV. Reduction of epoxy ketones by chromous salts. J. org. Chem. 19, 131-138.

4. Feuer, G., Golberg, L. and LePelley, J.R. (1965). Liver response tests I. Exploratory studies on glucose 6-phosphatase and other liver enzymes. Fd. Cosmet. Toxicol. 3, 235-249.

5. Miller, G.L. (1959). Protein determination for large numbers of samples. Analyt. Chem. 31, 964.

6. Blanford, A.T., Wittman, W., Stroupe, S.D., and Westphal, U. (1978). Steroid-protein interactions. XXXVIII. Influence of steroid structure on affinity to the progesterone-binding globulin. J. Steroid Biochem. 9, 187-201.

7. Drangova, R., Law, M.H.R. and Feuer, G. (1980). Specific progesterone receptor in hepatic microsomes of the female rat. Res. Commun. Chem. Pathol. Pharmacol. 29, 183-192.

Cytochrome P-450, Biochemistry, Biophysics and Environmental Implications, E. Hietanen, M. Laitinen and O. Hänninen editors

THE PREPARATION OF SUBSTRATE ANALOGS FOR STUDYING THE MECHANISM OF CYTOCHROME $P\text{-}450_{cam}$ CATALYZED REACTIONS

LAURI HIETANIEMI[1], PENTTI MÄLKÖNEN[2]
[1]Oy Star Ab, Box 33, SF-33721 Tampere 72 (Finland), [2]University of Joensuu, Department of Chemistry and Biological Sciences, Box 111, SF-80101 Joensuu 10 (Finland)

INTRODUCTION

In cooperation between the University of Joensuu and the Yale University in order to understand the mechanism of cytochrome $P\text{-}450_{cam}$, the task of the Joensuu group is to prepare suitable model compounds. Up till now we have prepared optically active dehydrocamphor and 6-endo-bromocamphor. These compounds have been useful in studying $P\text{-}450_{cam}$ dependent epoxidation and halogen oxidation reactions.

DISCUSSION

Figure 1 shows the scheme of the method developed previously by Mälkönen (1) for preparing dehydrocamphor from camphor. The method was applied this time, also. As you can see, nine synthetic stages are needed for a small change, that is, for preparing a double bond between carbons 5 and 6 in the camphor skeleton. A more applicable method for preparing dehydrocamphor has not so far been presented. The advantage of this method is that it gives an optically active pure final product.

First, reduction with sodium in ethanol is performed in the series. This is followed by protective acetylation of the obtained hydroxyl group. Protection is necessary for the oxidation with CrO_3. Protecting of the hydroxyl group is also required for the most critical reaction in the series, i.e. the Tzugajeff reaction. Protection with dihydropyran has shown to be suitable for this. It is removed at the same stage in the Tzugajeff pyrolysis. Finally, dehydroborneol is oxidized to the ketone using the Oppenauer oxidation.

We prepared 6-endo-bromocamphor by a new method starting from α-pinene (Figure 2) (2) in four stages: $KMnO_4$ oxidation, $NaBH_4$ reduction, $ZnBr_2$ rearrangement, and HNO_3 oxidation. The stereochemistry of bromine was verified by preparing a new compound, 6-endo-bromoborneol. Its stereochemistry at carbon 1 was first verified by NMR with europium shift reagents. The stereochemical structure at carbon 6 was then solved by the fact that

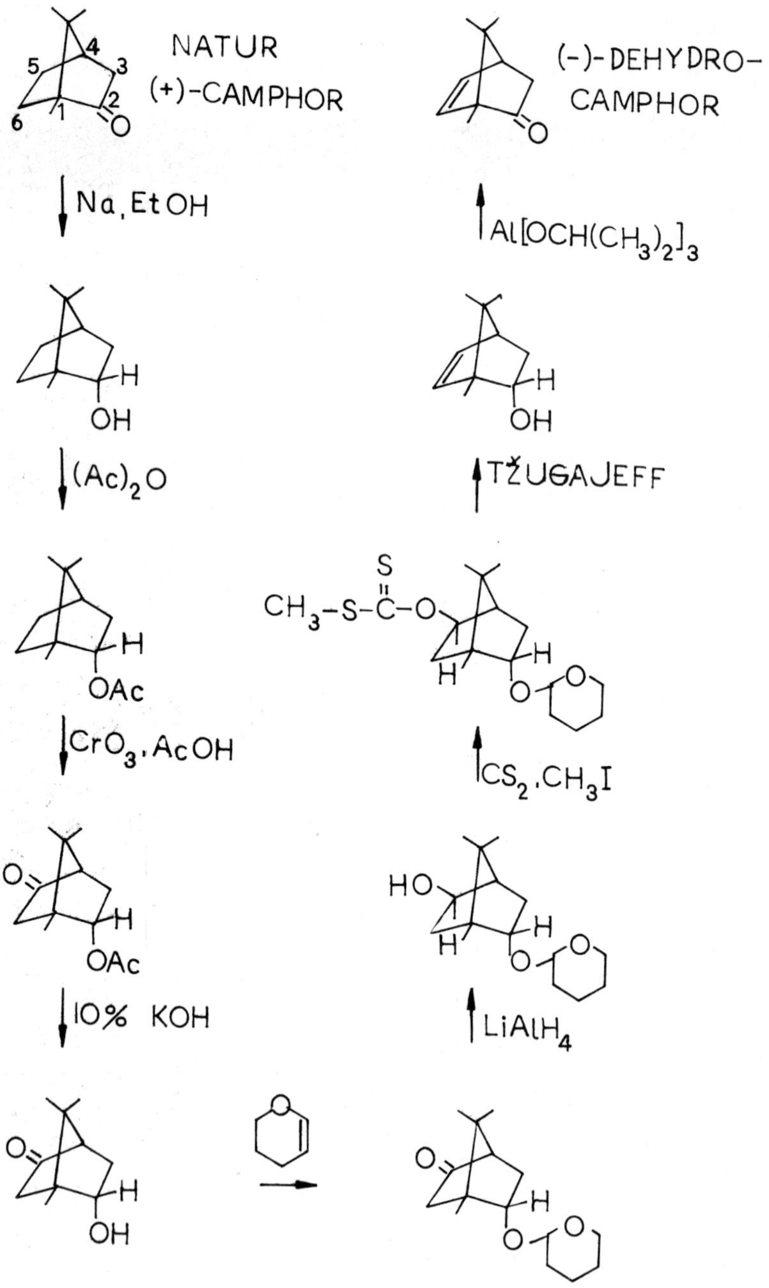

NATUR (+)-CAMPHOR

Na, EtOH

$(Ac)_2O$

CrO_3, AcOH

10% KOH

LiAlH$_4$

CS_2, CH_3I

$CH_3-S-C(=S)-O$

TŽUGAJEFF

$Al[OCH(CH_3)_2]_3$

(-)-DEHYDRO-CAMPHOR

Fig. 1. Preparation of optically active (-)-dehydrocamphor (1).

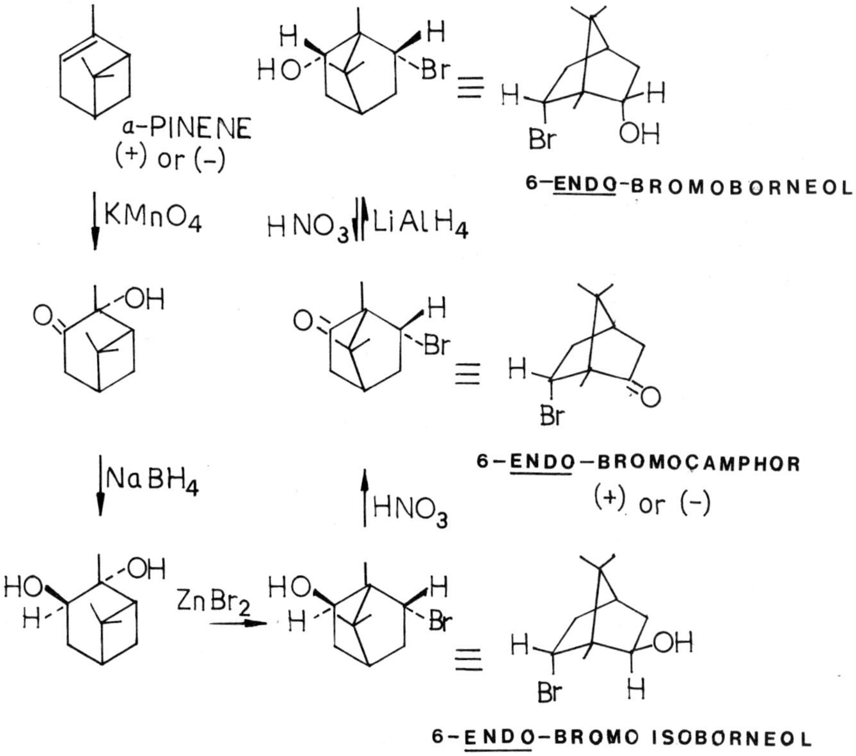

Fig. 2. Preparation of optically active 6-*endo*-bromocamphor (2).

this new compound had W-coupling of 2.2 Hz between protons 2 and 6 in a 250 MHz NMR spectrum. 6-*endo*-bromoisoborneol did not have this coupling.

It was interesting to compare the oxidation and reduction of these bromine compounds with camphor and isoborneol. Isoborneol can not be oxidized to the ketone with HNO_3, but corresponding 6-*endo*-bromoisoborneol was oxidized. Camphor can be reduced with lithium aluminum hydride to isoborneol, but the corresponding 6-bromocamphor surprisingly produced 6-bromoborneol. So the *exo*-hydroxyl is obtained from camphor, while the *endo*-hydroxyl is obtained from 6-*endo*-bromocamphor.

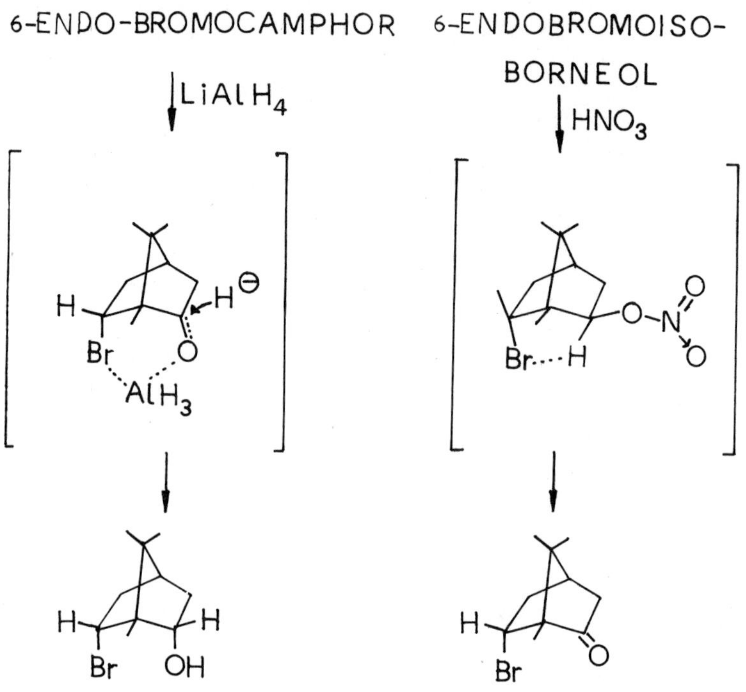

6-ENDO-BROMOCAMPHOR 6-ENDOBROMOISOBORNEOL

$LiAlH_4$ HNO_3

Scheme 1

In our opinion these surprising reactions canbe explained by the neighboring group effect caused by bromine (Scheme 1). Upon reduction the bromine atom of 6-*endo*-bromocamphor is coordinated with aluminium, and the hydride ion must come from the *exo* side. In HNO_3 oxidation of 6-*endo*-bromoisoborneol the bromine atom obviously contributes to the release of the 2-*endo* proton.

REFERENCES

1. Mälkönen, P.J. (1964) Ann. Acad. Scient. Fennicae AII, p. 7 - 61.
2. Hietaniemi, L.A. and Mälkönen, P.J. (1982) Finn. Chem. Lett. 3 - 4 (in Press).

© 1982 Elsevier Biomedical Press B.V.
Cytochrome P-450, Biochemistry, Biophysics
and Environmental Implications, E. Hietanen,
M. Laitinen and O. Hänninen editors

MULTI-SUBSTRATE MONOOXYGENASES — A FAMILY OF SIMILAR BUT DIFFERENT CATALYSTS

M. J. COON
Department of Biological Chemistry, Medical School, The University of Michigan, Ann Arbor, Michigan 48109 (U.S.A.)

Studies in this and other laboratories on purified cytochrome P-450 have shown that it occurs in many forms apparently differing in inducibility, chemical and physical properties, and substrate specificity. Speculation has ranged from the view that these forms represent a single cytochrome with varying properties because of the presence of non-peptide components (carbohydrate, lipid, or detergent) to the view that they represent an almost unlimited number of distinct gene products. In the latter case, one could imagine that each low molecular weight xenobiotic in the environment might cause the induction of a unique P-450 in animals, just as foreign macromolecules elicit the formation of specific antibodies. Various lines of evidence presented at recent international symposia support the concept that the various forms of P-450 are indeed isozymes — that is, enzymes differing to a greater or lesser extent in primary structure — and that the number of isozymes in a particular source such as the hepatic endoplasmic reticulum is at least as great as six and perhaps as large as ten or twenty. Highly important questions in this rapidly developing field are whether the upper limit will be an even larger number and how these isozymes differ from a biochemical point of view.

In the present brief paper some aspects of our knowledge of the similarities and differences among the many P-450's will be discussed. No attempt will be made to survey all of the known cytochromes of this type but rather to point out some interesting developments.

P-450's in a single organelle. The first mammalian cytochrome P-450's to be purified to electrophoretic homogeneity and extensively characterized were those from the hepatic endoplasmic reticulum. In view of the known multiplicity of inducers and substrates for the liver microsomal cytochromes, however, caution must be applied in estimating the total number of such isozymes. The data given in Table 1 show those rabbit P-450's which may be called distinct isozymes on the basis of work in various laboratories, as summarized elsewhere (1). In several instances the exchange of purified proteins among laboratories has simplified this list, in which only those cytochromes known to be electrophoretically homogeneous and to have

TABLE 1

P-450 ISOZYMES PURIFIED FROM RABBIT LIVER MICROSOMES

Isozyme	Inducer	Molecular weight	λ_{max} of carbonyl complex (nm)
2	Phenobarbital	48,000	451
3a	Ethanol	50,000	452
3b	Troleandomycin	52,000	450
3c		53,000	449
4	5,6-Benzoflavone	54,000	447
6	Isosafrole	58,000	448

distinct N- and C-terminal amino acid sequences are shown. The latter criterion is especially important in view of the fact that SDS-polyacrylamide gel electrophoresis may not distinguish between two proteins of highly similar minimal molecular weight or elucidate the relationship between a parent protein and its proteolytic product.

The six isozymes shown in the table differ electrophoretically and, accordingly, in their minimal molecular weights. They also differ in the λ_{max} values of the carbonyl adducts and in various other properties. It remains to be established whether all liver microsomal P-450's are inducible. Isozyme 2 is known to be drug-inducible, 3a ethanol-inducible as found in this laboratory (1-3), and 3b antibiotic-inducible as recently reported by Maurel and associates (4). Isozymes 4 and 6 have a number of inducers, only one of which is shown in each instance; 3c is presumed to be constitutive, but only because an inducer has not yet been found. Whether the level of some or all of these cytochromes is determined by hormonal control, as shown convincingly by Gustafsson and associates (5) for the rat, is not yet known. The list of isozymes in the table will probably be expanded when other possible rabbit cytochromes (6,7) are more thoroughly characterized and the explanation for separate steroid 7α- and 12α-hydroxylating activities in isozyme 4 preparations (8) is known.

P-450's in different strains. An interesting example of genetic heterogeneity was found by Johnson *et al*. (9), who reported the occurrence of both leucine and tyrosine residues at position 10 of P-450 isozyme 3b as well as the presence or absence of another peptide, and by our laboratory, which also reported variability (threonine, leucine, etc.) for residue 10 (1). This may possibly indicate strain differences in the population of New Zealand rabbits used. More recently, Walz *et al*. (10) have described differences in the P-450's present in liver microsomes from two strains of rats.

P-450's in different tissues. Although much remains to be learned about the similarities and differences between P-450's from different tissues from the same species, mention should be made that two isozymes of P-450 have been obtained in an electrophoretically homogeneous state from rabbit lung microsomes by Philpot and colleagues (11). Several lines of evidence showed that isozyme 2 of liver microsomal P-450 and lung microsomal form I are apparently identical, even though the latter protein is not inducible upon phenobarbital administration.

P-450's in different species. Our studies on the amino acid sequence of isozyme 2, as presented at this meeting (12), show extensive similarity between this phenobarbital-inducible P-450 from rabbit and the coresponding protein from rat, P-450_b, as recently deduced by Fujii-Kuriyama *et al*. (13) from the cDNA sequence. The sequence homology is approximately 80%. On the other hand, very little homology exists between the sequence of our isozyme 2 and that of P-450_{cam} as recently reported by Haniu *et al*. (14).

ACKNOWLEDGMENTS

Recent studies in this laboratory were supported by Grant AM-10339 from the National Institutes of Health and Grant PCM102324 from the National Science Foundation.

REFERENCES

1. Coon, M.J., Black, S.D., Koop, D.R., Morgan, E.T., and Tarr, G.E. (1982) in: Sato, R., and Kato, R. (Eds.) Microsomes, Drug Oxidations, and Drug Toxicity, Japan Scientific Societies Press, Tokyo, pp. 13-23.

2. Koop, D.R., Morgan, E.T., and Coon, M.J. (1982) in: Sato, R., and Kato, R. (Eds.) Microsomes, Drug Oxidations, and Drug Toxicity, Japan Scientific Societies Press, Tokyo, pp. 85-86.

3. Koop, D.R., Morgan, E.T., Tarr, G.E., and Coon, M.J. (1982) J. Biol. Chem., in press.

4. Bonfils, C., Dalet-Beluche, I., and Maurel, P. (1982) Biochem. Biophys. Res. Commun. 104, 1011-1017.

5. Gustafsson, J.-Å., Mode, A., Norstedt, G., Eneroth, P., and Hökfelt, T. (1982) in: Sato, R., and Kato, R. (Eds.) Microsomes, Drug Oxidations, and Drug Toxicity, Japan Scientific Societies Press, Tokyo, pp. 281-288.

6. Aoyama, T., Imai, Y., and Sato, R. (1982) in: Sato, R., and Kato, R. (Eds.) Microsomes, Drug Oxidations, and Drug Toxicity, Japan Scientific Societies Press, Tokyo, pp. 83-84.

7. Dieter, H.H., Muller-Eberhard, U., and Johnson, E.F. (1982) Biochem. Biophys. Res. Commun. 105, 515-520.

8. Wikvall, K., Bostrom, H., Hansson, R., and Kalles, I. (1982) in: Sato, R., and Kato, R. (Eds.) Microsomes, Drug Oxidations, and Drug Toxicity, Japan Scientific Societies Press, Tokyo, pp. 265-272.

9. Johnson, E.F., Dieter, H.H., and Muller-Eberhard, U. (1982) in: Sato, R., and Kato, R. (Eds.) Microsomes, Drug Oxidations, and Drug Toxicity, Japan Scientific Societies Press, Tokyo, pp. 35-43.

10. Walz, F. G., Jr., Vlasuk, G.P., Omiecinski, C.J., Bresnick, E., Thomas, P.E., Ryan, D.E., and Levin, W. (1982) J. Biol. Chem. 257, 4023-4026.

11. Wolf, C.R., Slaughter, S.R., Marciniszyn, J.P., and Philpot, R.M. (1980) Biochim. Biophys. Acta 624, 409-419.

12. Black, S.D., Tarr, G.E., and Coon, M.J. (1982), This Symposium.

13. Fujii-Kuriyama, Y., Mizukami, Y., Kawajiri, K., Sogawa, K., and Muramatsu, M. (1982) Proc. Natl. Acad. Sci. U.S.A. 79, 2793-2797.

14. Haniu, M., Armes, L.G., Tanaka, M., Yasunobu, K.T., Shastry, B.S., Wagner, G.C., and Gunsalus, I.C. (1982) Biochem. Biophys. Res. Commun. 105, 889-894.

Control and Biophysics of Cytochrome P-450 Function

© 1982 Elsevier Biomedical Press B.V.
Cytochrome P-450, Biochemistry, Biophysics
and Environmental Implications, E. Hietanen,
M. Laitinen and O. Hänninen editors

STABILITY, CONFORMATIONAL RIGIDITY AND LIFE-TIME OF MICROSOMAL REDOX ENZYMES IN SOLUBLE AND MEMBRANE-BOUND STATE

A.I.ARCHAKOV
Second Moscow Medical Institute, Moscow, USSR

INTRODUCTION

There are many points of view on the mechanism of phospholipid reactivation action on the solubilized biomembrane enzymes. The most popular of them is the opinion that the inactive conformational state of soluble enzyme changes in active catalytic state. The role of polar head and hydrophobic tail is discussed at protein-phospholipid interaction (1,2,3). Some years ago it was discovered in our lab that the incorporation of soluble purified cytochrome P-450 into phospholipid bilayer was not accompanied by great conformational changes in α-helix structure in spite of its reactivation. At the same time we observed the increase of conformational rigidity and stability of this enzyme in membrane-bound state. This effect did not depend on the type of phospholipid used for the preparation of liposomes. The polar head of phospholipid influences the rate of phospholipid-protein or protein-protein interaction only (1,4). It was suggested that the reactivation action of phospholipid bilayer on membrane enzymes is the result of an increase of their conformational rigidity and stability. To prove this suggestion the influence of phospholipid bilayer on the structure and functional activity of cytochromes P-450, b_5 and NADPH-cytochrome c reductase was studied.

Abbreviations used: P-450, cytochrome P-450-LM2; b_5, cytochrome d-b_5; FP, NADPH-specific flavoprotein; PL, phospholipid; MPL, the mixture of microsomal phospholipids; PCh, phosphatidylcholine; DMA, dimethylaniline; p-NA, p-nitroanisole; p-NPh, p-nitrophenol; PE, phosphatidylethanolamine; PS, phosphatidylserine; PI, phosphatidylinosite; SM, sphingomyelin; CuOOH, cumole hydroperoxide.

MATERIALS AND METHODS

Rabbit liver microsomal cytochromes P-450-LM2 (P-450), b_5 and NADPH-specific flavoprotein (FP) were obtained according to (5, 6,7). Monobilayer liposomes were prepared from different types of phospholipid. The α-helix content was measured by CD-method (8). The activity of NADPH-cytochrome c reductase, the rate of p-nitroanisole (p-NA) and aniline oxidation, stop-flow and spectrofluorimetricexperiments, the inactivation rate of reduced P-450 were determined as described earlier (9,10). The incorporation of proteins was performed into preliminary prepared monobilayer phospholipid liposomes (8).

RESULTS

The effectivity of protein incorporation into phospholipid (PL) bilayer prepared from the mixture of microsomal phospholipids (MPL-liposomes) was controlled by spectrofluorimetric method (Fig.1). The incorporation of b_5 and FP was accompanied by the increase of fluorescence of tryptophan residue of these proteins. On the contrary the incorporation of P-450 was not followed by the increase of tryptophane fluorescence. It means that tryptophan residue of soluble cytochrome P-450 is localized in hydrophobic microenviroment.

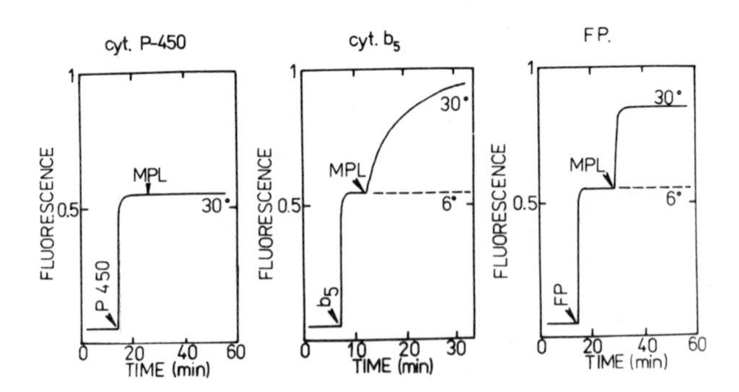

Fig. 1. Tryptophane fluorescence at incorporation of enzymes into membranes of MPL-liposomes

Liposomes with incorporated enzymes were sedimented at 150000g for 90 min from incubation mixture containing 100mM tris-HCl buffer, 300 mM NaCl (proteoliposomes). Nonincorporated enzymes and free liposomes were not sedimented under these conditions. Proteoliposomes which were prepared from MPL contained: PCh- 56%, PE - 24%, PI + PS - 12%, SM - 12% and 1.8 FP, or 1.8 P-450 or 1.6 b_5 per 1000 moles of PL.

Cytochromes P-450 and b_5 were incorporated into MPL- and PCh--liposomal membrane as well. On the contrary FP was incorporated into MPL-liposomal membrane only.

The data of the influence of protein incorporation into phospholipid bilayer on the α-helix content of isolated purified enzymes are given in Table 1. The incorporation of all biomembrane proteins into MPL-liposomes did not result in any significant changes of their α-helix content. The changes of conformational state of b_5 and P-450 incorporated into PCh-liposomes may be the result of other phospholipid microenviroment in comparison with MPL-proteoliposomes.

TABLE 1

α-HELIX CONTENT OF SOLUBLE AND MEMBRANE-BOUND MICROSOMAL ENZYMES (in per cent of aminoacids)

Protein	Soluble	Proteoliposomes	
		PCh	MPl
FP	20	-	20
P-450	60	55	57
b_5	37	50	41

Apparent catalytic constants of soluble and membrane-bound microsomal enzymes are given in Table 2.

Obtained results allowed us to conclude that the PL-bilayer did not influence redox-behaviour of cytochrome b_5 and FP. Their electron acceptor-donor properties are the same in soluble and membrane-bound state. In contrast to this, electron acceptor property in fast phase and the hydroxylase activity of membrane--bound P-450 are higher than that of soluble form.

Comparing these results with the data of CD-analysis of these proteins one can see that there is no dependence between the ac-

TABLE 2

FUNCTIONAL ACTIVITY OF SOLUBLE AND MEMBRANE-BOUND MICROSOMAL ENZYMES (apparent fast-order constants, min^{-1})

Indices	Soluble	MPL-proteoliposomes
Cytochrome P-450		
Reduction rate by dithionite		
fast phase	18,6	166
slow phase	2,4	1,6
CuOOH-dependent hydroxylation of		
p-NA	4,1	8,3
aniline	3,5	9,5
DMA	57	113
Cytochrome b_5		
Reduction by dithionite	900	720
ascorbate	0,6	0,4
FP		
NADPH-cytochrome c reductase activity	3100	3100

tivity of microsomal enzymes and changes of their conformational states. In the case of P-450 some decrease of α-helix content (from 60% down to 57%) was followed by a remarkable increase of its activity. In the case of b_5 some increase of α-helix content (from 37% up to 41%) did not result in any change of its electron acceptor properties. α-helix content and reductase activity of soluble and membrane-bound FP are the same.

Thus there is no direct correlation between the changes of α-helix content and functional activity of isolated soluble and membrane incorporated microsomal enzymes.

To investigate the mechanism of reactivation action of MPL--bilayer on P-450 its influence on stability, conformational rigidity and life-time of this hemoprotein was studied in comparison with b_5 and FP. The data of the influence of PL-bilayer on the stability of the three enzymes are given in Fig.2

It is seen that the incorporation of protein into PL-bilayer increases the stability of all the enzymes.

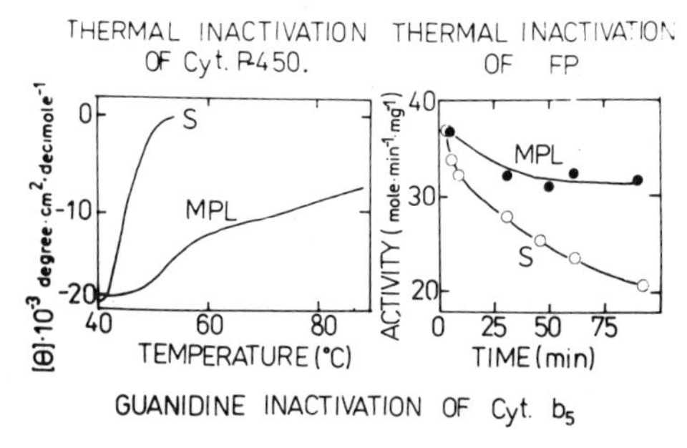

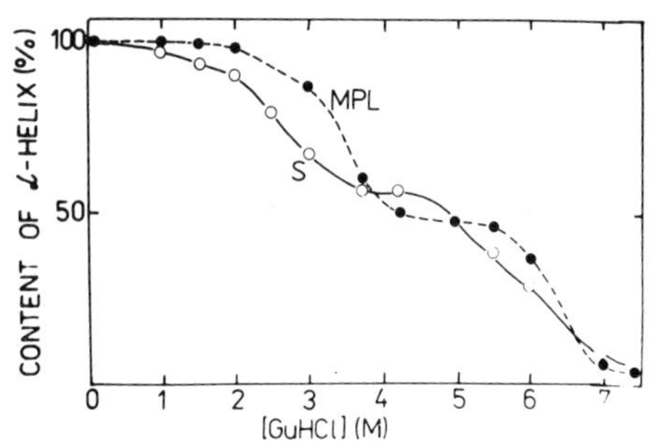

Fig. 2. The stability of P-450, b_5 and FP in soluble and membrane-bound state (S - soluble, MPL - P-450, incorporated into MPL-liposomes)

The influence of PL-bilayer on conformational rigidity is given in Table 3. In contrast to the general stabilizing action of PL-bilayer on the three microsomal enzymes the increase of conformational rigidity was observed in the case of P-450 only. Such difference between P-450 and b_5, FP may be explained by P-450-LM2 active centre localization in the hydrophobic space of PL-bilayer. Thus the reactivation action of PL-bilayer on P-450 may be the result of an increase of molecule conformational rigidity after its membrane incorporation.

To explain whether the enhancement of conformational rigidity of P-450 leads to the increase of its stability (enzyme life-time) at its catalytic action we studied the inactivation of P-450 at NADPH- and CuOOH-dependent oxidation reactions of p-NA in microsomes and soluble state (Fig.3)

It is seen that in all the cases the inactivation of P-450 is accompanied by the decrease of product formation in the same way. To estimate the constant of inactivation of P-450 we calculated the inactivation of P-450 in moles per moles of formed product and called it "life-time constant" (k_{lt}) (Table 4). This constant did not depend on the rate of hydroxylase reactions, the time of reaction, cosubstrate and substrate. k_{lt} of solubilized

TABLE 3

THE INDUCED CONFORMATIONAL RIGIDITY OF CYTOCHROMES P-450, b_5 AND FP

Indices	Content of α-helix			
	Soluble		MPL-proteoliposomes	
	% of amino-acids	% of change	% of amino-acids	% of change
Cytochrome P-450 (Fe^{3+})	60		57	
+1.8mM dithionite	70	+17	62	+9
+8.4mM diaminooctan	65	+ 8	57	0
Cytochrome b_5(Fe^{3+})	37		41	
+1.8mM dithionite	32	-14	36	-12
FP				
without NADPH	20		20	
with 1mM NADPH	20	0	20	0

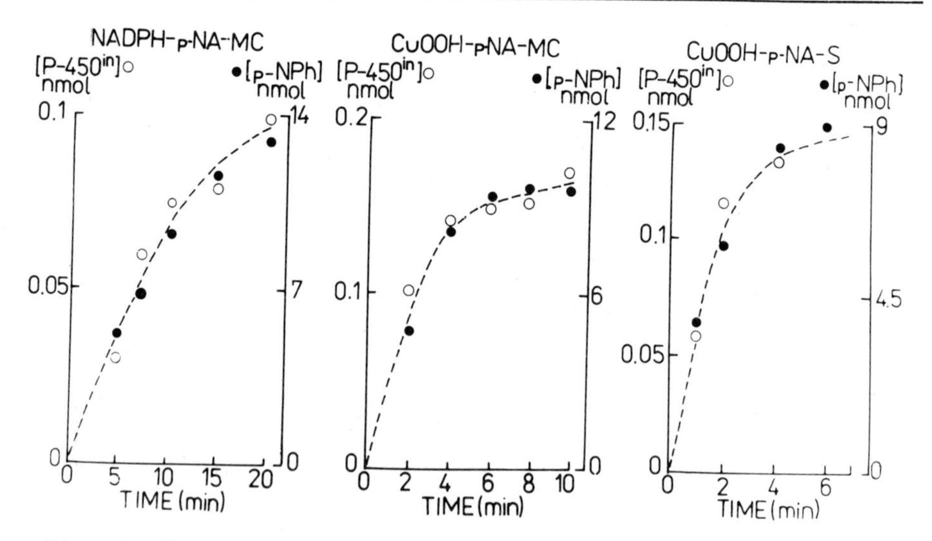

Fig. 3. The inactivation of soluble (S) and membrane-bound (microsomal - MC) P-450 at CuOOH- and NADPH-dependent reactions

TABLE 4

THE LIFE-TIME CONSTANT (k_{lt}) OF CYTOCHROME P-450 AT p-NA AND ANILINE OXIDATION (nmoles inactivated P-450 . 10^3 per nmole of formed product) AND FP AT NADPH-CYTOCHROME C REDUCTION REACTION (nmole FP.10^3 per nmole of reduced cytochrome c)

Conditions	State of enzyme	
	Soluble	Microsomal
Cytochrome P-450		
O-demethylation of p-NA		
NADPH	-	7
CuOOH	21	11
p-hydroxylation of aniline		
NADPH	-	7
CuOOH	16	7
FP	0.013	0.004

enzyme is higher than that for the membrane-bound P-450. It is interesting to point out that both cosubstrates NADPH and CuOOH have the same inactivation constant according to this calculation. Thus the value of inactivation is constant in calculation per mole of formed product. It characterizes the number of catalytic acts which may be performed by one molecule of enzyme before its inactivation and consists of 90-140 catalytic acts for membrane-bound state of hemoprotein in both NADPH and CuOOH-dependent reactions or 50-60 acts for soluble P-450 in CuOOH-dependent reaction. It is interesting to note that k_{lt} of NADPH--FP is about 3-order lower than that of P-450 (Table 4) i.e. FP is inactivated much slower than P-450 at catalytic process.

Thus the reactivation action of PL-bilayer on P-450 may be the result of the enhancement of its conformational rigidity. To prove this suggestion the thermodynamic parameters of p-NA O--demethylation reactions catalyzed by soluble and membrane-bound P-450 were measured. If that is true the increase of catalytic constant of membrane-bound enzyme should be accompanied by a smaller degree of change of its ΔS than that of the soluble one.

TABLE 5
CATALYTIC CONSTANTS ΔH AND ΔS OF p-NA DEMETHYLATION FOR CuOOH DEPENDENT REACTION OF SOLUBLE AND MEMBRANE-BOUND P-450

State of P-450	$k_{cat.}$, min^{-1}	ΔH, kcal·$mole^{-1}$	ΔS, e.u.
Soluble	4,8	13,2	-61,6
PCh-proteoliposomes	12,1	21,2	-34,2

It is seen that the enhance of catalytic constant at P-450 incorporation into PL-bilayer is followed by the increase of activation energy and is the result of ΔS decrease for the reaction catalysed by the membrane-bound hemoprotein. Therefore PL activation action on P-450 is the result of the increase of hemoprotein conformational rigidity with concomitant decrease of enthropy factor in catalytic process. FP and b_5 have the same activity and conformational rigidity in soluble and membrane-bound states. A more general action of PL-bilayer on microsomal enzymes is their stabilization against different denaturation agents. Catalytic life-time of P-450 is a constant value in calculation per mole of formed product under different experimental conditions.

ACKNOWLEDGEMENTS

The experiments were performed in collaboration with Drs. G.I. Bachmanova, A.V.Karyakin, I.I.Karuzina, V.Y.Uvarov, E.D.Skotzelyas. The author wishes to thank all of them.

REFRENCES

1. Ingelman-Sundberg M., Johansson I., Brunstrom G., Haaparanta I. and Rydstrom,J. (1980) in: Gustafsson, J.-A.et al. (Eds.), Biochemistry, Biophysics and Regulation of Cytochrome P-450, Elsevier/North Holland Biomedical Press, Amsterdam, pp. 299-306
2. Ruckpaul, K., Rein, U., Blanck, J., Ristau, O. and Coon M.J. (1980) Ibid, pp. 539-549
3. Kissel, M.A., Usanov, S.A., Metelitza, D.I. and Akhrem, A.A. (1979) Bioorganic Khimia,5, 1553-1558
4. Bachmanova, G.I. and Archakov, A.I. (1981) in: Microsomes,

Drug Oxidations and Drug Toxicities, 5th Inernational Symposium on Microsomes and Drug Oxidations, July 26-29, 1981, Tokyo, Japan, P-1-2, p.25

5. Karuzina, I.I., Bachmanova, G.I., Mengazetdinov, D.E., Myasoedova, K.N., Zhikhareva, V.O., Kuznetsova, G.P. and Archakov, A.I. (1979) Biokhimia, 44, 1149-1159
6. Spatz, L. and Strittmatter, P. (1971) Proc.Natl.Acad.Sci. USA, 68, 1042-1046
7. Dignam, J.D. and Strobel, H.W. (1977) Biochemistry, 16, 1116-1123
8. Archakov, A.I., Uvarov,V.Yu., Bachmanova, G.I., Sukhomudrenko, A.G., Myasoedova,K.N. (1981) Arch.Biochem.Biophys., 212, 378-384
9. Borodin, E.A., Dobretsov, G.E., Karasevich, E.I., Karuzina, I.I., Karyakin, A.V., Kuznetzova, G.P., Spirin, M.M. and Archakov,A.I. (1981) Biokhimia, 46, 1109-1118
10. Uvarov, V.Yu., Bachmanova, G.I., Mazurov, A.V. and Archakov, A.I. (1981) Bioorganic Khimia, 7, 621-627

© 1982 Elsevier Biomedical Press B.V.
Cytochrome P-450, Biochemistry, Biophysics
and Environmental Implications, E. Hietanen,
M. Laitinen and O. Hänninen editors

LIPID-PROTEIN INTERACTIONS AS DETERMINANTS OF THE ASSOCIATION OF CYTOCHROME b_5 AND CYTOCHROME P-450 REDUCTASE WITH CYTOCHROME P-450

BERNHARD BÖSTERLING AND JAMES R. TRUDELL
Department of Anesthesia, Stanford University School of Medicine, Stanford, California 94305 (U.S.A.)

INTRODUCTION

Studies aimed at elucidation of the mechanism of cytochrome b_5 (b_5) effects on cytochrome P-450-mediated oxidation using micelle-reconstituted systems of purified proteins have revealed a complex pattern of interactions. The variability in effects of b_5 has been suggested to be due in part to differences in the substrate metabolized or the type of cytochrome P-450 used in reconstitution experiments (1,2). However, it may also be reflective of differences in the molar ratios of NADPH-cytochrome P-450 reductase (reductase), cytochrome b_5, and cytochrome P-450 and/or the ratios of these proteins to phospholipid molecules in the micelle.

MATERIALS AND METHODS

Cytochrome P-450 LM-2 (P-450), NADPH-cytochrome P-450 reductase and cytochrome b_5 were purified as previously described (3,4,5). Egg phosphatidylcholine and phosphatidylethanolamine were prepared under an atmosphere of nitrogen. Reconstitution was accomplished by the previously described cholate dialysis technique (6-9). All measurements were carried out in 0.02 M potassium phosphate buffer, pH 7.5, containing 20% glycerol. Formaldehyde production was determined by the Nash reaction. A Cary 219 spectrometer was used for absorption measurements. Matched tandem cuvettes with inner compartments of 4-mm diameter each were used for difference spectroscopy measurements. All magnetic circular dichroism (MCD) spectra were recorded on a JASCO model J-40 spectropolarimeter.

RESULTS

The activity of benzphetamine N-demethylation by micelle-reconstituted systems was measured by the production of formaldehyde at various b_5 to P-450 ratios. The highest rate was observed in those systems containing the greatest amount of reductase in that this is the rate-limiting component. However, the greatest percentage increase in activity due to b_5 was observed in the system with the lowest ratio of reductase to P-450. In every case, strong inhibition was observed in micelles as the b_5 to P-450 ratio approached one. It was found that micelles containing b_5 exhibit either inhibition, activation, or no effect on activity as a function of the molar ratio of phospholipid to total protein. The efficiency of electron transport from reductase to P-450 was constant as a function of b_5 when the ratio of reductase to P-450 was 1. However, when the ratio of reductase to P-450 was 0.1:1, b_5 increased the efficiency from 40% to 80%.

In contrast to micelle-reconstituted systems, the increase of activity by b_5 of the vesicle-reconstituted systems of various reductase to P-450 ratios slowly reached a plateau value at a b_5 to P-450 ratio of 1. The observed difference in the effect of b_5 appears to reflect a difference between the reconstituted systems used. Since all protein-to-protein ratios were comparable, it seemed possible that b_5 has a particularly large molar requirement for phosphilipids.

If reductase or b_5 transfer electrons directly to cytochrome P-450, they may do so in a preformed complex of considerable lifetime or only during a rapid collision event. In that the MCD spectrum of cytochrome P-450 is a time weighted average of all possible states of its chromophore, it is possible to use this spectroscopic technique to distinguish between the two possibilities described above. In Fig. 1 it is seen that the MCD spectrum of vesicle co-reconstituted reductase and P-450 is considerably different than that of a mixture containing precisely equal amounts of separately vesicle-reconstituted proteins. This result suggests that the two co-reconstituted proteins are together in a complex for a significant percentage of the time.

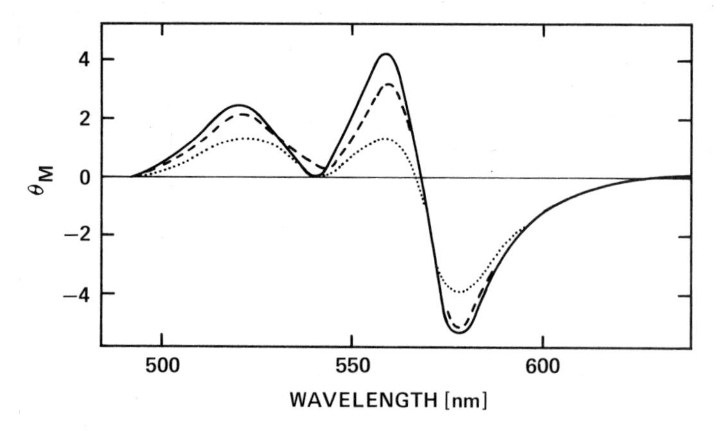

Fig. 1. Evidence for an association between cytochrome P-450 and NADPH-cytochrome P-450 reductase co-reconstituted in vesicles as measured by MCD. The molar elipticity was measured for three suspensions of identical protein and phospholipid concentrations: a suspension of co-reconstituted P-450 and reductase (dashed line); a 1:1 mixture of separately reconstituted P-450 and reductase (solid line); a suspension of the above co-reconstituted P-450 and reductase to which benzphetamine was added (dotted line).

Evidence for an interaction in a bilayer between b_5 and P-450 was obtained from the MCD-spectra of equimolar concentrations of a P-450 vesicle suspension and a solution of b_5 before and after mixing and equilibration. The difference corresponded to a decrease of 13% of the absolute intensity of the Soret band. No significant difference spectrum was obtained before and after mixing a suspension of vesicle-reconstituted reductase with a solution of b_5.

DISCUSSION

Studies of cytochrome P-450-mediated reactions in hepatic microsomal preparations have led to proposals that cytochrome b_5 can play a facilitory role in the transfer of a second electron in substrate oxidations (10,11). The addition of purified b_5 to NADPH-dependent micellar reconstituted systems has led to several reports of variable effects on substrate metabolism. Phospholipid requirements could change radically in systems containing three, rather than two, interacting protein components. In our micelle-reconstituted systems b_5 causes both activation and inhibition of the N-demethylation reaction, or has no measurable effect, depending upon the lipid to protein and protein to protein ratio. In vesicles, which contain a higher phospholipid:protein ratio than micelles, the inhibitory effects of b_5 are present only when the mole ratio of b_5 to P-450 exceeds unity. Maximal stimulation by b_5 and an increase in efficiency occurred at lower ratios of reductase to P-450 similar to those found in hepatic microsomes.

An interaction in a phospholipid bilayer of purified P-450 with reductase as well as with b_5 was demonstrated by changes in absorption and MCD spectra of the porphyrin prosthetic groups. Similar changes have been observed in micelle-reconstituted systems (12). It is important to demonstrate that the observed protein-protein association reactions are meaningful in that they also occur in the membrane of the endoplasmic reticulum. The present study supports such a projection in that an association in a membrane would not be expected to occur in the absence of specific contact sites. In the membrane of a vesicle, every protein molecule could be surrounded by several phospholipid layers and no protein would be forced to interact with another protein in order to circumvent exposure of hydrophobic surface areas to water. We would like to propose that electron transfer would occur after lateral and rotational diffusion of the individual dissociated proteins had lead to formation of complexes of specific orientation.

The most simple mechanism would be two linked reversible association reactions in the two-dimensional membrane.:

$$
\begin{array}{ccc}
 & b_5^{red} & \\
 & + & \\
R^{red} + & P\text{-}450^{ox} & \rightleftarrows \quad [RP\text{-}450] \\
 & \uparrow\downarrow & \qquad\quad \downarrow \\
 & [b_5P\text{-}450] & \longrightarrow \quad P\text{-}450^{red} + b_5^{ox} \\
 & & \qquad + \\
 & & \qquad R^{ox}
\end{array}
$$

ACKNOWLEDGEMENTS

We would like to thank Ms. Marie Bendix and Ms. Ruth Records for expert technical assistance. This research was supported by the National Institute of Occupational Safety and Health (OH 00978) and the Alexander von Humboldt-Stiftung.

REFERENCES

1. Sugiyama, T., Miki, N. and Yamano, T. (1979) Biophys. Res. Commun. 90, 715-720.
2. Kuwahara, S-I. and Omura, T. (1980) Biochem. Biophys. Res. Commun. 96, 1562-1568.
3. Haugen, D.A., Van der Hoeven, T.A. and Coon, M.J. (1975) J. Biol. Chem. 250, 3567-3570.
4. Yasukochi Y. and Masters, B.S.S. (1976) J. Biol. Chem. 251, 5337-5344.
5. Spatz, L. and Strittmatter, P. (1971) Proc. Natl. Acad. Sci. U.S.A. 68, 1042-1046.
6. Bösterling, B., Stier, A., Hildebrandt, A.G., Dawson, J.H. and Trudell, J.R. (1979) Mol. Pharmacol. 16, 332-342.
7. Trudell, J.R., Bösterling, B. and Trevor, A.J. (1981) Biochem. Biophys. Res. Commun. 102, 372-377.
8. Bösterling, B., Trudell, J.R., Trevor, A.J., and Bendix, M. (1982) J. Biol. Chem. 257 (in press).
9. Bösterling, B. and Trudell, J.R. (1982) J. Biol. Chem. 254 (in press).
10. Hildebrandt, A. and Estabrook, R.W. (1971) Arch. Biochem. Biophys. 143, 66-79.
11. Correia, M.A. and Mannering, G.J. (1973) Mol. Pharmacol. 9, 470-485.
12. French, J.S., Guengerich, F.P. and Coon, M.J. (1980) J. Biol. Chem. 255, 4112-4119.

© 1982 Elsevier Biomedical Press B.V.
Cytochrome P-450, Biochemistry, Biophysics
and Environmental Implications, E. Hietanen,
M. Laitinen and O. Hänninen editors

POSSIBILITY OF ACTIVATING SUBSTRATES BY CYTOCHROME P 450

L.M. WEINER[1], Ya.Yu. WOLDMAN[1] AND V.V. LYAKHOVICH[2]

[1]Institute of Chemical Kinetics and Combustion, Novosibirsk 630090 (USSR) and [2]Institute of Clinical and Experimental Medicin, Novosibirsk 630091 (USSR)

INTRODUCTION

In the studies concerned with the mechanism of the catalytic action of cytochrome P 450 main attention is usually focused on the processes of molecular oxygen activation (1). Meanwhile, some substrates of this enzyme containing quinone structures can be reduced to semi- and hydroquinones in the microsomal system (2-4). Obviously, the chemical properties of these substrates must essentially change upon reduction. Moreover, some semi- and hydroquinones can interact with molecular oxygen and superoxide radical producing oxidized products (5-7).

The objective of the present work was to examine the possibilities of reduction of some substrates and their analogs by reduced cytochrome P 450. The role of electron acceptors was provided by stable nitroxide radicals showing substrate properties toward cytochrome P 450 (8), and 1-piperidinoanthraquinone (1-PA), which can be oxidized on cytochrome P 450 to give (N-anthraquinonyl-1)-δ-aminovaleric acid (AAV) (9).

MATERIALS AND METHODS

Microsomes were prepared from livers of Wistar rats as described in (10). Isolation and purification of cytochrome P 450 and NADPH-cytochrome c reductase was performed according to the known procedures (11,12). Antibodies to the isolated cytochrome P 450 were obtained, purified and tested as described by Kamataki et al. (13). The concentration of microsomal protein and cytochrome P 450 was determined as described in (14,15). The activity of NADPH-cytochrome c reductase was found from the reduction rate of cytochrome c (ε_{550}=2.1 x $10^4 M^{-1}cm^{-1}$ (16)). Highly purified NADPH-cytochrome c reductase and cytochrome P 450 were inserted into liposomes from the overall microsomal lipid by the cholate-dialysis method (17).

The microsomal oxidation and reduction of 1-PA were examined under standard conditions: the reaction mixture contained 50 mM potassium phosphate buffer, pH = 7.4, 5 mM $MgCl_2$, 0.3-1.0 mg/ml microsomal protein, and T = 25°C. The microsomal oxidation of 1-PA was tested from the amount of accumulated AAV as described in (9). The radicals were reduced both in aerobic and anaerobic conditions. The typical reaction mixture was: 10^{-4}M radical, 5 x 10^{-4}M EDTA, 1 mg/ml microsomal protein, 10^{-3}M NADPH or NADH, 0.1 M tris HCl, pH=7.6, T=37°C.

The radical reduction was tested from the disappearance of the EPR spectrum of the radical (central component). To obtain anaerobic conditions, the following system was used: 30 units/ml glucose oxidase, 3000 units/ml catalase, and 100mM glucose.

The nitroxide radicals were kindly donated by A.B. Shapiro (Institute of Chemical Physics, Moscow) and L.B. Volodarskii (Institute of Organic Chemistry, Novosibirsk).

Spectral dissociation constants of substrates with microsomal cytochrome P 450 were determined by the procedure described by Schenkman et al. (18). The EPR spectra were taken on a Varian E-3 spectrometer in a capillar. A Hitachi-556 spectrophotometer was used for optical measurements.

RESULTS

The use of nitroxide radicals as electron acceptors in the microsomal system

Nitroxide radicals (R1-R3) bind to the microsomal cytochrome P 450 and induce in its optical spectra (registered in a differential form) the changes attributed to type I and II substrates. The chemical formulas of the radicals used, the types of their binding, and spectral dissociation constants, K_s, are listed in Table 1.

In the presence of NADPH the reduction of radicals 1-3 in the microsomal fraction was observed (NADPH did not reduce these radicals directly in the solution). The control experiments suggest that the disappearance of the EPR signal of the radical in microsomes is associated with the formation of diamagnetic hydroxyl amine rather than with the radical destruction:

TABLE 1
BINDING OF STABLE NITROXIDE RADICALS TO MICROSOMAL CYTOCHROME P 450

Radical		Type of Binding	Constant, K_s (10^{-6}M)
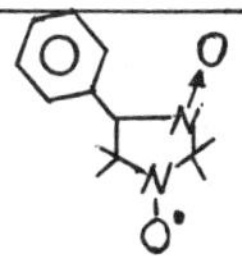	R1	I	28
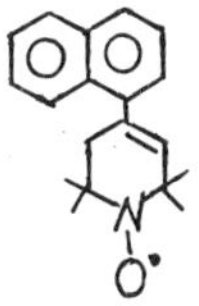	R2	I	8.3
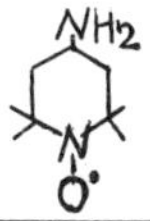	R3	II	27

the EPR signal reappeared after blowing the reaction mixture with oxygen. In anaerobic conditions the reduction rates of radicals increased. With NADH used as a reductant, the reduction rates of the radicals were lower by a factor of 2-3 compared to the rates in NADPH-experiments. This observation provides an evidence for the predominant reduction of the radicals in a NADPH-dependent microsomal chain.

The reduction rates of the radical analogs to the substrates in the microsomal system were compared to those in the individual components of the microsomal system (Table 2). Table 2 gives also the data on the effect of CO on the reduction rate of radicals R1 and R2 in the reconstructed system and in the intact microsomes.

Microsomal reduction and oxidation of 1-PA

In anaerobic conditions 1-PA is reduced to the corresponding hydroquinone (1-PAH). This follows from the disappearance of the absorption band of the initial quinone with a maximum at 520 nm and from the appearance of new bands at 390 and 430 nm, typical for 9,10-anthrahydroquinones and their derivatives (19).

In order to determine the role of cytochrome P 450 in

TABLE 2

REDUCTION RATES OF R1 and R2 RADICALS IN VARIOUS CONDITIONS $(M \times min^{-1} \times (unit\ act.)^{-1}) \times 10^8$

System	Radicals	
	1	2
Purified NADPH-cytochrome c reductase in solution	3.8	1.0
NADPH-cyt. c reductase inserted in liposomes	2.2	0.61
NADPH-cyt. c reductase inserted in liposomes + cytochrome P 450	4.9	2.5
+ CO	4.7	0.6
Intact microsomes	10.7	8.7
+ CO	7.2	5.8

The observed reduction rates of R1 and R2 were calculated per the activity of NADPH-cytochrome c reductase. The concentration of cytochrome P 450 in the microsomal system was 10 nmol per unit activity, and in the selforganized system - 6 nmol per unit activity of reductase.

reduction of 1-PA to 1-PAH, we carried out the following experiments: i) The kinetics of 1-PAH formation was measured in the presence of metyrapone which is a competitive inhibitor of cytochrome P 450, ii) The formation rates of 1-PAH in control and phenobarbital-induced microsomes were compared, iii) The effect of antibodies to cytochrome P 450 on the formation rate of 1-PAH was examined. The results obtained are compiled in Table 3.

Since many of quinone-containing compounds can be reduced in microsomes to semi- and hydroquinones by NADPH-cytochrome c reductase (2-4), we have compared the formation rates of 1-PAH in microsomes and in a highly purified NADPH-cytochrome c reductase (at the same activity of this latter). The formation rate of 1-PAH was found to be more than one order of magnitude higher in microsomes.

TABLE 3

THE RATE OF 1-PAH FORMATION IN MICROSOMES IN ANAEROBIC CONDITIONS

System	Rate (μM) of 1-PAH formation per min per mg of microsomal protein
Control microsomes	2
+ Metyrapone (10^{-3}M)	0.95
+ Antibodies to cyt. P 450	0.8
Phenobarbital microsomes	5.1

The rates were determined from the initial parts of the kinetic curve. The content of microsomal protein was about 0.7 mg/ml, [1-PA] = 10 μM, [NADPH] = 100 μM. The content of cyt. P 450 was: 0.9 nmol/mg protein in control microsomes; 2.5 nmol/mg in microsomes induced by phenobarbital. The concentration of antibodies was 3 mg/ml. Such concentration of antibodies suppressed metabolism of 1-PA by 50% (see Materials and Methods). The rate of 1-PAH accumulation was measured with a tandem cuvette (24).

DISCUSSION

The experiments with the substrates' radical analogs (Table 2) suggest that the reduced cytochrome P 450 can reduce these compounds to diamagnetic hydroxyl amines, i.e. the following reaction occurs:

$$\text{Cyt. P 450 (red.)} + \rangle N - \dot{O} \xrightarrow{e,H^+} \text{Cyt. P 450 (oxid.)} + \rangle N - OH .$$

Note that the reduction rate of R3 changes in the series: purified reductase —— reductase inserted into liposomes —— selforganized system. The observed decrease in the rate for the reductase inserted into liposomes is likely to be due to the inaccessibility of the radical-reducing site of the reductase during insertion into liposomes. At the same time the significant increase of the reduction rate of R3 in the selforganized system gives evidence in favor of cytochrome P 450 participation in the reduction of this radical analog characterized by the high affinity to cytochrome P 450 (Table 1). This proposal is supported by the effective (more than 70%) inhibition of the reduction of this compound in the presence of CO (Table 2).

The qualitative formation of 1-PAH in the microsomal system affords further evidence for the participation of cytochrome P 450 in reduction of its substrate. Inhibition of 1-PAH formation by cytochrome P 450 inhibitors (Table 3) indicates that the reduction is catalyzed by this enzyme. This follows also from the observed increase of the 1-PA reduction rate in phenobarbital-induced microsomes (Table 3) and a far lower reduction rate in experiments with highly purified NADPH-cytochrome c reductase. This conclusion accords well with the data obtained for R1 and R2 (Table 2), for which the reduction rate was higher in the intact microsomes than in the purified NADPH-cytochrome c reductase.

All the experimental data taken together suggest that the reduced cytochrome P 450 can reduce some compounds capable of accepting electrons. On the other hand, as reported in (5-7), some semiquinones and hydroquinones can react with the O_2^- radical in nonaqueous solutions yielding oxidized products. Accounting the fact that $O_2^{\dot{-}}$ can be produced in the microsomal system (20-23), it seemed of great interest to examine the possibility of oxidation of semi- and hydroquinones by $O_2^{\dot{-}}$ or oxygen.

This possibility was checked by conducting the photochemical reduction of 1-PA to 1-PAH in ethanol in anaerobic conditions (24). The produced hydroquinone was oxidized by oxygen. The oxidation product, (N-anthraquinonyl-1)- δ -aminovaleric aldehyde (AAVal) with a 20% yield and initial 1-PA were detected by means of the thin-layer chromatography and IR spectroscopy. However, the use of the oxidants other than O_2, and also the chemical generation of 1-PAH suggest that, most probably, the formation of AAVal is assistant neither by the substrate semi-reduced forms, nor by $O_2^{\dot{-}}$ (24).

To our opinion, this result does not necessarily preclude the possibility of participation of semi-reduced and reduced by cytochrome P 450 forms of some substrates in their oxidation on cytochrome P 450. The spatial proximity of thus activated substrates and activated oxygen in the enzyme active center may result in their more efficient oxidation. To elucidate this possibility, it seems necessary to use porphyrin systems, which model the reduced cytochrome P 450 and can reduce cytochrome P 450 substrates in the presence of oxygen or $O_2^{\dot{-}}$ radical (25).

REFERENCES

1. Hayaishi, O. (1969) Ann. Rev. Biochem., 38, 21-45.
2. Bachur, N.R., Gordon, S.L. and Gee, M.V. (1977) Mol.Pharm. 13, 901-910.
3. Bachur, N.R., Gordon, S.L. and Gee, M.V. (1978) Cancer Res. 38, 1745-1750.
4. Sinha, B.K. and Chignell, C.F. (1979) Chem. Biol. Interact. 28, 301-308.
5. Lee-Ruff, E., Lever, A.B. and Rigandy, J. (1976) Can. J. Chem. 54, 1837-1839.
6. Moro-Oka, Y. and Foote, C.S. (1976) J. Amer. Soc. 98, 1510-1512.
7. Pedersen, J.A. (1973) J. Chem. Soc. Perkin Trans. 11, 424-431.
8. Weiner, L., Eremenko, S., Popova, V., Sagdeev, R., Tsyrlov, I. and Lyakhovich, V. (1979) in: Proceedings of XX Congress AMPERE, Tallin, Springer Verlag, p. 520.
9. Weiner, L., Yamkovoi,V ., Tsyrlov, I., Pospelova, L.N. and Lyakhovich, V. (1979) Biochimiya 44, 634-639.
10. Tsyrlov, I.B., Zakharova, N.E., Gromova, O.A. and Lyakhovich, V.V. (1976) Biochim. Biophys. Acta 421, 44-56.
11. Imai, Y. (1976) J. Biochem. 80, 267-276.
12. Dignam, J.D. and Strobel, H.W. (1977) Biochemistry 16, 1116-1123.
13. Kamataki, T., Beecher, D.H. and Neal, R.A. (1976) Mol. Pharmacol. 12, 921-932.
14. Lowry, O.H., Ressebrongh, N.J., Farr, A.L. and Randall, R.J. (1951) J. Biol. Chem. 193, 265-275.
15. Omura, R. and Sato, R. (1964) J. Biol. Chem. 239, 2370-2378.
16. Phillips, A.H. and Langdom, R.G. (1962) J. Biol. Chem. 237, 2652-2658.
17. Yamakura, F., Kido, T. and Kimura, T. (1981) Biochim. Biophys. Acta 649, 343-354.

18. Schenkman, J.B., Remmer, H. and Estabrook, R.W. (1976) Mol. Pharmacol. 3, 113-123.

19. Davies, A.K., McKellar, Y.E. and Phillips, G.O. (1971) Proc. R. Soc. 323, 69-87.

20. Strobell, H.W. and Coon, M.J. (1971) J. Biol. Chem. 246, 7826-7830.

21. Richter, Ch., Azzi, A. and Wendel, A. (1977) J. Biol. Chem. 252, 5061-5065.

22. Rumyantseva, G.W. and Weiner, L.M. (1982) Biokhimiya 47, 921-930.

23. Ullrich, V. and Kuthan, H. in:Biochemistry, Biophysics and Regulation of Cytochrome P 450 (J.A. Gustafsson et al. eds.) Elsevier/North-Holland Biochemical Press, Amsterdam, pp. 267-272.

24. Weiner, L.M., Gritzan, N.P., Bazhin, N.M. and Lyakhovich, V.V. (1982) Biochim. Biophys. Acta 714, 234-242.

25. Mansuy, D. and Fontecave, M. (1982) Biochem. Biophys. Res. Commun. 104, 1651-1657.

Cytochrome P-450, Biochemistry, Biophysics and Environmental Implications, E. Hietanen, M. Laitinen and O. Hänninen editors

REGULATION OF THE CYCLIC FUNCTION OF LIVER MICROSOMAL CYTOCHROME P-450: ON THE ROLE OF CYTOCHROME b_5

JÜRGEN WERRINGLOER, SUNAO KAWANO* AND HARTMUT KUTHAN
Department of Toxicology, University of Tübingen, Wilhelmstr. 56, D-7400 Tübingen (F.R.G.) and Department of Biochemistry, The University of Texas Hlth. Sci. Ctr., Dallas, Texas 75235 (U.S.A.)

INTRODUCTION

The reaction cycle of liver microsomal cyt.P-450 has been recognized to be controlled by the electron transfer reactions responsible for the stepwise reduction of high spin ferric cyt.P-450 and of ferrous cyt.P-450 in its oxygenated state. The velocities of both these reactions were found to be regulated differentially in an inverse pH-dependent manner termed "counterpoise regulation" (1,2). In order to characterize the function of cyt.b_5 as an electron donor for oxy-cyt.P-450 quantitative analyses of the effect of NADH on the NADPH-dependent steady state levels of oxy-cyt. P-450 and of ferrous cyt.b_5 were carried out at varying proton concentrations.

MATERIALS AND METHODS

Microsomes were prepared from livers of phenobarbital-treated male Charles River, CD outbred albino, rats (3). The reaction media were composed of varying mixtures of 50 mM TRIS and 50 mM HEPES containing each 150 mM KCl, 10 mM $MgCl_2$, 2 mM 5'-AMP and 1 µM rotenone. The steady state levels of oxy-cyt.P-450 and of ferrous cyt.b_5 were analyzed by difference spectrophotometry. The extinction coefficients employed were as follows: a. Oxy-cyt.P-450 minus high spin ferric cyt.P-450: E*(440-500 nm)* = 42 mM^{-1} cm^{-1} (1); b. ferrous cyt.b_5 minus ferric cyt.b_5: E*(425-500 nm)* = 130 mM^{-1} cm^{-1}; this extinction coefficient was applied after correction for the spectral contribution of oxy-cyt.P-450, i.e. E*(425-500 nm)* = 21 mM^{-1} cm^{-1} (present investigation). The further methods employed were as described previously (1). All experiments were carried out at 25°C.

---

*Present address: The First Department of Internal Medicine, Osaka University Hospital, Fukoshima, Osaka (Japan)

RESULTS AND DISCUSSION

Analyses of the steady state concentrations of oxy-cyt.P-450 and of ferrous cyt.b_5 in the presence of NADPH as the sole electron donor indicate a progressive decrease of the levels of both these intermediates with increasing concentrations of protons (cf., Fig. 1). These results are consistent with the progressive attenuation with decreasing pH of the activity of the NADPH-cyt.P-450 reductase as discussed previously (1,2). The additional presence of NADH as an electron donor for the NADH-cyt.b_5 reductase was found to enhance markedly the steady state reduction of cyt.b_5, in particular at low pH. It should be noted, however, that a full reduction of cyt.b_5 is approached only at high pH, a phenomenon associated with the accumulation of cyt.P-450 in its oxygenated state and an attenuation of its monooxygenase function (cf., Fig. 1 and 2). The latter observations have been interpreted to suggest that the reduction of oxy-cyt.P-450 is progressively impaired with increasing pH due to a modification of its physico-chemical properties (1,2). This interpretation is consistent as well with the observed maximal levels of the steady state reduction of cyt.b_5 at high pH.

From the data shown in Figure 2 it is readily apparent that the "synergistic" action of NADH on the monooxygenase function of cyt. P-450 is enhanced with decreasing pH. This effect is highly sig-

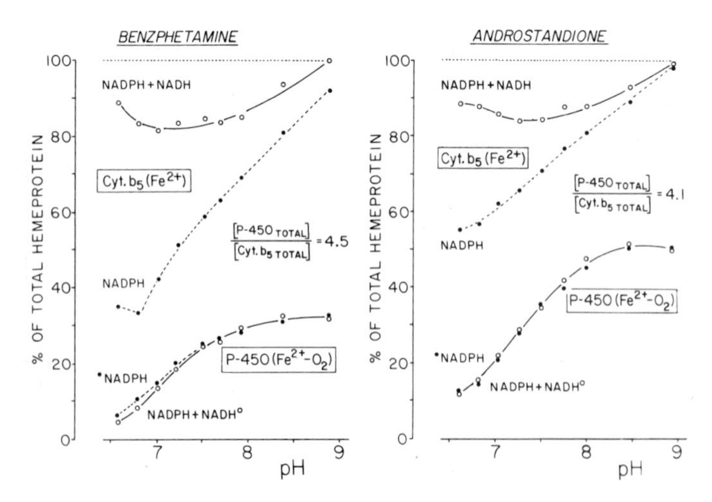

Fig. 1. Steady state levels of oxy-cyt.P-450 and of ferrous cyt.b_5 in the presence of benzphetamine and androstandione.

nificant at proton concentrations comparable to those of the cytoplasmic compartment of the liver cell, i.e. approximately pH 6.9 to 7 (2). The observed differences of the effect of NADH on the oxidative transformation of benzphetamine and androstandione remain to be investigated. These differences, however, are in agreement with those observed with regard to the effect of NADH on the molar ratios of the concentrations of oxy-cyt.P-450 and ferrous cyt.b_5 under identical conditions (cf., Fig. 3).

The marked effect of NADH on the NADPH-dependent steady state reduction of cyt.b_5 as well as its "synergistic" action on substrate hydroxylation (4) is contrasted by its failure to alter significantly the steady state levels of oxy-cyt.P-450 (cf., Fig. 1). This phenomenon may be interpreted to result from an increased availability of the NADPH-cyt.P-450 reductase for the reduction of high spin ferric cyt.P-450 in response to the enhanced participation of ferrous cyt.b_5 in the reduction of the hemeprotein in its oxygenated state. This interpretation is based upon the observed stimulation of the monooxygenase activity of cyt.P-450 in the additional presence of NADH (cf., Fig. 2).

In summary, the data presented substantiate the potential significance of the function of cyt.b_5 as an electron donor for oxy-cyt.P-450 in vitro. However, a variety of factors in addition to

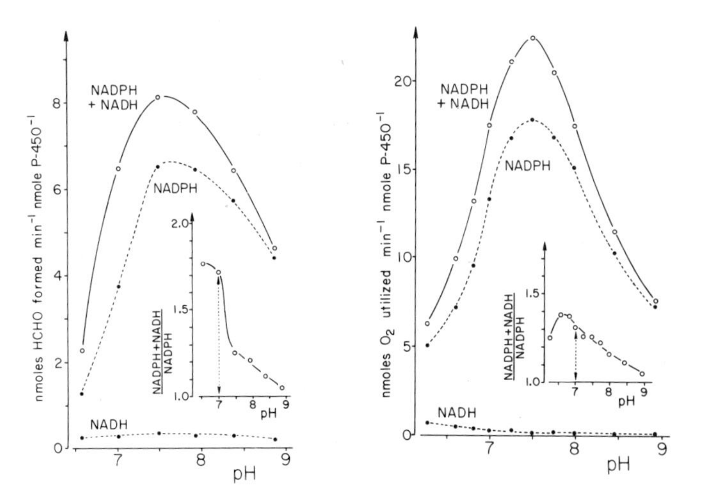

Fig. 2. The "synergistic" effect of NADH on the NADPH-dependent metabolism of benzphetamine (left) and androstandione (right).

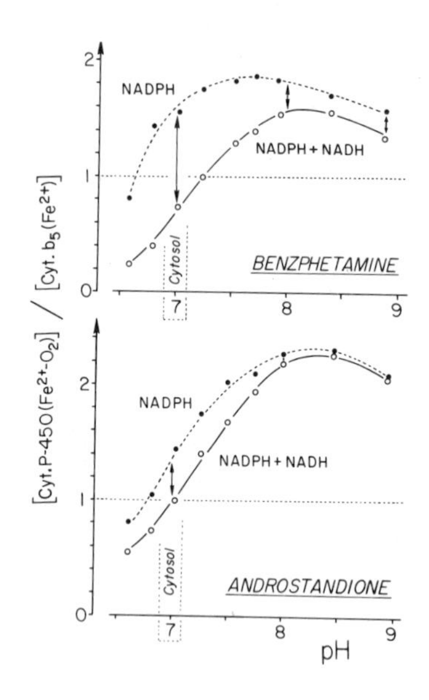

Fig. 3. The effect of NADH on the molar ratios of the NADPH-dependent steady state concentrations of oxy-cyt.P-450 and ferrous cyt.b_5 during the metabolism of bezphetamine and androstandione (calculated from the data shown in Fig. 1).

the concentration of protons have to be considered before assigning such a role to cyt.b_5 under intracellular conditions (5).

ACKNOWLEDGEMENTS

This work was supported in part by a grant from the National Institute of General Medical Sciences (# 16488). The participation of J.W. in this conference was supported by a travel grant from the BREUNINGER STIFTUNG GmbH, Stuttgart (F.R.G.) and is gratefully acknowledged.

REFERENCES

1. Werringloer, J. and Kawano, S. (1980) in: Gustafsson, J.A. et al. (Eds.), Biochemistry, Biophysics and Regulation of Cytochrome P-450, Elsevier/North Holland Biomedical Press, Amsterdam, pp. 359-362.

2. Werringloer, J. (1982) in: Sato,R. and Kato, R. (Eds.), Microsomes, Drug Oxidations and Drug Toxicity, Japan Scientific Society Press, Tokyo, in press.

3. Werringloer, J., Kawano, S. and Estabrook, R.W. (1979) Acta Biol. Med. Germ. 38, 163-175.

4. Hildebrandt, A.G. and Estabrook, R.W. (1971) Arch. Biochem. Biophys. 143, 66-79.

5. Noshiro, M., Ullrich, V. and Omura, T. (1981) Eur. J. Biochem. 116, 521-526.

Cytochrome P-450, Biochemistry, Biophysics and Environmental Implications, E. Hietanen, M. Laitinen and O. Hänninen editors

ACTIVE SITE MODIFICATION OF CYTOCHROME P-450

G.-R. JÄNIG, G. SMETTAN, J. FRIEDRICH, R. BERNHARDT, O. RISTAU and K. RUCKPAUL
Central Institute of Molecular Biology, Academy of Sciences of the GDR, 1115 Berlin-Buch, GDR

INTRODUCTION

The unique catalytic function of P-450 as terminal oxidase of the monooxygenatic system consists in its capability to activate molecular oxygen. The enzymatic specificity of P-450 is determined by its axial ligands and on the other hand by the three-dimensional structure provided by its individual protein. The oxygen activating capability of P-450 has been attributed to the strong electron donor properties of the negatively charged sulfur of cysteine residue acting as 5th axial ligand. Numerous experimental data have been accumulated indirectly evidencing the existence of a thiol group. The data about the nature of the ligand trans to the thiol group are rather speculative. The groups of Coon and Ullrich provided data which suggest an oxygen rather than a nitrogen to occupy the position of the 6th iron ligand.

In preceding experiments we have followed this idea and provided data indicating tyrosine as probable candidate for the 6th axial ligand (1). Photoaffinity labeling experiments of Dus revealed at least one tyrosine residue to be contained in the heme peptide of P-450 LM2 thus contributing to the active center (2).

To characterize the functional importance of tyrosine residues in P-450 LM2 we employed the technique of selective chemical modification using N-acetylimidazole which has been reported as relatively specific agent for the acetylation of hydroxyl groups of tyrosine under mild conditions.

MATERIAL AND METHODS

P-450 LM2 was isolated from liver microsomes of phenobarbital-treated male rabbits according to Haugen and Coon (3). The different preparations used showed an absorbance ratio of $A_{418\ nm}/$

$A_{278\ nm}$ from 1.23 to 1.71. Acetylation of tyrosyl groups of P-450 LM2 with N-acetylimidazole proceeds as a nucleophilic reaction and was carried out at 25° C in 0.05 M Tris-HCl buffer, 5 % (v/v) glycerol, pH 7.5, by adding different amounts of solid reagent corresponding to a 60- to 300-fold excess over tyrosine (4). The reaction was stopped after 60 min by gelfiltration on Sephadex G-25. The number of acetylated tyrosyl residues was calculated from the decrease in the ultraviolet absorbance at 278 nm ($\Delta\varepsilon$ = 1.16 $mM^{-1}cm^{-1}$) and corrected by the amount of apoprotein based on an absorbance ratio $A_{418\ nm}/A_{278\ nm}$ of 2.0 for the holoenzyme (5). Absorption spectra, the corresponding 2nd derivative spectra, K_s values and the kinetics of substrate binding were determined as described (6).

RESULTS

The modification of P-450 LM2 with N-acetylimidazole was monitored by optical and second derivative spectroscopy as well after removal of the modifyer. In dependence on the molar excess of the modifyer 2 classes of tyrosines can be differentiated. On average one fast reacting tyrosine residue per molecule P-450 LM2 is acetylated at a 60-fold excess of the reagent whereas concentrations up to a 300-fold excess are necessary for acetylation of two further tyrosines without or only small conversion to P-420. The residual 6 tyrosine residues of P-450 LM2 are not accessible. Additional modification has been excluded (i) by demonstrating a nearly complete accessibility of amino groups of acetylated P-450 LM2 for fluorescein isothiocyanate and (ii) by restoring the original form in the presence of 1 M hydroxylamine at pH 7.5.

The N-demethylase activity of the modified P-450 LM2 in a reconstituted system is almost completely lost when 3 tyrosine residues per enzyme monomer were modified. However, the activity remained unaltered when only one tyrosine residue had reacted. Obviously, the fast reacting tyrosine residue does not contribute to catalytic activity.

Binding studies (see Table 1) revealed a differentiation between aniline and benzphetamine. The binding affinity towards benzphetamine of differently acetylated P-450 LM2 remains unchanged whereas the affinity towards aniline is increased already after modification of one tyrosine. After acetylation of 3 tyrosines the spin equilibrium in the presence of benzphetamine exhibits a more pronounced shift to the high spin state than the control.

Table 1
Substrate binding to O-acetylated cytochrome P-450 LM2

Extent of	Benzphetamine		Aniline	
modification	K_s(mM)	ΔA_∞ ($mM^{-1}cm^{-1}$)	K_s(mM)	ΔA_∞ ($mM^{-1}cm^{-1}$)
0	0.15	14.2	8.2	23.7 ±2.6 (12)
0.6	0.15	16.2	4.0	
1.9	0.13	20.6	1.5	
> 2.4	0.14	19.8	1.7	

Kinetic studies on binding of aniline to P-450 LM2 with the acetylated fast reacting tyrosine revealed a decrease of the rate constant (fast phase) from $8.1 \cdot 10^3\ M^{-1}s^{-1}$ to $1 \cdot 10^3\ M^{-1}s^{-1}$. At higher extent of modification no further changes of the rate constants can be observed. Taking into account the K_s values and the kinetic data an about 40-fold decreased rate constant of the aniline dissociation as compared to the control has been calculated ($k_{-1} = 66\ s^{-1} \rightarrow 1.5\ s^{-1}$). The increased affinity of modified P-450 LM2 for aniline, therefore, is mainly caused by a slowed dissociation reaction. Acetylation of one tyrosine residue in P-450 LM2 does not change the rate constants of benzphetamine binding. However, the value of $2.5 \cdot 10^5\ M^{-1}s^{-1}$ for the fast phase is decreased by more than one order of magnitude at modification of one further tyrosine residue. This decrease of the fast phase is accompanied by a decrease of the slow phase with an almost unaltered phase distribution. The changes of the binding constant correspond to that of the dissociation reaction of the benzphetamine complex of modified P-450 resulting in almost unchanged K_s values.

To prove if tyrosines are involved in the reduction reaction the sodium dithionite reduction of modified P-450 LM2 was compared with the NADPH supported reduction via reductase and related to the unmodified control. The dithionite reduction remains almost unchanged. The half time of the reduction via reductase is slightly increased. A slowed reaction, however, is only observed if more than one tyrosine is modified (13 s as compared to 8 s for the control). These results indicate that the fast reacting tyrosine is functionally not involved in the reduction reaction. The functional role of the slower reacting tyrosine(s) is conceivable by assuming either that it is involved in P-450/reductase interactions or that it is part of the electron pathway.

Nitration of tyrosines of P-450 LM2 by tetranitromethane (Jänig, G.R. and Usanov, S.A., unpublished) results in the formation of nitrophenolate groups. The tyrosine absorption increased concomitant with a decrease and broadening of the absorbance in the Soret region and a significant conversion into P-420. The decrease of the Soret band is paralleled by an increase of the heme UV-band at 360 nm and a new band at about 460 nm. These bands are characteristic for a hyperspectrum which is formed by the heme in the presence of two negatively charged ligands (7). That means nitration increases the basicity of the phenolic hydroxyl group of a heme-linked tyrosine leading to a weakening of the thiolate iron bond at the 5th axial coordination position thus indicating the phenolic hydroxyl group of a tyrosine residue as the 6th axial heme iron ligand.

REFERENCES

1. Ruckpaul, K., Rein, H., Ballou, D.P. and Coon, M.J. (1980) Biochim. Biophys. Acta 626, 41-56.
2. Dus, K. (1980) in: Biochemistry, Biophysics and Regulation of Cytochrome P-450 (Gustafsson, J.-Å., Carlstedt-Duke, J., Mode, A., Rafter, J., Eds.) Elsevier/North-Holland Biomedical Press, Amsterdam, pp. 129-132.
3. Haugen, D.A. and Coon, M.J. (1976) J. Biol. Chem. 251, 7929-7939.
4. Riordan, J.F. and Vallee, B.L. (1972) Meth. Enzymol. 25, 500-506.
5. Shimizu, T., Nozawa, T., Hatano, M., Satake, H., Imai, Y., Hashimoto, C. and Sato, R. (1979) Biochim. Biophys. Acta 579, 122-133.
6. Ruckpaul, K., Rein, H., Blanck, J., Ristau, O. and Coon, M.J. (1980) in: Biochemistry, Biophysics and Regulation of Cytochrome P-450 (Gustafsson, J.-Å., Carlstedt-Duke, J., Mode, A., Rafter, J., Eds.) Elsevier/North-Holland Biomedical Press, Amsterdam, pp. 539-549.
7. Nastainczyk, W., Ruf, H.H. and Ullrich, V. (1976) Chem.-Biol. Interactions 14, 251-263.

Cytochrome P-450, Biochemistry, Biophysics
and Environmental Implications, E. Hietanen,
M. Laitinen and O. Hänninen editors

AFFINITY MODIFICATION OF CYTOCHROMES P450 AND P448

V.V. LYAKHOVICH[1], V.M. MISHIN[1], O.A. GROMOVA[1], V.I. POPOVA[2]
AND L.M. WEINER[2]
[1]Institute of Experimental and Clinical Medicine, USSR Academy of Sciences, Novosibirsk 630091 and [2]Institute of Chemical Kinetics and Combustion, USSR Academy of Sciences, Novosibirsk 630090 (U.S.S.R.)

INTRODUCTION

Affinity modification is a promising means to study the structure of enzyme active centers. In (1,2) the affinity modification of cytochrome $P450_{cam}$ was performed with the use of photoaffinity labels and compounds containing chemically active groups capable of covalent binding to the enzyme active center. We have proposed (3-5) to use the analogs of cytochrome P450 substrates containing alkylating groups and spin labels for affinity modification of cytochrome P450. This approach was used to study the structure of the active center of microsomal cytochrome P450 and inhibition of oxidative functions of this enzyme (4,5).

In the present work the alkylating analogs of substrates were used for investigation of the structure of the active center of microsomal cytochrome P450 induced by phenobarbital and 3-methylcholatrene as well as for the study of highly purified forms of this protein.

MATERIALS AND METHODS

Chemically pure aniline, naphthalene and sodium cholate (Sojuzkhimreaktiv) and NADPH (Reanal) were used in experiments. Radicals RI and RII were synthetized according to (6) and kindly given by L.B. Volodarskii (Institute of Organic Chemistry, Novosibirsk).

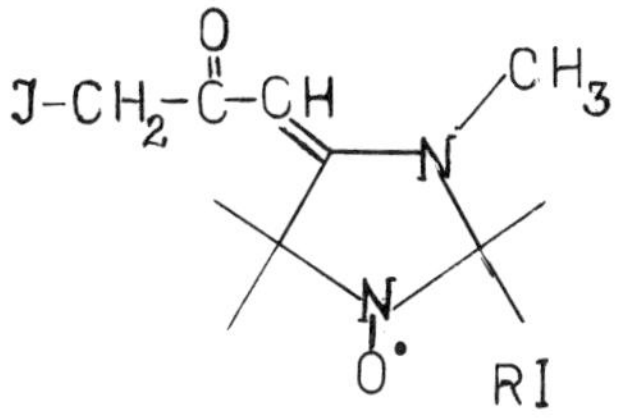

RI 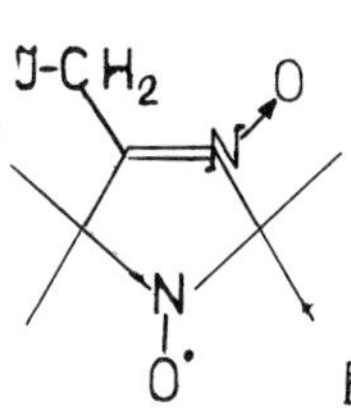 RII

Preparation of microsomes

Liver microsomes were prepared as in (7) from Wistar rats (120-140 g). The concentration of the microsomal protein and cytochrome P450 was determined as described in (8,9). Induction by 3-methylcholantrene and phenobarbital was performed as in (10). Highly purified cytochromes P450 and P448 were obtained using the combination of methods reported by Haugen et al. (11) and Imai (12). The content of cytochrome P450 (P448) was 17.5 nmol/mg of protein, the total yield being 15-20%.

Analysis methods

The microsomal oxidation of aniline and naphthalene was tested from the amount of accumulated p-aminophenol (13) and α-naphthol (14). The radicals were added to the reaction mixture dissolved in acetonitrile (10 μl per 1 ml reaction mixture). The spectral constants, K_s, of the radicals bound to cytochrome P450 were determined as described in (15). Modification of purified cytochrome P450 by radicals was carried out at T=303 K. Molar ratio cytochrome P450/radical was 1:1. Incubation time ranged 30 min to 1 hr. After incubation, the reaction mixture was inserted in the column with Sephadex LH-20 balanced by tris-HCl buffer, pH= 7.5, 0.1 M , to remove the nonbound radical. Then the protein was concentrated to attain the concentration of about 100 μM. In experiments on the determination of the distances between the radical $N\text{-}O^{\bullet}$ group and Fe^{3+}, glycerin (20-50%) was added to the modified protein.

Spectral measurements

Optical spectra were registered on a Hitachi-556 spectrophotometer. The binding of radical RI was examined with a tandem cuvette device to compensate both the turbidity changes and the absorbance of RI radical itself. The EPR spectra of RI and RII radicals were taken by a Varian E-109 spectrometer in a capillary tube at 303 K and in a quartz dewar at 77 K.

The distances between the covalently bound radicals and Fe^{3+} located in the cytochrome P450 active center were estimated by comparing the saturation curves of EPR spectra of the radical covalently bound to the protein and of the radical involved in the lipid (16). The comparison of the saturation curves (at 77 K) made it possible to determine the contribution of Fe^{3+} ion to the relaxation time T_1 of RI radical bound to the protein and to calculate the distance between Fe^{3+} and the nitroxide group of the radical via the formula (16):

$$T_1^{-1} = \frac{\mu^2\gamma^2}{6r^6\tau} \cdot \left\{ \frac{4}{5(\omega-\omega_i)^2} + \frac{24}{5(\omega+\omega_i)^2} + \frac{12}{5\omega^2} \right\}$$

where $\mu = 5.8 \times \mu_B$; $\mu_B = 0.93 \times 10^{-20}$ erg /G is the Bohr magneton; $\gamma = 1.77 \times 10^{-7}$ radian/s·G ; $\omega = 5.9 \times 10^{10}$ radian/s ; $\tau_{Fe^{3+}} = 3 \times 10^{-9}$ s; $\omega_i = 1.8 \times 10^{11}$ radian/s.

RESULTS

The radicals in study were found to bind to the microsomal cytochrome P450 (P448) as Type I or II substrates (Table 1).

TABLE 1

CHARACTERISTICS OF RI AND RII BINDING TO MICROSOMAL CYTOCHROME P450

Characteristics	RI			RII		
	Microsomes			Microsomes		
	Control	PB	3-MC	Control	PB	3-MC
Type of binding	I	II	II	I	I	I
λ_{max}, λ_{min} (nm)	403,422	425, (390-410)	425, 395	390,425	390, 428	402,424
K_s	100 μM	-	60 μM	300 μM	800 μM	300 μM

As seen in Table 1, in the case of induced microsomes, RI binds to cytochrome P450 (P448) as Type II substrate. For the phenobarbital-treated microsomes with RI, only the type of binding has been determined. We failed to find K_s because the minimum of the curve of the absorption spectrum was in the shoulder of a strong signal, which was not compensated even with the use of a tandem cuvette.

For the highly purified forms of cytochrome P450, the type I of RII binding to cytochrome P450 (λ_{min}=425, λ_{max}=408) and cytochrome P448 (λ_{min}=425, λ_{max}=403) was obtained. The same type of RI binding to cytochrome P450 (λ_{min}=430, λ_{max}=410) was observed. No spectrum corresponding to RI binding to cytochrome P448 was registered.

It was not possible in these cases to determine spectral constants (K_s) because of the weak spectral changes $\Delta A \div [R]$.

RI and RII reagents were found to be efficient inhibitors of the microsomal oxidation of aniline and naphthalene. The observed inhibition of the enzymatic activity reached usually 60-80% at the concentration of the inhibitor of about 1 mM. The use of the inhibitor concentrations higher than 1 mM is not recommended since it leads to partial conversion of cytochrome P450 to P420. Table 2 shows the concentrations of the inhibitor, which lead to the inhibition comprising 1/2 of the total inhibition effect. The ratios of the inhibitor/cytochrome P450(P448) concentrations, at which the half-inhibition was observed, are also listed in Table 2.

TABLE 2

INHIBITION OF OXIDATION OF ANILINE AND NAPHTHALENE BY RI AND RII RADICALS

Substrate	Probes	RI I_{50}	RI $\frac{[I]}{[P450]}$	RII I_{50}	RII $\frac{[I]}{[P450]}$
Aniline	Control	100 μM	42	100 μM	33
	PB	5 μM	4	20 μM	3
	3-MC	300 μM	94	70 μM	35
Naphthalene	Control	300 μM	63	100 μM	49
	PB	70 μM	16	65 μM	15
	3-MC	200 μM	117	80 μM	50

The distances between Fe^{3+} involved in the active center of cytochrome P450 and N-O$^{\bullet}$ group of the radicals covalently bound to the protein were determined at 77 K according to the procedure described in (16) (Tables 3 and 4).

TABLE 3 [a]

DISTANCES BETWEEN N-O˙ GROUP OF RI(RII) AND Fe^{3+} FOR MICROSOMES[b]

Microsomes	RI (Å)	RII (Å)
Control	11.5 (1.04)	10 (1.0)
PB	10 (1.8)	10.3 (0.3)
3-MC	9 (0.7)	8.5 (1.2)

TABLE 4 [a]

DISTANCES BETWEEN N-O˙ GROUP OF RI AND Fe^{3+} FOR PURIFIED ENZYMES[b]

Radical	P450	P448
RI	8.5 Å (1.8)	7.9 Å (1.7)

[a]Modification extent of cytochrome P450 is given in brackets.
[b]The experimental error was no more than 10%.

DISCUSSION

The inhibition of the oxidation of cytochrome P450 substrates by RI and RII was reported earlier to be due to the covalent binding of these radicals to the microsomal cytochrome P450 at the enzyme active center (4,5). When used in the concentrations 10-1000 μM, these radicals do not promote the conversion of cytochrome P450 to P420. They neither inhibit the activity of NADPH-cytochrome c reductase, no affect the reduction rates of cytochrome P450 (5).

As follows from Table 1, the type of RI binding changes for induced microsomes. These changes may be caused by the interaction of the RI carbonyl group with Fe^{3+}. Also, it can be related to the peculiarities of the structure of cytochrome P450 (P448) active center in induced microsomes.

The difference in the types of radical binding and in K_s values (Table 1) is not manifested in the observable inhibition of aniline and naphthalene oxidation by radicals (Table 2). For PB-microsomes, a much more efficient suppression of the microsomal activity is observed for both RI and RII on the both substrates of cytochrome P450. However, inhibition of the effect of 3-methylcholantrene microsomes is at the control level or even less pronounced (Table 2). As seen in Table 3, the distances between the radical N-O$^{\bullet}$ group and Fe^{3+} are longer for control micro somes than for induced ones. Assuming that RI and RII modify the same protein group at the active center, the difference in the distances indicates the different conformation of the active center of the cytochrome P450 forms in control and induced microsomes. The distances for the purified cytochrome P450 are 1-1.5Å less than those obtained for microsomes (Table 4). This fact suggests that the lipid surrounding may affect the conformation of the enzyme active center in the lipid phase (microsomes) and in water-glycerin matrices (purified enzyme). Hence, the affinity modification of cytochrome P450 by the radical analogs of substrates provides wide possibilities to study the structure and functions of the various forms of cytochrome P450.

REFERENCES

1. Swanson,R. and Dus,K. (1979) J.Biol.Chem., 254, 7238.
2. Dus,K. (1980) in: J.A.Gustafsson et al.(Ed.), Biochem., Biophys. and Regulation of Cytochrome P450, Elsevier/North-Holland Biomedical Press, Amsterdam, pp. 129-133.
3. Weiner,L., Eremenko,S. et al. (1979), Proc. XX Congress AMPERE, Tallin, p. 520.
4. Lyakhovich,V., Polyakova,N., Popova,V. et al. (1980), FEBS Lett., 115, 31.
5. Lyakhovich,V. and Weiner,L. (1980) in: J.A.Gustafsson et al. (Ed.) Biochem., Biophys. and Regulation of Cyt.P450, Elsevier/North-Holland Biomedical Press, Amsterdam, pp.283-290.
6. Volodarsky,L., Grigor'ev,I. and Sagdeev,R. (1980) in: Berliner,L.J.(Ed.) Biol.Magnetic Resonance, Plenum Press, N.Y., pp. 169-178.
7. Tsyrlov,I., Zakharova,N. et al. (1976) Biochim. Biophys. Acta, 421, 44.
8. Lowry,O., Rosenbrough,N. et al. (1951) J.Biol.Chem., 193, 265.
9. Omura,T. and Sato,R. (1964), Ibid., 239, 2370.
10. Weiner,L., Yamkovoi,V. et al. (1979) Biokhim.(USSR), 44, 634.
11. Haugen,D., Hoeven van der T.A. and Coon,M. (1975) J. Biol. Chem., 250, 3567.
12. Imai, Y. (1976) J. Biochem., 80, 267.
13. Imai, Y., Ito, A. and Sato, R. (1966) Ibid., 60, 417.
14. Achrem, A., Usanov, S. et al.(1975) Dokl.AN SSSR, 223, 1014.
15. Schenkman, J.B., Remmer, H. and Estabrook, R.W. (1976) Mol. Pharmacol., 3, 113.
16. Kulikov, A.V. (1976) Molek. Biol. (USSR), 10, 132.

Cytochrome P-450, Biochemistry, Biophysics
and Environmental Implications, E. Hietanen,
M. Laitinen and O. Hänninen editors

THE SIXTH LIGAND OF FERRIC CYTOCHROME P-450

JOHN H. DAWSON, LAURA A. ANDERSSON AND MASANORI SONO
Department of Chemistry, University of South Carolina, Columbia, South Carolina, 29208, U.S.A.

ABSTRACT

In order to identify the sixth ligand of low-spin ferric cytochrome P-450, we have performed a comprehensive investigation of biomimetic nitrogen, oxygen and sulfur donor complexes of the ferric enzyme. The nature of the sixth ligand of P-450 is of key importance in controlling the catalytic cycle, because dioxygen cannot coordinate to P-450 without dissociation of this ligand. Use of three complementary spectroscopic techniques, UV-visible absorption, magnetic circular dichroism and electron paramagnetic resonance spectroscopy, has enabled us to identify the class of ligand complexes that most closely reproduce *all* of the spectral properties of the native enzyme. In particular, marked similarities have been observed between the spectra of resting ferric P-450 and those of oxygen donor complexes; nitrogen and sulfur donors could be systematically excluded from consideration. Especially close correlation is seen with the spectra of amide and,particularly, alcohol oxygen donor adducts. These data strongly suggest that the sixth ligand of native ferric P-450 is an oxygen donor such as from an endogenous alcohol-containing amino acid, or water.

INTRODUCTION

The ability to activate molecular oxygen for insertion into organic substrate molecules is a key property of cytochrome P-450, a heme iron mono-oxygenase noted also for its unusual spectroscopic properties (1). Because P-450 has both beneficial and harmful physiological roles, ranging from steroid and drug metabolism to carcinogen activation (1), intense interest has been focussed upon its catalytic pathways. Of particular importance have been attempts to define the non-porphyrin, axial ligand(s) in the various reaction states of P-450, since accurate mechanistic interpretations require a thorough understanding of the basic structural features of an enzyme. Strong evidence has been obtained from EPR, UV-visible absorption, magnetic circular dichroism (MCD) and extended x-ray absorption fine structure (EXAFS) spectroscopy for the presence of cysteinate sulfur as an axial ligand in states 1 and 2 (Scheme 1) (2-9). However,

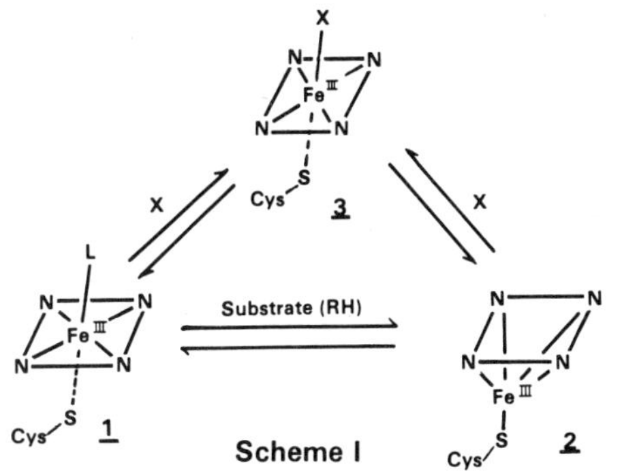

Scheme I

the identity of the ligand, L, *trans* to cysteinate in 1 has not been determined, although histidine (10,11) or an oxygen donor such as alcohol, water or an amide (4,12-14) have been suggested. Since dioxygen cannot coordinate to ferrous P-450 without dissociation of this ligand and reduction of the iron, its identity is of crucial relevance. Further, there may be chemical, electronic or steric interactions between dioxygen and the sixth ligand, such as positioning of the bound substrate for interaction with activated O_2 or hydrogen bonding to O_2. We have utilized UV-visible absorption, MCD and EPR spectroscopy to study and compare a series of biomimetic ligand complexes of P-450 to the native, resting enzyme. The combination of results obtained here have enabled us to sharply delineate the structural possibilities for 1; our results are most consistent with an oxygen donor ligand such as an endogenous alcohol-containing amino acid, or possible water.

MATERIALS AND METHODS

Cytochrome P-450-CAM was isolated and purified from *Ps. putida* to electrophoretic homogeneity (15,16). Substrate removal utilized the procedure of Peterson and co-workers (17). Ligand purity, complex generation and spectrophotometric procedures were as reported elsewhere (15,18,19). All P-450 ligand complexes were examined for the presence of inactive P-420 (1) by bubbling with carbon monoxide and treatment with $Na_2S_2O_4$. None of the species reported showed more than a few percent P-420. Ligands examined are listed in part in Table I.

RESULTS AND DISCUSSION

The primary assumption upon which this study is based is that a P-450 ligand adduct with the same sixth ligand type as in native six-coordinate P-450 will exhibit UV-visible absorption, MCD and EPR spectral properties that are highly similar to those of the native enzyme. Representative models for all reasonable amino acid coordinating types have been examined (Table I). In each case, spectra have been obtained under conditions that ensured complex homogeneity and lack of P-420 contamination. Ligand complexes which were particularly similar to native P-450 were prepared from both substrate-free low-spin, 1, and substrate-bound high-spin, 2, P-450 to produce the adduct, 3. Direct binding of oxygen donor ligands to the heme iron was confirmed by competition with well-known ligands such as cyanide (Figure 1).

TABLE I

Oxygen Donor Atom

Ligand Type: Alcohol   Amino Acid: Serine, Threonine, Tyrosine
Examples: Ethanol, 1-butanol, 1-pentanol, cyclohexanol, p-cresol, phenol

Ligand Type: Amide   Amino Acid: Asparagine or Glutamine; Peptide Amide
Examples: Benzamide, butyramide, N-methylpyrrolidinone, dimethylformamide

Ligand Type: Carboxylate Amino Acid: Glutamate or Aspartate
Examples: Formate, acetate, propionate

Non-Biomimetic Ligands: methyl formate, ethyl acetate, butyrolactone, diethylether, tetrahydrofuran, acetone, cyclohexanone

Nitrogen Donor Atom

Ligand Type: Amine   Amino Acid: Lysine
Examples: Methylamine, 1-butylamine, 1-octylamine

Ligand Type: Imidazole   Amino Acid: Histidine
Examples: Imidazole (Imid.), N-methylimid., 2-methylimid., 1,2-dimethylimid. N-phenylimid., 2-phenylimid., 4-phenylimid.,benzimid.

Ligand Type: Indole   Amino Acid: Tryptophan
Example: Indole

Sulfur Donor Atom

Ligand Type: Disulfide   Amino Acid: Cystine
Example: Dimethyl disulfide

Ligand Type: Thioether   Amino Acid: Methionine
Examples: Dimethyl sulfide, ethylene sulfide, tetrahydrothiophene

Ligand Type: Thiol(ate) Amino Acid: Cysteine(ate)
Examples: Benzyl mercaptan, p-chlorothiophenol, dithiothreitol, 2-mercaptoethanol, 1,3-propanedithiol, 1-propanethiol, thiophenol.

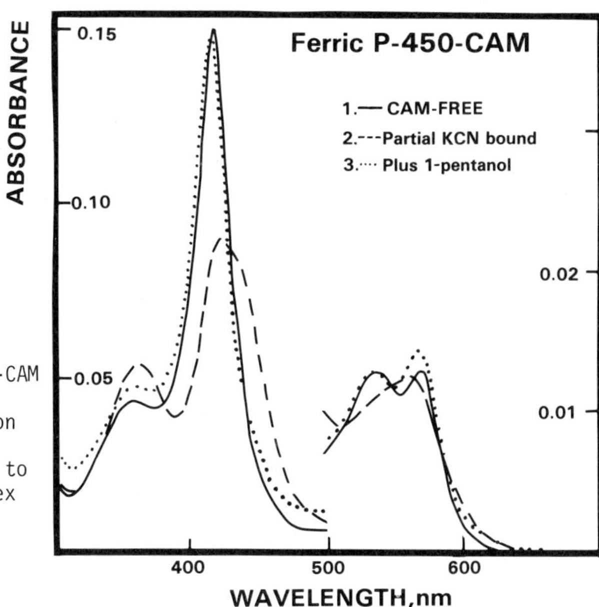

FIGURE 1. COMPETITION STUDY

Solid: substrate-free P-450-CAM

Dashed: partial KCN titration

Dotted: 1-pentanol addition to the P-450-KCN complex

Table II lists the key UV-visible absorption, MCD and EPR spectral properties of resting ferric P-450-CAM. The protein was shown to be essentially 100% low-spin and substrate-free by the absence of the 645 nm high-spin marker band in either the UV-visible absorption or MCD spectrum (15). Table III lists the same properties for the nitrogen, oxygen and sulfur donor complexes of ferric P-450. The various nitrogen donors, which consisted primarily of a series of substituted imidazoles (15), can clearly be eliminated as mimics for the sixth ligand L, after comparison of Tables II and III. These data thus remove histidine from consideration as a possible ligand, and suggest re-consideration of data interpreted to favor it (10,11). Sulfur can also be excluded as a possibility for the sixth ligand in P-450 state 1. The spectra of thioether and thiolate adducts can readily be distinguished from those of P-450. The latter complex has a bis-thiolate structure and exhibits a unique hyperporphyrin (split Soret) spectrum (4,18). The disulfide adduct has some spectral properties that are similar to those of P-450 state 1 (15,18); cystine cannot be the sixth ligand since some P-450's do not have sufficient sulfur-containing amino acids for such ligation to occur (20). Thiolate and thiol complexes interconvert as a function of pH (18). The spectral properties of the thiol bound form are quite similar to those of P-450 state 1; however the native enzyme does not undergo any pH dependent spectral changes. Thus the sixth ligand to P-450 state 1 cannot be a thiol (15,18). Substantial differences in the pKa of heme bound propanethiol were observed between P-450 ($pKa \approx 6.7$) and myoglobin ($pKa < 4.5$), thus suggesting that the heme iron of P-450 is considerably more electron-rich (18). This difference in electron density may be partly responsible for the different functions of the two proteins: oxygen activation by P-450 and reversible oxygen binding by myoglobin.

Comparison of the values in Table III to those of Table II readily reveals that the oxygen donor adducts most closely duplicate *all* of the spectroscopic properties of native P-450. This is particularly interesting when one considers the wide range of oxygen donors examined, from alcohols to ketones. Within this group, the best spectral correlation is seen with alcohol and amide adducts. Figures 1 and 2 show the close spectral similarity between a P-450-alcohol adduct and the native enzyme. The ligation of novel oxygen donors such as esters, ethers and ketones provides an explanation for the slight variations in the substrate-binding and spectral properties of P-450 observed in the presence of non-phosphate or mixed solvent systems containing oxygen functionalities (19) Precedent for oxygen donor ligation to the ferric form of oxygen-binding heme proteins is seen in the case of water and hydroxide ligation to ferric

TABLE II

LOW-SPIN FERRIC P-450-CAM SPECTRAL PROPERTIES

A. UV-visible Absorption

nm	356	417.0	536	569
$\varepsilon(mM \cdot cm)^{-1}$	32	115	10.6	11.1

B. Electron Paramagnetic Resonance

	g_1	g_2	g_3
	2.45	2.26	1.91
range[a]	2.39-2.46	2.23-2.30	1.90-1.93

C. Magnetic Circular Dichroism

nm	355	416	460	518	558	575
$\Delta\varepsilon/H$[b]	-11	0[c]	5.4	9.4	11.3	-18.2

a. Reference 10, EPR g values for P-450 from several sources.
b. $(M \cdot cm \cdot T)^{-1}$
c. Soret Crossover

TABLE III

NITROGEN, OXYGEN AND SULFUR DONOR COMPLEXES OF P-450

A. UV-visible Absorption

Nitrogen		Oxygen		Sulfur	
nm	ε[a]	nm	ε[a]	nm	ε[a]
354-360	37-47	356-362	34-38	358-374	36-67
421-425	98-111	416-420	98-116	417-461	57-98
536-541	10.9-12.6	533-539	10.2-12.2	536-557	10.5-11.6
568-578	8.8-9.5	566-571	10.2-12.4	567-570	7.5-10.3

B. Electron Paramagnetic Resonance

	Nitrogen	Oxygen	Sulfur
g_1	2.47-2.62	2.43-2.48	2.38-2.50
g_2	2.25-2.28	2.25-2.27	2.24-2.27
g_3	1.85-1.92	1.91-1.93	1.89-1.95

C. Magnetic Circular Dichroism

Nitrogen		Oxygen		Sulfur	
nm	$\Delta\varepsilon/H$[b]	nm	$\Delta\varepsilon/H$[b]	nm	$\Delta\varepsilon/H$[b]
360-368	(-9)-(-14)	350-359	(-9)-(-12)	358-368	(-10)-(-35)
419-426	0[c]	412-419	0[c]	416-461	0[c]
468-475	4.8-8.0	455-468	3.1-5.4	460-480	3.5-5.5
520-530	5.7-8.6	513-522	6.5-9.3	515-519	6.1-6.3
560-565	0.8-3.1	552-558	5.7-12.2	556	7.1-7.2
579-585	(-6.5)-(-10.7)	573-578	(-12.5)-(-19.5)	576-583	(-4.9)-(-15.6)

a. $(mM \cdot cm)^{-1}$
b. $(M \cdot cm \cdot T)^{-1}$
c. Soret crossover

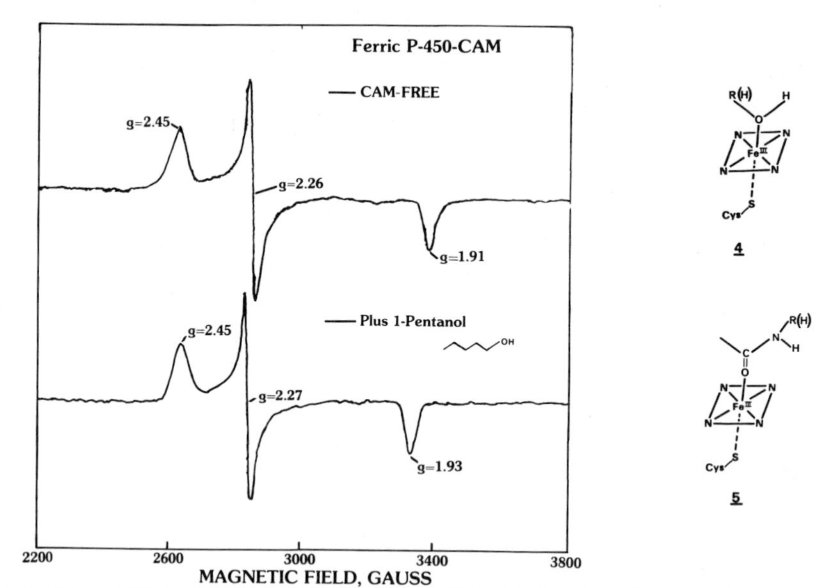

FIGURE 2: EPR SPECTRA OF FERRIC P-450-CAM AND P-450 OXYGEN DONOR ADDUCT
Top: Ferric P-450-CAM, 220 μM. Bottom: P-450 + 1-pentanol;[P-450-CAM]=220μM; [1-pentanol] = 642 μM.

myoglobin. However, aquometmyoglobin is substantially high-spin, whereas the oxygen donor adducts of P-450 are entirely low-spin. In addition, other ligands that yield high-spin complexes with myoglobin form low-spin adducts with P-450 (21). Thus oxygen donor ligation to P-450 is consistent with the known low-spin nature of the resting form. Additional indirect support for oxygen donor ligation to ferric resting P-450 comes from the observation that such ligands are not known to coordinate to ferrous iron porphyrins, a fact consistent with the observed five-coordinate nature of ferrous (deoxy) P-450.

CONCLUSION

UV-visible absorption, MCD and EPR spectroscopy have been used to characterize an extensive series of P-450 ligand complexes. The ligands were chosen as models for potential endogenous amino acid ligands and included nitrogen, oxygen and sulfur donors. Based on the empirical comparison of key spectral features seen in native, ferric low-spin P-450 to the spectral features of the ligand complexes, the *best* correlation was observed with oxygen donors, in particular alcohols, 4, but also amides, 5. As an amide ligand would lack a dissociable proton in close proximity to the metal, as required by magnetic resonance studies (12-14), an alcohol (or water) is the most likely candidate for the sixth ligand. The detailed spectral properties of the nitrogen and sulfur donor adducts are sufficiently different from those of the native enzyme to exclude them as candidates for the sixth ligand, L. Analysis of all available data leads to the conclusion that an endogenous alcohol-containing amino acid, or possibly water, is the sixth ligand to resting, ferric P-450.

ACKNOWLEDGEMENTS:

We thank Ed Phares, Mary V. Long, Susan E. Gadecki, Ian M. Davis and Dr. Grant W. Cyboron for their assistance in bacterial growth and/or protein purification and analysis amd Joseph V. Nardo and Edmund W. Svastits for assistance with data manipulation. This research was supported by a grant from the National Institutes of Health (GM-26730). The electromagnet for the CD spectrophotometer was purchased through a grant from Research Corporation.

REFERENCES

1. Sato,R., and Omura, T., eds. (1978) "Cytochrome P-450", Academic Press, N.Y.
2. Holm, R.H., Tang, S.C., Koch, S., Papaefthymiou, G.C., Foner, S., Frankel, R.R., and Ibers, J.A. (1976) Adv. Exp. Med. Biol. 74, 321-334.
3. Collman, J.P., and Sorrell, T.N. (1977) in ACS Symposium Series No. 44, "Drug Metabolism Concepts" (Jerina, D.M., ed.) pp 27-45, American Chemical Society, Washington, D.C.
4. Ruf, H.H., Wende, P., and Ullrich, V. (1979) J. Inorg. Biochem. 11, 189-204.
5. Dawson, J.H., Holm, R.H., Trudell, J.R., Barth, G., Linder, R.E., Bunnenberg,E., and Djerassi, C. (1976) J. Amer. Chem. Soc. 98, 3707-3709.
6. Collman, J.P., Sorrell, T.N., Dawson, J.H., Trudell, J.R., Bunnenberg, E., and Djerassi, C. (1976) Proc. Natl. Acad. Sci. U.S.A. 73, 6-10.
7. Chang, C.K., and Dolphin, D (1975) J. Amer. Chem. Soc. 97, 5948-5950.
8. Cramer, S.P., Dawson, J.H., Hodgson, K.O., and Hager, L.P. (1978), J. Amer. Chem. Soc. 100, 7282-7290.
9. Dawson, J.H., Andersson, L.A., Davis, I.M., and Hahn, J.E. (1980) in "Biochemistry, Biophysics and Regulation of Cytochrome P-450" (Gustafsson, J.A., Carlstedt-Duke, J., Mode, A., and Rafter, J., eds) pp. 565-572, Elsevier, Amsterdam.
10. Chevion, M., Peisach, J., and Blumberg, W.E. (1977) J. Biol. Chem. 252, 3637-3645.
11. Peisach, J., Mims, W.B., and Davis, J.L. (1979) J. Biol. Chem. 254, 12379-12389.
12. LoBrutto, R., Scholes, C.P., Wagner, G.C., Gunsalus, I.C., and DeBrunner, P.G. (1980) J. Amer. Chem. Soc. 102, 1167-1170.
13. Griffin, B.W., And Peterson, J.A. (1975) J. Biol Chem. 250, 6445-6541.
14. Philson, S.B., DeBrunner, P.G., Schmidt, P.G., and Gunsalus, I.C.(1979) J. Biol. Chem. 254, 10173-10179.
15. Dawson, J.H., Andersson, L.A., and Sono, M. (1982), J.Biol. Chem. 257, 3606-3617.
16. Andersson, Laura A., Ph. D. Thesis; to be submitted to the University of South Carolina, 1982.
17. O'Keeffe, D.H., Ebel, R.E., and Peterson, J.A. (1978) Methods Enzymol. 52, 151-157.
18. Sono, M., Andersson, L.A., and Dawson, J.H. (1982) J. Biol. Chem., in press.
19. Andersson, L.A., and Dawson, J.H., submitted for publication.
20. Dus, K., Goewert, R., Weaver, C.C., Carey, D., and Appleby, C.A. (1976) Biochem. Biophys. Res. Commun. 69, 437-445.
21. Sono, M., and Dawson, J.H., (1982) J. Biol. Chem. 257, 5496-5502.

© 1982 Elsevier Biomedical Press B.V.
Cytochrome P-450, Biochemistry, Biophysics
and Environmental Implications, E. Hietanen,
M. Laitinen and O. Hänninen editors

INACTIVATION OF KEY METABOLIC ENZYMES BY P450-LINKED MIXED FUNCTION OXIDATION SYSTEMS

C. Oliver, L. Fucci, R. Levine, M. Wittenberger and E. R. Stadtman
Laboratory of Biochemistry, National Heart, Lung, and Blood Institute,
National Institues of Health, Building 3, Room 222, Bethesda, MD 20205 U.S.A.

INTRODUCTION

It is well established that the rates of degradation of different enzymes vary greatly and are differentially affected by variations in the nutritional state of the cell. However, almost nothing is known about the mechanism by which a given enzyme is marked for degradation. In the course of our investigations on the turnover of various enzymes, it was observed that cell-free extracts of *Klebsiella aerogenes* catalyzed inactivation of glutamine synthetase (GS). On the assumption that the inactivation reaction might "mark" GS for subsequent degradation, the properties of the inactivation reaction were determined. It was found that the inactivation of GS in cell-free extracts of *K. aerogenes* is dependent upon the presence of NADPH and O_2, and is stimulated by $FeCl_3$ or ($CuSO_4$), and is inhibited by either catalase or chelating agents such as EDTA or *O*-phenanthroline (1). These results suggested that the inactivation is catalyzed by a Fe(III)-dependent mixed function oxidation system and involves the generation of H_2O_2. We report here that mixed function oxidation systems comprised of NADH-diaphorase or of highly purified preparations of NADPH-cytochrome P450 reductase and P450 (LM_2) are able to catalyze the inactivation of several other key enzymes. We show also that xanthine oxidase together with hypoxanthine will catalyze inactivation of GS and that this inactivation is greatly enhanced by the presence of either clostridial ferredoxin or putida redoxin. In either case, the further addition of $P450_c$ increases the rate of GS inactivation and decreases the sensitivity of the inactivation reaction to free radical scavengers.

MATERIAL AND METHODS

Highly purified preparations of rabbit liver microsomal P450 reductase and cytochrome P450 (LM_2) were generously supplied by M. J. Coon. I. C. Gunsalus kindly supplied highly purified preparations of redoxin, redoxin reductase and cytochrome $P450_c$ from *Pseudomonas putida*. Purified preparations of ferredoxin from *Clostridium sticklandii* and acetate kinase from *Escherichia coli* were gifts of T. C. Stadtman. Milk xanthine oxidase, bovine liver catalase, blood

TABLE 1

Inactivation of Glutamine Synthetase by Various Mixed-Function Oxidation Systems

Additions	Relative Rate of GS Inactivation							
	Diaphorase	Cytochrome reductase + P450(LM_2)	Redoxin reductase + Redoxin		Xanthine oxidase			
					+ Redoxin		+ Ferredoxin	
			$-P450_c$	$+P450_c$	$-P450_c$	$+P450_c$	$-P450_c$	$+P450_c$
None	100	100	100	100 (430)	100	100 (117)	100	100 (133)
Catalase	0	0	4	3	0	0	0	0
EDTA	0	0	0	0	0	0	0	0
$MnCl_2$	0	0	0	0	0	--	0	--
Inert gas	0	3	20	8	--	--	14	--
SOD	65	--	36	90	19	8	16	34
Mannitol	--	--	83	85	49	74	71	92
Histidine	100	--	50	85	29	38	63	71
Dimethylsulfoxide	--	--	50	95	40	90	52	90
$FeCl_3$	143	135	122	114				

Reaction mixtures were as described in Materials and Methods. Incubations were at 37°C in air. Where specified, the inert gas was argon. The relative rates of inactivation refer to the pseudo first-order rate constants with the values obtained in the absence of inhibitors set at 100. Numbers in parenthesis indicate how the rates of inactivation in the presence of P450 compare with those obtained in the absence of P450 = 100.

superoxide dismutase (SOD), dilauroyl phosphatidylcholine, rabbit muscle aldolase, *E. coli* alkaline phosphatase, pancreatic amylase, pancreatic carboxypeptidase A, rabbit muscle creatine kinase, rabbit muscle fructose 1,6-bisphosphatase, and glucose-6-phosphate dehydrogenase (from *Leuconostoc mesenteroides* and from yeast) were from Sigma. Alcohol dehydrogenase (from yeast and *L. mesenteroides*), microbial diaphorase, NADPH and NADH were from Boehringer-Mannheim.

Inactivation of glutamine synthetase. Inactivation of GS was measured as previously described (1). For inactivation by the diaphorase, cytochrome P450 reductase and redoxin reductase systems, the reaction mixtures contained 20 μM GS, 1 mM NADH or NADPH, 100 mM KCl, and as indicated, 10 mM mannitol, 100 mM histidine, 94 mM dimethylsulfoxide, 50 μM $FeCl_3$, 0.61 μM catalase. In addition, the diaphorase system contained 0.08 units of diaphorase, 10 mM $MgCl_2$, 50 mM Tris·HCl (pH 7.4), and where indicated, 4.7 μM SOD, 0.7 mM EDTA and 1.0 mM $MnCl_2$; the microsomal $P450(LM_2)$ system contained 0.35 μM cytochrome P450 reductase, 0.15 μM $P450(LM_2)$, 20 μg dilauroyl phosphatidylcholine, 50 mM Tris·HCl (pH 7.4), 10 mM $MgCl_2$, and where indicated, 0.1 mM EDTA and 4 mM $MnCl_2$. The putida redoxin systems contained 0.5 μM putida redoxin reductase, 5 μM putida redoxin, 5 mM $MgCl_2$ and 50 mM Hepes buffer (pH 7.4), and where indicated, 0.5 μM $P450_c$ and 0.7 mM EDTA. The same conditions were used in studies of the inactivation of other enzymes by diaphorase and the microsomal $P450(LM_2)$ systems except that in the inactivation of pyruvate kinase and phosphoglycerate kinase, the concentration of KCl and MgCl were reduced to 10 mM and 1.0 mM, respectively. In experiments with xanthine oxidase, the mixtures contained 2 μM GS, 75 mM Hepes (pH 7.4), 2.5 mM $MgCl_2$, 1 mM hypoxanthine, 10 μM $FeCl_3$ and, where indicated, 5 μM putida redoxin, 8.5 μM ferredoxin, 0.5 μM $P450_c$, 0.033 μM xanthine oxidase, 0.15 μM SOD, 0.5 mM histidine, 50 μM dimethylsulfoxide, 100 mM mannitol, 9 μM catalase. Incubations were at 37°C in air. Where specified, the inert gas was argon.

Reduction of iron(III). The reduction of iron by the xanthine oxidase and diaphorase systems was determined by the method of Weber *et al.* (2). For the cytochrome P450 system, reaction mixtures were gassed 2 minutes with N_2 and then stoppered. Then 25 μM $FeCl_3$ and 75 μM *O*-phenanthroline were added. After 30 minutes, the absorbance at 510 nm was measured. The concentration of Fe(III) was calculated based on an extinction coefficient of 1.13×10^4 liter mol^{-1} cm^{-1} at 510 nm (2).

RESULTS

As shown in Table 1, inactivation of GS is catalyzed by any one of several different mixed function oxidation systems; namely systems composed of (a) NADPH-cytochrome P450 reductase + P450 (LM_2) from rabbit liver microsomes, (b) NADH-redoxin reductase + redoxin $\pm$ $P450_c$ from *P. putida*, (c) NADH-diaphorase, (d) xanthine oxidase + clostridial ferredoxin $\pm$ $P450_c$, (e) xanthine oxidase + redoxin $\pm$ $P450_c$. In the xanthine oxidase coupled systems, hypoxanthine serves as the electron donor. The inactivation reaction is also catalyzed by non-enzymic systems composed of ascorbate, Fe(III) and O_2 or by Fe(II) + O_2 (1).

Site of inactivation. The inactivation of GS does not lead to a significant change in quaternary structure nor in the cleavage of peptide bonds (1,3). R. Levine (4) has demonstrated that the inactivation of GS is accompanied by the appearance of a small increase in absorbancy at 286 nm, and by the modification of just one of sixteen histidine residues in each subunit of the enzyme. The histidine modification leads to formation of a carbonyl containing derivative as judged by its reactivity with phenylhydrazine reagents, and, after acid hydrolysis to the formation of new acetic amino acid (unpublished data).

Susceptibility of inactivated glutamine synthetase to proteolysis. Our assumption that the inactivation reaction "marks" GS for proteolytic attack was confirmed by the demonstration that after inactivation GS is degraded by partially purified enzyme preparations of *E. coli*. As shown in Table 2, native GS is much more resistant to proteolytic attack than enzyme that has been inactivated by either the *P. putida* redoxin reductase-redoxin-$P450_c$ system or by the nonenzymic ascorbate system.

TABLE 2

EFFECT OF INACTIVATION ON PROTEOLYTIC DEGRADATION

Experiment	GS Preparation	GS Degraded in 1 Hour
		%
I	Native	9
	Inactive	61
II	Native	0
	Inactive	24

The assay conditions were described previously (1,5). In experiment I, the GS was inactivated by the nonenzymic ascorbate system. In experiment II, the GS was inactivated in the *P. putida* redoxin reductase-putida redoxin-$P450_c$ system.

Generality of the inactivation reaction. Of 23 enzymes tested, the following enzymes were readily inactivated by either the NADH-diaphorase system or the rabbit liver microsomal NADPH-cytochrome P450 system: alcohol dehydrogenase from *L. mesenteroides* or yeast; creatine kinase, glyceraldehyde-3-phosphate dehyrogenase, pyruvate kinase and lactate dehydrogenase from rabbit muscle; GS from *E. coli* or rat liver; and, phosphoglycerate kinase from yeast. Aspartate kinase III from *E. coli* was also inactivated by the diaphorase system; its inactivation by the P450 system has not been tested.

The following enzymes were not inactivated: acetate kinase, aldolase, alkaline phosphatase (*E. coli*) α-amylase, carboxypeptidase A, fructose-1,6-bisphosphatase, β-galactosidase, glucose-6-phosphate dehydrogenase, β-glucuronidase, hexokinase (yeast) lysozyme, malate dehydrogenase (heart).

As shown in Table 3, the inactivation of pyruvate kinase, GS and phosphoglycerate kinase by both mixed function oxidase systems are dependent upon NADH (or NADPH), O_2, and Fe^{3+}, and are inhibited by Mn^{2+}, catalase, and EDTA.

TABLE 3

PROPERTIES OF INACTIVATION REACTIONS

	Glutamine Synthetase		Phosphoglycerate Kinase		Pyruvate Kinase	
Addition	P450‡	Diaphorase	P450‡	Diaphorase	P450‡	Diaphorase
None*	100	100	100	100	100	100
Inert Gas	3	0	0	0	0	0
Catalase	0	0	0	0	0	0
SOD	--	65	86	100	100	85
$FeCl_3$	135	143	227	308	140	306
$MnCl_2$	0	0	11	4	0	23
EDTA	0	0	16	0	24	15
O-Phenanthroline	7	0	0	0	24	15
NaN_3	100	96	81	100	100	100

‡The P450 system consisted of microsomal P450 reductase + P450(LM_2) + NADPH.

*The composition of the various reaction mixtures were described in Materials and Methods. The numbers refer to the pseudo first-order rates of inactivation with the rate in the absence of inhibition set equal to 100.

It may be significant that all enzymes inactivated by the mixed function oxidation systems are either kinases or dehydrogenases; *i.e.*, they have substrate binding sites for either NADH, NADPH, or ATP. In addition, most of the inactivated enzymes are known to contain a histidine residue at or near the catalytic site. In any case, it is evident from these studies that in the presence of Fe(III) the cytochrome P450 and other mixed function oxidation systems are capable of inactivating a number of key enzymes in metabolism.

Regulation of the inactivation reaction. Other studies (5) have shown that the inactivation of GS by the cytochrome P450 mixed function oxidation system is affected by both the state of adenylylation (6) of the enzyme and by the substrates, ATP and glutamate. Substrates protect the unadenylylated enzyme from inactivation, but enhance the inactivation of the adenylylated enzyme. Similar effects have been observed with other mixed function oxidation systems (1,7). Either ATP or phosphoglycerate can protect phosphoglycerate kinase from inactivation by the NADH-diaphorase system (unpublished data). Protection of enzymes from inactivation by mixed function oxidation reactions by their substrates may therefore be a general phenomenon and might be the basis of an important physiological control mechanism. In the absence of its substrate an enzyme is nonfunctional and is therefore expendable; so it is destroyed by means of inactivation and subsequent proteolysis. Additional regulation could be achieved through the control of catalase and superoxide dismutase activities.

Physiological significance. Evidence that the inactivation reaction occurs *in vivo* was obtained by showing that the rate of inactivation of GS in washed cell suspensions of *E. coli* is dependent upon the presence of Fe(III) and of glucose (presumably to supply NADH and NADPH), and is inhibited by Mn(II) which, as shown in Table 1, is a potent inhibitor of all the mixed function oxidation systems tested. Furthermore, the rate of GS inactivation in cell suspensions varied inversely with the endogenous level of catalase, which could be altered experimentally by growth conditions or by a mutation leading to heme deficiency (8).

Mechanism. The fact that either ascorbate + Fe(III), or Fe(II) alone will catalyze inactivation of GS in the presence of O_2, and the fact that all inactivation reactions are inhibited by catalase, suggests that H_2O_2 and Fe^{2+} are intermediates in this reaction. This possibility is supported by the data in Table 4 showing that all mixed function oxidation systems tested are able to catalyze the reduction of Fe(III) (at least in the presence of orthophenanthroline), and also by the fact that GS is rapidly inactivated by a mixture of Fe(II) and H_2O_2 under anaerobic conditions (data not shown). Unclear, however, are the mechanisms by which Fe(III) is reduced and by which H_2O_2 is formed.

TABLE 4

REDUCTION OF Fe(III) BY VARIOUS MIXED-FUNCTION OXIDATION SYSTEMS

Mixed-Function Oxidase	Fe(III) Reduction* nmoles/30 min
Microsomal P450 system + NADPH	5.0
Redoxin + Redoxin Reductase + NADH	5.7
Redoxin + Redoxin Reductase + $P450_c$ + NADH	5.3
Xanthine Oxidase + Hypoxanthine	4.5
Diaphorase + NADH	26.5

*Iron reduction was measured as described in Methods. Reaction mixtures (1.0 ml) were as follows: the xanthine oxidase system contained 0.4 μM xanthine oxidase, 60 μM *O*-phenanthroline, 0.8 mM hypoxanthine, 20 μM $FeCl_2$, 2 mM $MgCl_2$ and 75 mM Hepes buffer (pH 7.4); the diaphorase system contained 100 μM NADH, 100 μM $FeCl_3$, 300 μM *O*-phenanthroline, 0.4 units of diaphorase, 10 mM $MgCl_2$, 100 mM KCl and 50 mM Tris·HCl (pH 7.4); the redoxin systems and the microsomal P450 systems contained 25 μM $FeCl_3$, 75 μM *O*-phenanthroline; otherwise conditions were the same as for the GS inactivation reactions (see Materials and Methods).

The attractive hypothesis that Fe(III) is reduced by superoxide anion (reactions 1, 2) and that Fe(II) reacts with H_2O_2 to generate hydroxyl radical (reaction 3) via the well known Haber-Weiss reaction (reaction 4) is not always supported by studies with superoxide dismutase and radical scavengers:

$$H + O_2 \rightarrow O_2^- + H^+ \quad (1)$$

$$Fe(III) + O_2^- \rightarrow Fe(II) + O_2 \quad (2)$$

$$Fe(II) + H_2O_2 \rightarrow Fe(III) + OH^\bullet + OH^- \quad (3)$$

$$\text{Sum:} \quad H_2O_2 + O_2^- \rightarrow OH^\bullet + OH^- + O_2 \quad (4)$$

For example, GS inactivation by either of the nonenzymic systems, or by the putida redoxin NADPH-linked $P450_c$ mixed function oxidation system is not inhibited by superoxide dismutase nor by any one of several free radical scavengers; *viz* dimethylsulfoxide, mannitol or thiourea.

However, superoxide dismutase and free radical scavengers do inhibit inactivation reactions catalzyed by the NADH-redoxin reductase-redoxin system (in absence of P450) and by the xanthine oxidase-ferredoxin systems (Table 1). Moreover, in the xanthine oxidase systems, H_2O_2 rather than O_2^- is probably involved in Fe(III) reduction, because catalase, but not superoxide dismutase, inhibits Fe(III) reduction by this system; see also (2,9). It is noteworthy that the addition of $P450_c$ to the mixed function oxidase system composed of redoxin reductase + redoxin or of xanthine oxidase + redoxin (or ferredoxin) leads to significant stimulation of the inactivation reaction and also to a decrease in the sensitivity of inactivation to radical scavengers. In contrast, the reduction of Fe(III) by the diaphorase system is insensitive to either superoxide dimustase or catalase. The fact that only one of sixteen histidine residues per subunit is modified, suggests that the inactivation reaction is a highly site-specific process. Perhaps Fe(II) binds to one of the several metal binding sites on GS (possibly to the sensitive histidine residue), and then peroxidation of the Fe(II)-enzyme complex leads to an activated oxygen species (OH', metalo-oxygen radical, etc.) which oxidizes the liganded histidine residue. A similar mechanism has been proposed for the H_2O_2-dependent inactivation of erythrocyte superoxide dismutase which is accompanied by the oxidation of enzyme-Cu^{1+} and the modification of one histidine residue at the catalytic site (10,11). The failure to observe effects of superoxide dismutase or radical scavengers under some conditions might be due to inaccessibility of these reagents to the enzyme site at which the activated oxygen is generated.

REFERENCES

1. Levine, R.L., Oliver, C.N., Fulks, R.M. and Stadtman, E.R. (1981) Proc. Natl. Acad. Sci. USA, 78, 2120.
2. Weber, M.M., Lenhoff, H.M. and Kaplan, N.O. (1955) J. Biol. Chem., 220, 93.
3. Levine, R.L. (1980) Fed. Proc., 39, (Abst.) 1681.
4. Levine, R.L. (1981) Fed. Proc., 40, (Abst.) 871.
5. Oliver, C.N., Levine, R.L. and Stadtman, E.R. (1981) in: Ornston, L.N. (Ed.), Experiences in Biochemical Perception, Academic Press, New York, pp. 233-249.
6. Stadtman, E.R., Chock, P.B. and Rhee, S.G. (1980) in: Mildner, P. and Ries, B. (Eds.), Enzyme Regulation and Mechanisms of Action, FEBS Symposia, Pergamon Press, Oxford, Vol. 60, pp. 57-68.
7. Oliver, C.N., Levine, R.L. and Stadtman, E.R. (1981) in: Holzer, H. (Ed.), Metabolic Interconversion of Enzymes, Springer-Verlag, Berlin, pp. 259-268.
8. Oliver, C.N. and Stadtman, E.R. (1981) Fed. Proc., 41, (Abst.) 872.
9. Ingelman-Sundberg, M. and Johansson, I. (1981) J. Biol. Chem. 256, 6321.
10. Hodgson, E.K. and Fridovich, I. (1975) Biochemistry, 14, 5294.
11. Sinet, Pierre-M. and Garben, P. (1981) Arch. Biochem. Biophys., 212, 411.

© 1982 Elsevier Biomedical Press B.V.
Cytochrome P-450, Biochemistry, Biophysics
and Environmental Implications, E. Hietanen,
M. Laitinen and O. Hänninen editors

CHOLESTEROL SIDE CHAIN CLEAVAGE CYTOCHROME P-450scc. AMINO-STEROIDS AS PROBES OF THE ACTIVE SITE STRUCTURE

LARRY E. VICKERY AND JOEL J. SHEETS
Department of Physiology and Biophysics, University of California,
Irvine, CA 92717 (U.S.A.)

INTRODUCTION

Cytochrome P-450scc catalyzes the side chain cleavage of cholesterol to produce pregnenolone (1). The reaction, which is the initial and rate-determining step in the biosynthesis of steroid hormones (2), is unusual in that it involves three mono-oxygenation cycles in carrying out the C-C bond cleavage (3,4): Cholesterol is first hydroxylated at C-22 and then at C-20 and the resulting glycol is cleaved to the 20-ketone (5,6). How the enzyme achieves these regio- and stereoselective (7,8) reactions is not understood. Studies on P-450scc (9,10), as well as other P-450 enzymes (11-16), provide evidence for the involvement of higher valence iron-oxo intermediates, and it is presumed that it is such an activated form of oxygen which "attacks" the substrate. Indeed, results with model systems have shown that certain iron-bound oxidants are capable of direct, stereospecific hydroxylations at an unactivated carbon (17-19). For this type of mechanism to be involved in the side chain cleavage of cholesterol by cytochrome P-450scc, the carbon to be attacked must be positioned close to the heme moiety of the enzyme. Our initial experiments have been directed toward determining the distance between the substrate binding site and the heme-iron catalytic site (20). As an approach to this problem, we have prepared steroid derivatives which have the potential to bind to each of these sites and have investigated their interaction with the enzyme. Our results suggest that cholesterol can bind with C-22 adjacent to an axial coordination position of the heme iron.

MATERIALS AND METHODS

Cytochrome P-450scc, adrenodoxin and adrenodoxin reductase were purified from bovine adrenocortical mitochondria as described previously (20,21). Cholesterol side chain cleavage activity was determined in 33 mM potassium phosphate, pH 7.2, containing 0.3% Tween-20 ("Assay Buffer"). Pregnenolone production was measured by radioimmunoassay and exhibited linear rates and a direct dependence on the concentration of P-450scc in all experiments (20,21). Spectral measurements were recorded in a Cary 17D spectrophotometer, and P-450 concentrations were calculated according to Omura and Sato (22).

RATIONALE

In developing a steroid derivative capable of interacting with both the substrate site and the heme, we have linked a series of primary aliphatic amines to 5-androstene-3β-ol (Figure 1). The androstene moiety could be expected to bind to the same site as the identical ring system present in cholesterol; the amine could form a bond with the heme-iron in the complex with the enzyme.

AMINE SUBSTRATE ANALOGS

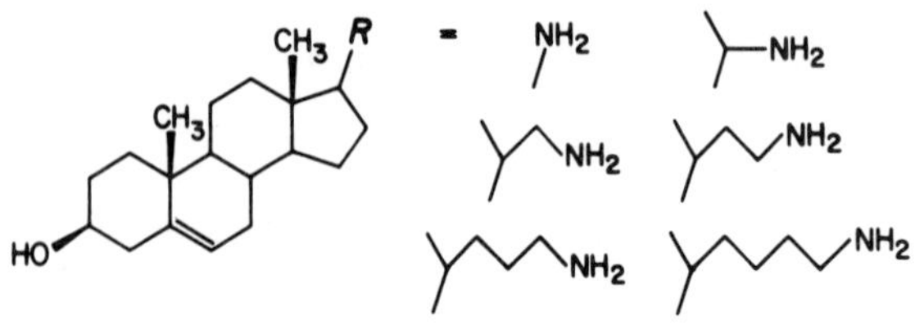

Fig. 1. Amine derivatives of 5-androsten-3β-ol.[1] The amine nitrogen is shown attached to carbons 17, 21, 22, 23, 24 and 25.

We would expect that a derivative with a properly positioned amine would 1) bind more tightly than either the steroid ring or the amine alone, 2) competitively inhibit cholesterol side chain cleavage, and 3) produce a characteristic absorption spectrum. In addition, derivatives in which the amine was too close or too far from the steroid ring system should exhibit weaker interactions, thus defining the distance between the substrate and catalytic sites.

RESULTS AND DISCUSSION

Inhibition studies. The initial testing of amine derivatives was carried out by assaying their ability to inhibit P-450-catalyzed cholesterol side chain cleavage (Fig. 2). With the amine in the 17β or 20α or β position, very little inhibition was observed indicating weak interaction with the enzyme.

---

[1]For preparation of 20- and 22-amine derivatives see Reference 20. The synthesis of other derivatives will be published separately (Sheets and Vickery, in preparation).

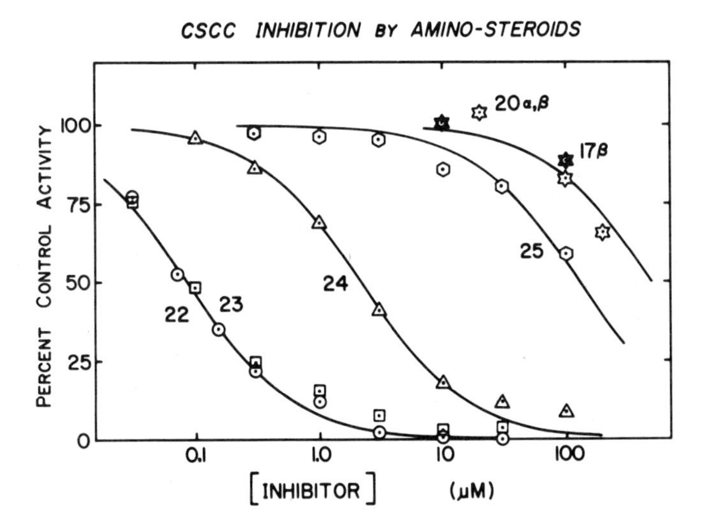

Fig. 2. Dose response curves for inhibition of cholesterol conversion to pregnenolone by cytochrome P-450scc. The cholesterol concentration was 70 μM in all assays. Curves shown were calculated assuming apparent binding constants of 0.1, 2.3, 130, and 500 μM.

Positioning of the amine on C-22 or C-23, however, yielded very potent inhibitors having I_{50} values of approximately 0.1 μM in the presence of 70 μM cholesterol. Shifting the amine to C-24 or C-25 led to a progressive decrease in inhibitory potency indicating a dependence on the distance of the amine from the steroid. Further studies were carried out with the 22-amine, 22-amino-23,24-bisnor-5-cholen-3β-ol (22-ABC), since this seemed to be the minimum distance for high affinity inhibition.

The question of whether 22-ABC causes inhibition by interacting with the substrate site was investigated by a kinetic analysis of inhibition of the cholesterol side chain cleavage reaction. As shown in Fig. 3A, cholesterol and 22-ABC binding are competitive. Because of the structural identity of the rings of the two compounds, we interpret this competition to mean that 22-ABC binds to the enzyme at the same site as cholesterol. Figure 3B shows that the inhibition is linear competitive and that the apparent K_i for 22-ABC is approximately 40 nM.

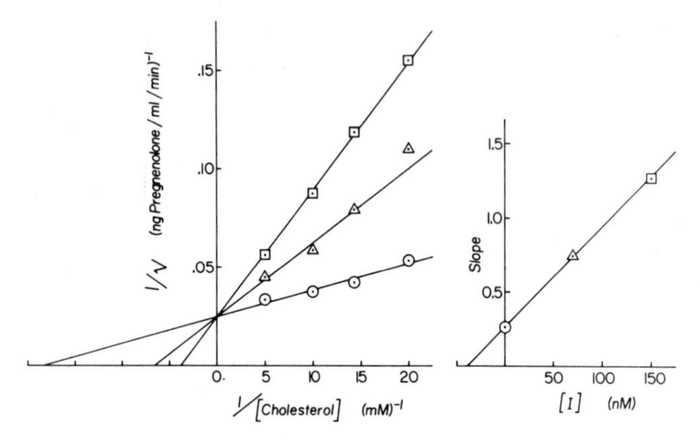

Fig. 3. Kinetic analysis of the mechanism of inhibition of P-450scc by 22-ABC. -O-, no inhibitor; -△-, 70 nM 22-ABC; -□-, 150 nM 22-ABC.

Spectral studies. The effect of 22-ABC on the heme of P-450scc was investigated by absorption spectroscopy. Fig. 4 shows the near-UV and visible spectra of the high spin cholesterol complex (Soret peak 393 nm), the low spin substrate-free form in 0.5% Tween-20 (417 nm), and the complex with 22-ABC (422 nm). The red-shifted Soret peak and the weak α band of the 22-ABC complex are typical of coordination of nitrogen ligands to cytochromes P-450 (23,24) which suggests bonding of the amine to the heme-iron. Futher evidence in support of this conclusion are the findings 1) that an identical spectrum can be produced by binding of isobutylamine to P-450scc and 2) that the 22-amide derivative binds only weakly and produces the 417 nm species (20).

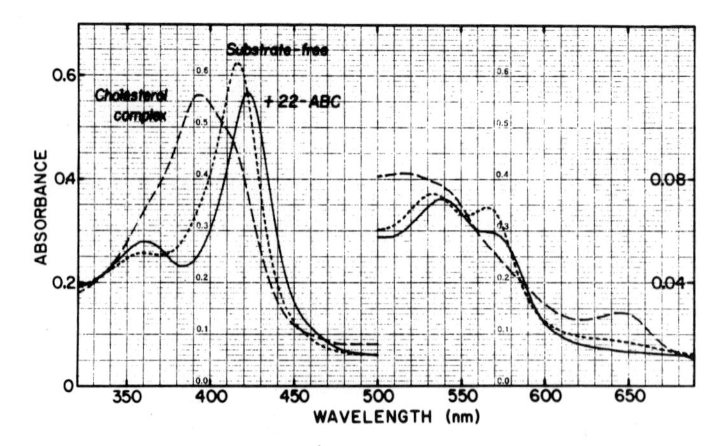

Fig. 4. Effect of 22-ABC on the spectral properties of cytochrome P-450scc.

Spectral titrations of ferri-P-450scc with 22-ABC were carried out in Assay Buffer to determine whether the 22-ABC complex formed was responsible for the observed enzyme inhibition. Figure 5 shows the difference spectra obtained. The spectral dissociation constant calculated, K_s = 42 nM, is very close to the K_i value of 40 nM indicating the role of the complex in inhibition.

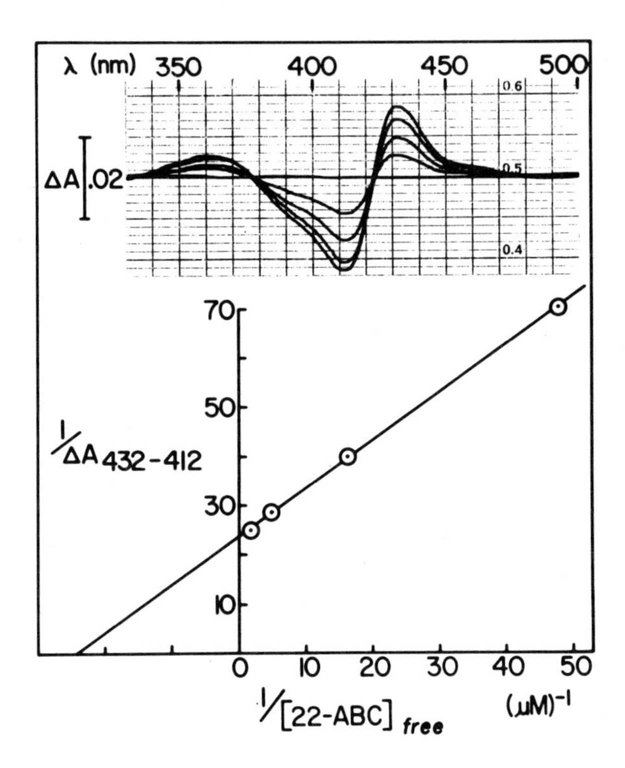

Fig. 5. Spectral titration of cytochrome P-450scc with 22-ABC in Assay Buffer. Path = 4 cm; 0.23 μM P-450scc; total concentrations of 22-ABC were 0.1, 0.2, 0.4, and 0.8 μM.

Structural and mechanistic implications. The results of these spectral studies, taken together with the results of the inhibition studies, suggest that 22-ABC is capable of binding both to the substrate binding site and to the heme-iron catalytic site to yield an inactive form of the enzyme. The structure of such a complex is given in Fig. 5. The heme group is assumed to be bound to the protein through non-covalent interactions of the porphyrin with non-polar amino acid side chains and through coordination of the iron in an axial position by another residue, probably cysteinate (23-25). The amine

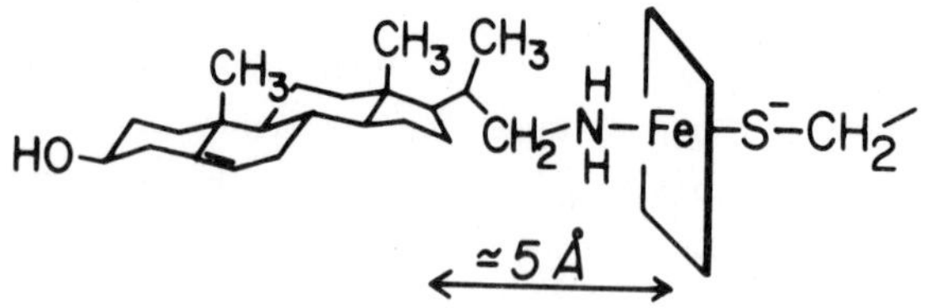

Fig. 6. Complex of 22-ABC with cytochrome P-450scc.

group of 22-ABC is shown coordinated to the heme-iron in the sixth position, *trans* to the endogenous fifth ligand. This bonding of the nitrogen to the iron places restrictions on the position of the steroid ring system in the complex. Using Kendrew skeletal molecular models and varying the conformation of the side chain, we have estimated limits of the distances between different regions of the steroid binding site and the iron catalytic site. C-17 of the steroid can approach to within approximately 3.5Å of the iron and can be positioned no farther than 5.5Å away. The oxygen atom of the 3β-hydroxyl group lies 11-16Å from the iron, depending upon the side chain orientation.

Assuming that the ring system of cholesterol binds in the same manner as that of 22-ABC, this would allow for C-22 of the side chain to lie immediately adjacent to the heme-iron, and an iron-bound oxidant generated during the P-450scc reaction cycle could directly hydroxylate that carbon. This positioning of C-22 of cholesterol closer than C-20 to the heme cataylic site may explain the preferred initial hydroxylation at C-22 rather than C-20. Thus the functions of the enzyme are to simultaneously generate the bound, reactive species of atomic oxygen and position the substrate to favor its regio- and stereoselective attack.

ACKNOWLEDGEMENTS

These studies were supported by U.S. National Institutes of Health Grant AM-30109, Research Career Development Award AM-1005 (L.E.V.), and Training Grant GMO-7311 (J.J.S.).

REFERENCES

1. Simpson, E.R. and Boyd, G.S. (1967) Eur. J. Biochem. 2, 275-285.
2. Stone, D. and Hector, O. (1954) Arch. Biochem. Biophys. 51, 457-469.
3. Shikita, M. and Hall, P.F. (1974) Proc. Natl. Acad. Sci. U.S.A. 71, 1441-1445.
4. Hall, P.F., Lewes, J.L. and Lipson, E.D. (1975) J. Biol. Chem. 250, 2283-2286.
5. Burstein, S., Middleditch, B.S. and Gut, M. (1975) J. Biol. Chem. 250, 9028-9037.
6. Hume, R. and Boyd, G.S. (1978) Biochem. Soc. (London) Trans. 6, 893-898.
7. Morisaki, M., Sato, S., Ikekawa, N. and Shikita, M. (1976) FEBS Lett. 72, 337-346.
8. Byon, C.-Y. and Gut, M. (1980) Biochem. Biophys. Res. Commun. 9, 549-552.
9. Van Lier, J.E. and Rousseau, J. (1976) FEBS Lett. 70, 23-27.
10. Van Lier, J.E., Rousseau, J., Langlois, R. and Fisher, G.J. (1977) Biochim. Biophys. Acta 487, 395-399.
11. Hamilton, G.A. (1974) in: Hayaishi, O. (Ed.), Molecular Mechanisms of Oxygen Activation, Academic, New York, pp. 405-451.
12. Rahimtula, A.D., O'Brien, P.J., Hrycay, E.G., Peterson, J.A., and Estabrook, R.W. (1974) Biochem. Biophys. Res. Commun. 60, 695-702.
13. Hrycay, E.G., Gustafsson, J.-A., Ingelman-Sundberg, M. and Ernster, L. (1975) Biochem. Biophys. Res. Commun. 66, 209-216.
14. Nordblom, G.D., White, R.E. and Coon, M.J. (1976) Arch. Biochem. Biophys. 175, 524-533.
15. Lichtenberger, F., Nastainczyk, W. and Ullrich, V. (1976) Biochem. Biophys. Res. Commun. 70, 939-946.
16. Groves, J.T., McClusky, G.A., White, R.E. and Coon, M.J. (1978) Biochem. Biophys. Res. Commun. 81, 154-160.
17. Groves, J.T. and Van der Puy, M. (1974) J. Am. Chem. Soc. 96, 5274-5275.
18. Groves, J.T. and Van der Puy, M. (1976) J. Am. Chem. Soc. 98, 5290-5297.
19. Groves, J.T. and McClusky, G.A. (1976) J. Am. Chem. Soc. 98, 859-861.
20. Sheets, J.J. and Vickery, L.E. (1982) Proc. Natl. Acad. Sci., U.S.A., in press.
21. Duval, J.F. and Vickery, L.E. (1980) Steroids 36, 473-481.
22. Omura, T. and Sato, R. (1964) J. Biol. Chem. 239, 2379-2385.
23. White, R.W. and Coon, M.J. (1982) J. Biol. Chem. 257, 3073-3083.
24. Dawson, J.H., Andersson, L.A. and Sono, M. (1982) J. Biol. Chem. 257, 3606-3617.
25. White, R.E. and Coon, M.J. (1980) Ann. Rev. Biochem. 49, 315-356.

© 1982 Elsevier Biomedical Press B.V.
Cytochrome P-450, Biochemistry, Biophysics
and Environmental Implications, E. Hietanen,
M. Laitinen and O. Hänninen editors

RESONANCE RAMAN DETECTION OF A Fe-S BOND IN CYTOCHROME $P450_{cam}$

P.M. CHAMPION[1], B.R. STALLARD[2], G.C. WAGNER[3] AND I.C. GUNSALUS[3]
[1]Department of Chemistry, Worcester Polytechnic Institute, Worcester, Massachusetts 01609, [2]Department of Chemistry, Cornell University, Ithaca New York 14853 and [3]Department of Biochemistry, University of Illinois, Urbana, Illinois 61801

In this study we employ resonance Raman spectroscopy as a structural probe of the active site of cytochrome $P450_{cam}$. We observe changes in the ground electronic state vibrational frequencies, upon isotopic substitution, due to the mass effect. In the present report we discuss results obtained with $P450_{cam}$ samples enriched in S^{34} (92%) by bacterial culture (1) and in Fe^{54} (96%) by heme reconstitution of apoprotein (2). Enriched sulfur was converted to sulfate ($^{34}SO_4^{2-}$) by the Schöniger method and used as the sole, limiting source of sulfur in bacterial growth. Sulfate uptake during culture was monitored with a trace of $^{35}SO_4^{2-}$. Isotope enrichments provided by the Oak Ridge National Laboratory were corrected for known sulfur or iron dilutions during protein preparations. The spectra of the enriched samples are compared directly to the spectra of native protein with nuclei of natural abundance (S^{32} and Fe^{56}).

In figure 1 we present the Raman spectrum of the oxidized enzyme-substrate complex obtained with 363.8nm excitation (see Champion et al. (3) for a more detailed presentation of the $P450_{cam}$ Raman spectra). The insert at upper right is the absorption spectrum of this complex in the Soret region; the vertical line denotes the position of the laser excitation. The Raman difference spectra (S^{32}-S^{34} and Fe^{56}-Fe^{54}) in the region near 350 cm^{-1} are displayed in the upper left of figure 1. The derivative shapes of the difference spectra are used to calculate the absolute Raman shifts by the method of Laane and Kiefer (4). (A small correction is made by computer simulation to account for the less than 100% isotope enrichment). We find a -4.9 ± 0.3 cm^{-1} downshift in the case of the S^{34} isotopic substitution and a $+2.5 \pm 0.2$ cm^{-1} upshift in the case of the Fe^{54} substitution. The results of these two experiments unambiguously demonstrate the existence of an Fe-S linkage in the oxidized enzyme-substrate complex of cytochrome $P450_{cam}$. Moreover, the 351 cm^{-1} mode is assigned to the (predominately) Fe-S stretching vibration, confirming an earlier suggestion (5).

We now utilize the general solution to the three body oscillator problem that has been discussed by Cross and Van Vleck (6) and by Wilson (7). If we

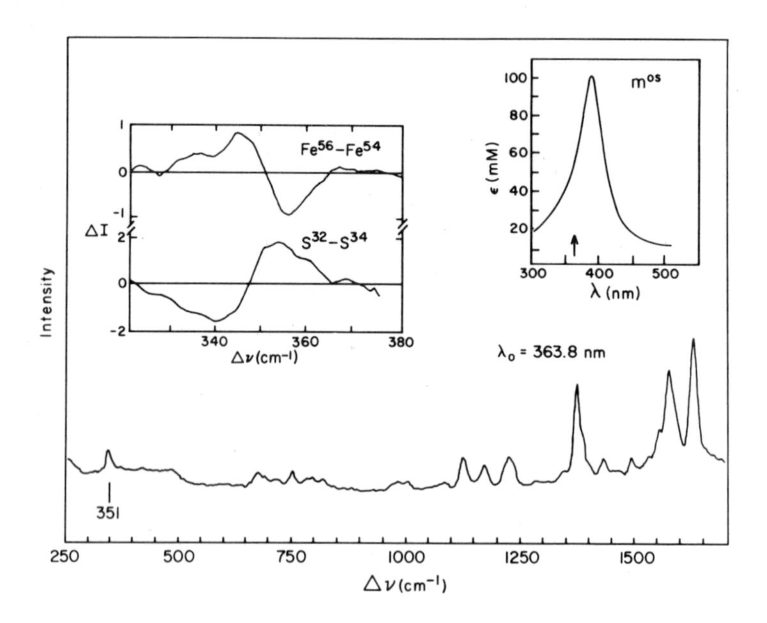

Fig. 1. The resonance Raman spectrum of cytochrome $P450_{cam}$ in the ferric camphor bound state is displayed in the lower curve. The absorption spectrum of the complex is shown in the upper right hand corner; the vertical line denotes the position of the resonant laser excitation. The Raman difference spectra in the region of the 351 cm^{-1} Raman mode are displayed in the upper left hand corner. The vertical scales for the difference spectra are given in units of 1000 counts/channel.

make the standard assumption that the bending force constant is much smaller than the stretching force constants (8) k_1 and k_2, we can write down the eigenfrequencies for the stretching vibrations as:

$$\omega^2 = \tfrac{1}{2}(k_1/\mu_1 + k_2/\mu_2) \pm \tfrac{1}{2}\{(k_1/\mu_1 + k_2/\mu_2)^2 - 4k_1k_2(1/M^2 + F(\theta)/m_S^{\,2})\}^{\frac{1}{2}} \quad (1)$$

where μ_1, k_1 and μ_2, k_2 are the reduced masses and force constants of the iron-sulfur and carbon-sulfur systems, respectively; m_S is the sulfur mass, and the quantity M^2 is given by:

$$M^2 = (m_{Fe}m_Cm_S)/(m_{Fe} + m_C + m_S) \quad (2)$$

The function $F(\theta)$ carries the dependence of the eigenfrequencies on the Fe-S-C bond angle, θ, and is given by:

$$F(\theta) = 4\cos^2(\theta/2)\sin^2(\theta/2) \quad (3)$$

The plus sign in eq. 1 corresponds to the predominantly C-S stretching vibration and the minus sign corresponds to the predominantly Fe-S stretching vibration which is under investigation here.

In figure 2 we have plotted the shifts (Δ_{Fe} and Δ_S) of the 351 cm^{-1} mode that are predicted from eq. 1 as the bond angle is varied from 90° to 180°.

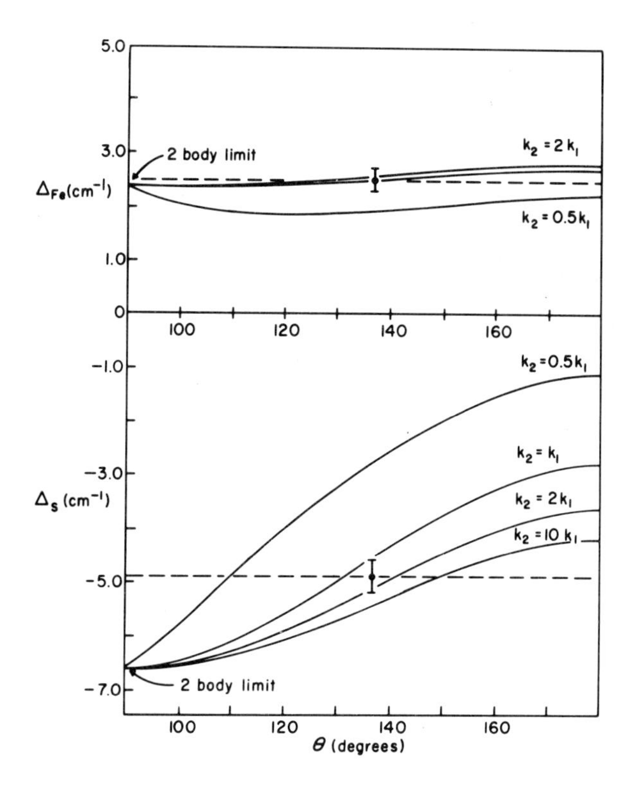

Figure 2. The theoretical predictions of the Fe^{56}-Fe^{54} (Δ_{Fe}) and S^{32}-S^{34} (Δ_S) Raman shifts using the three-body oscillator model discussed in the text.

A family of theoretical curves is shown which represents a variation in the ratio of the Fe-S and S-C force constants (k_1 and k_2, respectively). Normally, we would expect that k_2 is somewhat greater than k_1 since the C-S stretching vibration of cysteine (9) is found at 683 cm^{-1} and the Fe-S stretching vibration (10) is expected near 350 cm^{-1} as observed in this study (using these frequencies and the reduced masses, a simple calculation gives $k_2=1.63k_1$). However, since the bonding strengths in the Fe-S-C system of cytochrome P450 are not yet known with certainty, we have displayed curves showing the extremes $k_2 = 0.5k_1$ and $k_2 = 10k_1$ as well as the more reasonable values $k_2 = k_1$ and $k_2 = 2k_1$. Perhaps the most striking feature of figure 2 is the strong dependence of Δ_S on the bond angle. The shift, Δ_{Fe}, is much less sensitive to the angle due to the larger mass of the iron atom as well as its location at the end (not the middle) of the three body oscillator.

Superimposed on the theoretical curves in figure 2 are the experimentally observed shifts, shown as horizontal dotted lines with the errors shown explicitly. Clearly the observed shift, Δ_{Fe}, is quite well approximated by this model for any angle $\lesssim$ 150° and for any $k_2 \gtrsim k_1$. The result for the sulfur shift, Δ_S, is much more constraining, however. Inspection of the figure reveals that, for reasonable values of k_1 and k_2, the bond angle is confined to the range 125° - 145°.

Finally, we note that the resonance enhancement effect, quantitated as the laser excitation is tuned through an absorption band (i.e., measurement of a Raman excitation profile), can offer a great deal of information about both the nature of the excited molecular state(s) and the strength of the electron-nuclear coupling. If the laser frequency is such that an iron-sulfur charge transfer resonance condition is optimized, we expect enhancement of the scattering amplitudes of the iron-sulfur mode. In fact, now that such an iron-sulfur mode is definetly identified (e.g., the 351 cm^{-1} mode), a careful study of its Raman excitation profile will allow us to extract a great deal of information about the shape and absolute intensity of the charge transfer transition. Recent theoretical developments in this area have shown how the Raman excitation profile and the absorption bandshape (and intensity) are related in a general way (11). This opens up the possibility of using the Raman excitation profile of an iron-axial ligand mode to "deconvolute" the iron-ligand charge transfer band from within a spectrally congested region.

ACKNOWLEDGEMENTS

Supported by NIH grants AM30714, and AM00562.

REFERENCES

1. Gunsalus, I.C.; Wagner, G.C. Methods Enzymol. **1978**, 52, 166.
2. Wagner, G.C.; Perez, M.; Toscano, W.A.; Gunsalus, I.C. J. Biol. Chem. **1981**, 256, 6262.
3. Champion, P.M.; Gunsalus, I.C.; Wagner, G.C. J. Am. Chem. Soc. **1978**.
4. Laane, J.; Kiefer, W. J. Chem. Phys. **1980**, 72, 5305.
5. Felton, R.; Yu, N. in "The Porphyrins" Vol. 3 (ed. D. Dolphin) Academic Press: New York, 1978; p. 347.
6. Cross, P.; Van Vleck, J. J. Chem. Phys. **1933**, 1, 350.
7. Wilson, E. J. Chem. Phys. **1939**, 7, 1047.
8. Herzberg, G. "Infrared and Raman Spectra" Van Norstrand Reinhold Co.: New York, 1945; p. 173.
9. Garfinkel, D.; Edsall, J. J. Am. Chem. Soc. **1958**, 80, 3823.
10. Tang, S.; Spiro, T.; Antanaitis, C.; Holm, R.; Herskovitz, T.; Mortensen, L. Biochem. Biophys. Res. Commun. **1975**, 62, 1.
11. Champion, P.M.; Albrecht, A.C. Chem. Phys. Lett. **1981**, 82, 410.

© 1982 Elsevier Biomedical Press B.V.
Cytochrome P-450, Biochemistry, Biophysics
and Environmental Implications, E. Hietanen,
M. Laitinen and O. Hänninen editors

SUICIDE INACTIVATION OF REDUCED CYTOCHROME P-450 IN SOLUBLE AND MEMBRANE-BOUND STATES

G.I.BACHMANOVA, V.Yu.UVAROV, I.P.KANAEVA, A.I.ARCHAKOV
Second Moscow Medical Institute, Moscow, USSR

INTRODUCTION

It was shown earlier in our lab that the reduction of isolated soluble cytochrome P-450 by sodium dithionite is followed by its rapid inactivation. On the contrary this hemoprotein is inactivated much slower in membrane-bound state or in the presence of cytochrome b_5. Oxidized cytochrome P-450 is stable in soluble and membrane-bound state (1,2). In the present paper we studied the mechanism of inactivation of reduced cytochrome P-450 and the protective action of phospholipid bilayer and cytochrome b_5.

MATERIALS AND METHODS

Cytochromes P-450-LM2 (P-450), and b_5 were obtained from rabbit liver microsomes according to (3,4). Monobilayer liposomes from phosphatidylcholine (PCh) and microsomal phospholipids (MPL), the incorporation of P-450 and b_5, α-helix content of these hemoproteins and melting curves, the registration of reduced P-450 inactivation were carried out as described earlier (1,5).

RESULTS

To study the inactivation mechanism of P-450 reduced by sodium dithionite conversion rate of P-450 into P-420 was determined in the presence of different substances. (Fig.1). The inactivation rate decreased under anaerobic conditions, at the addition of ionole, catalase, cysteine, reduced gluthatione and did not change in the presence of SOD. Thus the inactivation of reduced P-450 at its autooxidation is the result of H_2O_2 or $^{\bullet}OH$ radical formation.

Abbreviations used: P-450, cytochrome P-450-LM2; b_5, cytochrome d-b_5; MPL, microsomal phospholipid; P-420, cytochrome P-420; DOC, deoxycholate; PL, phospholipid

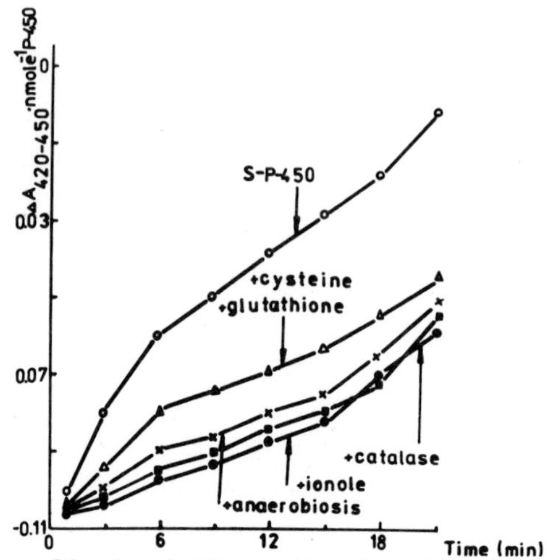

Fig. 1. The inactivation of soluble (S) reduced P-450

The effect of CO on the inactivation of soluble reduced P-450 proved most interesting (Fig.2). Carbon monooxide decreased the inactivation rate at continuous shaking of the incubation mixture The protective effect of CO was not observed without shaking and the inactivation rate of P-450 was the same in the absence or presence of CO. It means that P-450 is inactivated at its autooxidation due to the formation of the active form of oxygen, and in the case of the increase of O_2 concentration in the incubation mixture at its shaking the rate of P-450 inactivation increased and CO decreased the inactivation rate. In the case without shaking the limiting stage was O_2 diffusion and CO did not influence the inactivation rate of P-450.

It was suggested to find out whether the inactivation of reduced P-450 and its conversion into P-420 is followed by any change of apocytochrome P-450 secondary structure. Thus we investigated α-helix content of this hemoprotein by CD-method (Fig.3). There is no dependence between the inactivation of reduced P-450 and its conversion into P-420 and the change of α--helix content. The conversion of P-450 into P-420 does not result in any change of α-helix either in the case of DOC-treatment On the other hand cholate, pCMB, GuHCl and pH 10.5 induced the decrease of α-helix content. Thus cytochromes P-420 are characterized by a different content of α-helix. It is possible to assume that the forms of P-420 with insignificant changes of α-helix content may be reconverted into P-450(Table 1).

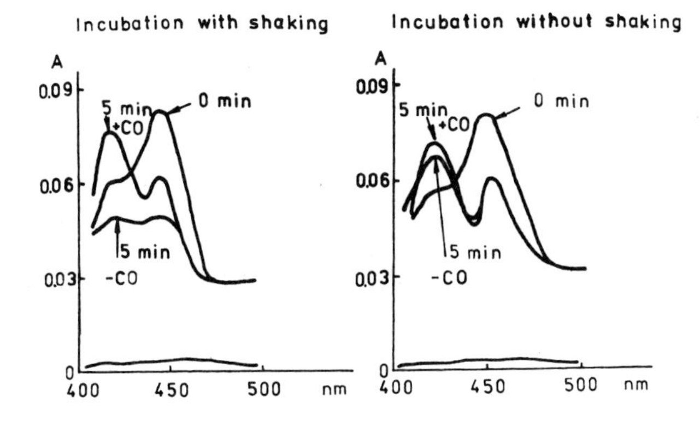

Fig. 2. The influence of CO on inactivation of reduced soluble P-450.

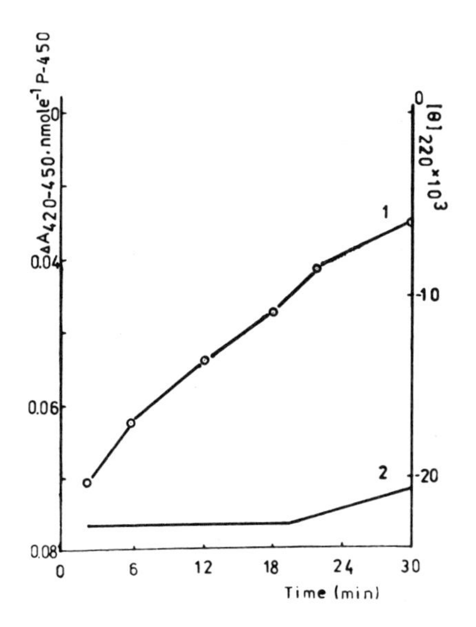

Fig. 3. The inactivation (1) and ellipticity change (2) of reduced soluble P-450.

TABLE 1.

α-HELIX CONTENT OF CYTOCHROMES P-420

Forms of P-450	α-helix content (% of aminoacids)
P-450 (Fe^{3+})	60
P-420	
+ 1% DOC	58
+ 1% cholate	47
+ 1 mM pCMB	44
+ 1.5 mM GuHCl	44
pH 10.5	20
P-450 (Fe^{2+})	70
P-420 (Fe^{2+})	65

Inactivation rate of reduced P-450 decreased after its incorporation into PL-bilayer. However α-helix content of soluble and membrane-bound state of P-450 was almost the same: 55% for PCh- 57% - for MPL-liposomes and 60% for the soluble one. Thus the decrease of inactivation rate of reduced membrane-bound P-450 cannot be explained by the increase of α-helix content. The stabilizing action of PL-bilayer may be the result of α-helix transfer from water space in the case of soluble P-450 into hydrophobic space of PL-bilayer after hemoprotein incorporation.

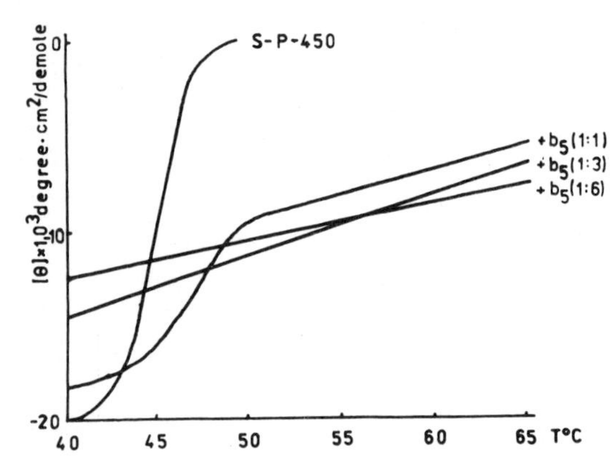

Fig. 4 **The thermal stability of soluble (S) P-450 in the presence of b_5 (Molar ratio of hemoproteins is shown in brackets).**

In the presence of b_5 the decrease inactivation rate of reduced soluble P-450 is observed as well. P-450-b_5 complex formation is accompanied by the decrease of α-helix content up to 5% in comparison with the calculated value. On the other hand P-450 was more stable in these complexes than the soluble one (Fig. 4). The mechanism of stabilizing action of b_5 is still unclear.

REFRENCES

1. Bachmanova, G.I., Karuzina, I.I., Mengazetdinov, D.E., Tveritinov, V.N., Pogodina, O.K., Chaitina, S.Z., Archakov, A.I. and Myasoedova, K.N. (1981) Biokhimia, 46, 280-285

2. Bachmanova, G.I., Pogodina, O.K., Kanaeva, I.P. and Archakov, A.I. (1980) in:Gustafsson, J-A. et al.(Eds.) Biochemistry, Biophysics and Regulation of Cytochrome P-450, Elsevier/North Holland Biomedical Press, Amsterdam, pp. 599-602

3. Karuzina, I.I., Bachmanova, G.I., Mengazetdinov, D.E., Myasoedova, K.N., Zhikhareva, V.O., Kuznetzova, G.P. and Archakov, A.I. (1979), Biokhimia 44, 1149-1159

4. Spatz,L. and Strittmatter,P. (1971) Proc,Natl.Acad.Sci.USA, 68, 1042-1046.

5. Archakov,A.I., Uvarov,V.Yu., Bachmanova,G.I., Sukhomudrenko A.G. and Myasoedova,K.N.(1981)Arch.Biochem.Biophys.212,378-384

Cytochrome P-450, Biochemistry, Biophysics
and Environmental Implications, E. Hietanen,
M. Laitinen and O. Hänninen editors

PLEUROMUTILIN DERIVATIVES AS A TOOL FOR INVESTIGATING THE SIZE AND THE SHAPE OF THE ACTIVE DOMAIN OF CYT P450.

INGE SCHUSTER, HEINZ BERNER, HELMUT EGGER
Sandoz Forschungsinstitut Ges.m.b.H., Brunner Str. 59, A-1235 Wien / Austria

INTRODUCTION

In spite of the numerous investigations on substrate interactions with Cyt P450 only very rough information exists about the structural features of the binding domain. Since X-ray crystallographic data will not be available within the next years indirect measurements might give valuable information on the orientation of molecules in the proximity of the paramagnetic center (1, 2, 3) and their distance to the heme iron. White *et al.* (2) concluded that the active center of Cyt P450 LM2 and LM4 should be large and thus small molecules are binding in a mobile rather nonspecific manner by hydrophobic interactions.

For investigations of size and shape of the active center a large substrate, like the antibiotic pleuromutilin, with its rigid structure appears to be particularly siutable. A set of derivatives with a different kind of substitution should enable us to learn at which points the compound is in contact with the protein moiety and how it is oriented towards the heme iron.

MATERIALS AND METHODS

Some 40 derivatives of pleuromutilin (Fig. 2) were prepared as previously described (4, 5, 6, 7, 8). Due to lack of space detailed structures and properties of the compounds will have to be reported in a separate paper. Systematic introduction of substituents at the positions 1, 2, 3, 4, 11, 12 and 14 of the pleuromutilin nucleus (i.e. hydroxyl-, alkyl- and halogen groups), along with hydrogenation and epimerization afforded the set of requisite substrates. Especially position 22 was substituted with a large number of lipophilic groups containing up to 16 C-atoms together with various functional groups.

Microsomes were prepared by sedimentation at 105,000 g / 60 min from livers of female guinea pigs (400 g) and from human livers derived from male and female brain dead organ donors. Titrations were performed in isotonic KCl/Tris buffer (pH 7.4) at 25^{0}C on a Perkin Elmer 557 spectrophotometer recording difference spectra between 360 and 500 nm. The compounds were added dissolved in DMSO, identical volumes of solvent were given to the reference sample. Cyclohexan, Metyrapon and SKF 525A were used as standards. From the analysis of the binding isotherms apparent affinities and the maximal portion of Cyt P450 interacting in TYPE I manner after (9) were calculated.

RESULTS AND DISCUSSION

Pleuromutilin derivatives are in general excellent substrates for liver Cyt P450 (10, 11), exhibiting apparent affinities up to 10^7 M^{-1} (pH 7.4, 25°C) and interacting with a high part of Cyt P450 in TYPE I manner. Obviously they possess overlapping specificities with Metyrapon and SKF 525A, since both compounds are expelled from their high affinity binding sites in the presence of pleuromutilin. No competition is observed with cyclohexane. The list of apparent affinities and portions of Cyt P450 converted to the high spin complex by the different pleuromutilin derivatives has to be given in a separate paper. General conclusions drawn from these results are listed below.

Human liver Cyt P450 differs from that of guinea pig liver with respect to the binding affinities and the numbers of interacting cytochromes: In human liver the binding constants are generally higher, thus indicating a still better fit in the active center than in that of guinea pig Cyt P450. The portion of Cyt P450 converted to the high spin complex is individually constant for all the compounds investigated and rather high for humans: Up to 40% of total Cyt P450 are involved in this interaction.

With guinea pig liver strong differences in the maximal heights of TYPE I spectra are observed for the different compounds: Pleuromutilin binds to about 15% of total Cyt P450 in TYPE I manner. Substitution of position 2 by polar residues (OH-groups) reduces the portion of interacting Cyt P450 to about 2%. Introduction of hydrophobic groups at the same position increases the number of interacting sites to about 30%. A similar increase is seen when large unpolar residues are located in position 12 and 14. Even a molecule consisting of two pleuromutilins linked together by a thioether bond interacts with higher affinity and a more than twofold number of sites than the monomer.

The strongly varying extent of TYPE I interactions is certainly not due to superposition by TYPE II binding and/or specific binding without a spin change: The additional binding of a "high extent TYPE I binder" shows no inhibition even in the presence of a high excess of a strongly interacting "low extent TYPE I binder".

A possible explanation for this effect is given in Fig. 1. Cyt P450 appears to be buried in the membrane to a very great part (12). Substrates are thought to gain access to the active center via a hydrophobic pocket (12). In human liver microsomes all the pleuromutilin binding forms of Cyt P450 possess similar good accessibility (B).

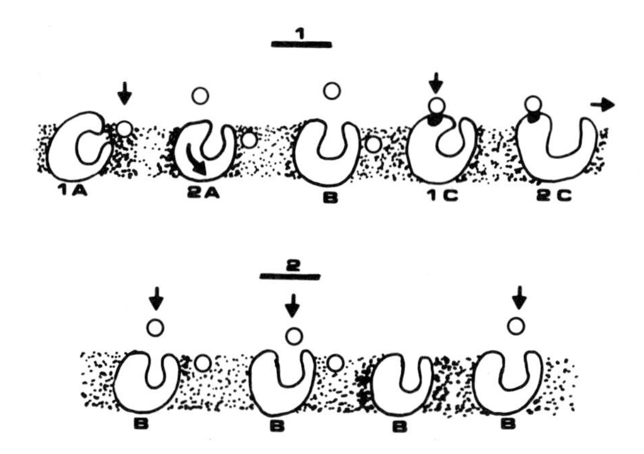

Fig. 1. Accessibility of the active site of Cyt P450 in liver microsomes from guinea pig (1) and man (2). Cyt P450 A: Approach of substrate via lipid interaction. Cyt P450 B: Freely accessible. Cyt P450 C: Substrate induces opening of access.

In guinea pig liver microsomes only a small part of Cyt P450's can be freely approached (B). Variations at some critical points of the lipophilic pleuromutilin molecule leads to structures which are either recognized by the flexible protein molecule as a signal to open the access to the active site (C) or which open up the path by interactions with the closely attached lipid moiety (A). From kinetic investigations of tiamulin Cyt P450 interactions we have suggested several years ago (10) that binding of the drug to the surrounding lipids induces changes in the ability of the protein to interact further in a TYPE I manner.

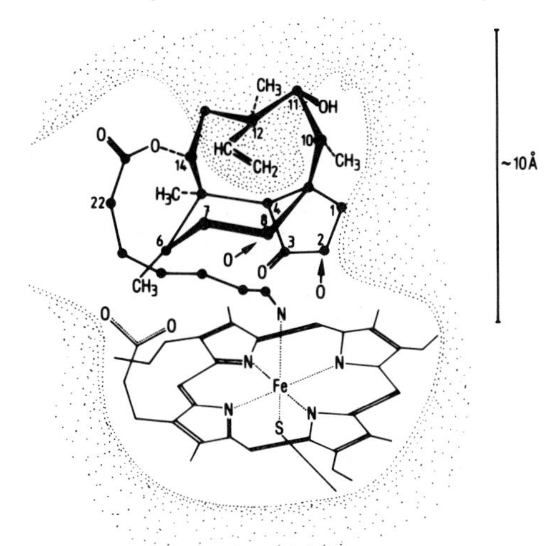

Fig. 2. Location of pleuromutilin derivatives in the active center of Cyt P450. Pleuromutilin = substituent in C-14 = $OCOCH_2OH$.

Substitution at various positions of the pleuromutilin molecule leads in both species to changes in the binding affinity which indicates that several subsites are involved in the interaction (Fig. 2). The 5- and 6-ring should be oriented towards the heme iron as indicated, since oxidative metabolism of several derivatives leads primarily to products hydroxylated in the 2 β and 8 α position. The 5 ring is in close contact to the protein at the position 1 and 2 but not at 3. Contributions are assumed to come from the 8 ring moiety especially at the positions 11 and 12. A wide empty space has to be thought to exist around position 14: Bulky and long substituents (dimeric pleuromutilin, side chains containing more than 10 - CH_2 groups, heterocycles) fit into the gap and increase the overall affinity. The side chain in C-14 bends down to the space between the 5 and 6 membered ring (Fig. 2). Derivatives containing amines and amides in the side chain separated from C-14 by about 7 carbons (heteroatoms), interact directly with the heme iron in a TYPE II interaction.

Considering the dimensions of the pleuromutilin molecule a minimal size and shape of the active domain of Cyt P450 can be calculated: The heme iron is located at the bottom of an at least 10 Å - 12 Å deep and wide groove which still widens at the site where the C-14 side chain expands (possibly the entering pathway of the molecule?).

REFERENCES

1. Novak, R.F., Vatsis, K.P. and Kaul, K.L. (1980) in: Biochemistry, Biophysics and Regulation of Cytochrome P450 (J.A. Gustafsson *et al.*, eds.).
2. White, R.E., Opiran, D.D. and Coon, M.J. (1981) in: Microsomes, Drug Oxidations and Chemical Carcinogenesis (M.J. Coon *et al.*, eds.). Academic Press, New York, 243-251.
3. Swanson, R.A. and Dus, K.M. (1979). J. Biol. Chem. 254, 7238.
4. Egger, H. and Reinshagen, H. (1976). J. Antibiotics 29, 923.
5. Egger, H. and Reinshagen, H. (1976). Ibid 29, 915.
6. Berner, H., Schulz, G. and Schneider, H. (1981). Tetrahedron 37, 915-919.
7. Berner, H., Schulz, G. and Schneider, H. (1980). Ibid 36, 1807-1811.
8. Berner, H., Vyplel, H., Hildebrandt, J. and Laber, G. (1981). 12th Int. Congr. Chemother., Florence.
9. Ebel, R.E., O'Keefe, D.H. and Peterson, J.A. (1978). J. Biol. Chem., 253, 3888.
10. Schuster, I. and Fleschurz, Ch. (1975). Eur. J. Biochem. 51, 511.
11. Schuster, I., Fleschurz, Ch. and Helm, I. (1977) in: Microsomes and Drug Oxidations (V. Ullrich *et al.*, eds.). Pergamon Press, Oxford. 51.
12. Gander, J.E. and Mannering, G.J. (1980). Pharmac. Ther. 10, 191-221.

Cytochrome P-450, Biochemistry, Biophysics
and Environmental Implications, E. Hietanen,
M. Laitinen and O. Hänninen editors

ACUTE ACTH STIMULATION OF CYTOCHROME P-450 DEPENDENT ADRENAL CORTICOSTEROIDOGENESIS: DISCOVERY OF A RAPIDLY INDUCED PROTEIN

RICK JAMES KRUEGER[1] AND NANETTE R. ORME-JOHNSON[1,2,*]
[1]Department of Chemistry, Massachusetts Institute of Technology and [2]The Mary Ingraham Bunting Institute, Radcliffe College, Cambridge, MA, U.S.A.

INTRODUCTION

The initial (1) and rate determining (2) step in adrenal corticosteroid synthesis is the cytochrome P-450_{scc} catalyzed oxidative cleavage of the cholesterol side chain to yield pregnenolone (3) and isocaproaldehyde. Metabolic control of this reaction is exerted via the pituitary polypeptide hormone ACTH (4), mediated by cAMP (5-7). Subsequent to ACTH binding to the plasma membrane of adrenal cells of the zona fasciculata-reticularis, the rate of pregnenolone formation is increased 5-10 fold at constant cytochrome P-450 concentration (8), but in a manner dependent on protein synthesis (5, 9-10). It was found that prior administration of a protein synthesis inhibitor prevented ACTH or cAMP stimulation, that subsequent administration of the inhibitor reversed the stimulation, and that removal of the inhibitor allowed the stimulation to occur again (5, 9-12). The ability of protein synthesis inhibitors to reverse ACTH or cAMP stimulation led to the postulation of a newly formed protein and its designation "labile protein factor" since, after stimulation, without continuing protein synthesis, the rate of steroid production decayed back to the unstimulated level. Several attempts have been made to detect such a protein, formed specifically subsequent to acute ACTH stimulation of adrenal cells, but, until now, no protein appearing synchronously with the increase in steroid product has been reported. We may note here that longer-term administration of ACTH (i.e., hours or days) in these systems produces effects which are probably unrelated to the initial acute steroidogenic response and occur with a much longer time course. We demonstrate here the synthesis of a specific protein in isolated rat adrenal cells, in response to acute ACTH or cAMP stimulation, whose ACTH dose response and time course of synthesis are closely similar to those for the increase in corticosterone production.

MATERIALS AND METHODS

Cell Preparation. Cells were isolated by collagenase digestion (13) of adrenals from female Holtzman rates and inculated in Krebs-Ringer bicarbonate containing 0.5% BSA and 0.2% glucose. Erythrosin B. Exclusion was used to estimate cell viability. Corticosterone production was determined fluorometrically.

Two dimensional electrophoresis. Electrophoresis was carried out essentially by the methods of O'Farrell (14). For first dimension isoelectric focussing, the ampholyte ratio was pH 3-10/pH 5-7: 1/4. For the second dimension, acrylamide gels were either 7.5-15% linear gradient or 10% uniform. Radioactivity was detected by fluorography and film spot densities were converted to cpm. ^{35}S-methionine incorporation into soluble pools and bulk protein was determined according to Rodriques and Yates (15).

RESULTS

Fig. 1 shows the production of a protein (i) in ACTH or dibuteryl-cyclic-AMP (Bt_2cAMP) stimulated cells but not in control cells or cells treated with cycloheximide before ACTH. n and p denote two proteins, which are present in unstimulated as well as stimulated cells and which have the same molecular weight (∿28,000) as but differ in charge from i.

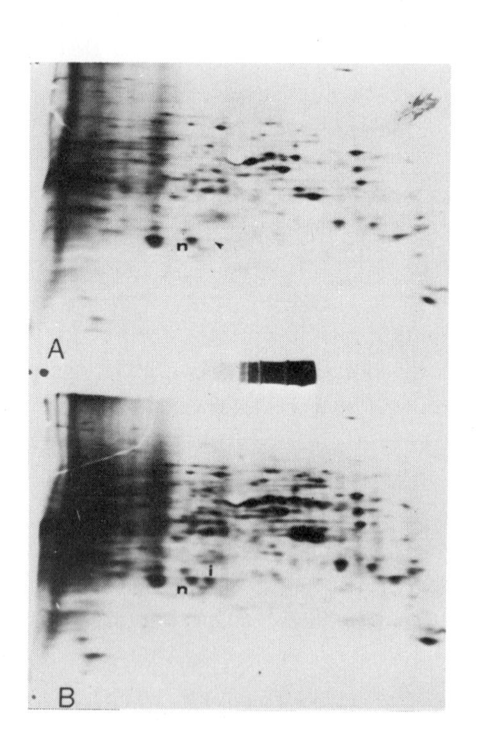

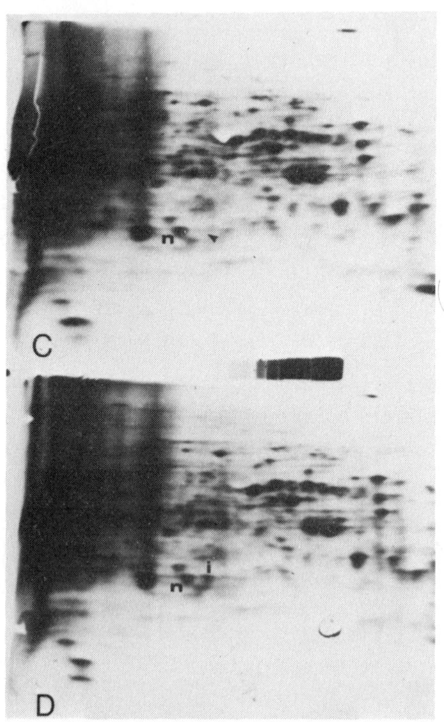

Fig. 1. Two dimensional gel electrophoresis of rat adrenal cortical cell proteins. Cells (40,000 viable cells/ml) were labeled with ^{35}S-methionine (80μ Ci/ml) for 30 min. At t = 2 min, cycloheximide (0.2mM) was added to the cells in A. At t = 0, ACTH (88m unit/ml) was added to the cells in A and B and Bt_2cAMP (2.0mM) was added to the cells in D. The cells in C are controls.

Fig. 2. Time course of production of the ACTH induced protein. Samples were prepared as described in Fig. 1. Data is shown for ^{35}S-methionine incorporation into the induced protein i for: Δ, control; O, ACTH stimulated; , Bt_2cAMP; and cycloheximide + ACTH (same values as for control). Data is shown for corticosterone production for: Δ, control; O; ACTH stimulated; and cycloheximide + ACTH (same values as for control).

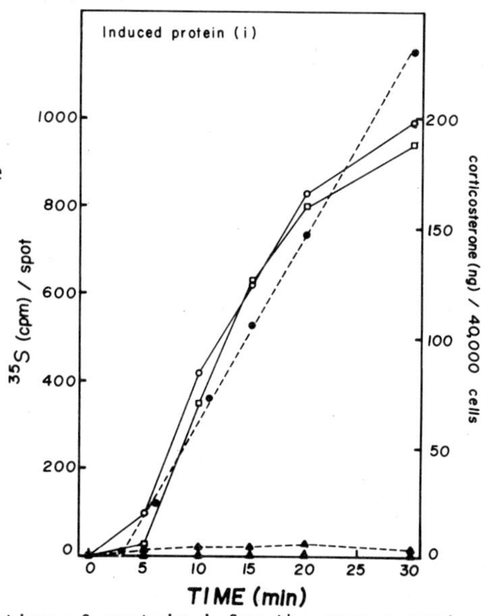

Fig. 2 shows the time course of production of protein i for the same experiments as in Fig. 1, compared to corticosterone production.

Fig. 3 shows the ACTH dose-response curve for the formation of the induced protein i of a non-induced protein n, and of corticosterone.

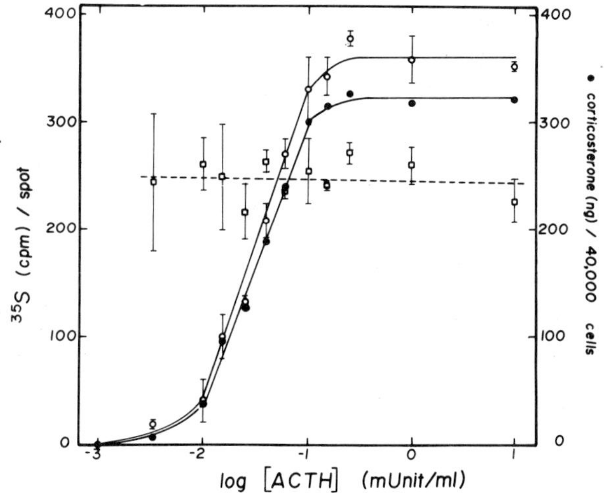

Fig. 3. ACTH dose-response for formation of induced protein and of corticosterone. Samples were prepared as for Fig. 1. Data is shown for ^{35}S-methionine incorporation into the induced protein i (O) and a non-induced protein n ( ). Data is also shown for corticosterone formation, O.

DISCUSSION

The production of a protein in adrenal cortical tissue in response to acute ACTH stimulation has been surmised from a variety of experiments using several protein synthesis inhibitors (5, 9-12). We have identified such a protein in isolated adrenal cells. This protein is not produced in unstimulated cells but is formed subsequent to acute ACTH (or Bt_2cAMP) stimulation and appears with the same ACTH does response and time course as the increase in corticosterone synthesis. Further, its formation is prevented by addition of cycloheximide before stimulation by ACTH or Bt_2cAMP. Addition of cycloheximide after stimulation causes the formation of the protein to cease at approximately the same time as the rate of steroid production begins to decrease.

At present, both the function and the source of this rapidly induced protein i are open to speculation. Indeed the participation of the protein in increasing the rate of corticosteroid formation is not demonstrated. All the present data would be consistent with the formation of this protein as a result of increased corticosterone concentration. However, this is the first direct demonstration of the cycloheximide sensitive formation of a protein in adrenal cells in response to acute ACTH or Bt_2cAMP stimulation with the same dose response and time course as that for corticosterone production.

ACKNOWLEDGEMENTS

This work was supported by USPHS Research Grant AM27381 (to N.R.O.-J.), by a fellowship (to N.R.O.-J.) funded by ONR Research Grant N00014-80-C-0084 to Radcliffe College and by USPHS Posdoctoral Fellowship AM06263 (to R.J.K.).

REFERENCES

1. Soloman, S., Levitan, P. and Lieberman, S. (1956) Rev. Can. Biol., 15, 282.
2. Dorfman, R. I. (1957) Cancer Rev., 17, 535-536.
3. Koritz, S. B. and Kumar, A. M. (1970) J. Biol. Chem., 245, 152-159.
4. Stone, D. and Hechter, O. (1954) Arch. Biochem. Biophys., 51, 457-469.
5. Haynes, R. C. Jr., Koritz, S. B. and Peron, F. G. (1959) J. Biol. Chem., 234, 1421-1423.
6. Grahame-Smith, D. G., Butcher, R. W., Ney, R. L. and Sutherland, E. W. (1957) J. Biol. Chem., 242, 5535-5541.
7. Davis, W. W. and Garren, L. D. (1968) J. Biol. Chem., 243, 5153-5157.
8. Brownie, A. C., Alfano, J., Jefcoate, C. R., Orme-Johnson, W. H. and Beinert, H. (1973) Ann. N. Y. Acad. Sci., 212, 344-360.
9. Ferguson, J. J. (1962) Biochim. Biophys. Acta, 57, 616-617.
10. Garren. L. D., Ney, R. L. and Davis, W. W. (1965) Proc. Nat'l Acad. Sci. U.S.A., 53, 1443-1450.
11. Schulster, D., Tait, S. A. S., Tain, J. F. and Mrotek, J. (1970) Endocrinology, 86, 487-502.
12. Crivello, J. F. and Jefcoate, C. R. (1978) Biochim. Biophys. Acta, 542, 315-329.
13. Ray, P. and Strott, C. A. (1978) Endocrinology 103, 1281-1288.
14. O'Farrell, P. H. (1975) J. Biol. Chem. 250, 4007-4021.
15. Rodriques, H. J. and Yates, J. T. (1980) Biochim. Biophys. Acta 596, 64-80.

© 1982 Elsevier Biomedical Press B.V.
Cytochrome P-450, Biochemistry, Biophysics
and Environmental Implications, E. Hietanen,
M. Laitinen and O. Hänninen editors

THE POTENTIAL ROLE OF CYTOCHROME b_5 AS AN EFFECTOR IN CYTOCHROME P-450-DEPENDENT DRUG OXIDATIONS

PETER HLAVICA
Institut für Pharmakologie und Toxikologie der Universität, Nussbaumstrasse 26, D-8000 München 2 (F.R.G.)

INTRODUCTION

Cytochrome b_5 and its apohemoprotein have been found to accelerate autoxidation of photoreduced cytochrome $P\text{-}450_{cam}$ associated with an increased rate of substrate hydroxylation (1). Similar observations were made with preparations of rabbit liver microsomal cytochrome P-450 (2). These results were interpreted to mean that cytochrome b_5 might act as an effector in cytochrome P-450-mediated drug oxidations. We, therefore, decided to look at the influence of cytochrome b_5 on the cumene hydroperoxide-supported oxidative metabolism of 4-chloroaniline in a system containing highly purified cytochrome P-450LM_2 incorporated into lipid vesicles.

MATERIALS AND METHODS

Detergent-solubilized cytochromes P-450 and b_5 from the livers of phenobarbital-treated rabbits were purified by published procedures (3,4). Incorporation of the hemoproteins into dilauroylglyceryl-3-phosphorylcholine (di12GPC) vesicles was performed by the cholate-gel filtration technique (5). Stopped-flow experiments were conducted with an Aminco-Morrow apparatus attached to an Aminco Dasar/DW-2 spectrophotometer system. Rates of interaction of 4-chloroaniline (4-CA) and cumene hydroperoxide (CHP) with ferric cytochrome P-450LM_2 were determined by following the increase in absorbance at 424 and 440 nm, respectively. The change in absorbance at 437 nm was monitored, when the reaction of the peroxide with the hemoprotein in the presence of the amine substrate was studied. The final concentrations of the reactants in the mixing chamber were as follows: 0.075 M KH_2PO_4/Na_2HPO_4, pH 7.4; 4 µM cytochrome P-450LM_2; 10 mM 4-CA, and/or 10 mM CHP; the temperature was maintained at 25°C. Peroxidatic N-hydroxylation of 4-CA was measured in a system composed as described above; incubation of the samples was carried out for 7 min at 37°C. 4-Chlorophenylhydroxylamine was determined after being oxidized to the corresponding nitroso derivative (6).

RESULTS AND DISCUSSION

Titration of the di12GPC-bound cytochrome P-450LM_2 with cytochrome b_5 results

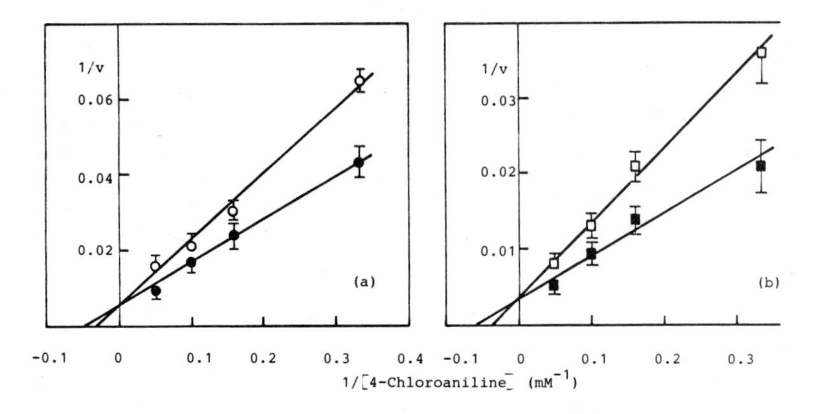

Fig. 1. Influence of cytochrome b5 on the CHP-dependent N-hydroxylation of 4-CA by cytochrome P-450LM2 after incorporation into di12GPC vesicles (a) or into intact microsomal membranes (b). The molar ratios of cytochrome b5 to cytochrome P-450LM2 were as follows: o, 0:1; •, 1.4:1; □, 0.46:1 (native phenobarbital-induced microsomal fraction); ■, 1.3:1. The initial rate (v) indicates pmol 4-chloronitrosobenzene formed/min per nmol of cytochrome P-450LM2.

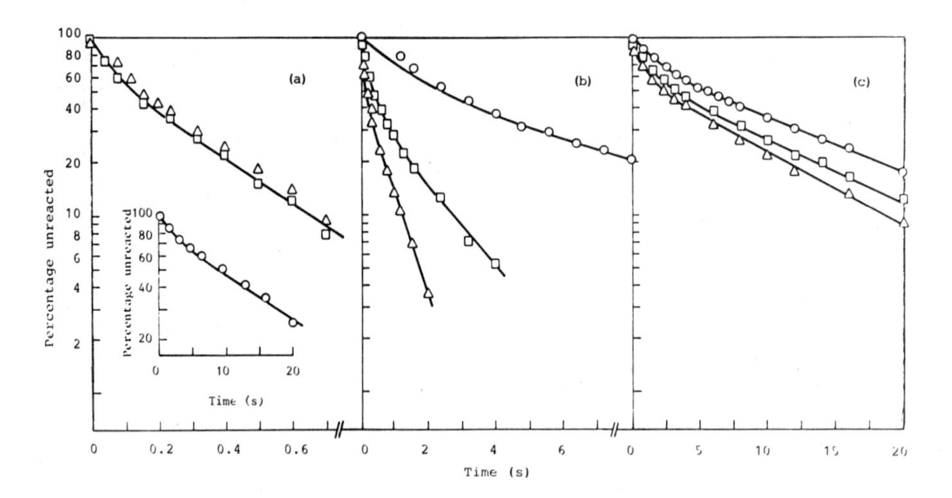

Fig. 2. Influence of cytochrome b5 on the kinetics of interaction of 4-CA and CHP with cytochrome P-450LM2 after incorporation into di12GPC vesicles. (a) Formation of the 424 nm complex; (b) formation of the 440 nm complex; (c) formation of the 437 nm complex. The molar ratios of cytochrome b5 to cytochrome P-450LM2 were as follows: o, 0:1; □, 1.3 or 1.4:1; △, 2.7 or 3.7:1.

in a type I difference spectrum characterized by a Soret band at 390 nm and a trough at 420 nm; this spectral perturbation indicates a transition of the spin

TABLE 1

INFLUENCE OF CYTOCHROME b_5 ON THE RATE CONSTANTS FOR INTERACTION OF 4-CA AND/OR CHP WITH CYTOCHROME P-450LM$_2$

Subscripts f and s refer to the fast and slow phases of the processes, respectively. The relative amplitude of each phase is given in parentheses. Cytochromes P-450LM$_2$ and b_5 were incorporated into dil2GPC vesicles.

$[b_5]/[P\text{-}450LM_2]$	Type of complex formed	k_f (s^{-1})	k_s (s^{-1})
0.00	424 nm	0.52 (13.5%)	0.06 (86.5%)
1.25		17.33 (27.5%)	3.06 (72.5%)
3.66		18.32 (24.5%)	2.81 (75.5%)
0.00	440 nm	0.69 (29.5%)	0.16 (70.5%)
1.43		3.47 (51.0%)	0.59 (49.0%)
2.65		6.13 (45.5%)	1.43 (54.5%)
0.00	437 nm	0.50 (28.5%)	0.07 (71.5%)
1.43		0.73 (34.5%)	0.09 (65.5%)
2.65		0.93 (44.5%)	0.09 (55.5%)

state of the oxidase from low to high. By means of CD spectrophotometry in the far u.v. region cytochrome b_5 was demonstrated to affect the α-helical content of cytochrome P-450LM$_2$. Such structural change of the pigment is associated with an alteration of the catalytic function of the oxidase, as evidenced by its ability to mediate CHP-dependent N-hydroxylation of 4-CA: when incorporated into artificial lipid membranes, cytochrome b_5 decreases the apparent K_m value for 4-CA without appreciably affecting V_{max} of the N-oxidation process (Fig. 1a). Similar observations are made, when cytochrome b_5 is incorporated into native microsomal membranes (Fig. 1b). The individual steps involved in the catalytic cycle were studied by means of the stopped-flow technique. Plots of absorbance change at 424 and 440 nm (Figs. 2a and 2b), representing interaction with the oxidase of 4-CA and CHP, versus time are biphasic. The rate constants deduced from such plots at various ratios of cytochrome b_5 to cytochrome P-450LM$_2$ are given in Table 1. The most prominent feature is that, as the molar ratio of the hemoproteins increases, both the rate of binding to the oxidase of 4-CA and CHP and the relative contribution of the fast phase of interaction are increased. However, the stability of the 440 nm complex appears to be decreased (data not shown). Most interestingly, the presence of 4-CA in the assay media decelerates formation of the oxy-complex in its substrate-bound form (followed at 437 nm;

TABLE 2

INFLUENCE OF CYTOCHROME b_5 ON THE ENERGY OF ACTIVATION REQUIRED FOR FORMATION OF THE 437 nm SPECIES

$[b_5]/[P\text{-}450LM_2]$	Type of hemoprotein reacting	Activation energy (kJ/mol)	
		Below critical temperature	Above critical temperature
0	fast	35.7	111.8
	slow	38.2	165.7
2.65	fast	23.9	79.5
	slow	63.8	63.8

Fig. 2c). Such phenomenon is easily understood, since 4-CA, by competing with CHP for the common heme binding site, partially displaces the oxygen donor from the receptor molecule. As there is a more pronounced stimulatory effect by cytochrome b_5 on 4-CA than on CHP binding (cf. Table 1), retardation of formation of the 437 nm intermediate becomes the more marked the stronger the ratio of cytochrome b_5 to cytochrome P-450LM$_2$ is raised. On the other hand, 4-CA stabilizes the ternary oxy-complex formed (data not shown). As summarized in Table 2, cytochrome b_5 acts by decreasing the energy of activation required for generation of the Fe^{3+}_S-O compound. As a resultant of the effects described above, cytochrome b_5, incorporated into lipid vesicles, decreases the optical dissociation constant (K_S) of the complex between 4-CA and cytochrome P-450LM$_2$ from 2.5 to 0.45 mM and that of the adduct of CHP from 0.5 to 0.24 mM.

In conclusion, our findings suggest that, in addition to its function as an electron carrier, cytochrome b_5 might modify the catalytic capacity of the cytochrome P-450 system by inducing a conformational change facilitating substrate interaction with the pigment.

REFERENCES

1. Lipscomb, J.D., Sligar, S.G., Namtvedt, M.J. and Gunsalus, I.C. (1976) J. Biol. Chem., 251, 1116.
2. Guengerich, F.P., Ballou, D.P., and Coon, M.J. (1976) Biochem. Biophys. Res. Commun., 70, 951.
3. Hlavica, P. and Hülsmann, S. (1979) Biochem. J., 182, 109.
4. Strittmatter, P., Fleming, P., Connors, M. and Corcoran, D. (1978) Methods Enzymol., 52, 97.
5. Ingelman-Sundberg, M. and Glaumann, H. (1980) Biochim. Biophys. Acta, 599, 417.
6. Herr, F. and Kiese, M. (1959) Naunyn-Schmiedebergs Arch. Pharmak. exp. Path., 235, 351.

Cytochrome P-450, Biochemistry, Biophysics
and Environmental Implications, E. Hietanen,
M. Laitinen and O. Hänninen editors

CHARACTERISATION OF Ah RECEPTOR MUTANTS AMONG BENZO(a)PYRENE-RESISTANT MOUSE HEPATOMA CLONES

CATHERINE LEGRAVEREND[1], RITA R. HANNAH[1], HOWARD J. EISEN[1], IDA S. OWENS[1], DANIEL W. NEBERT[1] AND OLIVER HANKINSON[2]
[1]Developmental Pharmacology Branch, NICHD, NIH, MD 20205 (U.S.A.) and [2]Laboratory of Biomedical and Environmental Sciences, University of California, Los Angeles, CA 90024 (U.S.A.)

INTRODUCTION

The Ah locus controls the induction by 3-MC B(a)A and TCDD of many drug metabolizing enzyme activities. Multiple forms of cytochrome P-450 are the products of structural genes "turned on" during the induction by PAH while a cytosolic receptor is the major product of regulatory genes (1).

Recently a number of Hepa-1 B(a)P resistant mutant clones have been isolated (2). By means of somatic-cell fusion experiments with the wild-type Hepa-1 parent, most of the mutants were shown to be recessive and belong to at least three different complementation groups (3), therefore reflecting mutations in at least three different genes.

The purpose of this study was to examine some of these mutant clones with regard to their Ah receptor content and translocation of TCDD-receptor complexes into the mucleus in an attempt to understand better the sequence of events occuring during the AHH induction process.

MATERIALS AND METHODS

A detailed description of all the experimental procedures has been published elsewhere (6) the description, cell culture conditions and abbreviations of the parent line and six mutants in this report and their designations in previous studies can be found in Ref 2 and 5 and in Table 1 of the present report.

Preparation of cytosolic fractions and nuclear extracts
Cells were lysed, following a hypotonic treatment, by hand homogenization in Potter-Elvehjem homogenizers with teflon pestles. The homogenate was centrifuged at 800 g for 10 min. The supernatant was centrifuged at 105,000 xg for 1 hr; the supernatant

fraction of this was the cytosolic fraction. Washed nuclei were suspended in .5M K cl Buffer pH 8.5. Following a 30' 105,000 xg centrifugation, the supernatant fraction was collected (nuclear extract).

Treatment of cytosal or nuclear extracts with (^{3}H) TCDD *in vitro* Cytosal or nuclear extracts from cells that never had been exposed to (^{3}H) TCDD in culture or from cells that had been exposed to (^{3}H)TCDD in culture, were treated with (^{3}H)TCDD *in vitro*. 1 ml of cytosol or nuclear extract (2 to 5 mg of protein) was treated *in vitro* with 1 or 10 nm (^{3}H)TCDD in the absence or presence of a 100 fold excess of nonlabelled TCDD for 1 h at 4°C.

Velocity sedimentation analysis .3 ml samples of cytosol or nuclear extracts were layered on 5 % to 20 % sucrose density gradients - the gradients were centrifuged at 2°C for 16 h at 235,000 xg in a Beckman SW 60 Ti rotor. .2 ml fractions were collected and counted in aquasol.

RESULTS

Kinetics of aryl hydroxylase induction by TCDD or Benzo(a)anthracene (Fig. 1) Three distinct patterns of AHH inducibility were seen among the six clones. (a) the TCDD and BA-inducible enzyme activity in c2 was 10 % to 20 % of that observed in the parent line. (b) the inducible activity in c3 and c6 was measurable and between 1 % and 10 % of the parent line's activity. (c) neither control nor the inducible hydroxylase activity was detectable in Cl, C4 and C5 lines.

Quantitation of *Ah* receptor in cytosolic and nuclear extracts of Hepa-1 and B(a)P-resistant mutants Treatment of the Hepa-lclc7 cytosol with (^{3}H)TCDD *in vitro* results in a distinct peak of about 9 S size on sucrose density gradients while the same treatment of the nuclear extract produces no peak. In the mutants no strict relationship was found between the quantity of cytosolic or nuclear receptor and AHH activities. (Fig. 2 and Table 1) After one hour incubation with (^{3}H)TCDD in culture cytosolic and nuclear receptor levels as well as cytosolic-to-nuclear receptor ratios are similar to that of Hepa-lclc7 in cl, c3, and c5 with a normal nuclear translocation of inducer-receptor complexes. In c2 and c6 nuclear translocation of the inducer-receptor complex is also unimpaired but less than 10 % of the wild type receptor level is found in the cytosol.

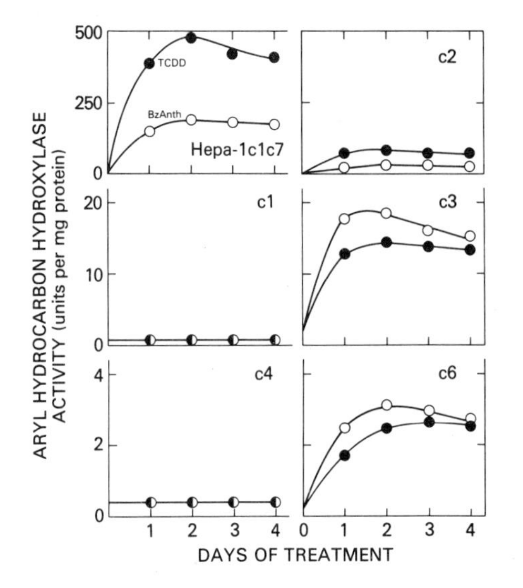

Fig. 1. Time course of AHH induction by 1.3 μM BA or 1.0 nM TCDD

Fig. 2. Characterisation of the Ah receptor in the cytosol and nucleus following 1.0 nM (^{3}H) TCDD exposure in culture.

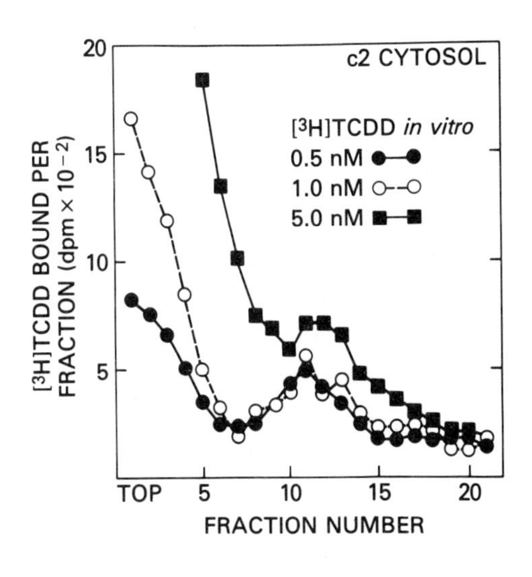

Fig. 3. Saturation of the c2 cytosolic receptor with different concentrations of (^{3}H TCDD) in vitro.

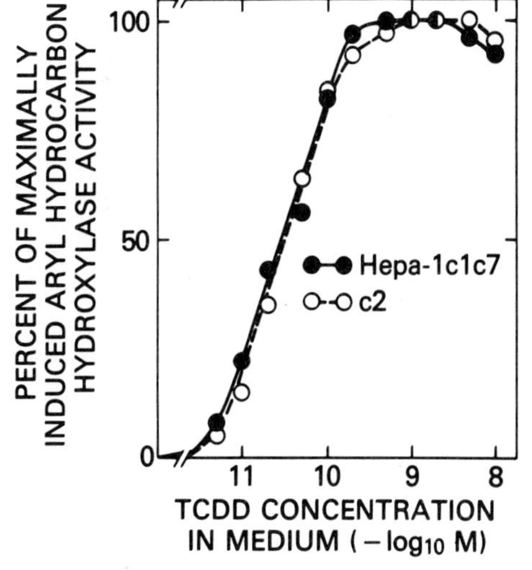

Fig. 4. Dose-response curve for AHH activity induced by TCDD.

TABLE 1

QUANTIFICATION OF THE Ah RECEPTOR AND MAXIMALLY INDUCIBLE AHH ACTIVITIES IN THE Hepa-1c1c7 PARENT LINE AND SIX CLONES

Cell line[a]	Ah receptor concentrations[b]		Maximally Inducible AHH[c]	
	Cytosol, in culture	Nuclei, in culture	B(a)A	TCDD
	(fmol/10^6 cells)		(Units/mg prot.)	
Hepa-1c1c7	11.9±1.9 (N=4)	6.0±0.4 (N=4)	210	520
c1	7.6	2.0	<0.4	<0.4
c2	1.2, 2.2	0.5	42	110
c3	18.3	3.4	22	16
c4	15.4±6.6 (N=4)	0.3±0.09 (N=3)	<0.4	<0.4
c5	6.0	3.4	<0.4	<0.4
c6	0.5	0.4	4	3

[a]Mutant clones c1 and c5 are in the same complemention group. Mutants c2 and c6 are in the same complementation group. Clones c1, c2, and c4 are known to be in different complementation groups. Clone c3 is a dominant mutant (4).

[b]Values are expressed as the means ± standard deviations. The numbers in parentheses denote the number of separate experiments performed. Following exposure of the cells to 1.0 nM [^{3}H]TCDD in the growth medium, cytosolic and nuclear extracts were prepared and, after dextran-charcoal adsorption, were examined on sucrose density gradients. The total radioactivity in gradient fractions 6 through 15 and fractions 6 through 12 (cf. Fig. 2) was equated with femtomoles of Ah receptor in the cytosol and nuclei, respectively.

[c]The maximally inducible AHH activity typically was observed after 48 h of exposure of the cultures to the inducer (Fig. 1). These values reflect at least two, and in some cases more than ten, separate induction-kinetics experiments carried out during different weeks. It was determined that µM B(a)A was no more effective than 1.3 µM B(a)A and that 10 nM TCDD was no more effective than 1.0 nM TCDD in inducing the AHH activity.

The possibility that this defect in the c2 receptor content could reflect a lack of saturation was ruled out *in vitro* and *in vivo*. At concentrations greater than .5 or 1.0 nM (^3H)TCDD *in vitro* (Fig. 3), background radioactivity increased without affecting significantly the size of the *Ah* receptor peak. The dose-response curve and ED_{50} of maximally induced AHH activity were identical for c2 and Hepa-1c1c7 (Fig. 4).

c4 has as much total *Ah* receptor as the Hepa-1c1c7 parent, but nuclear translocation is completely blocked.

Clones c1, c3, and c5 can be regarded as having essentially normal cytosolic and nuclear receptor charecteristics. Nevertheless, these clones have very low or non detectable AHH activity. The primary defect in these mutants therefore may be in the P_1-450 structural genes or other genes responsible for AHH.

Clones c2 and c6 could be regarded as r^-, or receptor-deficient with about 10 % of wild-type receptor levels, yet normal nuclear translocation kinetics, and about 10 % to 20 % of wild-type AHH inducibility.

Clone c4 could be regarded as nt^-, deficient in nuclear translocation. c4 has wild type levels of the *Ah* receptor which however fails to accumulate in the nucleus and the mutant lacks any detectable AHH.

The ease with which these benzo(a)pyrene-resistant clones of the Hepa-1c1c7 parent line were developed and the heterogenicity of the mutants thus far characterised demonstrate the potential usefulness of these established cell cultures lines. With these mutant clones we hope to elucidate important aspects of the mechanism of induction of cytochrome P_1-450 and its corresponding induction of AHH activity by PAH.

ABBREVIATIONS

AHH, aryl hydrocarbon (benzo(a)pyrene hydroxylase (EC 1.14.14.1); 3.MC, 3-methylcholanthrene; B(a)A, benzo(a)anthracene; B(a)P, benzo(a)pyrene; TCDD, 2, 3, 7, 8. -tetrachlorodibenzo-p-dioxin; (^3H)TCDD, 2, 3, 7, 8. (1,6.^3H)tetrachlorodibenzo-p-dioxin.

REFERENCES

1. Nebert, D.W., Negishi, M., Lang, M.A., Hjelmel, L.M., and Eisen, H.J. (1982) Advanc. Genet., in press.

2. Hankinson, O. (1979) Proc. Nat. Acad. Sce. U.S.A. 76, 373-376.

3. Hankinson, O. (1981) Somat. Cell Genet. 7. 373-388.

4. Hankinson, O. (1980) in Microsomes, Drug, Oxidation, and Chemical Carcinogenesis (Coon, M.J., Conney, A.H., Esrabrook, R.W., Gelborn, H.V., Gillette, J.R., and O'Brien, P.J., eds.) vol. II, pp. 1149-1152, Academic Press, New York.

5. Hannah, R.R., Nebert, D.W., and Eisen, H.J. (1981) J. Biol. Chem. 256, 4584-4590.

6. Legraverend, C., Hannah, R.R., Eisen, H.J., Owens, I.S., Nebert, D.W., and Hankinson, O. (1982) J. Biol. Chem., in press.

© 1982 Elsevier Biomedical Press B.V.
Cytochrome P-450, Biochemistry, Biophysics
and Environmental Implications, E. Hietanen,
M. Laitinen and O. Hänninen editors

PREPARATION AND PROPERTIES OF MANGANESE-SUBSTITUTED CYTOCHROME $P450_{CAM}$

MICHAEL H. GELB[1], WILLIAM A. TOSCANO, JR.[2] AND STEPHEN G. SLIGAR[1]
[1]Department of Molecular Biophysics and Biochemistry, Yale University, New Haven, Connecticut 06511 (U.S.A.) and
[2]Department of Toxicology, Harvard University School of Public Health, Boston, Massachussetts 02115 (U.S.A.)

INTRODUCTION

The homogeneous catalysis of hydrocarbon oxygenation in organic solution by artificial metalloporphyrins in the presence of appropriate oxidants have been reported. Chloro(tetraphenyl-porphinato)manganese(III) (Mn(III)TPPCl) and chloro(tetra-phenylporphinato)chromium(III) (Cr(III)TPPCl) in the presence of iodosobenzene, as a source of active oxygen, form higher valence metal-oxo intermediates that can oxidize a variety of hydrocarbons (1a,b,c). The reaction pathway for this model system is illustrated with Mn(III)TPPCl:

1. Mn(III)TPPCl + iodosobenzene $\rightarrow$ O=Mn(V)TPPCl + iodobenzene
2. O=Mn(V)TPPCl + RH $\rightarrow$ Mn(III)TPPCl + ROH

Reaction 1 is the transfer of an oxygen atom from iodosobenzene to the metal, giving a metal-oxo species in which the metal ion is oxidized by 2 electrons. Reaction 2 is the transfer of the active oxygen atom to the substrate with the regeneration of the catalyst. This last reaction occurs in two steps, involving first the 1 electron reduction of the metal-oxo species to give a substrate radical intermediate followed by oxygen transfer to give the alcohol product (1b,c). Iodosobenzene has been shown in several P450 systems to promote substrate oxygenation reactions when added to the ferric form of the enzyme (2a,b,c), suggesting that the model systems cited above may accurately mimic cytochrome P450 monoxygenase chemistry. Unfortunately, higher valence states of the heme cofactor of cytochrome P450 have not been convincingly characterized in the oxygen dependent reaction, presumably because of the rapid breakdown of such intermediates with the concomitant generation of product. Indeed, Fe(III)TPPCl and iodosobenzene have also been shown to effectively catalyze hydrocarbon oxygenation reactions, although unlike the reaction with Mn(III)TPPCl and Cr(III)TPPCl, no quasi-stable intermediate was observed (3). More recently, the higher valence states of Fe(III)TPPCl have been partially characterized at low temperature (4); however, it is not yet clear what the precise electronic structure of the active oxidants are. While there are reports of Fe(IV) and Fe(VI) compounds, the Fe(V) oxidation state has yet to be observed. The possibility of forming an Fe(V)-oxo species, analogous to Mn(V)-oxo, must be considered; however, a more likely route is to store the oxidation equivalents in the porphyrin macrocycle (5).

We have recently approached the understanding of the mechanism of cytochrome P450-dependent oxygenations by preparing and characterizing the manganese-substituted enzyme (Mn-P450). The observation of a metal-oxo intermediate and the ability of such a species to proceed through a substrate oxygenation reaction pathway would lend credence to the proposal of the ability of the Mn(III)TPPCl catalysts to model substrate oxidation by cytochrome P450.

PREPARATION AND SPECTRAL CHARACTERIZATION OF Mn-P450

Mn-P450 was prepared by the acidified-acetone method of Yu and Gunsalus (6) and treated with Mn(III)protoporphyrin IX acetate as described for hemin addition (6). The enzyme was purified as described by Wagner et al. (7). The manganese/protein ratio of the purified enzyme was measured to be 0.91 $\pm$ 0.05. The purity of the enzyme was monitored by the A^{378}/A^{280} ratio which had a final value of 1.0 $\pm$ 0.06. Remaining native P450 was quantitated by CO difference spectroscopy and constituted only 1-2 % of the total protein.

In figure 1 the optical spectra of Mn(III)-P450 and Mn(II)-P450 are compared to the free metalloporphyrin spectra. The optical spectral data for previously prepared mangansese-substituted hemoproteins along with the data for Mn-P450 is presented in Table 1. Mn(III)-P450 is unique among the manganese-substituted proteins in that no charge transfer band is seen near 470 nm; it possibly appears at higher wavelength since an additional band is seen in the visible region of the spectra (500-626 nm). The spectrum of Mn(II)-P450, obtained by dithionite reduction of Mn(III)-P450, shows a normal metalloporphyrin spectrum characteristic of Mn(II)-porphyrins (9).

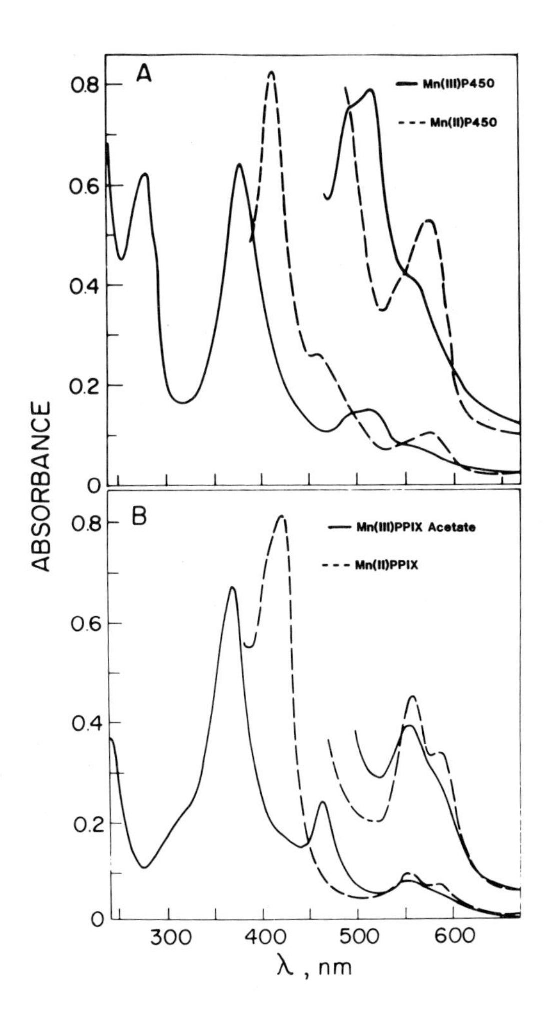

Fig. 1. Optical absorption spectrum of: A, 6 μM Mn(III)-P450 in 100 mM KP_i, pH 7.0 (solid line) and Mn(II)-P450 (dashed line); B, aquo(acetato)manganese(III)protoporphyrin IX (solid line) and the Mn(II) porphyrin (dashed line).

TABLE 1
LIGHT ABSORPTION MAXIMA OF MANGANESE PROTOPORPHYRIN IX-SUBSTITUTED HEMOPROTEINS[a]

Protein	Absorption Maxima (nm)	
	Soret	Additional Bands
Mn(III)-P450	378	496, 516, 560
Mn(III)-cytochrome c peroxidase	377	483, 568, 605
Mn(III)-horseradish peroxidase	373	482, 564, 584
Mn(III)-myoglobin	373-377	471, 556
Mn(III)-protoporphyrin IX	370	465, 556
Mn(II)-P450	412	548, 576
Mn(II)-cytochrome c peroxidase	439	563, 595
Mn(II)-horseradish peroxidase	412	509, 624
Mn(II)-myoglobin	438-440	560, 595
Mn(II)-protoporphyrin IX	420	547-556, 577-588
Mn(II)-NO-P450	447	363, 558, 588
Mn(II)-NO-cytochrome c peroxidase	427	543, 572
Mn(II)-NO-hemoglobin	433	538, 580

[a]With the exception of Mn-P450, the data come from references (8a,b,c).

Oxygenation of a solution of Mn(II)-P450 at -20°C had little effect on the spectrum, but rapid oxidation back to Mn(III)-P450 was observed on warming without observing an oxy-complex, as is the case with Mn(II)-hemoglobin and myoglobin (10).

The spectrum of the nitric oxide complex of Mn(II)-P450, obtained by adding NO to an anaerobic solution of Mn(II)-P450, is dramatically different from the spectra of NO complexes of other manganese-substituted proteins (Table 1, Figure 2). Mn(II)-NO-P450 shows a hyper-metalloporphyin spectrum with a red-shifted Soret band at 447 nm and a near-UV band at 364 nm as opposed to a single Soret band seen near 430 nm for Mn(II)-NO-cytochrome c peroxidase and -hemoglobin (Table 1). In fact, the spectrum of Mn(II)-NO-P450 is remarkably similar to its isoelectronic species, Fe(II)-CO-P450 (11). The NO complex of ferrous cytochrome P450 has been shown to display a hyper- spectrum (12) and model studies in solution have confirmed that ligation of a thiolate anion, rather than a thiol, to the heme—iron is necessary in order to mimic the observed spectral properties of nitrosyl cytochrome P450 (13). These results suggest that the thiolate is a ligand for the porphyrin-bound manganese in Mn-P450.

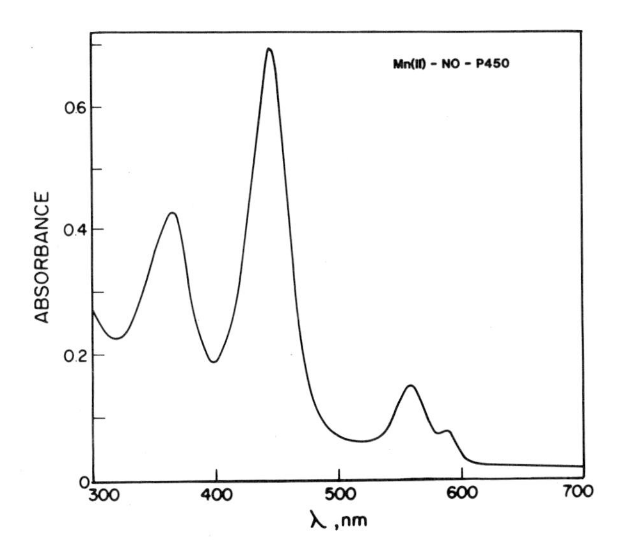

Fig. 2. Optical absorption spectrum of the nitric oxide complex of Mn(II)-P450 (6.4 μM).

CATALYTIC PROPERTIES OF Mn-P450

Mn-P450 was not effective in the promotion of NADH oxidation in the reconstituted hydroxylase system consisting of flavoprotein, putidaredoxin, NADH, and camphor. The redox potential for the Mn(III)/Mn(II) couple in Mn-P450 was measured and found to have a value of -260 mV. As expected (14), this value is considerably more negative than for Fe-P450 (-173 mV (15)). The redox potential for P450-bound putidaredoxin is -196 mV (15) and hence electron flow from reduced iron-sulfur protein to Mn(III)-P450 is not favored and could give rise to an absence of pyridine nucleotide oxidation.

Addition of iodosobenzene to Mn(III)-P450 immediately produces a new spectral species (Figure 3), showing a striking similarity to the Mn(V)TPP-oxo complex (insert to figure 3). Thus the Soret band of Mn(III)-P450 is red-shifted upon addition of iodosobenzene (378 to 410 nm) and the visible bands (450-600 nm) are broadened. It seems likely that oxygen transfer from oxidant to protein has occurred to generate a Mn(V)-oxo-P450 complex. Addition of iodosobenzene to Mn-P450 also leads to substantial porphyrin destruction as judged by the rapid loss of all porphyrin absorption bands (Figure 3).

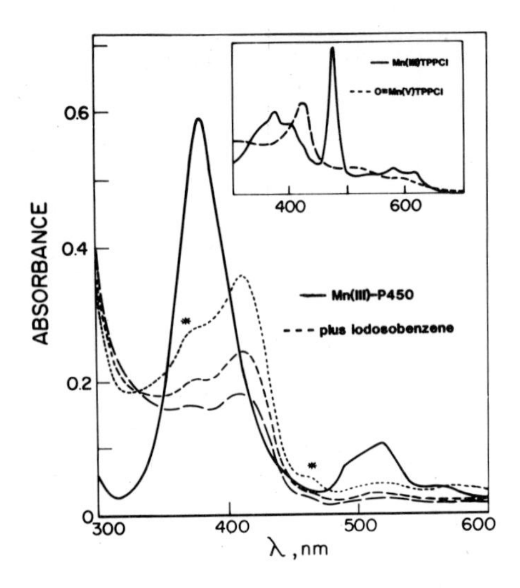

Fig. 3. Addition of iodosobenzene to a 5.7 μM solution of Mn(III)-P450 (solid line) in 100 mM KP_i, pH 7.0 to produce the Mn(V)-oxo complex (dashed line). Three spectra taken at 1 min intervals are shown to illustrate the destruction of porphyrin. Asterisked peaks refer to the small amount of free Mn-porphyrin generated during the reaction. Insert: The absorbance spectrum of Mn(III)-TPPCl (solid line) and the Mn(V)-oxo complex (dashed line) formed by iodosobenzene addition.

Oxidation of Mn(III)TPPCl with iodosobenzene also leads to irreversible porphyrin destruction (1b), although the reaction is much less rapid than with Mn(III)protoporphyrin IX due to steric hindrance to oxidant attack by the <u>meso</u>-substituted phenyl groups of TPP. Indeed, addition of iodosobenzene to free Mn(III)-protoporphyrin IX acetate leads to immediate porphyrin degradation, thus demonstrating the ability of the protein to protect the cofactor in Mn-P450 from oxidative attack. The results of turnover studies of camphor and 5,6-dehydrocamphor with Mn-P450/iodosobenzene are given in Table 2.

TABLE 2
IODOSOBENZENE-SUPPORTED OXYGENATION REACTIONS WITH CYTOCHROME $P450_{cam}$

Enzyme	Substrate	Product Yield (nmol product/nmol enzyme)
Fe(III)-P450	camphor	4.1
Fe(III)-P450	dehydrocamphor	4.3
Mn(III)-P450	camphor	less than 0.02
Mn(III)-P450	dehydrocamphor	2.3

We have previously demonstrated that native $P450_{cam}$ will catalyze the epoxidation of dehydrocamphor to give 5,6-exo-epoxycamphor (16). Although the Fe-P450/iodosobenzene system was active in both hydroxylation and epoxidation reactions, the Mn-P450/iodosobenzene system supported only the epoxidation (Table 2). The reasons for the inability of Mn-P450 to catalyze camphor hydroxylation in the presence of iodosobenzene are not yet clear. It is possible that the steps involved in oxygenation of a C-H bond are slow compared to the steps involved in olefin epoxidation; if the difference in these rates is of the same magnitude as the cofactor destruction rate, no hydroxylation will be observed. Alternatively, electronic differences in the epoxidation and hydroxylating metal-oxo species may exist. Unlike with Mn-P450, no epoxidation was observed with Mn(III)-protoporphyrin IX acetate and iodosobenzene. In this case, it is likely that the very rapid oxidation of the unprotected porphyrin prevents substrate oxygenation from taking place.

ACKNOWLEDGEMENTS

We are grateful to Drs. Heimbrook, Gunsalus, Debrunner, and Orme-Johnson for informative discussions. Supported by NIH grants GM24976, GM28897 and AM00778.

REFERENCES

1. a, Groves, J.T. and Kruper, W.J., Jr. (1979) J. Am. Chem. Soc., 101, 7613; b, Hill, C.L. and Schardt, B.C. (1980) ibid., 102, 6374; c, Groves, J.T., Kruper, W.J., Jr. and Haushalter, R.C. (1980) ibid., 102, 6375.

2. a, Lichtenberger, F., Nastainczyk, W. and Ullrich, V. (1976) Biochem. Biophys. Res. Commun. 70, 939; b, Gustafsson J-A, Rondahl, L. and Bergman, J. (1979) Biochemistry, 18, 865; c, Heimbrook, D.C. and Sligar, S.G (1981) Biochem. Biophys. Res. Commun., 99, 530.

3. Groves, J.T., Nemo, T.E. and Meyers, R.S. (1979) J. Am. Chem. Soc., 101, 1032.

4. Groves, J.T., Haushalter, R.C., Nakamura, M., Nemo, T.E. and Evans, B.J. (1981) ibid., 103, 2884.

5. Dolphin, D. (1981) Israel J. Chem., 21, 67.

6. Yu, C-A. and Gunsalus, I.C. (1974) J. Biol. Chem., 249, 107.

7. Wagner, G.C., Perez, M., Toscano, W.A., Jr. and Gunsalus, I.C. (1981) ibid., 256, 6262.

8. a, Yonetani, T. and Asakura, T. (1969) ibid., 243, 4580; b, Yonetani, T., Yamamoto, H., Erman, J.E., Leigh, J.S., Jr. and Reed, G.H. (1972) ibid., 247, 2447; c, Hoffman, B.M., Gibson, Q.H., Bull, C., Crepeau, R.H., Edelstein, S.J., Fisher, R.G. and McDonald, M.J. (1975) Ann. N.Y. Acad. Sci., 244, 174.

9. Boucher, L.J. (1968) J. Am. Chem. Soc., 90, 6640.

10. Hoffman, B.M. (1979) in: Dolphin, D. (Ed.), The Porphyrins, Academic Press, New York, vol. 7, pp. 3405-3410.

11. Hanson, L.K., Eaton, W.A., Sligar, S.G., Gunsalus, I.C., Gouterman, M. and Connell, C.R. (1976) J. Am. Chem. Soc., 98, 2672.

12. Ebel, R.E., O'Keeffe, D.H. and Peterson, J.A. (1975) FEBS Lett., 55, 198.

13, Stern, J.O. and Peisach, J. (1976) FEBS Lett., 62, 364.

14. Boucher, L.J. and Garber, H.K. (1970) Inorg. Chem., 9, 2644.

15. Sligar, S.G. and Gunsalus, I.C. (1976) Proc. Natl. Acad. Sci. U.S.A., 73, 1078.

16. Gelb, M.H., Malkonen, P.J. and Sligar, S.G. (1982) Biochem. Biophys. Res. Commun., 104, 853.

© 1982 Elsevier Biomedical Press B.V.
Cytochrome P-450, Biochemistry, Biophysics
and Environmental Implications, E. Hietanen,
M. Laitinen and O. Hänninen editors

LOCALIZATION OF FUNCTIONAL RESIDUES OF CYTOCHROME P-450 LM2

R. BERNHARDT, G. ETZOLD, H. STIEL*, W. SCHWARZE, G.-R. JÄNIG AND K.RUCKPAUL
Central Institute of Molecular Biology, Academy of Sciences, 1115 Berlin-Buch, GDR; *Institute of Optical Spectroscopy, 1199 Berlin-Adlershof, GDR

INTRODUCTION

Due to lacking amino acid sequence of P-450 LM2 and its lacking three-dimensional structure little is known about amino acid residues which are involved in recognition, interaction and electron transfer processes between cytochrome P-450 and NADPH-cytochrome P-450 reductase. An approach to describe in more detail the structure of P-450 LM2 and the molecular basis of the electron transfer between the reductase and P-450 is the application of more indirect methods such as chemical modification, photoaffinity and fluorescence labeling. We combined the advantages of fluorescence labeling with those of selective chemical modification to get more insight into the structure/function relationships of the P-450 LM2 molecule and into the nature and localization of amino acid residues.

Previous studies with fluorescein isothiocyanate (FITC) suggested that the N-terminus of P-450 LM2 may be involved in interactions between P-450 and reductase (1). Its emission optimum at about 525 nm allows to observe energy transfer from the excited FITC molecule to the prosthetic group heme due to spectral overlapping. This way the distance between the labeled group(s) and the active center of the enzyme can be determined.

MATERIAL AND METHODS

Preparation and labeling of the samples were carried out as previously described (1). Dansylation of P-450 and identification of the N-terminal amino group was performed as described (2). If the N-terminus is occupied by FITC, dansylation cannot proceed and therefore a fluorescent methionine is not formed. The efficiency of energy transfer (E_T) between FITC and heme was estimated according to the formula of Förster (3):

$$E_T = 1 - \frac{\tau}{\tau_o} = 1 - \frac{q}{q_o} \qquad (1)$$

where τ, τ_o denote the lifetimes of the fluorescence in the presence and in the absence of the acceptor; q, q_o are the quantum yields in the presence and in the absence of the acceptor, respectively. The distance R between donor and acceptor was then estimated according to:

$$R = R_o \left(\frac{1}{E_T} - 1 \right)^{\frac{1}{6}} \qquad (2)$$

where R_o is the so-called "critical Förster-distance" at which transfer efficiency is 50 %. Fluorescence decay and polarization measurements were carried out with the commercially available laser impulse spectrophotometer LIF-200 (4).

RESULTS AND DISCUSSION

Extending previous studies (1) (i) the distance between the fluorescent group (FITC) and the heme of P-450 LM2 was calculated by means of energy transfer measurements, (ii) the functional importance of labeled groups, and (iii) the localization of the labeled group with respect to the membrane surface were analyzed.

(i) Reliable results on molecular distances require the exclusion of conformational changes of FITC-labeled P-450 LM2 at least in the immediate heme environment. An almost unchanged P-450 content, maintained spectral binding constants for aniline and benzphetamine and an undisturbed CHP- or H_2O_2-dependent N-demethylase activity (1) evidence that the introduction of FITC does not significantly impair the native structur of the P-450 molecule. Estimation of the distance between the chromophore FITC and the heme as acceptor of FITC-fluorescence was carried out according to equations (1) and (2). Due to varying amounts of apoprotein different P-450 LM2 preparations were modified by FITC and the values extrapolated to 0 and 100 % content of apoprotein (Fig. 1) τ was estimated to be 1.1 ns and τ_o equal to 3.3 ns. A similar ratio q/q_o was obtained by measuring the quantum yields of the fluorescence in static fluorescence measurements. From these values and a R_o value equal to 39.8 Å the distance between the label and the heme was calculated to be 36 Å as mean distance

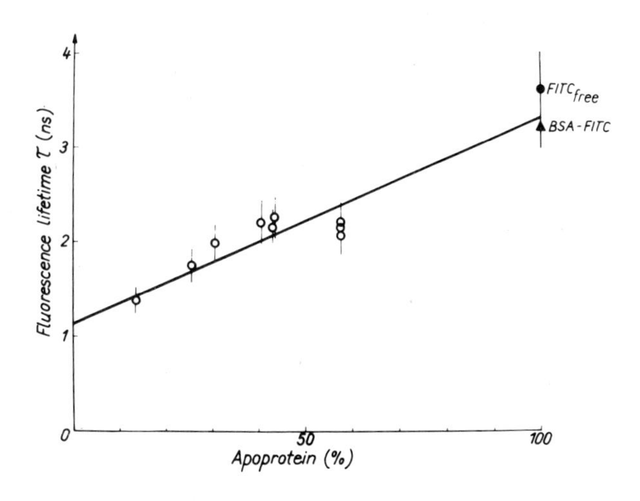

Fig. 1 Dependence of the fluorescence lifetime on the apoprotein content of P-450 LM2

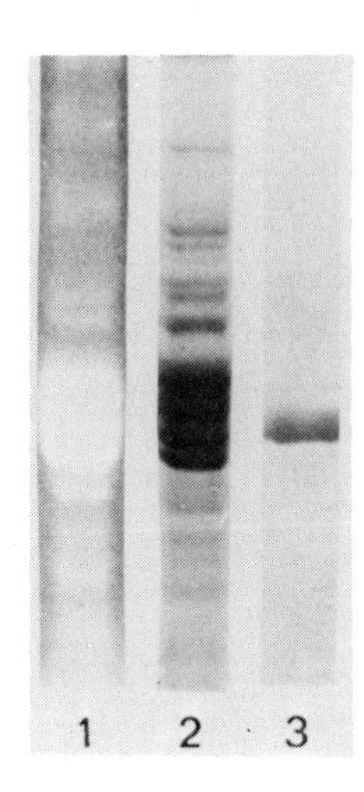

Fig. 2 Fluorescence traces (1), protein staining (2) of rabbit liver microsomes and P-450 LM2 (3) as studied by SDS-PAGE

between all FITC molecules and all hemes in the oligomeric P-450 LM2 which has to be assumed in aqueous solution (5).

(ii) Labeling of the N-terminus only or together with a further as yet not identified group of P-450 LM2 with FITC leads to a decrease in the overall activity of 25 % and 50 %, respectively, caused by disturbed electron transfer from reductase to heme. The decreased reduction rate is explicable by assuming: first, intramolecular disturbances of the electron transfer within the P-450 molecule itself. This seems to be of less importance due to the unchanged conformation of the P-450 molecule in the heme environment indicated by the parameters mentioned before. Secondly, interactions between both proteins may be disturbed either by masking of groups directly involved in the interaction or by steric hindrance of the site of contact due to the rather bulky FITC molecule. Electrostatic effects may play an important role as FITC introduces two negative charges into the protein molecule (6). Possibly similar charge pair interactions between P-450 and reductase occur as in the cytochrome b_5 system (7).

(iii) The electrostatic interactions postulated require the localization of the labeled group(s) at the cytoplasmic side of the endoplasmic reticulum. Incubation of microsomal P-450 with FITC (Fig. 2) and SITS (4-acetamino-4'-isothiocyanostilbene-2,2'-disulfonic acid), a well known surface label, elution of P-450 LM2 after SDS-PAGE from the gel and determination of the N-terminus revealed a modified α-amino group which is not further accessible for dansylation[1].

Summarizing it can be stated that the fluorescence label FITC under special conditions (pH 7.45, 35-fold molar excess over protein) can be attached specifically to the N-terminus of P-450 LM2 allowing the determination of the distance between the N-terminal bound chromophore and heme in the active center. In addition, functional analysis after introduction of the FITC molecules into P-450 LM2 revealed that the N-terminus and the up to now not identified further residue(s) may be directly involved in the interaction between P-450 and reductase by electrostatic forces or are at least localized near the binding site for reductase. A localization of the α-amino group at the cytoplasmic side of the endoplasmic reticulum as necessary prerequisite for such interactions has been demonstrated.

REFERENCES

1. Bernhard,R., Ngoc Dao, N.T., Friedrich, J., Ristau, O., Jänig, G.-R. and Ruckpaul, K. in: Biochemistry, Biophysics and Regulation of Cytochrome P-450 (J.A. Gustafsson et al. eds.) Elsevier/North-Holland Biomedical Press, 1980, 595-598
2. Etzold, G., Büttner, D. and Welfle, H. (1980) Acta biol. med. germ. 39, 1143-1146.
3. Förster,. T. (1959) Discuss. Faraday Soc. 27, 7-17
4. Scholz, M., Teuchner, K., Näther, M., Becker, W. and Dähne,S. (1978) Acta Physica Polonica A 54, 823-831.
5. Behlke, J., Jänig, G.-R. and Pfeil, D. (1979) Acta biol. med. germ. 38, 389-397.
6. Kalousek, I., Jandova, D. and Vodrazka, Z. (1980) Int. J. Biol. Macromol. 2, 284-288.
7. Dailey, H.A. and Strittmatter, P. (1979) J. Biol. Chem. 254, 5388-5396.

[1] The authors thank Mrs. D. Büttner for the determination of the N-terminal amino acid.

© 1982 Elsevier Biomedical Press B.V.
Cytochrome P-450, Biochemistry, Biophysics
and Environmental Implications, E. Hietanen,
M. Laitinen and O. Hänninen editors

SPIN SHIFT, REDUCTION RATE AND N-DEMETHYLATION CORRELATION IN CYTOCHROME P-450 LM USING A SERIES OF BENZPHETAMINE ANALOGUES

H. REIN, J. BLANCK, M. SOMMER[1], O. RISTAU, K. RUCKPAUL AND W. SCHELER
Central Institute of Molecular Biology, Academy of Sciences of the GDR, 1115 Berlin-Buch, GDR; [1]Institute of Pharmacol. and Toxicol., Friedrich-Schiller-University, 6900 Jena, GDR

INTRODUCTION

P-450 catalyzed reactions proceed in a sequential reaction cycle initiated by the binding of the substrate. Binding of type I substrates shifts the spin equilibrium to the high spin state thus favouring the reduction of the heme iron, i.e. the second reaction step. Previously we have shown that the substrate induced high spin shift correlates with the rate constant of the reduction (1). In that analysis substrates of different chemical structure were used which by P-450 were differently converted (hydroxylation, N-demethylation) thus restricting a comparative analysis of the relations between single reaction steps with the overall reaction. Consequently in the present paper the effect of a series of benzphetamine analogues on the spin equilibrium and the reduction rate of the heme iron was analyzed and the data obtained compared with the uniform N-demethylation of these substrates.

MATERIAL AND METHODS

For the studies liver microsomes from phenobarbital induced male rats were used. The high spin shift was calculated from ΔA_{max} values obtained from difference spectra of the enzyme substrate complexes using a difference extinction coefficient of $\Delta\varepsilon = 55.0\ M^{-1}cm^{-1}$ for the low spin band (2). The reduction kinetics were measured anaerobically in the presence of CO by use of a stopped flow spectrophotometer and the kinetic curves were analyzed by a multiexponential nonlinear estimation procedure. The N-demethylase activity was determined with slight modifications according to Werringloer. K_s, V_{max} and K_m values were derived from Lineweaver-Burk plots.

Tab. 1

Substrates	1 ΔA_{max} (417-387 nm) $[mM^{-1} \cdot cm^{-1}]$	2 % high spin shift	3 % high spin content 42%*, 20°C	4 K_s $[\mu mol/l]$	5 k_1 $10^{-1}[s^{-1}]$ 3.4 ± 0.7**	6 K_m $[\mu M]$	7 V_{max} nmol HCHO/ nmol P-450·min
① $C_6H_5-CH_2-CH_2-N(CH_3)-CH_2-CH_2-C_6H_5$	19.0 ± 0.9	17.3	59.3	35.4 ± 5.7	5.9 ± 1.0	101.1 ± 3.0	3.24 ± 0.1
② $Cl-C_6H_4-CH_2-N(CH_3)-CH_2-C_6H_4-Cl$	20.0 ± 0.5	18.2	60.2	16.7 ± 1.6	4.4 ± 0.3	88.3 ± 5.2	3.76 ± 0.4
③ $C_6H_5-CH_2-CH_2-N(CH_3)-CH_2-C_6H_5$	26.2 ± 0.6	23.8	65.8	44.0 ± 3.5	5.8 ± 0.4	163.9 ± 10.6	6.04 ± 0.5
④ $C_6H_5-CH_2-CH(CH_3)-N(CH_3)-CH_2-C_6H_4-Cl$	32.0 ± 0.8	29.1	71.1	33.9 ± 2.6	5.7 ± 0.7	140.2 ± 4.9	6.64 ± 0.2
⑤ $C_6H_5-CH_2-CH(CH_3)-N(CH_3)-CH_2-C_6H_5$	32.5 ± 0.6	29.5	71.5	34.2 ± 2.1	6.0 ± 0.6	172.4 ± 10.1	7.04 ± 0.4
⑥ $C_6H_5-CH_2-C(CH_3)_2-N(CH_3)-CH_2-C_6H_5$	37.6 ± 0.7	34.4	76.4	22.9 ± 1.5	6.0 ± 0.6	139.9 ± 4.1	7.16 ± 0.2
⑦ $(Cl)C_6H_4-CH_2-N(CH_3)-CH_2-C_6H_4(Cl)$	39.6 ± 0.3	36.0	78.0	7.2 ± 2.5	7.1 ± 0.7	214.4 ± 16.0	10.2 ± 1.0
⑧ $C_6H_5-CH(CH_3)-N(CH_3)-CH(CH_3)-C_6H_5$	45.0 ± 0.6	40.9	82.9	11.2 ± 1.6	7.4 ± 0.6	210.6 ± 12.0	13.5 ± 1.1

*intrinsic high spin content of P-450 and ** reduction rate constant, both in the absence of a substrate.

RESULTS

The series of tertiary amines which are related to benzphetamine induce type I difference spectra with significantly different ΔA_{max} values. In Table 1 (column 1) the series of substrates 1-8 is listed in the order of increasing ΔA_{max} values. The corresponding spin shifts are contained in column 2. The total high spin content of microsomal P-450 in the presence of the different substrates has been calculated on the basis of a high spin content of 42 % of the microsomal P-450 (using $\varepsilon_{645\ nm}$ = 2.14 $mM^{-1}cm^{-1}$ derived from the high spin form of P-450 CAM) (column 3). In column 4 the calculated apparent spectral binding constants K_s are shown which do not correlate with the ΔA_{max} values. Column 5 exhibits the physiologically relevant fast rate constants k_1 of the anaerobic NADPH reduction of P-450 at 20 °C in the presence of the substrates used. The rate constants differ by about 1-2 orders of magnitude from the two slow phase constants k_2 and k_3. The fast phase was found to amount up to about 60 % of total reduction independent of the different substrates. The order of increasing spin shift is followed by the reduction rate constants which increase from 4.4-7.4 $10^{-1}s^{-1}$. In column 6 and 7 K_m and V_{max} of the N-demethylation of the substrates measured at 37 °C are shown. Again from compound 1 to 8 the capability to be N-demethylated is increased (see increasing V_{max} values from 1 to 8). The order of increase indicates that the rate of N-demethylation is correlated with the spin state and the reduction rate as well. A calculation of the validity of correlation revealed values of r = 0.81 and r = 0.94 for the correlationship between the spin shift and the reduction rate and the N-demethylation, respectively.

DISCUSSION

Extending our previous findings that the rate of the P-450 reduction is correlated with the substrate induced high spin shift (1), the present studies on a series of benzphetamine analogues revealed that likewise the overall reaction correlates with the spin equilibrium shift. The simultaneous existence of the high spin and low spin conformation in P-450 means that the measured reduction rate k is a function of the individual rate constants of both conformations.

$$\bar{k} = \alpha k_{high\ spin} + (1-\alpha) k_{low\ spin} \quad 1$$

The validity of the equation is based on the rapid equilibrium of both conformations ($\tau < 10^{-4}$ s) (2). The high spin conformation is favourable reduced due to its more positive redox potential, therefore the reduction rate $\bar{k}$ is determined by the high spin fraction (α). The observed linear increase of the reduction rate with increasing high spin fraction evidences a rapid pre-equilibrium control of the reduction according to equation 1.

With respect to the intrinsic high spin content of 42 % due to the microsomal isozyme pattern the range of substrate regulation (spin shift) thus accounts for maximal 58 %. Interestingly the reduction rate ($3.4\ 10^{-1} s^{-1}$) of the intrinsic high spin content of 42 % fits to the observed correlation. If so, the reduction kinetics of the low spin conformation approaches almost zero.

The correlation between spin shift and substrate conversion implies that the overall reaction may be regulated by the rapid spin equilibrium via the reduction reaction. The reduction rate, however, exceeds significantly the substrate turnover, thus the first reduction cannot be rate limiting but a subsequent step. Considering this result it can be derived that the substrates besides their capability to shift the spin equilibrium to the high spin state induce substrate specific conformations of P-450 which are reflected in the different ΔA_{max} values. The substrate specific conformations should differ in the redox potential and this way regulate the conversion rate. That implies the second electron transfer to be rate limiting. This conclusion is impaired by the isozyme pattern of microsomes. Therefore studies on homogeneous reconstituted systems are under way.

REFERENCES

1. Rein, H., Ristau, O., Misselwitz, R., Buder, E. and Ruckpaul, K. (1979) Acta Biol. Med. Germ. 38, 187-200.
2. Ristau, O., Rein, H., Greschner, S., Jänig, G.-R. and Ruckpaul, K. (1979) Acta Biol. Med. Germ. 38, 177-186

© 1982 Elsevier Biomedical Press B.V.
Cytochrome P-450, Biochemistry, Biophysics and Environmental Implications, E. Hietanen, M. Laitinen and O. Hänninen editors

THE ACTIVE SITE STRUCTURE OF CYTOCHROME P-450 AS DETERMINED BY EXTENDED X-RAY ABSORPTION FINE STRUCTURE SPECTROSCOPY

JOHN H. DAWSON*, LAURA A. ANDERSSON*, KEITH O. HODGSON† AND JAMES E. HAHN†
*Department of Chemistry, University of South Carolina, Columbia, South Carolina, 29208 and †Department of Chemistry, Stanford University, Stanford, California, 94305

A thorough understanding of the mechanism of action of cytochrome P-450 necessarily requires a detailed foundation of structural information. Extended x-ray absorption fine structure (EXAFS) spectroscopy is an experimental technique that has the capability to provide such information, in particular metal-ligand bond distances and ligand identities. We report herein the application of EXAFS spectroscopy to substrate-free ferric and to substrate-bound ferric, ferrous and ferrous-CO P-450-CAM. Direct evidence for a sulfur donor atom as the fifth ligand to the heme iron in all of these states has been obtained. In each case, the iron-sulfur bond distance is equal to, or shorter than, analogous Fe-S bonds in model iron porphyrin thiolate complexes of known structure. As known thiol-sulfur: iron heme bond distances are noticeably longer, the observed sulfur donor atom is assigned to be a cysteinate. The iron-nitrogen (porphyrin) distances indicate five-coordination in substrate-bound ferric and ferrous P-450, and six-coordination in substrate-free ferric and in ferrous-CO P-450. The ligand identity, state of ligand protonation and metal-ligand bond distances reported in this paper establish important fundamental molecular details about the active site iron coordination environment of P-450 that are critical to a complete understanding of the P-450 reaction cycle and mechanism of oxygen activation.

INTRODUCTION

Cytochrome P-450 is a heme iron mono-oxygenase that is noted both for its unique spectral properties, chiefly the 450 nm UV-visible absorption maximum of its ferrous-CO complex, and for its capability to activate molecular oxygen for insertion into organic substrates (1). Despite extensive study, the detailed metal coordination structure of the active site of P-450 remains a matter of controversy (2). We have undertaken a study of the various ferric and ferrous states of P-450 utilizing extended x-ray absorption fine structure (EXAFS)[1] spectroscopy in order to provide the structural foundation required for an understanding of the mechanism of oxygen activation by P-450.

The identity of the non-porphyrin, axial ligand(s) to the heme iron of P-450 is a matter of considerable importance. As shown in Figure 1, four isolable states (1-4) have been characterized in the P-450 reaction cycle. The low-spin, six-coordinate ferric form (1) becomes high-spin, five-coordinate (2) upon substrate binding. Spectral investigations of P-450 with UV-visible absorption,

---

[1]The abbreviations used are: EXAFS, extended x-ray absorption fine structure; TPP, dianion of meso-tetraphenylporphyrin; PPIXDME, dianion of protoporphyrin (IX)dimethyl ester.

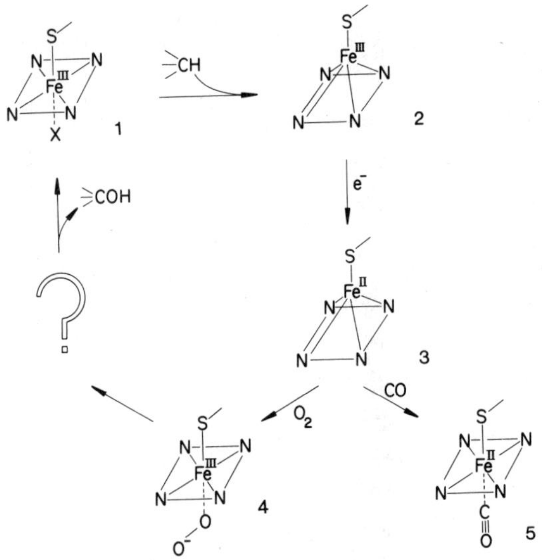

Figure 1. CATALYTIC CYCLE OF P-450. The postulated structures of the iron site for the isolable intermediates in the P-450 catalytic cycle.

magnetic circular dichroism and EPR spectroscopy and comparisons to synthetic model porphyrin systems and/or myoglobin ligand adducts have provided convincing evidence of thiolate ligation in both states 1 and 2 (3-6). Preliminary examination of mammalian P-450 with EXAFS demonstrated that a sulfur donor is one of the axial ligands to P-450 state 1, with an iron-sulfur bond distance of ca. 2.2 Å (7,8). Recent evidence suggests that the ligand trans to thiolate in P-450 state 1 is an oxygen donor, such as an endogeneous alcohol-containing amino acid or water (6,9,10). Comparison of the spectroscopic properties of ferrous-CO P-450 (5) to those of model porphyrin complexes again suggests that a thiolate ligand is trans to carbon monoxide (5,11,12). However, the ligand assignments for ferrous, 3, and oxygenated, 4, P-450 are less clear. Thus, by examination of the EXAFS of P-450-CAM, we hoped to directly establish the ligand assignments for states 1, 2, and 5, as well as to determine the ligand in state 3. EXAFS spectroscopy is a technique of particular utility for the study of metalloproteins (13) since it can determine the coordination sphere of the metal including the number, identity and distance of the ligands surrounding the x-ray absorbing species. In favorable circumstances, EXAFS can determine absorber-ligand distances to an accuracy of ± 0.02 Å (13). Because EXAFS is the only technique other than x-ray diffraction that is capable of determining bond distances, it has great advantages over other spectroscopic methods. Thus, EXAFS examination can provide the structural data required as a basis for understanding the mechanism of oxygen activation by P-450.

MATERIALS AND METHODS

Cytochrome P-450-CAM was purified from *Ps. Putida* grown on d-camphor as reported elsewhere (9)[2]. A typical preparation of the camphor-bound form ($A_{391}/A_{280} \cong 1.55$, electrophoretically homogeneous) was concentrated to 3-4 mM in 20 mM potassium phosphate, 5-8 mM d-camphor, 100 mM KCl, pH 7.4. Camphor was removed by gel filtration (14). All samples were examined in the ferrous-CO form by UV-visible absorption spectroscopy; negligible contamination by inactive cytochrome P-420 (1) was seen before or after exposure to x-rays. EXAFS spectra were obtained at 2°C with sealed cells in a helium flushed sample box. Samples were checked by UV-visible absorption spectroscopy directly after EXAFS experiments using short pathlength cells and/or anaerobically diluted protein. In no case was there any loss of spectral integrity. Given the sensitivity of P-450 to conversion to P-420 and/or auto-oxidation when reduced, the above results indicate that each protein sample was in the desired form throughout the period of spectral examination. P-450 EXAFS data were collected at the Stanford Synchrotron Radiation Laboratory using radiation monochromatized with a double Si[III] crystal monochromator. Energy calibration was based on the first inflection point (7111.2 eV) of an iron foil spectrum. Data were obtained as fluorescence excitation spectra; 12-24 scans using five NaI detectors were averaged and then analyzed as described elsewhere (7,8,15). The absorption is corrected for the background, normalized to a per-iron basis and converted to $\underline{k}$ space[3] using an E_0 of 7130 eV, the threshold energy for liberating a core electron. Extensive experience with iron-porphyrin EXAFS has shown that the assumption of a constant E_0 does not introduce a detectable inaccuracy in the determination of the absorber-scatter distance, R, as long as the same E_0 value is used for all compounds (7,13,15).

RESULTS AND DISCUSSION

Figure 2 is the EXAFS data for the four P-450 samples, plotted as a function of $\underline{k}$. Initial examination reveals a complex pattern, with "beats" in the

---

[2]Complete details of bacterial growth and protein purification procedures will be reported: L.A. Andersson, PH.D. thesis, to be submitted to the University of South Carolina.

[3]The photoelectron wavevector $\underline{k}$ is defined by $\underline{k} = \sqrt{2\, me\, (E-E_0) / \hbar}$.

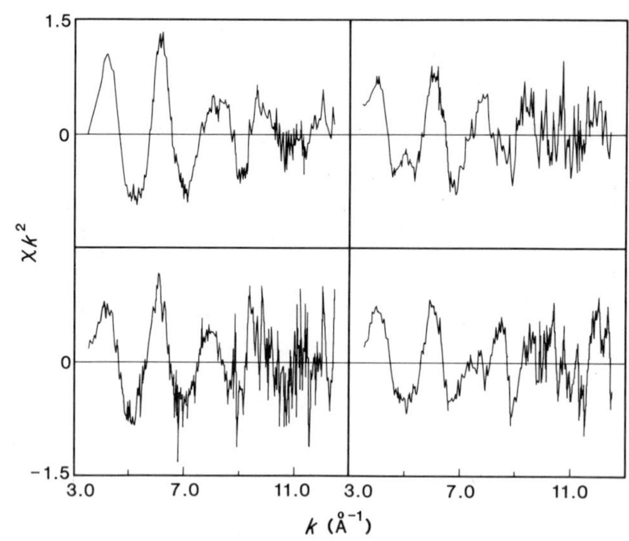

Figure 2. EXAFS SPECTRA OF P-450-CAM. The EXAFS spectra (weighted by k^2) for low-spin ferric (1) (upper left), high-spin ferric (2) (lower left), high-spin ferrous (3) (upper right) and ferrous-CO (5) (lower right) P-450-CAM.

amplitude; the beat pattern reveals that there are several different shells of atoms contributing to the Fe EXAFS. The Fourier-transform (not shown) of the EXAFS data yields the phase-shifted radial distribution function around the absorbing atom (13). Transforms alone cannot allow determination of the nature of axial ligands unless all data are well-resolved. Curve fitting is the most sensitive technique for analysis of EXAFS data, utilizing the measured amplitude and phase of the EXAFS of model compounds of known structure. This procedure is based on an assumption of chemical transferrability. Extensive testing with model porphyrins of known structure indicates that by use of empirical curve fitting procedures, an accuracy of ca. ± 0.02 Å in distance determination and 1 atom in 3-4 in coordination number determination is obtained (7,13,15) with accuracies comparable to the reproducibility of the data. Fe(III)(TPP)(imidazole)$_2$Cl, Fe (II)(TPP), Fe(diethyldithiocarbamate)$_3$, Fe-(acetylacetonate)$_3$, and HFe(CO)$_4$[bis(triphenylphosphine)imminium cation] were used as models for Fe-N, Fe-C (alpha and meso), Fe-S, Fe-O, and Fe-C(CO) and Fe-O(CO) parameters, respectively (15). EXAFS is insensitive to small changes in the atomic number of the scattering atom. Thus C,N,O, or F first shell donor atoms would be indistinguishable and could all be modeled using our "N" parameters. So, a first shell N could actually be C,N,O, or F, although only

O or N are biologically reasonable. Similarly, only S (of Si, P, S, and Cl) is a biologically reasonable donor atom. A measure of the accuracy and precision of EXAFS results obtained using these parameters can be seen in Table I of reference 7 and Table I of reference 15. The curve fitting results for low-spin ferric P-450 are shown in Figure 3; the results for all states of P-450 examined are collected in Table I.

Curve fitting of the P-450 EXAFS data clearly indicates the presence of a sulfur donor atom in all states of P-450 examined. *In all cases, an acceptable fit could only be obtained by including a sulfur atom.* Interatomic distances and coordination numbers for P-450 are reported in Table I; structural interpretations of the P-450 bond distances and numbers follow.

TABLE I

Structural details for P-450-CAM and relevant porphyrin model compounds[a]

	Fe-N(porphyrin)		Fe(axial)		
	R(Å)	N[b]	R(Å)	N[b]	Reference
low-spin Fe(III)					
P-450 state 1	2.00	5.0	2.22	0.6	this work
1(mammalian P-450)	2.00	4.8	2.19	0.8	7
$Fe(TPP)(SC_6H_5)_2$[c]	2.008	4	2.336	1	25
$Fe(TPP)(HSC_6H_5)(SC_6H_5)$[c]	not reported		2.27[d]	1	17
high-spin Fe(III)					
P-450 state 2	2.06	5.2	2.23	0.8	this work
$Fe(PPIXDME)(SC_6H_4NO_2)$[c]	2.064	4	2.324	1	4
high-spin Fe(II)					
P-450 state 3	2.08	3.0	2.34-2.38	0.6	this work
$Fe(TPP)SC_2H_5)$[c]	2.096	4	2.360	1	18
low-spin Fe(II)					
P-450 state 5	1.98	3.3	2.32	1.0	this work
$Fe(TPP)(SC_2H_5)(CO)$[c]	1.993	4	2.352	1	18

a. All parameters are from EXAFS measurements except where noted and were obtained from curve fitting.
b. The number (N) of ligands at the distance indicated.
c. Data from crystal structure determination.
d. The Fe-S(thiolate) bond distance; Fe-S(thiol) distance equals 2.43 Å.

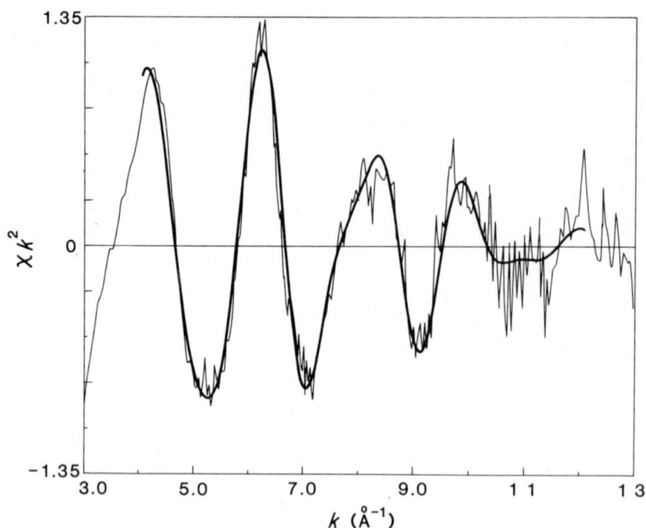

Figure 3. CURVE FITTING RESULTS FOR LOW-SPIN FERRIC P-450-CAM. The least-squares fit (dark line) to the EXAFS data (light line). Three waves [N,S,C(alpha,meso)] were used. Fitting was over a range of $\underline{k}$=4-12 Å^{-1}. The best fits to the other P-450 data are qualitatively similar. Numerical results of all P-450 curve fitting are summarized in Table I.

Although relatively few compounds that directly model the P-450 active site have been structurally characterized, the structural properties of iron porphyrins in general are very well known (16). Low-spin six-coordinate ferric porphyrins have Fe-N(porphyrin) distances of about 1.99 Å, while the high-spin five-coordinate ferric Fe-N(porphyrin) distance is approximately 2.065 Å. Ferrous porphyrins have Fe-N(porphyrin) distances that are about 0.02 Å longer. Thus, the data presented for P-450 in Table I are indicative of six-coordination for P-450 states $\underline{1}$ and $\underline{5}$ and five-coordination for states $\underline{2}$ and $\underline{3}$. Comparison of the P-450 Fe-S distances for states $\underline{1}$,$\underline{2}$,$\underline{3}$ and $\underline{5}$ to various model porphyrin complexes (Table I) is strongly suggestive of thiolate ligation to P-450 in all of the states examined, since a coordinated thiol would undoubtedly have longer bond lengths. For example, Fe(III)TPP(HSC_6H_5)(SC_6H_5) has an Fe-S(thiol) distance of 2.43 Å, and an Fe-S(thiolate) distance of 2.27 Å (17). While the data are compelling for states $\underline{1}$,$\underline{2}$ and $\underline{5}$, the somewhat lower quality of the EXAFS data for P-450 state $\underline{3}$ introduces some uncertainty. However, the Fe-S distance determined for $\underline{3}$ is closely similar to that reported for the

$Fe(II)TPP(SC_6H_5)$ anion (18), thus the EXAFS data are most consistent with thiolate ligation in this state. The presence of thiolate ligation to the heme iron of P-450, directly demonstrated in this and previous (7,8,15) work, has significance for the catalytic activity of the enzyme. It has been suggested that thiolate ligation results in an increased electron density at the heme iron of P-450, as compared to the iron of the histidine-ligated oxygen transport proteins of hemoglobin and myoglobin (3,19,20). Since P-450 not only binds but also activates dioxygen, the electronic structure at the active site is a critical feature of the catalytic process. Transient, high-valent metal-oxo (M=O) species have commonly been suggested to be the key intermediates in P-450 catalysis (3,21-24). An electron-rich thiolate ligand might serve to stabilize the normally unfavorable oxidation state of iron in such species and might also aid in cleaving the O-O bond enroute to formation of the metal-oxo stage.

CONCLUSION

Cytochrome P-450-CAM has been examined in its ferric low-spin (1) and high-spin (2), ferrous high-spin (3) and ferrous-CO (5) states by extended x-ray absorption fine structure spectroscopy. Curve fitting techniques used in EXAFS data analysis are capable of accuracy of approximately ± 0.02 Å and ± 25% in coordination number with Fe-porphyrin systems. *Such a procedure conclusively demonstrates the presence of a sulfur ligand to the heme iron of P-450 states 1, 2, 3 and 5.* Fe-N(porphyrin) bond distances indicate six-coordination for states 1 and 5, and five-coordination for states 2 and 3. Further, analysis of Fe-S bond lengths clearly indicates thiolate rather than thiol sulfur donor ligation for 1, 2 and 5, and strongly suggests it for 3. The structural results presented herein provide the basic foundation for future model and mechanistic studies of P-450.

ACKNOWLEDGEMENTS

We thank Ed Phares, Mary V. Long, Susan E. Gadecki, Ian M. Davis, and Dr. Grant W. Cyboron for their assistance in bacterial growth and/or protein purification and analysis. Dr. Robert A. Scott, Steven D. Conradson, Man Sung Co and Jon Sessler assisted with data collection and analysis. This research was supported by grants from the National Science Foundation (PCM 79-11518) to JHD and (PCM 79-04915) to KOH. Synchrotron radiation beam time was provided by the Stanford Synchrotron Laboratory which is supported by the National Science Foundation and the National Institutes of Health (RR 101749) in cooperation with the Stanford Linear Accelerator Center and the Department of Energy. JHD acknowledges initial support by the American Cancer Society. JEH is an NSF predoctoral fellow.

REFERENCES

1. Sato, R., and Omura, T. eds. (1978) "Cytochrome-P450", Academic Press, N.Y.
2. Peterson, J.A. (1979) in "Oxygen: Biochemical and Clinical Aspects", (Caughey, W.S., ed.) pp 227-262, Academic Press, N.Y.
3. Dawson, J.H., Holm, R.H., Trudell, J.R., Barth, G., Linder, R.E., Bunnenberg, E., Djerassi, C., and Tang, S.C. (1976) J. Amer. Chem. Soc. 98, 3707-3709.
4. Tang, S.C., Koch, S., Papaefthymiou, G.C., Foner, S., Frankel, R.B., Ibers, J.A., and Holm, R.H. (1976) J. Amer. Chem. Soc. 98, 2414-2434.
5. Collman, J.P., and Sorrell, T.N. (1979) in "Drug Metabolism Concepts" (Jerina, D.M., ed.) American Chemical Society Symposium Series No. 44, pp. 27-45, American Chemical Society, Washington, D.C.
6. Ruf, H.H., Wende, P., and Ullrich, V. (1979) J. Inorg. Biochem. 11, 189-204.
7. Cramer, S.P., Dawson, J.H., Hager, L.P., and Hodgson, K.O. (1978) J. Amer. Chem. Soc. 100, 7282-7290.
8. Dawson, J.H., Andersson, L.A., Davis, I.M., and Hahn, J.E. (1980) in "Biochemistry, Biophysics and Regulation of Cytochrome P-450" (Gustaffson, J.A., Carlstedt-Duke, J., Mode, A., and Rafter, J., eds.) pp. 565-572, Elsevier, Amsterdam.
9. Dawson, J.H., Andersson, L.A., and Sono, M. (1982) J. Biol. Chem. 257, 3606-3617.
10. White, R.E., and Coon, M.J. (1982) J. Biol Chem. 257, 3073-3083.
11. Chang, C.K., and Dolphin, D. (1975) J. Amer. Chem. Soc. 97, 5948-5950.
12. Collman, J.P., Sorrell, T.N., Dawson, J.H., Trudell, J.R., Bunnenberg, E., and Djerassi, C. (1975) Proc. Natl. Acad. Sci. U.S.A. 73, 6-10.
13. Cramer, S.P., and Hodgson, K.O. (1979) Progress in Inorganic Chemistry 25, 1-39.
14. O'Keeffe, D.H., Ebel, R.E., and Peterson, J.A. (1978) Methods in Enzymol. 52, 151-157.
15. Hahn, J.E., Hodgson, K.O., Andersson, L.A., and Dawson, J.H. (1982) J. Biol. Chem. 257, in press.
16. Scheidt, W.R., and Reed, C.A. (1981) Chemical Reviews, 81, 543-555.
17. Collman, J.P., Sorrell, T.N., Hodgson, K.O., Kulshrestha, A.K., and Strouse, C.E. (1977) J. Amer. Chem. Soc. 99, 5180-5181.
18. Caron, C., Mitschler, A., Riviere, G., Richard, L., Schoppacher, M., and Weiss, R. (1979) J. Amer. Chem. Soc. 101, 7401-7402.
19. Sono, M., and Dawson, J.H. (1982) J. Biol. Chem. 257, 5496-5502.
20. Sono, M., Andersson, L.A., and Dawson, J.H. (1982) J. Biol. Chem. 257, in press.
21. Ullrich, V. (1979) Topics in Current Chemistry 83, 67-103.
22. White, R.E., and Coon, M.J. (1980) Ann. Rev. Biochem. 49, 315-356.
23. Gunsalus, I.C., and Sligar, S.G. (1979) Adv. Enzymol. 47, 1-44.
24. Groves, J.T. (1979) Adv. Inorg. Biochem. 1, pp. 119-145.
25. Byrn, M.P., and Strouse, C.E. (1981) J. Amer. Chem. Soc. 103, 2633-2635.

Cytochrome P-450, Biochemistry, Biophysics and Environmental Implications, E. Hietanen, M. Laitinen and O. Hänninen editors

MICROSOMAL ELECTRON TRANSPORT. THE RATE OF THE ELECTRON TRANSFER TO OXYGENATED CYTOCHROME P450

HANS H. RUF AND V. EICHINGER
Physiological Chemistry, University of the Saarland,
D-6650 Homburg-Saar (Federal Republic of Germany)

INTRODUCTION

Monooxygenation by cytochrome P450 requires the sequential transfer of two electrons. For a kinetic description of the reaction cycle, it is essential to measure the rate of these electron transfers. The "first" electron, for the reduction of ferric cytochrome P450, can be measured straightforward under anaerobic conditions with CO. The fastest phase of the obtained heterogeneous kinetics is due to the reduction of the enzymatically active substrate-bound cytochrome P450 (1) with a pseudo-first order rate constant of about 2 s^{-1}. Since this rate is faster than the turnover of benzphetamine N-demethylation of less than 1 s^{-1}, a second rate-limiting step is presumed, namely the transfer of the "second" electron for the reduction of the oxygenated cytochrome P450.

The rate of the second electron, however, cannot be measured in microsomes directly since appropriate starting conditions for such a kinetic experiment cannot be achieved. The instability of oxygenated cytochrome P450 as well as a pool of electrons in the other electron transfer proteins in microsomes make this experiment infeasible.

With isolated cytochromes P450 and b_5 incorporated in phospholipid vesicles electron transfer from cytochrome b_5 to oxygenated cytochrome P450 has been demonstrated (2). In the more native system, in microsomes, the rate of the second electron has been estimated from the absorption at 440 nm due to oxygenated cytochrome P450 during the hydroxylating steady state (3, 4). In contrast to kinetic studies, steady state measurements cannot resolve heterogeneous reactions of the multiple forms of cytochrome P450.

We could resolve the pre-steady state kinetics of the formation of oxygenated cytochrome P450 after starting the hydroxylation of benzphetamine. From these data the rate of the second electron was calculated to be about equal to the rate of the first electron. Using this method, we could demonstrate electron transfer from additionally incorporated cytochrome b_5 to oxygenated cytochrome P450.

MATERIALS AND METHODS

Microsomes were prepared from phenobarbital-pretreated rats by differential centrifugation. The rate of the first electron was measured under anaerobic conditions with CO as described (1). The reactions were started by mixing equal volumes of microsomes and pyridine nucleotides in a modified Dionex (Durrum) D-103 stopped flow apparatus with an optical pathlength of 2 cm. The reactions were followed by a dual-wavelength photometer (to be described elsewhere) which was connected to a computer for data storage and analysis. The kinetic traces were analyzed by a multiexponential fit program (5) under the assumption of simultaneous pseudo-first order reactions which was reasonable since the reductases and even cytochrome b_5 was reduced more rapidly than cytochrome P450.

We used the wavelengths (sample/reference) 450/470 nm for measuring the first electron (ferrous P450.CO), 442/470 nm for oxygenated P450 (there is no interference with cytochrome b_5) and 424/409 nm for reduced cytochrome b_5 where the interference of P450.CO was removed by two-components analyses.

RESULTS AND DISCUSSION

When microsomes with benzphetamine as substrate were rapidly mixed with NADPH under aerobic conditions the pre-steady state kinetics showed a decrease of the ferric substrate-bound cytochrome P450 at 650/670 nm with a concomitant increase of oxygenated cytochrome P450 at 442/470 nm (fig. 1, aerobic traces).

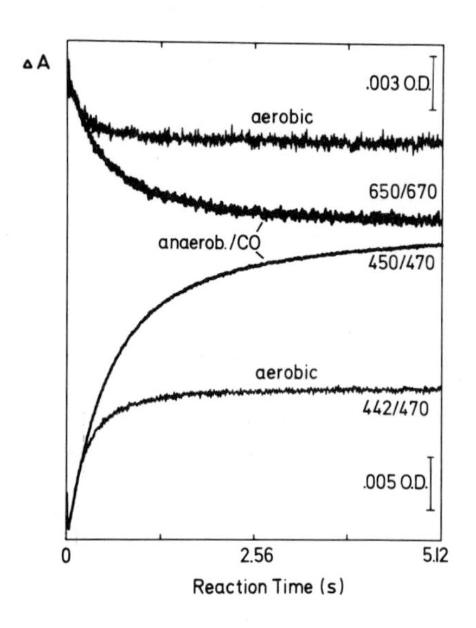

Fig. 1. Comparison of aerobic and anaerobic kinetics of microsomal cytochrome P450. Microsomes were mixed with equal volumes of NADPH yielding the final concentrations: protein 0.39 mg/ml, P450 0.89 µM, benzphetamine 1 mM, NADPH 0.1 mM in 0.05 M HEPES pH 7.7, 0.125 M KCl, 0.01 M $MgCl_2$ and 1 mM EDTA. For the anaerobic conditions the solution was gassed with CO, and O_2 was removed by a glucose oxidase/catalase system (1). The trace at 450/470 nm was scaled as if it had the same extinction coefficient as P450.O_2 at 442/470 nm. The traces were averages from four to eight runs.

This indicated that there were two states accumulating during the catalytic cycle, namely ferric substrate-bound and oxygenated cytochrome P450. Thus, the rate-limiting steps are the transfers of the two electrons. This led to the simple kinetic model

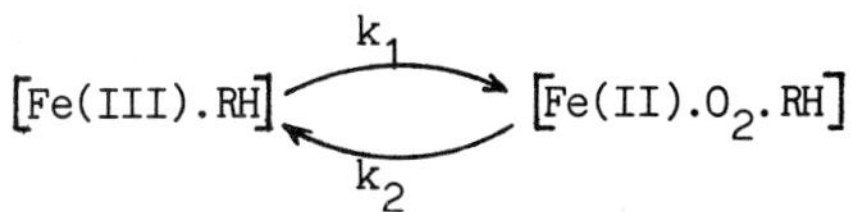

where k_1 and k_2 were pseudo-first order rate constants for the transfer of the first and second electron, respectively. Then an exponential presteady state kinetic could be derived with $k_{obs} = k_1 + k_2$, $A_{obs} = A_{tot}.k_1/(k_1 + k_2)$ and during the steady state a turnover of $k_1.k_2/(k_1 + k_2)$. Data fitting of the observed kinetics revealed two exponential phases where more than 65% of the reaction proceeded in the rapid phase which was used to calculate the rate of the second electron (table 1).

TABLE 1

ELECTRON TRANSFER RATES TO MICROSOMAL CYTOCHROME P450[a,b]

Wavelengths	442(450)/470	650/670	nm
k_1 (CO)	2.2	2.8 ± 0.3	s^{-1}
k_{obs} (air)	4.8 ± 0.2	5.0 ± 0.4	s^{-1}
$k_2 = k_{obs} - k_1$	2.6 ± 0.2	2.2 ± 0.7	s^{-1}
A_{tot} (CO)	(400)	5.2 ± 0.7	* 10^{-3}
A_{obs} (air)	(37 ± 1)	3.2 ± 0.2	* 10^{-3}
$k_2 = k_1(A_{tot}/A_{obs} - 1)$	-	1.8 ± 0.8	s^{-1}

[a]The same conditions as figur 1.
[b]For explanations see the text.

The rate constants k_2 calculated both from the rates and the amplitudes were consistent within the error limits supporting the kinetic model. When k_2 was corrected for the autoxidation of P450.O_2 (less than 0.5 s^{-1}, H. Kuthan,to be published) k_2 was about equal to k_1 yielding about half of the active P450 in the oxygenated state. It is noteworthy that at pH 7, approximately the intracellular pH, k_2 was slightly increased and k_1 strongly decreased (data not shown).

With this method the role of cytochrome b_5 in the transfer of the second electron was studied. Isolated cytochrome b_5 was incorporated into microsomes. Functional incorporation was achieved when the following NADH-dependent reaction rates increased proportionally to cytochrome b_5 content: reduction of cytochrome c and 7-ethoxycoumarin-O-deethylation. With additional cytochrome b_5 the rate of the second electron was increased, especially when both NADPH and NADH were used to ensure the rapid initial reduction of cytochrome b_5 (table 2).

TABLE 2

EFFECT OF CYTOCHROME B_5 ON THE RATE OF THE "SECOND' ELECTRON

Cyt.b_5/cyt.P450	0.21[c]	0.23[c]	0.67[b]	0.90[b]	
k_2	3.2 ± 0.5	2.9 ± 0.4	8.2 ± 1.3	9.6 ± 1.5	s^{-1}

[a]The conditions correspond to fig. 1 except both NADH and NADPH were used.
[b]Isolated cytochrome b_5 was incubated for 20 min at 22°C with microsomes. After centrifugation (1 h at 100 000 x g) the pellet was resuspended and used for the experiments.
[c]The controls were treated as in [b] but without additional cytochrome b_5.

Under the assumption of pseudo-first order reactions the extrapolation to the absence of cytochrome b_5 yielded a rate constant of 1 s^{-1} which was tentatively assigned to the transfer of the second electron from the NADPH-dependent reductase. Thus, more than half of the second electrons for the hydroxylation of benzphetamine were transfered via cytochrome b_5 in native microsomes at pH 7.7.

Further work will try to resolve the pseudo-first order rate constants with protein-protein interactions. Special emphasis will be put on studies at physiological intracellular conditions regarding pH and ionic strength.

ACKNOWLEDGMENTS

We are indebted to Drs. H.Graf and J.Poensgen for the preparation of cytochrome b_5. This work was supported by the Deutsche Forschungsgemeinschaft, SFB 38-L1-B.

REFERENCES

(1) Ruf, H.H., (1980) in: Gustafsson, J.A., et al. (Eds), Biochem.Biophys. and Regulation of Cytochrome P450, Elsevier/North-Holland Biomedical Press, Amsterdam, pp. 355-358

(2) Bonfils, C., Balny, C. and Maurel, P. (1981) J.Biol.Chem., 256, 9457-9465

(3) Noshiro, M. and Ullrich, V. (1981) Eur.J.Biochem., 116, 521-526

(4) Werringloer, J. (1981) in: Sato, R. and Kato, R. (Eds.) Microsomes, Drug Oxidations and Drug Toxicity, Japan Scientific Press, Tokyo, pp.

(5) Provencher, S.W. (1976) J.Chem.Phys. 64, 2772-2777

Cytochrome P-450, Biochemistry, Biophysics and Environmental Implications, E. Hietanen, M. Laitinen and O. Hänninen editors

THE SPONTANEOUS RELEASE OF HEME FROM HEMEPROTEINS

MICHAEL L. SMITH, PER-INGVAR OHLSSON AND KARL GUSTAV PAUL
Department of Physiological Chemistry, University of Umeå, S-901 87 Umeå, Sweden

The transfer of heme has been observed from *Aplysia* metmyoglobin to horse apomyoglobin (apo Mb) (1), from horse metmyoglobin to horseradish isoperoxidase C (HRP C) (2), and from metleghemoglobin to HRP C (3). The facile release of heme from tryptophan 2,3-dioxygenase has been noted (4). The spontaneous loss of heme from human hemoglobins may result from mutations of amino acid residues lining the heme pocket, e.g. Hb Köln (β98 val → met) (5).

We have undertaken a systematic study of this problem in an attempt to understand some of the factors controlling spontaneous heme release and to ascertain which hemeproteins exhibit this property. Spontaneous heme release may be related to the "openess" of the active site and/or to the strength of the amino acid - heme - van der Waals interactions. Biologically seen, the exchange of hematin may be a form of metabolic control for an organism at the post-ribosomal level or a first step in protein degradation.

EXPERIMENTAL

All reactions were carried out in 50 mM potassium phosphate, pH 7.0, 1 mM EDTA (standard buffer) at 25 ^{o}C unless otherwise noted. Typically, the reaction was initiated by the addition of 100-200 μl of apoMb (sperm whale, Sigma type III purified by ion-exchange chromatography) to about 2 ml, in a quartz cuvette, of a 5-10 μM solution of the hemeprotein under study. Just prior to initiating the reaction an identical quantity of standard buffer was added to give the same protein concentration in a reference cuvette. The reaction was followed by repeated scans in the Soret region triggered by a "homemade" automatic timer. For HRP A, cytochrome C peroxidase (CCP) and met Mb the reactions were followed for at least 24 hours, which was at least the reaction half life for all proteins except met Mb.

Rate constants were determined by the method of linear least squares. For for release of heme by metlegHb att data pairs (> 20) were used. For the reactions of HRP A and CCP which gave obvious "biphasic" kinetics, a plot was made of the first order dependence upon holoprotein concentration verses time, then points from the most linear portions of the plot were used to determine the pseudo first order rate constants $^{1}k_1$ and $^{2}k_1$.

Apomyoglobin was determined to be 99.8 % heme free by addition of cyanide and observation of the spectrum in the Soret region. HRP A was prepared as described previously (6). CCP was a gift from Drs. T. Yonetani and M. Ikeda-Saito, University of Pennsylvania. Leg Hb_a was a gift from Dr. Gunhild Sievers, University of Helsinki.

RESULTS

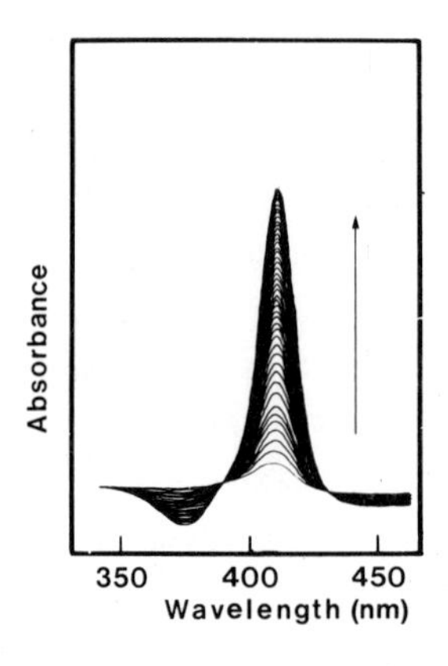

Fig. 1. Reaction of HRP A (9.2 μM) and apo Mb (127 μM) followed for 24 hours.

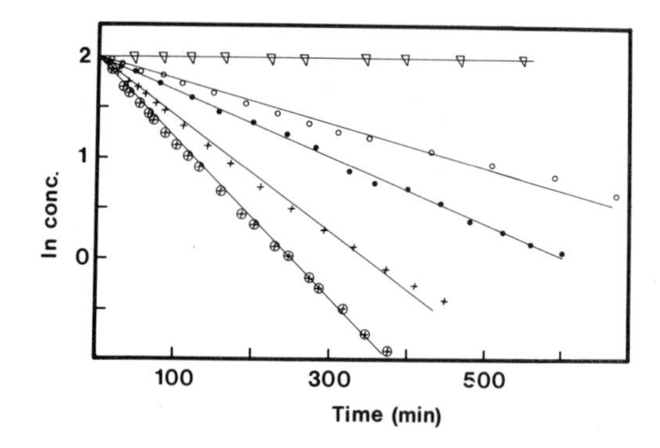

Fig. 2. Normalized first order plots of the release of heme from met legHb (about 8 μM) to apo Mb (127 μM) in standard buffer alone (⊕); with 0.1 M NaCl(x); 0.1 M NaF(●); 0.5 M NaF (o); 0.05 M NaCN (▽).

The reaction shown in Fig. 1 is typical of all those studied so far. The pair of isosbestic points at 389 and 429 nm excludes the existence of a long lived intermediate or protein precipitation. The addition of sodium cyanide allows identification of the hemeprotein product as cyanomet Mb by its characteristic spectrum. For the proteins Leg Hb, HRP A and CCP the apparent first order rate constants were found to be independent of apo Mb concentration as long as an excess of Mb is present.

For Leg Hb the transfer of heme was first order with respect to Leg Hb over the entire reaction (3 to 4 half lives) and could be described by a single rate constant (Table 1). For HRP A and CCP two first order reactions were observed. A relatively fast reaction comprising 20-30 % of the reaction was followed by a slower release of heme. Met Mb releases heme slowest of the four proteins which were found to spontaneously release heme. No reaction was observed for HRP C or lactoperoxidase.

The rate for Leg Hb was found to be highly dependent upon iron ligation and slightly dependent upon ionic strength (Table 2). The reaction rate is cut to half by the conversion to the fluoride form and is completely inhibited when the protein is in the cyanomet form (Fig. 2). Both rate constants for the release of heme from HRP A were found to be linearly dependent upon pH (Table 3), increasing with increased hydroxide concentration.

TABLE 1

RATE CONSTANTS FOR THE SPONTANEOUS RELEASE OF HEME FROM VARIOUS Fe(III) HEMEPROTEINS

Protein	First order $^1k_1$ (min^{-1})		Rate constants $^2k_1$ (min^{-1})	Samples
Leghemoglobin a (Leg Hb_a)	-		$6.7 \pm 1.3 \times 10^{-3}$	(3)
Horseradish isoperoxidase A_2 (HRP A)	$2.2 \pm 0.5 \times 10^{-3}$	(7)	$4.9 \pm 1.1 \times 10^{-4}$	(7)
Cytochrome c peroxidase (CCP)	1.4×10^{-3}	(2)	$1.7 \pm 0.5 \times 10^{-4}$	(4)
Myoglobin (Mb)	-		1.1×10^{-5}	(1)[a]
Horseradish isoperoxidase C_2 (HRP C)	no reaction			
Lactoperoxidase (LP)	no reaction			

[a]Using an excess of apohorseradish peroxidase C

CONCLUSIONS

1. The apparent rate law for the initial spontaneous release of heme from HRP A was found to be

$$\text{rate} = (^1k + {}^2k)\,[\text{HRP A}]\,[OH^-] \ldots$$

between pH 7 and 9. For the later reaction stage (> 20 % completion) the rate law becomes simply

$$\text{rate} = {}^2k\,[\text{HRP A}]\,[OH^-] \ldots$$

TABLE 2

RATE CONSTANTS OF THE SPONTANEOUS RELEASE OF HEME FROM VARIOUS DERIVATIVES OF LEG HEMOGLOBIN

Salt	(M)	First order rate constant $^{2}k_1$ (min)	Samples
Only buffer		$6.7 \pm 1.3 \times 10^{-3}$	(3)
NaCl	(0.1)	5.3×10^{-3}	(2)
NaF	(0.1)	3.0×10^{-3}	(2)
NaCN	(0.05)	no reaction	(2)

TABLE 3

pH DEPENDENCE OF THE RATE CONSTANTS OF SPONTANEOUS RELEASE OF HEME FROM HORSERADISH PEROXIDASE A

pH	First order rate constants	
	$^{1}k_1$ (min^{-1})	$^{2}k_1$ (min^{-1})
7.0	$2.2 \pm 0.5 \times 10^{-3}$	$4.9 \pm 1.1 \times 10^{-4}$
8.0	8.6×10^{-3}	9.3×10^{-4}
9.0 [a]	3.4×10^{-2}	1.7×10^{-3}

[a] in 50 mM potassium borate, with 1 mM EDTA at 25 ^{o}C

2. The apparent rate law for the spontaneous release of heme from leg Hb was found to be

$$\text{rate} = k\ [\text{leg Hb}]\ \ldots$$

at pH 7, during the entire reaction course. Salt (NaCl) slightly inhibits the reaction, and heme ligands can moderately (F^-) or totally (CN^-) inhibit the reaction.

3. The reaction rates are independent of apomyoglobin concentration from 5- to 120-fold excess. This means that the spontaneous release of heme is a function of the donor protein and independent of protein - protein interactions. The rate dependence upon hydroxyl ions and inhibition by iron ligands may mean that aquation at the iron may play a role in spontaneous heme release.

4. The first order rate constants, ${}^{2}k_{1}$, are interpreted as expressing the minimum frequencies at which these proteins unfold to allow increased access of the aqueous environment to the heme. The rate constant may well depend upon the fraction of heme exposed in the active holoprotein and/or the flexibility of the polypeptide lining the "heme pocket".

5. The more rapidly released heme from HRP A and CCP may be adventitiously bound heme though the possibility exists that for both proteins a population of molecules is present which more rapidly releases heme but only very slowly interconverts with the major portion of the protein molecules.

ACKNOWLEDGEMENTS

This study was supported by grants from Statens medicinska forskningsråd (3X-6522) and Magn. Bergvalls stiftelse.

REFERENCES

1. Rossi-Fanelli, A. and Antonini, E. (1960) J. Biol. Chem. 235, PC 4.
2. Rosenquist, U. and Paul. K.G. (1964) Acta Chem. Scand. 18, 1802.
3. Ellfolk, N., Pertilä, U. and Sievers, G. (1973) Acta Chem. Scand. 27, 3601.
4. Tokuyama, K. (1968) Biochim. Biophys. Acta 151, 76.
5. Wajcman, H., Byckova, V., Haidas, S. and Labie, D. (1971) FEBS Lett. 13, 145.
6. Paul, K.G. and Stigbrand, T. (1970) Acta Chem. Scand. 24, 3607.

Cytochrome P-450, Biochemistry, Biophysics
and Environmental Implications, E. Hietanen,
M. Laitinen and O. Hänninen editors

THIOL COMPOUNDS AS LIGANDS OF CYTOCHROME P-450

ELKA ELENKOVA[1], BORIS ATANASOV[1], OTTO RISTAU[2], CHRISTIANE JUNG[2], HORST REIN[2] and KLAUS RUCKPAUL
[1]Institute of Organic Chemistry, Bulgarian Academy of Sciences, Sofia, 1113 (Bulgaria) and [2]Central Institute of Molecular Biology, Academy of Sciences of the GDR, 1115 Berlin-Buch (GDR)

INTRODUCTION

Thiol compounds are widely used as protective agents in the preparation of P-450 and also for therapeutical purposes, e.g. using their capability to support effects of antitumor drugs. The knowledge of the interaction of thiols with P-450 may be a first step in understanding the molecular mechanism of these effects.

This report describes the interaction of cysteine, 1,4-dithioerythritol and 2,3-dimercaptopropanol with P-450. The influence of these compounds on the absorption spectrum of P-450 and their binding properties are studied. The enzymatic activity of P-450 is inhibited by these compounds (1).

MATERIAL AND METHODS

All studies were performed with solubilized P-450, which was prepared from phenobarbital induced rabbit liver microsomes according to (2). The concentration of P-450 was determined from the CO spectrum using $\varepsilon = 91\ mM^{-1}cm^{-1}$ (3). L-cysteine was purchased from Fluka (Switzerland). 1,4-dithioerythritol (erythro-1,4-dimercapto-2,3-butandiol) was obtained from Ferak (Berlin) and 2,3-dimercaptopropanol as Sulfactin^R from Homburg (FRG).

The absorption spectra were recorded using the spectrophotometer UV-300 (Shimadzu, Japan) at 25°C. The titration was carried out in 0.1 M phosphate buffer, pH 7,4 containing 20 % glycerol (v/v). The binding constants and different classes of binding sites were calculated by a computer with statistical processing of the optical data.

RESULTS

The Soret band of P-450 at 418 nm is about 5 nm red-shifted in the presence of cysteine. The difference spectrum formed in the presence of cysteine exhibits a maximum at 423 nm and a minimum at 407 nm characteristic of a type II spectrum. Computer processing of the titration curves revealed two classes of binding sites (Fig. 1). The Hill coefficient of approximately 1 (Table 1) excludes any cooperative effect between both classes of binding sites. The Soret band of P-450 in the presence of 1,4-dithioerythritol is likewise 5 nm red-shifted. Different from cysteine in the presence of 1,4-dithioerythritol an unusual difference spectrum with P-450 is formed, exhibiting peaks at 465, 423 and 370 nm. The analysis of the titration curves again revealed two classes of binding sites (Fig. 1) with about one order of magnitude higher affinity than cysteine and also without cooperative effect (Table 1).

In the presence of an excess of 2,3-dimercaptopropanol P-450 exhibits bands at 460, 418 and 375 nm (Fig. 2). Due to spectral characteristics of P-450 which are induced by 2,3-dimercaptopropanol and 1,4-dithioerythritol these spectra are interpreted as 'hyper' porphyrin spectra (4).

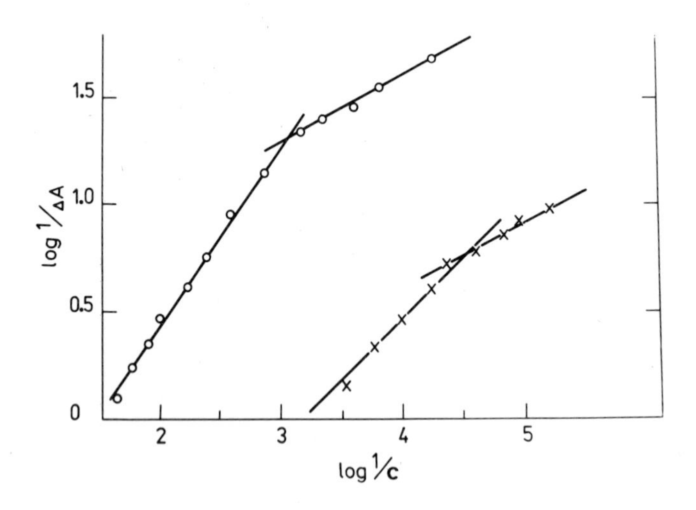

Fig. 1 Hill-plot of the titration of solubilized P-450 with cysteine o-o and 1,4-dithioerythritol x - x.

TABLE 1

Binding constants of thiol P-450 complexes

Ligand	Binding site	Hill coefficient	pK	Correlation coefficient
Cysteine	I	1.27	3.74	0.986
	II	1.03	1.14	0.999
1,4-dithio-	I	1.10	4.59	0.989
erythritol	II	1.06	3.42	0.997

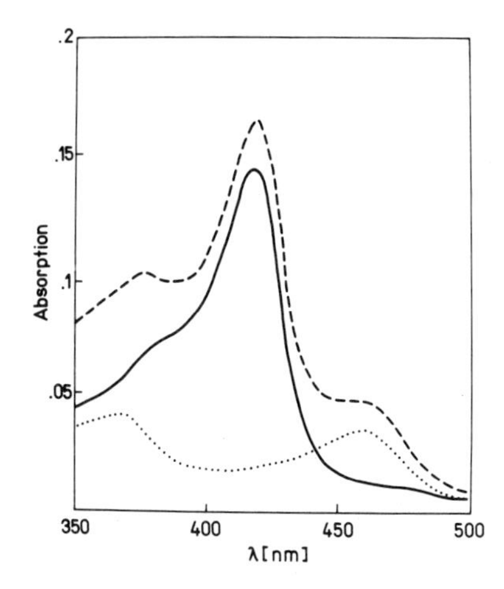

Fig. 2 - - - Absorption spectra of P-450 (2μM) in the presence of 2,3-dimercaptopropanol (100 μM) (i)
... hemin n-butylmercaptide (ii) obtained from (5)
--- the calculated spectrum based on the difference between (i) and (ii), which is very similar to the spectrum of the P-450-cystein complex and to the corresponding model complexes with thiol-thiolate as axial ligands.

DISCUSSION

The breaks in the titration curves have been attributed to two classes of binding sites. They may be ascribed to P-450 isozymes with different affinity which are contained in the partially purified P-450.

From model complexes of P-450 it is known that thiol compounds react with the heme iron (4,5). The formation of type II difference spectra of P-450 in the presence of thiol compounds, described in this paper indicates that these compounds are bound to the heme iron. According to spectra of respective model complexes (4, 5), the unusual optical spectra obtained from P-450 in the presence of 1,4-dithioerythritol and 2,3-dimercaptopropanol can be interpreted as 'hyper'porphyrin spectra evidencing thiolate as heme iron ligand. 1,4-dithioerythritol induces only a small fraction of 'hyper'porphyrin spectrum whereas 2,3-dimercaptopropanol produces a higher fraction which is estimated to amount up to 50 %. The reason for the formation of a susbstrate dependent varying portion of 'hyper'porphyrin spectral characteristics in the spectrum originates from the capability of the respective substrate to form a thiolate ($-S^-$) which is mixed with the usual low spin spectrum of the thiol (-SH) form to the final spectrum. In the ligands used obviously an equilibrium between the thiol and the thiolate forms exists resulting in a mixed spectrum of 'hyper'porphyrin ($-S^-$) and low spin spectrum emerging from the thiol ligand (-SH).

REFERENCES

1. Elenkova, E., Atanasov, B., Staikova, M. and Goranov, J. (1981) Dokl. Bolg. Acad. Nauk, 36, 33-35.

2. Lu, A.Y.H., Kuntzman, R., Jacobson, M. and Conney, A.H. (1972) J. Biol. Chem., 247, 1727-1734.

3. Omura, T. and Sato, R. (1967) Methods in Enzymology, 10, 536-561.

4. Nastainczyk, W., Ruf, H.H. and Ullrich, V. (1976) Chem. Biol. Interactions, 14, 251-263.

5. Nastainczyk, W., Ruf, H.H. and Ullrich, V. (1975) Eur. J. Biochem., 60, 615-620

Cytochrome P-450, Biochemistry, Biophysics
and Environmental Implications, E. Hietanen,
M. Laitinen and O. Hänninen editors

RESONANCE RAMAN STUDY OF IRON COORDINATION IN THE FERRIC LOW SPIN CYTOCHROME P-450

PAVEL ANZENBACHER[1], ZDENĚK ŠÍPAL[1], BOHUSLAV STRAUCH[2] AND JACEK TWARDOWSKI[3]
[1]Department of Biochemistry and [2]Department of Inorganic Chemistry, Charles University, Albertov 2030, 128 40 Prague 2 (Czechoslovakia) and [3]Department of Animal Physiology, Jagellonian University, Karasia 6, 30060 Cracow (Poland)

INTRODUCTION

Resonance Raman spectroscopy has been recently shown to be the unique tool for studying the stereochemistry of hemoproteins (1,2). RR* spectra of cytochrome P-450 were interpreted in favour of mercaptide ligation (2-5) and pentacoordination of the heme iron in the ferric high spin P-450 (4).

Because of the vibrational nature of RR spectra, it is possible to extend the field of their application to the study of the identity of heme axial ligands (6,7). This is particularly interesting for the P-450 heme iron, where the nature of the sixth ligand in the ferric low spin state remains controversial (8).

MATERIALS AND METHODS

Model compounds with N-Fe-S and O-Fe-S heme axial ligation were prepared according to (9). From a wide variety of models, these of visible, EPR and Mössbauer spectral characteristics closest to P-450 ones were chosen. N-methylimidazole adduct of ferriprotoporphyrin IX dimethyl ester p-nitrobenzene thiolate (N-Fe-S) was prepared by addition of the methylimidazole base to the thiolate compound in toluene solution under nitrogen atmosphere; RR spectra of this solution and of 1:1 mixture of solid substance with KBr were subsequently taken. The N,N-dimethylformamide (DMF) adduct of the same compound (O-Fe-S) was prepared also in solution by dissolving the thiolate com-

*Abbreviations: RR, resonance Raman, P-450, cytochrome P-450, N-Fe-S, (N-methylimidazole)(ferriprotoporphyrin IX dimethyl ester)(p-nitrobenzene thiolate), O-Fe-S, (N,N-dimethylformamide)(ferriprotoporphyrin IX dimethyl ester)(p-nitrobenzene thiolate)

pound in 1:3 mixture of dichloromethane and DMF.

Raman spectra were excited with the 488.0 nm line of CR 2 and Spectra Physics 164 Ar^+ lasers and recorded on Jeol JRS S1 and Cary 82 spectrometers. The samples were placed in a home-built (JRS S1) or Oxford Instruments CF 100 (Cary 82) cryogenic unit and kept under cold nitrogen. The stability of all samples was continuously watched. Typical conditions were: slit width 7 cm^{-1}, scan rate 36 $cm^{-1} \cdot min^{-1}$, sensitivity, 2000 pulses per second, power 60 mW at the sample. Raman data of P-450 were taken from previously published reports (3-5, 10,11).

RESULTS

Among various possible ligands of heme iron in the low spin ferric P-450, nitrogen and oxygen are the most serious candidates for the sixth coordination position (8). Whereas the NMR and visible spectral data are in good agreement with a hypothesis of oxygen coordination (8,12,13), the EPR and magnetic CD data reveal to the imidazole nitrogen (14,15).

The positions of the most prominent peaks in the RR spectra of various ferric low spin P-450 preparations of different origin together with RR data for both model compounds are presented in Table 1.

TABLE 1

RESONANCE RAMAN BAND POSITIONS OF VARIOUS FERRIC LOW SPIN P-450 AND MODEL COMPOUNDS

Compound	Band positions /cm^{-1}/				Reference
P-450 rat	1370	1505	1580	1638	10
P-450 LM 2	1371	1502	1585	1638	5
P-450 cam	1372	1502	1581	1635	3
P-450 scc	1368	1500	1582	1637	11
O-Fe-S	1373	1500	1579	1637	this work
N-Fe-S	1376	1503	1585	1632	this work

rat, LM 2, cam and scc explains the origin of P-450: rat microsomal, rabbit microsomal preparation LM 2, bacterial (grown on camphor) and bovine adrenal mitochondrial (catalyzing cholesterol side chain cleavage).

For both models, high quality RR spectra were obtained which allowed clear resolution of RR peaks with approximately 2 cm^{-1} accuracy (an example of RR spectra of the O-Fe-S is displayed in Fig. 1). The algebraic sum of differences between O-Fe-S and all P-450 RR band positions is 36 cm^{-1}, whereas for the N-Fe-S compound is 62 cm^{-1}. This fact clearly shows that the oxygen atom as the sixth ligand of heme iron cannot be ruled out and remains as a very serious candidate for the sixth co-ordination position in the low spin ferric P-450.

The importance of this statement lies in (1) the nature of RR spectroscopy, which is by its origin a vibrational one and is, consequently, very sensitive to the nature of vibrating atoms and (2) the fact, that both the model compounds used are of very similar structure and, therefore, RR band positions reflect the changes of the sixth ligand.

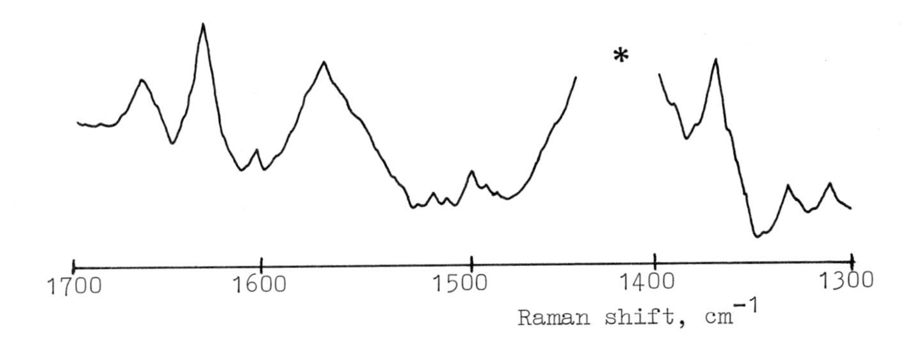

Fig. 1. RR spectrum of the O-Fe-S model compound. For experimental conditions, see text. The region marked by asterisk is influenced by solvent.

ACKNOWLEDGEMENT

This work was supported partly by agreement between Charles University, Prague and Jagellonian University, Cracow.

REFERENCES

1. Spiro, T.G. (1975) Biochim. Biophys. Acta,416, 169-189.
2. Felton, R.H. and Yu, N.-T. (1978) in: Dolphin, D. (Ed.), The Porphyrins, Academic Press, New York, pp. 374-393.
3. Champion, P.M., Gunsalus, I.C. and Wagner, G.C. (1978) J. Am. Chem. Soc., 100, 3743-3751.

4. Anzenbacher, P., Šípal, Z., Strauch, B., Twardowski, J. and Proniewicz, M. (1981) J. Am. Chem. Soc., 103, 5928-5929.

5. Ozaki, Y., Kitagawa, T., Kyogoku, Y., Imai, Y., Hashimoto-Yutsudo, C. and Sato, R. (1978) Biochemistry, 17, 5826-5831.

6. Rönnberg, M., Österlund, K. and Ellfolk, N. (1980) Biochim. Biophys. Acta, 626, 23-30.

7. Teraoka, J. and Kitagawa, T. (1980) J. Phys. Chem., 84, 1928-1935.

8. White, R.E. and Coon, M.J. (1980) Ann. Rev. Biochem., 49, 315-356.

9. Tang, S.C., Koch, S., Papaefthymiou, G.C., Foner, S., Frankel, R.B., Ibers, J.A. and Holm, R.H. (1976) J. Am. Chem. Soc., 98, 2414-2434.

10. Anzenbacher, P., Šípal, Z., Chlumský, J. and Strauch, B. (1980) Studia biophysica, 78, 73-74.

11. Shimizu, T., Kitagawa, T., Mitani, F., Iizuka, T., and Ishimura, Y. (1981) Biochim. Biophys. Acta, 670, 236-242.

12. Philson, S.B., Debrunner, P.G., Schmidt, P.G. and Gunsalus, I.C. (1979) J. Biol. Chem., 254, 10173-10179.

13. Ruf, H.H., Wende, P. and Ullrich, V. (1979) J. Inorg. Biochem., 11, 189-204.

14. Chevion, M., Peisach, J. and Blumberg, W.E. (1977) J. Biol. Chem., 252, 3637-3645.

15. Shimizu, T., Iizuka, T., Shimada, H., Ishimura, Y., Nozawa, T. and Hatano, M. (1981) Biochim. Biophys. Acta, 670, 341-354.

© 1982 Elsevier Biomedical Press B.V.
Cytochrome P-450, Biochemistry, Biophysics
and Environmental Implications, E. Hietanen,
M. Laitinen and O. Hänninen editors

COMPARATIVE STUDIES OF SUPEROXIDE RADICAL GENERATION IN MICROSOMES AND RECONSTITUTED MONOOXYGENASE SYSTEMS

S.K.SOODAEVA, E.D.SKOTZELYAS, A.A.ZHUKOV AND A.I.ARCHAKOV
Second Moscow Medical Institute, Moscow, USSR

INTRODUCTION

It is well known that superoxide radical ($O_2^{\bar{\cdot}}$) is generated in microsomes at NADPH oxidation (1-4). Recent investigations showed that the main site of $O_2^{\bar{\cdot}}$ generation at hydroxylation reactions is cytochrome P-450 (5-7). The rate of $O_2^{\bar{\cdot}}$ generation has lately been measured by means of succinylated cyt.$\underline{c}_{suc}$* (7). There are some difficulties, however, in quantitive estimation of the rate of $O_2^{\bar{\cdot}}$ formation by this method. Rate constants for $O_2^{\bar{\cdot}}$ dismutation and its interaction with cyt.$\underline{c}_{suc}$ are comparable and, besides, there is a possibility of direct reduction of cyt.$\underline{c}_{suc}$ by FP. Taking this into account, we tried to determine experimental conditions for an adequate registration of $O_2^{\bar{\cdot}}$ formation in microsomes and reconstituted monooxygenase systems using cyt.$\underline{c}_{suc}$.

MATERIALS AND METHODS

Cyt.$\underline{c}_{suc}$ was prepared by the method of (8). Two moles of succinyc anhydride per each free lysine ε-aminoresidue of cyt.$\underline{c}$ were used. Microsomal fraction was prepared from phenobarbital induced rabbits. Cytochrome P-450, FP were prepared by the methods described elsewhere (9,10). Reconstitution was performed by the method of (11), FP activity was measured monitoring the reduction of cyt.$\underline{c}$. The true rate of $O_2^{\bar{\cdot}}$ formation was obtained from cyt.$\underline{c}_{suc}$ SOD-sensitive reduction rate using the coefficient calculated as the ratio of V_{max} for cyt.$\underline{c}$ reduction and cyt.$\underline{c}_{suc}$ reduction rate in xanthine/xanthine oxidase sytem.

RESULTS

To choose the optimal conditions the rate of $O_2^{\bar{\cdot}}$ -dependent cyt.$\underline{c}_{suc}$ reduction was measured in xanthine/xanthine oxidase system under different conditions (Table 1).

* Abbreviations used: FP,NADPH-specific flavoprotein; cyt.$\underline{c}$ and cyt.$\underline{c}_{suc}$, native and succinylated cytochrome $\underline{c}$ correspondingly.

TABLE 1

THE INFLUENCE OF BUFFER, Mg^{2+} AND IONIC STRENGTH ON THE REDUCTION OF CYT.C IN XANTHINE/XANTHINE OXIDASE SYSTEM

Conditions	Native	Succinylated
	(nmoles reduced cytochrome·ml^{-1}·min^{-1})	
25 mM Tris-HCl buffer, pH 7,8	18	0
+ 10mM $MgCl_2$	11	1,8
50 mM K-phosphate buffer, pH 7,8	14	3,2
100 mM K-phosphate buffer, pH 7,8	14	2,0

Reaction medium (final volume 3 ml) contained: buffer, 50 μM cyt.c, 0,5 mM xanthine, 20 μg/ml xanthine oxidase, 500 U/ml catalase at 37°C

The data show that cyt.c_{suc} is not reduced in this system in 25 mM Tris-HCl buffer in the absence of $MgCl_2$. A higher rate of reduction is observed in 50 mM K-phosphate buffer.

Table 2 presents the data concerning $O_2^{\bar{\cdot}}$ generation in microsomes.

TABLE 2

THE INFLUENCE OF BUFFER, Mg^{2+} AND IONIC STRENGTH ON $O_2^{\bar{\cdot}}$ GENERATION IN MICROSOMES

Conditions	Reduction rate (nmole $O_2^{\bar{\cdot}}$·mg^{-1}·min^{-1})	
	SOD-sensitive	SOD-insensitive
25 mM Tris-HCl buffer,pH 7,8	0	0
+ 10 mM $MgCl_2$	14,0	2,4
50 mM K-phosphate buffer, pH 7,8	0	0
+ 10 mM MgCl	8,8	0,5
100 mM K-phosphate buffer, pH 7,8	14,3	0
+ 10 mM $MgCl_2$	14,3	0

Reaction medium (final volume 3 ml) contained: buffer, 0,5 mg/ml microsomal protein, 50 μM cyt.c_{suc}, 0,3 mM NaCN, 1 mM NaN_3, 0,3mM NADPH at 37°C

This data show that $O_2^{\bar{}}$ production rate is maximal and completely SOD-sensitive in the case of 100 mM K-phosphate buffer.

The results obtained allow us to suggest that the interaction of cyt.$\underline{c}_{suc}$ with both $O_2^{\bar{}}$ and FP depends on the charge of the protein molecules and thus may be affected by cations.

TABLE 3

COMPARATIVE STUDY OF NADPH- AND NADH-DEPENDENT $O_2^{\bar{}}$ GENERATION IN MICROSOMES

Cosubstrate	Control	+CO[a]
	(nmole $O_2^{\bar{}} \cdot min^{-1} \cdot mg^{-1}$)	
1. NADPH	12,1	5,3
2. NADH	3,0	3,0

Reaction medium (final volume 3 ml) contained: 100 mM K-phosphate buffer, pH 7,8; 50 μM cyt.$\underline{c}_{suc}$, 0.5 mg/ml microsomal protein, 0.3 mM NaCN, 1 mM NaN_3, 0.3 mM NADPH or 1 mM NADH at 37℃

[a]Reaction medium was equilibrated by gas mixture (80% CO and 20% O_2) for 2 min.

TABLE 4

SUPEROXIDE GENERATION IN MICROSOMAL AND RECONSTITUTED SYSTEMS

System	Components	$O_2^{\bar{}}$ generation (nmole · $ml^{-1} \cdot min^{-1}$)
1. Microsomes		14,7
2. Soluble	FP	0
	FP+P-450	1,5
	P-450	0
3. Reconstituted with Emulgen 913	FP	1,2
	FP+P-450	3,2
	P-450	0
4. Incorporated into PCh-liposomes	FP	0
	FP+P-450	5,8
	P-450	0

Reaction medium (final volume 3ml) contained: 100 mM K-phosphate buffer, pH 7.8; 50 μM cyt.$\underline{c}_{suc}$; 0.16 U/ml FP, 1nmole/ml cytochrome P-450, in systems 1-3 and 0.5 nmole/ml cytochrome P-450 in system 4. System 1 contained also 0.3 mM NaCN and 1mM NaN_3. Emulgen 913 concentration in system 3 was 53 g/ml.

NADPH-dependent $O_2^{\bar{\cdot}}$ production was inhibited by the typical inhibitor of cytochrome P-450 carbon monooxide, whereas NADH-dependent production was CO-insensitive (Table 3), thus supporting the suggestion that cytochrome P-450 is tne main site of $O_2^{\bar{\cdot}}$ production in microsomes (2,7) at NADPH-oxidation. At NADH-dependent oxidation it may be cytochrome b_5.

To compare $O_2^{\bar{\cdot}}$ generation in microsomal and reconstituted systems different types of reconstituted systems were used (Table4). The data illustrate that the system of phospholipid incorporated carriers is most similar to the microsomal one and cytochrome P-450 is the main site of $O_2^{\bar{\cdot}}$ generation in reconstituted systems as well.

REFRENCES

1. Aust, S.D., Roerig, D.L. and Pederson, T.C. (1972) Biochem. Biophys.Res.Commun.,47,1133-1137
2. Auclair, Chr., De Prost,D. and Hakim,J.(1978) Biochem. Pharmacol.,27,355-358
3. Debey,P. and Balny,C. (1973) Biochimie,55,329-332
4. Sasame, H.A., Mitchell, J.R. and Gillette, J.R.(1975) Fed.Proc 34, 729-(abstr.2886).
5. Richter, C., Azzi,A., Weser,U. and Wendel,A.J. (1977) J.Biol. Chem.,252,5061-5066
6. Estabrook,R.W., Kawano,S., Werringloer, J., Kuthan,H., Tsuji, H., Graf,M. and Ullrich V.(1979) Acta Biol.Med.Germ.,38,423-432
7. Ullrich, V. and Kuthan, H. (1980) in:J.-A. Gustafsson et al. eds.Biochemistry, Biophysics and Regulation of Cytochrome P-450, Elsevier/North-Holland Biomedical Press, Amsterdam, pp.267-272
8. Takemori,S., Wada, K., Ando,K., and Hosokawa,M (1962) J.Biochem., 52, 28-37.
9. Dignam,J.D. and Strobel,H.W. (1977) Biochemistry,16,1116-1123
10. Karuzina,I.I., Bachmanova,G.I., Mengazetdinov,D.E., Myasoedova,K.N., Zhichareva, V.O., Kuznetsova, G.P. and Archakov,A.I. (1979) Biokhimia,4,1149-1159
11. Archakov,A.I., Uvarov,V.Yu., Bachmanova,G.I., Sukhomudrenko, A.G. and Myasoedova,K.N. (1981) Arch.Biochem.Biophys.,212, 378-384

Cytochrome P-450, Biochemistry, Biophysics
and Environmental Implications, E. Hietanen,
M. Laitinen and O. Hänninen editors

USE OF VESICLES OF RECONSTITUTED HUMAN CYTOCHROME P-450 IN STUDIES OF FREE RADICAL PRODUCTION BY REDUCTIVE METABOLISM OF HALOCARBONS

B. BÖSTERLING, J.R. TRUDELL AND A.J. TREVOR
Department of Anesthesia, Stanford University School of Medicine, Stanford, California, 94305, U.S.A.

INTRODUCTION

It has been proposed that covalent binding of reactive metabolites to liver membrane constituents may be responsible for the hepatoxicity of halocarbons, such as carbon tetrachloride or halothane (1). If halothane (2-bromo-2-chloro-1,1,1-trifluoroethane) is the substrate, reductive metabolism results in increased binding of metabolites to phospholipids of the endoplasmic reticulum, reduction of content of glutathione, increase in content of conjugated dienes in the fatty acid chains, loss of structural integrity of the endoplasmic reticulum, and liver necrosis. It has been suggested that this metabolite could be a free radical (2). However, the molecular structures of such reactive metabolites have not been determined and there has been no characterization of adducts formed or of their mechanism of binding to tissue molecules. Such knowledge could be crucial to our further understanding of the biochemical events leading to hepatic cell damage.

To circumvent the problem of heterogeneity of microsomal phospholipids, we have reconstituted purified cytochrome P-450 from human liver, NADPH-cytochrome P-450 reductase, and cytochrome b_5 in phospholipid vesicles that contained egg phosphatidylethanolamine and dioleoylphosphatidylcholine (DOPC) as the sole phosphatidylcholine. These lipids were shown to act as structural components of the vesicle, to support catalytic activity of the reconstituted enzyme system, and to serve as an effective target for reactive metabolites that bind covalently to the double bonds in fatty acyl chains (3-6).

MATERIALS AND METHODS

Human cytochromes P-450 were purified 12-fold from liver microsomes (7). NADPH-cytochrome P-450 reductase was purified from liver microsomes of phenobarbital-pretreated rabbits to an activity of 32 μmol/min per mg of protein (8). Cytochrome b_5 was purified from the same microsomes as the reductase to a purity of over 90% (9). Egg phosphatidylethanolamine (egg PE) was prepared from fresh eggs under N_2.

Reconstitution of the purified proteins into phospholipid vesicles was achieved by a modification of the slow cholate dialysis method (10). The vesicle suspension was deoxygenated, an NADPH-generating system was introduced and then 0.4 μl/ml CCl_4 or halothane was added after which the mixture was stirred for 1 hr at 30°C. The extracted phospholipids were applied to a Lichrosorb Si-100 HPLC column. The phosphatidylcholine fraction was subjected to transesterification and the resulting methyl esters were separated on Lichrosorb reverse-phase C-18 column. The major radioactive fraction was rechromatographed on a reverse-phase Lichrosorb C-8 HPLC column. The single radioactive peak was subjected to analysis by mass spectrometry. Direct inlet electron impact mass spectra were measured on a Varian CH-7 mass spectrometer at 20 eV. Desorption electron impact mass spectra were measured on a Ribermag R10-10B.

RESULTS AND DISCUSSION

In an initial series of experiments, identification of the reactive metabolite from halothane (6) bound to fatty acid chains of phospholipids in microsomes was attempted by subjecting the phosphatidylcholine fraction purified by preparative HPLC to transesterification. Unfortunately a broad distribution of radioactive metabolites bound to fatty acid methyl esters was observed. In a previous study we demonstrated that, when the 1-chloro-2,2,2-trifluoroethyl radical is generated by UV photolysis of halothane, it adds to either position 9 or position 10 of the double bond in methyl oleate (4). We suggest that a metabolically produced radical will add to double bonds in a similar fashion, and the broad distribution of radioactivity is consistent with random addition of a metabolite to any of the double bonds in the many different unsaturated fatty acids found in microsomes.

To circumvent the isolation problem encountered in microsomes, we used a reconstituted phospholipid vesicle system in which the single double bond of oleoyl presented the only target for free radical additon in the phosphatidylcholine fraction. Anaerobic incubation of ^{14}C- halothane or $^{14}CCl_4$ with a suspension of either human or rabbit cytochrome P-450 reconstituted in phospholipid vesicles with NADPH-cytochrome P-450 reductase and cytochrome b_5 resulted in similar binding of radioactive metabolites in the egg PE and DOPC fractions (5,6). When the DOPC fraction which was purified by preparative HPLC was subjected to transesterification and then applied on the reverse-phase HPLC column, a single radioactive fraction was obtained that was subjected to capillary gas chromatography coupled to a mass spectrometer. The mass spectrum is consistent with addition of a CF_3CHCl radical to the double bond of oleic acid followed by abstraction of a hydrogen radical from a neighboring donor. The use of desorption chemical ionization mass spectrometry with ammonia as reagent gas allowed the identification of the single adduct that was produced during reductive metabolism of CCL_4 as a mixture of 9- and 10-(trichloromethyl)-stearate methyl esters.

When NADPH-cytochrome P-450 reductase was reconstituted in DOPC and egg PE, no binding of halothane metabolites was observed. The proposed 9,10-cyclopropane-substituted methyl stearate that would have resulted from addition of the 2,2,2-trifluroethyl carbene or the CCl_2 carbene (11,12) was not observed. The absence of the product of addition of a carbene to a double bond in our study suggests that either the cytochrome P-450-carbene complex is so stable that it does not dissociate once formed, that the lifetime for the carbene is too short to diffuse to a double bond, or that the peak in absorption at 470 nm is not due to a carbene complex.

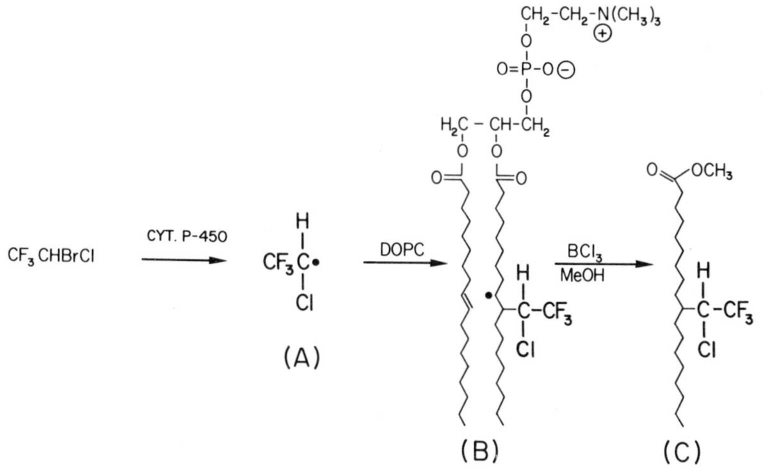

The molecular structure can be explained by the reaction scheme shown above. When halothane is reduced with one electron it can lose a bromide ion and form the 1-chloro-2,2,2-trifluorethyl radical (A). This free radical could then add to either end of the double bond in the two oleoyl chains in DOPC (B). This mixture can be transesterified to yield methyl oleate and a mixture of 9- and 10-(1-chloro-2,2,2-trifluoroethyl)-stearate methyl esters.

CONCLUSION

It is likely that this pathway of reductive metabolism will be a general one for halocarbons in incubation in vitro of reconstituted cytochrome P-450 under anaerobic conditions and in exposure in vivo under conditions that cause very low hepatic oxygen concentrations.

ACKNOWLEDGEMENTS

We are indebted to Ellis N. Cohen, M.D. for much support and helpful suggestion. We would like to thank Ms. Marie Bendix for expert technical assistance and the Stanford Cardiac Anesthesia and Cardiac Transplantation teams for their help in obtaining the human liver specimens. This research was supported by the National Institute of Occupational Safety and Health (OH 00978) and the Alexander von Humboldt-Stiftung.

REFERENCES

1. Cohen, E.N. and Hood, N. (1969) Anesthesiology 31, 553-558.
2. Stier, A. (1968) Anesthesiology 29, 453-454.
3. Trudell, J.R., Bösterling, B. and Trevor, A.J. (1981) Biochem. Biophys. Res. Commun. 102, 372-377.
4. Bösterling, B., Trevor, A.J. and Trudell, J.R. (1982) Anesthesiology (in press).
5. Trudell, J.R., Bösterling, B. and Trevor, A.J. (1982) Proc. Natl. Acad. Sci. (U.S.A.) 79, 2678-2682.
6. Trudell, J.R., Bösterling, B. and Trevor, A.J. (1982) Mol. Pharmacol. 21 (in press).
7. Bösterling, B. and Trudell, J.R. (1980) Microsomes, Drug Oxidations, and Chemical Carcinogenesis, pp. 115-118, Academic Press, New York.
8. Yasukochi, Y. and Masters, B.S.S. (1976) J. Biol. Chem 251, 5537-5544.
9. Strittmatter, P. and Rogers, M.J. (1975) Proc. Natl. Acad. Sci. (U.S.A.) 72, 2658-2661.
10. Bösterling, B., Stier A., Hildebrandt, A.G., Dawson, J.H. and Trudell, J.R. (1979) Mol. Pharmacol. 16, 332-342.
11. Mansuy, D., Nastainczyk, W. and Ullrich, V. (1974) Naunyn-Schmiedebergs Arch. Pharmacol. 285, 315-324.
12. Nastainczyk, W., Ullrich, V. and Sies, H. (1978) Biochem. Pharmacol. 27, 387-392.

Cytochrome P-450, Biochemistry, Biophysics
and Environmental Implications, E. Hietanen,
M. Laitinen and O. Hänninen editors

STRUCTURAL AND MECHANISTIC DIFFERENCES IN QUINONE INHIBITION OF MICROSOMAL DRUG METABOLISM: INHIBITION OF NADPH-CYTOCHROME P-450 REDUCTASE ACTIVITY

EVAN D. KHARASCH AND RAYMOND F. NOVAK
Department of Pharmacology, Northwestern University Medical and Dental Schools
Chicago, Illinois 60611 U.S.A.

INTRODUCTION

Inhibition of microsomal oxidative drug metabolism by quinone-containing compounds is a well known phenomenon, first reported by Gillette *et al.* for menadione inhibition of antipyrene demethylation (1). It has also been reported that benzo(a)pyrene 1,6-,3,6-, and 6,12-quinones inhibit the microsomal mixed function oxidation of benzo(a)pyrene and *trans*-7,8-dihydro-7,8-dihydroxybenzo(a)pyrene in a noncompetitive manner (2). The mechanism of inhibition of drug metabolism which occurs is that of an electron shunt which diverts electrons from cytochrome P-450 via an accelerated flow of reducing equivalents through NADPH-cytochrome P-450 reductase to the quinone and molecular oxygen (3).

1,4-Bis(2-((2-hydroxyethyl)amino)-ethylamino)-9,10-anthracenedione (HAQ), and the ring hydroxylated derivative, 1,4-dihydroxy-5,8-bis(2-((2-hydroxyethyl)-amino)-ethylamino)-9,10-anthracenedione (DHAQ) are two new quinone-containing antineoplastic agents (Fig. 1). We have shown that HAQ decreases microsomal oxidative drug metabolism by inhibiting P-450 reductase activity (4). This study examines the effect of ring hydroxylation on quinone inhibition of microsomal drug metabolism.

Figure 1

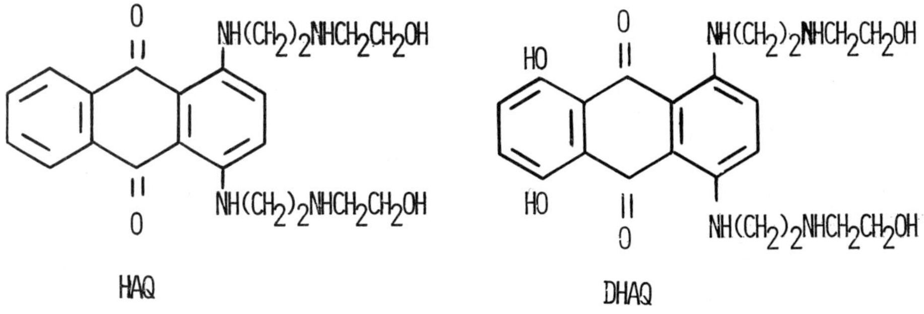

HAQ

DHAQ

METHODS

Hepatic microsomes were prepared from phenobarbital- and β-naphthoflavone-induced rabbits. Microsomal incubations were performed at 37°C in an oscillating water bath using a reaction mixture which contained 0.1 M potassium phosphate buffer pH 7.5, 1 mM NADPH, 10-300 μM *p*-nitroanisole or 50-800 μM dimethylaniline, and 1 mg microsomal protein in a final volume of 1.0 ml. Reactions were initiated after a 2 min preincubation period by the addition of NADPH and terminated after 10 min by the addition of 0.25 ml cold 20% trichloroacetic acid. Aryl hydrocarbon hydroxylase activity was determined under similar conditions using a 1.0 ml reaction mixture containing 50 mM potassium phosphate buffer pH 7.5 , 360 μM NADPH, 1-100 μM benzo(a)pyrene, and 1 mg protein. The reaction was terminated after 10 min by the addition of 1.0 ml cold acetone. Reaction rates were linear over the time period employed. Drug metabolism was assayed in the presence of HAQ and DHAQ as described (4). NADPH-cytochrome P-450 reductase was prepared according to the procedure of French and Coon (5).

RESULTS

HAQ noncompetitively inhibited aryl hydrocarbon hydroxylase activity in β-naphthoflavone-induced microsomes with a $K_i \approx 3.0$ mM. Similar noncompetitive inhibition of N,N-dimethylaniline ($K_i = 2.6$ mM) N-demethylase and *p*-nitroanisole O-dealkylase ($K_i = 2.9$ mM) activity by HAQ was observed in phenobarbital-induced microsomes (Table 1). Basal and substrate-stimulated NADPH oxidation was also inhibited, whereas HAQ failed to inhibit cumene hydroperoxide-supported metabolism which does not depend on NADPH cytochrome P-450 reductase activity. HAQ appears to be a relatively poor substrate for the flavoprotein-catalyzed redox cycling pathway as compared to most quinones. Table 2 provides data for NADPH oxidation and superoxide formation by HAQ in the presence of P-450 reductase. HAQ fails to stimulate NADPH oxidation to a significant degree. In contrast, stimulation of NADPH oxidation by DHAQ is ∿ 10-fold greater than that of HAQ. In addition, HAQ also inhibited the activity of P-450 reductase towards menadione, cytochrome *c*, and several quinone-containing antineoplastic agents (5).

The ring-hydroxylated derivative, DHAQ, also inhibited microsomal metabolism. Table 3 compares the effect of HAQ and DHAQ on aryl hydrocarbon hydroxylase activity in microsomes from phenobarbital-induced rabbits. Activity was diminished 20 and 31%, by 1 and 3 mM HAQ, respectively, while DHAQ was a more potent inhibitor, decreasing activity 41 and 56% respectively, at identical concentrations. In contrast to HAQ, DHAQ stimulated NADPH oxidation and superoxide formation by P-450 reductase (Table 2), suggesting that DHAQ is a substrate for flavoprotein-catalyzed redox cycling.

TABLE 1

HAQ INHIBITION OF DRUG METABOLISM

Substrate[a]	Product Formation (nmoles/min/mg protein)		
	Control	1 mM HAQ	3 mM HAQ
DMA[b]	6.5	4.5	3.2
p-NA[b]	9.0	6.8	4.4
BP[c]	0.93	0.62	0.45

[a]Saturating concentrations of all three substrates were employed: N,N-dimethylaniline (DMA), 0.8 mM: *p*-nitroanisole (*p*-NA), 0.3 mM: and benzo(a)pyrene (BP), 100 μM.
[b]Phenobarbital-induced microsomes.
[c]β-naphthoflavone-induced microsomes.

TABLE 2

QUINONE ACTIVATION BY NADPH CYTOCHROME P-450 REDUCTASE

Quinone	NADPH oxidation[a] (nmoles/min/nmole)	Superoxide Formation[b]
HAQ	36 ± 2	150 ± 18
DHAQ	327 ± 23	429 ± 39

[a]Assayed in 0.1 M phosphate buffer pH 7.7, 100 μM EDTA, 100 μM NADPH, 100 μM quinone.
[b]Measured as the rate of superoxide dismutase-inhibitable acetylated cytochrome *c* (21 μM) reduction.

DISCUSSION

These data suggest that HAQ inhibits electron transfer by microsomal NADPH-cytochrome P-450 reductase, diminishing electron flow to cytochrome P-450 and thereby inhibiting substrate metabolism. In contrast, DHAQ appears to inhibit substrate hydroxylation by the conventional electron shunt mechanism, accelerating the flow of electrons through the reductase to the quinone and molecular oxygen thereby diverting reducing equivalents from the cytochrome.

Thus, it appears that HAQ is an inhibitor of P-450 reductase. The ring hydroxylation of HAQ to DHAQ confers redox-cycling properties such that the

TABLE 3
QUINONE INHIBITION OF MICROSOMAL ARYL HYDROCARBON HYDROXYLASE ACTIVITY[a]

Quinone	Product (nmoles/min/mg)	% Inhibition
None	0.64 ± .02	-
1 mM HAQ	0.51 ± .02	20
3 mM HAQ	0.44 ± .01	31
5 mM DHAQ	0.49 ± .01	23
1 mM DHAQ	0.38 ± .01	41
3 mM DHAQ	0.28 ± .01	56

[a]Measured using 100 μM benzo(a)pyrene and phenobarbital-induced microsomes.

drug acts by a classical electron-shunt mechanism. The importance of aromatic ring hydroxylation adjacent to the quinone moiety for redox cycling behavior has recently been confirmed with the mono-hydroxylated derivative of HAQ, which displays activity similar to that of DHAQ.

ACKNOWLEDGEMENTS

This work was supported in part by NIH grant GM27836 to RFN and NIH Training Grant GM02763 to the Department of Pharmacology.

REFERENCES

1. Gillette, J.R., Brodie, B.B., and LaDu, B.N. (1957) J. Pharm. Exp. Therap. 119, 532-540.
2. Shen, A.L., Fahl, W.E., Wrighten, S.A., Jefcoate, C.R. (1979) Canc. Res. 39, 4123-4129.
3. Netter, K.J., Pharmac. Ther. (1980) 10, 515-535.
4. Kharasch, E.D. and Novak, R.F., (1982) Mol. Pharmacol. in press.
5. French, J.S., M.J. Coon (1979) Arch. Biochem. Biophys. 195, 565-577.
6. Kharasch, E.D. and Novak, R.F. (1981) Biochem. Pharmacol. 30, 2881-2884.

Cytochrome P-450, Biochemistry, Biophysics
and Environmental Implications, E. Hietanen,
M. Laitinen and O. Hänninen editors

CYTOCHROME P45011β AND P450scc SYSTEMS IN ADRENOCORTICAL MITOCHONDRIA: IMMUNOCYTOCHEMICAL EVIDENCE FOR A HETEROGENEOUS DISTRIBUTION AMONG THE MITOCHONDRIA WITHIN A SINGLE CELL.

FUMIKO MITANI[1], YUZURU ISHIMURA[1], SHINICHI IZUMI[2], NORIYUKI KOMATSU[2], AND KEIICHI WATANABE[2]
[1]Department of Biochemistry, School of Medicine, Keio University, Shinjuku-ku, Tokyo 160 and [2]Department of Pathology, School of Medicine, Tokai University, Isehara, Kanagawa 259-11 (Japan)

INTRODUCTION

We have previously shown by using the horseradish-labeled antibody method that adrenodoxin (Ad) and adrenodoxin reductase (AdR), components of cytochrome P450-dependent monooxygenase systems in adrenal cortex, are localized in the matrix side of the inner membrane of the mitochondria (1). In this study, we examined localizations of the terminal enzymes of the systems, i.e. cytochrome P450scc (P450scc) and cytochrome P45011β (P45011β), in bovine adrenocortical cells by the same technique. The results indicated that both P-450s are localized at the same site as that of Ad and AdR and that the distribution of these P-450s is heterogeneous among the mitochondria within a single adrenocortical cell. Furthermore physiological states of the animals appears to influence the distribution pattern.

MATERIALS AND METHODS

P450scc, P45011β, AdR and their antibodies (Fab' fragments of IgG) labeled with horseradish peroxidase were prepared according to the published methods (1,2). Details for the immunocytochemical manupulations have also been described (1,3). For the experiments with rat, anti-AdR which cross-reacted with the rat enzyme was used to study the distribution pattern of AdR under various physiological conditions.

RESULTS AND DISCUSSION

Fig. 1a is an immuno-electron microscopic view of the cells in the zona fasciculata of bovine adrenal cortex to show the localization of P45011β. The immunostaining (dark deposits) was found to be associated only with the inner membrane of mitochondria and

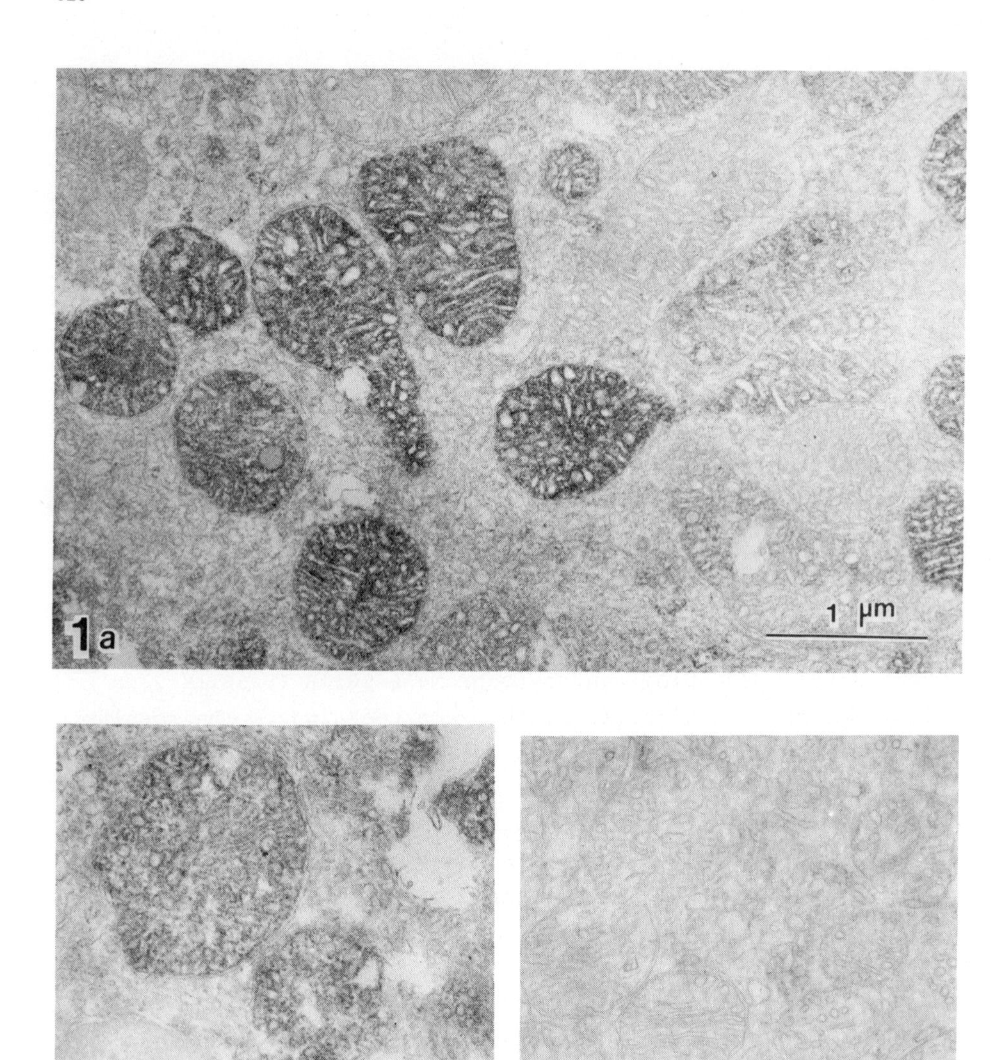

Fig. 1. Immunocytochemical localization of cytochrome P450s in bovine zona fasciculata cells by HRP-labeled antibody (Fab') method: 1a, Staining for P45011β; 1b, Staining for P450scc; 1c, Control staining with anti-mouse kidney basement membrane.

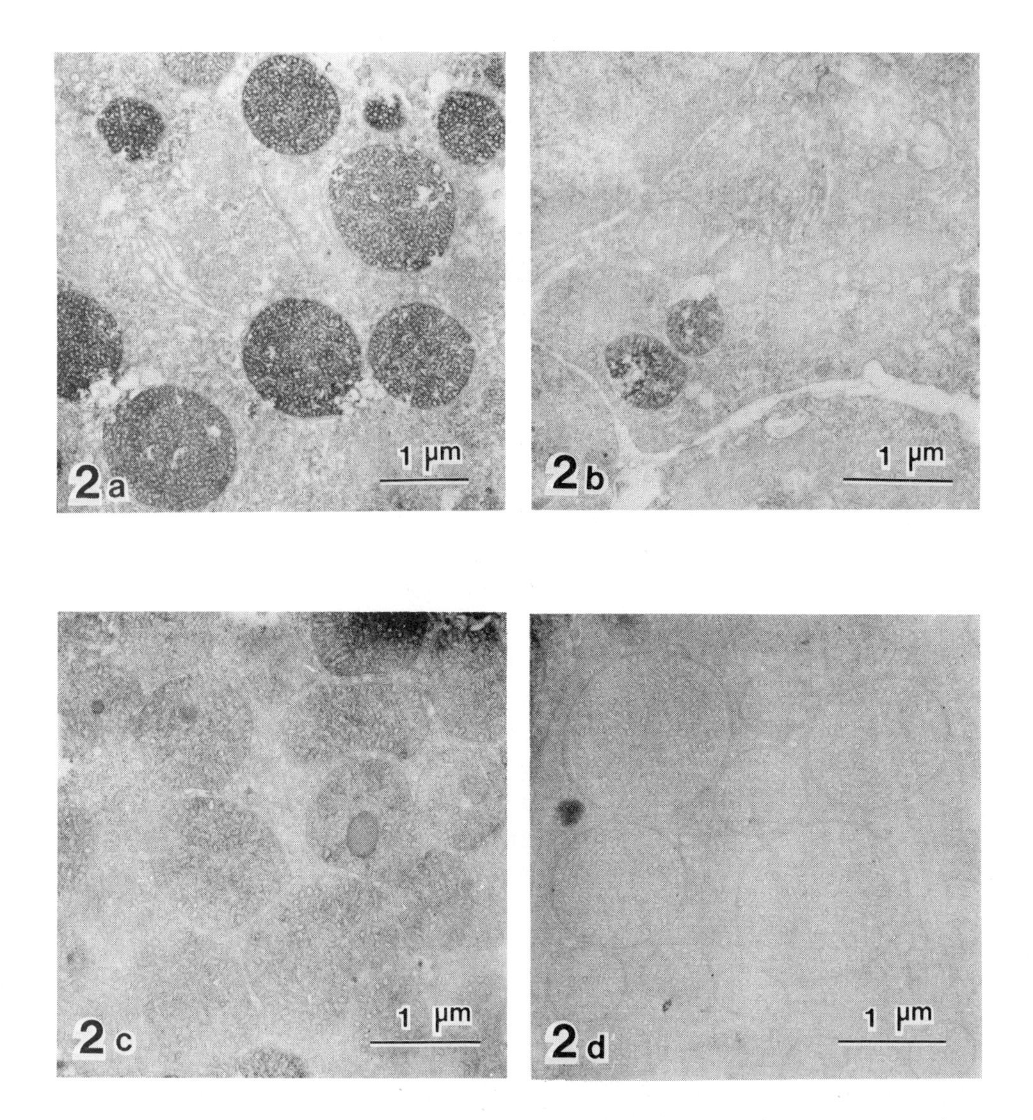

Fig. 2. Immunocytochemical stainings for adrenodoxin reductase in rat adrenal cortex: 2a, In zona fasciculata cells from untreated rats; 2b, In zona glomerulosa cells from untreated rats; 2c, In zona fasciculata cells from ACTH-treated (for 1 week) rats; 2d, Control staining with HRP-labeled normal rabbit IgG (Fab' fragment) in zona fasciculata cells from untreated rats.

not with the outer membrane, spaces between the inner and outer membranes, cytosol nor other intracellular organelles. Almost identical results were obtained with P450scc as shown in Fig. 1b. No such a specific immunochemical staining was observed with anti-mouse kidney basement membrane used as a control (Fig. 1c).

Figs. 1a and 1b also show that about a half of the mitochondria were found always negatively stained for both P450scc and P45011β. Similar results have been obtained for both Ad and AdR (1). These results should not be due to technical artifacts such as the difficulty of the labeled antibody penetration, for essentially the same results were obtained even when the isolated mitochondria were stained for each P450. Furthermore, changes of physiological (hormonal) states of the animals dictated the staining patterns of the systems as described below.

In untreated rats, AdR-positive mitochondria were about a half of total mitochondria in a cell from zona fasciculata (Fig. 2a) and less than a half in a cell from zona glomerulosa (Fig. 2b). However, when rats were administered with ACTH for a week, almost all of the mitochondria in the cells became positive for AdR as shown in Fig. 2c. In contrast, when rats were hypophysectomized for 2 weeks, the population of positively stained mitochondria was greatly reduced and was not restored to normal level until the animals were re-treated with ACTH for another week (4). All these results strongly suggest that there exists a functional heterogeneity among the mitochondria in the adrenocortical cell and that the degrees of such a heterogeneity reflect the physiological state of the animals.

REFERENCES

1. Mitani, F., Ishimura, Y., Izumi, S., and Watanabe, K. (1979) Acta Endocrinol., 90, 317.
2. Katagiri, M., Takemori, S., Itagaki, E. and Suhara, K. (1978) Methods Enzymol., 52, 124.
3. Mitani, F., Shimizu, T., Ueno, R., Ishimura, Y., Izumi, S., Komatsu, N. and Watanabe, K. (1982) J. Histochem. Cytochem. in press.
4. Izumi, S., Komatsu, N., Watanabe, K., Mitani, F. and Ishimura, Y. in preparation.

© 1982 Elsevier Biomedical Press B.V.
Cytochrome P-450, Biochemistry, Biophysics
and Environmental Implications, E. Hietanen,
M. Laitinen and O. Hänninen editors

ACTH MEDIATED INDUCTION OF CYTOCHROMES P-450 IN BOVINE ADRENAL CORTEX

M.R. WATERMAN[1], R. E. KRAMER[2], B. FUNKENSTEIN[2], J. L. McCARTHY[3], V. BOGGARAM[1], AND E. R. SIMPSON[1,2], [1] Dept. of Biochemistry and [2]Cecil H. and Ida Green Center for Reproductive Biology Sciences, University of Texas Health Science Center, Dallas, Texas 75235 and [3]Dept. of Biology, Southern Methodist University, Dallas, Texas 75275.

INTRODUCTION

The fasiculata-reticularis zone of the adrenal cortex contains four major forms of cytochrome P-450 which catalyze hydroxylation steps necessary for the conversion of cholesterol to cortisol. Cholesterol side chain cleavage cytochrome P-450 (P-450_{scc}) and 11β-hydroxylase cytochrome P-450 (P-$450_{11\beta}$) are localized in the inner mitochondrial membrane and are reduced by NADPH via adrenodoxin reductase and adrenodoxin, which are located in the mitochondrial matrix. 17α-Hydroxylase cytochrome P-450 (P-$450_{17\alpha}$) and 21-hydroxylase cytochrome P-450 (P-450_{C-21}) are localized in the endoplasmic reticulum and are reduced by NADPH via NADPH-cytochrome P-450 reductase which is also located in the endoplasmic reticulum.

Using confluent monolayer cultures of bovine adrenocortical cells, this laboratory has undertaken the investigation of the mechanism of induction by ACTH of various forms of cytochrome P-450 and their respective reducing systems. ACTH is well-known to have both an acute and a chronic effect on the production of cortisol by the adrenal cortex. Thus an understanding of the mechanism of action of this hormone on the synthesis of the enzymes involved in steroid hydroxylation will provide further insight into the regulation of the conversion of cholesterol to glucocorticoids.

MATERIALS AND METHODS

Monolayer cultures of bovine adrenocortical cells were prepared as previously described (1). Cells were maintained in the absence or presence of 1 μM ACTH for various periods of time. P-450_{scc} and P-$450_{11\beta}$ enzyme activities were determined in intact cells in the presence of appropriate inhibitors and precursors by measuring pregnenolone and cortisol levels, respectively, in culture medium, by means of radioimmunoassay. P-$450_{17\alpha}$ enzyme activity was determined in post-mitochondrial supernatants derived from cultured cells by measuring steroid products by HPLC using progesterone as a substrate. Enzyme synthesis in cultured cells was measured either by pulse-radiolabeling cells at various times following initiation of ACTH treatment or by isolating RNA from such cells followed by translation in the reticulocyte lysate translation system. In either instance, specific immunoisolation of newly synthesized proteins was performed followed by SDS-PAGE and quantitation of the resultant autoradiograms (1).

RESULTS

In Table 1 is shown the stimulation of P-450_{scc} and P-$450_{11\beta}$ activities measured in cell cultures following initiation of ACTH treatment.

TABLE 1
INDUCTION OF P-450_{scc} AND P-$450_{11\beta}$ ACTIVITIES BY ACTH

Enzyme Activity	Maximal Fold Induction	Time of Maximal Induction (hr)
Cholesterol→ Pregnenolone (P-450_{scc})	2.1	36
11-Deoxycortisol→ Cortisol (P-$450_{11\beta}$)	2.6	48
Deoxycorticosterone→ Corticosterone (P-$450_{11\beta}$)	2.0	48

In Table 2 is shown the induction of synthesis of P-450_{scc}, P-$450_{11\beta}$ and adrenodoxin measured by both pulse-radiolabeling cultured cells and translation of RNA isolated from cultured cells. In each case comparison was made between ACTH treated and untreated cells at different times following initiation of ACTH treatment.

TABLE 2
ACTH INITIATED INDUCTION OF SYNTHESIS OF MITOCHONDRIAL COMPONENTS OF THE STEROID HYDROXYLASE PATHWAY

	CELL LABELING		RNA TRANSLATION	
Protein	Maximal Fold Induction	Time of Maximum	Maximal Fold Induction	Time of Maximum
P-450_{scc}	9	36 hr.	6	36 hr.
P-$450_{11\beta}$	13	36 hr.	5	36 hr.
Adrenodoxin	2	36 hr.	5	36 hr.

It can be seen that synthesis of each of these components reached an optimal value 36 hours following initiation of ACTH treatment. This result was obtained in both cell labeling and RNA translation experiments. Preliminary studies indicate that the ACTH-mediated induction of the synthesis of adrenodoxin reductase also reached an optimal value at 36 hours.

In addition, this laboratory has undertaken investigation of the effect of ACTH on the microsomal components participating in steroid hydroxylation. P-$450_{17\alpha}$ activity was examined in post-mitochondrial supernatants prepared from cells cultured in the presence or absence of ACTH using progesterone as substrate and measuring 17OH-progesterone (17OH-P) by HPLC. A marked stimulation by ACTH of enzymatic activity was observed which optimized at 36 hours. When cells are cultured in the absence of ACTH, the rate of 17α-hydroxylation was 3.4 nmol 17OH-P formed /hr/mg protein. Following exposure of the cells to ACTH for 36 hr, this activity was increased to 72.4 nmol 17OH-P formed/hr/mg protein. This 20-fold stimulation of enzymatic activity in response to ACTH is the largest we have observed for any of the steroid hydroxylase systems studied. Synthesis of P-450_{C-21} has been examined by pulse-radiolabeling of cells and has been found to increase 12-fold in response to ACTH treatment. In this instance, the optimal rate of synthesis was observed 24 hours following initiation of ACTH treatment. Preliminary results on induction of synthesis of NADPH-cytochrome P-450 reductase indicate that its rate of synthesis reached a maximum as early as 12 hours after initiation of ACTH treatment.

As we have reported previously, all the mitochondrial components of this pathway are synthesized in the cytoplasm as higher molecular weight precursors which are presumably processed proteolytically upon insertion into the mitochondrion (2). We have found that P-450_{C-21} was synthesized in a cell-free translation system as the mature form in agreement with results obtained by Omura and his colleagues (3). In addition, we have found that NADPH-cytochrome P-450 reductase is also synthesized as the mature form in a cell free translation system. A similar observation has been made for NADPH-cytochrome P-450 reductase of rat liver (4).

DISCUSSION

We conclude from the above results that treatment of bovine adrenocortical cells in culture with ACTH leads to induction of synthesis and/or stimulation of activity of all the forms of cytochrome P-450 involved in the conversion of cholesterol to cortisol. The induction of synthesis of the mitochondrial components of this steroid hydroxylase pathway (cytochrome P-450_{scc}, cytochrome P-$450_{11\beta}$, adrenodoxin, adrenodoxin reductase) appears to be coordinately regulated. However, the induction of synthesis and/or stimulation of activity of the microsomal components of this steroid hydroxylase pathway (cytochrome P-450$_{17\alpha}$, cytochrome P-450$_{C-21}$, NADPH cytochrome P-450 reductase) does not appear to be coordinately regulated. The mechanism(s) involved in the regulation of the synthesis of these enzymes is currently under investigation as is the relationship of the activation of these enzymes to the physiological function of the adrenal cortex.

ACKNOWLEDGEMENTS

The skillful technical assistance of Janette Tuckey, Grace Carlson and Albert Dee is gratefully acknowledged. This research was supported in part by USPHS Grants AM 28350 and GM 27151 and Grant I-624 from The Robert A. Welch Foundation.

REFERENCES

1. DuBois, R. N., Simpson, E. R., Kramer, R. E. and Waterman, M. R. (1981) J. Biol. Chem., 256, 7000.

2. Kramer, R. E., DuBois, R. N., Simpson, E. R., Anderson, C. M., Kashiwagi, K., Lambeth, J. D., Jefcoate, C. R. and Waterman, M. R. (1982) Arch. Biochem. Biophys., 215, 478.

3. Nabi, N., Kominami, S., Takemori, S. and Omura, T. (1980) Biochem. Biophys, Res. Commun., 97, 687.

4. Gonzalez, F. J. and Kasper, C. B. (1980) Biochemistry, 19, 1970.

Cytochrome P-450 Tissue Specificity and Toxicity

© 1982 Elsevier Biomedical Press B.V.
Cytochrome P-450, Biochemistry, Biophysics
and Environmental Implications, E. Hietanen,
M. Laitinen and O. Hänninen editors

CYTOCHROME P-450 AND PROSTAGLANDIN SYNTHETASE CATALYZED ACTIVATION OF PARACETAMOL AND p-PHENETIDINE

PETER MOLDÉUS, BO ANDERSSON, ROGER LARSSON AND BJÖRN LINDEKE[1]
Department of Forensic Medicine, Karolinska Institutet, S-104 01 Stockholm (Sweden) and [1]Department of Organic Pharmaceutical Chemistry, Faculty of Pharmacy, BMC, University of Uppsala, S-751 23 Uppsala (Sweden)

INTRODUCTION

Paracetamol (acetaminophen) is an analgesic and antipyretic drug which, in large overdoses, can cause liver necrosis in man and laboratory animals (1,2). Renal damage caused by this drug has also been reported (3). In the liver paracetamol is metabolized to a reactive metabolite, presumably the N-acetyl-p-benzoquinoneimine, by a cytochrome P-450 dependent oxidation. The mechanism of this reaction is, however, still unclear.

Another enzyme capable of metabolizing paracetamol to a reactive metabolite is prostaglandin synthetase (PGS). This has been demonstrated with microsomes from both ram seminal vesicles (RSV) (4,5) and rabbit kidney medulla (5,6,7). A possible role for this type of activation in the nephrotoxicity of paracetamol has been suggested.

Paracetamol is a primary metabolite of phenacetin, a drug known to cause serious damage to the kidneys in both animals and man. Phenacetin itself is not activated by PGS but another of its primary metabolites, p-phenetidine, is. This activation results in the formation of products which have been shown to be genotoxic (8).

This paper summarizes some recent findings and ideas concerning the cytochrome P-450 and prostaglandin synthetase catalyzed activation of paracetamol and p-phenetidine.

METHODS

Hepatocytes were isolated from phenobarbital treated rats by collagenase perfusion (9). Microsomes from RSV and rat liver were isolated as described previously (10,11).

Hepatocyte incubations were performed in Krebs-Hepes buffer, pH 7.4. Incubations with RSV and rat liver microsomes were performed in a 0.1 M phosphate buffer, pH 8.0 and 7.4, respectively, supplemented with 1.0 mM EDTA.

Paracetamol glutathione conjugate formation was determined as described earlier (12). Irreversible binding to protein was determined according to the method of Jollow *et al.* (13). Reduced and oxidized glutathione were determined by HPLC, according to the method of Reed *et al.* (14).

RESULTS AND DISCUSSION

Much attention has been focused on the mechanism of cytochrome P-450 dependent activation of paracetamol. It is generally agreed that the ultimate reactive metabolite of the drug is the N-acetyl-p-benzoquinoneimine. The mechanism by which it is formed, however, has not yet been elucidated. Three possible pathways for its formation are shown in Fig. 1. Originally it was postulated that cytochrome P-450 catalyzed an N-oxidation of paracetamol to N-hydroxyparacetamol, which in turn was dehydrated to form the N-acetyl-p-benzoquinoneimine (15). There are, however, several reports now (16,17) which repudiate the N-hydroxy metabolite as an intermediate.

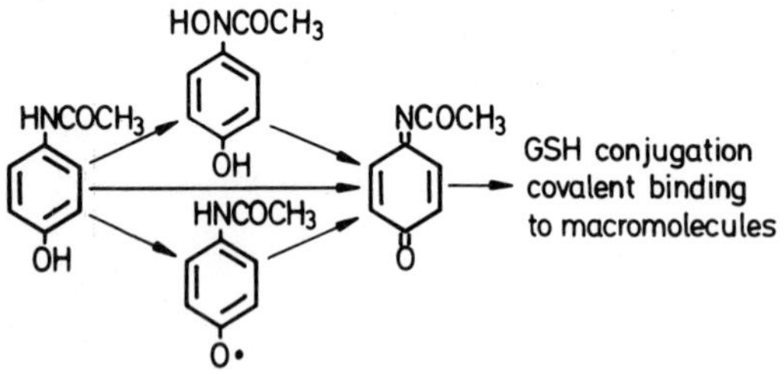

Fig. 1. Cytochrome P-450 catalyzed activation of paracetamol.

Since cytochrome P-450, in addition to acting as an oxygenase, may also function as a peroxidase a radical intermediate of paracetamol has been suggested (18). Such a mechanism would then involve a reaction of paracetamol with a ferryloxyradical complex of cytochrome P-450, resulting in a hydrogen abstraction to form a semiquinone or nitrenium radical. This paracetamol radical could then easily by oxidized by a rapid second electron transfer to produce the N-acetyl-p-benzoquinoneimine and a hydrated ferric cytochrome P-450 complex (18). Even though the idea of this pathway is attractive, no radical species of paracetamol has so far been detected. In isolated hepatocytes for instance, oxidation of GSH would be expected if a paracetamol radical was formed. As shown in Fig. 2 there was, however, only negligable GSSG formation when paracetamol was incubated with isolated hepatocytes. The very rapid depletion of GSH is, instead, directly correlated with the formation of a paracetamol glutathione conjugate. Thus, a radical of paracetamol is either not formed in the system during the cytochrome P-450 dependent reaction or it never escapes from the active site of the enzyme.

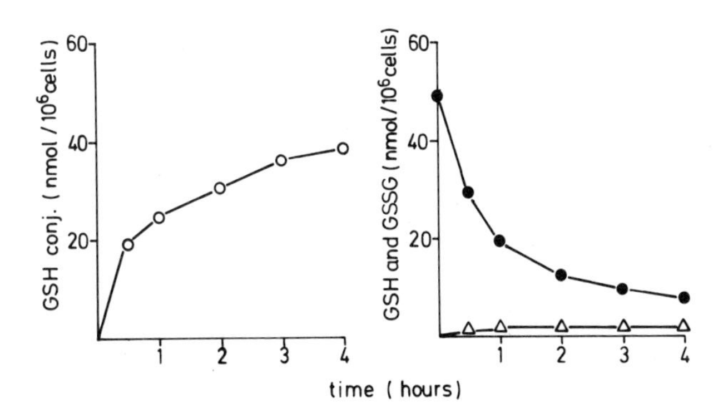

Fig. 2. Paracetamol glutathione conjugate formation, GSH and GSSG levels in hepatocytes from phenobarbital treated rat incubated in the presence of paracetamol. Paracetamol glutathione conjugate (o), GSH (●), GSSG (Δ).

Alternatively, the N-acetyl-p-benzoquinoneimine may be formed directly as suggested by Nelson *et al.* (18). In this case a perferryl form of cytochrome P-450 would react with paracetamol to give a ferric oxyamide complex which could readily decompose to give the N-acetyl-p-benzoquinoneimine.

That a radical of paracetamol can indeed be formed in peroxidase reactions although not involving cytochrome P-450, has been shown with both PGS and horse radish peroxidase (HRP) (5,18). In the PGS catalyzed reaction, for which RSV microsomes and arachidonic acid (AA) were used, formation of a paracetamol radical was indicated by a rapid oxidation of GSH in the presence of paracetamol (Fig. 3,5). A direct identification of a paracetamol radical could be achieved in the HRP/hydrogen peroxide reaction by EPR spectroscopy (18).

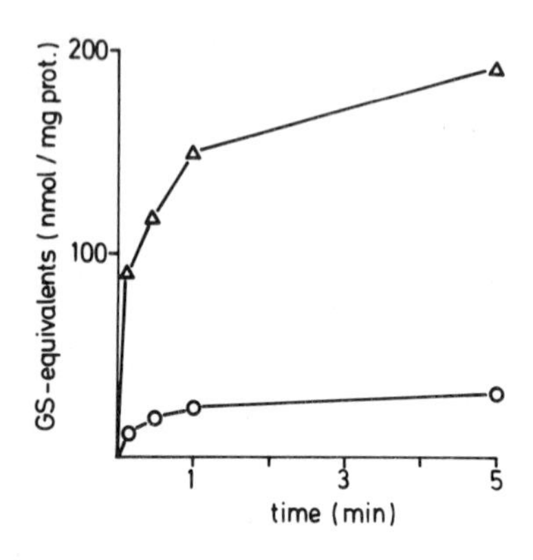

Fig. 3. Paracetamol glutathione conjugate formation and paracetamol dependent oxidation of GSH in an incubation with RSV microsomes (1 mg/ml), AA (300 μM) and GSH (1 mM).
GSSG (Δ), paracetamol glutathione conjugate (o).

The PGS catalyzed activation of paracetamol by RSV microsomes is very rapid and results, in the presence of GSH, not only in GSSG formation but also in the formation of a paracetamol glutathione conjugate (Fig. 3). This conjugate is apparently identical with that formed by liver microsomes in the presence of NADPH and GSH which thus indicates the formation of the N-acetyl-p-benzoquinoneimine also in the PGS catalyzed reaction. The quinoneimine may be formed either by further oxidation of the paracetamol radical or by disproportionation of two radicals.

A PGS catalyzed activation of paracetamol has been observed also using microsomes from rabbit kidney medulla (5,6,7). Since the PGS reaction has a very high affinity for paracetamol and also seems to be more active than the cytochrome P-450 dependent activation in the kidney, it may indeed be of importance for the nephrotoxicity of paracetamol.

TABLE 1

COVALENT BINDING TO PROTEIN OF p-PHENETIDINE IN DIFFERENT EXPERIMENTAL SYSTEMS

Experimental system	nmol/min per mg protein
RSV micr. + AA (100 μM)	14.7±0.4
Pb micr.[1] + CHP[2] (1 mM)	1.2±0.2
Pb micr.[1] + NADPH (gen. syst.)	0.2±0.05

p-Phenetidine concentration was 100 μM.
[1]Pb micr. = Microsomes from phenobarbital treated rats
[2]CHP = Cumene hydroperoxide

Paracetamol is a major metabolite of phenacetin and may thus contribute to the nephrotoxicity of the drug. Another primary metabolite of phenacetin which may be of importance for this toxicity since it is a substrate for PGS, (8, Table 1), is p-phenetidine. The PGS catalyzed metabolism of p-phenetidine, which is considerably faster than the cytochrome P-450 dependent one, results in the formation of protein binding products (Table 1) which have also been shown to induce single strand breaks in DNA (8). Several products from the PGS catalyzed reaction, have been separated on T.L.C. Subsequent masspectrometric investigations showed that the structures of some of the isolated products are compatible with trimers and dimers of p-phenetidine, conceivably formed through coupling reactions of initially formed p-phenetidine radicals.

In conclusion, the metabolic activation of paracetamol and p-phenetidine may be catalyzed both by cytochrome P-450 and peroxidases like PGS and HRP. Whereas the peroxidase catalyzed reaction seems to result in the formation of radical intermediates, no such intermediates have so far been detected during the cytochrome P-450 catalyzed reactions.

The PGS catalyzed activation of paracetamol and p-phenetidine may also be of relevance to the nephrotoxicity of paracetamol and phenacetin.

ACKNOWLEDGEMENTS

Supported by the Swedish Medical Research Council (grants Nos 5645-03A and 5918-02A).

REFERENCES

1. Proudfoot, A.T. and Wright, N. (1970) Br. Med. J., 3, 557-559.
2. Mitchell, J.R., Jollow, D.J., Potter, W.Z., Davis, D.C., Gillette, J.R. and Brodie, B.B. (1973) J. Pharmacol. Exp. Therap., 187, 185-194.
3. Boyer, T.D. and Rouff, S.L. (1977) J. Am. Med. Ass., 2187, 440-441.
4. Moldéus, P. and Rahimtula, A. (1980) Biochem. Biophys. Res. Commun., 96, 469-475
5. Moldéus, P., Andersson, B., Rahimtula, A. and Berggren, M. (1982) Biochem. Pharmacol., 31, 1363-1368.
6. Mohandas, J., Duggin, G.G., Horvath, J.S. and Tiller, D.J. (1981) Toxicol. Appl. Pharmacol., 61, 252-259.
7. Boyd, J.A. and Eling, T.E. (1981) J. Pharmacol. Exp. Therap., 219, 659-664.
8. Andersson, B., Nordenskjöld, M., Rahimtula, A. and Moldéus, P. (1982) Molec. Pharmacol., in press.
9. Moldéus, P., Högberg, J. and Orrenius, S. (1978) Meth. Enzym., 52, 60-68.
10. Egan, R.W., Paxton, J. and Kuehl, F.A. (1976) J. Biol. Chem., 251, 7329-7335.
11. Ernster, L., Siekevitz, P. and Palade, G.E. (1962) J. Cell Biol., 15, 541-562.
12. Moldéus, P. (1978) Biochem. Pharmacol., 27, 2859-2863.
13. Jollow, D.J., Mitchell, J.R., Potter, W.Z., Davis, D.C., Gillette, J.R. and Brodie, B.B. (1973) J. Pharmacol. Exp. Therap., 187, 195-202.
14. Reed, D.J., Babson, J.R., Beatty, P.W., Brodie, A.E., Ellis, W. and Potter, D.W. (1980) Anal. Biochem., 106, 55-62.
15. Hinson, J.A., Pohl, L.R. and Gillette, J.R. (1979) Life Sci., 24, 2133-2138.
16. Nelson, S.D., Forte, H.J. and Dahlin, D.C. (1980) Biochem. Pharmacol., 29, 1617-1620.
17. Mitchell, J.R., Jollow, D.J., Gillette, J.R. and Brodie, B.B., (1973) Drug Metab. Disp., 1, 418-423.
18. Nelson, S.D., Dahlin, D.C., Rauckman, E.J. and Rosen, G.M. (1981) Molec. Pharmacol., 20, 195-199.

Cytochrome P-450, Biochemistry, Biophysics
and Environmental Implications, E. Hietanen,
M. Laitinen and O. Hänninen editors

COMPARISON OF CARCINOGEN METABOLISM IN DIFFERENT ORGANS AND SPECIES

HERMAN AUTRUP AND ROLAND C. GRAFSTROM
Carcinogen Macromolecular Interaction Section, Laboratory of Human Carcinogenesis, National Cancer Institute, Bethesda, Maryland 20205 (U.S.A.)

INTRODUCTION

Most chemical carcinogens in the environment require metabolic activation to exert their mutagenic and carcinogenic effect. While the liver is considered to be the major site for metabolic conversion of xenobiotics, it may play a less important role in the activation of environmental carcinogens. The significance of extrahepatic metabolism, especially in tissues at the portals of entry of carcinogens, is of considerable interest as these tissues represent the major sites of human cancer. The development of models for culturing human target tissues (1) has made it possible to study the activation of chemical carcinogens in intact human tissues (2). Furthermore, they also allow comparative studies between human and experimental animals at the same level of biological organization. We (2-4) and others (7-11) have used these models to study activation of chemical carcinogens as determined by metabolic profile, binding to cellular macromolecules and to identify the nature of the carcinogen-DNA adducts.

MATERIALS AND METHODS

Tissues. Non-tumorous specimens of bronchus, colon, and esophagus from adult donors with or without cancers were collected at the time of surgery and autopsy. Specimens were transported to the laboratory in L-15 culture medium at 4°C, and the explants cultured in chemically defined medium for 1-7 days prior to carcinogen exposure (1). Rat tissues were obtained from male CD rats (4-5 weeks), and culture conditions identical to those of human tissues were used.

Incubation with carcinogens. To minimize the effect of exogenous agents, such as drugs and dietary constitutents, the tissues were cultured for 7 days and were then incubated with radioactively labeled [^{3}H] aflatoxin B_1 (AFB) or [^{3}H]benzo(a)pyrene (BP) at a final concentration of 1.5 M for 24 hrs. DNA was isolated from the surface epithelium of the explants and used for both quantitation and for identification of carcinogen-DNA adducts (5,12). Metabolites released into the media were extracted and analyzed by chromatographic procedures (5).

RESULTS AND DISCUSSION

Cultured bronchial, colonic, and esophageal tissues from both rats and humans metabolized BP and AFB to metabolites that bind to DNA. Autoradiographic methods as well as cell culture of epithelial and fibroblasts from the same individual has shown that the metabolic capability of epithelium is several fold higher than in the stroma (2). The highest mean binding level for both carcinogens was seen in human bronchus and esophagus, the binding level of BP generally being higher than that of AFB (Table 1). However, no

TABLE 1

COMPARISON OF DNA BINDING LEVELS OF BENZO(a)PYRENE (BP) AND AFLATOXIN B_1 (AFB) IN DIFFERENT HUMAN TISSUES

Organ	BP	AFB	Binding of BP > AFB[b]	Correlation Coefficient[b]
Bronchus	25[a](<1-160);26 cases	16(<1-92);28 cases	18/25	-0.20
Colon	8(<1-46);27 cases	8(<1-46);24 cases	14/22	-0.08
Esophagus	27(<1-104);11 cases	11(<1-35);11 cases	7/9	0.02

[a]Pmol per 10 mg DNA; mean (range); number of cases.
[b]Data from cases where the binding level of both carcinogens were measured.

correlation between the two binding levels was observed. We have previously reported a positive correlation between the binding levels of BP and 7,12-dimethylbenz(a)anthracene in human bronchus and between BP and 1,2-dimethylhydrazine in human colon (13,14). The lack of correlation between AFB and BP is probably due to the different forms of cytochrome P-450 being involved in formation of the ultimate carcinogen. Using a reconstituted activation system, it has been shown that the 3-methylcholanthrene-induced form of cytochrome P-450 from rat liver mostly converted AFB to aflatoxin M_1 (15), a metabolite that is not further metabolized. This form was also responsible for the formation of the ultimate carcinogenic form of BP and BP phenols (16). The binding level of BP in human tissues was higher than in rat, while the opposite was seen for AFB. The mean binding level of BP to human bronchus DNA was similar to the level in hamster DNA, a species susceptible to the lung carcinogenic action of BP. Among human tissues the highest level of binding for both carcinogens was seen in the bladder (Table 2). A wide interindividual variation in the binding levels was noted in all organs. The magnitude

TABLE 2

BINDING LEVEL OF BP AND AFB TO DNA IN CULTURED HUMAN TISSUES[a,b]

Tissue	Culture[c]	BP	AFB	Interindividual Variation in BP
Bladder	E	6.4	1.7	55
Bronchus	E	1.0	0.5	75
Colon	E	0.3	0.1	150
Duodenum	E	0.6	0.4	50
Endometrium	E	0.1	n.t.[d]	70
Esophagus	E	0.7	0.4	99
Keratinocytes	C	2.9	n.t.	--
Liver (fetal)	E	n.t.	0.7	--

[a]Refs. 3,6-8, 11.
[b]The results are expressed as number of modification per 10^6 nucleotides. Concentration of carcinogens used was $1.5 \mu M$. Incubation time was 24 hrs.
[c]E, Explants; C, Cells.
[d]n.t. = not tested.

of this variation ranges from 50- to 150-fold and is similar to that found in pharmacogenetic studies (17). This variation in human tissues complicates a quantitative comparison between animal and human data.

The adducts formed in cultured tissues and cells between the ultimate carcinogen and DNA is quite similar in all human organs and in tissues from experimental animals in which the compounds are carcinogenic. However, the adduct pattern is generally different from the pattern when liver microsomes are used as the activating system in combination with native DNA. In tissue explants, guanosine is the major target in DNA for the reaction with both AFB and BP. The activated forms of BP -- BP-7,8-diol-9,10-epoxides (BPDE) -- react predominantly with the exocyclic 2-amino group (5), while aflatoxin B_1-2,3-oxide reacts mostly with the N-7 position (Table 3) (12). Minor amounts of adduct formed between BPDE and adenine have only been observed in rat tissues (5).

Both BP and AFB are metabolized into organosoluble and water-soluble metabolites. The major metabolites of BP in both humans and rats were BP tetrols and BP 9,10-diol (Table 4), an observation different from cell-free systems where BP phenols are the major metabolites. However, in intact tissues the primary metabolites may serve as substrate for conjugation reactions. No qualitative differences in the metabolic profile of BP were observed between

TABLE 3

RELATIVE DISTRIBUTION OF AFLATOXIN B_1 DNA ADDUCTS IN DIFFERENT HUMAN EPITHELIAL CELL TYPES

Organ		Initial	III	II	Diol	I	Ratio I/II
Bronchus	Human	4.1	8.7	33.6	n.d.	41.3	1.23
	Rats	3.6	14.7	70.1	3.2	11.2	0.16
Colon	Human	73.0	16.8	7.1	n.d.	3.0	0.43
	Rats	3.4	14.5	63.4	4.1	14.4	0.23
Esophagus	Human	1.3	16.2	70.0	1.2	7.0	0.10
	Rats	3.4	10.1	46.6	2.8	37.0	0.80
Liver	Human	6.7	8.6	35.5	n.d.	47.8	1.35

DNA was isolated by phenol extraction from cells or explants treated with ^3H aflatoxin B_1, hydrolyzed and adducts separated on a μBondapack C_{18} column with 18% ETOH, 20 mM NH_4Ac, pH 5.1 at 1 ml/min.

TABLE 4

METABOLISM OF BP TO ORGANOSOLUBLE METABOLITES BY HUMAN AND RAT TISSUES[a,b]

Metabolite	Rat			Human		
	Colon	Bladder	Bronchus	Colon	Bladder	Bronchus
Tetrols	15.5	20.0	13	4.8	19.7	21
9,10-diol	20.6	20.0	12	19.8	21.1	28
7,8-diol	8.4	13.5	6	9.9	10.8	5
9-hydroxy	4.7	2.0	8	2.0	3.0	4
3-hydroxy	4.0	10.5	8	27.6	10.8	2

[a]Ref. 3 and 10.

[b]After incubation of the explant with [^{3}H] BP for 24 hrs, the media were extracted with ethylacetate/acetone, and the metabolites were separated by high pressure liquid chromatography.

different human organs and the corresponding rat tissues although quantitative differences do exist. This is probably due to different relative amounts of cytochrome P-450's in various organs and animal species. Another possibility is that the prostaglandin synthetase pathway appears to be involved in the oxidation of BP metabolites in cultured animal tracheal tissues. In our studies with human bronchus, we found a positive correlation between the amount of BP tetrols, BP 7,8-diol, and the level of binding to DNA (3). The major metabolite of AFB in all tissues was aflatoxin M_1, but smaller amounts of AFQ_1 and AFP_1 were also detected by high pressure liquid chromatography analysis of the media. However, organosoluble metabolites only represent 5-10% of the total metabolites.

CONCLUSIONS

Understanding of the metabolic activation of environmental carcinogens by human tissues is an important factor in the evaluation of the human risk to these compounds. By using explant culture methodology, we have shown that human tissues can 1) activate several classes of environmental carcinogens into metabolites that bind to DNA and 2) the metabolism of BP and AFB was qualitatively similar to that found in experimental animals in which these compounds are carcinogenic. The results emphasize the advantages of using intact cellular systems instead of a subcellular system for the study of carcinogen metabolism as the cellular system possess the capability for both activation and deactivation and, therefore, resemble more closely the *in vivo* situation.

ACKNOWLEDGEMENT

This project is the result of a collaboration between our laboratory, Dr. A. M. Jeffrey, Institute of Cancer Research, Columbia University, Dr. J. M. Essigmann, Department of Nutrition and Food Science, MIT, and Dr. B. F. Trump, Department of Pathology, University of Maryland School of Medicine.

REFERENCES

1. Harris, C. C., Trump, B. F., and Stoner, G. D. (1980) in Harris, C. C., Trump, B. F., and Stoner G. D. (Eds), Methods in Cell Biology, Vol. 21A and 21B, Academic Press, New York.
2. Autrup, H. (1982) Drug Metab. Rev., 13, 603.
3. Autrup, H., Grafstrom, R. C., Brugh, M., Lechner, J. F., Haugen, A., Trump, B. F., and Harris, C. C. (1982) Can. Res., 42, 934.
4. Autrup, H., Grafstrom, R. C., Christensen, B., and Kieler, J. (1981) Carcinogenesis, 2, 763.
5. Autrup, H., Wefald, F. C., Jeffrey, A. M., Tate, H., Schwartz, R. D., Trump, B. F., and Harris, C. C. (1980) Int. J. Cancer, 25, 293.
6. Harris, C. C., Trump, B. F., Grafstrom, R. C., and Autrup, H. (1982).
7. Daniel, F. B., Stoner, G. D., Sandwisch, D. W., Schenck, K. M., Hoffman, C. A., Schut, H. A. J., and Patrick, J. R. (submitted).

8. Dorman, B. H., Genta, V. M., Mass, M. J., and Kaufman, D. G. (1981) Cancer Res., 41, 2718.

9. Stampfer, M. R., Bartholomew, J. C., Smith, H. S., and Bartley, J. C. (1981) Proc. Natl. Acad. Sci. USA, 78, 6251.

10. Stoner, G. D., Daniel, F. B., Schenck, K. M., Schut, H. A. J., Goldblatt, P. J., and Sandwisch, D. W. (1982) Carcinogenesis, 3, 195.

11. Theall, G., Eisinger, M., and Grunberger, D. (1981) Carcinogenesis, 2, 581.

12. Autrup, H., Essigmann, J. M., Croy, R. G., Trump, B. F., Wogan, G. N., and Harris, C. C. (1979) Cancer Res., 39, 694.

13. Harris, C. C., Autrup, H., Stoner, G., Yang, S. K., Leutz, J. C., Gelboin, H. V., Selkirk, J. K., Connor, R. J., Barrett, L. A., Jones, R. T., McDowell, E. M., and Trump, B. F. (1977) Cancer Res., 37, 3349.

14. Autrup, H., Schwartz, R. D., Smith, L., Trump, B. F., and Harris, C. C. (1980d) Carcinogenesis, 1, 375.

15. Yoshizawa, H., Uchimaru, R., Kamataki, T., Kato, R., and Ueno, Y. (1982) Cancer Res., 42, 1120.

16. Gozukara, E. M., Guengerich, F. P., Miller, H., and Gelboin, H. V. (1982) Carcinogenesis, 3, 129.

17. Atlas, S. A., Vessell, E. S., and Nebert, D. W. (1980) Cancer Res., 36, 4619.

Cytochrome P-450, Biochemistry, Biophysics and Environmental Implications, E. Hietanen, M. Laitinen and O. Hänninen editors

PURIFICATION AND PROPERTIES OF CYTOCHROME P-450MC FROM RAT LUNG MICROSOMES AND ITS ROLE TO ACTIVATION OF CHEMICAL CARCINOGENS

MINRO WATANABE, IKUKO SAGAMI, TATSUYA ABE AND TETSUO OHMACHI
Research Institute for Tuberculosis and Cancer, Tohoku University, 4-1 Seiryo-machi, Sendai 980 (Japan)

INTRODUCTION

Hepatic cytochrome P-450s in rats were purified and their precise properties were widely examined. But, there was only one paper on the separation and purification of microsomal cytochrome P-450 (P-450) from rat lung, showing 0.66 nmoles per mg protein as a specific content of the purified form (1). Lung is apparently exposed to several xenobiotics present in urban air and also is seemingly one of the target organ to carcinogenic hydrocarbon in rodent and human species. It was well observed that the content of P-450 in the lung was very low, compared to that in the liver from the corresponding species. On the other hand, there were different properties of benzo[a]pyrene hydroxylation between lung and liver microsomes from mice (2), guinea pigs (3,4) and rats (5). Therefore, this paper aims to describe the characteristics of the highly purified form of lung P-450MC from 3-methylcholanthrene-treated rats.

MATERIALS AND METHODS

Purification of P-450. Pulmonary and hepatic P-450MC from 3-methylcholanthrene-treated Buffalo rats were purified by the modified method of published procedures (6,7). Hepatic P-450PB and NADPH-cytochrome P-450 reductase (f_{PT}) from phenobarbital-treated Buffalo rats were purified by published procedures (6,8). Either hepatic P-450MC or P-450PB was a major fraction of the microsomal P-450s, which consists of 56,000 and 53,000 Mr., respectively.

Analytical procedure. Assay of benzo[a]pyrene hydroxylation and 7-ethoxycoumarin O-dealkylation activities was performed as measuring the formation of 3-hydroxybenzo[a]pyrene and of 7-hydroxycoumarin, respectively, in the reconstituted systems, containing P-450, f_{PT}, dilauroylphosphatidylcholine, NADPH and $MgCl_2$ (9). Benzphetamine N-demethylation activity was expressed by the reduction rate of NADPH in the reconstituted system. Sodium dodesyl sulfate-polyacrylamide gel electrophoresis (SDS-PAGE) was performed with the modification of the method of Fairbanks et al (10), and polypeptide bands were detected by staining with silver nitrate (11). Limited proteolysis of cytochromes with either α-chymotrypsin, papain or *S. AUREUS* V8 protease was

TABLE 1

PURIFICATION OF PULMONARY P-450MC FROM 3-METHYLCHOLANTHRENE-TREATED RATS

Fraction	Protein (mg)	Pulmonary P-450MC Total (nmoles)	Specific Content (nmoles/mg protein)	Yield (%)	Purification (fold)
Microsomes	1334	53.1	0.04	100	1.0
Cholate extract	1202	56.9	0.05	107	1.3
Aminooctyl Sepharose	77.5	37.5	0.48	71	12
Hydroxyapatite	10.0	19.2	1.92	36	48
DEAE-Cellulose	1.4	9.4	6.63	18	166
CM-Sephadex	0.3	3.3	12.5	6	313

carried out according to the procedures of Cleveland et al (12).

RESULTS

Purification. Pulmonary P-450MC was purified approximately 313 fold, as shown in Table 1. The purified samples containing 12.5 nmoles per mg of protein were essentially free of f_{PT} and NADH-cytochrome b_5 reductase activities. SDS-PAGE of the purified pulmonary P-450MC gave one major band, showing 54,000 Mr., which was clearly different from hepatic P-450MC (56,000 Mr.) (13).

Spectral properties. Difference spectrum of the reduced hemoprotein-carbon monoxide complex expressed an absorption maximum at 448 nm, and absolute absorption spectrum of the oxidized form of this hemoprotein suggests to be low spin state of heme iron, as previously observed in hepatic P-450MC in the rats.

Electrophoretic profile of digested forms. Significant difference in the peptide pattern of P-450MC after the partial proteolysis with either *S. AUREUS* V8 protease or papin was detected between lung and liver, as shown in Figure 1. On the other hand, no difference was observed after the digestion with α-chymotrypsin, even though each sample of the P-450MC, which was purified as a main band on the first electrophoresis using 7.5% polyacrylamide, was again applied to the second electrophoresis using 12.5% polyacrylamide for examining the peptide pattern of the digested P-450MC. These data indicate that primary amino acid sequences of pulmonary P-450MC are partially different from those of hepatic P-450MC, suggesting the presence of tissue specificity in the P-450MC molecules.

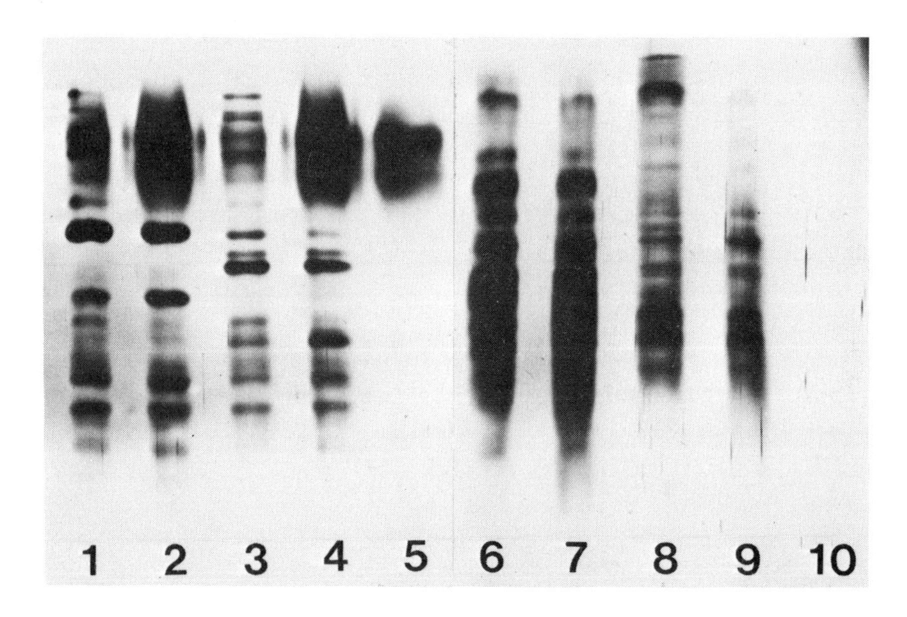

Fig. 1. Peptide patterns on SDS-PAGE after digestion of P-450MC with either V8 protease or papain. Track No. 1,2 Liver (11 μg) + Protease (0.5, 2.0 μg), 3,4) Lung (11 μg) + Protease (0.5, 2.0 μg), 5) Protease only (1.0 μg), 6,7) Liver (13 μg) + Papain (0.01, 0.05 μg), 8,9) Lung (13 μg) + Papain (0.01, 0.05 μg), 10) Papain only (0.05 μg). The gel is stained with silver nitrate.

Catalytic activity in the reconstituted system. The purified form of pulmonary P-450MC catalyzed apparently benzo[a]pyrene hydroxylation and 7-ethoxycoumarin deethylation as the hepatic P-450MC did, as shown in Table 2. The

TABLE 2

CATALYTIC ACTIVITIES OF THREE FORMS OF PURIFIED CYTOCHROME P-450 IN THE RECONSTITUTED ENZYME SYSTEM IN THE RAT

Substrate	Turnover Number[a]		
	Pulmonary P-450MC	Hepatic P-450MC	Hepatic P-450PB
Benzo[a]pyrene	11.9	17.1	1.0
7-Ethoxycoumarin	23.5	40.0	10.1
Benzphetamine	6.3	8.8	257

[a]nmoles of product formed/min/nmole of P-450

pulmonary P-450MC shows to be slightly lower activity in the hydroxylation reaction than the hepatic P-450MC, but apparently a lower activity of benzphetamine N-demethylation in the pulmonary P-450MC was observed, compared to the activity in the hepatic P-450PB. When considering the significance of the role of P-450MC in the activation system of chemical carcinogen, it remains to be clarified whether or not pulmonary P-450MC has a major role in polycyclic hydrocarbon carcinogenesis in the rats.

ACKNOWLEDGEMENTS

This research was supported, in part, by a Grant-in Aid for Cancer Research from the Ministry of Education, Science and Culture, Japan, and by a grant from the Japan Tabacco and Salt Public Corporation.

REFERENCES

1. Jernström, B., Capdevila, J., Jakobsson, S. and Orrenius, S. (1975) Biochem. Biophys. Res. Commun. 64, 814.
2. Watanabe, M., Konno, K. and Sato, H. (1978) Gann 69, 1.
3. Bilimoria, M.H., Johnson, J., Hogg, J.C. and Witschi, H.P. (1977) Toxicol. Appl. Pharmacol. 41, 433.
4. Abe, T. and Watanabe, M. (1982) Biochem. Pharmacol. 31, in press.
5. Wiebel, F.J., Leutz, J.C., Diamond, L. and Gelboin H.V. (1971) Arch. Biochem. Biophys. 144, 78.
6. West, S.B., Huang, M-T., Miwa, G.T. and Lu, A.Y.H. (1979) Arch. Biochem. Biophys. 193, 42.
7. Harada, N. and Omura, T. (1981) J. Biochem. Tokyo 89, 237.
8. Taniguchi, H., Imai, Y., Iyanagi, T. and Sato, R. (1979) Biochim. Biophys. Acta 550, 341.
9. Tamura, Y., Abe, T. and Watanabe, M. (1981) J. Toxicol. Sci. 6, 71.
10. Fairbanks, G., Steck, T.L. and Wallack, D.F.H. (1971) Biochemistry 10, 2606.
11. Merril, C.R., Switzer, R.C. and Van Keuren, M.L. (1979) Proc. Natl. Acad. Sci. U.S. 76, 4335.
12. Cleveland, D.W., Fischer, S.G., Kirschner, M.W. and Laemmli, U.K. (1977) J. Biol. Chem. 252, 1102.
13. Watanabe, M., Takahashi-Sagami, I., Tamura, Y. and Abe, T. (1982) in: Sato, R. and Kato, R. (Ed.), Microsomes, Drug Oxidation and Drug Toxicity, Japan Scientific Societies Press, Tokyo, in press.

Cytochrome P-450, Biochemistry, Biophysics and Environmental Implications, E. Hietanen, M. Laitinen and O. Hänninen editors

REGULATION AND ROLE OF ARYL HYDROCARBON HYDROXYLASE IN RAT ADRENAL AND GONADS

MARGOT BENGTSSON, JOHAN MONTELIUS, EINAR HALLBERG, LOUISE MANKOWITZ AND JAN RYDSTRÖM
Department of Biochemistry, Arrhenius Laboratory, University of Stockholm, S-106 91 Stockholm, Sweden

INTRODUCTION

The carcinogenicity of various polycyclic aromatic hydrocarbons (PAH) is dependent on both the structure of the PAH and the type of target organ (1). Particularly strong carcinogens are methylated derivatives (2,3), e.g., 7,12-dimethylbenz(a)anthracene (DMBA) which is especially potent in generating mammary cancer (4). DMBA is also adrenocorticolytic, i.e., it causes necrosis of the rat adrenal (4). In gonads DMBA efficiently destroys oocytes (5) and spermatogonia (4). In both adrenal (6) and ovary (7) aryl hydrocarbon hydroxylase (AHH) appears to be unrelated to the major steroid hydroxylases involved in the synthesis of corticosterone, testosterone and estradiol. The pituitary has been shown to exert a regulatory function on AHH in adrenal (8) as well as in testis (9).

MATERIALS AND METHODS

Adrenal glands were removed from at least 5 female Sprague--Dawley rats (>180 g) after decapitation and kept frozen until used. Subcellular fractionation and assay of benz(a)pyrene (BP) metabolism and cytochrome P-450 content were carried out as described earlier (5). Metabolism of estradiol was assayed by replacing (^{14}C)-BP with 50 μM (^{14}C)-estradiol (57 mCi/mmol), followed by HPLC analysis essentially as described previously (6).

Estrus cycle phases were determined by vaginal smears and the rats were divided into three groups, i.e., the proestrus, estrus and metestrus/diestrus phase groups. After decapitation the ovaries from at least 5 rats from each group were removed, pooled and fractionated immediately.

3-(1,2,3,4-tetrahydro-1-oxo,2naphthyl)-pyridine (SU-9055) was a gift from Ciba-Geigy (Basel, Switzerland).

RESULTS

As shown Table 1 estradiol was metabolized by rat adrenal microsomes at a rate similar to those of DMBA and BP (5). That the AHH involved in the estradiol metabolism most likely is identical to that metabolizing DMBA and BP is indicated by the fact that carbon monoxide and DMBA were potent inhibitors of estradiol metabolism. Likewise, the 17-hydroxylase and AHH inhibitor SU-9055 (5) also inhibited estradiol metabolism. In contrast, the epoxide hydrolase inhibitor trichloropropene oxide (TCPO) was without effect.

TABLE 1

METABOLISM OF ESTRADIOL BY RAT ADRENAL MICROSOMES

Conditions were as described in Materials and Methods. The concentration of estradiol was 50 μM. Additions were: 100 μM DMBA; 500 μM SU-9055; 100 μM TCPO; and bubbling with carbon monoxide for 1 min.

Addition	Activity (pmoles/min/mg prot)	Inhibition (%)
none	110.3	-
carbon monoxide	47.0	57
SU-9055	57.5	48
DMBA	42.7	61
TCPO	111.5	-

The adrenal does not synthesize detectable amounts of estrogens (5). However, it contains steroid receptors for both estradiol (10) and testosterone (11). Thus, it is conceivable that the physiological role of adrenal AHH may involve inactivation of estradiol by converting it to e.g. estriol. Indeed, preliminary results (not shown) indicate that the primary product of estradiol metabolism by adrenal microsomes is estriol, a relatively weak estrogen. The fact that estradiol metabolism is accompanied by covalent binding of metabolites to protein at a rate of 2.8 pmoles/min/mg protein, suggests that estradiol metabolism may be of importance in adrenal carcinogenesis (cf. ref. 12).

Conversions of steroids catalyzed by AHH *in vitro* may reflect a physiological role of AHH in steroid metabolism (cf. ref 13). This possibility is supported by the regulation of liver AHH by growth hormone (13), adrenal AHH by adrenocorticotropic hormone (8) and testicular AHH by luteinizing (LH) and follicle stimulating hormone (FSH) (10). However, neither of these effects have been demonstrated to occur under short term physiological conditions.

Ovary contains a microsomal AHH, the activity of which is inducible by PAH (14). As shown in Table 2 both microsomal AHH and cytochrome P-450 content in the rat ovary were markedly regulated by the estrus cycle. Maximal AHH activity occurred during the proestrus phase whereas the activities during the estrus and metestrus/diestrus phases were several-fold lower. These changes were parallelled by similar relative changes in microsomal cytochrome P-450 content. To our knowledge this is the first time a cyclic physiological regulation of AHH has been demonstrated. The results suggest that ovarian AHH is directly controlled by LH and/or FSH, a possibility which was tested by direct administration of pregnant mare's serum gonadotropins (PMSG) to the rats. Indeed,

TABLE 2

EFFECT OF THE ESTRUS CYCLE ON MICROSOMAL AHH ACTIVITY AND CYTOCHROME P-450 CONTENT IN THE RAT OVARY

Conditions were as described in Materials and Methods.

Activity[a]	Estrus cycle phase		
	proestrus	estrus	metestrus/diestrus
cytochrome P-450 (nmoles/mg protein)	0.045 (1)	0.009 (1)	0.019 (1)
AHH (pmoles/min/mg protein)	3.50 ± 1.18 (3)	0.67 ± 0.45 (3)	1.06 ± 0.56 (3)

[a]Numbers within parenthesis denote number of experiments.

the AHH activity was increased to about 9.4 pmoles/min/mg protein, i.e., approximately a 2.6 and 14-fold stimulation as compared to the AHH activities of the proestus and estrus phases, respectively. Which of the two pituitary hormones that is active remains to show. Experiments are now in progress which will clarify the mechanism of regulation of ovarian AHH by the pituitary, as well as the physiological role of AHH in the ovary.

ACKNOWLEDGEMENTS

This work was supported by the Swedish Cancer Society and the Swedish Council for Planning and Coordination of Research.

REFERENCES

1. Searle, C.E. (1976) Chemical Carcinogenesis, ACS monograp 173, American Chemical Society, Washington.
2. Hecht, S.S., Loy, M., Mazzarese, R. and Hoffman, D. (1978) in Gelboin, H.V. and Ts´o, O.P. Polycyclic Hydrocarbons and Cancer, Academic Press, New York, p. 119.
3. DiGiovanni, J. and Juchau, M.R. (1980) Drug Metab. Rev. 11, 61.
4. Huggins, C.B. (1979) Experimental Leukemia and Mammary Cancer. Induction, Prevention, Cure. The University of Chicago Press, Chicago and London.
5. Mattison, D.R. and Thorgeirsson, S.S. (1979) Cancer Res. 39, 3471.
6. Montelius, J. Papadopoulos, D., Bengtsson, M. and Rydström, J. (1982) Cancer Res. 42, 1479.
7. Bengtsson, M., Montelius, J., Mankowitz, L. and Rydström, J. (1982) submitted.
8. Guenthner, T.M., Nebert, D.W. and Menard, R.H. (1979) Mol. Pharmacol. 15, 719.
9. Lee, J.P., Suzuki, K., Mukhtar, H. and Bend, J.R. (1980) Cancer Res. 40, 2486.
10. Cutler, G.B., Barnes, K.M., Sauer, M.A. and Loriaux, D.L. (1978) Endocrinology 102, 252.
11. Gustafsson, J.-Å. and Pousette, Å. (1975) Biochemistry 14, 3094.
12. Tsibris, J.C.M., Eppert, J.E., Williams, A.G., Spellacy, W.N. and McGuire, P.M. (1978) in Gelboin, H.V. and Ts´o, O.P. Polycyclic Hydrocarbons and Cancer, Academic Press, New York, p. 361.
13. Gustafsson, J.-Å., Eneroth, P., Hansson, A., Hökfelt, T., Lefevre, A., McGeoch, C., Mode, A., Norstedt, G. and Skett, P. (1980) in Gustafsson, J.-Å., Carlstedt-Duke, J., Mode, A. and Rafter, J. Biochemistry, Biophysics and Regulation of Cyto chrome P-450, Elsevier/North-Holland, Amsterdam and New York, p. 171.
14. Mattison, D.R. and Thorgeirsson, S.S. (1978) Cancer Res. 38, 1368.

Cytochrome P-450, Biochemistry, Biophysics
and Environmental Implications, E. Hietanen,
M. Laitinen and O. Hänninen editors

DESTRUCTION OF CYTOCHROME P-450 BY A RUBBER ANTIOXIDANT

ANTTI ZITTING
Department of Industrial Hygiene and Toxicology, Institute of Occupational Health, Haartmaninkatu 1, SF-00290 Helsinki 29, Finland

INTRODUCTION

N-isopropyl-N'-phenyl-p-phenylenediamine (IPPD) is a common antioxidant in black rubber at concentrations of 0.5 - 1.5%.

IPPD

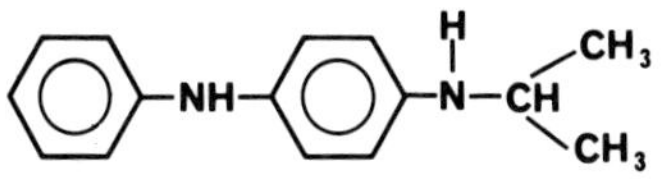

N-isopropyl-N'-phenyl-phenylene-diamine

It has been recently demonstrated that IPPD oxidizes and denaturates hemoglobin as well as causes hemolysis and Heinz body formation in red cells (1). IPPD can react with free radicals and form less reactive species inhibiting the propagation of autooxidation of rubber. The effects on hemoglobin prompted a study of the action of IPPD on microsomal cytochrome P-450.

MATERIALS AND METHODS

Ca^{++}-aggregated microsomes were prepared from the livers of adult male Wistar rats. IPPD (Santoflex^{R}IP, Monsanto, U.S.A.)was incubated with microsomes (0.7 mg protein/ml) in air-saturated 0.1 M phoshate buffer (pH 7.4) at 37°C. IPPD was added as a 0.1 M stock solution in acetone. To the control incubations, the similar amounts of acetone were added. Cytochrome P-450 contents were assayed according to Omura and Sato (2).

The anaerobic incubations (37°C) were performed in closed tubes after the microsomal suspensions were first deoxygenated with nitrogen bubbling for 15 min at 0°C. NADPH and ascorbic acid were added to the incubations in phosphate buffer (0.1 M, pH 7.4).

Ethoxycoumarin deethylase activity and the amount of malondialdehyde were determined according to Aitio (3) and Högberg *et al.* (4), respectively.

RESULTS AND DISCUSSION

The content of cytochrome P-450 and the activity of ethoxycoumarin deethylase decreased significantly when microsomes were incubated with IPPD at the concentration of 20 μM or more (Fig. 1 and 2).

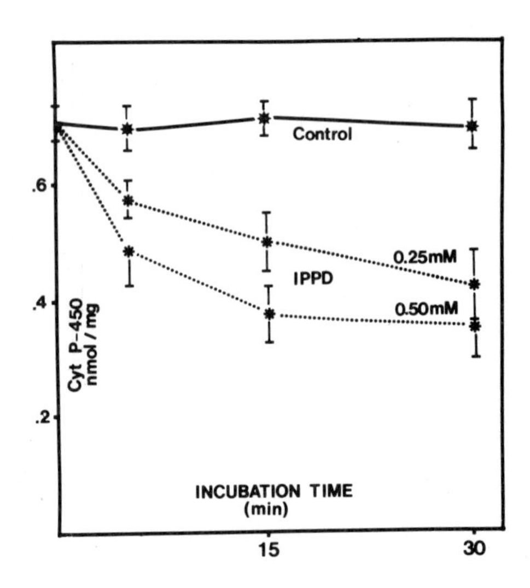

Fig. 1. Decrease of microsomal cytochrome P-450 content caused by IPPD.

The absorbance of the carbon monoxide complex of cytochrome, however, did not increase (Fig. 3). The loss of the absorption peak at 450 nm was independent of the presence of NADPH (0 - 1 mM).

IPPD effectively inhibited the NADPH-dependent microsomal lipid peroxidation which was measured as a production on malondialdehyde (Fig. 4.).

The absence of oxygen from the incubation mixtures as well as the presence of ascorbic acid (5 μM) completely protected cytochrome P-450 against IPPD-induced destruction (100 μM) during 60-minute incubations.

The obtained results suggest that IPPD does not destruct cytochrome P-450 "suicidally", i.e. through the formation of a covalent

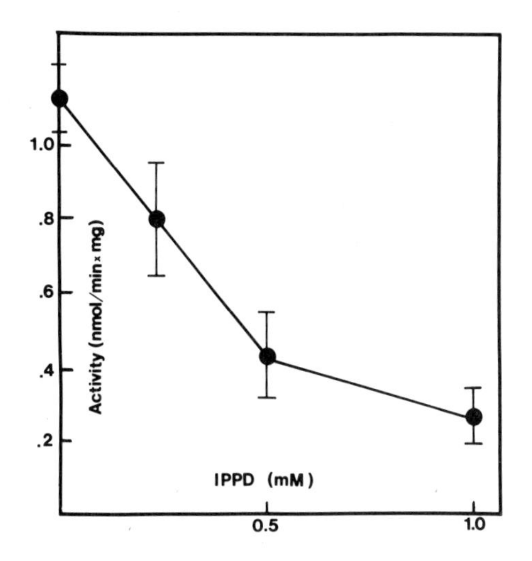

Fig. 2. Decrease of microsomal ethoxycoumarin deethylase activity after 30-minute incubation with IPPD.

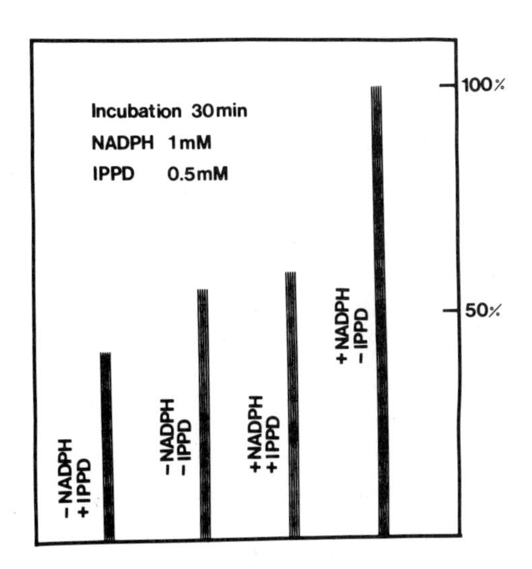

Fig. 3. Effect of IPPD on microsomal malondialdehyde production.

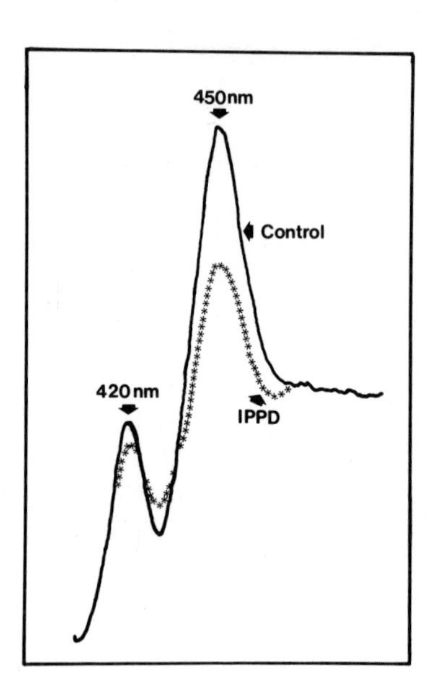

Fig. 4. Effect of IPPD on carbon monoxide compex spectrum of microsomal cytochrome P-450.

adduct with the cytochrome heme after metabolic activation, because NADPH is not needed for the reduction of the peak at 450 nm. The loss through lipid peroxidation is also unprobable - on the contrary, IPPD decreases the production of malondialdehyde.

The fact that oxygen is needed for the activity of IPPD and the inhibitory effect of ascorbic acid point to a mechanism in which IPPD reacts with oxygen or intermediates of oxidative processes. The produced derivative of IPPD could then destroy cytochrome or react with ascorbic acid which inactivates it.

REFERENCES

1. Williamson, D., Winterbourn, C.C., Swallow, W.H. and Missen, A.W. (1981) Hemoglobin, 5, 73.
2. Omura, T. and Sato, R. (1964) J. Biol. Chem., 239, 2370.
3. Aitio, A. (1978) Anal. Biochem., 85, 488.
4. Högberg, J., Orrenius, S. and Larsson, R.E. (1975) Eur. J. Biochem., 50, 595.

Cytochrome P-450, Biochemistry, Biophysics
and Environmental Implications, E. Hietanen,
M. Laitinen and O. Hänninen editors

MULTIPLE FORMS OF INSECT CYTOCHROME P-450: ROLE IN INSECTICIDE RESISTANCE

MOISES AGOSIN
Department of Zoology, University of Georgia, Athens, Ga. 30602, USA

ABSTRACT

The microsomal fraction from several insect species contains an electron-transport system which is strikingly similar to that of mammalian liver (1). Low levels of microsomal monooxygenase activity are usually associated with susceptibility to insecticides; whereas, high activity correlates with varying degrees of resistance. However, considerable debate has centered on the apparent lack of correlation between levels of cytochrome P-450 and *in vivo* oxidation of insecticides. This lack of correlation may now be explained as due to the existence of multiple cytochrome P-450 species which may differ in catalytic and other properties (2). Measurement of cytochrome P-450 by its CO-difference spectrum represents the total hemoprotein pool which may underestimate large amounts of specific hemoprotein forms. This may be particularly relevant during induction by chemical insecticides. Several insect cytochrome P-450 species have been partially purified in our laboratory and their properties are briefly summarized.

INTRODUCTION

Two major cytochrome P-450 systems have been reported in insects, microsomal and mitochondrial. The microsomal system is involved in the monooxygenation of xenobiotics, including insecticides with an apparently endless array of chemical specificities. Endogenous substrates such as juvenile hormones are also metabolized. The microsomal cytochrome P-450 system may represent the most important detoxication mechanism available to insects when exposed to a foreign chemical. Detoxication reactions are quite similar in insects and mammals; although in Phase II, conjugation is catalyzed by glucosyl rather than by glucuronyl transferases. The mitochondrial system metabolizes polyhydroxylated ketosteroids, such as ecdysteroids, which are involved in the regulation of insect development and metamorphosis. A mitochondrial cytochrome P-450 which hydroxylates α-ecdysone into β-ecdysone — the active hormone — has been characterized in *Manduca sexta*. Mitochondria contain an iron-sulfur protein tentatively identified as ferrodoxin which appears to function as adrenodoxin in mammals. However, a dual localization for ecdysteroid metabolism has been postulated (3).

Most of the work on insect microsomal cytochrome P-450 has been conducted with the hemoprotein associated to microsomal membranes, and few reports on the purification and reconstitution of the system are available (3). As in mammals, substrate specificity resides in the cytochrome P-450 fraction as phospholipid and NADPH-cytochrome P-450 reductase are interchangeable (2). The insect cytochrome P-450 is a single polypeptide whose levels are modulated through induction by drugs, hormones and a variety of environmental chemicals including those present in the diet. As in mammals, several isozymes have been reported; and differences in substrate specificity and regio- and stereo-specificity of the various forms seem to be involved in the type of monooxygenase activity observed toward a given chemical under *in vivo* and *in vitro* conditions. Studies on insect cytochrome P-450 are hindered by the usually small size of the organisms which often makes isolation of individual tissues impossible and by low specific content of the hemoprotein(s). In spite of this, insects are an interesting model which may contribute to the elucidation of important questions such as: a) Is it possible to extrapolate *in vitro* monooxygenase activity with either microsomes or reconstituted systems to *in vivo* conditions? b) What is the role of the microsomal system in insecticide metabolism and hence in resistance and cross-resistance to insecticides? c) How is activity of cytochrome P-450 controlled and what is the role of various cytochrome P-450 forms? Because of rapid advances made in the field of mammalian hemoproteins, we tend to extrapolate results and overemphasize the role that this system may play in *in vivo* reactions. Maybe this is the time to approach this problem with a broader perspective.

LEVELS OF CYTOCHROME P-450 AND MONOOXYGENASE ACTIVITY

A survey of reactions catalyzed by insect cytochrome P-450 reveals the high versatility of the system. They include aromatic hydroxylation (naphthalene, carbaryl, Baygon); alycyclic hydroxylation (dihydroaldrin, dihydroisodrin, rotenone); aliphatic hydroxylation (DDT, landrin, pyrethroids); *O*-dealkylation (metoxychlor, Baygon); *N*-dealkylation (various carbamates, parathion); cyclic ether cleavage (1,3-dioxoles); ester cleavage (parathion, diazinon); *N*-methyl hydroxylation (*N*-methyl carbamates); sulfooxidation (mesurol); desulfuration (Schradan, gluthion, parathion); epoxidation (aldrin, isodrin, heptachlor, juvenoids); dehalogenation (γ-hexachlorocyclohexane); and reduction (γ-hexachlorocyclohexane) (3). The contents of cytochrome P-450 in insects ranges from about 0.05 to 0.5 nmol/mg microsomal protein; although in certain specific tissues, such as *Periplaneta americana* fat body, it may reach up to 0.8 nmol/mg protein. Low levels of microsomal monooxygenase activity frequently are associated with susceptibility to insecticides. Whereas, high monooxygenase activity correlates with varying degrees with resistance. The phenomenon has been mostly studied in houseflies where higher levels of monooxygenation toward a given insecticide are often paralleled by high LD_{50} values for the same insecticide or a chemically unrelated compound. However, as seen in Table I, although high levels of activity appear to be translated into increased resistance toward Baygon, the correlation among resistance and levels and turnover numbers of cytochrome P-450 is far from perfect. The Rutgers strain of houseflies metabolizes four times more Baygon than the susceptible strain; whereas, the Fc strain metabolizes eleven times more; however, the LD_{50} for Baygon are 28.5- and 7.7-fold higher, respectively, than in the susceptible NAIDM strain. A similar situation occurs with diazinon. However, cytochrome P-450 inhibitors increase the LD_{50} values in both cases. *In vitro* to *in vivo* extrapolation is difficult because of competing enzymes. Furthermore, patterns in a species may not be generalized. Thus, a carbofuran derivative is ring-hydroxylated in houseflies, but in mammals is excreted as a conjugate plus a *N*-methylol derivative (Fig. 1). In extreme cases substrates that are metabolized to a single product by cytochrome P-450 in reconstituted systems are not metabolized at all by cytochrome P-450 in systems that closely resemble *in vivo* conditions, such as hepatocytes in suspension (4).

TABLE I

MONOOXYGENASE ACTIVITY OF HOUSEFLY MICROSOMES AND INSECTICIDE RESISTANCE

Strain	Cyt. P-450		Baygon hydroxylated		LD_{50}, μg/fly	
	nmol/mg protein	nmol/ fly	nmol/nmol P-450	nmol/ fly	Baygon	diazinon
NAIDM (S)	0.18	0.049	1.0	0.049	0.35	< 0.1
Rutgers (R)	0.47	0.116	1.7	0.197	> 10	7.1
Fc (R)	0.26	0.100	5.48	0.548	2.7	4.3

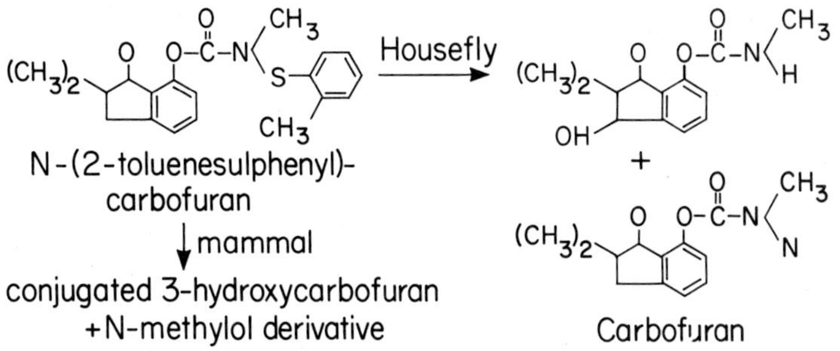

Fig. 1. *In vivo* metabolism of a carbofuran derivative in mammals and insects.

EFFECT OF INDUCTION

The effect of induction has been minimized because the correlation among levels of cytochrome P-450, catalytic activity and resistance is in many instances nonconclusive. Phenobarbital increases slightly the amount of cytochrome P-450 in the NAIDM strain of housefly and very markedly levels of epoxidase; while the opposite occurs in the resistant Rutgers strain (Table II). Phenobarbital on the other hand increases cytochrome P-450 in the Fc strain which is paralleled by an increase in Baygon hydroxylation, a phenomenon which does not occur in the Rutgers strain. Again, although naphthalene increases cytochrome P-450 to similar levels as phenobarbital, the increase in Baygon hydroxylation is much higher. However, while induction by naphthalene in the Fc strain is accompanied by increases in LD_{50}s toward Baygon, this does not occur with phenobarbital; although a small degree of tolerance is afforded by phenobarbital treatment in the normally susceptible NAIDM flies. It is now apparent that the extent of induction by various chemicals is a function of the inducer used, the dosage, the species and strain of insects, developmental stage, sex and diet. When all these factors are considered, selective induction may increase resistance under experimental conditions as it occurs with naphthalene.

TABLE II

EFFECT OF INDUCERS ON CYTOCHROME P-450 AND MONOOXYGENASE ACTIVITY

Strain	Inducer	nmol P-450/ mg protein	Aldrin epoxidation	Baygon hydroxylation	LD_{50} Baygon (μg/fly)
		(Percentage of Control Values)			
NAIDM (S)	Phenobarbital	135	417[a]	---	0.76[b]
Rutgers (R)	Phenobarbital	254	125	92	---
Fc (R)	---	100	---	100	2.7
Fc (R)	Phenobarbital	170	---	145	1.75
Fc (R)	Naphthalene	173	---	200	6.15

[a]Data from Moldenke & Terriere (7)
[b]Data from Yu & Terriere (8)

MULTIPLE CYTOCHROME P-450S AND RESISTANCE

The observation, that susceptible insects usually have lower amounts of cytochrome P-450 than resistant ones coupled to the fact that in many instances high monooxygenase activity may be linked to the phenomenon of resistance, suggested the possibility that susceptible insects have a cytochrome P-450 different from that of resistant insects. Initial observations indicated that susceptible strains (NAIDM and CSMA) had a cytochrome P-450 that absorbs maximally at 452 nm in the CO-difference spectrum; whereas, resistant strains had either a cytochrome P-450 (Fc strain) or P-448 (Rutgers strain). It was also speculated that these differences in spectral properties could be translated into differences in substrate specificities and turnover rates which would account for the differences in tolerance to insecticides. However, results obtained by induction (Table II) remained unexplained until the concept of multiple cytochrome P-450s in insects was considered (3). Induction results in shifts in the maximum absorbance in the CO-difference spectrum not only in resistant strains but also in susceptible ones (Table III). Additional spectral evidence suggested that induction gave rise to hemoproteins which were different than those found in noninduced insects (5). It was found that induction was paralleled by increases in microsomal peptides of varying molecular weights and the pattern of peptide induction depended on the inducer and strain used (Table III). All strains examined show multiple peptides in a range of 40,000 to 60,000 m.w. Predominantly peptides in the Fc and Rutgers strains are 43,000; 44,000; and 53,000 m.w. Induction of NAIDM houseflies with phenobarbital caused increases in peptides of 45,000; 48,000; and 53,000 m.w.; whereas, in the Fc strain the 53,000 peptide decreased consistently. Two of the cytochrome P-450s were partially purified from the Fc strain, a P-450 and a P-448 species (1). The cytochrome P-450 species had a m.w. of 45,000 but cytochrome P-448 corresponded to an aggregate. Similarly three species of cytochrome P-450 were partially purified from the susceptible NAIDM strain with absorbance maxima in the CO-difference spectrum of 450, 452 and 453 nm and m.w. of 43,000; 48,000; and 53,000, respectively (6). This work established that insect microsomes contain multiple species of cytochrome P-450, and it seems that at least six different species may be present in housefly microsomes (3). It is obvious that the spectral studies conducted with insect microsomes have been made with mixtures of the

TABLE III

TYPE OF CYTOCHROME P-450 FOUND IN DIFFERENT HOUSEFLY STRAINS AFTER INDUCTION (3)

Strain	Inducer	Absorption maximum of CO-reduced complex, nm
NAIDM (S)	---	452
	α-pinene	451
	β-naphthoflavone	452
	Phenobarbital	449
Fc (R)	---	450
	Phenobarbital	448
	Naphthalene	448
Rutgers (R)	---	448

cytochrome P-450 hemoproteins and that the differences between strains are related to proportions of individual forms and nature of the species of hemoprotein present. Estimation of cytochrome P-450 content by the CO-difference spectrum represents the hemoprotein pool level which may lead to an underestimation of specific constitutive or induced species. This may explain the lack of correlation in many cases among levels of cytochrome P-450 in control as well as in induced insects, catalytic activities and resistance. It may not be surprising that resistance to a given insecticide brought about by monooxygenation may be afforded by a cytochrome P-450 species which occurs in much larger amounts in resistant and/or induced insects than suspected.

DROSOPHILA CYTOCHROME P-450S

Preliminary work (2) and data of Table IV are indicative of difficulties to demonstrate multiple cytochrome P-450s in insects. Although the purification factors are remarkably high, the purity of the hemoproteins is not as desirable due to its low initial specific content. This, coupled to low yields of hemoproteins from about 400 gm insects, makes purification quite impractical unless better fractionation procedures are devised. Nevertheless, these results clearly support multiple forms of cytochrome P-450 in Drosophila, which have different molecular weights and spectral properties (2). However, an additional complication arises when spectral properties of these hemoproteins are examined at

TABLE IV

PURIFICATION OF DROSOPHILA CYTOCHROME P-450

Fraction	Cyt. P-450 nmol	Specific content nmol/mg protein	Yield, %	Purification factor
I Microsomes	220	0.066	100	1
II 0.6% cholate	171	0.15	78	2
III Octylamino-sepharose 4B	59	3.3	13	50
IV Hydroxyl apatite				
0.05 M $PO_4^=$	5.9	5.3	2.7	80
0.10 M $PO_4^=$	10.6	5.9	4.8	89
0.20 M $PO_4^=$	5.9	6.8	2.7	103
V Octylamino-sepharose 4B				
$P\text{-}450_I$	2.5	6.4	1.1	96
$P\text{-}450_{II}$	5.3	7.1	2.2	108
$P\text{-}450_{III}$	5.4	8.2	2.4	124
VI Phenyl sepharose				
$P\text{-}450_I$	1.7	8.0	0.8	121
$P\text{-}450_{II}$	3.7	9.7	1.7	147
$P\text{-}450_{III}$	0.6	11.7	0.3	177

purification stage V. It was found that the species show an absolute CO-reduced complex with a maximum absorbance at about 450 nm and a second band which coincides with the reduced form of the hemoproteins at about 416 nm. The second band decreases with time as the CO-reduced complex increases until no further changes are observed after 40 to 70 min. At this stage it was not clear whether the second band corresponded to a fraction of cytochrome P-450 that is incompletely reduced as it was suggested for purified cytochrome P-450 from houseflies (1), to contaminating pigments, to formation of cytochrome P-420 or to the presence of cytochrome P-450-adduct complexes which would be unavailable for carbon monoxide binding. This problem was reexamined with cytochrome P-450_{III} at purification stage VI. The CO-difference spectrum, which reveals little if any cytochrome P-420 after 40 min, increases with time quite substantially (Fig. 2). The reduction of the hemoprotein appears to occur rapidly with an immediate shift from 415 to 417 nm; but the binding of carbon monoxide in the absolute spectrum, initially obscured by the contribution of the reduced form, increases with time while the reduced form concomitantly decreases until no further changes are observed after 70 min. Although more extensive purification of these hemoproteins may be needed to resolve this problem, it may appear from

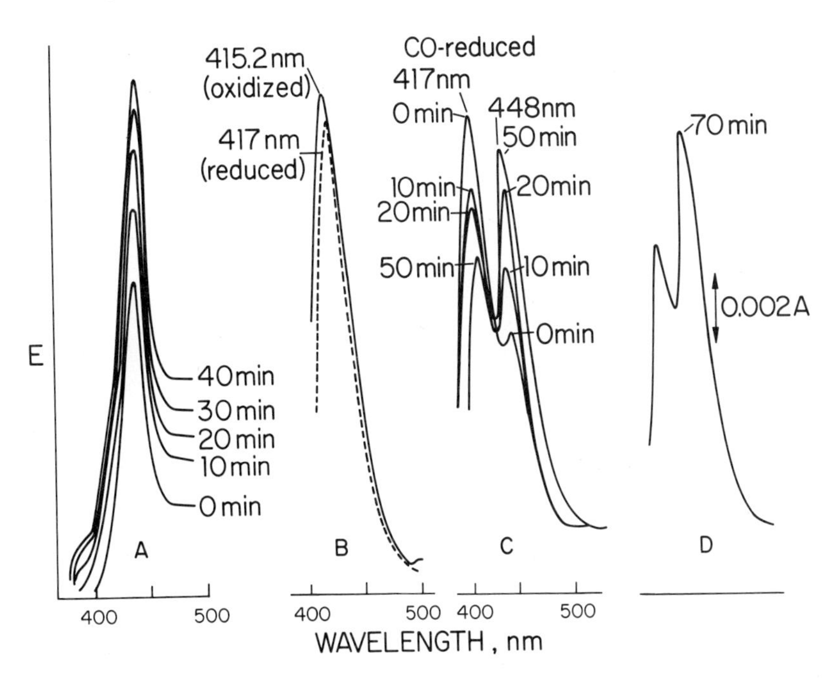

Fig. 2. Spectral characteristics of purified cytochrome P-450_{III} from Drosophila: a) CO-difference spectrum, b) absolute oxidized and reduced spectra, c) absolute CO-reduced spectra, d) absolute CO-reduced spectrum after 70 min.

these observations that, at least in Drosophila and possibly in housefly and other insect species, cytochrome P-450 may contain tightly bound endogenous substrates which may make them unavailable for catalysis in varying extents. We do not have information on the nature of the endogenous substrates that may bind to cytochrome P-450, but a good guess may point to steroids of the ecdysteroid type. The endogenously bound substrates are not displaced by either ferricyanide or 7-ethoxycoumarin. From the absolute CO-reduced spectra, it was estimated that about two-thirds of cytochrome P-450_{III} is available for CO-binding compared to only one-third in cytochrome P-450_{I} and P-450_{II}.

The three cytochrome P-450 species show different substrate specificities (Table V). Cytochrome P-450_{I} and $_{II}$ have a high turnover number for aldrin epoxidation but cytochrome P-450_{I} has low activity toward 7-ethoxycoumarin as opposed to cytochrome P-450_{II}. Cytochrome P-450_{III} has low activity for 7-ethoxycoumarin and aldrin. These differences in turnover numbers clearly justify the assumption that resistance to a given insecticide may be related to the presence of specific cytochrome P-450 species whose levels have to be determined by procedures other than the CO-difference spectrum. Whether the cytochrome P-450-adduct complexes in insects play a role in modulating the activity of these hemoproteins toward insecticides and other substrates remains to be established. In conclusion, the role of cytochrome P-450 in insecticide resistance may be better defined only when the precise number of constituted and induced species is known; their number may be larger than expected.

TABLE V

MONOOXYGENASE ACTIVITY OF DROSOPHILA CYTOCHROME P-450S IN RECONSTITUTED SYSTEMS. RECONSTITUTION WAS ESSENTIALLY AS DESCRIBED (2) UTILIZING NADPH-CYTOCHROME P-450 REDUCTASE PURIFIED FROM RAT LIVER

Cytochrome P-450	Substrate metabolized, nmol/nmol P-450/min	
	7-ethoxycoumarin	Aldrin
I	1.96	13.6
II	14.02	11.1
III	0.31	2.26

ACKNOWLEDGEMENT

Supported in part by NIH grant, AI 17133.

REFERENCES

1. Agosin, M. (1976) Mol. Cell. Biochem. 12, 33-44.
2. Naquira, C., White, R.A.Jr. and Agosin, M. (1980) in: Gustaffson, J.A. (Ed.), Biochemistry, Biophysics and Regulation of Cytochrome P-450, Elsevier/North-Holland Biomedical Press, Amsterdam, pp. 105-108.

3. Agosin, M. (1982) in: Kerkut, G.A. *et al* (Eds.), Comprehensive Insect Physiology, Biochemistry and Pharmacology, Pergamon Press, New York/London, in press.

4. Morello, A., Repetto, Y. and Agosin, M. (1980) Drug Met. Disp. 8, 309-312.

5. Capdevila, J., Morello, A., Perry, A.S. and Agosin, M. (1973) Biochemistry 12, 1445-1451.

6. Capdevila, J. and Agosin, M. (1977) in: Ullrich, V. (Ed.), Microsomes and Drug Oxidations, Pergamon Press, Oxford/New York, Vol. I, #19, pp. 144-151.

7. Moldenke, A.F. and Terriere, L.C. (1981) Pestic. Biochem. Physiol. 16 (3), 222-230.

8. Yu, S.J. and Terriere, L.C. (1973) Pestic. Biochem. Physiol. 3(2), 141-148.

Cytochrome P-450, Biochemistry, Biophysics and Environmental Implications, E. Hietanen, M. Laitinen and O. Hänninen editors

ARYL HYDROCARBON HYDROXYASE ACTIVITY IN RAT BRAIN MITOCHONDRIA

MUKUL DAS, PRAHLAD K. SETH AND HASAN MUKHTAR[1]

Industrial Toxicology Research Centre, P.O.Box 80, Lucknow 226001, India and [1]VA Medical Center, Cleveland, Ohio 44106, U.S.A.

INTRODUCTION

Aryl hydrocarbon hydroxylase(AHH) and other cytochrome P-450 (P-450) dependent enzymes once thought to be of microsomal origin only have now also been detected in the nuclear and mitochondrial membranes of liver (1,2). Although the physiological role of mitochondrial P-450 and its dependent enzymes have been thouroughly studied in steroidogenic organs, little is known about their role in non-steroidogenic tissues. Presence of mitochondrial P-450 dependent enzymes in chicken kidney (3) and small intestinal mucosa of rat (4) have been reported recently. Significant activity of AHH in rat brain mitochondria was noticed by us while studying the subcellular distribution of the enzyme in this organ (5). In the present study some properties of brain mitochondrial AHH have been investigated.

MATERIALS AND METHODS

Male Wistar albino rats (100-120g and 200-300g), male guinea pigs (225-265g) and male Swiss mice (15-21g) derieved from ITRC animal breeding colony and raised on commercial pellet diet and water ad libitum were used in the present study.

The AHH activity was estimated in isolated brain mitochondria as described previously (5). The purity of mitochondria was ensured with the help of the marker enzymes of microsomes (glucose -6-phosphatase) and mitochondria (succinic dehydrogenase and monoamine oxidase). The preparation showed only 2-6% contamination of mitochondria in microsomes and viceversa.

RESULTS

The NADH-dependent AHH activity in brain mitochondria of the selected mammalian species was found to be 3 and 12-45 times higher than that of NADPH-dependent mitochondrial and microsomal enzyme activity respectively (Table 1). The Km value of NADH-dependent mitochondrial AHH was 4-6 times lower than the NADPH-dependent mitochondrial or microsomal enzyme (Table 2). Adminis-

TABLE 1

MITOCHONDRIAL AND MICROSOMAL AHH IN BRAIN OF DIFFERENT MAMMALIAN SPECIES

Animals	Mitochondrial AHH[a]		Microsomal AHH[a]
	NADH-dependent	NADPH-dependent	NADPH-dependent
Rat	12.16 ± 0.48	4.66 ± 0.53	1.09 ± 0.04
Guinea pig	20.75 ± 0.91	6.89 ± 0.56	0.61 ± 0.04
Mice	18.65 ± 0.41	6.67 ± 0.32	0.42 ± 0.06

a - p moles 3-OH BP/min/mg protein

Data represent mean ± S.E. of four animals of body weight : Rats (260 ± 20g), Guinea pigs (240 ± 15g) and Mice (18 ± 3g).

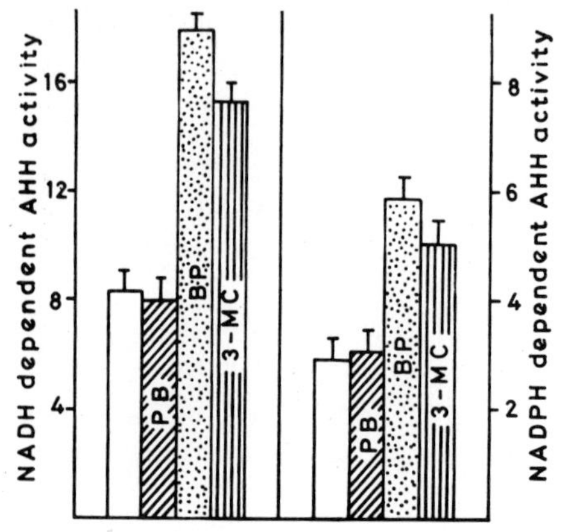

Fig. 1. Effect of PB(80mg/Kg),BP(30 mg/Kg) and 3-MC(30mg/Kg) given i.p. for 5 days on brain mitochondrial aryl hydrocarbon hydroxylase.

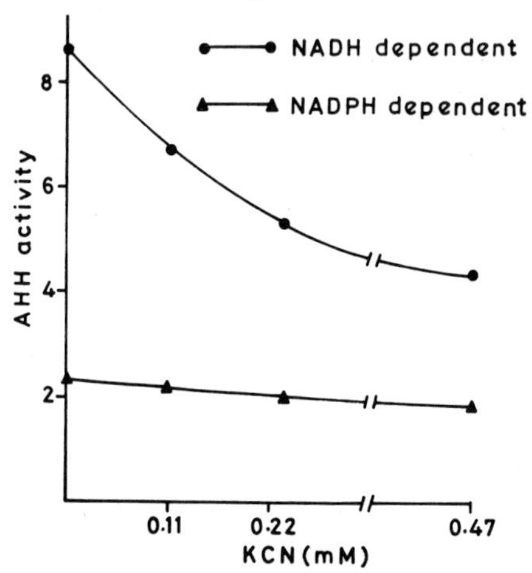

Fig. 2. Sensitivity of brain mitochondrial aryl hydrocarbon hydroxylase towards cyanide.

tration of 3-methylcholanthrene(3-MC) and BP induced the NADH- and NADPH-dependent mitochondrial AHH activity while phenobarbital (PB) had no such effect (Fig. 1). The effect of inhibitors on control and 3-MC induced brain NADH-dependent mitochondrial AHH is summarized in Table 3. Metyrapone, 1-benzylimidazole, SKF-525A and α-napthoflavone(α-NF) inhibited control and 3-MC induced NADH-dependent mitochondrial AHH in a concentration dependent manner. α-NF caused a greater inhibition of 3-MC induced enzyme (Table 3).

TABLE 2

KINETIC CONSTANTS OF MITOCHONDRIAL AND MICROSOMAL AHH OF RAT BRAIN

Organelle	Apparent Km µM	Vmax (p moles 3-OH BP/min/mg protein)
Mitochondria		
NADH-dependent AHH	1.18	15.41
NADPH-dependent AHH	4.22	3.52
Microsomes		
NADPH-dependent AHH	6.66	1.62

Data from a typical experiment, repeated 3 times with identical values, are presented.

TABLE 3

INHIBITION OF CONTROL AND 3-MC STIMULATED RAT BRAIN NADH-DEPENDENT MITOCHONDRIAL AHH

Inhibitors		Control (p moles 3-OH BP/min/mg protein)	3-MC induced
None		8.53	16.32
1-Benzylimidazole	10^{-4}M	3.12	7.32
	10^{-3}M	1.15	4.16
SKF-525A	10^{-5}M	5.63	8.14
	10^{-4}M	3.41	6.12
	10^{-3}M	2.82	4.19
Metyrapone	10^{-4}M	4.19	11.12
	10^{-3}M	3.21	8.32
α-Naphthoflavone	10^{-6}M	6.12	7.12
	10^{-5}M	5.32	4.81
	10^{-4}M	3.19	2.15

Rats (100-120g) were injected with 3-MC as indicated in Fig. 1. The values represent mean of two experiments.

KCN caused a concentration dependent inhibition of NADH-dependent brain mitochondrial AHH while NADPH-dependent enzyme remained unaffected even at its highest concentration, 0.47mM (Fig. 2).

DISCUSSION

This is perhaps the first report on the presence and properties of AHH in brain mitochondria. This organelle showed a greater enzyme activity in comparison to the corresponding microsomal AHH, while an inverse relation between the two organelles was observed in liver, kidney and intestinal mucosa (2-4). The low Km value of NADH-dependent brain mitochondrial AHH suggests a higher affinity of BP towards this enzyme and therefore a higher metabolism of BP is expected in brain mitochondria than that in microsomes. Hepatic mitochondrial DNA has been shown to be a major cellular target for a dihydrodiol epoxide derivative of BP (6) which initiates the process of carcinogenisis. In this context the presence of AHH in brain mitochondria observed in this study assumes a great significance. Inhibition of NADH-dependent brain mitochondrial AHH by KCN and its no effect on NADPH-dependent mitochondrial enzyme suggests the possibility of two forms of P-450 in brain mitochondria.

The brain mitochondrial AHH seems to fulfill the requirements (induction, inhibition and cyanide sensitiveness) for its classification as mixed function oxidase (7). Thus the P-450 localized in brain mitochondria may be helpful in the disposal of various CNS acting drugs and neurotoxins.

REFERENCES

1. Mukhtar, H., Elmamlouk, T.H. and Bend, J.R. (1979) Arch. Biochem. Biophys. 192, 10-21.
2. Uemura, T. and Chiesara, E. (1976) Eur. J. Biochem. 66, 293-307.
3. Ghazarian, J.F., Jefcoate, C.R., Knutson, J.C., Orme Johnson, W.H. and De Luca, H.P. (1974) J. Biol. Chem. 249, 3026-3033.
4. Jones, D.P., Grafstorom, R. and Orrenius, S. (1980) J. Biol. Chem. 255, 2383-2390.
5. Das, M., Seth, P.K. and Mukhtar, H. (1981) J. Pharmacol. Exp. Ther. 216, 156-161.
6. Backer, J.M. and Weinstein, I.B. (1980) Science 209, 297-299.
7. Mason, H.S. (1957) Adv. Enzymol. 19, 79.

Cytochrome P-450, Biochemistry, Biophysics
and Environmental Implications, E. Hietanen,
M. Laitinen and O. Hänninen editors

ENERGY REGULATION AND DRUG METABOLISM IN HEPATIC AND INTESTINAL MICROSOMES OF RAT AND CHICK

YASMEEN ZUBAIRY[1], SANJAY GOVINDWAR[2], MADHUSUDAN SONI[3], MANVENDRA KACHOLE AND SITARAM PAWAR
Biochemistry Division, Chemistry Department,
Marathwada University, Aurangabad 431 004, India

Factors responsible for tissue specificity of cytochrome P_{450} mediated reactions were investigated in hepatic and intestinal microsomes of rat and chick. A system comprising cell wall, cytosol, mitochondria and microsomes was used to evaluate contribution of each fraction. The variation in the magnitude of reactions and the levels of electron transport constituents appear to be a response controlled by compartmentalization of the modifier or substrate. Omission of cell wall and substitution of other constituents after eliciting changes in energy metabolism were found to function equally in reactions of all microsomal preparations. The tissue specific metabolism of substrates is depending largely on the cell wall mediated energy regulation inside the cell and other reactions in the cytoplasm.

INTRODUCTION

In an attempt to quantitate the effects of caffeine on reactions of microsomal mixed function oxidase system and the recently stressed factors (1) of energy supply, we had designed a system comprising hepatic cell wall, cytosol, mitochondria and microsomes. At a concentration around 10 mM, caffeine could alter the energy supply through its interactions with cell wall and cytosolic enzymes as well as microsomal electron transport components (2). The system was used in the present studies to investigate energy factors responsible for tissue specificity of cytochrome P_{450} mediated reactions in hepatic and intestinal microsomes of rats and chicks.

METHODS

Cell wall, cytosol, mitochondria and microsomes were isolated (2) from livers and musocal lining from male albino rats and sonali pearls chicks

---

Supported in part by grants (1) PGS 1/RF/81-82/74965, (2) RD-25/1277-78 and (3) 45/125/81-BS

treated with phenobarbital (80 mg/kg body weight, ip., 3 days) or benzo(a)pyrene (20 mg/kg body weight, ip., 2 days) or respective vehicles. The composition of the system and assay procedures have been described elsewhere (2).

RESULTS AND DISCUSSION

Preparations of microsomes from livers and intestine adjusted to equal cytochorome P_{450} contents carried out aminopyrine N-demethylase reaction at similar rates. The stoichiometric values (Table 1) were equal in both tissues and were altered by same factors due to inducers. However, the inductions of intestinal systems were similar in case of phenobarbital and lower in benzo(a)pyrene treatment. The differences in inducibilities seem to be a factor associated with compartmentalization of the drug.

TABLE 1
STOICHIOMETRIC RELATIONSHIPS FOR AMINOPYRINE N-DEMETHYLASE

Animals	Tissue	Treatment		
		Control	Phenobarbital	Benzo(a)pyrene
Rat	Liver	0.612 ± 0.010	0.962 ± 0.015	0.622 ± 0.008
	Intestine	0.581 ± 0.012	0.901 ± 0.016	0.575 ± 0.005
Chick	Liver	0.400 ± 0.008	0.650 ± 0.012	0.380 ± 0.006
	Intestine	0.380 ± 0.009	0.620 ± 0.010	0.350 ± 0.010

Values are the means ± SE of four experiments

The microsomes were allowed to carry out N-demethylation in presence of respective cytosol and fructose 1,6 diphosphate. The rates of reactions with fructose 1,6 diphosphate were similar to those in presence of NADPH + NADH, showing that fructose 1,6 diphosphate in presence of cytosol and mitochondria served as an efficient source of energy (Table 2).

In presence of glucose 6 phosphate, mitochondria, cytosol, NADP and NAD, rat intestinal microsomes showed a lower activity, whereas, in chick intestinal microsomes, glucose 6 phosphate could as well serve as an efficient source of energy.

Glycogen, cyclic AMP, cytosol, mitochondria, NADP, NAD and ATP were used in another reaction mixture. With this system glycogen was the

TABLE 2

AMINOPYRINE N-DEMETHYLATION IN RAT AND CHICK MICROSOMES FROM LIVER AND INTESTINE IN PRESENCE OF DIFFERENT ENERGY SOURCES.

Microsomes	Energy	Cytosol	Mitochondria	Rat			Chick		
				C	PB	BP	C	PB	BP
Hepatic	1			100	100	100	100	100	100
	2			90	100	100	88	90	87
	3			85	90	94	83	82	82
	4			78	85	75	77	75	75
	5			76	85	73	76	70	73
Intestinal	1			100	100	100	100	100	100
	2			85	80	90	97	100	95
	3			65	69	85	84	90	86
	4			58	63	85	76	80	77
	5			58	60	80	75	80	78
Hepatic	4	H	H	100	100	100	100	100	100
	4	I	H	58	60	54	66	64	56
	4	H	I	64	68	60	54	60	58
	4	I	I	30	36	28	50	54	42
Intestinal	4	I	I	100	100	100	100	100	100
	4	H	I	112	135	127	122	130	120
	4	I	H	136	140	132	137	141	132
	4	H	H	180	197	175	200	212	190

Activities % of respective controls;
C - Untreated,
PB-Phenobarbital, treated,
BP-Benzo(a)pyrene treated,
H - Hepatic,
I - Intestinal

1 - NADPH + NADH
2 - F16DP + NADH + Cytosol
3 - G6P + NAD + NADP + Cytosol + Mitrochondria
4 - Glycogen + CAMP + ATP + NAD + NADP + Cytosol + Mitochondria
5 - Glycogen + ATP + NAP + NADP + Cytosol + Mitochondria + Cell Wall + Epinephrine

ultimate source of energy. The hepatic microsomes from rat and chick as well as intestinal microsomes from chick carried out aminopyrine N-demethylation with slightly lower efficiency. Rat intestinal microsomes were very poor in metabolizing aminopyrine in this assay system.

Exchange of mitochondria between hepatic and intestinal preparations increased the rate of aminopyrine N-demethylase significantly in both intestinal preparations from rat and chick (Table 2). Further exchange of cytosol brought the activities close to those of hepatic microsomes, indicating thereby the efficiencies of microsomes could be influenced by mitochondria as well as cytosol.

Phenobarbital treatment and benzo(a)pyrene treatment to rats and chicks further characterized the impairment. The increase in the efficiency on replacement of intestinal mitochondria and cytosol by hepatic preparations was lower in phenobarbital treatment and followed a pattern similar to untreated rats in benzo(a)pyrene treatment.

The tissue specific responses of the cell wall to receptors in circulation bringing about the changes in adenyl cyclase and therefore cyclic AMP can be overruled in the system by incorporation of cyclic AMP in the assay mixture or by addition of preincubated cell wall, epinephrine and ATP. Both the systems are found to satisfy the requirements for glycogen breakdown and supply of energy for drug metabolism.

The system needs detailed explorations on the critical concentrations of cyclic AMP, NADPH/NADH synergisms, competition for reducing equivalents, effects of $p0_2$, $pC0_2$, etc. for further information on tissue specific functioning of cytochrome P_{450}.

REFERENCES

1. Thurman, R.G. and Kauffman, F.C. (1980) Pharmacol. Reviews, 31, 229-251.

2. Govindwar, S.P., Siddiqui, A.M., Soni, M.G. and Kachole, M.S. Ind. J. Med. Res. (In Press).

© 1982 Elsevier Biomedical Press B.V.
Cytochrome P-450, Biochemistry, Biophysics
and Environmental Implications, E. Hietanen,
M. Laitinen and O. Hänninen editors

WHAT ARE THE SIGNIFICANT TOXIC METABOLITES OF STYRENE?

HARRI VAINIO[1], FRANCESCO TURSI[2] AND GIORGIO BELVEDERE[1,2,3]
[1]Department of Industrial Hygiene and Toxicology, Institute of Occupational Health, Haartmaninkatu 1, SF-00290 Helsinki 29 (Finland) and [2]Istitute di Ricerche Farmacologiche "Mario Negri", Via Eritrea, 62-20157 Milano (Italy)

INTRODUCTION

Styrene is used extensively in the production of synthetic rubbers and plastics as a modifier or solvent of polyester resins. Studies in experimental animals and in man indicate that the major route of the metabolism proceeds via styrene-7,8-oxide (phenyloxirane) (cf. Fig. 1). However, recently also other active intermediates have been proposed in the metabolism of styrene. These include styrene-3,4-oxide, a product of oxidation at the aromatic ring (1,2) and phenylacetaldehyde (3). In this short review we attempt to discuss the various metabolic pathways of styrene, and the existing evidence on the importance of different reactive intermediates in the toxicity of styrene.

Cytochrome P-450 dependent metabolism

Styrene is biotransformed in vivo to a great variety of metabolites (Fig. 1) that are mainly excreted in the urine. The urinary metabolites of styrene that account for more than 50 % of the dose, i.e., mandelic acid and phenylglyoxylic acid, are oxidative products of styrene glycol that is formed in the liver by hydration of styrene-7,8-oxide. This metabolic intermediate is not detectable in vivo, but has been identified in *in vitro* incubations of styrene with rat liver microsomes (4).

Styrene-7,8-oxide formation is a key step in the metabolism of styrene *in vivo* and it has been shown to bind to DNA and protein (5) and to be genotoxic (6,7). For these reasons methods have been established to study the kinetic and the mechanisms of styrene-7,8-oxide formation and its detoxication to the glycol (8,9). Styrene is oxidized to styrene-7,8-oxide by liver microsomal cytochrome P-450 dependent monooxygenase(s) and it is hydrated to the

---

[3]recipient of an EEC-fellowship

TABLE 1

KINETIC PARAMETERS OF STYRENE MONOOXYGENASE AND STYRENE-7,8-OXIDE HYDROLASE IN LIVER AND BLOOD

Species	System	K_m (mM), V_{max} (nmol/min/mg protein)			
		Monooxygenase		Hydrolase	
		K_m	V_{max}	K_m	V_{max}
Rat	microsomes	0.23	2.71	0.93	10.50
Mouse	microsomes	0.039	3.81	0.73	5.52
Guinea pig	microsomes	0.18	5.41	0.75	8.97
Rabbit	microsomes	0.076	4.18	0.51	9.98
Man	erythrocytes	15.9	253.2 (nmol/30 min/ml)	-	-

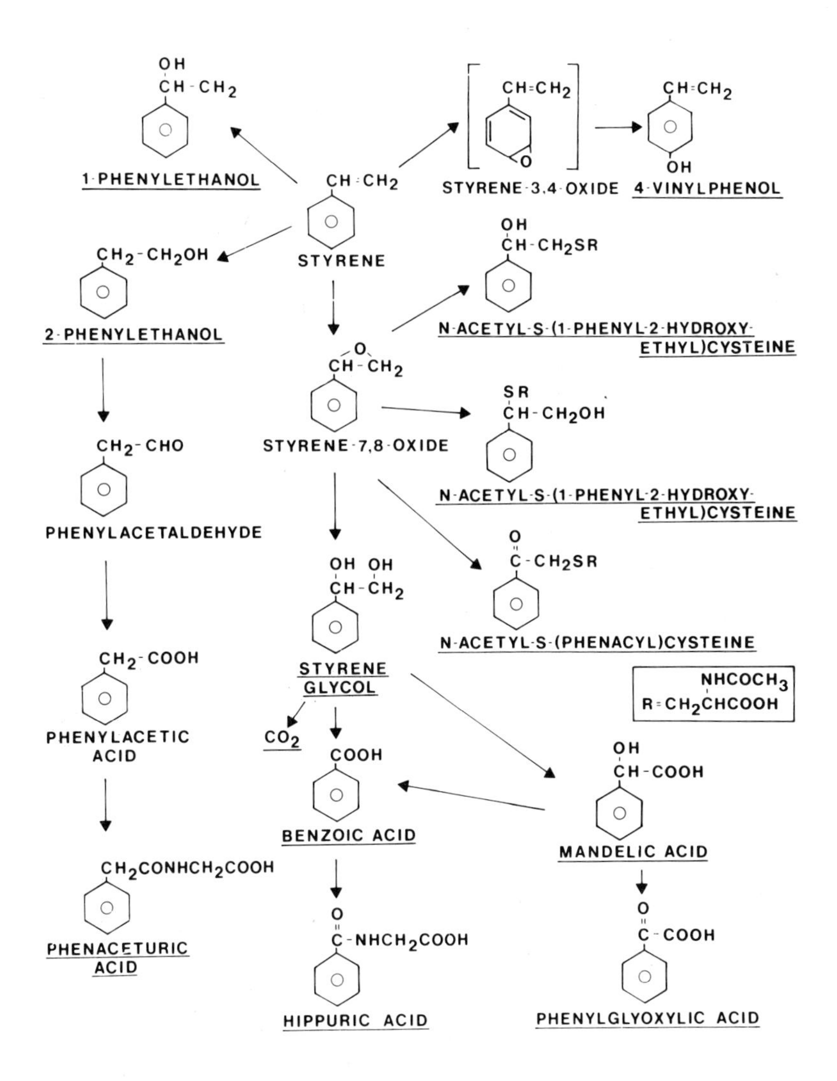

Fig. 1. Styrene metabolism. Underlined metabolites are excreted in the urine.

glycol by epoxide hydrolase. The affinity (K_m) and the specific activity (V_{max}) for these two enzymes have been determined in liver microsomes of different species (Table 1). The affinity of styrene for the monooxygenase had broad interspecies variability, whereas the specific activity was more homogeneous, both parameters showed an interspecies similarity for the hydrolase that showed activities similar to the monooxygenase but much lower affinities (10).

4-Vinylphenol is a urinary metabolite of styrene that accounts for less than 0.5 % of the dose in the urine of animals and man (1,11). The half-life of 3,4-oxide in buffer solutions is only a few seconds (12), this is probably why it has never been identified after styrene metabolism *in vivo* and *in vitro*.

Non-cytochrome P-450 dependent metabolism

The development of more sensitive methods (9,13) allowed a detailed study of the activity of styrene monooxygenase in different biological systems. Washed human erythrocytes and whole blood were found to oxidize styrene to styrene-7,8-oxide in the absence of NADPH and other cofactors required for metabolic reactions catalysed by microsomal enzymes (Table 2). This reaction was not

TABLE 2

STYRENE OXIDE FORMATION IN HUMAN ERYTHROCYTES

System	Styrene glycol[a] nmol/30 min/ml
Erythrocytes	130.5 ± 6.0
+ Cofactors[b]	204.0 ± 13.0
+ CO[c]	29.2 ± 1.0
+ Superoxide dismutase (18 μg/ml)	173.2 ± 5.7
+ Catalase (50 μg/ml)	158.3 ± 16.2
+ Tryptophane (2 mM)	104.0 ± 5.2
+ Mannitol (20 mM)	126.0 ± 6.3
+ Dimethyl sulfoxide (280 mM)	150.0 ± 8.0
Erythrocytes membranes + cofactors	n.d.[d]
Erythrocytes $-O_2$[e] + cofactors	n.d.
Methemoglobin + cofactors	n.d.
Blood	164.5 ± 6.5
Blood + cofactors	172.5 ± 5.5

Styrene concentration in the incubation mixture was 50 mM. The data represent the mean ± S.E.

a) The styrene oxide enzymatically formed was chemically converted to styrene glycol.

b) The cofactors mixture consisted of: NADPH 0.87 mM, NADH 0.25 mM, nicotinamide 2.7 mM and $MgCl_2$ 5 μM.

c) CO was bubbled into the incubation mixture for 1 min with a flow of 50 ml/min before styrene addition.

d) Styrene glycol values, after styrene incubation with erythrocytes membranes (1 mg/of protein/ml), were the same as for blank samples consisting of PBS styrene and cofactors. n.d. not detectable.

e) Solutions were purged with N_2 and samples incubated under vacuum.

f) Purified human methemoglobin (Sigma) was dissolved in PBS and added to the incubation mixture at the same concentration of hemoglobin present in samples containing erythrocytes.

catalysed by erythrocytes membranes or methemoglobin. Styrene oxidation was not inhibited by superoxide dismutase or catalase and by scavangers of ·OH radicals such as tryptophane, mannitol and dimethylsulfoxide, indicating the free reactive oxygen intermediates such as $O_2^{-\cdot}$, H_2O_2 and ·OH were not involved in this reaction. The inhibition of styrene-7,8-oxide formation in these cells caused by CO and the linear relationship of its formation with the molar fraction of oxyhemoglobin (14), indicate this agent as the likely catalyst of the reaction. Styrene-7,8-oxide formation in human erythrocytes showed a Michaelis-Menten kinetic, but the affinity of styrene for oxyhemoglobin was 80 times lower than that for rat liver microsomal monooxygenase (Table 1). The V_{max} of the reaction has been estimated to be 253.2 nmol/30 min/ml of erythrocytes and a similar activity has been observed in rat blood (unpublished results). The time course of the reaction, however, is not linear (Fig. 2), but it can be estimated that about 100 nmol

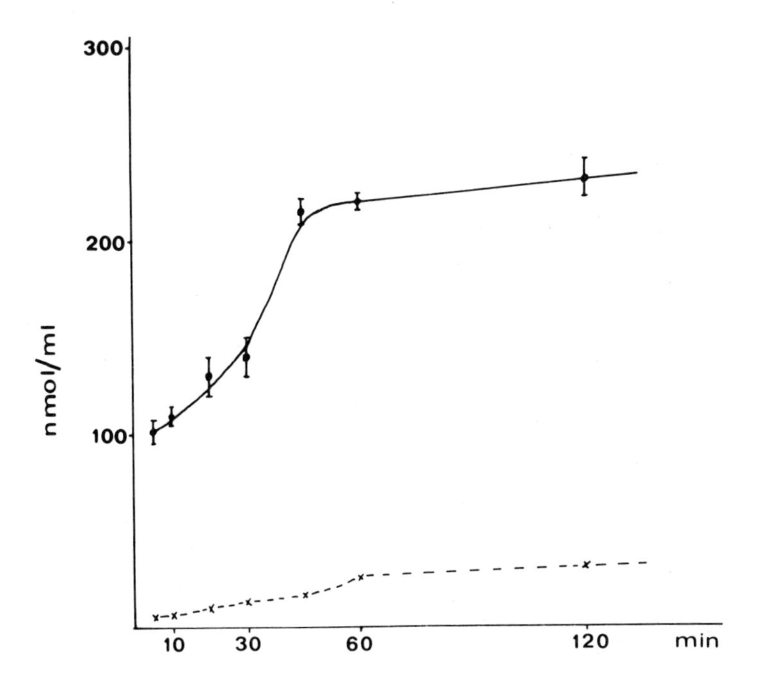

Fig. 2. Time course of styrene-7,8-oxide formation in human erythrocytes. Red blood cells were washed reconstituted to the original blood volume and incubated with styrene 50 mM (o) and 0.8 mM (x). Brackets represent the mean ± S.E.M.

of epoxide are formed in one ml of blood in one minute with a saturating concentration of styrene. This amount of styrene oxide is about 30 times higher than the specific activity of styrene monooxygenase in rat liver microsomes.

Styrene was also oxidated to styrene-7,8-oxide by xanthine oxidase in the presence of xanthine and Fe^{3+} (Table 3). This reaction was inhibited by superoxide dismutase and catalase and by scavangers of ·OH radicals (15) indicating that also free reactive oxygen intermediates might be involved in styrene oxidation to styrene-7,8-oxide. It thus appears that styrene can be oxidized to styrene-7,8-oxide in different biological systems by different mechanisms. The relevance of these alternative metabolic pathways for styrene activation in relation to its toxic effects are now under investigation.

TABLE 3

STYRENE OXIDATION TO STYRENE OXIDE DURING REACTION OF XANTHINE (X) WITH XANTHINE OXIDASE (XO) IN THE PRESENCE OF Fe^{3+}

Additions[a]	Styrene glycol[b] nmol	% of control
X + XO + Fe^{3+}	12.80 ± 0.74 (11)[c]	100
- Fe^{3+}	4.33 ± 0.65* (8)	33
+ Superoxide dismutase 0.93 μg/ml	9.33 ± 0.65* (11)	73
+ Superoxide dismutase 93.0 μg/ml	6.50 ± 0.46* (6)	51
+ Catalase 6.6 μg/ml	2.40 ± 0.24* (6)	23
XO + H_2O_2 (0.1 mM)	n.d.[d]	-

a) Reaction mixtures contained 50 mM styrene (dissolved in acetonitrile), 50 M xanthine, 0.50 μM xanthine oxidase, 0.15 mM EDTA, 50 μM $Fe_2(SO_4)_3$, 50 mM potassium phosphate, and the indicated additions at pH 7.4 and at 37°C. The data are the mean ± S.E.

b) Styrene oxide formed enzymatically was chemically converted to styrene glycol

c) In brackets the number of determinations

d) Styrene glycol values were the same as for blank samples consisting of buffer and styrene. n.d. not detectable

* $p < 0.01$

Reactive metabolites and their toxicity

Styrene is a typical indirect mutagen which needs metabolic activation to be able to bind covalently to nucleophilic biological macromolecules or to be mutagenic in various mutagenicity assays (16). Inhalation of styrene by mice resulted in the formation of sister chromatid exchanges in the alveolar macrophages, bone marrow and regenerating liver cells (17). The role of styrene-7,8-oxide as the toxic intermediate has been proposed by many authors. However, only recently, its role has been questioned in favour of styrene-3,4-oxide (18).

Styrene-7,8-oxide can bind with nucleic acid bases (19). It is mutagenic in bacteria and fungi in the absence of an exogenous metabolic activation system, and induces recessive lethal mutations in Drosophila (16,20). In cultured mammalian cells, styrene-7,8-oxide induced mutations, sister chromatid exchanges, chromosomal aberrations and DNA repair (21). Styrene-7,8-oxide has also been shown to be carcinogenic in rats after peroral administration (22).

Styrene-3,4-oxide, a putative percursor of 4-vinylphenol, is a highly unstable vinyl-conjugated benzene oxide. It has been shown to be a more cytotoxic and a more potent mutagen in two strains of Salmonella typhimurium in respect to styrene-7,8-oxide without any metabolic activation (12). It has not, however, been studied in any other test system. If present *in vivo* in sufficient quantities, styrene-3,4-oxide may also be of toxicological importance.

Phenylacetaldehyde

Phenylacetaldehyde, which can be formed via the intramolecular rearrangement of styrene-7,8-oxide or oxidation of 2-phenylethanol, has shown no mutagenic activity in any of the studies done. Thus, it appears highly improbable that it has any great impact as a toxic metabolite of styrene.

CONCLUSIONS

Various experimental assay systems suggest that styrene after a metabolic activation has a mutagenic potential. The discrepancies observed in studies published so far would seem to arise from the variations in activating and inactivating metabolic reactions in different tissues and organisms rather than on the inability of the reactive intermediates to bind to intracellular macromolecules.

Styrene-7,8-oxide is formed from styrene by monooxygenase(s) in microsomes or in erythrocytes by oxyhemoglobin. Styrene-3,4-oxide, a putative intermediate in 4-vinylphenol formation, is another candidate as the toxic intermediate of styrene. The amount of 4-vinylphenol excreted in urine after inhalation exposure to styrene, suggest, however, that it may be of minor importance as far as the genotoxicity of styrene is concerned. Furthermore, it is only styrene-7,8-oxide which is being formed by e.g. erythrocytes in blood.

There are also some pieces of evidence that favour the extrahepatic non-cytochrome P-450 catalyzed metabolic activation of styrene to styrene-7,8-oxide: (i) The same amount of sister chromatid exchanges has been observed in regenerating rat liver hepatocytes, in bone marrow cells and in alveolar macrophages after inhalation exposure to styrene (17). (ii) The hepatotoxicity of styrene in rats is not inhibited *in vivo* by SKF-525A and it is not increased by phenobarbital (23) and (iii) sister chromatid exchange - frequency increased in human lymphocytes in culture only after addition of red blood cells (24).

Taking all the data available together, it appears as if styrene-7,8-oxide still is the most relevant candidate as the toxic intermediate of styrene in mammals.

REFERENCES

1. Pantarotto, E., Fanelli, R., Bidoli, F. Morazzoni, P., Salmona, M. and Szczawinska, K. (1978) Scand. J. Work Environ. Health, 4, Suppl. 2, 67.
2. Watabe, T., Isobe, M., Sawahata, T., Yoshikawa, K., Yamada, S. and Takabatake, E. (1978) Scand. J. Work Environ. Health, 4, Suppl. 2, 142.
3. Delbressine, L.P.C., Ketelaars, H.C.J., Seuter-Berlage, F. and Smeets, F.L.M. (1980) Xenobiotica, 10, 337.
4. Leibman, K.C. and Ortiz, E. (1970) J. Pharmacol. Exp. Ther., 173, 242.
5. Van Anda, J., Smith, B.R., Fouts, J.R. and Bend, J.R. (1979) J. Pharm. Exp. Ther., 211, 207.
6. Meretoja, T., Vainio, H., Sorsa, M. and Härkönen, H. (1977) Mutat. Res., 56, 193.
7. Linnainmaa, K., Meretoja, T., Sorsa, M. and Vainio, H. (1978) Mutat. Res., 56, 277.

8. Belvedere, G., Pachecka, J., Cantoni, L., Mussini, E. and Salmona, M. (1977) J. Chrom. 118, 387.

9. Duverger-Van Bogaert, M., Noel, G., Rollman, B., Cumps, J., Roberfroid, M. and Mercier, M. (1978) Biochem. Biophys. Acta, 526, 77.

10. Belvedere, G., Cantoni, L., Facchinetti, T. and Salmona, M. (1977) Experientia, 33, 708.

11. Pfäffli, P., Hesso, A., Vainio, H. and Hyvönen, M. (1981) Toxicol. Appl. Phamacol., 60, 85.

12. Watake, T., Hirotzuka, A., Aizawa, T., Sawahata, T., Ozawa, N., Isobe, M. and Takebatake, E. (1982) Mutat. Res., 93, 45.

13. Gazzotti, G., Garattini, E. and Salmona, M. (1980) Chem. Biol. Interact., 29, 189.

14. Tursi, F., Samaia, M., Salmona, M. and Belvedere, G. (1982) Experientia, submitted.

15. Belvedere, G. and Tursi, F. (1982) Tox. Lett., in press.

16. Vainio, H., Norppa, H., Hemminki, K. and Sorsa, M. (1982) in: Snyder, R., Parke, D.V., Kocsis, J.J., Gibson, G. and Witmer, C. (eds.), Biological Reactive Intermediates - II, Part A., Plenum Publishing Corporation, pp. 257-274.

17. Conner, M.K., Alarie, Y. and Dombroske, R.L. (1979) Toxicol. Appl. Pharmacol., 50, 365.

18. Beije, B. and Jenssen, D. (1982) Chem. Biol. Interact., 39.

19. Hemminki, K., Paasivirta, J., Kurkirinne, T. and Virkki, L. (1980) Chem. Biol. Interact., 30, 259.

20. Donner, M., Sorsa, M. and Vainio, H. (1979) Mutat. Res., 67, 373.

21. Norppa, H. (1981) Academic dissertation, Helsinki, Institute of Occupational Health and University of Helsinki, 56 pp.

22. Maltoni, C., Failla, G. and Kassapidis, G. (1979) Med. Lavoro, 5, 358.

23. Chakrabarti, S. and Brodeur, J. (1981) J. Tox. Env. Health, 8, 599.

24. Vainio, H., Norppa, H. and Belvedere, G. (1982) 13th International Cancer Congress, Seattle, September 8-15, 1982, Abstract.

© 1982 Elsevier Biomedical Press B.V.
Cytochrome P-450, Biochemistry, Biophysics
and Environmental Implications, E. Hietanen,
M. Laitinen and O. Hänninen editors

REGIOSELECTIVITY OF CAFFEINE METABOLISM IN VIVO AND PERIPHERAL BLOOD LYMPHOCYTES IN DIFFERENT SPECIES

G. BELVEDERE[1], F. TURSI[1], L. JRITANO[2], F. GALLETTI[2] AND M. BONATI[2]
[1]Laboratory for Enzyme Research and [2]Laboratory of Clinical Pharmacology, Istituto di Ricerche Farmacologiche "Mario Negri", via Eritrea, 62 - Milano, Italy

INTRODUCTION

Human peripheral blood lymphocytes have been shown to metabolize benzo(a)pyrene (BP), benzo(a)pyrene-7,8-dihydrodiol and 17β-estradiol (1-3). The interest in the use of human lymphocytes in the study of xenobiotics metabolism arose mainly for the accessibility of these cells, thus the possibility to investigate interindividual differences in carcinogens metabolism.

Kellerman et al. (4-5) showed a trimodal distribution of the inducibility of aryl hydrocarbon hydroxylase (AHH) and a correlation between bronchogenic carcinoma and the inducibility of AHH in lymphocytes. However this correlation has not been confirmed (6) probably because the measure of AHH, which detects only BP phenols formation, was not sufficient to evidentiate differences in the activation pathway of BP to the more mutagenic and toxic BP epoxides and diol epoxides (7). An investigation of BP metabolism by a more selective high pressure liquid chromatographic method has shown a different regioselectivity in the metabolism of BP in human lymphocytes and monocytes in respect to rat liver microsomes and also a difference between the two populations of leukocytes (2,8). These differences may be caused by different forms of cytochromes P-450 involved in BP metabolism. Highly purified forms of cytochrome P-450 isolated from rabbit liver showed in fact a different regioselectivity in BP metabolism (9). In human lymphocytes, however, the activity of other enzymes such as epoxide hydratase, that is present in these cells (10) and possibly glutathione S-transferase, may represent a conflicting factor in interpreting BP metabolism. Caffeine (1,3,7-trimethylxanthine, TMX) is a substrate that such as BP is metabolized by cytochrome P-448 and whose metabolism is greatly induced by polycyclic aromatic hydrocarbons (11-12). The metabolic pathway of TMX is constituted by oxidative and demethylative steps and does not involve enzymatic systems other than mixed function oxidases (13). For this reason and for its safety TMX is more suitable than BP for the investigation of the metabolic regioselectivity in different species including man.

In this study the regioselectivity of lymphocytes and hepatic mixed function oxidases on TMX metabolism in three species is reported. The results indicate a similar regioselectivity in man and substantial differences in rat and rabbit.

MATERIALS AND METHODS

$(2-^{14}C)$Caffeine was generously supplied by NSDA (National Soft Drink Association, Washington D.C., USA). ILSI (International Life Sciences Institute, Washington D.C., USA) kindly supplied TMX metabolites. Reagents (LiChrosolv, Merck, Darmstadt GFR) were UV grade. Specifications on schedule sampling, materials and methods and _in vivo_ data concerning humans and animals, after oral TMK administration are as previously described (14-17).

Peripheral blood lymphocytes from man, rabbit and rat were isolated as previously reported (18). The cells ($10\text{-}20\ 10^6$) were incubated in 1 ml of PBS (phosphate-buffered saline) (pH 7.8) containing: 0.87 mM NADPH, 0.25 mM NADH, 2.7 mM nicotinamide, 5 μM $MgCl_2$ and 0.1 mM $(2-^{14}C)$caffeine with a specific activity of 7.8 μCi/μmol. Samples were incubated for 1 hr at 37°C with shaking and the reaction was terminated by adding 1 ml of ice cold acetone.

Lymphocytes incubation samples, after the extraction procedure previously described (14) were analyzed by the HPLC method of Voelter et al. (19) substantially modified. By using a Perkin-Elmer (Norwalk, CT, USA) series 2/2 liquid chromatograph equipped with a LC 55 spectrophotometer (254 nm) and a reverse phase column (Hibar RP-8,7 μm, Merck), the samples were eluted with a linear gradient starting from 0% of acetonitrile in 10 mM tetrabutylammonium hydroxide in water (pH 7.5), up to a final concentration of 10% of acetonitrile within 24 min at a flow rate of 2 ml/min. Fractions corresponding to individual peaks of TMX and its metabolites were collected in vials and assayed by scintillation counting according to a previous report (20).

RESULTS AND DISCUSSION

Caffeine metabolism in lymphocytes (Tables 1-2) showed formation of theobromine, paraxanthine, theophylline and trimethyluric acid for the three species tested. Only in rat formation of the four metabolites was observed, while rabbit and man lymphocytes did not produce theophylline. Also theobromine formation was lacking in man leukocytes.

TABLE 1

CAFFEINE METABOLISM IN PERIPHERAL BLOOD LYMPHOCYTES ($pmol/hr/10^6$ cells) AND PLASMA PEAK LEVELS (μg/ml) IN DIFFERENT SPECIES

Species	Theobromine	Paraxanthine	Theophylline
Man			
lymphocytes	nd	0.50 ± 0.17	nd
plasma[a]	0.41 ± 0.03	2.54 ± 0.62	0.38 ± 0.06
Rabbit			
lymphocytes	1.80 ± 0.63	0.56 ± 0.20	nd
plasma[b]	0.84 ± 0.31	4.98 ± 1.30	1.26 ± 0.20
Rat			
lymphocytes	0.66 ± 0.23	0.46 ± 0.16	1.96 ± 0.69
plasma[b]	0.89 ± 0.06	0.95 ± 0.22	0.73 ± 0.07

Mean ± S.E., n = 4.
[a] 5 mg/kg p.o.
[b] 10 mg/kg p.o.
nd: ≤ 0.1 $pmol/hr/10^6$ cells.

The pharmacokinetic of TMX was followed after an oral dose of 10 mg/kg in animals and 5 mg/kg in human volunteers. Plasma peaks concentrations of primary metabolites are also reported in Table 1. This parameter was chosen to compare TMX metabolism *in vivo* and lymphocytes in order to avoid interferences due to the further *in vivo* dimethylxanthines metabolism (13-14). These metabolites showed similar peak concentrations in rats, but paraxanthine was predominant in man and rabbits (Table 1). Since trimethyluric acid is a final metabolite of TMX and it is excreted in the urine as such, its urinary excretion and formation in lymphocytes were used to compare the metabolic rate in both systems; a good agreement was found for both animals and humans (Table 2).

TABLE 2

TRIMETHYLURIC ACID URINARY EXCRETION AND METABOLIC FORMATION IN PERIPHERAL BLOOD LYMPHOCYTES

Species	Trimethyluric acid	
	Urines (% of dose)	Lymphocytes (pmol/hr/10^6 cells)
Man[a]	0.5	0.54
Rabbit[b]	4.0	1.02
Rat[b]	20.0	2.28

Urines were collected for 48 hr.
[a]Caffeine 5 mg/kg p.o.
[b]Caffeine 10 mg/kg p.o.

A close similarity of the metabolic pattern was observed in man, also considering trimethyluric acid excretion (Tables 1-2). However the regioselectivity of TMX metabolism *in vivo* and lymphocytes in rat and rabbit was highly different although a dose dependent kinetic of TMX metabolism was observed in rat (20). The difference between the metabolic pathway of caffeine *in vivo* and lymphocytes in rat is even more distinct considering that the major *in vivo* metabolite of TMX, 6-amino-5-formylmethylamino-1,3-dimethyl uracil (16-17) is undetectable in lymphocytes. This metabolite is about 1% of TMX dose in man and rabbit (16-17).

Multiple forms of cytochrome P-450 have been isolated from liver and extrahepatic tissues microsomes of untreated or induced animals (21-23). The alteration of the relative rate of warfarine metabolites formation in microsomes by using different inducing agents has implicated the involvement of multiple forms of cytochrome P-450 (24). This was confirmed by studies with highly purified forms of cytochrome P-450 that showed a different regioselectivity (25-26). The relative ratio of TMX metabolites was found to be greatly altered in rat liver microsomes induced with phenobarbital or 3-methylcholanthrene (16,27) and *in vivo* in beagle dogs by β-naphtoflavone pretreatment (28).

Lymphocytes have been used as a non invasive system to study carcinogens metabolism in man, therefore it would be highly desirable to know if similar forms of cytochrome P-450 are present in this system and in the liver. The data here reported indicate that different forms of cytochrome P-450 could be responsible for TMX metabolism in the liver and lymphocytes of rat and rabbit, while a similar enzymatic system may be present in man.

ACKNOWLEDGEMENTS

This work was partially supported by a Grant from the International Life Science Institute, Washington D.C., USA; and a Grant on Clinical Pharmacology and Rare Diseases from CNR (National Research Council, Rome, Italy).

REFERENCES

1. Whitlock Jr., J.P., Cooper, H.L. and Gelboin, H.V. (1972) Science, 177, 618.
2. Okano, P., Miller, H.N., Robinson, R.C. and Gelboin H.V. (1979) Cancer Res., 39, 3184.
3. Muijsson, I.E., Coomes, M.L., Cantrell, E.T., Anderson, D.E. and Busbee, D.L. (1975) Biochem. Genet., 13, 501.
4. Kellerman, G., Luyten-Kellerman, M. and Shaw, C.R. (1973) Am. J. Hum. Genet., 25, 327.
5. Kellerman, G., Shaw, C.R. and Luyten-Kellerman, M. (1973) N. Engl. J. Med., 289, 934.
6. Paigen, B., Gurtoo, H.L., Minowada, J., Hauten, L., Vincent, R., Paigen, K., Parker, N.B., Ward, E. and Hayner, N.T. (1977) N. Engl. J. Med., 297, 346.
7. Huberman, E., Sachs, L., Yang, S.K. and Gelboin, H.V. (1976) Proc. Natl. Acad. Sci. USA, 73, 607.
8. Yang, S.K., Selkirk, J.K., Plotkin, E.P. and Gelboin, H.V. (1975) Cancer Res., 35, 3642.

9. Deutsch, J., Leutz, J.C., Yang, S.K., Gelboin, H.V., Chiang, Y.L., Vatsis, K.P. and Coon, M.J. (1978) Proc. Natl. Acad. Sci. USA, 75, 3123.

10. Glatt, H., Kaltenbach, E. and Oesch, F. (1980) Cancer Res., 40, 2552.

11. Aldridge, A., Parsons, W.D. and Neims, A.H. (1977) Life Sci., 21, 967.

12. Parsons, W.D. and Neims, A.H. (1978) Clin. Pharm. Exp. Ther., 24, 40.

13. Burg, A.W. (1975) Drug Met. Rev., 4, 199.

14. Bonati, M., Latini, R., Galletti, F., Young, J.F., Tognoni, G. and Garattini, S., Clin. Pharm. Exp. Ther., in press.

15. Bonati, M., Latini, R., Marzi, E., Cantoni, R. and Belvedere, G. (1980) Tox. Lett., 7, 1.

16. Latini, R., Bonati, M., Marzi, E. and Grattini, S. (1981) Tox. Lett., 7, 267.

17. Garattini, S. (1980) Nutrition Rev., 38, 196.

18. Boyum, A. (1968) J. Clin. Lab. Invest., suppl. 97.

19. Voelter, W., Zech, K., Arnold, P. and Ludwig, G. (1980) J. Chrom., 199, 345.

20. Latini, R., Bonati, M., Castelli, D. and Garattini, S. (1978) Tox. Lett., 2, 267.

21. Coon, M.J., Ballou, D.P., Haugen, D.A., Krezoski, S.O., Nordbloom, G.D. and White, R.E. (1977) in: Ullrich, V. (Ed.), Microsomes and Drug Oxidation, Academic Press, p. 82.

22. Comai, K. and Gaylor, J.L. (1973) J. Biol. Chem., 248, 2947.

23. Guengerich, F.P. (1977) J. Biol. Chem., 252, 3970.

24. Pohl, L.R., Porter, W.R. and Trager, W.F. (1977) Biochem. Pharmacol., 26, 109.

25. Fasco, M.J., Vatsis, K.P., Kaminski, L.S. and Coon, M.J. (1978) J. Biol. Chem., 253, 7813.

26. Kaminski, L.S., Fasco, M.J. and Guengerich, F.P. (1980) J. Biol. Chem., 255, 85.

27. Arnaud, M.J. and Welsch, C. (1980) in: Coon, M.J., Conney, A.H., Estabrook, R.W., Gelboin, H.V., Gillette, J.R. and O'Brien, P.J. (Eds.), Microsomes, Drug Oxidation and Chemical Carcinogenesis, Academic Press, p. 813.

28. Aldridge, A. and Neims, A.H. (1979) Drug Met. Disp., 7, 378.

Cytochrome P-450, Biochemistry, Biophysics and Environmental Implications, E. Hietanen, M. Laitinen and O. Hänninen editors

ENZYMATIC REDUCTION OF LIPOSOME-EMBEDDED CYTOCHROME $P450_{scc}$ FROM BOVINE ADRENOCORTICAL MITOCHONDRIA. EFFECT OF CHOLESTEROL CONCENTRATION IN MEMBRANE

TOKUJI KIMURA, FUMIYUKI YAMAKURA AND BEHLING CHENG
Department of Chemistry, Wayne State University, Detroit, Michigan 48202

INTRODUCTION

The activity of cholesterol side chain cleavage reaction in adrenocortical mitochondria is regulated by the influence of ACTH by as yet undefined mechanism. The current hypothesis strongly suggests that the availability of substrate cholesterol to $P450_{scc}$ molecules in membrane is a crucial part of the regulatory mechanisms of the overall steroidogenic reactions (1). In order to elucidate this important problem, we have previously examined conditions for the preparation of a stable $P450_{scc}$-containing liposomes, revealing dioleoylglycerophosphocholine (DOPC) as a best phospholipid for this purpose (2). Subsequently, we found that the most of the heme site is located on the outside monolayer of bilayer membrane as judged by tryptic digestion and titration of impermeable p-chloromercuriphenylsulfonic acid. As for membrane-free $P450_{scc}$, trypsinolysis of the heme protein resulted in the formation of two major core fragments which had molecular weights of 30,000 and 26,000 (3).

In this study, we have tested the effect of membrane-cholesterol concentration on the reducibility of liposomal $P450_{scc}$ by NADPH, adrenodoxin reductase, adrenodoxin in the presence of carbon monoxide. In the presence of cholesterol, the reduction reactions had a double phase. The fast rates of reduction markedly increased as the cholesterol concentration increased, whereas the slow rates were independent from the cholesterol concentration. From these results, the half-maximum concentration of cholesterol was 3.7 mol % for the $P450_{scc}$ reduction by NADPH, the dissociation constant of $P450_{scc}$-cholesterol complex was 1.9 mol % cholesterol, and the slow rate constant was 4×10^{-4} sec^{-1} regardless of the cholesterol concentrations.

MATERIAL AND METHOD

$P450_{scc}$ was prepared from bovine adrenocortical mitochondria as previously described (1). The preparation of $P450_{scc}$-containing liposomes was based on the previously reported method (2). Adrenodoxin and adrenodoxin reductase were purified according to the method described before (4). The kinetical studies were carried out under anaerobic conditions by use of a Cary 118 spectrophotometer at 22° C.

RESULTS AND DISCUSSION

In agreement with our observations on the trypsinolysis of $P450_{scc}$ and the titration with p-chloromercuriphenylsulfonic acid, liposomal $P450_{scc}$ was reduced by externally added adrenodoxin reductase, adrenodoxin, and NADPH in the presence of cholesterol, indicating that the most of the heme proteins is located on the outside monolayer of the bilayer vesicles.* As previously demonstrated by us (5), phospholipids are low-spin inducers and cholesterol is a high-spin inducer for $P450_{scc}$. At 7.3 mol % cholesterol, liposomal $P450_{scc}$ displayed a considerable amount of the oxidized low-spin form, although cholesterol was largely excess relative to $P450_{scc}$ in membrane. Upon enzymatic reduction, approximately 70% of the total heme was reduced. At 39.5 mol % cholesterol, the high-spin form was spectroscopically detected by a small peak at 393 nm, and more than 90% of $P450_{scc}$ was reduced enzymatically.

We have then studied kinetically the reduction reaction of liposomal $P450_{scc}$ by adrenodoxin reductase, adrenodoxin, and NADPH. As shown in Fig. 1, the reductive process follows by fast and slow phases. Both rates approximately fit to a pseudo first order reaction. The fast rates increased as the membrane-cholesterol concentration increased, whereas the slow rates were independent from the cholesterol concentrations. The reciprocal plots of the fast rates against the cholesterol concentration were linear, and the intercept provided a half-maximal concentration of cholesterol ($Kmax/_2$). The value was 3.7 mol % cholesterol. From the slow rates obtained from prolonged reactions, the first order rate constants were calculated. The rates were constant within experimental errors at the range of up to 36 mol % cholesterol. The extrapolation of the slow phase to zero time provided the content of the high-spin form of $P450_{scc}$, which is originally present in the reaction mixture at the presence of variable amount of cholesterol. The reciprocal plots of the content of the high-spin form against the cholesterol provided the dissociation constant of $P450_{scc}$-cholesterol complex. The value was calculated to be 1.9 mol % cholesterol.

---

* The tryptic profile of the membrane-free protein was essentially the same as when $P450_{scc}$ was embedded into either DOPC-liposomes or vesicles prepared from phospholipids of adrenocortical mitochondria. The molecular shape of $P450_{scc}$ was implicated from our trypsinolysis experiments as follows: the heme molecule consists of two ordered core domains and a short connecting loop similar to the one of immunoglobins.

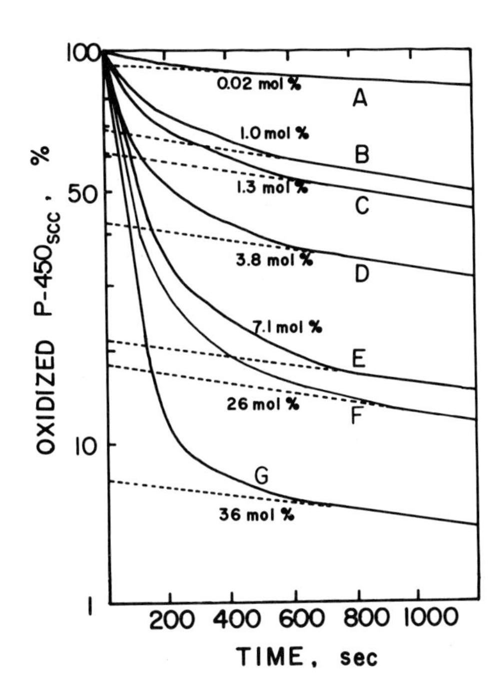

Fig. 1. Effect of DOPC-liposomal cholesterol concentration on NADPH-dependent reduction of $P450_{scc}$

Purified $P450_{scc}$ was embedded into DOPC-liposomes which contain variable amounts of cholesterol. Curves A to F contain cholesterol at 0.02, 1.0, 1.3, 3.8, 7.1, 26, and 36 mol %, respectively.

The enzymatic reduction was carried out under anaerobic conditions in a reaction mixture containing 30 nmoles of NADP, 500 nmoles of glucose-6-phosphate, 5 units glucose-6-phosphate dehydrogenase, 0.06 nmoles of adrenodoxin, 0.01 nmole of adrenodoxin reductase, 0.7 nmoles of variable cholesterol-containing liposome-embedded $P450_{scc}$, and carbon monoxide in a total volume of 1.0 ml. The reduction was followed spectrophotometrically at 448 nm.

TABLE I

CHOLESTEROL CONTENTS IN BOVINE ADRENAL CORTEX SUBMITOCHONDRIAL FRACTIONS

Fractions	Cholesterol
	mol %
Mitochondria (n = 7)	9.6 ± 2.4
Inner membrane (n = 5)	6.2 ± 0.7
Outer membrane (n = 4)	12.6 ± 1.3
Mitoplast (n = 4)	8.2 ± 1.5

Cholesterol was determined by the method of cholesterol oxidase (9) after extraction by $CHCl_3$:MeOH (2:1).

As shown in Table I, the inner membrane of bovine adrenocortical mitochondria has approximately 6 mol % cholesterol. This implies that isolated bovine mitochondria are partially stimulated by ACTH action. Upon full stimulation by this hormone, the cholesterol concentration in the inner membrane must increase far more than 3.7 mol %, if cholesterol uniformally distributes in the membrane. Alternatively, one has to assume a specific transfer mechanism of cholesterol to $P450_{scc}$ across membrane and a facilitated lateral diffusion of cholesterol. In a separate experiment, the maximum extent of cholesterol binding to $P450_{scc}$ was observed at approximately 20 mol % by titrating $P450_{scc}$ by liposomal cholesterol (5). Lambeth et al (6) recently reported the dissociation constant of $P450_{scc}$-cholesterol complex to be 9.1 mol % in DOPC-vesicles, based on their titration experiments. Our kinetical estimation gave a smaller value than those obtained by titration experiments. It appears that the optical titration provides an overestimation in view of the half-maximum concentration calculated from the fast rate reactions. This view may be supported by the fact that at 7.3 mol % cholesterol the majority of $P450_{scc}$ is optically low-spin but the form is reduced readily by the enzyme system. The cholesterol-bound form, which displays optically a low-spin form in membrane, may be an important species in the catalysis.

Based on the present results, together with existing knowledge on this field (1), we would point out four possibilities of the ACTH action on adrenal cortex mitcchondria in terms of cholesterol availability. First, the transfer from cholesterol-rich outer membrane to cholesterol-poor inner membrane. Secondly, the transmembrane movement of cholesterol across the inner membrane. Thirdly, the cholesterol molecules in aqueous matrix phase to the hydrophobic heme procket in the cytochrome molecule. The fourth possibility is the lipid phase lateral diffusion of cholesterol in the matrix side layer of the inner membrane, where $P450_{scc}$ locates (7). This last possibility may not be likely from the two facts indicating the exposure of heme to aqueous medium (2) and the rapid binding of hydroxy-cholesterols to the heme (8).

Our recent experiments on the depletion of mitochondrial cholesterol by malate-supported steroidogenesis indicated that the cholesterol pool in the inner membrane decreased more than that of the outer membrane. In addition, all of the cholesterol in the inner membrane was not utilized by the hydroxylase system. These results suggest that in the inner membrane more than two cholesterol pools for steroidogenesis may exist.

ACKNOWLEDGEMENTS

This study was supported by a Research Grant from the National Institutes of Health (AM-12713-13).

REFERENCES

1. T. Kimura (1981) Mol. Cell. Biochem., 36, 105-122.
2. F. Yamakura and T. Kimura (1981) Biochim. Biophys. Acta, 649, 343-354.
3. F. Yamakura and T. Kimura (1982) Biochem. International, in press.
4. T. Kimura, J. Parcells, and H. P. Wang (1978) Methods Enzymol, 52, 132-142.
5. T. Kido and T. Kimura (1981) J. Biol. Chem., 256, 8561-8568.
6. J. D. Lambeth, S. E. Kitchen, A. A. Faroqui, R. Tuckey, and H. Kamin (1982) J. Biol. Chem., 257, 1876-1884.
7. P. F. Churchill and T. Kimura (1979) J. Biol. Chem. 254, 10443-10448.
8. T. Kido, F. Yamakura, and T. Kimura (1981) Biochim. Biophys. Acta, 666, 370-381.
9. A. C. Deacon and P. J. G. Dawson (1979) Clin. Chem., 25, 976-984.

Cytochrome P-450, Biochemistry, Biophysics
and Environmental Implications, E. Hietanen,
M. Laitinen and O. Hänninen editors

SUICIDE INACTIVATION OF RAT LIVER CYTOCHROME P-450 IN VIVO AND IN VITRO

JAMES HALPERT, BIRGITTA NÄSLUND AND INGVAR BETNER
Department of Medical Nutrition, Karolinska Institute, Huddinge Hospital F69, S-141 86 Huddinge, Sweden.

INTRODUCTION

The antibiotic chloramphenicol has previously been shown to act in vitro as a suicide substrate of the major phenobarbital-induced form of rat liver cytochrome P-450, inactivating the enzyme by virtue of the covalent modification of the protein rather than of the heme moiety (1). The inactivation is accompanied by the covalent binding of 1.5 nmol metabolite(s) per nmol cytochrome P-450. Approximately 50% of the bound chloramphenicol exists as an adduct of chloramphenicol oxamic acid and the ε-amino group of one or more lysine residues in the cytochrome P-450, whereas the remainder of the bound material is very labile, and appears to be derived from a different active metabolite than the putative acyl chloride which gives rise to the lysine adduct (2-5). In vitro, chloramphenicol is a very effective suicide substrate for cytochrome P-450, requiring approximately five turnovers to inactivate the enzyme (2).

In humans chloramphenicol is known to influence the plasma levels and half-lives of other therapeutic agents (6), and in vivo administration of chloramphenicol to experimental animals has been shown to increase barbiturate-induced sleeping time (7-9) and to inhibit various cytochrome P-450 dependent monooxygenase activities assayed in vitro (9-10). However, in none of the previous studies has the precise mechanism of the in vivo inhibition of cytochrome P-450 mediated reactions been elucidated, although inhibition of substrate binding to the enzyme has been suggested as a possible explanation (9).

In the present investigation we have sought to examine the mechanism by which administration of chloramphenicol to phenobarbital-treated rats causes inhibition of cytochrome P-450 dependent monooxygenase activity and to relate the inhibition to the covalent binding of chloramphenicol metabolites to cytochrome P-450.

RESULTS

Effect of in vivo treatment with chloramphenicol on rat liver microsomal enzymes

As seen in table 1, chloramphenicol at a dose of 100 mg/kg caused significant inhibition of the microsomal metabolism of 7-ethoxycoumarin and 1,1,2,2-tetrachloroethane and of the endogenous NADPH oxidase activity, but had no effect on the level of cytochrome P-450 detectable as its carbon monoxide complex or on the NADPH-cytochrome c reductase activity. Since these determinations utilizing individual microsomes from the chloramphenicol-treated and control groups clearly showed that the inhibition observed was a specific effect of the chloramphenicol treatment, subsequent experiments aimed at elucidating the precise nature of the inhibition were performed using pooled microsomes from each group. Furthermore, since rather similar degrees of inhibition of the metabolism of tetrachloroethane and ethoxycoumarin were observed, only the latter substrate was used in the remainder of the investigation.

TABLE I

EFFECT OF CHLORAMPHENICOL ON RAT LIVER MICROSOMAL ENZYMES

Microsomal Constituent	Control	Chloramphenicol[a)]
Cytochrome P-450 (nmol/mg protein)	2.56 ± 0.26	2.55 ± 0.27[b)]
NADPH-cytochrome c reductase (U/mg)	0.59 ± 0.05	0.67 ± 0.05[b)]
NADPH oxidase (nmol/min/nmol P-450)	6.72 ± 1.22	4.85 ± 0.56
7-ethoxycoumarin deethylase (nmol/min/nmol P-450)	2.81 ± 0.10	1.58 ± 0.08
1,1,2,2-tetrachloroethane metabolism[c)] (nmol/min/nmol P-450)	1.84 ± 0.22	0.87 ± 0.11

Unless otherwise noted, each value represents the mean ± standard deviation of duplicate determinations on individual microsomes from 3-4 rats per group.

a) Chloramphenicol, 100 mg/kg i.p. in 0.5 ml propylene glycol, 1 hr prior to sacrifice; controls received vehicle only. Both groups had been treated for 5 days with 0.1% phenobarbital in the drinking water and starved overnight prior to treatment.

b) Not significantly different ($P>0.05$) from respective control value.

c) Assayed as described in reference 11; single determinations per rat.

Effect of chloramphenicol on kinetics of 7-ethoxycoumarin deethylation and on ethoxycoumarin binding to cytochrome P-450

Three experiments were carried out in which the steady-state kinetics of ethoxycoumarin metabolism were examined at concentrations between 10 and 240 μM.

In all cases the Vmax was found to be 48-50% lower in the microsomes from the chloramphenicol-treated group, whereas the Km was unaffected. In agreement with these results, chloramphenicol was found to have no effect on the Ks or on the ΔAmax for the type I difference spectrum induced by ethoxycoumarin, suggesting inhibition of some step subsequent to substrate binding.

Effect of chloramphenicol on cytochrome P-450 reduction

The chloramphenicol-mediated decrease in endogenous NADPH oxidase activity in the absence of any effect on the level of dithionite-reduced cytochrome P-450 or on the cytochrome c reductase activity (Table I) suggested that some modification of the P-450 had occurred which inhibited its ability to accept electrons from the reductase. In accordance with this hypothesis, both the steady-state level of enzymatically reducible cytochrome P-450 under aerobic conditions and the rate of enzymatic P-450 reduction under anaerobic conditions were found to be decreased by 30% in the microsomes from the chloramphenicol-treated rats.

Effect of chloramphenicol on cumene hydroperoxide dependent 7-ethoxycoumarin deethylase activity

In order to short-circuit the steps involving an interaction between the cytochrome and the reductase, the metabolism of ethoxycoumarin in the presence of 1 mM cumene hydroperoxide was examined by direct fluorimetry. At room temperature the reaction was found to be linear with time for up to 25 seconds and linear with protein up to 1.0 nmol cytochrome P-450 per ml. Under these conditions there was no difference in the cumene hydroperoxide supported metabolism of ethoxycoumarin between the chloramphenicol-treated and control groups, whereas the NADPH dependent activity assayed under the same conditions showed the expected inhibition.

Covalent binding of chloramphenicol metabolites in vivo

After it was established that a dose of 100 mg/kg chloramphenicol caused a ca 50% loss of monooxygenase activity assayed in vitro, the same dose of radiolabeled compound (1 mCi/mmol) was administered to two phenobarbital-treated rats, and microsomes were prepared one hour later. The microsomes contained 3.3 nmol ^{14}C/mg protein, which corresponded to 1.7 nmol ^{14}C/nmol P-450. Alkaline hydrolysis of the labeled microsomes released 70% of the ^{14}C, which was found to exist mainly as oxalic acid and chloramphenicol oxamic acid as shown by chromatography on a Bio-Rad P-2 column. In all respects these results are comparable to those obtained when intact microsomes or a reconstituted system are labeled in vitro with chloramphenicol and leave little doubt that the greater part if not all of the ^{14}C associated with the microsomes represented covalently-bound adducts of chloramphenicol metabolites.

By solubilization with sodium cholate and chromatography on octylamino Sepharose (12), a cytochrome P-450 fraction was purified from the ^{14}C microsomes. This fraction contained 0.75 nmol ^{14}C/nmol P-450 and accounted for approximately one-half of the total protein-bound radioactivity recovered. Digestion with pronase followed by chromatography on Bio-Rad P-2 and by HPLC on C_{18} in 25% methanol revealed that approximately one-third of the radioactivity associated with the P-450 fraction was in the form of N-ε-chloramphenicol oxamyl lysine (2). When the P-450 fraction was chromatographed on DEAE Sephacel (12), approximately equal ratios of $^{14}C/A_{417}$ were found in all four P-450 fractions. The level of binding associated with the homogeneous B_2 isozyme was 0.7 nmol ^{14}C/nmol P-450. Since this level of binding results from a dose of chloramphenicol which causes 50% inhibition of enzymatic activity, the stoichiometry of inactivation of 1.5 nmol ^{14}C/nmol P-450 previously established in vitro (1) appears to be valid in vivo.

DISCUSSION

The results of the present investigation leave little doubt that the inhibitory effects of chloramphenicol administration in vivo on rat liver mixed-function oxidase activity assayed in vitro are mediated by the suicide inactivation of cytochrome P-450. The same metabolic pathways which give rise to covalently-bound adducts of chloramphenicol and P-450 in vitro (2-5) appear to be operative in vivo and to inactivate the enzyme in a similar fashion. The loss of monooxygenase activity may be related to an impaired interaction between the cytochrome P-450 and the reductase.

REFERENCES

1. Halpert, J. and N~al R.A. (1980) Mol. Pharmacol. 17, 427.
2. Halpert, J. (1981) Biochem. Pharmacol. 30, 875.
3. Halpert, J. (1982) Mol. Pharmacol. 21, 166.
4. Pohl, L.R., Nelson, S.D., and Krishna, G. (1978) Biochem. Pharmacol. 27, 491.
5. Pohl, L.R. and Krishna, G. (1978) Biochem. Pharmacol. 27, 335.
6. Christensen, L.K. and Skovsted, L. (1969) Lancet 2, 1397.
7. Dixon, R.L. and Fouts, J.R. (1962) Biochem. Pharmacol. 11, 715.
8. Alvin, J. and Dixit, B.N. (1974) Biochem. Pharmacol. 23, 139.
9. Adams, H.R., Isaacson, E.I., and Masters, B.S.S. (1977) J. Pharmacol. Exp. Ther. 203, 388.
10. Reilly, P.E.B., and Ivey, D.E. (1979) FEBS Lett. 97, 141.
11. Halpert, J. (1982) Drug Metab. Disp., in press.
12. Guengerich, F.P. and Martin, M.V. (1980) Arch. Biochem. Biophys. 205, 365.

Cytochrome P-450, Biochemistry, Biophysics
and Environmental Implications, E. Hietanen,
M. Laitinen and O. Hänninen editors

DIETARY LIPIDS AS MODIFIERS OF MONOOXYGENASE INDUCTION

EINO HIETANEN[1], MARKKU AHOTUPA[1], ARNO HEIKELÄ[1] AND MATTI LAITINEN[2]
[1]Department of Physiology, University of Turku, 20520 Turku 52 (Finland) and
[2]Department of Physiology, University of Kuopio, 70101 Kuopio 10 (Finland)

INTRODUCTION

The monooxygenase enzymes metabolizing xenobiotics and endogenous compounds are located in the endoplasmic reticulum of cells. Both nutritional compounds and environmental agents may alter the structure of the endoplasmic membranes and have consequently effects on the membrane function (1,2).

Dietary lipids change the membrane composition and also modify the drug metabolizing activity in liver (3-5). Dietary lipids have also been found to modify the experimental cancer formation although the mechanisms are unsolved (6-8). We have studied the effects of dietary cholesterol and neutral lipids on the inducibility of monooxygenases by xenobiotics.

MATERIALS AND METHODS

Weanling male Wistar rats (aged 3 weeks) were divided into four groups: one was fed cholesterol free pelleted diet, another 2 % cholesterol diet, both for 4 weeks, third group fat free and fourth group 45 % fat diet for 2 weeks. Diets were from ICN Life Sciences group. During the last week of feeding each dietary group was divided further into 4 subgroups: first served as a control, second received daily phenobarbital (Pb) in drinking water for 1 week at an estimated dose of 100 mg/kg, third received carbon tetrachloride (CCl_4) as a single s.c. dose of 1.5 g/kg 6 days before sacrification and fourth both Pb and CCl_4.

Liver microsomes were isolated by a centrifugation at 105 000 g x 60 min. The phospholipid content was measured as inorganic phosphorous (9) and cholesterol content as described previously (10,11). The cytochrome P-450 concentration (12), NADPH cytochrome c reductase (13), aryl hydrocarbon hydroxylase (14), ethoxycoumarin O-deethylase (15,16) and PPO hydroxylase (17,18) activities were determined as described previously. Student's t test was used for statistical comparisons and means ± SEM are shown.

RESULTS

Cholesterol feeding increased markedly microsomal phospholipid and cholesterol contents while high fat diet decreased phospholipid content slightly from that found in the fat free group (Table 1). In the cholesterol free group

Pb or CCl_4 didn't change phospholipid content of microsomes. Pb increased phospholipid content of liver microsomes in rats fed high lipid diets. CCl_4 decreased microsomal phospholipid content in rats fed fat free and high fat diets (Table 1).

TABLE 1

HEPATIC MICROSOMAL PHOSPHOLIPID AND CHOLESTEROL CONTENTS OF RATS FED LIPID DIETS

Rats in each dietary group were administered either by phenobarbital (Pb), carbon tetrachloride (CCl_4) or both in combination (Pb + CCl_4). Controls of dietary groups were compared to each other, respectively. Otherwise comparisons are made in each group separately to respective control. Statistical significancies are shown as follows: o: $P<0.1$, *: $P<0.05$, **: $P<0.01$ and ***: $P<0.001$. Means ± SEM are shown.

Diet		N	Phospholipid (μmol P/g w wt)	Cholesterol (μmol/g w wt)
Chol. free	Control	7	4.2 ± 0.4	0.66 ± 0.04
	Pb	7	5.2 ± 0.3	0.88 ± 0.13
	CCl_4	8	5.0 ± 0.2	0.92 ± 0.07**
	Pb + CCl_4	7	5.3 ± 0.3	1.07 ± 0.01***
High chol.	Control	8	6.7 ± 0.3***	0.99 ± 0.07**
	Pb	6	9.9 ± 1.4*	0.87 ± 0.12
	CCl_4	8	6.3 ± 0.2	1.33 ± 0.20
	Pb + CCl_4	7	7.4 ± 0.6	1.29 ± 0.24
Fat free	Control	5	5.2 ± 0.1	Not determined
	Pb	5	5.7 ± 0.1*	
	CCl_4	5	4.2 ± 0.2**	
	Pb + CCl_4	5	4.9 ± 0.3	
High fat	Control	4	4.4 ± 0.5	
	Pb	4	7.8 ± 0.6***	
	CCl_4	4	3.0 ± 0.6 o	
	Pb + CCl_4	4	6.1 ± 0.8	

The hepatic cytochrome P-450 concentration was about 2-fold in rats fed either high cholesterol or high fat diet as compared to groups fed lipid free diets (Table 2). Pb increased the hepatic cytochrome P-450 content more in rats fed high lipid than lipid free diets whether it was given alone or in compination with CCl_4 (Table 2). The NADPH cytochrome c reductase activity was rather nonresponsive to dietary manipulations (Table 2). The aryl hydrocarbon hydroxylase (AHH) did not either change due to dietary modifications (Table 2). The ethoxycoumarin O-deethylase (EtOH) activity doubled by the high lipid diets (Table 2). The inducibility by Pb was rather similar in each dietary group. Both the 2 % cholesterol and 45 % fat diets increased the PPO hydroxylase activity (Table 2). Pb increased the activity in all other groups except for the

TABLE 2

THE HEPATIC CYTOCHROME P-450 CONCENTRATION, NADPH CYTOCHROME C REDUCTASE, ARYL HYDROCARBON HYDROXYLASE, ETHOXYCOUMARIN O-DEETHYLASE AND PPO HYDROXYLASE ACTIVITIES OF RATS FED VARIOUS LIPID DIETS

Diet	Cyt.P-450[a] nmol x g^{-1}	Cyt c red.[a] nmol x min^{-1} x g^{-1}	AHH[a] pmol x min^{-1} x g^{-1}	EtOH[a] nmol x min^{-1} x g^{-1}	PPO[a] units x min^{-1} x g^{-1}
Chol free					
Control	4.2 ± 0.5	2.3 ± 0.3	252 ± 35	2.5 ± 0.2	34.7 ± 3.5
Pb	5.8 ± 0.6	2.8 ± 0.5	142 ± 16	10.7 ± 1.2***	48.0 ± 10.1
CCl_4	4.1 ± 0.5	2.2 ± 0.3	132 ± 23	2.5 ± 0.3	29.4 ± 3.7
Pb + CCl_4	6.8 ± 0.7*	2.8 ± 0.3	161 ± 6	8.4 ± 0.6***	40.9 ± 3.0
2 % chol.					
Control	8.5 ± 1.3**	2.9 ± 0.4	344 ± 44 o	5.9 ± 0.4***	38.7 ± 3.5**
Pb	16.8 ± 3.6*	4.6 ± 0.8 o	245 ± 52	19.9 ± 2.0***	76.7 ± 15.3 o
CCl_4	7.1 ± 0.5	2.6 ± 0.3	237 ± 12	4.9 ± 0.2	39.2 ± 4.2
Pb + CCl_4	13.6 ± 1.9**	3.1 ± 0.2	209 ± 45	14.5 ± 1.6***	54.6 ± 8.0
Fat free					
Control	4.0 ± 0.2	1.7 ± 0.1	137 ± 19	3.1 ± 0.2	22.8 ± 1.9
Pb	8.8 ± 0.9***	2.6 ± 0.1***	107 ± 11	10.7 ± 0.9***	41.8 ± 2.1***
CCl_4	4.0 ± 0.3	1.4 ± 0.2	83 ± 15	2.3 ± 0.2	17.2 ± 1.3
Pb + CCl_4	7.0 ± 0.6**	2.3 ± 0.4	94 ± 11	9.9 ± 1.0***	35.7 ± 3.8*
45 % fat					
Control	7.9 ± 0.6***	2.8 ± 0.4*	171 ± 19	5.9 ± 0.5***	37.0 ± 1.5**
Pb	20.0 ± 1.3***	4.6 ± 0.4*	208 ± 18	26.8 ± 2.4***	79.5 ± 14.6*
CCl_4	6.3 ± 1.4	2.5 ± 0.2	185 ± 39	5.6 ± 0.6	32.5 ± 2.4
Pb + CCl_4	18.0 ± 2.7**	4.2 ± 0.2*	165 ± 7	23.9 ± 2.3***	60.0 ± 9.7*

[a] g tissue wet weight; For other explanation see Table 1.

cholesterol free group.

DISCUSSION

Dietary lipids have been found to be essential in maintaing the inducibility of cytochrome P-450 dependent monooxygenase activities (4,19). Lipids may also modify the effects of other compounds affecting monooxygenase activities (20). Well in accordance with our data Marshall and McLean (4) found that the cytochrome P-450 contents of livers from rats fed the fat free diet was only about 36 % of that found in rats fed standard diet when the cytochrome content was measured in phenobarbital induced rats. However, our present study revealed that the inducibility by phenobarbital was actually nearly as high in rats fed fat free diet as in those fed high fat diet although the activity level was much lower. Furthermore, although both in livers of rats fed cholesterol free and fat free diets the cytochrome P-450 levels were about equal the inducibility of cytochrome P-450 was much less in those rats fed the cholesterol free diet in comparison to those fed fat free diet suggesting that cholesterol might be even more important constituent to enzyme induction than other dietary lipids. When studying monooxygenases using various substrates it is evindent from the present study that the dietary lipids influence rather specifically on certain substrates in a way that the AHH activity is much less influenced than that of ethoxycoumarin in a rather parallel way to PPO.

ACKNOWLEDGEMENTS

This study was supported by grants from NIH (ROI ES 01684) and J. Vainio Foundation (Finland).

REFERENCES

1. Campbell, T.C. (1977) Clin. Pharmacol. Ther., 22,699.
2. Hietanen, E. (1981) in: Rechcigl, M., Jr. (Ed.), CRC Handbook of Nutritional Requirements in a Functional Context, CRC Press, Boca Raton, pp. 361-395.
3. Agradi, E., Spagnuolo, C. and Galli, C. (1975) Pharmacol. Res. Commun., 7, 469.
4. Marshall, W.J. and McLean, A.E.M. (1971) Biochem. J., 122,569.
5. Parke, D.V. (1978) Wrld. Rev. Nutr. Diet., 29,96.
6. Nigro, N.D., Singh, D.U., Campbell, R.L. and Pak, M.S. (1975) N. Natl. Cancer Inst., 54,439.
7. Gori, C.B. (1977) J. Am. Diet. Ass., 71,375.
8. Jakoby, H.P. and Boumann, C.A. (1946) Am. J. Cancer, 39,338.
9. Bartlett, C.R. (1959) J. Biol. Chem. 234,466
10. Abell, L.L., Levy, B.B. and Kendall, F.E. (1952) J. Biol. Chem., 195,357.
11. Anderson, J.T. and Keys, A. (1956) Clin. Chem., 2,145.
12. Omura, T. and Sato, R. (1964) J. Biol. Chem., 239,2379.
13. Phillips, A.H. and Langdon, R.G. (1962) J. Biol. Chem., 237,2652.
14. Wattenberg, L.W., Leong, J.L. and Strand, P.J. (1962) Cancer Res., 22,1120.
15. Aitio, A. (1978) Anal. Biochem., 85,488.
16. Ullrich, V. and Weber, P. (1972) Hoppe-Seyler's Z. Physiol. Chem., 353,1171.
17. Ahokas, J.T. (1976) Res. Commun Chem. Path. Pharm., 13,439.
18. Cantrell, E.T., Abrew-Greenberg, M., Guyden, J. and Boubee, D.L. (1975) Life Sci., 17,317.
19. Stier, A. (1976) Biochem. Pharmac., 25,109.
20. Joly, I-G., Hetu, C., Marier, P. and Villeneuve, J.P. (1976) Biochem. Pharmac., 25,1995.

Cytochrome P-450, Biochemistry, Biophysics and Environmental Implications, E. Hietanen, M. Laitinen and O. Hänninen editors

EFFECTS OF AGE ON HEPATIC MONOOXYGENASE SYSTEM IN NORMAL AND TOLUENE-INDUCED FEMALE RATS

KAIJA PYYKKÖ AND HEIKKI VAPAATALO
Departments of Clinical and Biomedical Sciences, University of Tampere, Box 607, SF-33101 Tampere 10, Finland.

INTRODUCTION

Toluene is one of the most used aromatic solvents in industry. It is metabolized in liver by a cytochrome P-450 dependent enzyme system. Toluene induces microsomal drug-metabolizing enzymes, but the induction pattern seems to be different from that of phenobarbital or methylcholantrene (1). We have demonstrated earlier (2), that in male rats both control and toluene-induced activities of liver microsomal monooxygenases vary with age. In most cases the minimum inducibility was at the age of puberty. In the present study, the depelopment of liver microsomal enzymes was determined from control and toluene-exposed female rats, in order to detect possible sex differences.

MATERIALS AND METHODS

Female Sprague-Dawley rats 10, 20, 30, 40, 50, 60 and 90 days of age at the beginning of the experiment were used. Toluene (2 M in corn oil) was given orally by gastric tube, 20 mmol/kg daily for 3 consecutive days between 9 and 10 a.m. Controls received the same quantity of vehicle (10 ml/kg). On the fourth morning, the rats were decapitated and their livers removed. Microsomes were isolated by Sepharose 2B gel filtration (3). The concentration of cytochrome P-450 (P450) and b_5 (B5), and the activities of aniline hydroxylase (AH), aryl hydrocarbon hydroxylase (AHH), aminopyrine N-demethylase (APDM), 7-ethoxycoumarine O-deethylase (EDE) and NADPH - cytochrome c reductase (NCR) were determined by the methods described in the references (3-8).

RESULTS AND DISCUSSION

The age curves of microsomal enzymes in the livers of normal and toluene-treated female rats are shown in Figures 1 and 2. In young control rats, the cytochrome concentrations and enzyme activities, with the exception of AH, increased until the age of

20 or 30 days. Thereafter, P450 and B5 reduced slowly and AH, AHH, APDM and EDE sharply with age. In adult rats, the enzyme levels remained lower than in immature rats. The NCR curve differed from the others; it showed no decrease at the age of 30 days.

Toluene induced P450 and the monooxygenases in all age groups, but the shape of the age-activity curves of different enzymes after toluene induction varied markedly. The highest induced P450 and monooxygenase levels were found in immature rats at the age of 20 or 30 days, and the lowest ones at the age of 50 days.

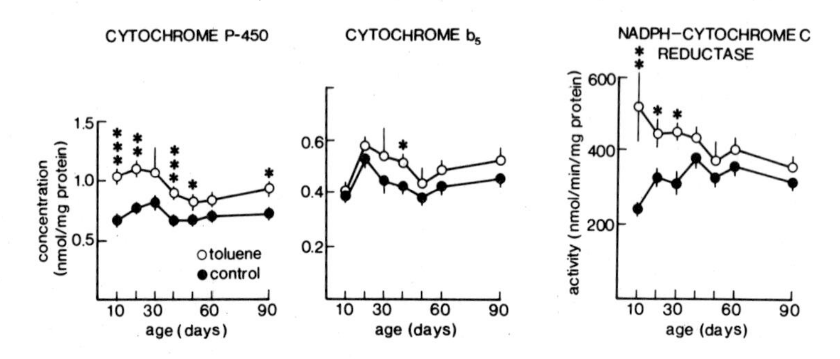

Fig.1. The effect of age on the concentration of cytochromes and the activity of NADPH - cytochrome c reductase in the liver microsomes of toluene exposed and control female rats.
*** p<0.001, ** p<0.01 and * p<0.05 compared to the controls.

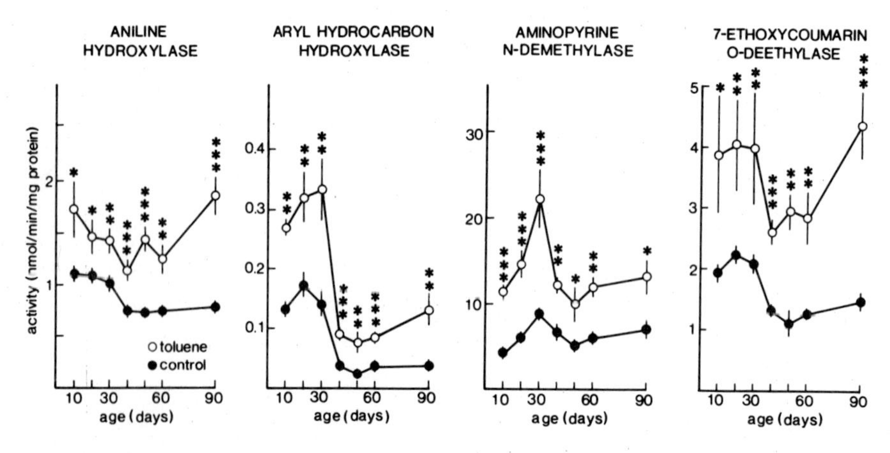

Fig.2. The effect of age on the activities of the monooxygenases in the liver microsomes of toluene exposed and control female rats. Asterisks as in Fig. 1.

AH and EDE activities increased again at the age of 90 days. However, the inducibility of AH, AHH and EDE increased, whereas that of APDM and P450 decreased during the age period studied. NCR was induced only in immature females, and B5 hardly significantly in any of the age groups.

The differences of the developmental curves in both uninduced and toluene-induced activities of monooxygenases reflect the multiplicity of P450 forms, the types and proportions of which vary with the age of the animal. Morover, the inducibility of various monooxygenase activities seems to develop separately.

In Table 1, the values of 30 and 60 days old female rats are compared to our earlier results from male rats of the same age (2). The enzyme activities of immature females and males, 30 days of age, were quite similar, but after sexual maturation, at the age of 60 days, AHH and APDM were considerably lower in control female rats. The levels of almost all enzymes after toluene induction were clearly lower in female than male rats; females showed weaker responses to the inducer.

The sex differences in the activities of drug-metabolizing enzymes of rat liver are known to be dependent on the substrate used. AHH and APDM are typically sex-dependent monooxygenases, and AH a sex-independent monooxygenase (9). Male rats, in general show a higher drug-metabolizing activity than females. Sex differences are explained to be determined by the androgen level and androgen-

TABLE 1.

THE RATIO BETWEEN FEMALE AND MALE RAT MICROSOMAL ENZYME LEVELS AT IMMATURE (30 DAYS) AND MATURE (60 DAYS) AGE*

	immature		mature	
	control	toluene exposed	control	toluene exposed
AH	1.1	1.0	1.1	0.6
AHH	0.9	1.1	0.3	0.4
APDM	0.9	1.1	0.7	0.5
EDE	1.1	1.1	1.2	0.6
P450	1.0	1.0	0.9	0.7
B5	1.0	1.0	1.3	0.9
NCR	0.9	1.0	1.2	0.8

*The data of male rats from Pyykkö, 1982 (2).

induced imprinting at birth (10). More recently, Gustafsson and coworkers (11) have presented a central regulatory system in the control of hepatic steroid and drug metabolism. The pituitary gland of female animals has been postulated to secrete a "feminizing factor", which maintains the female-type metabolism in the liver. The common regulatory mechanism in the metabolism of steroid hormones and foreign compounds with various specific and unspecific details could explained well the developmental changes during sexual maturation and the sex differences of various drug-metabolizing monooxygenases.

ACKNOWLEDGEMENT

This work was supported by grants of the Academy of Finland and the Foundation for Drug Research, Finland.

REFERENCES

1. Pyykkö, K. (1980) Biochim. Biophys. Acta 633, 1-8.
2. Pyykkö, K. (1982) in: Sato, R. (Ed.), Microsomes, Drug Oxidations, and Drug Toxicity, Japan Scientific Societies Press, Tokyo, in press.
3. Tangen, O., Johnsson, J. and Orrenius, S. (1973) Anal. Biochem. 54, 597-603.
4. Omura, T. and Sato, R. (1964) J. Biol. Chem. 239, 2370-2378.
5. Imai, Y., Ito, A. and Sato, R. (1966) J. Biochem. 60, 417-428.
6. Nebert, D. W. (1978) Methods in Enzymology, 52, 226-240.
7. Aitio, A. (1978) Anal. Biochem. 85, 488-491.
8. Strobel, H. W. and Dignam, J. D. (1978) Methods in Enzymology, 52, 89-96.
9. Kato, R. and Gillette, J. (1965) J. Pharmacol. Exp. Ther. 150, 279-284.
10. Gustafsson, J.-Å., Gustafsson, S. A., Ingelman-Sundberg, M., Pousette, Å., Stenberg, Å. and Wrange, Ö. (1974) J. Steroid Biochem. 5, 855-859.
11. Skett, P. and Gustafsson, J.-Å. (1979) Rev. Biochem. Toxicol. 1, 27-52.

Cytochrome P-450, Biochemistry, Biophysics and Environmental Implications, E. Hietanen, M. Laitinen and O. Hänninen editors

INTERACTION OF IMIDAZOLE WITH RABBIT LIVER MICROSOMAL CYTOCHROME P-450: INDUCTION AND INHIBITION OF ACTIVITY

KAREN KAUL HAJEK AND RAYMOND F. NOVAK
Department of Pharmacology, Northwestern University Medical and Dental Schools, Chicago, Illinois 60611, U.S.A.

INTRODUCTION

Previous work in our laboratory has demonstrated the utility of ^1H FT NMR spectroscopy in studying interactions of the substrates acetanilide and N,N-dimethylaniline with purified isozymes of rabbit liver microsomal cytochrome P-450. Such studies provide information on the distances of approach, the relative orientation, and dynamic accessibility of substrates to the binding sites of these hemeproteins (1). Recent studies have demonstrated that the nitrogenous base imidazole interacts comparably with the paramagnetic heme iron atom of purified rabbit liver microsomal cytochrome P-450$_{LM_2}$ or LM_4, producing an enhancement in the T_1 relaxation rates of the H(2) and H(4,5) protons of imidazole (2). The paramagnetic contribution to the T_1 relaxation rate enhancement, T_{1p}^{-1}, may be utilized, in conjunction with the spectral dissociation constant (K_D) values and the Solomon-Bloembergen equation, to obtain estimates of the distance of approach of the H(2) and H(4,5) protons of imidazole to the paramagnetic heme iron atom of the cytochrome. These distance estimates suggest direct coordination of imidazole to the heme iron atom of cytochrome P-450$_{LM}$, and agree with the observations that (a) imidazole prevents the interaction of acetanilide or N,N-dimethylaniline with purified cytochromes P-450$_{LM_2}$ or LM_4 and (b) imidazole noncompetitively inhibits the metabolism of acetanilide, N,N-dimethylaniline, and *p*-nitroanisole in uninduced, phenobarbital- or β-naphthoflavone-induced microsomal preparations (3-5).

Certain compounds which inhibit microsomal drug metabolism may also act as inducers of hepatic microsomal cytochrome P-450. Wilkinson *et al.* have demonstrated the inhibition of drug metabolism *in vivo* by various imidazole-containing compounds (6). Our present work suggests that imidazole and phenylimidazole act as inducers of hepatic microsomal cytochrome P-450 in rabbits, as evidenced by elevated hepatic microsomal cytochrome P-450 levels, gel electrophoresis, and enhanced catalytic activity towards the substrate N,N-dimethylnitrosamine.

MATERIALS AND METHODS

Phenobarbital-inducible P-450$_{LM_2}$ and β-naphthoflavone-inducible P-450$_{LM_4}$ were purified from rabbit liver microsomes according to published procedures (7). Difference spectra were recorded using a Cary 219 spectrophotometer, equipped with a Lauda water bath operating at 25°C.

Male New Zealand white rabbits (2-2.5 kg) were injected with imidazole (200 mg/kg, i.p., 5 d) in 0.1 M potassium phosphate buffer, pH 7.5, or phenylimidazole (100 mg/kg, i.p., 4d) in corn oil. Animals were fasted 12 hrs prior to sacrifice, and liver microsomes were prepared using accepted procedures. SDS-polyacrylamide gel electrophoresis was performed according to the method of Laemmli (8). The formation of formaldehyde from dimethylnitrosamine was measured after a 30-minute incubation at 37°C using the method of Nash (9). Incubation mixtures consisted of 1.0 mg microsomal protein, 3.0-5.0 mM dimethylnitrosamine, and 1.0 mM NADPH in a total volume of 1 ml.

RESULTS AND DISCUSSION

The spectral dissociation constant (K_D) values obtained for the imidazole-cytochrome P-450 complexes are shown in Table 1. Binding of imidazole to the purified isozymes was found to be biphasic, with K_D values of 5.5×10^{-6} and 6.1×10^{-5} M for P-450$_{LM_2}$ or 6.3×10^{-6} and 1.6×10^{-4} M for P-450$_{LM_4}$.

TABLE 1

SUMMARY OF UV-VISIBLE DIFFERENCE SPECTRAL RESULTS OBTAINED FOR IMIDAZOLE WITH PURIFIED CYTOCHROMES P-450$_{LM_2}$ AND P-450$_{LM_4}$

Isozyme[a]	Minimum	Maximum	K_{D_1} (M)	K_{D_2} (M)
LM_2	412	432	$5.5 \pm 1.3 \times 10^{-6}$	$6.1 \pm 1.6 \times 10^{-5}$
LM_4	393	430	$6.3 \pm 1.8 \times 10^{-6}$	$1.6 \pm 0.8 \times 10^{-4}$

[a]Present at 1 μM in 0.1 M (LM_2) or 0.3 M (LM_4) phosphate buffer, pH 7.5.

The smaller K_D value which represents the higher affinity binding site, and previously obtained paramagnetic relaxation rate (T_{1p}^{-1}) values, were used with the Solomon-Bloembergen equation to estimate the distances of approach of the H(2) and H(4,5) protons of imidazole to the heme iron atom of cytochrome P-450$_{LM_2}$ and LM_4. For low-spin hemeproteins and primarily dipolar interactions, the Solomon-Bloembergen equation reduces to:

$$R(\text{Å}) = 540\left(T_{1M}\left(\frac{3\tau_c}{1+\omega_I^2\tau_c^2}\right)\right)^{1/6} \qquad \text{Eq. 1}$$

where $T_{1M} = \alpha T_{1p}$. T_{1M} represents the relaxation time of a nucleus bound in proximity to a paramagnetic metal ion, α is the mole fraction of substrate in complex, ω_I is the nuclear precession frequency, and τ_c is the correlation time, which lies in the range of 10^{-11} to 10^{-10} sec for most ferric hemeproteins. The estimated distances of approach for the H(2) and H(4,5) protons of imidazole to the iron atom using equation 1, are presented in Table 2. The H(2) protons are

TABLE 2

DISTANCES OF SEPARATION OF THE H(2) AND H(4,5) PROTONS OF IMIDAZOLE FROM THE PARAMAGNETIC HEME IRON ATOM OF CYTOCHROME P-450LM$_2$ OR LM$_4$

	P-450LM$_2$		P-450LM$_4$	
τ_c (sec)	H(2) Å	H(4,5) Å	H(2) Å	H(4,5) Å
10^{-11}	3.9	4.1	3.7	3.9
10^{-10}	5.6	5.9	5.3	5.6

positioned slightly nearer the heme iron atom of cytochrome P-450LM$_2$ than the H(4,5) protons, with distance estimates of 3.9 and 4.1 Å ($\tau_c = 10^{-11}$ sec) or 5.6 and 5.9 Å ($\tau_c = 10^{-10}$ sec) for the H(2) and H(4,5) protons of imidazole, respectively. Similar results were obtained with cytochrome P-450LM$_4$ with distances of 3.7 and 3.9 Å ($\tau_c = 10^{-11}$ sec) or 5.3 and 5.6 Å ($\tau_c = 10^{-10}$ sec) for the H(2) and H(4,5) protons, respectively. Correlation time values of $\sim 10^{-11}$ sec have been obtained for a number of low-spin hemeproteins. Thus, these results suggest that imidazole may directly coordinate to the heme iron atom of cytochrome P-450LM$_2$ or LM$_4$. Furthermore, these distance estimates are in agreement with the observations that (a) imidazole prevents the interaction of acetanilide, which binds 5-8 Å from the catalytic heme iron atom, or N,N-dimethylaniline with either P-450LM$_2$ or LM$_4$, and (b) imidazole is a noncompetitive inhibitor of acetanilide *p*-hydroxylase, N,N-dimethylaniline N-demethylase and *p*-nitroanisole O-dealkylase activities in uninduced, phenobarbital- and β-naphthoflavone-induced microsomes (3-5).

Pretreatment of rabbits with imidazole or phenylimidazole resulted in an increase in content of cytochrome P-450, from 1.3 to 1.8 nmoles/mg microsomal protein for imidazole and 1.0-1.2 nmoles/mg protein for phenylimidazole as compared to 0.7-0.8 nmoles/mg protein in uninduced animals (Table 3). β-Naphthoflavone and phenobarbital pretreatment increased P-450 levels to 1.2-2.2 and 2.4-3.2 nmoles/mg protein, respectively. The SDS polyacrylamide gel electrophoretic pattern of imidazole-induced microsomes included bands of protein with enhanced intensity occurring at the approximate positions of P-450LM$_2$, LM$_3$, and LM$_4$.

TABLE 3

CYTOCHROME P-450LM CONTENT AND DIMETHYLNITROSAMINE N-DEMETHYLASE ACTIVITY IN UNINDUCED, IMIDAZOLE-, PHENOBARBITAL-, AND β-NAPHTHOFLAVONE-INDUCED MICROSOMES

Pretreatment	Cytochrome P-450 nmole/mg protein	Formaldehyde Production	
		nmoles/min/ mg protein	nmoles/min/ nmole P-450LM
None	0.7 - 0.8	0.7	0.9
Imidazole	1.3 - 1.8	2.5	2.0
Phenylimidazole	1.0 - 1.2	2.1	2.0
β-naphthoflavone	1.2 - 2.2	0.4	0.4
Phenobarbital	2.4 - 3.2	0.4	0.2

The metabolism of N,N-dimethylnitrosamine by imidazole-induced microsomal suspensions differs from that observed for other microsomal preparations, as shown in Table 3. Imidazole- and phenylimidazole-induced microsomes produced 2.5 and 2.1 nmoles HCHO/min/mg protein, respectively, as compared to 0.7, 0.4 and 0.4 nmoles HCHO/min/mg protein for uninduced, β-naphthoflavone- and phenobarbital-induced microsomes, respectively. When expressed per nanomole P-450LM, these differences become even more apparent. In addition, the binding of imidazole to imidazole-induced microsomes is monophasic with $K_D = 2.8 \pm 0.8 \times 10^{-6}$ M.

ACKNOWLEDGEMENTS

Supported by NIH grant GM27836 (RFN) and the Monsanto Fund Fellowship (KKH).

REFERENCES

1. Novak, R.F., and Vatsis, K.P. (1982) Mol. Pharmacol. 21, 701-709.
2. Kaul, K.L., Novak, R.F., and Vatsis, K.P. (1980) Fed. Proceed. 39, 661.
3. Novak, R.F., Vatsis, K.P., and Kaul, K.L. (1981) in Biochemistry, Biophysics, and Regulation of Cytochrome P-450 (Gustafsson, J.A., Ruckpaul, K., Coon, M.J., Estabrook, R.W. and Gunsalus, I.V., eds.) pp. 583-586, Elsevier North Holland Biomedical Press, Amsterdam.
4. Kaul, K.L., and Novak, R.F. (1980) Fed. Proceed. 39, 2367.
5. Hajek, K.K., Cook, N.I., and Novak, R.F. (1981) Pharmacologist 23, 5.
6. Wilkinson, C.F., Hetnarski, K., and Yellen, T.O. (1972) Biochem. Pharmacol. 21, 3187-3192.
7. Coon, M.J., van der Hoeven, T.J., Dahl, S.B., and Haugen, D.A. (1978) in Methods in Enzymology, 52, Part C, 109-117, Academic Press, New York.
8. Laemmli, U.K. (1970) Nature 227, 680-685.
9. Nash, T. (1953) Biochem. J. 55, 416-421.

Cytochrome P-450, Biochemistry, Biophysics
and Environmental Implications, E. Hietanen,
M. Laitinen and O. Hänninen editors

SPECTRAL STUDIES OF CYTOCHROME P-450 IN ISOLATED RAINBOW TROUT LIVER CELLS

TOMMY ANDERSSON AND LARS FÖRLIN
Department of Zoophysiology, University of Göteborg, Box 25059, 400 31 Göteborg, Sweden

INTRODUCTION

It has been known for some time that fish possess the enzymes involved in biotransformation of xenobiotic chemicals e.g. cytochrome P-450 and conjugating enzymes, and that the former is inducible by polyaromatic hydrocarbons. Although considerable information on cytochrome P-450 mediated metabolism in fish has been gathered, little is known about its intracellular regulation.

In mammals isolated hepatocytes have been used in studies on regulation of cytochrome P-450 linked processes in the intact cell. In this paper we have used the freshly isolated rainbow trout liver cells and microsomes from control and β-naphtoflavone (BNF) treated trout to study optical absorption characteristics of cytochrome P-450.

MATERIAL AND METHODS

Fish: Cultured rainbow trout, *Salmo gairdnerii*, were kept in basins with aerated, filtered and recirculating water at a temperature of 10°C. BNF (100 mg/kg) were administrated via i.p. injections in peanut oil.

Isolation of hepatocytes: The procedure was essentially as described by Walton and Cowey (1979). The liver was perfused with calcium free Salmo Ringer (2) containing 0.5 mM EGTA for 10 minutes. Perfusion with collagenase and hyaluronidase medium was performed with a calcium containing Salmo Ringer for 30 min. and a final incubation of the cellsuspension with the digestive enzymes in 5min. After filtration, the cells were sedimented and washed twice with Salmo Ringer. Cellsuspensions were always examined for trypan blue exclusion. A high precentage (>90 %) of unstained cells were routinely observed.

Difference spectra were recorded in an Aminco DW-2a UV-VIS spectrophotometer.

RESULTS AND DISCUSSION

In isolated rat hepatocytes it has been suggested that only the substrate bound form of cytochrome P-450 can accept endogenous reducing equivalents and thus interact with CO (3). In the present study neither the introduction of CO nor the subsequent addition of ethylmorphine (EM) to the sample cuvette produced any difference absorption spectra in isolated trout cells (Fig. 1).

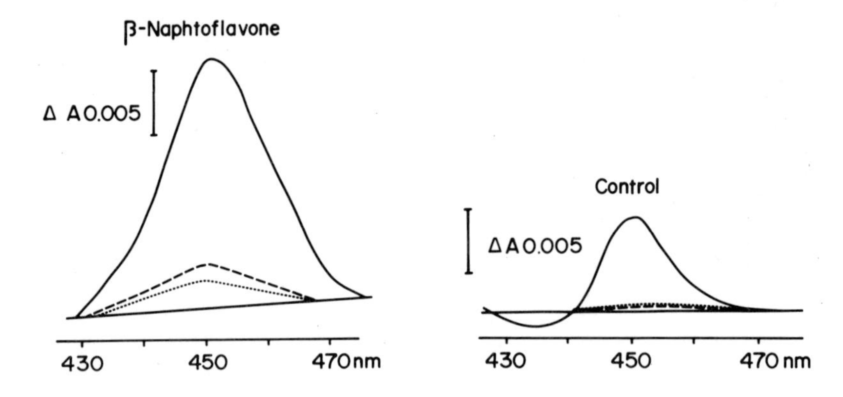

Fig. 1. Difference spectra of isolated liver cells from control and β-naphtoflavone treated trout. Each cuvette contained $4x10^6$ cells per ml. CO was bubbled for 30 s. into the sample cuvette and the difference spectrum recorded (······). Ethylmorphine was then added to the sample cuvette to a concentration of 1 mM (-‒-). Finally sodium dithionite was added to the same cuvette.

On the other hand, in hepatocytes from BNF treated trout the CO-bubbling and the subsequent addition of EM to the sample cuvette resulted in spectral changes with an absorption peak at about 450 nm (Fig. 1). These results show that a fraction of the cytochrome P-450 was endogenously reduced in the isolated liver cells from BNF treated fish. Finally the maximal spectral changes was obtained when the sample cuvette was reduced with sodium dithionite suggesting the major fraction of cytochrome P-450 in the isolated trout hepatocytes to be present in the oxidized non-substrate bound form.

When 7-ethoxycoumarin (EC) was added to the sample cuvette, containing microsomes from trout, a difference spectrum was recorded with a peak at 415 nm and through at 432 nm (Fig. 2). This spectral change was different from that obtained when EM was added (peak at about 390 and a through at about 420 nm), but almost in accordance with EC difference spectra recorded in rat microsomes (5). Type I spectral change has been suggested to be a manifestation of enzyme substrate complex due to similaritis between the apperent half maximal activities (K_m) and appearent half maximal spectral change (K_s) (4). In rat microsomes the appearent K_m and K_s values of EC was found identical (5). In the present study the calculated appearent K_s value for EC different spectra from rainbow trout microsomes was in the same order of magnitude as the the reported microsomal appearent K_m value for EC-O-deethylase of the same species (1). This indicates that the EC induced spectral change in trout may show the enzyme substrate

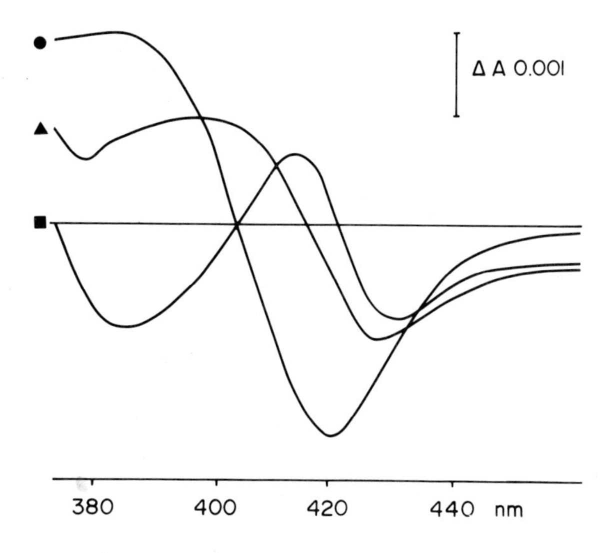

Fig. 2. Difference spectra upon addition of substrate to cytochrome P-450. 7-ethoxycoumarin (EC) caused spectral shift in microsomes (■) and hepatocytes (▲) from β-naphtoflavone treated fish. EC induced spectral shift were compared to spectral change caused by ethylmorphine (EM) in microsomes from control trout (●). EC and EM additions to the sample cuvette were 0.1 and 2 mM respectively. 2 mg microsomal protein and $1x10^7$ cells per ml were used.

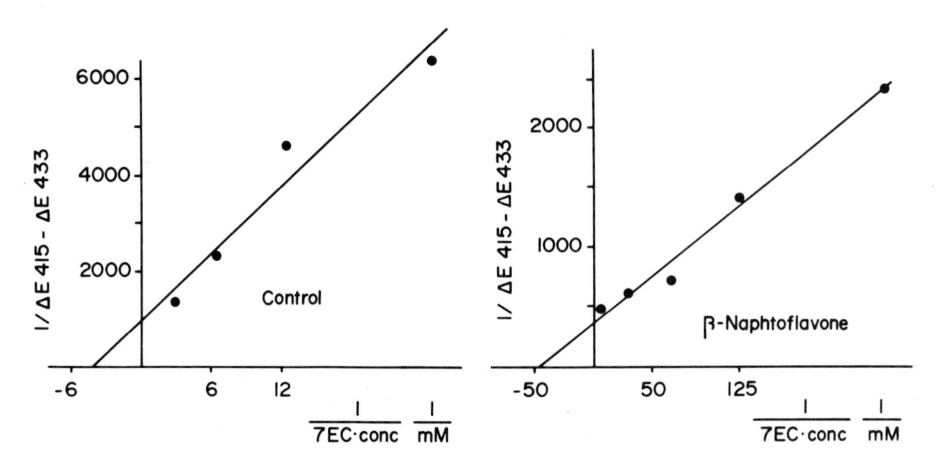

Fig. 3. Spectral dissociation constant (K_s) for spectral interactions of 7-ethoxycoumarin with microsomes from control and β-naphtoflavone (BNF) treated trout. The K_s obtained were 0.23 and 0.022 mM in microsomes from control and BNF treated trout respectively.

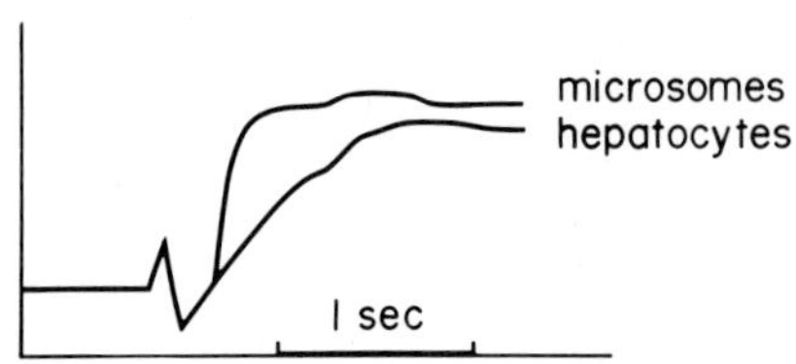

Fig. 4. Rate of formation of substrate cytochrome P-450 complex in microsomes and hepatocytes. The reaction was started by plunging 2 μl 1 mM 7-ethoxycoumarin into the cuvette. 2 mg microsomal protein and 1×10^7 cells per ml were used.

complex and thus to be a type I spectral change. The appearent k_s value was markedly lower in microsomes from BNF treated trout than in control (Fig. 3). This may indicate that the ratio between different cytochrome P-450 species has changed and/or that a novel type of cytochrome P-450, with a high binding specificity for EC, has been induced in BNF treated fish.

In isolated trout hepatocytes the EC difference spectra had a peak at about 408 nm and a through at about 428 nm (Fig. 2). The appearence of this spectral change was compared with that obtained with microsomes and used to study the time course for the uptake of EC into the isolated liver cells (Fig. 4). Although the formation of spectral change in hepatocytes was fast, and took place within seconds, it was slower than spectral change observed in microsomes.

ACKNOWLEDGEMENTS

This study was supported by grants from Magn. Bergvalls Foundation and the National Swedish Environment Protection Board.

REFERENCES

1. Elcombe, C.R. and Lech, J.J. (1979) Toxicol. Appl. Pharmacol. 49, 437.
2. Lockwood, A.P.M. (1961) Comp. Biochem. Physiol. 2, 241.
3. Moldeus, P., Grundin, R., vonBahr, C. and Orrenius, S. (1973) Biochem. Biophys. Res. Commun. 55, 937.
4. Schenkman, J.B., Sligar, S.G. and Cinti, D.L. (1981) Pharm. Ther. 49, 437.
5. Walton, M.S. and Cowey, S.B. (1979) Comp. Biochem. Physiol. 62B, 75.

© 1982 Elsevier Biomedical Press B.V.
Cytochrome P-450, Biochemistry, Biophysics
and Environmental Implications, E. Hietanen,
M. Laitinen and O. Hänninen editors

USE OF A SITE-SPECIFIC NEPHROTOXIN FOR THE INVESTIGATION OF CYTOCHROME P-450 MULTIPLICITY AND LOCALISATION IN RAT AND MOUSE KIDNEY

C.R. WOLF[1], J.B. HOOK[2] AND E.A. LOCK[3]
[1]Pharmacological Institute, University of Mainz, D-6500 Mainz, BRD. [2]Center of Environmental Toxicology, East Lansing, USA. [3]Imperial Chemical Industies, Central Toxicology Lab., Alderly Park, Cheshire, UK.

INTRODUCTION

The multiplicity of cytochrome P-450 (P-450) has been extensively investigated in the liver (1). It is now also clear that multiple forms of P-450 are present in other organs. Pulmonary and adrenal monooxygenase systems have been relatively well charactrised (2,3), however, there is still little known about cytochrome P-450 multiplicity in the kidney.

Hexachloro-1,3-butadiene (HCBD) is a potent nephrotoxin and causes a specific lesion in the S_3 and S_2 and S_3 segments of the proximal tubules in rats and mice respectively (4 and unpublished observations). The specificity of this effect has been used here to investigate P-450 multiplicity and distribution in the kidneys of these two species.

MATERIALS AND METHODS

Male Wistar-derived rats (200 g) and Swiss-derived mice (25 g) were used. HCBD was dosed i.p. in corn oil (200 mg/kg for rats and 50 mg/kg for mice) 24 h before the preparation of kidney microsomal preparations. Control animals received corn oil only. In some experiments animals were pretreated with β-naphthoflavone (βNF), 100 mg/kg/day, i.p., for three days before HCBD administration. In this case control animals received βNF only.

Microsomal metabolism was determined using 0.2 → 2.0 mg of kidney microsomal protein suspended in 1.0 ml, 0.066 M, Tris-HCl buffer, pH 7.4, containing 0.33 mM EDTA and NADPH (1 mM). The P-450 substrates used were aldrin, 0.2 mM, 7-ethoxycoumarin (7-EC), 1 mM, 7-ethoxyresorufin (7-ERF), 1 μM, and lauric acid, 150 μM, samples were incubated at 37 °C for periods up to 30 min. Metabolite identification for aldrin was according to Wolf et al. (5). The methods used for the other substrates are described in ref. 6. Components of the P-450-dependent monooxygenase system were quantitated by procedures described in the literature (6).

TABLE 1. MONOOXYGENASE LEVELS AND ACTIVITIES IN RAT AND MOUSE KIDNEY MICROSOMES FROM UNTREATED AND β-NAPHTHOFLAVONE-TREATED ANIMALS

	P-450 (nmol/mg)	P-450 reductase (units)	Cyt b_5 (nmol/mg)	Substrate Metabolism (nmol/min/mg protein: Aldrin epoxidation	Ethoxy-coumarin deethylation	Ethoxy-resorufin deethylation	Lauric acid hydroxylation: 11-OH	12-OH	Total
RAT									
Control	0.13±.03	50.1±19.4	0.18±.04	0.05 ±.02	0.028± .005	ND	1.59±.45	2.10±.94	3.69±1.64
β-NF	0.34±.02*	46.2± 3.5	0.34±.04*	0.08 ±.04	2.3 ± .2*	3.2 ±.9*	0.92±.35	2.34±.19	3.26± .48
MOUSE									
Control	0.18±.04	110 ±13.1	0.38±.01	0.024±.015	0.032± .004	ND	0.50±0.1	3.46±.35	3.90± .43
β-NF	0.13±.02	167 ± 9.9*	0.24±.03*	0.007±.0	0.018± .001*	0.036±.01*	0.2 ±0.1	0.90±.21	1.1 ± .20*

Control = untreated animals, βNF = animals treated with βNF, 100 mg/kg for 3 days prior to use.
* significantly different from control, $p < 0.01$.

TABLE 2. EFFECT OF HCBD-TREATMENT *IN VIVO* ON MONOOXYGENASE COMPONENTS AND SUBSTRATE METABOLISM IN RAT AND MOUSE KIDNEY MICROSOMAL PREPARATIONS

	% Control: P-450	P-450 reductase	Cyt b_5	Aldrin epoxidation	Ethoxy-coumarin deethylation	Ethoxy-resorufin deethylation	Lauric acid hydroxylation: 11-OH	12-OH	Total
RAT									
HCBD[1]	46±10*	76±30	104±24	33±12*	68±12*	ND	83±23	91±41	88±39
βNF+HCBD[2]	66± 8*	113±23	94± 3	38±13*	147±13*	137±19	105±27	81±15	87± 4
MOUSE									
HCBD	15±12*	65± 6*	70± 1*	0	28± 9*	ND	24±18*	23± 6*	23± 8
βNF+HCBD	17± 5*	83± 4*	83± 9	23± 7*	67±16*	116±22	40±20*	33± 7*	33± 6

ND = not detectable, 1 = HCBD animals treated with HCBD 200 mg/kg, i.p., 24 h before use. 2 = Animals treated with β-naphthoflavone (100 mg/kg, i.p. 3 days) prior to administration of HCBD (200 mg/kg i.p.)
* Significantly different from control $p < 0.01$. Control values are shown in Table 1.

RESULTS AND DISCUSSION

A time dependent reduction in renal microsomal cytochrome P-450 concentration was observed following the administration of HCBD to rats (Fig. 1). Maximum destruction (approx 70 %) was observed after 12 h. The cytochrome loss appeared

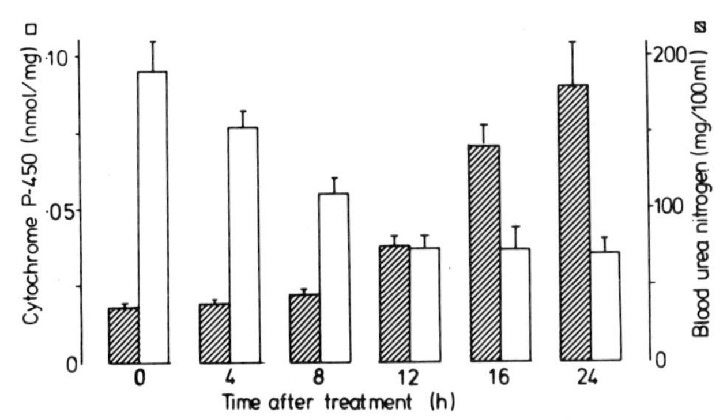

Fig. 1. Time course for HCBD-mediated destruction of rat renal cytochrome P-450.

to be an early event in the toxicity as it occurred prior to the increase in blood urea concentration (taken as a measure of overt kidney dammage). The restriction of the HCBD-induced lesion to the S_3 segment of the proximal tubule suggests that the majority of renal P-450 in rat is localised within this area.

These findings are in agreement with those of Dees et al. (7). In mice treated with HCBD nearly all the cytochrome was destroyed (Table 2) suggesting that most of the P-450 is in the cells of the S_2 and/or S_3 segments.

In the rat the P-450 concentration was increased significantly following ßNF-treatment and this effect was accompanied by a very large increase in 7-EC and 7-ERF metabolism (see also 8) but aldrin and lauric acid metabolism remained essentially unchanged. These results are indicative of multiple forms of P-450 in induced rat kidney. In the mouse ßNF treatment resulted in a decrease in nearly all activities measured. 7-ERF metabolism was the only enzyme activity which increased significantly. It is of interest that the ratio of 11-hydroxy to 12-hydroxy lauric acid metabolites in the mouse kidney was 1:9 compared to 1:1.3 in the rat.

The HCBD-mediated reduction in rat kidney P-450 concentration was paralelled by a 67 % reduction in the metabolism of aldrin. 7-EC metabolism was reduced by 32 % and lauric acid metabolism remained essentially unchanged (Table I). These differences suggest the presence of at least three forms of P-450 in control kidney. The data obtained with ßNF-treated rats (Table I) also suggests the presence of a minimum 3 forms; that metabolise aldin, 7-ERF and lauric acid. The difference between the enzymes active in aldrin and lauric acid metabolism was further exemplified by differences in their suceptability to inhibition by SKF 525A (Fig. 2). Neither enzyme was inhibited by 1 mM metyrapone.

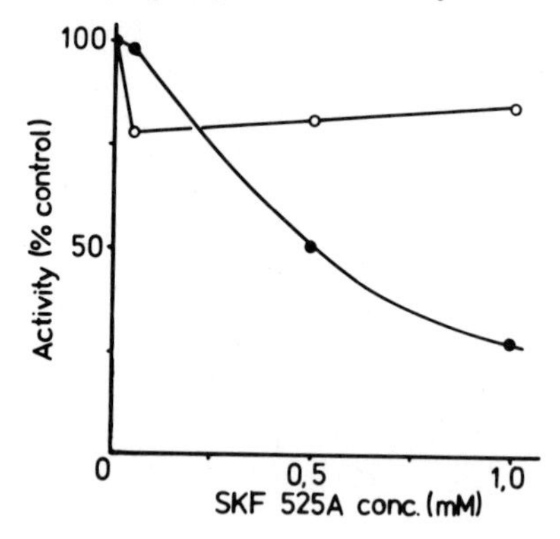

Fig. 2. Effect of SKF 525A on renal microsomal metabolism of aldrin and lauric acid. ● = Aldrin metabolism o = lauric acid metabolism (total metabolites).

In the mouse the large reduction in P-450 following HCBD administration was paralelled by a reduction in all activities measured. In mice treated with ßNF prior to HCBD, as in the case of the rat, the 7-ERF activity was not reduced, inspite of the large cytochrome loss. It would appear that the enzymes induced by ßNF are either more resistent to destruction and/or are localised in different areas of the kidney.

REFERENCES

1. Johnson, E.F. (1979) in: Hodgson, E., Bend, J.R. and Philpot, R.M. (Eds.) Reviews in Biochemical Toxicology Vol. 1, Elsevier/North-Holland Biomedical Press, Amsterdam, pp. 1-26.
2. Philpot, R.M. and Wolf, C.R. (1981) in: Hodgson, E., Bend, J.R. and Philpot, R.M. (Eds.) in Reviews in Biochemical Toxicology Vol. 3, Elsevier/North-Holland Biomedical Press, Amstrdam, pp. 51-76.
3. Suhara, K., Gomi, T., Sato, H., Hagahi, E., Tahemori, S. and Katagiri, M. (1978) Arch. Biochem. Biophys., 225, 2606.
4. Lock, E.A. and Ishmael, J. (1982) Adv. Pharmacol. Ther. 5, 87.
5. Wolf, T., Dene, F., Wander, H. (1979) Drug Metab. Dispos., 7, 301.
6. In Methods in Enzymology Vol. LII (1978) Academic Press/New York.
7. Dees, J.H., Masters, B.S.S., Muller-Eberhard, U. and Johnson, E.F. (1982) Cancer Res., 42, 1423.
8. Hook, J.B.,Elcombe, C.R., Rose, M. and Lock, E.A. (1982) submitted.

Cytochrome P-450, Biochemistry, Biophysics
and Environmental Implications, E. Hietanen,
M. Laitinen and O. Hänninen editors

STUDIES ON THE METABOLISM AND TOXICITY OF 1-NAPHTHOL IN ISOLATED HEPATOCYTES

Martyn T. Smith, Mary D'Arcy Doherty, John A. Timbrell and Gerald M. Cohen

Toxicology Unit, The School of Pharmacy, Brunswick Sq., London WC1N 1AX (U.K.)

A variety of toxic chemicals, including naphthalene, benzene and benzo(a)-pyrene, are thought to form reactive intermediates of quinonoid structure. These reactive intermediates are probably formed from the secondary metabolism of phenols generated during primary cytochrome P-450-dependent metabolism. For example, naphthalene is converted by the hepatic cytochrome P-450 monooxygenase to a primary epoxide which rearranges non-enzymatically to 1-naphthol. The 1-naphthol may then be further metabolised to dihydroxynaphthalene, which may, in turn, be relatively easily oxidized to naphthoquinones (1). The naphthoquinone metabolites could mediate both naphthalene and 1-naphthol toxicity either by covalently binding to important cellular macromolecules (2) or by redox cycling to produce large amounts of $O_2^{\bar{\cdot}}$ and oxidative stress (3). In the present investigation we have used isolated hepatocytes to determine which of these two mechanisms is likely to be the most important in the toxicity of 1-naphthol, and hence naphthalene, to liver cells.

METHODS

Hepatocytes were isolated from the livers of male Sprague-Dawley rats using the collagenase perfusion method described in (4). Incubations were performed in modified Krebs-Henseleit buffer, pH 7.4 (4) at 37°C under 95% O_2/5%CO_2.

Cell viability was determined as the exclusion of Trypan Blue. GSH was determined as acid-soluble thiols using Ellman's reagent. The metabolism of 1-naphthol was studied by thin layer chromatographic analysis of the extracellular medium. The total amount of irreversible covalent binding was determined using |^{14}C|1-naphthol essentially as described in (2), but following lysis of the cells by repeated freezing and thawing.

RESULTS AND DISCUSSION

Naphthoquinone metabolites of 1-naphthol could be metabolized via a one-electron reduction pathway, leading to semiquinone and oxygen radical formation, or via a two-electron-reduction to the hydroquinone (dihydroxynaphtha-

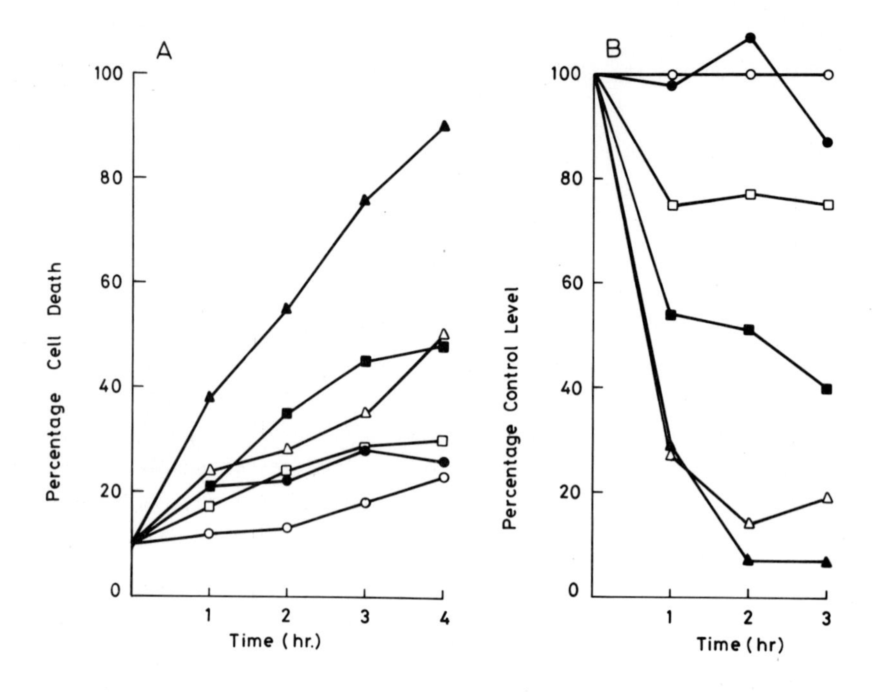

Fig. 1. Effect of dicoumarol on 1-naphthol induced cytotoxicity (A) and GSH loss (B). Control, o,●; + 0.1 μM 1-naphthol, □,■; + 0.25 μM 1-naphthol, Δ,▲; 30 μM dicoumarol also added, ●■▲. The means of 2-3 experiments are shown.

TABLE 1

EFFECT OF DICOUMAROL ON 1-NAPHTHOL CONJUGATE FORMATION IN ISOLATED HEPATOCYTES

Time (hr)	% 1-naphthol present as			
	glucuronide		sulphate	
	-dicoumarol	+dicoumarol	-dicoumarol	+dicoumarol
0	0	0	0	0
1	60.4	69.4	26.1	23.2
2	64.8	71.6	25.8	21.0
3	65.8	74.1	21.7	20.5

Initial 1-naphthol concentration was 100 μM. The results are corrected for blanks incubated at 37° but with no hepatocytes.

lene). The two-electron-reduction pathway is catalyzed by DT-diaphorase (NAD(P)H: quinone-acceptor oxidoreductase) and may protect against the potentially toxic one-electron-reduction pathway by competing with it for substrate. Support for this idea comes from the fact that dicoumarol inhibition of DT-diaphorase in isolated hepatocytes greatly potentiates the cytotoxicity of menadione (2-methyl 1,4-naphthoquinone) (5). Thus, if 1-naphthol toxicity is mediated by formation of 1,2- or 1,4-naphthoquinone one might also expect dicoumarol to potentiate the cytotoxicity of I-naphthol, since both 1,2- and 1,4-naphthoquinone have the same affinity for DT-diaphorase as menadione (6). Fig. 1A shows that dicoumarol does indeed potentiate the cytotoxicity of 1-naphthol in isolated hepatocytes. However, this does not necessarily show that I-naphthol is toxic to hepatocytes via the formation of quinones. For example, dicoumarol could interfere with the conjugation of 1-naphthol, thus increasing the free I-naphthol concentration. The results given in Table 1 show that dicoumarol does not significantly alter the amount of I-naphthol glucuronide or sulphate conjugates formed during the metabolism of I-naphthol. Dicoumarol could be acting in a variety of other ways, for example as a mild uncoupling agent. However, Fig. 1B shows that dicoumarol also increased 1-naphthol dependent GSH loss in isolated hepatocytes. Thus it seems very likely that a greater amount of reactive intermediates are formed when dicoumarol is present and it is difficult to see how dicoumarol could be acting if quinone metabolites are not involved. Interestingly, however, the presence of dicoumarol did not increase the

TABLE 2
EFFECT OF DICOUMAROL ON THE AMOUNT OF $|^{14}C|$1-NAPHTHOL IRREVERSIBLY BOUND TO ISOLATED HEPATOCYTES

Incubation conditions	Amount of $|^{14}C|$1-naphthol irreversibly bound after 2 hr (nmoles/mg protein)
37°, -dicoumarol	1.35±0.08
37°, +dicoumarol	1.22±0.09
4°, -dicoumarol	0.32±0.02

Results expressed as mean ± S.D. of 3 samples. Initial 1-naphthol concentration was 100 μM.

amount of reactive intermediates of 1-naphthol irreversibly bound to hepatocyte protein over a 2 h incubation period (Table 2). This result suggests that the total amount of covalent binding is not related to 1-naphthol toxicity in hepatocytes. It is therefore very likely that the mechanism of I-naphthol toxicity is related to the formation of active oxygen species and the creation of an oxidative stress. Dicoumarol could therefore be potantiating the cytotoxicity of 1-naphthol by inhibiting DT-diaphorase and making more naphthoquinone metabolite available for redox cycling at other flavoproteins. The importance of active oxygen species in 1-naphthol and naphthalene toxicity is presently under further investigation in our laboratory.

ACKNOWLEDGEMENTS

This work was supported by grants from the Cancer Research Campaign and the Nuffield Foundation.

REFERENCES

1. Daly, J., Inscoe, J. and Axelrod, J. (1965) J. Med. Chem. 8, 153
2. Hesse, S. and Mezger, M. (1979) Mol. Pharmacol. 16, 667
3. Powis, G., Svingen, B.A. and Appel, P. (1981) Mol. Pharmacol. 20, 387
4. Moldéus, P., Högberg, J. and Orrenius, S. (1978) Meths. Enzymol. 52, 60
5. Thor, H., Smith, M.T., Hartzell, P., Bellomo, G., Jewell, S. and Orrenius, S. (1982) J. Biol. Chem., submitted for publication
6. Ernster, L., Danielsson, L. and Ljunggren, M. (1962) Biochim. Biophys. Acta, 58, 171

Cytochrome P-450, Biochemistry, Biophysics and Environmental Implications, E. Hietanen, M. Laitinen and O. Hänninen editors

TOXIC AND NONTOXIC PATHWAYS DURING METABOLISM OF MENADIONE (2-METHYL-1,4-NAPHTHOQUINONE) IN ISOLATED HEPATOCYTES

Hjördis Thor, Martyn T. Smith, Pia Hartzell and Sten Orrenius
Department of Forensic Medicine, Karolinska Institutet, S-104 01 Stockholm (Sweden)

INTRODUCTION

The metabolism of quinones by flavoenzymes can occur by either one- or two-electron reduction routes, which differ greatly in cytotoxicity. NADPH-cytochrome P-450 reductase catalyzes the one-electron reduction to semiquinone radicals (1), which can rapidly reduce dioxygen, forming the superoxide anion radical, $O_2^{\bar{\cdot}}$, and regenerating the quinone (2). Dismutation of $O_2^{\bar{\cdot}}$ and production of other highly reactive oxygen species (e.g. $OH^{\cdot}$ and $^1\Delta gO_2$) quickly leads to conditions of oxidative stress and toxicity as redox cycling of the quinone continues (3). Quinones can also undergo two-electron reduction, forming hydroquinones without production of semiquinone intermediates (4). This reaction is catalyzed by DT-diaphorase (4), and may serve an important protective function for the cell by competing with the single-electron pathway (5).

In this study the potential protective role of DT-diaphorase was investigated using menadione and isolated hepatocytes.

METHODS

Cell and cell fraction preparation. Male Sprague-Dawley rats (200-220 g) were used, after pretreatment with either phenobarbital (PB) or 3-methylcholanthrene (3-MC). Hepatocytes were isolated by collagenase perfusion of the liver and incubated as previously described (5). Cell viability was determined by Trypan Blue exclusion (5). Microsomal and cytosolic fractions of rat liver were isolated as described by Ernster *et al* (6).

Assays. Oxygen consumption was measured at 37° with a Clark oxygen electrode and Antimycin A (25 μM) to inhibit mitochondrial O_2 uptake (5). Superoxide anion production was measured using acetylated cytochrome c (7). Hepatocyte glutathione (GSH) level was measured according to Saville (5).

RESULTS

Evidence for the two different pathways in cell fractions is shown in Fig. 1. In microsomal fraction, acetylated cytochrome c reduction was terminated by the addition of superoxide dismutase, but unaffected by dicoumarol, a specific DT-diaphorase inhibitor (4) (Fig. 1A). Dicoumarol did block acetylated cytochrome c reduction in the DT-diaphorase-containing cytosol (Fig. 1B). In this fraction superoxide dismutase was without effect, indicating that the DT-diaphorase-mediated menadione reduction proceeds without measurable $O_2^{\bar{}}$ production. In this case, acetylated cytochrome c is being reduced by the hydroquinone product, menadiol. Menadione metabolism by the one- and two-electron routes in intact hepatocytes is also measurable by this method, as shown in Fig. 2.

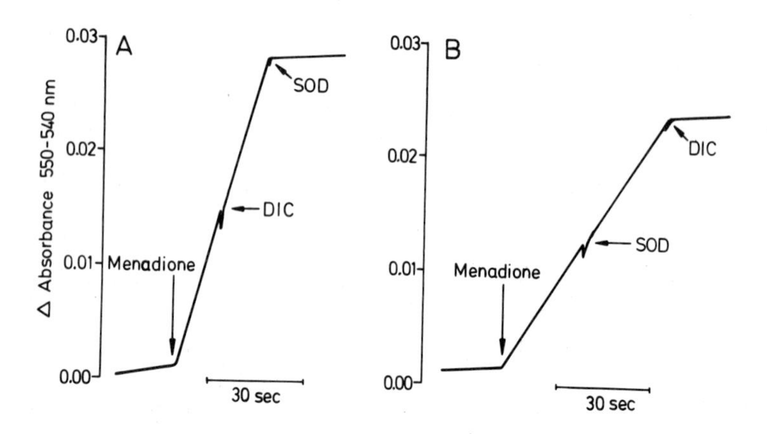

Fig. 1. Reduction of acetylated cytochrome c. A. Microsomes (1 mg protein/ml) in Tris-HCl buffer, pH 7.4, with 1 mM NADPH. B. Cytosol (1 mg protein/ml) in Tris-HCl buffer, pH 7.4, with 1 mM NADH. Menadione, 50 μM; Dicoumarol, 30 μM; Superoxide dismutase, 0.2 mg/ml.

The rate of O_2 uptake can be used as an indicator of menadione metabolism by the one-electron reduction pathway. Fig. 3 shows that hepatocytes from PB-treated rats consume significantly more O_2 compared with hepatocytes with induced DT-diaphorase (3-MC treatment). Addition of dicoumarol to PB-induced cells dramatically enhances menadione dependent O_2 consumption, indicating that even in hepatocytes with induced cytochrome P-450 reductase a large proportion of menadione metabolism occurs via the two-electron pathway.

The potential toxicity of the one-electron pathway is clearly shown in Fig. 4. In the PB-treated cells, the production of reactive oxygen products causes a precipitous drop in cellular GSH levels, and a marked loss of viability after 1 hour. In the 3-MC-treated cells, where the two-electron reduction pathway predominates, these effects are far less severe.

Taken together these results suggest preferential metabolism of menadione by the two-electron pathway reduction route, indicating a significant protective

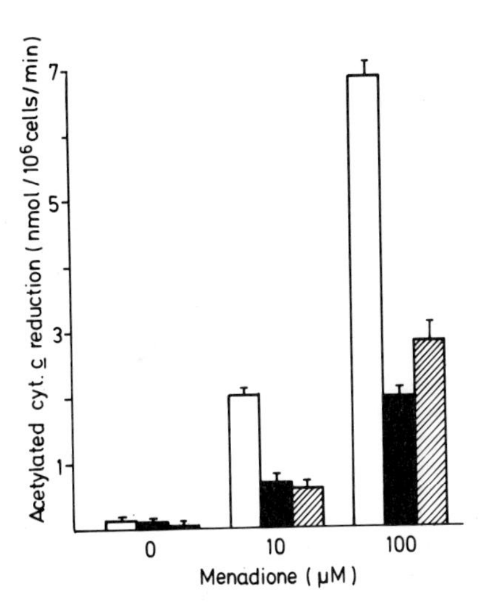

Fig. 2. Reduction of acetylated cytochrome c during metabolism of menadione in hepatocytes isolated from PB-induced rats. Dicoumarol (▮) 30 μM and superoxide dismutase (▨) 0.2 mg/ml.

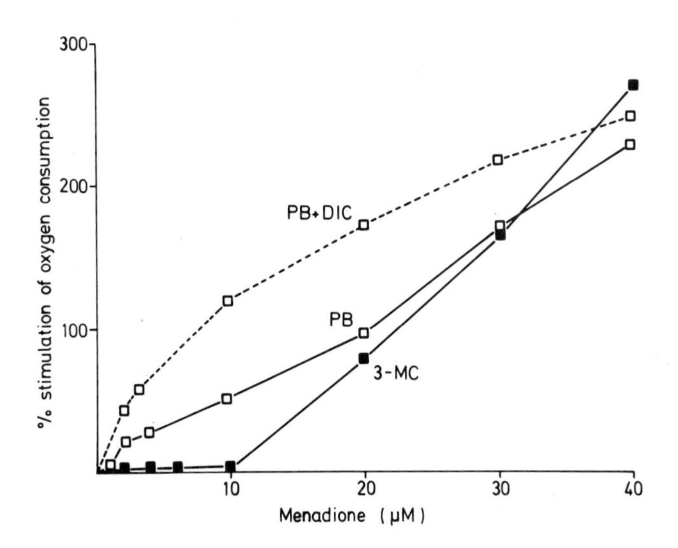

Fig. 3. O_2-consumption due to menadione metabolism in hepatocytes isolated from PB- or 3-MC treated rats. Dicoumarol, 30 μM.

role for DT-diaphorase. Quinone toxicity may therefore depend to a large extent on the relative affinity of the individual quinonoid structure for the enzymes catalyzing the two different reduction pathways.

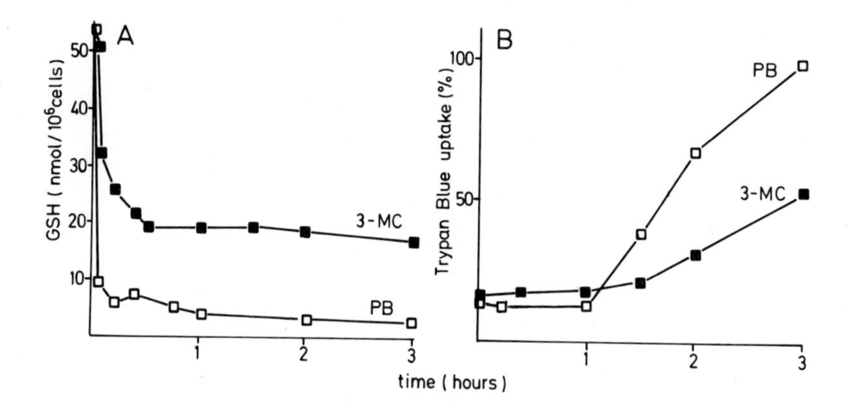

Fig. 4. A. Glutathione level. B. Trypan Blue uptake.
Relationship between GSH concentration (A) and Trypan Blue uptake (B) during metabolism of menadione in isolated hepatocytes from PB- and 3-MC-treated rats.

ACKNOWLEDGEMENTS

This work was supported by grants from the Swedish Medical Research Council (no. 03X-2471) and the Swedish Council for Planning and Coordination of Research.

REFERENCES

1. Iyanagi, T. and Yamazaki, I. (1970) Biochim. Biophys. Acta, 216, 282-294
2. Patel, K.B. and Willson, R.L. (1973) J. Chem. Soc. Faraday Trans. 69, 814-825
3. Beauchamp, C. and Fridovich, I. (1970) J. Biol. Chem. 245, 4641-4646
4. Ernster, L. (1967) Methods Enzymol. 10, 309-317
5. Thor, H., Smith, M.T., Hartzell, P., Bellomo, G., Jewell, S. and Orrenius, S. (1982) J. Biol. Chem., submitted
6. Ernster, L., Siekevitz, P. and Palade, G.E. (1962) J. Cell Biol. 15, 541-562
7. Azzi, A., Montecucco, C. and Richter, C. (1975) Biochem. Biophys. Res. Commun. 65, 597-607

Cytochrome P-450, Biochemistry, Biophysics
and Environmental Implications, E. Hietanen,
M. Laitinen and O. Hänninen editors

INDUCTION OF HEPATIC MONOOXYGENASES BY n-HEPTANE AND CYCLOHEXANE INHALATION

JORMA JÄRVISALO, HARRI VAINIO AND TUULA HEINONEN
Department of Industrial Hygiene and Toxicology, Institute of Occupational Health, Haartmaninkatu 1, SF-00290 Helsinki 29, Finland

INTRODUCTION

Numerous chemicals seem to be able to change the pattern of biotransformation enzymes of the liver (e.g. 1). In the case of benzo(a)pyrene monooxygenase it has been proposed that the induction process is controlled by a complex genetic mechanism (2). The induction of benzo(a)pyrene monooxygenase has also been studied in primary cultures of hepatocytes where UV-illumination in the presence of certain nutrients is also able to induce the enzyme activity (3). On the ground of such experiments, Paine (4) suggested that the induction of benzo(a)pyrene monooxygenase involves generation of oxygen or other biological radicals which for their part are direct or indirect effectors in the induction process.

To test the hypothesis of involvement of reduced oxygen species in the induction of drug biotransformation enzymes *in vivo* we exposed male rats to three concentrations of n-heptane and cyclohexane in ambient air. The microsomal metabolism of n-heptane has been shown to be uncoupled, leading to liberation of oxygen radicals from microsomes during n-heptane metabolism; cyclohexane on the other hand is metabolized in a coupled way, with no liberation of such radicals (5,6).

MATERIALS JA METHODS

Conditions of exposure. Adult male Wistar rats were exposed to 100, 500 and 1500 ppm of n-heptane and to 300, 1000 and 2000 ppm of cyclohexane in a 1 m^3 dynamic exposure chamber. The animals were otherwise kept on a commercial diet (Astra-Ewos, Södertälje, Sweden), they were deprived of food during the exposure hours. The exposure period was 6 h a day, 5 days a week, for one or two weeks.

TABLE 1

EFFECT OF EXPOSURE OF RATS TO n-HEPTANE (A) AND CYCLOHEXANE (B) ON LIVER DRUG BIOTRANSFORMATION ENZYMES AFTER THE FIRST WEEK OF THE EXPERIMENT

Exposure group	Cytochrome P-450 (nmol/mg)	Ethoxycoumarin deethylase (nmol/min/mg)	Ethoxyresorufin deethylase (pmol/min/mg)	UDP-glucuronosyl-transferase (nmol/min/mg)
A. n-HEPTANE				
Controls	0.74 ± 0.02	0.84 ± 0.13	70 ± 17	1.07 ± 0.22
100 ppm	0.81 ± 0.06*	0.97 ± 0.10	65 ± 12	1.02 ± 0.06
500 ppm	0.81 ± 0.03**	1.18 ± 0.12**	130 ± 20***	1.26 ± 0.19
1500 ppm	0.73 ± 0.03	1.32 ± 0.17***	110 ± 20**	1.24 ± 0.26
B. CYCLOHEXANE				
Controls	0.71 ± 0.07	0.74 ± 0.08	74 ± 11	1.03 ± 0.19
300 ppm	0.67 ± 0.09	0.82 ± 0.10	79 ± 12	1.37 ± 0.49
1000 ppm	0.69 ± 0.05	0.85 ± 0.06	107 ± 13**	1.47 ± 0.08***
2000 ppm	0.58 ± 0.07*	0.93 ± 0.14*	93 ± 40	2.24 ± 0.42***

The asterices indicate a statistically significant difference (Student's t-test): *$p < 0.05$; **$p < 0.01$; ***$p < 0.001$. The number of animals in each group was five.

For practical reasons, the groups exposed to either n-heptane or cyclohexane could not be exposed simultaneously; consequently, a comparison group was sham-exposed for each study group (Table 1). The animals were decapitated on the last day of the exposure, the livers were removed and stored at $-70^{o}C$ till analysis. After thawing, the livers were homogenized, and the microsomal fraction was obtained with the Ca^{2+}-aggregation method (7). The following determinations were performed from the microsomes: the activity of 7-ethoxycoumarin O-deethylase (8), 7-ethoxy resorufin O-deethylase (9) and paranitrophenol UDP-glucuronosyltransferase (10) (also measured after treatment of the microsomes with TRITON X-100 (11)), the content of cytochrome P-450 (12) and protein (13). The free sulfhydryl groups of the liver were determined with the method of Saville (14).

RESULTS AND DISCUSSION

The changes in the measured parameters of the liver biotransformation system after the first experimental week are given in Table 1. Both of the hydrocarbons affected the activity of the two hydroxylation enzymes only moderately. N-heptane caused only a small increase in the content of cytochrome P-450 in the microsomes and the highest air concentration of cyclohexane decreased the measurable P-450 content of the microsomes. The increase in the activity of the two ethylases caused by the two hydrocarbons seemed to be dose-dependent, no qualitative or quantitative differences between the inductive property of the two compounds were evident on the basis of the data. In contrast to n-heptane exposure, the exposure of the animals to cyclohexane caused a statistically significant increase in the activity of UDP-glucuronosyltransferase measured from intact microsomes. No difference , however, was evident between any of the exposed groups and the control group in the microsomes treated with TRITON X-100 (data not shown). This seems to indicate that a membranolytic effect of cyclohexane rather than induction was behind the apparent enhancement of the activity of UDP-glucuronosyltransferase by cyclohexane.

Neither of the two hydrocarbons at any exposure level had a significant effect on the liver content of free sulfhydryl groups (data not shown).

The changes in the liver drug biotransformation enzymes found after the second week of the experiment were both qualitatively and quantitatively very similar to those of the first week (data not shown).

Evidently, the difference between the inductive property of n-heptane and cyclohexane with respect to liver biotransformation enzymes is negligible, at least when measured with the present parameters. Ethoxyresorufin deethylase was chosen for the present study because it has been shown to have a close correlation with aryl hydrocarbon hydroxylase (9) whereas ethoxycoumarin deethylase is more typically induced by inducers of phenobarbital type (15). Provided that the metabolism of n-heptane is similarly uncoupled *in vivo* as found with microsomes *in vitro* (6), the present data seem to propose that oxygen radicals produced by microsomal drug metabolism are of minor importance in the regulation of the liver monooxygenase pattern. On the other hand, both n-heptane and cyclohexane are similar types of inducers as are many other simple hydrocarbons (for styrene and toluene, see ref. 15,16).

The environment with its multiple chemical exposures and the genetic mechanisms together constantly regulate the pattern of the drug biotransformation enzymes of the liver and also of other tissues. Such an environmental regulation is probably one mechanism behind the very great variation of human response to toxic, including carcinogenic, exposures.

REFERENCES

1. Estabrook, R.W. and Lindenlaub, E. (Eds.). Induction of Drug Metabolism, Schattaur Verlag, Stutgart, pp. 1-645.
2. Nebert, D.W. (1980) J. Natl. Cancer Inst., 64, 1279.
3. Paine, A.J. (1977) Chem.-Biol. Interact., 16, 309.
4. Paine, A.J. and Francis, J.E. (1980) Chem.-Biol. Interact., 30, 343.
5. Frommer, V., Ullrich, V., Staudinger, H. and Orrenius, S. (1972) Biochem. Biophys. Acta, 280, 487.
6. Staudt, H., Lichtenberger, F. and Ullrich, V. (1974) Eur. J. Biochem., 46, 99.
7. Aitio, A. and Vainio, H. (1976) Acta Pharmacol. Toxicol., 39, 555.

8. Aitio, A. (1978) Anal. Biochem., 85, 488.

9. Burke, M.D. and Mayer, R.T. (1974) Drug Metab. Disp., 6, 583.

10. Hänninen, O. (1968) Ann. Acad. Sci. Fenn. Ser. A II No 142, 1-92.

11. Vainio, H. and Aitio, A. (1974) Acta Pharmacol. Toxicol., 34, 130.

12. Omura, T. and Sato, R. (1964) J. Biol. Chem., 239, 2370.

13. Lowry, O.H., Rosebrough, N.J., Farr, A.L. and Randal, R.J. (1951) J. Biol. Chem., 193, 265.

14. Saville, B. (1958) Analyst, 83, 670.

15. Vainio, H., Järvisalo, J. and Taskinen, E. (1979) Toxicol. Appl. Pharmacol., 47, 7.

16. Elovaara, E., Savolainen, H., Pfäffli, P. and Vainio, H. (1979) Arch. Toxicol., suppl. 2, 345.

© 1982 Elsevier Biomedical Press B.V.
Cytochrome P-450, Biochemistry, Biophysics
and Environmental Implications, E. Hietanen,
M. Laitinen and O. Hänninen editors

THE RELATIVE INDUCTION OF THE HEPATIC DRUG-METABOLIZING ENZYME SYSTEM BY MEDROXYPROGESTERONE ACETATE IN RATS

HANNU SAARNI
Department of Pharmacology, University of Oulu, Oulu, SF-90220
Oulu 22 (Finland)

INTRODUCTION

Medroxyprogesterone acetate (MPA) is a synthetic progesterone which has also some glucocorticoid-like (1) and androgenic (2) properties. It induces hepatic drug metabolism both in rats (3) and in man (4). The induction produced by MPA resembles in many respects that of phenobarbital and pregnenolone-16α-carbonitrile. Female rats seem to be more sensitive to its effects than male rats (3).

The aim of this study was to investigate the inducing ability of MPA in female rats after chemically-produced liver injury, which is associated with impaired drug-metabolizing enzyme activities and diminished protein synthesis (5).

MATERIALS AND METHODS

Female Wistar rats (approx. 160 g) were divided into nine groups of six rats each. Three groups were pretreated with carbon tetrachloride (CCl_4)(1 ml/kg body weight s.c.) on the first and third day of each week for four weeks, and other three groups with dimethylnitrosamine (DMN)(10 µl/kg body weight i.p.) on the first three days of each week. Three groups received no treatment during that time. After four weeks the animals in the groups C (control), CCl_4 and DMN were killed by decapitation, the animals in the other groups received MPA (100 mg/kg body weight i.p.)(groups MPA, CCl_4-MPA and DMN-MPA) or MPA vehicle (groups V, CCl_4-V and DMN-V) for one week. The hepatic concentration of cytochrome P-450 (6) and the activities of NADPH-cytochrome c reductase (7) and epoxide hydrolase, by a modified method (8), were determined in the hepatic microsomal fraction. The activities of benzo(a)pyrene hydroxylase (9) and aminopyrine N-demethylase (10) were assayed in the 10,000 x g supernatant fraction of the liver, and glutathione S-transferase activity was measured in the liver homogenate (11).

RESULTS AND DISCUSSION

The relative changes in the hepatic parameters are shown in Fig. 1. The extent of liver injury was evaluated by comparing the CCl_4 and DMN-treated rats with the control animals, and the effect of MPA therapy was judged by comparing the MPA-treated rats with the corresponding vehicle animals, which showed spontaneous regeneration after chemical liver injury.

DMN proved more hepatotoxic than CCl_4 with the doses used. This was confirmed by histological findings and supported further by the elevated collagen content. A 5-fold increase in hepatic neutral lipid content was found in the CCl_4 rats, while the hepatic phospholipid and protein contents were decreased in the DMN rats.

The liver injury produced by DMN was accompanied by impaired activity of the enzymes involved in the hepatic microsomal monooxygenase system and a marked increase in epoxide hydrolase activity. Only glutathione S-transferase activity, measured in the whole-liver homogenate, remained practically unchanged in both groups.

MPA treatment caused a significant increase in all the parameters measured in the intact rats as compared with the vehicle-treated rats. The increase in the absolute values for the enzymes in the CCl_4 and DMN animals upon MPA treatment correlated with the extent of liver damage, but when the activities were compared with the corresponding vehicle rats, which showed spontaneous regeneration, the relative induction of each enzyme was essentially the same as in the intact rats. This means that MPA was able to enhance the production of new enzyme molecules even in the rats with chemical liver injury. Moreover, MPA stimulated this production to the same extent as in the intact rats in the situation where normal regeneration had already begun the process.

The relative cytochrome P-450 content of the damaged liver decreased to almost the same extent as the activities of the monooxygenase enzymes (benzo(a)pyrene hydroxylase and aminopyrine N-demethylase) and NADPH-cytochrome c reductase, possibly due to the destruction of cytochrome P-450 molecules and other components in the hepatic endoplasmic reticulum. In these animals MPA treatment increased the relative activity of the monooxygenases and also NADPH-cytochrome c reductase more than 2-fold, but only a slight increase in cytochrome P-450 content could be found. A correspond-

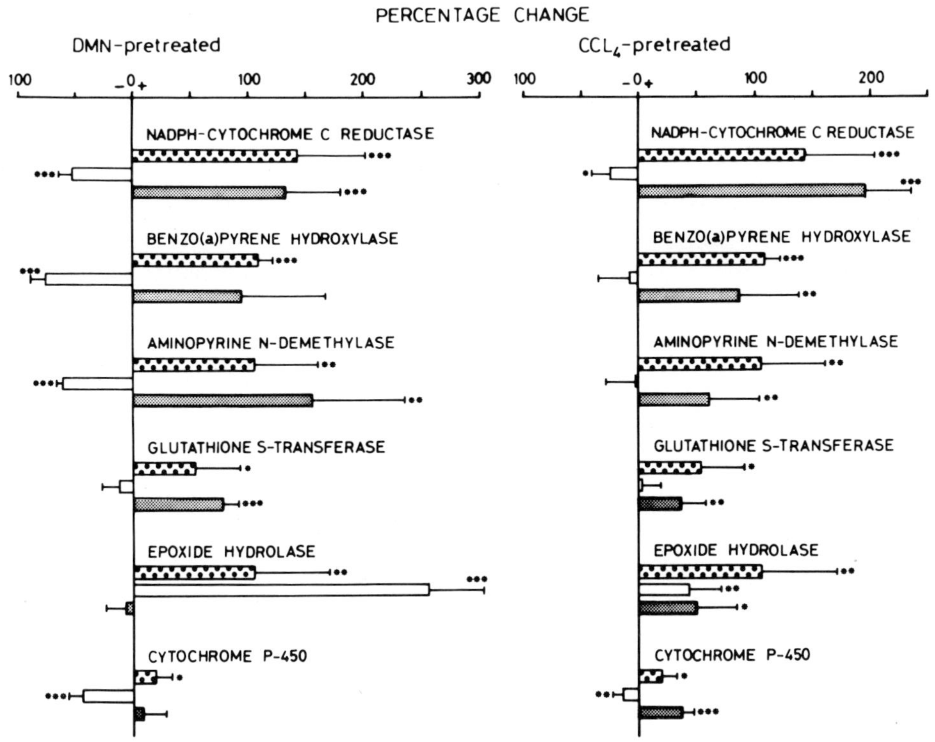

PERCENTAGE CHANGE

DMN-pretreated

CCL_4-pretreated

100 –0+ 100 200 300

100 –0+ 100 200

NADPH-CYTOCHROME C REDUCTASE

BENZO(a)PYRENE HYDROXYLASE

AMINOPYRINE N-DEMETHYLASE

GLUTATHIONE S-TRANSFERASE

EPOXIDE HYDROLASE

CYTOCHROME P-450

Fig. 1. Relative changes in hepatic parameters induced by MPA, CCl_4 and DMN in the intact rats, and by MPA in the CCl_4 or DMN-pretreated animals.

MPA-treated group vs. vehicle group

CCl_4 or DMN-treated group vs. control group

MPA-treated CCl_4 or DMN group vs. corresponding vehicle-treated group

The values are expressed per mg of protein. The differences between the treated and the control or vehicle groups are significant at the levels (+) $P < 0.05$, (++) $P < 0.01$ and (+++) $P < 0.001$.

ing change in the cytochrome P-450 content was found in the gel electrophoresis pattern of the microsomal proteins. If we assume that MPA increased the monooxygenase activities by enhancing the production of the whole cytochrome P-450 pool in the liver, then corresponding changes ought to be found in the relative cytochrome P-450 content and enzyme activities. The data support the assumption that MPA increased the quantity of cytochrome species associated with benzo(a)pyrene and aminopyrine metabolism more than it did the whole liver cytochrome P-450 pool. It is also possible that, in addition to its enhancement of protein synthesis, MPA also increased the availability of reducing equivalents in the monooxygenase system by increasing NADPH-cytochrome c reductase activity.

ACKNOWLEDGEMENTS

The excellent technical assistance of Ms. Ritva Saarikoski and Ms. Eija Saarikoski is greatly appreciated.

REFERENCES

1. Glenn, E.M., Richardson, L. and Bowman, B.J. (1959) Metabolism, 8, 265.
2. Bullock, L.P. and Bardin, C.W. (1977) Ann. N. Y. Acad Sci., 286, 321.
3. Saarni, H., Ahokas, J.T., Kärki, N.T., Pelkonen, O. and Sotaniemi, E.A. (1980) Biochem. Pharmacol., 29, 1155.
4. Ratio, A., Sotaniemi, E.A., Pelkonen, R.O. and Luoma, P.V. (1980) Clin. Pharmacol. Ther., 28, 629.
5. Sherlock, S. (1979) Gut, 20, 634.
6. Omura, T. and Sato, R. (1964) J. Biol. Chem., 239, 2370.
7. Masters, B.S.S., Williams, C.H. and Kamin, H. (1967) in: Estabrook, R.W. and Pulman, M.E. (Eds.), Methods in Enzymology, Academic Press, New York, p. 565.
8. Oesch, F., Jerina, D.M. and Daly, J. (1971) Biochim. Biophys. Acta, 227, 685.
9. Nebert, D.W. and Gelboin, H.V. (1968) J. Biol. Chem. 243, 6242.
10. Nash, T. (1953) Biochem. J., 55, 416.
11. Norling, A., Arranto, A.J. and Sotaniemi, E.A. (1982) Br. J. Clin. Pharmacol. (accepted).

Cytochrome P-450, Biochemistry, Biophysics
and Environmental Implications, E. Hietanen,
M. Laitinen and O. Hänninen editors

INTERACTION OF OLEYL ANILIDE, A COMPONENT IN DENATURATED SPANISH OIL, WITH HEPATIC AND PULMONARY CYTOCHROME P-450

IVAR ROOTS AND ALFRED G. HILDEBRANDT
Institute of Clinical Pharmacology, Klinikum Steglitz, Freie Universität Berlin, D-1000 Berlin 45 and Max von Pettenkofer-Institute, Federal Health Office, Berlin 33

INTRODUCTION

Oleyl anilide is among the chemicals being suspected of having caused the toxic-allergic syndrome after ingestion of aniline-denaturated rape-seed oil in Spain 1981 (1). To explain the pathological processes, a theory has been advanced that a peroxidation process is initiated, possibly with oleyl anilide behaving as catalyst or coenzyme, liberating free radicals that cause e.g. pulmonary damage or initiate immunological reactions.

Cytochrome P-450 could be involved in such toxification processes. It might react as an oxidase with oleyl anilide producing H_2O_2, similarly as known for its interaction with hexobarbital. Or, it might activate oleyl anilide to toxic metabolites. Therefore, experiments were designed to prove an interaction of oleyl anilide with microsomal cytochrome P-450 from lung and liver of various animal species. The significance of these animal studies was tested in human liver microsomes.

METHODS

Liver and lung microsomes were prepared according to standard procedures from rabbit (male and female, 3 - 4 kg body weight), male rat (150 - 250 g) and guinea pig (300 - 400 g), and from human surgical liver biopsy specimen. Partly, animals were pretreated with phenobarbital (PB) or methylcholanthrene (MC).

H_2O_2-assay (2): 50 mM Tris-HCl buffer, pH 7.5, 150 mM KCl, 10 mM $MgCl_2$, 8 mM D,L-isocitrate trisodium salt, 10 µl (20 mU) isocitrate dehydrogenase (IDH) per ml of the incubate, 3000 mU/ml catalase, 50 mM methanol containing various concentrations of oleyl anilide or hexobarbital. Microsomal protein was usually 1.5 mg/ml, with lung microsomes 2 mg/ml, with human liver microsomes 0.5 - 1 mg/ml. Reaction was started by addition of 400 µM NADPH; T = 37°C. Five samples were withdrawn within 9 min (animal liver), or two at 0.25 and 15 min (lung and human liver). Formaldehyde was measured according to Nash (3).

N-Demethylation of aminopyrine: As above, without catalase, but supplemented with 0.2 mM aminopyrine.

3-Hydroxylation of benzo(a)pyrene (4): 50 mM Tris-HCl buffer, pH 7.5, 150 mM

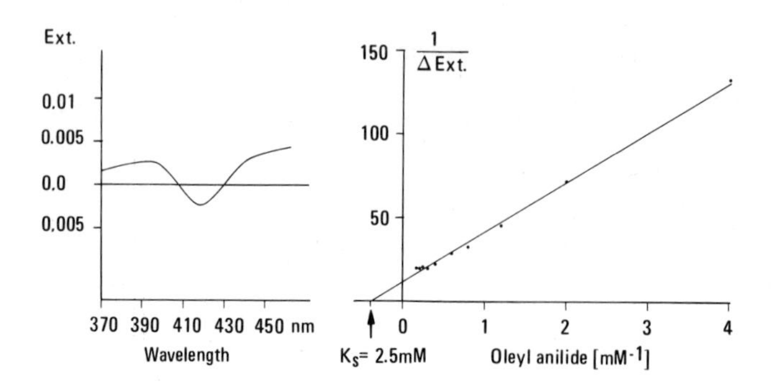

Fig. 1. Type-I binding spectrum with oleyl anilide (2.5 mM) in pulmonary microsomes (2 mg/ml) from male rabbit (control). Right: Binding of oleyl anilide to hepatic microsomes (1.5 mg/ml) from phenobarbital treated rat.

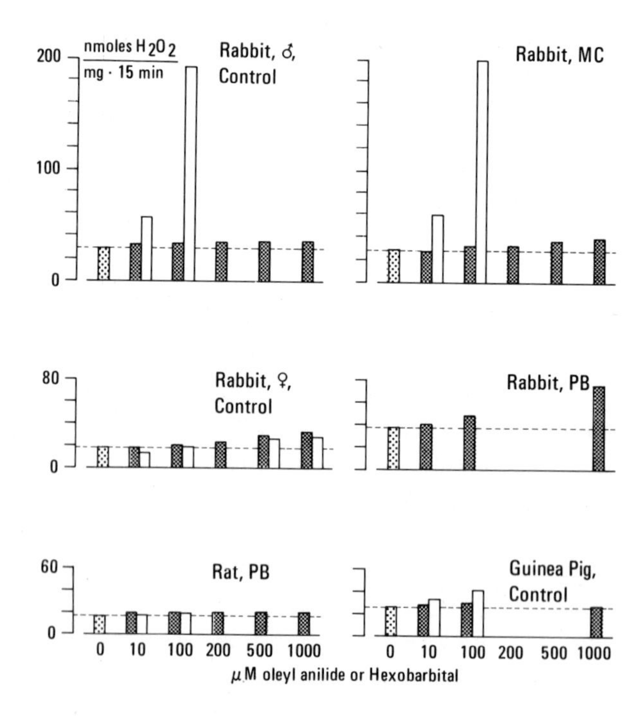

Fig. 2. Microsomal H_2O_2-production in lung microsomes as influenced by oleyl anilide (dark columns) or hexobarbital (light columns).

KCl, 10 mM $MgCl_2$, 8 mM D,L-isocitrate trisodium salt, 20 mU per ml IDH, 75 µg/ml of microsomal protein, 80 µM (finally) benzo(a)pyrene (added with 40 µl acetone per ml incubate), oleyl anilide (0 - 1000 µM finally, added with 2 µl of methanol per ml incubate). Samples were taken at 2.5, 5, 7.5, and 10 min (animal) or at 0.25 and 9 min (human liver).

Lipid peroxidation: ADP/Fe (1 mM/40 µM) stimulated lipid peroxidation was evaluated by measuring formation of malondialdehyde by the thiobarbituric acid method.

RESULTS

In contrast to aniline, oleyl anilide results a type-I spectrum in liver microsomes from rabbit, rat, or guinea pig and also in pulmonary microsomes from untreated rabbit (Fig. 1). The spectrum is reversible by metyrapone. K_s-values range from 0.5 to 7 mM, varying with species, pretreatment, organ, and factors yet to be investigated.

As shown in Figure 2, H_2O_2-formation in pulmonary microsomes is increased in presence of oleyl anilide up to 15 - 100 %. However, the surprisingly high capacity of oxidase function in some collectives could not be exhausted to the same degree as with hexobarbital.

As toxic effects in patients are not restricted to lung but strike many organs (1, 5), the examination was extended to liver microsomes. Figure 3 demonstrates an oleyl anilide interaction with cytochrome P-450 by increased H_2O_2-formation with relatively high concentrations of the toxin. Again, the effects are mostly smaller as reported for hexobarbital (6, 7). In contrast, a remarkable response

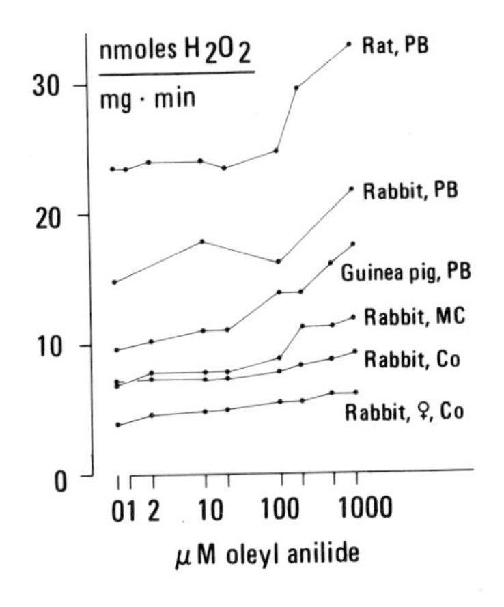

Fig. 3. Effect of oleyl anilide on H_2O_2 formation in liver microsomes of various animal species.

(10 - 117 %) is observed in human liver microsomes with low concentrations of oleyl anilide, significantly exceeding the effect of isomolar hexobarbital (Fig. 4). This might express an especially high affinity of oleyl anilide to human cytochrome P-450. With high concentrations, inhibition by oleyl anilide predominates.

Oleyl anilide exerts little influence on aminopyrine N-demethylation (Fig. 5), but clearly inhibits 3-hydroxylation of benzo(a)pyrene (Fig. 6) in most collectives. Interestingly, concentrations below 10 μM seem to activate the latter reaction in the rat, an effect already known for e. g. flavone (8).

The differential effects on both monoxygenase reactions are obtained in human liver microsomes as well. Considerable inhibition of 3-OH-benzo(a)pyrene formation with oleyl anilide concentration as low as 2 μM (Fig. 7) contrasts to an almost unaffected N-demethylation even with high concentrations (not shown).

Lipid peroxidation is inhibited by high concentrations of oleyl anilide - the extent varying with species and pretreatment (Fig. 8).

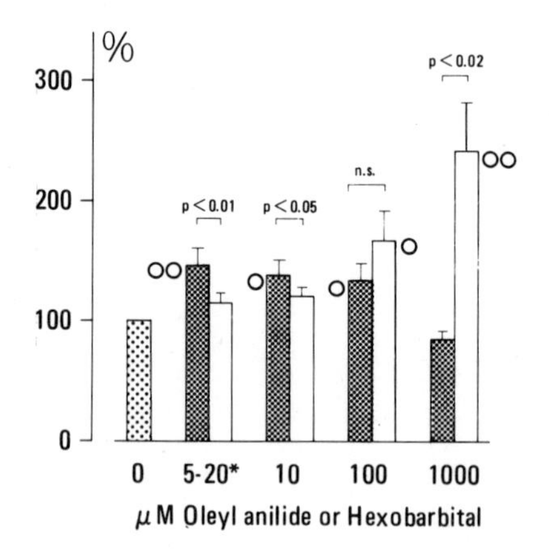

Fig. 4. Stimulatory and inhibitory effect of oleyl anilide (dark columns) on H_2O_2-formation in human liver microsomes, as compared to the hexobarbital effect (light columns). Control activity: 1.5 - 4.5 nmoles $H_2O_2 \cdot min^{-1} \cdot mg^{-1}$, n = 7 - 8.
*Measured at maximum effect of oleyl anilide; °P< 0.05, °°P< 0.02 (as compared to control activity).

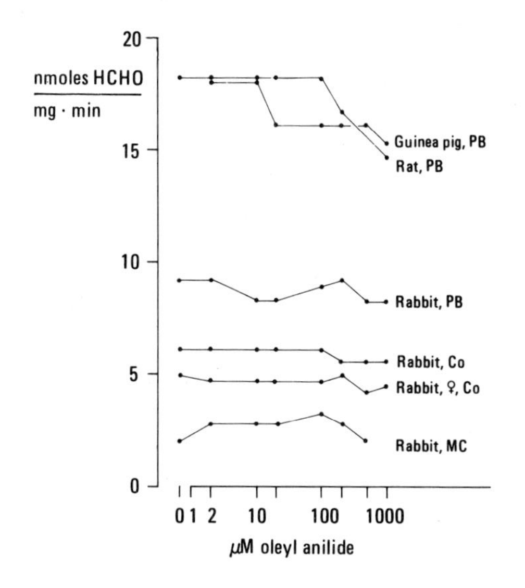

Fig. 5. Effect of oleyl anilide on aminopyrine N-demethylation in liver microsomes.

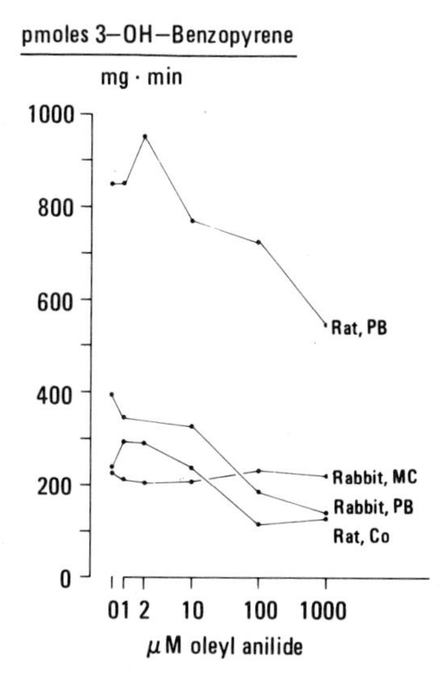

Fig. 6. 3-Hydroxylation of benzo(a)-pyrene in liver microsomes in the presence of oleyl anilide.

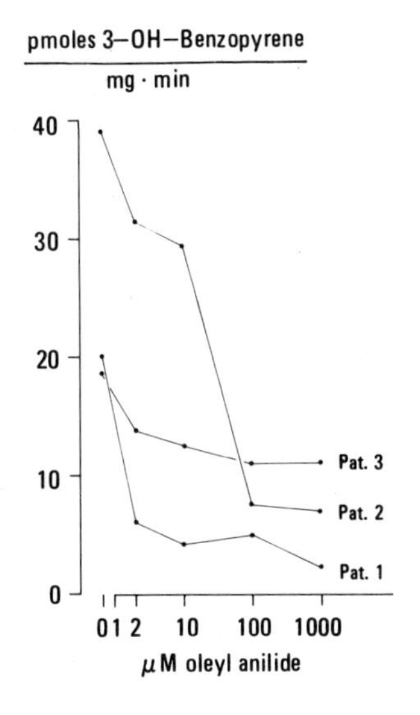

Fig. 7. Inhibition of 3-hydroxy-benzopyrene formation in human liver microsomes by oleyl anilide.

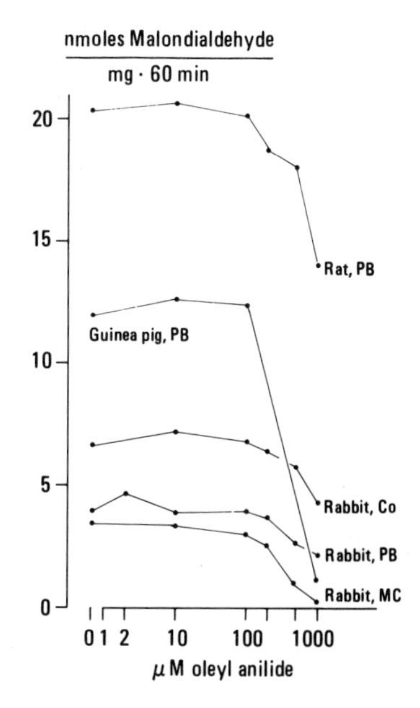

Fig. 8. Effect of oleyl anilide on lipid peroxidation in liver microsomes. Malondialdehyde production within 1 h is shown.

CONCLUSION

The data provide evidence for an interaction of oleyl anilide with cytochrome P-450 in lung and liver of several animal species and man. Not all forms of cytochrome P-450 seem to be affected equally as shown by the selective inhibition of benzo(a)pyrene 3-hydroxylation as compared to N-demethylation of aminopyrine and by the differential effects of enzyme inducers.

The relevance of these observations depends on the occurence of comparable concentrations in vivo. This should be achieved in man, as reactions in human liver microsomes exhibited a high response towards oleyl anilide in the 10^{-6} to 10^{-5} M range.

Metabolism of oleyl anilide itself via cytochrome P-450 seems likely, though not shown here directly. One might speculate that formation of reduced oxygen species or of reactive intermediates is important for the onset of the Spanish oil epidemic. As with other chemicals that are metabolically activated, the individual's balance between toxification and detoxication processes might become critical in different tissues.

ACKNOWLEDGEMENTS

We greatfully acknowledge the technical assistence of Mrs. E. Berg and Mrs. G. Minarek as well as the help of Mr. C. Bergs. Liver biopsy samples were kindly supplied by Prof. R. Häring and Dr. A. Hirner, Surgical Clinic.

REFERENCES

1. Tabuenca, J. M. (1981) Lancet II, 567
2. Hildebrandt, A. G. and Roots, I. (1975) Arch. Biochem. Biophys. 175, 385
3. Nash, T. (1953) Biochem. J. 55, 416
4. Nebert, D. W. and Gelboin, H. V. (1968) J. Biol. Chem. 243, 6242
5. Gilsanz, V. (1982) Lancet I, 335
6. Heinemeyer, G., Nigam, S., Hildebrandt, A. G. (1980) Arch. Pharmacol. 314, 201
7. Roots, I., Laschinsky G., Hildebrandt, A. G., Heinemeyer, G., Nigam, S. (1980) in: Coon, M. J., Conney, A. H., Estabrook, R. W., Gelboin, H. V., Gillette, J. R., O'Brien, P. J. (Eds.) Microsomes, Drug Oxidations, and Chemical Carcinogenesis, Vol. I, Academic Press, New York, pp. 375 - 378
8. Hildebrandt, A. G., Bergs, C., Heinemeyer, G., Schlede, E., Roots, I., Abbas-Ali, B., Schmoldt, A. (1982) in: Snyder, R. et al. (Eds.) Biological Reactive Intermediates-II, Part A, Plenum Publishing Corp., New York, pp. 179 - 198

Cytochrome P-450, Biochemistry, Biophysics
and Environmental Implications, E. Hietanen,
M. Laitinen and O. Hänninen editors

HYDRATION OF BENZO(a)PYRENE 4,5-OXIDE BY CADAVER TISSUES

M.G. PARKKI, M. AHOTUPA AND S. TOIKKANEN*
Department of Physiology and *Department of Pathological Anatomy,
University of Turku, Kiinamyllynkatu 10 SF-20520 Turku 52, Finland

INTRODUCTION

Epoxide hydrolase hydrates epoxides of very diverse structures. The function of epoxide hydrolase is necessary for the metabolic formation of the ultimate carcinogenic forms of polycyclic aromatic hydrocarbons which contain sc. bay-region (1). Recently epoxide hydrolase has been shown to be one of the preneoplastic antigens which originate during exposure of experimental animals to hepatocarcinogens (2).

In order to study the possible existence of epoxide hydrolase in human cancer tissues, methods for tissue preparation of cadaver tissues were searched. Also an attept to find epoxide hydrolase activity in formalin fixed tissue samples and paraffin bloks was committed.

METHODS

Biopsies from liver, kidney, lung, intestine and spleen were taken at autopsy 2 to 4 days after the exitus. The cadavers were stored at $4^{o}C$. The biopsies were stored at $-70^{o}C$ until tissue preparation and enzyme activity determination. The homogenization was performed in 0.25 mol/l sucrose solution 1:5 wet weight either with Ultra Turrax^{R}, Janke & Kunkel or by powdering tissues freezed in liquid nitrogen with Micro-Dismembrator^{R}, B. Braun. The homogenate obtained and 10000 g supernatant were compared. Deparaffination of paraffin blocs of hamster liver samples was performed by eluting thin sheets of the bloc for 3 days in xylene followed by ethanol washing with decreasing concentrations of alcohol-water solutions. Homogenation was performed with Ultra-Turrax. Homogenate was used as the enzyme preparation in these samples. The epoxide hydrolase activity was measured using benzo(a)pyrene 4,5-oxide (NCI Chemical Carcinogen Reference Standard Repository, NIH, Bethesda, Maryland) (3).

RESULTS AND DISCUSSION

The two homogenization methods gave essentially similar activities when measured in the homogenate. However, the activity in the 10000 g supernatant was at the level measured in the homogenate only in the tissue preparation obtained with Ultra-Turrax homogenation. Thus the 10000 g supernatant of Ultra-Turrax homogenated tissues was used routinely. When tissue samples stored 1 wk in 4 % formalin solution buffered at pH 7 by phosphate, were studied the activity ranged from 100 % to 30 % of the level of native preparations indicating, that no reproducible estimation of the tissue content of epoxide hydrolase in formalin-fixed samples can be made. Also the epoxide hydrolase activity in tissue preparations made from 3 years old paraffin blocs gave only marginal activity compared to the activity in fresh tissue preparation.

The activities of hydration of benzo(a)pyrene 4,5-oxide in cadaver tissues are given in the table:

Hydration of benzo(a)pyrene 4,5-oxide, nmol/min g liver wet weight

Liver	Kidney	Lung	Intestine	Spleen	
37.4 ± 4	7.7 ± 0.4	4.0 ± 0.4	1.0 ± 0.1	2.1 ± 0.2	(n=11)
100 %	20 %	10 %	3 %	6 %	

Cholangiocarcinoma	Surrounding liver tissue
22.2	39

The rather small interindividual differences in the enzyme activity suggest that no significant degradation due to autolysis has been occurred. The material included a 40 years old male who had died of cholangiocarcinoma. Cholangiocarcinoma originates from the epithelium of bile ducts. The epoxide hydrolase activity in this tissue is surprisingly high when taking into account, that the nonparenchymal cells of liver have rather low drug-metabolizing enzyme activities.

The present study suggest, that measurement of the catalytic activity of epoxide hydrolase can be used to quantitate epoxide hydrolase in cadaver tissues but other methods like antibody staining must be used when formalin fixed samples or paraffin blocs are studied.

ACKNOWLEDGEMENTS

Benzo(a)pyrene 4,5-oxide was received from the NCI Chemical Carcinogen Reference Standard Repository, NIH, Bethesda, Maryland, 20205.

Grants: NIH, U.S.A. ROI-01684.

REFERENCES

1. Jerina, D.M., Yagi, H., Lehr, R.E., Thakker, D.R., Schaefer-Ridder, M., Karle, J.M., Levin, W., Wood, A.W., Chang, R.L. and Conney, A.H. (1978) in: Ts'o, P.O.P. and Gelboin, H. (Eds.) Polycyclic hydrocarbons and cancer, Chemistry, molecular biology and environment. Academic Press New York, pp. 173-188.
2. Levin, W., Lu, A.Y.H., Thomas, P.E., Ryan, D., Kizer, D.E. and Griffin, M.J. (1978) Proc. Nat. Acad. Sci. U.S.A. 75, 3240-3243.
3. Schmassmann, H.U., Glatt, H.R. and Oesch, F. (1976) Anal. Biochem. 74,94-104.

Cytochrome P-450, Biochemistry, Biophysics
and Environmental Implications, E. Hietanen,
M. Laitinen and O. Hänninen editors

TRIACETYLOLEANDOMYCIN AS INDUCER OF P-450 LM_{3b} FROM RABBIT LIVER MICROSOMES

CLAUDE BONFILS, ISABELLE DALET-BELUCHE, CHRISTIAN DALET AND PATRICK MAUREL
INSERM U 128, B.P. 5051, 34033 MONTPELLIER Cedex (France).

INTRODUCTION

Recently, we showed that TAO (triacetyloleandomycin), a macrolide antibiotic, induced in rabbit liver microsomes a LM_3 form of P-450 which we denoted P-450 LM_3 (TAO), (1). This form was purified and according to electrophoretic, spectral and enzymic properties appeared to resemble P-450 LM_{3b}, recently isolated by Coon and associates from control rabbit liver microsomes, (2). Since, up to now, LM_3 forms were considered as non inducible, it was therefore of interest to determine whether P-450 LM_3 (TAO) was a form already present in control animals (for instance LM_{3b}) and stimulated by TAO or a new isozyme specifically induced by TAO. Answering such a question is fundamental since it is not clear whether there exists a limited number of forms of P-450 or as many forms as different inducers (3). A detailed biochemical study was undertaken to further characterize P-450 LM_3 (TAO). Some of our recent findings, reported in this communication, demonstrate, on the basis of peptide mapping and immunological reactivity, that P-450 LM_3 (TAO) and LM_{3b} are in fact the same protein.

MATERIALS AND METHODS

P-450 LM_3 (TAO) (specific content 16 mg/ml) was prepared from TAO treated rabbits as indicated in ref. (1). Forms LM_2, LM_4 and NADPH reductase were prepared according to published procedures (4) and (5). Peptide mapping and Ouchterlony double immunodiffusion were carried out as described (6). Antibodies against LM_2, LM_3 (TAO) and LM_4 were prepared from sera of immunized female sheeps. Immunizations were carried out by three series (at three weeks intervals) of multiple injections of 1 nmole of each LM in 1 ml mixed with an equal volume of Freund's complete adjuvant. Authentic standards P-450 LM_{3b} and LM_{3c} were kindly provided by Drs. Coon and Koop (University of Michigan).

RESULTS AND DISCUSSION

In Table I are presented data concerning several hydroxylating activity of forms LM_2, LM_3 (TAO) and LM_4 using the reconstituted system. LM_3 (TAO) appears to be the most active isozyme towards these substrates. In particular, both 6β and 16α hydroxylation of testosterone are indeed much higher with LM_3

than with both LM_2 and LM_4. Similar results were reported recently (2). On the other hand no detectable activity was observed with LM_3 in ethoxycoumarin deethylation.

TABLE I

HYDROXYLATION ACTIVITY OF RECONSTITUTED SYSTEM

Number given represents turnover in nmol of product formed per minute, per nmol P-450. Respective concentrations of P-450, reductase and dilauroylphosphatidylcholine were 0.5 μM, 0.5 μM and 45 μg/ml. Reactions were carried out in 0.1 M Tris HCl (7.4) at 37 °C. Chlorcyclizine (1 mM) hydroxylation was determined by the method of Nash, Testosterone (200 μM) by HPLC (7), estradiol (200 μM) according to (8).

Substrates	LM_2	LM_3 (TAO)	LM_4
Chlorcyclizine	1.0	3.3	1.4
Testosterone (6βOH)	0.16	1.65	0.42
Testosterone (16αOH)	0.23	0.86	0.12
Estradiol (2-OH)	0.13	0.30	0.13

In order to know whether P-450 LM_3 (TAO) was a new cytochrome specifically induced by TAO or a LM_3 form already present in the liver of control animal, we decided to compare it, by peptide mapping and immunological study, with LM_{3b} and LM_{3c} prepared from untreated rabbit liver and kindly provided by Drs. Coon and Koop.

The results are presented in Fig. 1 and 2. Whereas peptide maps of LM_2, LM_3 (TAO), LM_4 and LM_{3c} were quite different from one another (results not shown here) no difference could be detected when LM_3 (TAO) and LM_{3b} were similarly compared, Fig. 1. Moreover, it is clear from Fig. 2 that LM_3 (TAO) and LM_{3b} exhibit complete immunological identity. On the other hand no cross reactivity occurs between both LM_3 (TAO) and LM_{3b} and the other isozymes.

In addition to electrophoresic, enzymic and spectral evidences provided in a previous report, these results strongly suggest that LM_3 (TAO) and LM_{3b} are in fact the same protein. This is the first example of xenobiotic mediated induction of a P-450 LM_3 form in rabbit liver microsomes. Macrolide antibiotics such as TAO and erythromycin (which according to SDS-PAGE experiments also induces a LM_3 form with same apparent mobility as LM_{3b}) could thus be considered as new inducers. The finding that P-450 LM_{3b}, a form apparently involved in the

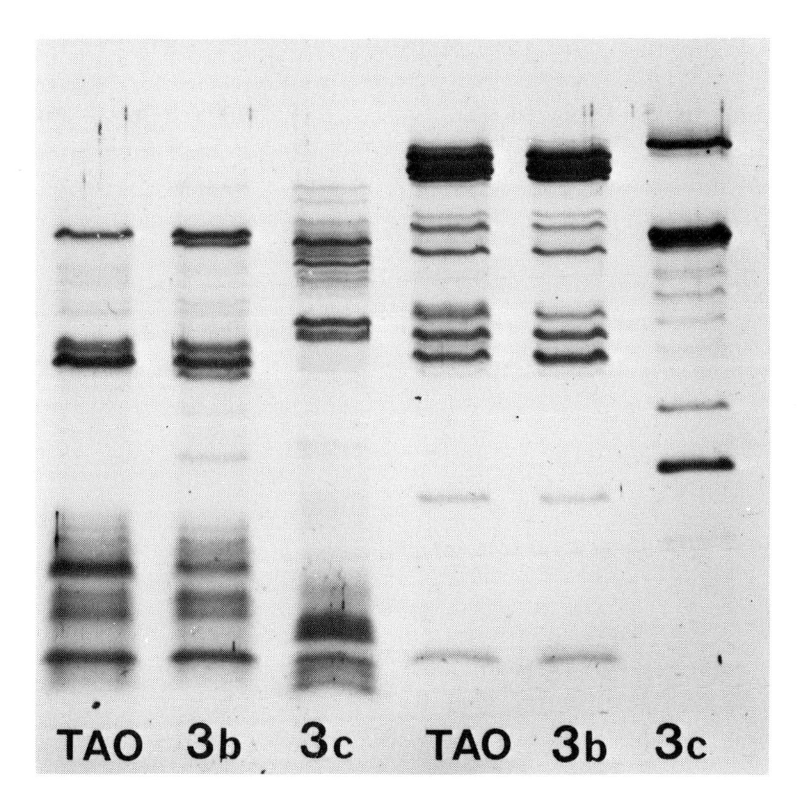

Fig. 1. Peptide mapping in SDS-PAGE of LM_3 (TAO), LM_{3b} and LM_{3c}. Protease V 8 (left) and α-chymotrypsine (right).

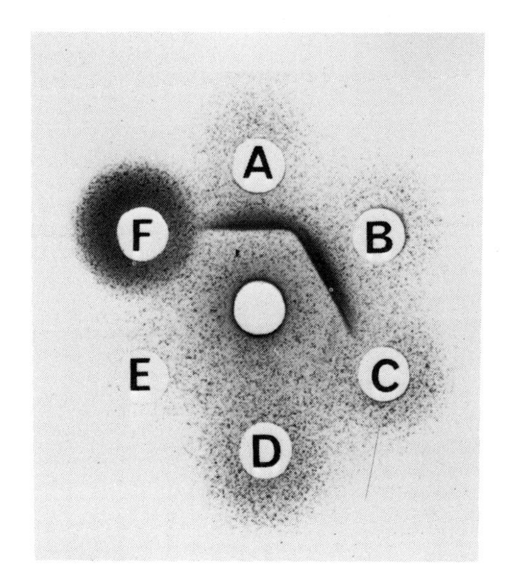

Fig. 2. Ouchterlony double diffusion experiments. Wells contain : central anti LM_3 (TAO), A LM_{3b} ; B LM_3 TAO ; C LM_4 ; D LM_{3c} ; E LM_2 ; F buffer.

hydroxylation of endogeneous substrates, can be induced by and is highly active towards xenobiotic (2), (9), suggests that a strict discrimination by any form of P-450 between metabolism of physiological substrates and disposition of xenobiotics is not likely to occur. Further studies are under current investigation in this laboratory to determine whether other forms of P-450 are actually induced by TAO, in addition to LM_{3b}.

ACKNOWLEDGEMENTS

We are grateful to Professor M.J. Coon and Dr. D.R. Koop (University of Michigan) for mailing us their preparations of LM_{3b} and LM_{3c} used as authentic standards in this work. Thanks also are due to Dr. Brugier, PFIZER-France for kindly providing TAO.

REFERENCES

1. Bonfils, C., Dalet-Beluche, I. and Maurel, P. (1982) Biochem. Biophys. Res. Commun. 104, 1011-1017.
2. Koop, D.R., Persson, A.V. and Coon, M.J. (1981) J. Biol. Chem. 256, 10704-10711.
3. Nebert, D.W. (1979) Mol. and Cell. Biochem. 27, 27-46.
4. Haugen, D.A. and Coon, M.J. (1976) J. Biol. Chem. 251, 7929-7939.
5. French, J.S. and Coon, M.J. (1979) Arch. Biochem. Biophys. 195, 565-577.
6. Cleveland, D.W., Fischer, S.G., Krischner, M.W. and Laemli, U.K. (1977) J. Biol. Chem. 252, 1102-1106.
7. Dalet, C. and Maurel, P. (manuscript in preparation).
8. Gelbke, H.P. and Knuppen, R. (1972) J. Chromatogr. 71, 465-471.
9. Johnson, E.F. (1980) J. Biol. Chem. 255, 304-309.

Cytochrome P-450, Biochemistry, Biophysics and Environmental Implications, E. Hietanen, M. Laitinen and O. Hänninen editors

INTERACTIONS OF POLYCYCLIC HYDROCARBONS WITH THE ESTRADIOL RECEPTOR IN RAT ADRENAL

LOUISE MANKOWITZ AND JAN RYDSTRÖM
Departement of Biochemistry, Arrhenius Laboratory, University of Stockholm, S-106 91 Stockholm, Sweden

INTRODUCTION

7,12-dimethylbenz(a)anthracene (DMBA) is a potent carcinogen and adrenocorticolytic polycyclic aromatic hydrocarbon (PAH), whereas other carcinogenic PAH, e.g., benz(a)pyrene (BP), is without effect on the adrenal. Acute adrenal necrosis in rat produced by DMBA is dependent on ACTH and is facilitated by estrogens. The mechanism of the generation of of the necrosis remains to be elucidated but is believed to involve the generation of reactive intermediates in the adrenal (1). However, rat adrenal is known to contain an estrogen receptor (2) and it is possible that reactive DMBA metabolites act by combining with this receptor as previously shown for the uterine estradiol receptor (3).

MATERIALS AND METHODS

Adrenals were removed from Sprague-Dawley rats after decapitation and the microsomal fraction was prepared essentially as described previously (4) and stored below -20°C. Metabolism of DMBA and BP and separation of metabolies by HPLC were carried out as previously described (1).

Adrenal cytosol were obtained from adrenals from female rats after homogenization in 0.01M Tris-HCl (pH 7.3), 1.5 mM EDTA, 1 mM dithiothreitol (TED) and centrifugation for 1 hour at 100.000 x g. Incubations for competition experiments were made by mixing the supernatant (200 µl) with 5 µl of (^{3}H)-estradiol in TED (final concentration 0.150 nM) and 5 µl of ethanol solutions of DMBA, BP or metabolites. After 2 hours of incubation on ice 200 µl of a dextran-coated charcoal mixture (3% charcoal, 0.3% dextran w/v) (DCC) was added. After 10 min of incubation on ice and 10 min centrifugation at 1000 x g, an aliquot of 300 µl was removed from each tube and placed into 10 ml of scintillation fluid. Unspecific binding was measured in incubations with a 100-fold excess of unlabelled estradiol.

(^{14}C)-DMBA (98.7 mCi/mmol), (^{14}C)-BP (29.7 mCi/mmol) and (^{3}H)-estradiol(137.1 Ci/mmol) were obtained from NEN chemicals (D-6072 Dreireich, West Germany). Dextran T-70 was obtained from Pharmacia Fine Chemicals (Uppsala, Sweden). Other biochemicals were of analytical grade.

RESULTS

In agreement with other investigators (2) the presence of an estradiol receptor was demonstrated in rat adrenal cytosol as indicated by high affinity binding of tritiated estradiol. The binding was saturable with a maximal binding of approximately 3 fmol/mg protein and a K_d value of approximately 0.05 nM (fig. 1). Heat treatment abolished binding completely (not shown).

As shown in fig. 2 up to 15 μM of HPLC-purified DMBA did not displace tritiated estradiol bound to the receptor significantly. However, a mixture of DMBA metabolites, generated by adrenal microsomal cyt. P-450,was capable of displacing estradiol essentially to 100%. A 50% inhibition was observed with a 1000-fold excess of pooled metabolites. Fig. 3. shows that BP, like DMBA did not displace estradiol from binding to the receptor whereas BP metabolites, generated in the same way as the DMBA metabolites, were more potent in displacing estradiol from the receptor. In this case a 10.000- fold excess was necessary for 50% inhibition. As compared to the DMBA metabolites those of BP were thus about 10 times less efficient. In an attempt to investigate the efficiency of various DMBA metabolites, with respect to binding to the estradiol receptor, DMBA metabolites were separated on HPLC and incubated at a concentration of 0.1 μM, i.e., a 700-fold excess, with adrenal cytosol as described above. As shown in Table 1 there was a difference in binding capacity between various DMBA metabolites. Metabolites 1 and 2, tentatively identified as dihydrodiols, showed 64% and 81% inhibition respectively. The more hydrophobic metabolites 3 and 4, tentatively identified as hydroxymethyl derivatives and/or phenols, showed a considerably lower inhibition, between 11-16%. These results indicate that the preferential binding of DMBA metabolites to the estradiol receptor may be involved in the mechanism of the toxic action of DMBA in rat adrenal and possibly other organs.

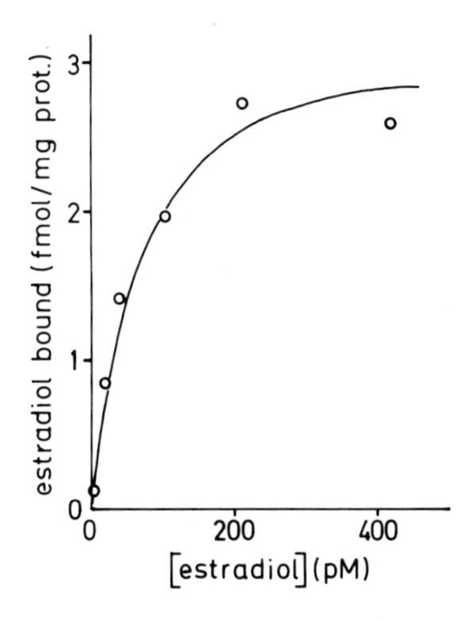

Fig. 1. Binding of estradiol to the estradiol receptor, as a function of the concentration of estradiol. Binding was assayed with (^{3}H)-estradiol as described in Materials and Methods.

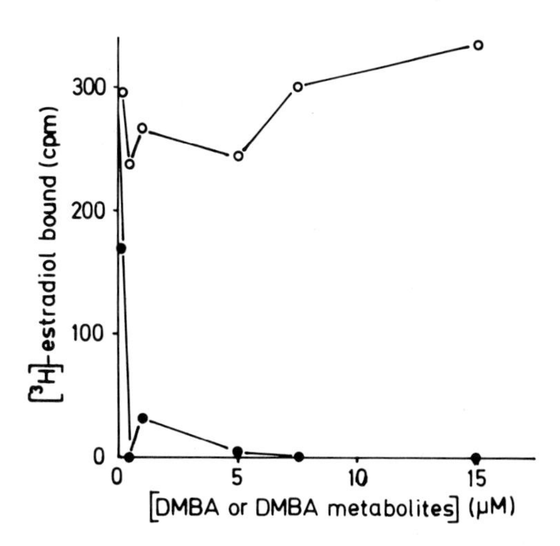

Fig. 2. Binding of (^{3}H)-estradiol to the estradiol receptor in the presence of various concentrations of DMBA (open circles) and DMBA metabolites (closed circles), respectively.

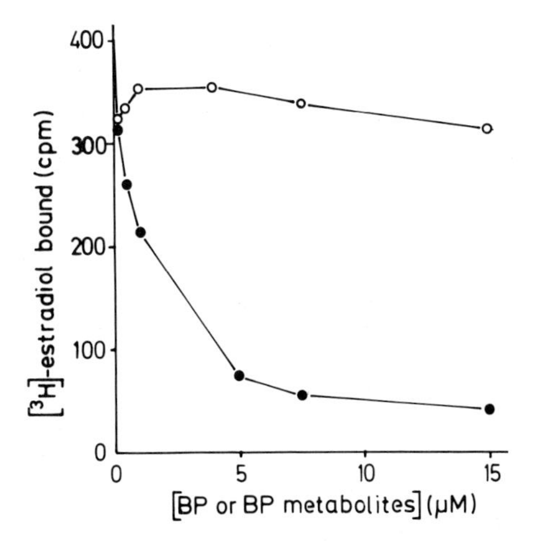

Fig. 3. Binding of (^{3}H)-estradiol to the estradiol receptor in the presence of various concentrations of BP (open circles) and BP metabolites (closed circles), respectively.

The possibility that DMBA and BP metabolites generated _in vivo_ may be translocated to the nuclei when bound to the receptor and exert some type of estrogen effect remains to be shown and is presently being investigated.

TABLE 1

EFFECT OF DMBA METABOLITES ON THE BINDING OF (^{3}H)-ESTRADIOL TO THE ESTRADIOL RECEPTOR IN RAT ADRENAL CYTOSOL

Conditions were as described in Materials and Methods.

Metabolite	Derivative	Inhibition (%)
1	dihydrodiol	64
2	"	81
3	hydroxymethyl	16
	and/or	
4	phenols	12
5	DMBA	11

ACKNOWLEDGEMENT

This work was supported by the Swedish Cancer Society.

REFERENCES

1. Montelius, J. Papadopoulos, D. Bengtsson, M. and Rydström, J. (1982) Cancer Res. 42, 1479.
2. Muller, R.E. and Wotiz, H.H. (1978) J. Biol. Chem. 253, 740.
3. Morreal, C.E. Schneider, S.L. Sinha, D.K. and Bronstein, R.E. (1979) JNCI 62, 1585.
4. Ogle, T.F. (1977) J. Steroid Biochem. 8, 1033.

Cytochrome P-450, Biochemistry, Biophysics
and Environmental Implications, E. Hietanen,
M. Laitinen and O. Hänninen editors

EFFECT OF DITHRANOL ON THE HEPATIC MICROSOMAL DRUG-METABOLIZING SYSTEM IN RATS.

ANNELI KARPPINEN AND NIILO T. KÄRKI
Department of Pharmacology, University of Oulu, Kajaanintie 52,
90220 Oulu 22 (Finland).

INTRODUCTION

Dithranol (anthralin) or 1,8,9-trihydroxyanthracene has been in clinical use for over 60 years for the topical treatment of psoriasis. Although intensively studied, the basic questions concerning its mode of action are still unanswered. It has been found in animal studies to be a tumour promoter (1) and very toxic on the liver and nervous system (2), but for psoriatics it has proved to be a safe and effective remedy (3) without producing changes in liver or kidney function (4) or increasing tumour incidence (5), only problems being irritation and staining of the skin.

Dithranol can be oxidized via a free radical chain mechanism into inactive 1,8-dihydroxyanthraquinone and coloured dimeric products (6). In view of its polycyclic aromatic hydrocarbon structure it has been claimed to be metabolized by aryl hydrocarbon hydroxylase (AHH), one of its metabolites being that responsible for the irritating property (7,8). The relationship between dithranol and the drug-metabolizing enzyme system is largely unknown, however.

MATERIALS AND METHODS

Animals. Male Wistar and Sprague-Dawley rats weighing 200 - 350 g were used. Water and standard pelleted rat diet were provided *ad libitum*.

Methods. Dithranol (ICN Pharmaceuticals) in corn oil (2 ml/kg rat weight) was injected intraperitoneally at 8-9 a.m. Because of its instability the dithranol solution was prepared just before use. The control rats received corn oil alone. The rats were killed by decapitation. Different dithranol doses, varying from 0.005 to 50 mg/kg, were tested and toxic signs compared with microsomal parameters. A single bolus injection of 0.15 or 1.5 mg/kg was selected for time-curve experiments.

The rat liver microsomes were isolated by ultracentrifugation, and the activities of AHH and aminopyrine demethylase were measured according to Nebert and Gelboin (9) and Cochin and Axelrod (10) respectively. Microsomal cytochrome P-450 content was determined by method of Greim *et al.* (11) and protein by that of Lowry *et al.* (12).

Statistical significance. A Student's t-test was used, and results expressed as means ± S.D.

RESULTS

Dithranol toxicity. LD_{50} for dithranol in a single-dose i.p. injection was about 3 mg/kg. The severity of the visible toxic effects, disturbance of balance and breath, bleeding of the nose and weakness were comparable with the severity of the autopsy findings, yellowish, swollen intestines, a greyish brittle liver covered with fibrous material, haemorrhagia in the lungs and ascites, when a dose of 0.5 mg/kg or more was used. The rats appeared normal after 0.005 or 0.05 mg dithranol/kg, although some inflammatory changes were seen in autopsy. The effects were similar in both Wistar and Sprague-Dawley rats.

Dose effect of dithranol on microsomal parameters. The rats were killed 48 hours after the i.p. injection of dithranol. The low doses gave no response in microsomal parameters in the Wistar rats, but significantly depressed values were seen at high doses in the Sprague-Dawley rats, as shown in Table I. No significant difference existed between the liver weights of the dithranol and control rats when calculated in relation to body weight. The body weights were 10 - 15 % lower in the dithranol groups of Sprague-Dawley rats, and these rats also showed interindividual variations in toxicity, AHH activity and cytochrome P-450.

Changes in microsomal parameters within one week of a single dithranol injection. Male Wistar rats weighing 220 ± 20 g had about 15 % lower liver:body weight ratios after 1 day with a dose of 0.15 mg/kg ($P < 0.01$) and after 1 and 2 days with 1.5 mg/kg ($P < 0.001$ and $P < 0.01$ respectively). They were about 10 % lighter than the controls on the second day ($P < 0.05$). The symptoms and toxic signs found at autopsy became less serious after the third day. No visible pathological changes were noted in the rats injected with 0.15 mg/kg by the seventh day.

The microsomal protein quantities differed from the control yields only on the seventh day, being 90 % of this value with the smaller dose ($P < 0.05$) and 72 % with the larger one ($P < 0.01$). Dithranol significantly reduced the cytochrome P-450 values from the first to the seventh day. The values were 73, 63, 55, 92 and 72 % and 88, 73, 80, 100 and 90 % of the controls after 1, 2, 3, 5 and 7 days when using doses of 1.5 mg/kg and 0.15 mg/kg respectively. The numbers of animals in each group were same as given in Figure 1.

The changes in AHH and aminopyrine demethylase activities are shown in Figure 1. AHH activity was still reduced by about a half at the end of the experiment.

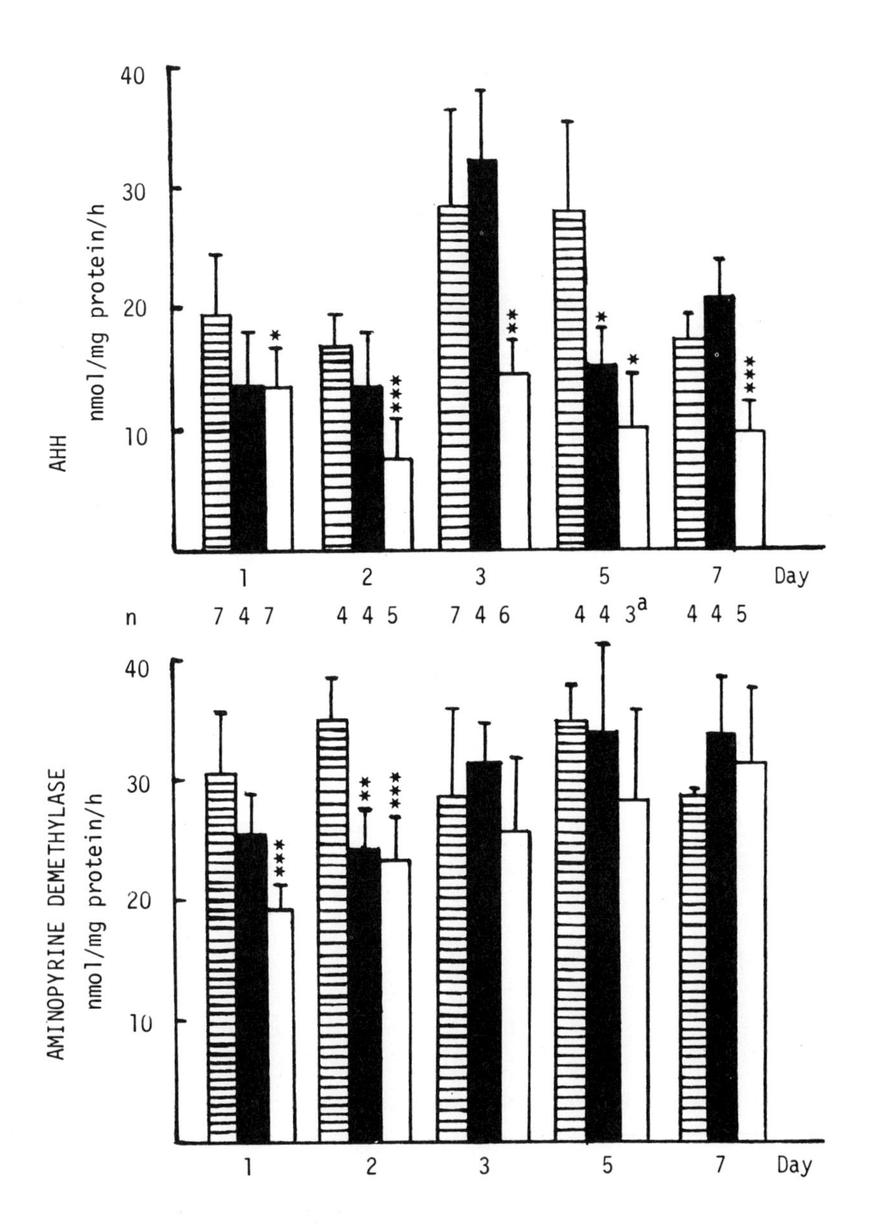

Fig. 1. Activities of AHH and aminopyrine demethylase after an intraperitoneal injection of dithranol in corn oil at a dose of 0.15 mg/kg ■ or 1.5 mg/kg □ in male Wistar rats, which were killed 1, 2, 3, 5 and 7 days afterwards. The controls ▤ received corn oil alone.
n indicates the number of animals in one group.

[a]One died in group on the third day.
* $P < 0.05$, ** $P < 0.01$, *** $P < 0.001$.

TABLE I.

EFFECT OF DIFFERENT DOSES OF DITHRANOL ON MICROSOMAL PROTEIN, CYTOCHROME P-450 CONTENT, AHH AND AMINOPYRINE DEMETHYLASE ACTIVITY.

The male Wistar rats (3/group) weighed 300 - 330 g and the male Sprague-Dawley rats (4/group) 270 ± 10 g. All the animals were killed 48 hours after the intraperitoneal injection of Dithranol in corn oil.

Dithranol Dose mg/kg	Microsomal Protein mg/g liver	Cytochrome P-450 nmol/mg prot.	AHH nmol/mg protein/h	Aminopyrine Demethylase
Wistar				
- control	17.8	0.90	27.2	29.3
0.005	17.9	0.98	28.9	28.6
0.05	18.8	0.93	29.7	25.1
0.5	15.0*	0.62**	26.8	23.9
Sprague-Dawley				
- control	16.6 ± 1.3	0.45 ± 0.03	23.3 ± 2.3	35.7 ± 2.9
0.5	14.6 ± 0.3*	0.50 ± 0.07	13.3 ± 9.0	26.6 ± 1.4**
1.5[a]	13.3 ± 0.8*	0.47 ± 0.07	19.1 ± 9.8	32.5 ± 1.9
3.0	14.3 ± 1.0*	0.38 ± 0.09	11.5 ± 7.4	28.1 ± 3.2*
4.0[a]	17.8 ± 2.7	0.40 ± 0.17	12.1 ±10.2	23.2 ± 4.5**

[a]One died. *$P < 0.05$, **$P < 0.01$.

DISCUSSION

The LD_{50} and toxic signs for the Wistar and Sprague-Dawley rats were similar to those found earlier for white mice (2): 4 - 5 mg/kg i.p. and fatty degeneration of the liver along with various pathological changes in the central nervous system and intestines. The tissue-drying effect of dithranol, because of its laxative property, seemed not to play any important part in the results, although the weight loss began earlier and was more marked with the high doses and in those animals which died or were very sick. The hepatic weight loss was rapid and was soon recovered. The decrease in microsomal protein and cytochrome P-450 recorded in day 7 might indicate some subacute damage.

Very low quantities of dithranol or its metabolites are reported to have penetrated from skin of young white pigs following topical application, the highest radioactivity being found in the liver (13). Low doses of dithranol produced no biochemical changes in the parameters measured. This may indicate a threshold value in a dose-response relationship in the liver, and could be one explanation for the lack of any noticeable toxicity in psoriatics.

The inhibition of AHH and aminopyrine demethylase activities and cytochrome P-450 differed during the follow-up period of one week. Using one 1.5 mg/kg i.p. injection of dithranol the maximal inhibitory effect on AHH activity was shown in 2 days, with no recovery by the seventh day, while normal values for aminopyrine demethylase were restored in the third day, the minimum for cytochrome content being reached at the same time. The degree of inhibition was dose-dependent. No induction of any of the hepatic microsomal components of drug metabolism measured here could be found at least under these experimental conditions.

REFERENCES

1. IARC MONOGRAPH (1977) Eval. Carcinog. Risk Chem. Man 13, 1.
2. Von Ippen, H. (1959) Dermatologica 119, 211.
3. Champion, R.H. (1981) Br. Med. J. 282, 343.
4. Gay, M.W., Moore, W.J., Morgan, J.M. and Montes, L.F. (1972) Arch. Dermatol. 105, 213.
5. Von Ippen, H. (1981) Br. J. Dermatol. 105 (Suppl. 20), 72.
6. Mustakallio, K.K. (1979) Acta Dermatovener. 59 (Suppl. 85), 125.
7. Chapman, P.H., Rawlins, M.D. and Shuster, S. (1979) Lancet 1, 297.
8. Kersey, P., Chapman, P., Rogers, S., Rawlins, M. and Shuster, S. (1981) Br. J. Dermatol. 105 (Suppl. 20), 64.
9. Nebert, D.W. and Gelboin, H.V. (1968) J. Biol. Chem. 243, 6242.
10. Cochin, J. and Axelrod, J. (1959) J. Pharmac. Exp. Ther. 125, 105.
11. Greim, H., Schenkman, M., Klotzbucher, M. and Remmer, H. (1970) Biochem. Biophys. Acta 201, 21.
12. Lowry, O.H., Rosebrough, N.J., Farr, A.L. and Randall, R.J. (1951) J. Biol. Chem. 193, 265.
13. Hooper, G. and Ayres, P.J. (1979) Clin. Exper. Dermatol. 4, 315.

Cytochrome P-450, Biochemistry, Biophysics
and Environmental Implications, E. Hietanen,
M. Laitinen and O. Hänninen editors

UNALTERED METABOLISM OF m-XYLENE IN THE PRESENCE OF ETHYLBENZENE

EIVOR ELOVAARA, KERSTIN ENGSTRÖM[1] AND HARRI VAINIO
[1]Institute of Occupational Health, SF-00290 Helsinki 29
Regional Institute of Occupational Health, SF-20500 Turku 50,
Finland

INTRODUCTION

The biotransformation of xylene isomers (1) and ethylbenzene (2) is a cytochrome P-450 dependent process. The technical xylene, which is a mixture of xylene isomers and ethylbenzene, is widely used in industry. The biological effects of ethylbenzene, either alone or in solvent mixtures, are poorly known as compared to the data e.g. on xylene. In a recent work it was proposed that oxidation of m-xylene at the aromatic nucleus is diminished in the presence of ethylbenzene (3). The aim of this study was to elucidate the metabolic interactions of m-xylene and ethylbenzene by screening their specific urinary metabolites in the rat.

MATERIALS AND METHODS

Twenty male Wistar rats (276 $\pm$ 21 g) were divided into 4 groups. The rats were exposed for 5 days (6h/day) in dynamic exposure chambers (1 m^3) at the concns. of 0, 25+75, 100+300, or 200+600 ppm of ethylbenzene (99 %, Fluka Ag, Buchs) and m-xylene (laboratory grade, Merck, Darmstadt) in air, respectively (4). Water was given _ad libitum_ but the rat chow only between the exposure periods. Three rats of each group were housed in metabolic cages to collect pooled urine samples throughout the experiment. The known urinary metabolites of m-xylene and ethylbenzene were measured gas-chromatographically (5).

Rats were decapitated after termination of the fifth days' exposures. Perirenal fat samples were prepared for the g.l.c. analysis of m-xylene and ethylbenzene (4). Liver were reserved at -70° until analysis. Calcium-aggregated microsomes were prepared from the livers (4) and enzyme assays performed as previously described (4). Reduced GSH concns. were determined from deproteinated liver homogenates (4). The statistical evaluation of the results were carried out with Students' t-test.

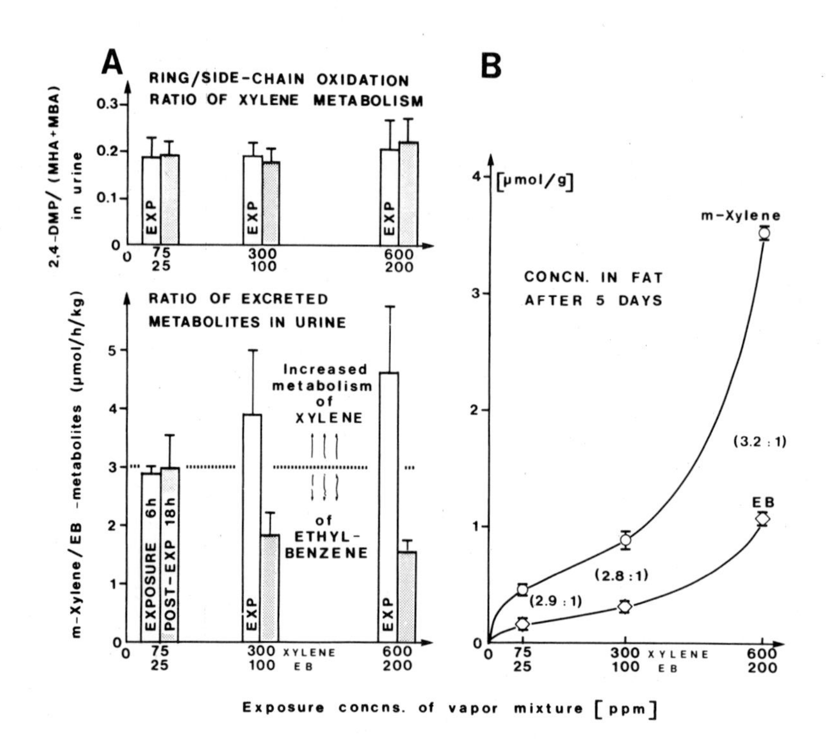

FIG. 1 Dose-dependent effects of intermittent exposure of 5 days to vapor mixtures of m-xylene and ethylbenzene.

A. Excretion of urinary metabolites during and after daily treatments (data given are mean + S.D. of 5 days and representative of all voided and pooled urine of 3 rats per a group).

B. Accumulation of solvents in fat (the mean ± S.D. of 5 rats).

TABLE 1

EFFECTS OF INHALATION EXPOSURES TO m-XYLENE (X) AND ETHYLBENZENE (EB) VAPOR MIXTURES (6 h PER DAY, FOR 5 DAYS) ON THE DRUG-METABOLIZING SYSTEM OF THE LIVER

Exposure level (ppm)	Liver/Rat (g/kg)	GSH (μmol/g w.wt)	NADPH-cytochrome c reductase[a)]
Control	35 ± 2	5.8 ± 0.3	95 ± 3
25+75 (EB+X)	32 ± 4	5.1 ± 0.5*	110 ± 26
100+300 "	36 ± 3	5.3 ± 0.8	87 ± 9
200+600 "	37 ± 4	5.2 ± 0.5*	98 ± 18
	Cyt. P-450 (nmol/mg)	Ethoxycoumarin O-deethylase[a)]	UDPglucuronyl transferase[a)]
Control	0.67 ± 0.06	1.21 ± 0.39	1.03 ± 0.20
25+75 (EB+X)	0.71 ± 0.02	1.32 ± 0.22	0.99 ± 0.15
100+300 "	0.76 ± 0.06*	1.36 ± 0.23	1.35 ± 0.14**
200+600 "	0.77 ± 0.10*	1.67 ± 0.24*	1.91 ± 0.24***

[a)] nmol/min per mg microsomal protein

Each figure is the mean for 5 rats ± S.D.

Difference from the control: $^{*}P \leq 0.05$, $^{**}P \leq 0.01$, $^{***}P \leq 0.001$

RESULTS AND DISCUSSION

Accumulation curves (Fig. 1) show no differences between the uptake of m-xylene or ethylbenzene into perirenal fat after a 5-day exposure period as judged by the unchanged relative ratios of solvents found in fat. The load of solvents in fat bears an exponential relationship to their concentrations in inhaled air. Presumably, the unlinearity is dependent on the cumulation of the solvents due to repetitive 6-h treatments. Hence, with the highest exposure concentrations studied the biotransformation capacity of the liver was apparently exceeded.

There are two distinct pathways of microsomal biotransformation of m-xylene (1). Methylbenzylalcohol (MBA) and methylhippuric acid (MHA) were measured in urine as indicators of the major route implicating side-chain oxidation whereas 2,4-dimethylphenol stands for the oxidation at the aromatic nucleus. To answer the question, does the presence of ethylbenzene impair m-xylene metabolism via the latter pathway, the ratio between ring and side-chain oxidations was evaluated (Fig. 1). Proportionally, the metabolism via ring oxidation remained constant at different concentrations of ethylbenzene. The rate of ethylbenzene metabolism was estimated from the total amount of its major urinary metabolites, i.e., hippuric acid, mandelic acid, phenylglyoxylic acid and 1-phenylethanol (2). The ratio between the m-xylene per ethylbenzene metabolites excreted in urine show at the lowest concentration tested (25+75) ppm the same ratio as in inhaled air (Fig. 1). However, with increasing dose the metabolism of xylene is preferred during the actual exposure periods at the expence of ethylbenzene metabolism. Consequently, this delay in ethylbenzene metabolism asserts itself through an inverse ratio during the post-exposure period (Fig. 1).

Only the highest dose which apparently saturates the biotransformation capacity of the liver resulted in small but statistically significant increase of the cytochrome P-450 linked drug metabolism (Table 1). The reduced glutathione levels in liver were only slightly reduced by the inhalation exposures, even though an almost linearly dose-dependent increase of up to 12-fold in urinary thioether excretion was measured.

In conclusion, the hepatic metabolism of ethylbenzene is delayed in the presence of m-xylene the more the higher the treatment dose. However, we failed to show any changes in the hepatic fate of m-xylene during its metabolism in the presence of ethylbenzene.

REFERENCES

1. Dean, B.J. (1978) Mutation Res., 47, 2, pp. 75-97.
2. Kiese, M. and Lenk, W. (1974) Xenobiotica, 4, 6, pp. 337-343.
3. Angerer, J. and Lehnert, G. (1979) Int. Arch. Occup. Environ. Health, 43, pp. 145-150.
4. Elovaara, E., Collan, Y., Pfäffli, P. and Vainio, H. (1980) Xenobiotica, 10, 6, pp. 435-445.
5. Engström, K., Husman, K. and Rantanen, J. (1976) Int. Arch. Occup. Environ. Health, 36, pp. 153-160.

Cytochrome P-450, Biochemistry, Biophysics and Environmental Implications, E. Hietanen, M. Laitinen and O. Hänninen editors

MITOCHONDRIAL MEDIATION OF ETHYLMORPHINE-INDUCED ALTERATIONS IN HEPATOCYTE Ca^{2+} HOMEOSTASIS

SARAH A. JEWELL, GIORGIO BELLOMO, HJÖRDIS THOR AND STEN ORRENIUS
Department of Forensic Medicine, Karolinska Institutet, S-104 01 Stockholm (Sweden)

INTRODUCTION

The Ca^{2+} ion plays a crucial regulatory role in many different cell functions. The maintenance of intracellular Ca^{2+} homeostasis is therefore highly regulated, involving sequestration within the mitochondrial and endoplasmic reticulum, and binding to proteins such as calmodulin (for reviews, see (1) and (2)). The mitochondria contain up to 80% of cellular Ca^{2+} (1), depending primarily upon the pyridine nucleotide redox level (3) and the respiration-generated transmembrane potential (4). Studies performed by Sies *et al.* with perfused rat liver (5) have shown that substrates of the hepatic cytochrome P-450 system cause a release of Ca^{2+} into the perfusate, although the cellular sites from which the Ca^{2+} was released could not be identified. Here we report that high rates of ethylmorphine metabolism by isolated hepatocytes cause a large mobilization of intracellular Ca^{2+} which is completely prevented by impairing the ability of the mitochondria to mediate intracellular Ca^{2+} homeostasis.

METHODS

Hepatocytes were isolated as described previously (6) from male Sprague-Dawley rats (180-200 g) given water (± 1 mg/ml sodium phenobarbital for 1 week) and food *ad libitum*. The cells were incubated in Krebs-Hepes buffer, pH 7.4, at 37° (7) for 30 minutes prior to substrate addition; in some cases this preincubation was carried out in the presence of 20 µM Ruthenium Red or 30 µM dicoumarol (entire 30 minutes), or 0.5 mM metyrapone (added 5 minutes prior to substrate). Ca^{2+} measurements were performed according to a method developed recently in this laboratory (8), involving the addition of 10 µM FCCP (carbonyl cyanide p-trifluoromethoxyphenylhydrazone) to release mitochondrial Ca^{2+}, and the cation ionophore A23187 (15 µM) to release the total cellular Ca^{2+}. Ca^{2+}

released from the cells was measured spectrophotometrically (wavelength pair 685-675 nm) by changes in the absorbance of Arsenazo III, a metallochromic indicator (cf. 8).

RESULTS

The FCCP-releasable, mitochondrial Ca^{2+} content consistently represented 60-70% of the total releasable Ca^{2+}, both after 30 min preincubation in Krebs-Hepes buffer (t=0) and 30 min later (t=30) (See Figs 1 and 2). Incubation with ethylmorphine caused a dramatic, dose-dependent increase in the Ca^{2+} levels of both the mitochondrial and total Ca^{2+} contents (Figs 1 and 2); the prevention of this effect by metyrapone indicates that the Ca^{2+} increase occurs as a result of metabolism by the cytochrome P-450 system rather than by direct effect of the substrate itself. Changes in intracellular Ca^{2+} were not observed with up to 4 mM aminopyrine or hexobarbital, which undergo oxidation by the P-450 system less readily than ethylmorphine (9), nor were any changes observed when these three substrates were added to suspensions of control (i.e. uninduced) hepatocytes (not shown). Studies in which the hepatocytes were incubated with ethylmorphine in the presence of $^{45}Ca^{2+}$, filtered, and total radioactivity measured by liquid scintillation indicate that the increase in measurable Ca^{2+} may be due to a mobilization of Ca^{2+} from previously inaccessible intracellular stores, rather than an uptake from the extracellular medium (not shown).

Since the mitochondria contain a large percentage of cellular Ca^{2+} and play an important role in the interdependent functioning of the various compartments (1), experiments were undertaken to determine the extent of mitochondrial involvement in the ethylmorphine-induced Ca^{2+} mobilization. Preincubation of the hepatocytes with 20 μM Ruthenium Red, a specific inhibitor of the uptake portion of normal mitochondrial Ca^{2+} cycling (10), caused loss of mitochondrial (and thus some total) Ca^{2+} during preincubation (Figs 1 and 2); subsequent addition of ethylmorphine did not cause any observable Ca^{2+} mobilization whatsoever. These results were repeated when mitochondrial Ca^{2+} regulation was impaired in a different way, by preincubation with 30 μM dicoumarol, a mild uncoupler which causes Ca^{2+} release by collapse of the mitochondrial membrane potential (11) (Figs 1 and 2).

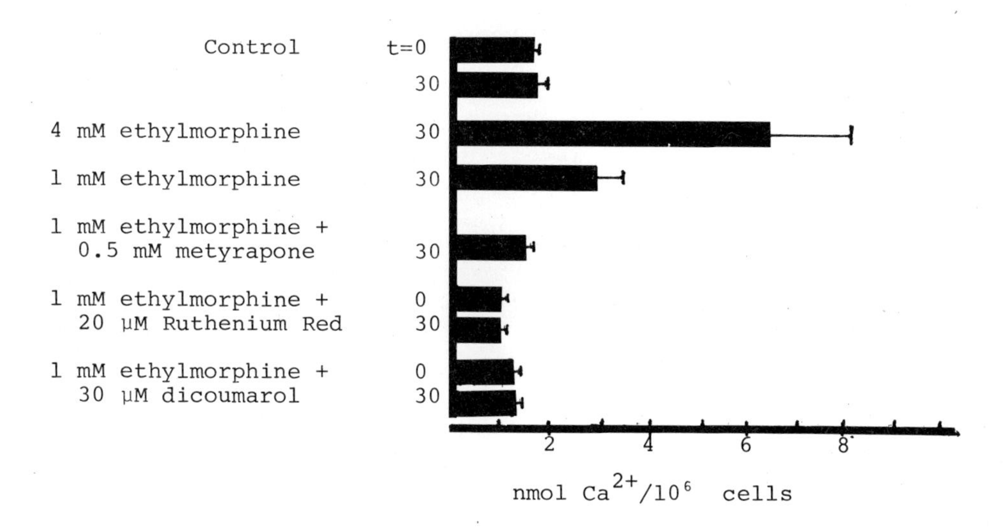

Fig. 1. Mitochondrial Ca^{2+} content in phenobarbital-induced hepatocytes. Values represent mean FCCP-releasable Ca^{2+} ± S.E.M. (n=3) for 0 or 30 minutes.

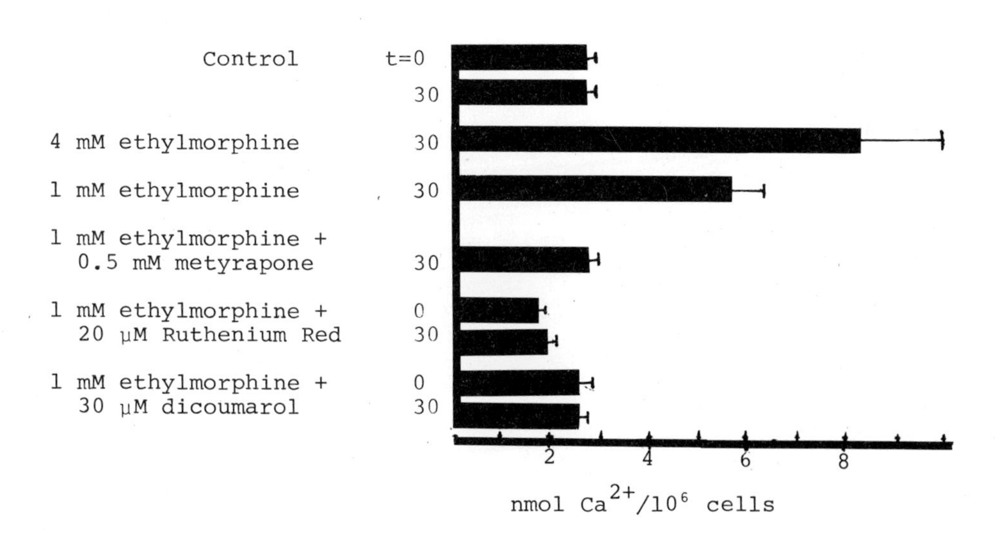

Fig. 2. Total Ca^{2+} content in phenobarbital-induced hepatocytes. Values represent mean A23187-releasable Ca^{2+} ± S.E.M. (n=3) for 0 or 30 minutes.

Thus, the mitochondria apparently play a major role in ethylmorphine-induced alterations in Ca^{2+} homeostasis. The fact that ethylmorphine caused a release of intracellular Ca^{2+} from perfused liver mentioned earlier (5), may be due to the fact that the perfusion medium contained only micromolar Ca^{2+}, thus destabilizing the normal intracellular control processes. Any significant perturbations in normal intercompartmental Ca^{2+} balance would be expected to have numerous consequences for cell function. The effects of a variety of substrates on intracellular Ca^{2+} homeostasis are currently under active investigation in our laboratory.

ACKNOWLEDGEMENTS

This work was supported by grants from the Swedish Medical Research Council, (no. 03X-2471), the Swedish Council for the Planning and Coordination of Research and the Nobel Foundation.

REFERENCES

1. Williamson, J.R., Cooper, R.H. and Hoek, J.B. (1981) Biochim. Biophys Acta 639, 243-295
2. Cheung, W.L. (1980) Science 207, 19-27
3. Lehninger, A.L., Vercesi, A. and Bababunmi, E.A. (1978) Proc. Natl. Acad. Sci. USA 75, 1690-1694
4. Beatrice, M.C., Palmer, J.W. and Pfeiffer, D.R. (1980) J. Biol. Chem. 255, 8663-8671
5. Sies, H., Graf, P. and Estrela, J.M. (1981) Proc. Natl. Acad. Sci. USA 78, 3358-3362
6. Moldéus, P., Högberg, J. and Orrenius, S. (1978) Methods Enzymol. 51, 60-71
7. Högberg, J. and Kristoferson, A. (1977) Eur. J. Biochem. 74, 77-82
8. Bellomo, G., Jewell, S.A. and Orrenius, S. (1982) Proc. Natl. Acad. Sci. USA in the press
9. Jones, D.P., Thor, H., Andersson, B. and Orrenius, S. (1978) J. Biol. Chem. 253, 6031-6037

© 1982 Elsevier Biomedical Press B.V.
Cytochrome P-450, Biochemistry, Biophysics
and Environmental Implications, E. Hietanen,
M. Laitinen and O. Hänninen editors

THE INDUCIBILITY OF HEPATIC CYTOCHROME P-450 CONTENT AND MONOOXYGENASES BY 2,2,2-TRICHLORO-1-(3,4-DICHLOROPHENYL)ETHYL ACETATE IN THE RAT

EERO MÄNTYLÄ[1], EINO HIETANEN[1], MARKKU AHOTUPA[1] AND HARRI VAINIO[2]
[1]Department of Physiology, University of Turku, SF-20520 Turku 52 and
[2]Institute of Occupational Health, SF-00290 Helsinki 29 (Finland)

INTRODUCTION

A new insecticide, 2,2,2-trichloro-1-(3,4-dichlorophenyl)ethyl acetate (Penfenate) is widely used as a DDT substitute but data is lacking about its effects on drug metabolizing enzymes. Like many other insecticides DDT (1,1,1-trichloro-2,2-bis(p-chlorophenyl)ethane) elevates hepatic polysubstrate monooxygenase activities and cytochrome P-450 content (1, 2, 3). Further DDT is a suspected carcinogen (4). In this study the effect of Penfenate on polysubstrate monooxygenase activities and hepatic cytochrome P-450 content was studied in rats.

MATERIALS AND METHODS

Eight weeks old male Wistar rats (155-220 g) were used. The rats were fed commercial pelleted diet _ad libitum_ and they had free access to tap water. Penfenate (Bayer A.G., Leverkusen, FRG) was dissolved in corn oil either 20 or 100 mg/ml. A single i.p. dose (100 or 500 mg/kg) of Penfenate was given to the animals and they were killed at different time points. When a dose (100 or 500 mg/kg) was given to rats daily for five days the animals were sacrified 24 hours after the last dose. Control animals received equal amounts of the vehicle. Liver, kidney and 40 cm segment of small intestine starting from pylorus were immersed in 0.25 mol/l sucrose solution ($0^{o}C$). The mucosa of the small intestine was scraped off with an ampoule file. The tissues were homogenized in 4 vol of 0.25 mol/l sucrose solution with a Potter-Elvehjem glass-Teflon[R] homogenizer. From liver and kidney Ca^{2+}-aggregated microsomes (5, 6) were prepared. The postmitochondrial (12 000 x g_{max}) supernatant was prepared from intestinal mucosa homogenate.

The hepatic cytochrome P-450 content was measured by the method of Omura and Sato (7). Ethoxycoumarin deethylation was measured by a modification (8) of the method of Ullrich and Weber (9). PPO (2,5-diphenyloxazole) hydroxylase activity was determined as described earlier by Cantrell _et al._ (10). In the statistical evaluation of the results Student's t-test was used. Means and S.E.M.'s are shown in figures and tables.

RESULTS

The single high Penfenate dose (500 mg/kg) increased the hepatic cytochrome P-450 content 1.3-1.7-fold when measured 1 or 3 days after the treatment. With the low dose (100 mg/kg) a slight decrease was found after one day (Fig. 1A). The high dose enhanced hepatic PPO hydroxylase activity 1.6-fold and ethoxycoumarin deethylase activity 2.5-fold in 3 days (Fig. 1B and 2A). In kidney and intestine there was a slight decrease in ethoxycoumarin deethylase activity after one day (Fig. 2B and C). The low dose had no effect on the monooxygenation activities.

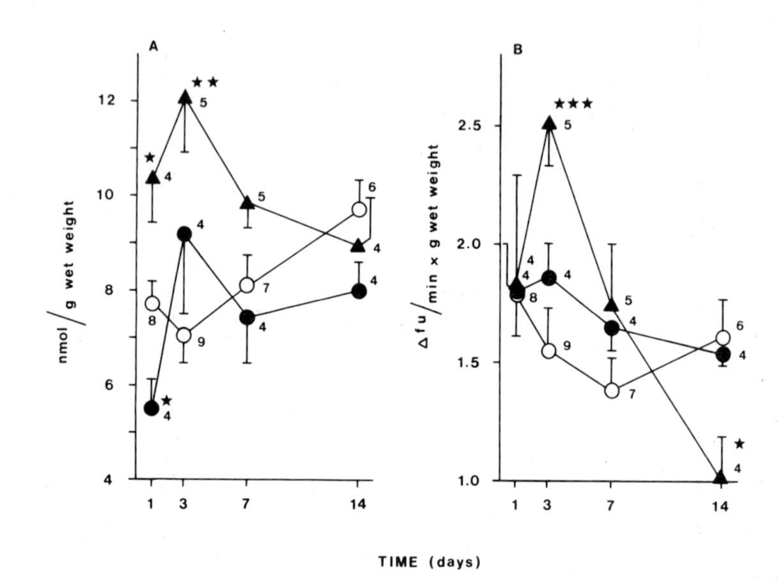

Fig. 1. Hepatic cytochrome P-450 content (A) and PPO hydroxylase activity (B). o-o = control, ●-● = Penfenate 100 mg/kg, ▲-▲ = Penfenate 500 mg/kg, * = 2p < 0.05, ** = 2p < 0.01, *** = 2p < 0.001, fu = fluorescence units. The number of the animals is indicated.

When Penfenate (500 mg/kg) was given to the rats for five consecutive days the hepatic cytochrome P-450 content increased 1.4-fold, PPO hydroxylase activity 1.6-fold and ethoxycoumarin deethylase activity 2.0-fold. The high dose decreased the intestinal ethoxycoumarin deethylase activity. The low repeated dose elevated the hepatic cytochrome P-450 content 1.4-fold (Table 1).

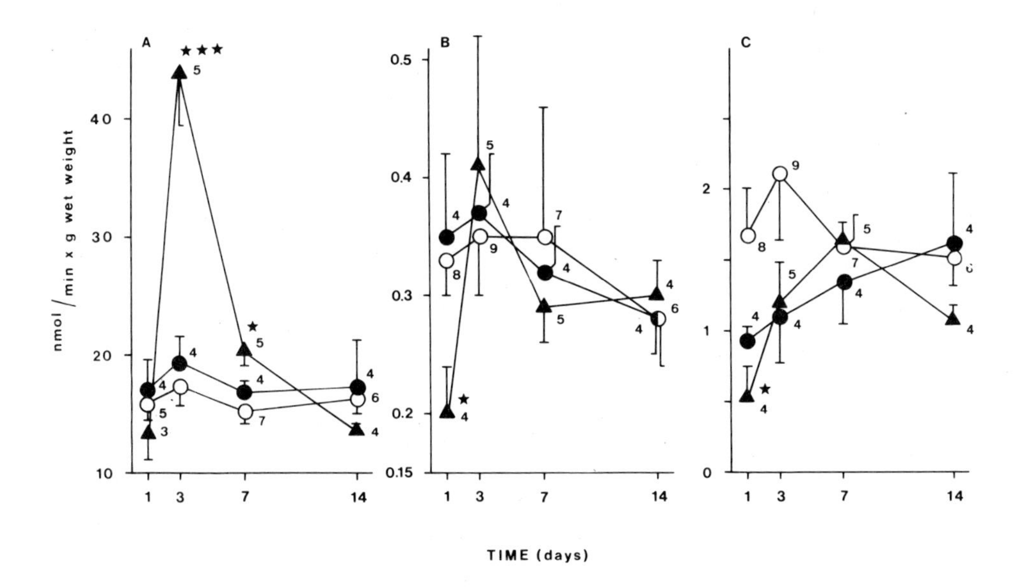

Fig. 2. Ethoxycoumarin deethylase activity in rat liver (A), kidney (B) and intestine (C). o-o = control, ●-● = Penfenate 100 mg/kg, ▲-▲ = Penfenate 500 mg/kg, * = 2p < 0.05, *** = 2p < 0.001. The number of the animals is indicated.

TABLE 1

THE EFFECT OF FIVE CONSECUTIVE DAILY I.P. DOSES (100 OR 500 mg/kg) OF PENFENATE ON THE AMOUNT OF CYTOCHROME P-450 AND ON POLYSUBSTRATE MONOOXYGENASE ACTIVITIES

	Cytochrome P-450 (nmol/g wwt)	PPO hydroxylase (Δfu/min x g wwt)	Ethoxycoumarin deethylase (nmol/min x g wwt)		
	Liver	Liver	Liver	Kidney	Intestine
Control	7.12±0.60[a]	1.48±0.22	15.7±2.1	0.26±0.06	1.38±0.18
100 mg/kg	9.95±0.64*	1.71±0.06[b]	15.9±3.9	0.42±0.05	1.45±0.20
500 mg/kg	10.20±1.81	2.30±0.17*	30.6±4.2*	0.22±0.04	0.88±0.07*

[a]Number of animals is 4-7, except [b]n=2.

DISCUSSION

Penfenate differs from DDT as the latter exerts its maximal effect 2 weeks after a single i.p. dose (11), while the effect of Penfenate was more rapid. This difference can be explained by the longer retention and slower catabolism of DDT than Penfenate (12). There was after 7 days no detectable amounts of Penfenate or its principal metabolite (2,2,2-trichloro-1-(3,4-dichlorophenyl) ethanol) in liver or adipose tissue with daily oral dose (50 mg/kg) in contrast to DDT (12). When Penfenate is fed to chicken it does not accumulate into any tissue. Further in chicken liver homogenate Penfenate is metabolized almost completely in 20 minutes (13).

Some evidence exists that Penfenate is neither mutagenic nor teratogenic (12). The acute toxicity of Penfenate in the rat is quite low, oral LD_{50} > 10 000 mg/kg and cutaneous LD_{50} > 1000 mg/kg (12). However, as Penfenate causes significant enhancement in the activities of hepatic polysubstrate monooxygenases it may also exert noxious effects via altering the relative amounts of reactive intermediates. The fact that Penfenate is often used in a combination with other insecticides must also be considered.

ACKNOWLEDGEMENTS

This study was supported by grants from J. Vainio Foundation (Finland) and NIH (R01 ES 01684).

REFERENCES

1. Hart, L.G. and Fouts, J.E. (1963) Proc. Soc. Exp. Biol. Med., 114, 288-292.
2. Juchau, M.R., Gram, T.E. and Fouts, J.R. (1966) Gastroenterology, 51, 213-218.
3. Vainio, H. (1974) Chem.-Biol. Interact., 9, 7-14.
4. Kimbrough, R.D. (1979) Ann. N.Y. Acad. Sci., 320, 415-418.
5. Kamath, S.A. and Narayan, K.A. (1972) Anal. Biochem., 48, 59-61.
6. Aitio, A. and Vainio, H. (1976) Acta Pharmacol. Toxicol., 39, 555-561.
7. Omura, T. and Sato, R. (1964) J. Biol. Chem., 239, 2370-2378.
8. Aitio, A. (1978) Anal. Biochem., 85, 488-491.
9. Ullrich, V. and Weber, P. (1972) Hoppe Seylers Z. Physiol. Chem., 353, 1171.
10. Cantrell, E.T., Abreu-Greenberg, M., Guyden, J. and Busbee, D.L. (1975) Life Sci., 17, 317-322.
11. Parkki, M.G., Marniemi, J. and Vainio, H. (1977) J. Toxicol. Environ. Health, 3, 903-911.
12. Behrenz, W., Büchel, K.H. and Meiser, W. (1977) Pflanzenschutz-Nachrichten Bayer, 30, 237-248.
13. Anonymous (1979) Acta Entomol. Sin., 22, 390-395.

Cytochrome P-450, Biochemistry, Biophysics
and Environmental Implications, E. Hietanen,
M. Laitinen and O. Hänninen editors

EFFECTS OF FLUORENONE AND TRINITROFLUORENONE ON THE MONOOXYGENASE, EPOXIDE HYDROLASE AND GLUTATHIONE S-TRANSFERASE ACTIVITIES IN RABBITS

EINO HIETANEN[1] AND HARRI VAINIO[2]
[1]Department of Physiology, University of Turku, SF-20520 Turku 52 (Finland)
and [2]Institute of Occupational Health, Haartmanink. 1, SF-00290 Helsinki 29
(Finland)

INTRODUCTION

The wide-spread use of photocopying machines has lead to the expansing exposure to the chemicals used. A common photoconducting agent is a polynuclear hydrocarbon trinitrofluorenone (2,4,7-trinitrofluoren-9-one) which is a derivative of the parent compound fluorenone (fluoren-9-one). Suspicions on the carcinogenicity of trinitrofluorenone were raised already in 1960'ies (1). Recently, data has been also obtained on the mutagenicity of trinitrofluorenone both in prokaryotic and eukaryotic assays.

As the toxicity of polycyclic hydrocarbons is modulated by their metabolism catalysed by the cytochrome P-450 dependent enzyme system, we tested the effects of trinitrofluorenone and fluorenone on the hepatic and intestinal biotransformation enzymes.

MATERIALS AND METHODS

Adult male Californian rabbits were used (N=15). The rabbits were administered by trinitrofluorenone (m.wt 315) as a subcutaneous (s.c.) dose of 1 mmole/rabbit dissolved in dimethylsulfoxide or by fluorenone (m.wt = 180) as a s.c. dose of 3 mmoles/rabbit. Both compounds were given every 3rd day for 2 weeks. Controls received respective amount (3 ml DMSO). After 2 weeks the rabbits were killed and the liver was excised and a 20 cm segment of the small intestine aborally from the pylorus and the intestinal mucosa was scraped off by an ampoule file. Both the liver and intestine were homogenized in a 4 x wet weight volume of 0.25 mol/l sucrose with a Potter-Elvehjem type glass homogenizer with a Teflon pestle. After homogenization the postmitochondrial supernatants were prepared by the centrifugation at 10 000 g_{av} x 15 min whereafter microsomes were separated by a centrifugation at 105 000 g_{av} x 60 min. The microsomes were resuspended in 0.25 mol/l sucrose at a concentration of 1 g tissue equivalent to 1 ml suspension.

The protein (2) and phospholipid phosphorous (3) contents were determined as described previously. The cytochrome P-450 concentration of liver microsomes was determined according to Omura and Sato (4) and NADPH cytochrome c reductase both in the hepatic and intestinal microsomes as described by Phillips and Langdon (5). The aryl hydrocarbon hydroxylase (AHH) was determined fluorometrically (6) and ethoxycoumarin O-deethylase (7) and ethoxyresorufin O-deethylase (8) as described previously. The epoxide hydrolase was measured using radio-labelled styrene oxide as a substrate (9). The glutathione S-transferase was measured in the postmicrosomal supernatant fraction (10). Student's t-test was used for statistical treatments. Means ± SEMs are shown.

RESULTS

Rabbits having either trinitrofluorenone or fluorenone weighed less than controls (Table 1). No difference was present in the liver weights or relative liver weights. The hepatic and intestinal microsomal protein contents were also unchanged due to the treatments (Table 1). Neither there were any changes in the microsomal phospholipid contents.

TABLE 1

WEIGHTS, PROTEIN AND PHOSPHOLIPID CONTENTS

The body and liver weights and tissue microsomal protein and phospholipid contents after fluorenone or trinitrofluorenone administration

Exposure	Body weight (g)	Liver weight (g)	Liver/ body wt ratio (%)	Microsomal protein (mg/g) Liver	Intestine	Phospholipids (μmol Pi/g) Liver	Intestine
Control (4)	4013±192	100±8	2.5±0.1	24.4±0.8	21.3±1.1	8.2±0.5	5.6±0.6
Trinitro-fluorenone (5)	3250±88**	85±8	2.5±0.2	24.0±0.7	21.0±1.1	8.2±0.5	5.9±1.5
Fluorenone (6)	3453±79*	92±5	2.7±0.1	24.8±0.7	22.8±0.8	8.7±0.7	6.0±1.0

*: $P < 0.05$; **: $P < 0.01$

The NADPH cytochrome c reductase activity did not change in the treated rabbit liver or intestine as compared to respective DMSO treated controls (Fig. 1). Also the hepatic cytochrome P-450 concentration was unchanged. The aryl hydrocarbon hydroxylase activity decreased in those rabbit livers treated with fluorenone while trinitrofluorenone did not cause any changes (Fig. 1). The ethoxycoumarin and ethoxyresorufin O-deethylase activities were unchanged while epoxide hydrolase activity decreased both by fluorenone and trinitrofluorenone (Fig. 2). The glutathione S-transferase activity did not change due to the administration of gluorenone or trinitrofluorenone (not illustrated).

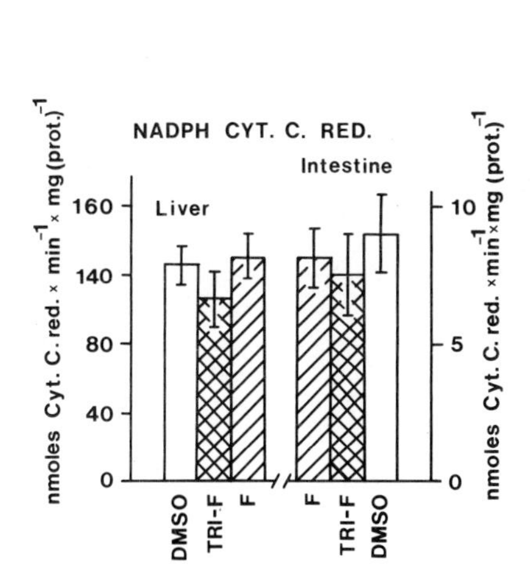

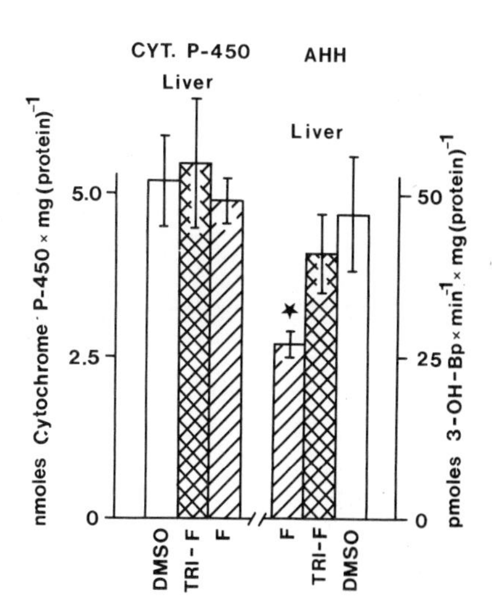

Fig. 1. The hepatic and intestinal NADPH cytochrome c reductase activities, hepatic cytochrome P-450 concentration and aryl hydrocarbon hydroxylase activity of rabbits exposed to fluorenone (F, shaded column) or trinitrofluorenone (TRI-F, cross-hatched columns) and in controls (DMSO, open columns). Means ± SEMs are given. For other explanations see Table 1.

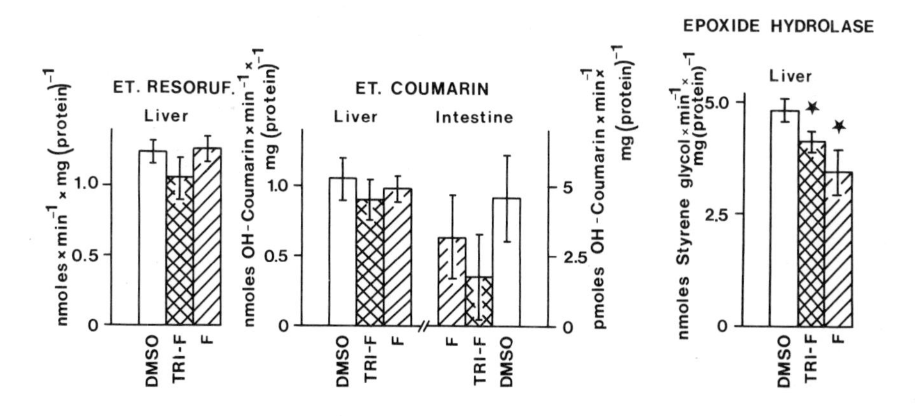

Fig. 2. The hepatic ethoxyresorufin, hepatic and intestinal ethoxycoumarin O-deethylase and hepatic epoxide hydrolase activities in the rabbits exposed to fluorenone or trinitrofluorenone for two weeks. For other explanations see Fig. 1 and Table 1.

DISCUSSION

The present data suggest that fluorenone decreases specifically aryl hydrocarbon hydroxylase activity of the monooxygenases while both fluorenone and trinitrofluorenone decrease epoxide hydrolase activity. Whether fluorenone decreases aryl hydrocarbon hydroxylase activity by attacking towards specific cytochromes or whether the metabolites of this compound bind to the same catalytic sites as those of benzo(a)pyrene, the substrate for the aryl hydrocarbon hydroxylase, remains to be solved. Both trinitrofluorenone and fluorenone decreased epoxide hydrolase activity which might suggest possible toxicity of these compounds in long term exposure. However, the toxicity data on both of the compounds tested is scanty and further studies are warrented to assess possible their safety in long-term exposure.

ACKNOWLEDGEMENTS

This study was supported by grants from NIH (ROI ES 01684) and J. Vainio Foundation (Finland).

REFERENCES

1. Huggins, C. and Yang, N.C. (1962) Science 137, 257.
2. Gornall, A.G., Bardawill, C.J. and David, M.M. (1949) J. Biol. Chem., 177, 751.
3. Bartlett, C.R. (1959) J. Biol. Chem. 234, 466.
4. Omura, T. and Sato, R. (1964) J. Biol. Chem., 239, 2379.
5. Phillips, A.H. and Langdon, R.G. (1962) J. Biol. Chem., 237, 2652.
6. Wattenberg, L.W., Leong, J.L. and Strand, P.J. (1962) Cancer Res., 22, 1120.
7. Aitio, A. (1978) Anal. Biochem., 85, 488.
8. Prough, R.A., Brake, M.D. and Mayer, R.T. (1978) in: Heischer, S. and Packer, L. (Eds.), Methods in Enzymology, vol. 3, Academic Press, New York, pp. 372-377.
9. Oesch, F., Jerina, D.M. and Daly, J. (1971) Biochim. Biophys. Acta 227, 685.
10. James, M.D., Fouts, J.R. and Bend, J. (1976) Biochem. Pharmacol. 25, 187.

Cytochrome P-450, Biochemistry, Biophysics and Environmental Implications, E. Hietanen, M. Laitinen and O. Hänninen editors

EFFECT OF EXPERIMENTAL DIABETES ON HEPATIC MONOOXYGENASES

EINO HIETANEN[1], RAINER RAURAMAA[2] AND MATTI LAITINEN[2]
[1]Department of Physiology, University of Turku, 20520 Turku 52 (Finland) and
[2]Department of Physiology, University of Kuopio, 70101 Kuopio 10 (Finland)

INTRODUCTION

Although diabetes is a common disease in man and is often involved with the use of many therapeutic drugs in addition to hypoglycemic agents rather little is known on the drug metabolism in diabetics. Salmela et al. (1) found in diabetic humans that changes in drug metabolism are not necessarily caused by the changes in the enzymes metabolizing drugs per se but rather due to the histological alterations in the liver. We compared the effects of insulin treatment and physical training on the hepatic cytochrome P-450 contents, NADPH cytochrome c reductase activity and the monooxygenase activities in the rat.

MATERIALS AND METHODS

Weanling male Wistar rats (4 weeks old) were produced diabetic by the intravenous injection of streptozotocin (Upjohn Co, Kalamazoo, MI) as a single dose of 75 mg/kg under anesthesia (2). Controls received respective amount of saline-citrate buffer. Rats were left diabetic for 3 months whereafter both diabetic and control rats were divided into three subgroups: one was trained 45 min in a treadmill at speed 16.2 m/min with a slope of 12 degrees twice a day (2), another group received insulin (zinc-protamin) as a dose of 31 units/day s.c. and a third subgroup was as a sedentary control group. The therapy or training was continued for two weeks whereafter livers were excised, centrifuged first at 10000 g x 15 min and thereafter at 100000 g x 60 min to prepare microsomes. Simultaneously blood was drawn for glucose determinations (glucose oxidase method). The microsomal protein was determined as described by Gornall et al. (3) and phospholipid phosphorous as described by Bartlett (4). The cytochrome P-450 concentration (5), NADPH cytochrome c reductase (6), aryl hydrocarbon hydroxylase (7), ethoxycoumarin O-deethylase (8) and ethoxyresorufin O-deethylase (9) activities were determined as described previously. Student's t test was used for statistical comparisons and means $\pm$ SEM are shown.

RESULTS

The blood glucose (non-fasted) values were about 3-fold in diabetic rats as compared to controls indicating the existence of diabetes (Table 1). No changes were found in the hepatic microsomal protein contents. The microsomal phospholipid contents were also equal in both diabetic and control animals without any treatment (Table 1). The insulin treatment did not change the phospholipid contents of liver microsomes from the control rats but decreased it significantly in diabetic rats (Table 1). The microsomal cytochrome P-450 concentration did not change due to diabetes or any of the treatments (Table 1). Also the NADPH cytochrome c reductase activity was unchanged in diabetic rats as compared to respective controls whether these were sedentary, trained or insulin treated (Table 1).

TABLE 1

BLOOD GLUCOSE, HEPATIC MICROSOMAL PHOSPHOLIPID AND CYTOCHROME P-450 CONTENTS AND NADPH CYTOCHROME C REDUCTASE ACTIVITY

Rats were diabetic for 3 months whereafter they had 2 weeks training or insulin treatment.

Rat Group	N	Blood Glucose (mmol/l)	Phospholipid (μmol Pi/g w. wt)		Cytochrome P-450 (nmol x mg prot.$^{-1}$)	NADPH Cyt. c Red. (nmol x min^{-1} x mg prot.$^{-1}$)
Sedentary						
Control	8	7.7 ± 1.0	9.80 ± 0.36		0.46 ± 0.02	88.1 ± 3.8
Diabetes	6	20.1 ± 2.3	10.20 ± 0.58		0.51 ± 0.02	92.9 ± 5.7
Insulin-treated						
Control	8	8.4 ± 0.3	9.00 ± 0.35	$P<0.05$	0.47 ± 0.02	84.8 ± 5.7
Diabetes	6	20.6 ± 4.8	7.50 ± 0.52		0.49 ± 0.03	87.3 ± 1.9
Trained						
Control	8	7.5 ± 0.7	9.5 ± 0.4		0.41 ± 0.04	107.0 ± 4.4
Diabetes	6	22.8 ± 2.1	8.9 ± 0.5		0.46 ± 0.03	112.1 ± 3.5

The polysubstrate monooxygenase activity was measured towards benzo(a)pyrene, ethoxyresorufin and ethoxycoumarin (Table 2). The aryl hydrocarbon hydroxylase was not changed due to diabetes. The ethoxyresorufin O-deethylase activity increased in the liver microsomes from diabetic rats while insulin treatment returned the activity to the control level. In trained non-diabetic rats the ethoxyresorufin O-deethylase activity was higher than in sedentary controls while in diabetic rats training did not change the activity (Table 2). The ethoxycoumarin O-deethylase activity was somewhat higher in diabetic rats than in controls although the difference was not statistically significant (Table 2). Insulin treatment decreased the activity in diabetics and further decrease was found in trained diabetic rats (Table 2).

TABLE 2

THE MONOOXYGENASE ACTIVITY TOWARDS VARIOUS SUBSTRATES IN LIVER MICROSOMES OF CONTROL AND DIABETIC RATS AFTER INSULIN TREATMENT OR PHYSICAL TRAINING

Rat Group	Benzo(a)pyrene[a]		Ethoxyresorufin[a]		Ethoxycoumarin[a]	
Sedentary						
Controls	0.29 ± 0.03		0.25 ± 0.03	P<0.05	0.84 ± 0.08	
Diabetics	0.22 ± 0.03		0.43 ± 0.07	P<0.05	1.09 ± 0.18	
Insulin-treated						
Controls	0.28 ± 0.01	P<0.05	0.29 ± 0.02		0.72 ± 0.09	P<0.05
Diabetics	0.28 ± 0.01		0.28 ± 0.01		0.70 ± 0.14	
Trained						
Controls	0.22 ± 0.02		0.33 ± 0.02	P<0.05	0.71 ± 0.09	
Diabetics	0.23 ± 0.02		0.41 ± 0.03		0.65 ± 0.14	

[a]nmoles x min^{-1} x mg $(protein)^{-1}$

DISCUSSION

The present study demonstrated that in liver microsomes from diabetic rats the polysubstrate monooxygenase activity is increased towards ethoxyresofurin and ethoxycoumarin but not towards benzo(a)pyrene. These changes were reversible by insulin and partly also by physical training. Previously Eacho and Weiner (10) found increased microsomal p-nitroanisole O-demethylase and aniline hydroxylase activities in diabetic rats with simultaneous increase in cytochrome P-450 contents. However, in hepatocytes they found a decrease in the metabolic rate of both substrates which changes were reversible by insulin. They also found that the conjugation of p-nitroanisole was faster in diabetic rat hepatocytes which might partly explain the discrepancy found between hepatocytes, capable also for conjugation, and microsomes (11). When they calculated the sums of conjugated and non-conjugated metabolites of p-nitroanisole also in hepatocytes the actual demethylation rate increased in diabetic rats (11). However, the enhanced glucuronidation rate was not due to the increased enzyme synthesis but due to the increased production of uridine diphosphoglucuronic acid (11). Al-Turk et al. (12) found enhance deethylation of ethoxycoumarin in the liver and intestine of both male and female rats but not in the lung. Also in this study the insulin treatment reversed the changes.

Possibly diabetes might influence on the structure of the microsomal membranes as seen in changes of membrane composition in addition to effects on the cytochrome P-450 concentration. The influence of diabetes on cytochrome P-450 seems to be rather substrate specific as no changes were found in the aryl hydrocarbon hydroxylase activity. In addition to insulin also physical training reversed partly the diabetes-induced changes in the monooxygenase activities.

ACKNOWLEDGEMENTS

This study has been supported by grants from NIH (R01 ES 01684) and J. Vainio Foundation (Finland).

REFERENCES

1. Salmela, P.I., Sotaniemi, E.A. and Pelkonen, O. (1980) Diabetes, 29, 788.
2. Rauramaa, R., Kuusela, P. and Hietanen, E. (1980) Horm. Metab. Res., 12, 591.
3. Gornall, A.G., Bardawill, C.J. and David, M.M. (1949) J. Biol. Chem., 177, 751.
4. Bartlett, C.R. (1959) J. Biol. Chem., 234, 466.
5. Omura, T. and Sato, R. (1964) J. Biol. Chem., 239, 2379.
6. Phillips, A.H. and Langdon, R.G. (1962) J. Biol. Chem., 237, 2652.
7. Wattenberg, L.W., Leong, J.L. and Strand, P.J. (1962) Cancer Res., 22, 1120.
8. Aitio, A. (1978) Anal. Biochem., 85, 488.
9. Prough, R.A., Brake, M.D. and Mayer, R.T. (1978) in: Heischer, S. and Packer, L. (Eds.), Methods in Enzymology, vol. 3, Academic Press, New York, pp. 372-377.
10. Eacho, P.I. and Weiner, M. (1980) Drug Metab. Disp., 8, 385.
11. Eacho, P.I., Sweeny, D. and Weiner, M. (1981) J. Pharmacol. Exptl. Ther., 218, 34.
12. Al-Turk, W.A., Stohs, S.J. and Roche, E.B. (1980) Drug Metab. Disp., 8, 44.

Cytochrome P-450, Biochemistry, Biophysics and Environmental Implications, E. Hietanen, M. Laitinen and O. Hänninen editors

METABOLISM OF POLYCYCLIC AROMATIC HYDROCARBONS BY CULTURED RAT AND HUMAN ADRENAL CELLS

EINAR HALLBERG[1], ANDERS RANE[2] AND JAN RYDSTRÖM[1]
[1]Dept. of Biochemistry, University of Stockholm, Stockholm, Sweden and [2]Dept. of Clinical Pharmacology at the Karolinska Inst., Huddinge Hospital, Huddinge, Sweden.

INTRODUCTION

Some methylated polycyclic aromatic hydrocarbons (PAH) are far more potent carcinogens than their parent compounds (1). For example, 7,12-dimethylbenz(a)anthracene (DMBA) is highly carcinogenic (2) and also causes selective necrosis in the rat adrenal cortex (2) and testis (2), whereas benz(a)anthracene is essentially inactive in these respects (2).

Generation of necrosis in the two inner zones of the adrenal cortex requires the presence of adrenocorticotrophic hormone (ACTH) (2), a pituitary peptide hormone, which plays a key role in the regulation of adrenal function. The adrenal cortex has a high content of cytochrome P-450 (3) and comprises a number of different steroid hydroxylases (3) as well as aryl hydrocarbon hydroxylase (AHH) (4). Although both types of hydroxylases are regulated by ACTH (cf.5), the relationship betwen them remains to be elucidated (but see M. Bengtsson et al in this volume).

In order to study the influence of ACTH on adrenal metabolism and toxicity of DMBA and the possible role of hepatic metabolism of DMBA in this context, primary cultures of rat adrenal cells (RAC) and human fetal adrenal cells (HFAC) were used as model systems.

MATERIALS AND METHODS

Isolation procedure. Isolation of cells from human fetal or rat adrenals was carried out essentially as described previously (6). The experiments were performed with cells either in suspension or in primary culture.

Assays. ACTH, dissolved and stored as described (6), was added directly to the medium. Corticosterone in the case of RAC or cortisol in the case of HFAC secreted into the medium was then estimated fluorimetrically (6).

Conversion of HPLC-purified (^{3}H)-DMBA or (^{14}C)-DMBA and analysis of metabolites, and covalent binding of metabolites to protein was assayed as described previously (7).

RESULTS

RAC and HFAC were kept in primary cultures for up to two weeks. Maintenance of in vivo function was indicated by a several-fold increase in corticosteroid formation in response to ACTH.

DMBA was metabolized by RAC at a rate of 3 - 30 pmol/min$\cdot 10^{6}$ cells with a K_m of 0.5 μM (fig 1). The specific activity at a concentration of DMBA of 0.5 μM or 10 μM was not significantly changed by ACTH (not shown). The presence of effective conjugation systems is indicated by the fact that 20 - 40 % of the products remained in the water phase after ethyl acetate extraction.

HPLC analysis of the ethyl acetate phase is shown in fig. 2. The two major products were identified as a phenol and a dihydrodiol,

TABLE 1

EFFECT OF VARIOUS INHIBITORS ON DMBA METABOLISM, AND COVALENT BINDING OF METABOLITES TO PROTEIN IN RAC

RAC were incubated in duplicates with 10 μM DMBA. Ellipticine was added at a concentration of 10 μM, all other inhibitors were added at a concentration of 100 μM.

Conditions	AHH activity (% of control)	Hydrophilic products (% of control)	Covalent binding (% of control)
No addition	100[a]	100[b]	100[c]
+ SU-9055	12	24	8
+ Ellipticine	10	23	0
+ α-naftoflavone	4	18	0
+ β-estradiol	61	78	69
+ Progesterone	25	31	18
+ Androstenedione	37	52	14
+ Testosterone	62	129	69

[a] 33.0 pmol/min$\cdot 10^{6}$cells ; [b] 7.1 pmol/min$\cdot 10^{6}$cells ;
[c] 1.9 pmol/min$\cdot 10^{6}$cells

probably 3-OH-DMBA and 8,9-diOH-DMBA, respectively. No change in metabolite pattern could be observed when RAC were incubated in the presence of ACTH (not shown).

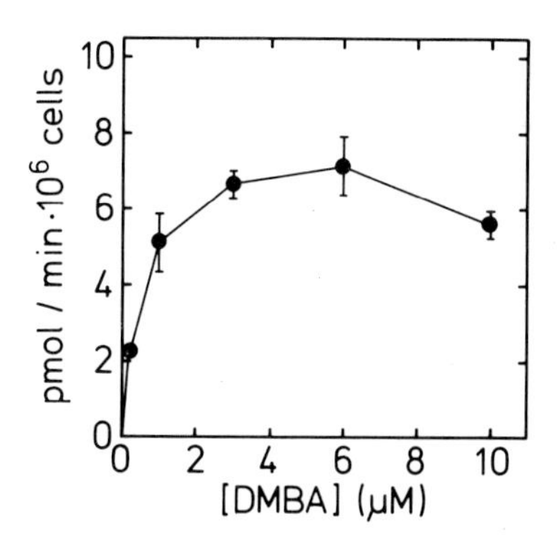

Fig. 1. AHH activity in RAC as a function of the concentration of DMBA. Freshly isolated RAC were incubated for 60 min with DMBA dissolved in DMSO (final concentration 2 %). The AHH activity was estimated by the organic/aqueous distribution method described in (7).

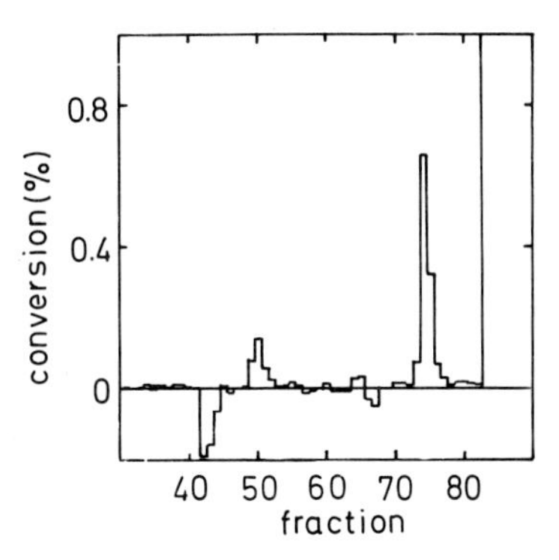

Fig. 2. HPLC pattern of DMBA metabolites obtained with RAC. The HPLC pattern was corrected for RAC-independent conversion. The incubation conditions were as described in fig 1. The incubations were extracted and analyzed on a reversed phase column using a linear methanol gradient as described in (7).

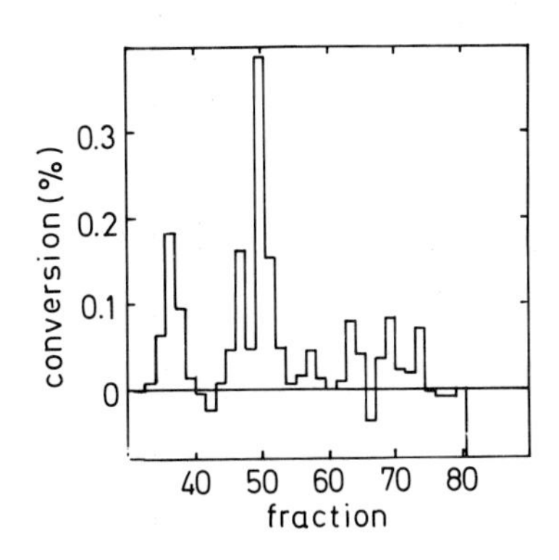

Fig. 3. HPLC pattern of DMBA metabolites obtained with HFAC. 24 hours old primary cultures of HFAC were incubated for 20 hours with DMBA. Incubation conditions were as in fig 1 and assay procedure as described in fig 2.

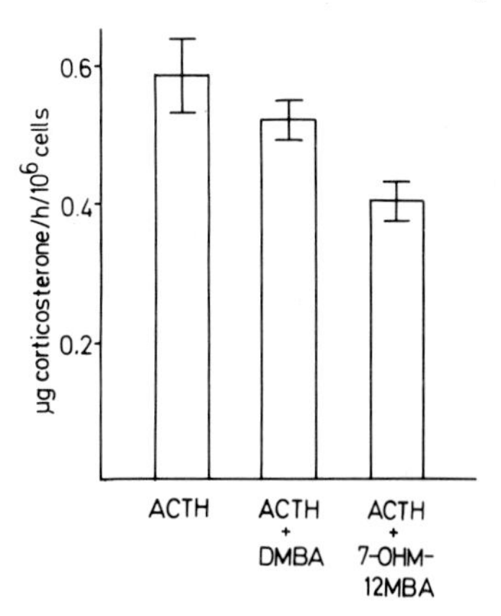

Fig. 4. ACTH-induced corticosterone synthesis. Primary cultures of RAC were incubated for 24 hours in the presence and absence of 20 μM DMBA or 7-OHM-12-MBA dissolved in DMSO (final concentration 2%). Incubations were made in quadruplicates.

Using HFAC the AHH activity was of the order of 1.5 pmol/min·10^6 cells. Microsomes isolated from human fetal adrenals showed a maximum conversion rate of 3.3 pmol/min·mg with a K_m of 17 μM. The metabolite pattern was somewhat different as compared to those obtained with RAC or rat adrenal microsomes (fig. 3).

The AHH activity in RAC was effectively inhibited by various nonsteroid inhibitors, i.e., SU-9055, ellipticine and α-naftoflavone and to a lesser extent by the endogeneous steroids estradiol, progesterone, androstenedione and testosterone (Table 1). The covalent binding of metabolites to protein followed the AHH activity as did the formation of hydrophilic products, i.e. predominantly conjugates, except in the case where estradiol and testosterone were added, which gave an increase in the conversion.

The toxicity of DMBA and 7-OHM-12-MBA, the latter being the major hepatic metabolite (8), was estimated by the change in ACTH-induced corticosterone biosynthesis of RAC (Fig. 4). Concentrations of DMBA up to 20 μM were essentially without effect, whereas 20 μM 7-OHM--12-MBA decreased the corticosterone output significantly. These results indicate that, in vivo, hepatic metabolism of DMBA is required for adrenal toxicity of DMBA.

ACKNOWLEDGEMENT

This work was supported by the Swedish Cancer Society, the Swedish Medical Research Council and the Expressen Prenatal Research Foundation.

REFERENCES

1. Hecht, S.S., Loy, M., Mazzarese, R. and Hoffman, D. in Gelboin, H.V. and Ts´o, O.P. (1978) Polycyclic Hydrocarbons and Cancer 1, Academic Press, New York, p. 119.
2. Huggins, C.B. (1979) Experimental Leukemia and Mammary Cancer. The University of Chicago Press, Chicago and London.
3. Gunsalus, I.C., Pederson, T.C. and Sligar, S.G. (1979) Annu. Rev. Biochem. 44, 377.
4. Dao, T.L. and Yogo, H.(1964) Proc. Soc. Exprtl. Biol. Med. 116, 1048,
5. Guenthner, T.M., Nebert, D.W. and Menard, R.H. (1979) Mol. Pharmacol. 15, 719.
6. Hallberg, E. and Rydström, F. (1981) Acta Chem. Scand. B 35,145.
7. Montelius, J., Papadopoulos, D., Bengtsson, M. and Rydström, J. (1982) Cancer Res. 42, 1479.
8. Yang, S.K. and Dover, W.V. (1979) Proc. Natl. Acad. Sci. USA 72, 2601.

Cytochrome P-450, Biochemistry, Biophysics
and Environmental Implications, E. Hietanen,
M. Laitinen and O. Hänninen editors

METABOLISM OF EXOGENOUS AND ENDOGENOUS COMPOUNDS IN THE RAT VENTRAL PROSTATE CHARACTERIZATION OF THE ENZYMES INVOLVED USING ANTIBODIES

TAPIO HAAPARANTA[1,2], JAMES HALPERT[1], HANS GLAUMANN[2] AND JAN-ÅKE GUSTAFSSON[1]
[1]Department of Medical Nutrition and Department of Pathology, Karolinska Institute, Huddinge University Hospital, S-141 86 Huddinge, Sweden

INTRODUCTION

Prostatic cancer is a common malignant disease among men in Europe and the United States. Both extrinsic (environmental and viral) and intrinsic (genetic and hormonal) agents may play a role in the etiology of the disease. A correlation between cadmium exposure and increased incidence of prostatic cancer has been suggested on the basis of epidemiological data. Epidemiologic studies have also demonstrated increased cancer incidence among rubber workers, indicating that chemical factors may influence the development of prostatic cancer.

Given this background the ability of the prostate gland to metabolize xenobiotics becomes of interest. It has recently been shown that drug metabolizing enzymes in the microsomal fraction of the rat prostate are very sensitive to inducers of cytochrome P-450. These enzymes may be active in the production of reactive metabolites of carcinogens in the prostate itself.

MATERIALS AND METHODS

Male Sprague-Dawley rats weighing 400 g were used throughout the study. The B_2-fractions of P-450 from PB and BNF treated animals as well as NADPH-cytochrome P-450 reductase were purified from rat liver essentially as described by Guengerich and Martin (1). Antibodies against the purified proteins were raised in rabbits. The immunological contents of the enzymes were determined by incubation of nitrocellulose blots of SDS-polyacrylamide gels with the primary antibody and subsequent incubation of the blot with ^{125}I-protein A (2). After autoradiography the labeled protein bands were cut out and counted for radioactivity. Quantitation was performed by running varying amounts of the purified antigens in parallel with the samples on the same immunoblot. Standard curves were constructed and the radioactivity was proportional to the amount of antigen.

Abbreviations used are: TCDD, 2,3,7,8-tetrachlorodibenzo-p-dioxin; AHH, aryl hydrocarbon hydroxylase; BP, benzo(a)pyrene; PB, phenobarbital; BNF, β-naphthoflavone; P-450, cytochrome P-450; P-450 BNF-B_2 β-naphthoflavone induced form of P-450; P-450 PB-B_2, phenobarbital induced form of P-450.

RESULTS AND DISCUSSION

All enzyme activities assayed were much lower in prostatic microsomes as compared to liver microsomes (Table 1). NADPH-cytochrome c reductase activity in prostatic microsomes was not significantly changed by BNF or PB treatment whereas two-fold induction was observed in liver microsomes. These enzyme activities were well correlated with the immunological content of rat liver NADPH-cytochrome P-450 reductase. The good correlation between the enzyme activity and amount of protein in the prostate and the liver suggests that the prostatic P-450 reductase has the same turnover as the corresponding liver enzyme. In addition, as could be determined from the immunoblots, the prostatic P-450 reductase had the same apparent molecular weight as the liver form. The prostatic NADPH-cytochrome c reductase activity was inhibited up to 65% by antibodies against liver P-450 reductase. In a similar experiment, the same antibodies inhibited liver microsomal activity up to 75%.

As regards the 7-ethoxyresorufin O-deethylase and AHH activities, the degree of induction after BNF treatment is much higher in the prostate whereas phenobarbital had almost no effect in the prostate. Prostatic aminopyrine N-demethylase activity was only slightly induced after BNF or PB treatment. When rats were treated with TCDD, the same type of induction and substrate specificity was seen as after BNF treatment. Ellipticine (0.01 mM), SU-9055 (0.5 mM), β-naphthoflavone (0.5 mM), α-naphthoflavone (0.5 mM), metyrapone (0.5 mM), and aminopyrine (0.5 mM)

TABLE I

EFFECT OF β-NAPHTHOFLAVONE (BNF) AND PHENOBARBITAL (PB) TREATMENT ON THE DRUG METABOLIZING ENZYMES IN PROSTATIC AND LIVER MICROSOMES

	Prostatic microsomes			Liver microsomes		
	Control	BNF	PB	Control	BNF	PB
NADPH-cyt P-450 reductase[a] (immunological content)	0.02	0.02	0.02	0.2	0.2	0.3
NADPH-cyt c reductase[b] (enzyme activity)	0.026	0.038	0.023	0.27	0.28	0.42
Cytochrome P-450[a] (spectral content)	0.02	0.06	n.m.	0.69	0.87	1.7
Cyt B_2-BNF P-450[a] (immunological content)	<0.0004	0.05	<0.0004	0.05	1.8	0.16
7-ethoxyresorufin O-deethylase[c]	0.04	300	0.11	94	8900	205
AHH[c]	1.4	290	n.m.	54	3400	n.m.

[a]nmol/mg [b]nmol/min/mg [c]pmol/min/mg n.m. not measured

showed the same pattern of inhibition of AHH and 7-ethoxyresorufin O-deethylase activities in both BNF and TCDD induced microsomes. The inhibition was for ellipticine 96%; SU-9055 76%; β-naphthoflavone 62%; α-naphthoflavone 96%; metyrapone 12% and for aminopyrine 27%.

Fig. 1 depicts the effect of different antibodies on 7-ethoxyresorufin O-deethylase activity in prostatic microsomes. As can be seen, the P-450 BNF-B_2 antibodies inhibited the activity almost completely whereas anti-P-450 reductase antibodies inhibited to approximately half the control activity; anti-P-450 PB-B_2 IgG and preimmune IgG had no inhibitory effect. It was also of interest to investigate if prostatic microsomes could hydroxylate steroids. Of 11 physiological steroids only 5α-androstane-3α,17β-diol was efficiently converted to polar metabolites. The identities of these metabolites have not yet been determined. The reaction was inhibited by metyrapone, α-naphthoflavone, β-naphthoflavone and SU-9055 suggesting that this reaction is also catalyzed by cytochrome P-450. When comparing the specific immunological content of P-450 BNF-B_2 in the liver and the prostate it is evident that the induction of the P-450 dependent monooxygenase present in the prostate can almost exclusively be accounted for by the BNF inducible form. The P-450 BNF-B_2 antibodies against the liver enzymes cross-react strongly with the prostatic enzyme suggesting great similarity between the liver and the prostatic enzymes. Furthermore, as judged from the immunoblots, the apparent molecular weights of these enzymes were similar.

The amount of the BNF inducible P-450 form present in the untreated prostatic microsomes was below the detection limit of approximately 0.4 pmol/mg microsomal protein. However, after treatment of the rats with BNF, this form represented about 0.3% of the microsomal protein or 50 pmol/mg. Spectral measurement of the P-450 content in prostatic microsomes gave a specific content of 61 pmol/mg. This

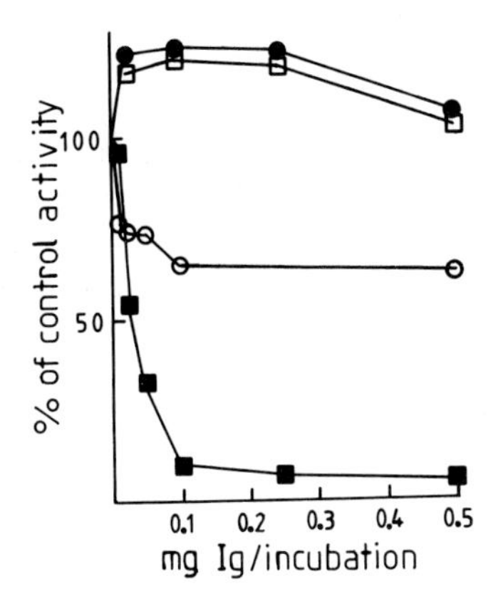

Fig. 1. Effect of antibodies against P-450 BNF-B_2, P-450 PB-B_2, P-450 reductase and preimmune IgG on 7-ethoxyresorufin O-deethylase activity in prostatic microsomes. The experiment was performed essentially as described by Guengerich (3).

■—■, anti P-450 BNF-B_2.

□—□, anti P-450 PB-B_2.

○—○, anti P-450 reductase

●—●, preimmune IgG.

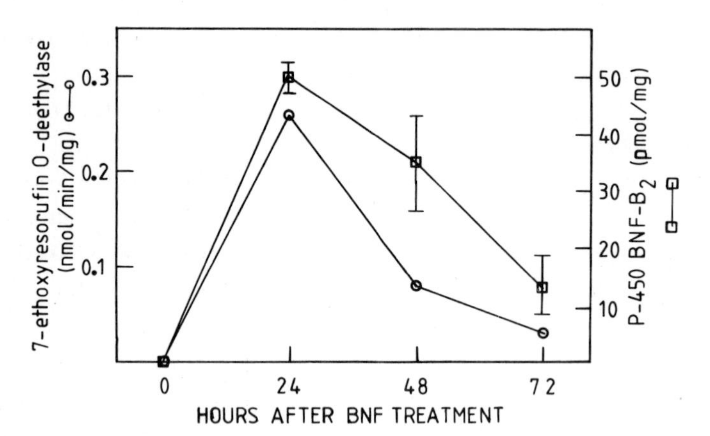

Fig. 2. Time-course of 7-ethoxyresorufin O-deethylase activity and the specific immunological content of P-450 BNF-B_2 after a single i.p. injection of BNF (8 mg/100 g).

value could however be an unprecise estimation of the actual content, because the prostatic microsomal fraction is not as pure as the liver preparation. Other cytochromes present might greatly influence the spectral determinations. This is especially true for the P-450 content of control microsomes which is close to the sensitivity limit of the Aminco-Chance spectrophotometer.

Fig. 2 demonstrates the time-response curve of 7-ethoxyresorufin activity and the specific content of P-450 BNF-B_2 in prostatic microsomes after a single i.p. injection of BNF (8 mg/100 g). The rate of induction of immunoreactive P-450 BNF-B_2 clearly follows the rate of the induction of the enzyme activity. In uninduced microsomes the amount of immunoreactive protein is below the detection level.

In conclusion this paper reports the presence of a P-450 form present after BNF induction in the rat ventral prostate which crossreacts with antibodies against the rat liver P-450 BNF-B_2. The induction of 7-ethoxyresorufin-O-deethylase activity in the prostate after induction can be attributed to this protein. Antibodies to rat liver P-450 reductase crossreact with prostatic P-450 reductase and the apparent molecular weights of these proteins are similar.

REFERENCES

1. Guengerich, F.P. and Martin, M.V. (1980) Arch. Biochem. Biophys. 205, 365.
2. Towbin, H., Staehelin, T. and Gordon, J. (1979) Proc. Nat. Acad. Sci. 76, 4350.
3. Guengerich, F.P., Wang, P. and Mason, P.S. (1981) Biochemistry 20, 2379.

ACKNOWLEDGEMENTS

This research was supported by grants from the Swedish Cancer Society and the Swedish Council for Planning and Coordination of Research.

Cytochrome P-450, Biochemistry, Biophysics
and Environmental Implications, E. Hietanen,
M. Laitinen and O. Hänninen editors

REDOX CYCLING OF GLUTATHIONE IN ISOLATED HEPATOCYTES

Lena Eklöw, Peter Moldéus and Sten Orrenius
Department of Forensic Medicine, Karolinska Institutet, Box 60400,
S-104 01 Stockholm (Sweden)

INTRODUCTION

Several drug substrates, including ethylmorphine and aminopyrine, stimulate H_2O_2 formation due to autooxidation of cytochrome P-450 in hepatocytes from phenobarbital-treated rats. A comparative study of H_2O_2 metabolism in hepatocytes indicates that catalase is predominant in the metabolism of H_2O_2 generated within the peroxisomes, whereas glutathione (GSH) peroxidase plays a major role in the decomposition of H_2O_2 formed in the endoplasmic reticulum (1). The metabolism of H_2O_2 and various organic hydroperoxides by GSH peroxidase is associated with glutathione oxidation. The glutathione disulfide (GSSG) formed is then rereduced by GSSG reductase, and GSH is formed at the expense of NADPH. However, an efflux of GSSG, leading to GSH depletion, is often observed as a consequence of stimulated GSH peroxidase activity. To study this further, we have used isolated hepatocytes and an inhibitor of GSSG reductase, BCNU (1,3-bis(2-chloro-ethyl)-1-nitrosourea) (2). The experimental conditions where chosen such that cells with normal GSH level, but inhibited GSSG reductase, were obtained (3). The cells were then challenged with the oxidizing agents diamide and t-butylhydroperoxide and the levels of GSH and GSSG were monitored.

METHODS

Hepatocytes were isolated and incubated as in (4). GSH and GSSG were measured by the fluorometric method of Hissin and Hilf (5) and $NADP^+$ and NADPH concentrations were assayed by the method described by Klingenberg (6). Glutathione reductase activity was determined by monitoring the oxidation of NADPH at 340 nm.

To inhibit GSSG reductase, the hepatocytes were incubated with 50 μM BCNU in a modified Krebs-Henseleit buffer, pH 7.4, supplemented with 25 mM Hepes and an amino acid mixture (7). After 30 minutes, the cells were centrifuged at 500 rpm, resuspended in an amino acid-containing medium and incubated for additional two hours, allowing the GSH level, which was decreased by the BCNU treatment, to rise to normal. Before the experiments were started, the incubation medium was changed to modified Krebs-Henseleit buffer, supplemented with 25 mM Hepes. The inhibitory effect of BCNU was maximal at 50 µM, but never reached more than approximately 90% inhibition.

RESULTS AND COMMENTS

During the incubation of hepatocytes with inhibited GSSG reductase, in absence of oxidizing agents, there was a slow decrease in cellular GSH level and an increase in GSSG concentration in the medium. These changes were, however, not significantly different from those seen with hepatocytes with fully active GSSG reductase. Moreover, the NADPH/$NADP^+$ ratio remained unchanged in both control and BCNU-treated cells during the entire incubation period.

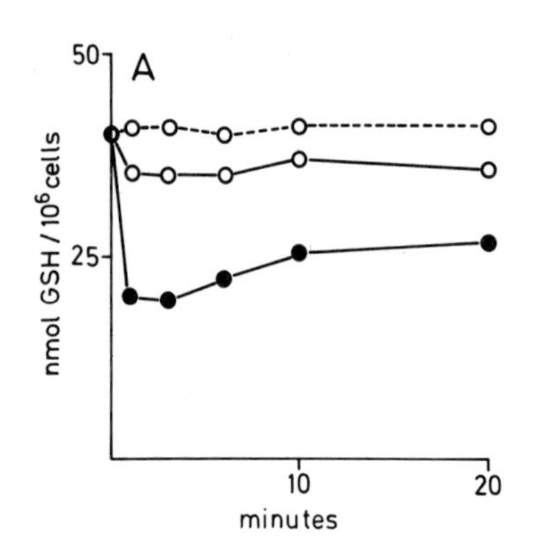

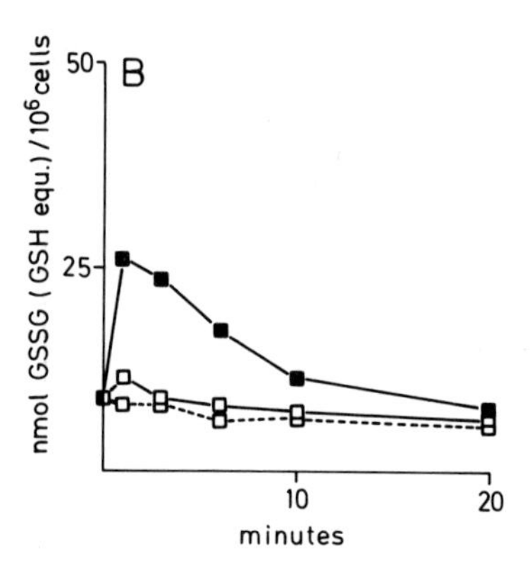

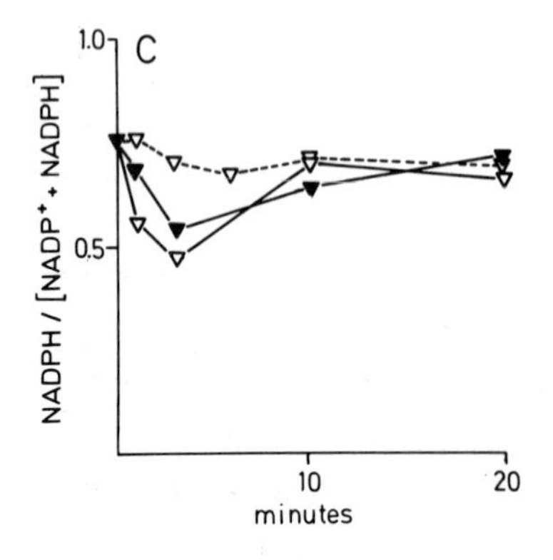

Fig. 1. Effect of 0.2 mM Diamide on the intracellular GSH and GSSG levels and on the NADPH/[$NADP^+$+NADPH] ratio in isolated hepatocytes with inhibited or uninhibited GSSG reductase: (A) GSH level; (B) GSSG level; (C) NADPH/[$NADP^+$+NADPH] ratio; (o---o, □---□, ▽---▽) control; (o——o, □——□, ▽——▽) 0.2 mM Diamide; ●——●, ■——■, ▼——▼) 0.2 mM Diamide and inhibited GSSG reductase.

Diamide, an agent that oxidizes GSH to GSSG (8), was then added to the hepatocytes. When GSSG reductase was fully active, nearly all of the formed GSSG was reduced back to GSH (Fig. 1A); this occurred so rapidly that it was not measurable in this type of experiment indicating the high capacity of noninhibited GSSG reductase. However, hepatocytes with only 5% remaining GSSG reductase activity due to BCNU pretreatment lost about 10 nmol more GSH than the cells with fully active reductase, leading to an accumulation of GSSG in the BCNU-treated cells (Fig. 1B). Diamide also caused a 30% decrease in the NADPH/NADP$^+$ ratio, which was less pronounced when GSSG reductase was inhibited (Fig. 1C). In both cases, the NADPH/NADP$^+$ ratio returned to normal level after 10 minutes.

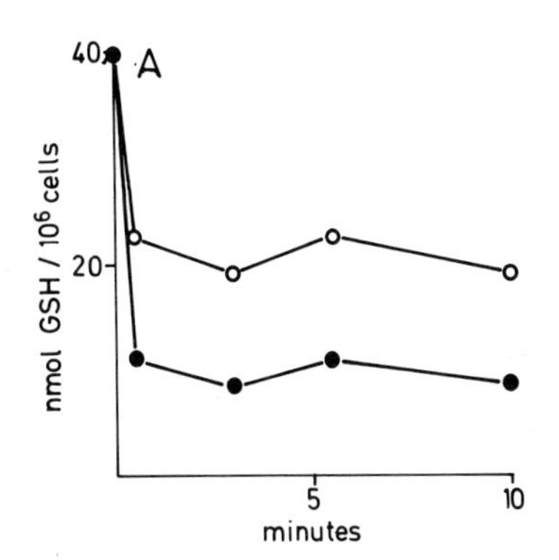

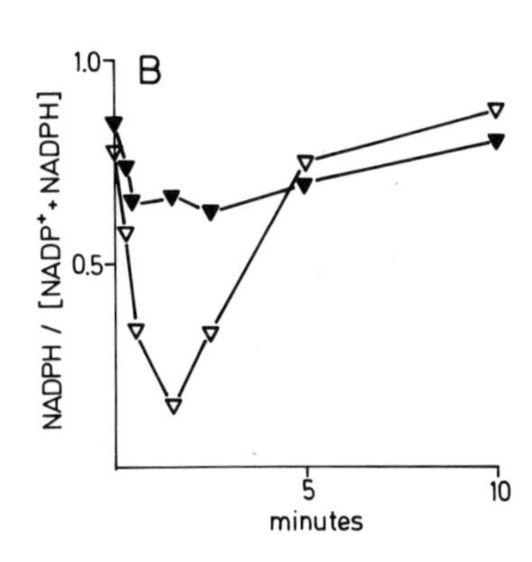

Fig. 2. Effect of 0.1 mM t-butylhydroperoxide metabolism on the GSH level and the NADPH/[NADP$^+$+NADPH] ratio with inhibited or uninhibited GSSG reductase: (A) GSH level; (B) NADPH/[NADP$^+$+NADPH] level; (o——o, ▽——▽) 0.1 mM t-butylhydroperoxide; (●——●, ▼——▼) 0.1 mM t-butylhydroperoxide and inhibited GSSG reductase.

The hepatocytes were also incubated with t-butylhydroperoxide, which is reduced by GSH peroxidase at the expense of GSH, and caused a dramatic drop in intracellular GSH (Fig. 2A). As expected, the GSH loss was more pronounced when GSSG reductase was inhibited, but under no conditions did the GSH level return to normal, since a fraction of the GSSG formed was excreted from the cells and not rereduced. Further, addition of t-butylhydroperoxide (0.1 mM) to hepatocytes with inhibited GSSG reductase caused a more pronounced decrease in the NADPH/$NADP^+$ ratio than diamide (0.2 mM) (Fig. 2B); this returned to normal after five minutes. These results indicate that GSSG reductase has a very high capacity to rereduce the GSSG formed in the presence of either diamide or t-butylhydroperoxide, and that a lowered NADPH/$NADP^+$ ratio can be rate-limiting for GSSG reduction.

However, an enhanced rate of GSSG release by isolated hepatocytes can also be seen when there are no apparent changes in the cellular NADPH/$NADP^+$ ratio, e.g. during incubation of hepatocytes with ethylmorphine (4). Moreover, with this substrate, inhibition of GSSG reductase does not affect either the decrease in GSH level or GSSG release. This finding indicates that the GSSG formed in hepatocytes may not always be available for the GSSG reductase, a possibility currently under investigation.

ACKNOWLEDGEMENT

This study was supported by a grant (03X-2471) from the Swedish Research Council.

REFERENCES

1. Jones, D.P., Eklöw. L., Thor, H. and Orrenius, S. (1981) Arch. Biochem. Biophys. 210, 505-516
2. Babson, J.R., Abell, N.S. and Reed, D.J. (1981) Biochem. Pharmacol. 30, 2299-2304
3. Eklöw, L. and Moldéus, P. (1982) in manuscript
4. Eklöw, L., Thor, H. and Orrenius, S. (1981) FEBS Letters, 127, 125-128
5. Hissin, P.J. and Hilf, R. (1976) Anal. Biochem. 74, 214-226
6. Klingenberg, M. (1970) Methoden der Enzymatischen Analyze (Bergmeyer, H.U.. ed), pp 1975, Verlag Chemie, Weinheim
7. Waymouth, C. and Jackson, R.B. (1959) J. Nat. Cancer Inst. 22, 1003-1017
8. Kosower, E.M., Correa, W., Kinon, B.J. and Kosower, N.S.(1972) Biochim. Biophys. Acta, 264, 39-44

Cytochrome P-450, Biochemistry, Biophysics
and Environmental Implications, E. Hietanen,
M. Laitinen and O. Hänninen editors

RESPONSE OF MICROSOMAL PEPTIDES IN *DENDROCTONUS TEREBRANS* AND RAT LIVER TO ALPHA-PINENE AND OTHER INDUCERS

ROBERT A. WHITE, JR. AND MOISES AGOSIN
Department of Zoology, University of Georgia, Athens, Georgia 30602

INTRODUCTION

Electrophoretic analysis of peptides has been used to evaluate mixed function oxidase differences between animals treated with various inducers (1,2). The characteristic peptide patterns may reflect the capacity to preferentially metabolize certain types of substrates and, therefore, could be useful indicators of the presence of similar enzyme functions in various zoological taxa or experimental groups.

We have previously demonstrated microsomal electrophoretic peptide differences in different insect strains and between treatment groups in the same strain (1). In the present report we demonstrate microsomal electrophoretic differences between two ecologically-related beetle species (*Dendroctonus terebrans* and *D. frontalis*) and between rats treated with alpha-pinene and various other inducers. Additionally, *D. terebrans* microsomes showed a 50-fold higher metabolic activity towards alpha-pinene than did either control or treated rat liver microsomes. Alpha-pinene-treated rat microsomes had greater metabolic reactivity toward alpha-pinene than the rat microsomes from control or the other induced groups.

METHODS

Beetles were obtained from naturally-infested *Pinus* trees and utilized within 4-24 hours. Adult *D. terebrans* were treated with alpha-pinene vapors as described before (1). Rats were treated as previously reported (3). All SDS-PAGE gels (1) and incubations (3) were carried out using microsomes prepared as before (3).

RESULTS AND DISCUSSION

The microsomal electrophoretic pattern is quite similar in the beetle (Fig. 1A, 1B) species with the major peptide at 47,000 daltons and primary differences in the 54-76,000 dalton range, somewhat similar to alpha-pinene-induced housefly peptides (1).

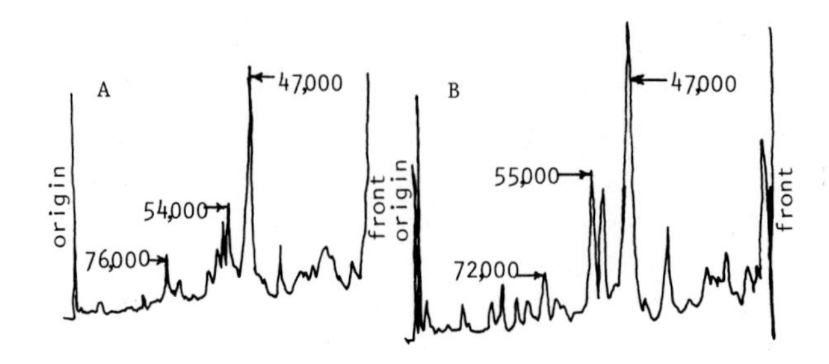

Fig. 1. SDS-PAGE scans of (A) *Dendroctonus terebrans* and (B) *D. frontalis*.

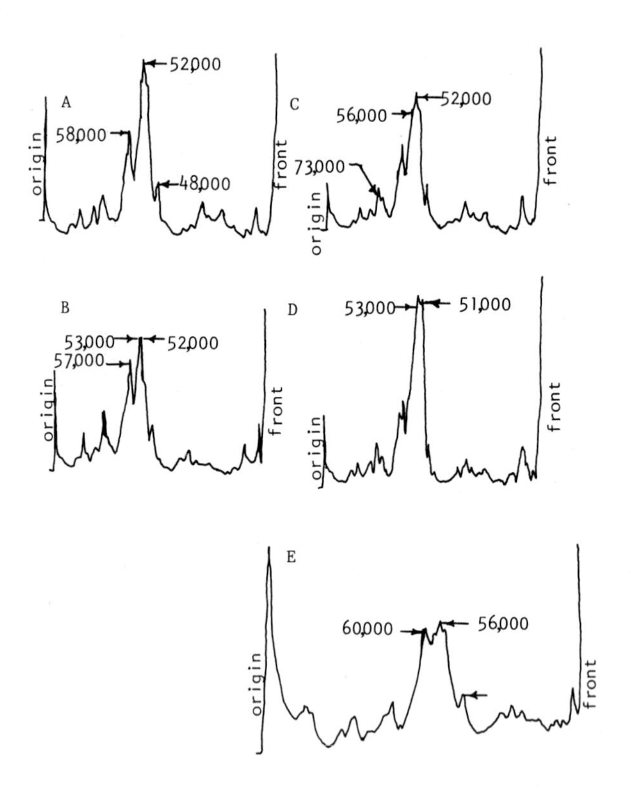

Fig. 2. SDS-PAGE scans of rat liver microsomes from (A) nontreated controls, (B) phenobarbital, (C) beta-naphthoflavone, (D) alpha-pinene, and (E) 3-methyl-cholanthrene-treated rats.

Rat microsomal peptide patterns (Fig. 2A-2E) were markedly different from the beetle patterns. Nontreated controls (A) had a major peptide at 52,000 with large secondary ones at 48,000 and 58,000 daltons; phenobarbital (B) induced peptides between 44-78,000 daltons, especially 57-58,000; beta-naphthoflavone (C) induced a 56,000 shoulder and increases between 70-73,000; alpha-pinene (D) predominantly induced peptides in the 50-54,000 dalton range; and 3-methylcholanthrene (E) induced peaks at 48,000; 55-64,000 and 75,000 daltons.

In spite of increases in levels of peptides presumably corresponding to cytochrome P-450 in rat liver treated with the above inducers, only alpha-pinene and beta-naphthoflavone induction resulted in an increase in catalytic activity towards alpha-pinene. On the other hand, microsomes from *D. terebrans* were at least 50-fold more active towards alpha-pinene than any of the rat liver microsomes. It is suggested that these differences are due to the presence of highly specific cytochrome P-450(s) in *Dendroctonus* which possibly are fully induced through the ingestion of terpene-containing foodstuffs by these phytophagous insects.

REFERENCES

1. Stanton, R.H., Plapp, F.W. Jr., White, R.A. and Agosin, M. (1978) Comp. Biochem. Physiol. 61B, 297-305.

2. Atlas, S.A., Boobis, A.R., Felton, J.S., Thorgeirsson, S.S. and Nebert, D.W. (1977) J. Biol. Chem. 252, 4712-4721.

3. White, R.A., Franklin, R.T. and Agosin, M. (1979) Pestic. Biochem. Physiol. 10, 233-242.

Cytochrome P-450, Biochemistry, Biophysics
and Environmental Implications, E. Hietanen,
M. Laitinen and O. Hänninen editors

METABOLISM OF CHEMICAL CARCINOGENS IN NUDE MOUSE FIBROBLAST CULTURES AND ITS EFFECT ON VIRAL TRANSFORMATION

MARIITTA LAAKSONEN[1], RAUNO MÄNTYJÄRVI[1] AND OSMO HÄNNINEN[2]
[1]Department of Clinical Microbiology and [2]Department of Physiology, University of Kuopio, 70101 Kuopio 10, Finland

INTRODUCTION

Virus transformation assay in mammalian cells provides a system in which the contribution of a chemical, as well as that of the virus, to the transformation process can be assessed. One of the factors having an effect on this kind of test is the metabolic activation of chemicals in the target cells. The levels of metabolizing enzymes vary *e.g.* between mouse strains. Earlier studies in our laboratory have shown that the skin of NMRI *nu/nu* (nude) mouse contains carcinogen-metabolizing enzymes (1,2). Fibroblast cultures prepared from nude mouse skin have also proved sensitive for transformation by carcinogenic chemicals (2). Furthermore, the tumorigenicity of morphologically transformed cells should be easy to test by transplantation in the parental nude mouse.

In this work we have studied the activity of the cytochrome P-450 -linked microsomal enzyme aryl hydrocarbon hydroxylase (E.C. 1.14.14.2) in secondary cultures of nude mouse skin fibroblasts. This communication also presents results of adapting these cells for testing carcinogenicity of chemicals by using 3-methylcholanthrene as a reference carcinogen and by using focus formation by a DNA tumor virus, SV_{40} as the test parameter.

MATERIALS AND METHODS

Animals. Homozygous normal (+/+) and homozygous *nu/nu* NMRI nude mice were obtained from Dr. O. Mäkelä, University of Helsinki. *Nu*/+ heterozygous mice were used for inter-crosses or for mating with *nu/nu* males to produce the *nu/nu* mice used in the experiments. Mice were produced in the Laboratory Animal Center, University of Kuopio.

Cell cultures. Primary fibroblast cultures were prepared by trypsinization from minced skin preparations of newborn *nu/nu* mouse as described by Laaksonen *et al.* (2).

BHK 21 (baby Syrian hamster kidney), VERO and CV-1 cell lines were purchased from Flow Laboratories.

Enzyme assay. Aryl hydrocarbon hydroxylase (AHH; E.C. 1.14.14.2.) activity was measured from secondary nude mouse fibroblast cultures as described by Nebert and Gelboin (3). Cell protein was determined by the method of Lowry *et al.* (4) using bovine serum albumin as a standard.

Virus. SV_{40} was replicated in CV-1 cultures. Virus released from infected cells was partially purified by pelleting through 20 % (w/v) sucrose solution followed by a density gradient centrifugation in CsCl. Purified virus was stored at -80°C. Virus concentration of the preparation used in the present experiments was 10^{11} plaque forming units (PFU)/ml.

Transformation assays. Secondary fibroblast cultures were prepared by plating trypsinized cells in plastic dishes. After 24 hours' cultivation the cells were infected with SV_{40} for 2 h at a multiplicity of 10^{5} PFU/cell, washed with phosphate-buffered saline, pH 7.0, and refed with complete medium. After another 24 hours' cultivation period they were treated with a chemical for three days, and after that they were fed weekly for 25 days. At the end of the assay the cells were fixed in methanol (Merck AG, Darmstadt, GFR), stained with Giemsa (Merck), and the transformed foci were counted.

The proportion of cells surviving the treatment was determined by counting colonies in dishes seeded with 10^{3} cells which were treated as in the experiments and stained five days after the treatment. The transformation frequency is expressed as the number of foci per 10^{6} surviving cells. This calculation is based on the assumption that cell survival is not a function of the cell number/dish under these conditions.

Test chemicals 3-methylcholanthrene and anthracene were purchased from Sigma Chemical Company (St. Louis, Mo, USA). Stock solutions of chemicals were made in dimethylsulfoxide (DMSO; Merck) at concentrations of 1 mg/ml or 10 mg/ml.

RESULTS AND DISCUSSION

The carcinogen-metabolizing enzyme complex was present in the nude mouse fibroblast cultures at a specific activity higher than in BHK 21 (5) and also higher than in VERO cells tested for comparison (Table 1).

The treatment for three days with 3-methylcholanthrene after the virus infection caused a very remarkable, concentration dependent increase in viral transformation (Table 2). Under these conditions the chemical alone did not induce transformation. In contrast to 3-methylcholanthrene a non-carcinogenic compound, anthracene, had no effect on viral transformation. In other experiments a short pretreatment with 3-methylcholanthrene also caused a significant increase in the focus formation by the virus (results not shown).

TABLE 1

ARYL HYDROCARBON HYDROXYLASE ACTIVITY IN SECONDARY FIBROBLAST CULTURES FROM NMRI NU/NU (NUDE) MOUSE

Cell culture	AHH[a] pmol/mg protein/min
Nude mouse	4.20 ± 1.60
BHK 21	0.93 ± 0.23
VERO	2.45 ± 0.93

[a]Cell cultures in the logarithmic growth phase were analyzed. All values are averages ± SD from four determinations.

Target cell density has an effect on the transformation frequency. The decreased cell density caused by the toxic effect of 3-methylcholanthrene was not, however, sufficient to explain the observed increase in transformation. Experiments under way should exclude this possibility more definitely. It is more difficult to estimate the possible selection for transformation-sensitive cells by the chemical treatment. There are several ways by which the carcinogenic metabolites of 3-methylcholanthrene may affect the frequency

TABLE 2

EFFECT OF 3-METHYLCHOLANTHRENE AND ANTHRACENE ON SV_{40} INDUCED TRANSFORMATION OF NMRI NU/NU (NUDE) MOUSE FIBROBLASTS

Chemical (μM)	Survival fraction[a]	No. foci / No. dishes	Transformation frequency[b]	Enchancement ratio
3-Methyl-cholanthrene				
3.5	0.5	50/8	500	3.1
7.0	0.4	96/8	1200	7.5
Anthracene				
14.0	0.7	19/8	136	0.9
28.0	0.6	20/8	167	1.0
Control				
DMSO	1.0	32/8	160	1.0

[a]Proportion of surviving cells compared to the cells plated in growth medium (10^3 cell/dish).

[b]Transformation frequency was determined by plating 2.5×10^4 cells in 30 mm dishes and is expressed as the number of foci per 10^6 surviving cells.

of viral transformation. Both genetic and epigenetic mechanisms may be involved. The importance of the virus for the transformation was suggested by the finding that all cell lines tested from the combination treatment experiments were T antigen positive, *i.e.* they carried viral genetic material. The cell lines also grew in soft agar and produced fibrosarcomas when transplanted in nude mice as a demonstration of their tumorigenicity (results not shown).

Further experiments with a variety of precarcinogenic and carcinogenic chemicals as well as structurally related non-carcinogenic compounds are required for a more definitive assessment of the test system described in this report. These experiments are now in progress.

ACKNOWLEDGEMENTS

This work has been supported by the Finnish Foundation for Cancer Research. We would like to thank Mrs. Päivi Kivistö for her skillfull technical assistance.

REFERENCES

1. Laaksonen, M., Mäntyjärvi, R.A. and Hänninen, O. (1980) in: Spiegel, A. and Mühlbock, O. (Ed.), 7th ICLAS Symp., Utrecht 1979, Gustav Fischer Verlag, Stuttgart, New York, pp. 389-393.
2. Laaksonen, M., Mäntyjärvi, R. and Hänninen, O. (1982) in: Snyder, R., Parke, D.V., Kocsis, J.J., Jollow, D.J., Gibson, C.G. and Witmer, C.M. (Ed.), Biological Reactive Intermediates-II, Chemical Mechanism and Biological Effects, Plenum Press, New York, pp. 1307-1317.
3. Nebert, D.W. and Gelboin, H.V. (1968), J. Biol. Chem., 243, 6242.
4. Lowry, O.H., Rosebrough, A.L., Fare, A.D. and Randall, R.J. (1951), J. Biol. Chem., 193, 265.
5. Huberman, E., Yamasaki, H. and Sachs, L. (1976), Int. J. Cancer, 18, 76.

Cytochrome P-450, Biochemistry, Biophysics and Environmental Implications, E. Hietanen, M. Laitinen and O. Hänninen editors

ASPECTS ON THE DETERMINATION OF THE OXIDATION-REDUCTION POTENTIAL OF LACTOPEROXIDASE

PER-INGVAR OHLSSON* AND KARL GUSTAV PAUL
Department of Physiological Chemistry, University of Umeå, S-901 87 Umeå, Sweden

INTRODUCTION

To determine low oxidation-reduction potentials, like those of peroxidases, it is necessary to reduce to an absolute minimum the oxygen concentration in the solution and the oxygen adsorbed on the surfaces of the reaction vessel. Usually oxygen is replaced by the somewhat heavier argon, by passing or bubbling the gas over or through the reaction solution for a number of hours, often in combination with stirring and pressure reduction to reduce the time needed. These more or less standard procedures were not applicable in our attempts to determine the oxidation-reduction potential of lactoperoxidase (LP, E.C.1.11.1.7). The present article shows some of the problems involved, and presents a cell for reductive titrations using hydrogen dissolved in platinum black as reductant.

MATERIALS AND METHODS

Lactoperoxidase was prepared from cow's milk as previously described (1). The first fraction to emerge from a DEAE-Sephadex® A 50 column was used.

The redox cell used for the simultaneously spectrophotometric and potentiometric measurements under strictly anaerobic condition was a modified Dixon--cell (2). It was equipped with a bright platinum electrode and as reference a saturated calomel electrode. The platinum electrode, 5 x 10 mm, was fused to a 20 x 0.8 mm platinum wire, inserted in a gold-plated holder soldered to the potentiometer wire. The wire was cemented in the stopper with a two--component expanding cement (Casco, Stockholm). The platinum electrode was mechanically activated by "ironing" with the round end of a glass rod (3), soaked for 10 minutes in concentrated nitric acid followed by extensive washing with distilled water, and finally heated to glow in a reducing ethanol flame (4). The semimicro calomel electrode (Radiometer K4112, Copenhagen) was connected to the cuvette solution via a saturated potassium chloride-agar (4 % w/v, Noble, Difco) bridge (5) in a 150 mm long Tygon®-tubing, o.d. = 3.2 mm, i.d. = 1.6 mm. The tube with the bridge was drawn through a 3.1 mm hole drilled in the cuvette stopper. This reference system when calibrated against a hydrogen electrode (3) gave a potential of 249.9 mV in 100 mM sodium phosphate, pH 7.00, 25 ^{o}C.

Potentials were registered with a potentiometer (Radiometer PHM 4 or PHM 64, Copenhagen) and spectra with a Beckman ACTA® III spectrophotometer. This was calibrated against a holmiumoxide filter and a potassium dichromate solution in 0.005 M sulphuric acid for wavelength accuracy and absorbance linearity, respectively. The standard thermostatted cuvette holder was equipped with a magnetic stirring device allowing intermittent stirring with a magnetic bar, a "flea", in the cell. Commercially available argon, helium and hydrogen (AGA, Stockholm) with an oxygen content of 1 - 3 ppm was further purified by passage through an Oxy-Trap® (Alltech Associates, Ill.) column in series to give a concentration of O_2 < 0.1 ppm.

The gas was conducted to the cell through a softdrawn copper tube, o.d. = = 3.2 mm. The end of the tube was silver brazed to a coil of stainless steel, o.d. = 0.8 mm. The end of the coil was firmly pressed (2) into the inlet tube of the cuvette stopper. The gas was passed through or over the solution depending on the type of gas. The gas outlet from the cell ended 10 mm under water to give a slight over-pressure in the cell. Gas flow was kept constant at 0.1 - 0.3 ml/min.

For reduction sodium dithionite or hydrogen dissolved in platinum black was used. Platinization of activated bright platinum foil was carried out as described by Bates (3) but with a somewhat prolonged depositing time to get a thicker layer of platinum black. The platinum black electrode, 5 x 10 mm, was then saturated with hydrogen by electrolysis of 3 M sulphuric acid with the platinum black electrode as cathode (6).

Mediators with potentials suitable for this work were 9.10-anthraquinon-2-sulphonate (AOS), midpotential at pH 7.00 ($E_{m,7.00}$) = -225 mV at 25 ^{o}C (7, 8) and 2-methyl-3-hydroxy-1,4-naptoquinone(phthiocol) $E_{m,7.00}$= -168 mV at 25 ^{o}C as measured with the above equipment (Fig. 1, Table 1) in 20 mM sodium cacodylate buffer pH 7.00. Mediators and solutions were protected against light exposure.

Chemicals. Fresh sodium dithionite (Merck, Darmstadt) was divided under argon into 2 ml tubes, which were sealed with paraffin wax and stored dark and dry at -20 ^{o}C. AOS (BDH, England) was recrystallized from hot water and airdried at room temperature. Other chemicals were of analytical grade. The water was bidistilled in glass vessels. All experiments were made at 25 ^{o}C.

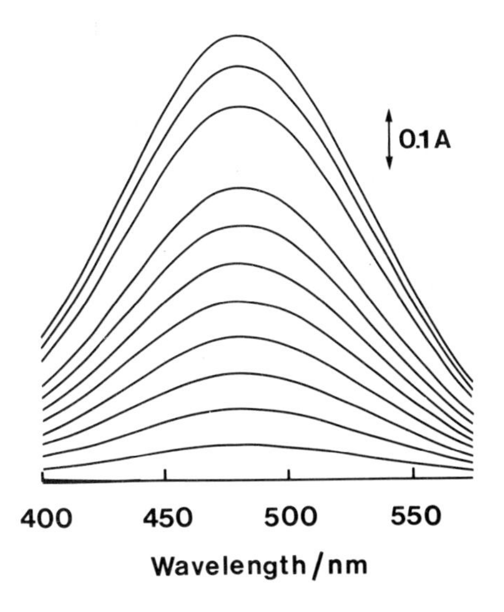

Fig. 1. Titration of 3.05×10^{-4} M phthiocol in 20 mM sodium cacodylate pH 7.0 with sodium dithionite. 25 °C.

TABLE 1

POTENTIOMETRIC TITRATION OF $3.05 \cdot 10^{-4}$ M PHTHIOCOL IN 20 mM SODIUM CACODYLATE pH 7.0 WITH SODIUM DITHIONITE. 25 °C. n = 2.

Curve no.	E_h mV	Percent reduced	$E_{m, 7.0}$ mV	
1	-	0	-	
2	-135.2	7.1	-168.3	
3	-147.1	16.1	-168.3	
4	-159.8	34.4	-168.1	
5	-164.4	42.9	-168.1	
6	-168.7	51.6	-167.9	
7	-172.8	59.8	-167.7	
8	-177.5	68.2	-167.7	
9	-183.0	76.5	-167.8	
10	-189.4	84.3	-167.9	
11	-199.9	92.41	-167.8	Mv -168.0 ± 0.2

RESULTS AND DISCUSSION

Deaeration of LP solution by bubbling Ar or N_2 caused precipitation, which increased at higher rates of gas flow. The rate of precipitation was followed in terms of absorbance changes at 500 nm. He, H_2, and noble gases with molecular weights above Ar gave no or little effect (Table 2).

Stirring of an LP solution, too, caused precipitation. It was irreversible and depended on the rotational speed of the magnetic bar and also on the material covering the bar (Fig. 2). Also here the precipitation was followed in terms of absorbance changes at 500 nm. Increasing the hydrophilicity by boiling the sodium glass covered bar in 3 M KOH for 15 minutes reduced the rate of precipitation to less than one tenth of that with the standard teflon--covered bar. Size and shape was the same for all the bars used.

TABLE 2

GAS BUBBLING EFFECTS ON LP MEASURED AS ΔA AT 500 nm. FLOW RATE 2 ML/MIN. LP 15.3 μM.

Gas	$\Delta A \times s \times 10^6$
H_2	-
He	-
N_2	2.0
Ar	3.8
Kr	0.9
Xe	-

At concentrations less than 4 mM Ca^{2+} did not affect the precipitation, but over 4 mM precipitation occurred without stirring (Fig. 3). A molecule, similar to LP, undergoes a calcium-dependent self-association giving mainly tetramers in 20 mM $CaCl_2$ (9). The precipitation of LP at Ca^{2+} over 4 mM is prevented by chelators like diethylene triamine pentaacetic acid but the precipitate does not redissolve.

An inert H_2 or He atmosphere and intermittent stirring by means of a bar in KOH-treated glass minimized the rate of precipitation. Under these conditions oxidation-reduction potentials could be determined. Optical determinations with phthiocol as indicator gave $E_{m,7.00}$ = -191 mV and the potentiometrically directly determined potential using phthiocol as mediator -188 mV. In both procedures hydrogen dissolved in platinum black functioned as reductant. Details will be published elsewhere.

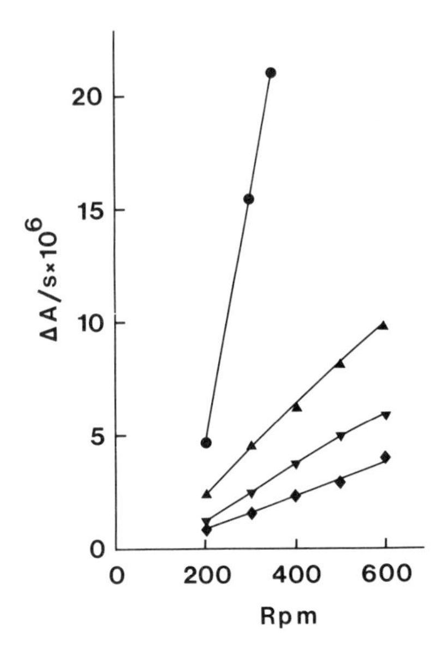

Fig. 2. Effects of magnetic stirring bar surface on the rate of precipitation measured as ΔA at 500 nm. LP 15.3 μM. —●— teflon, —▲— siliconized glass, —▼— sodium glass, and —◆— sodium glass boiled in 3 M KOH.

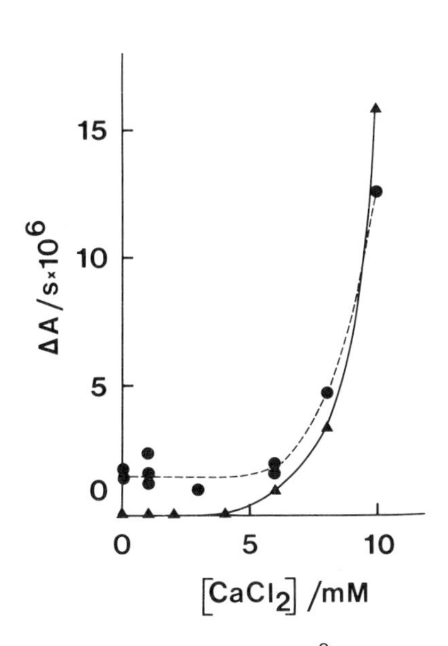

Fig. 3. Effects of Ca^{2+} on the rate of precipitation measured as ΔA at 500 nm. LP 15.3 μM. —▲— without stirring, —●— stirring with hydrophilic bar, 300 rpm.

ACKNOWLEDGEMENT

This study was supported by grant from Statens medicinska forskningsråd (13X-5423).

REFERENCES

1. Paul, K. G., Ohlsson, P.-I. and Henriksson, A. (1980) FEBS Letters, 110, 200-204.
2. Dixon, M. (1971) Biochim. Biophys. Acta, 226, 241-258.
3. Bates, R.S. (1973) Determination of pH. Theory and practice, Wiley-Interscience, New York.

4. Paul, K. G. and Ohlsson, P.-I. (1978) Acta Chem. Scand., B32, 395-404.
5. Clark, W.M. (1960) Oxidation-reduction potentials of organic systems, Williams and Wilkins, Baltimore.
6. Michaelis, L. and Rona, P. (1930) Praktikum der Physikalischen Chemie, Springer, Berlin.
7. Conant, J.B. and Fieser, L.F. (1922) J. Am. Chem. Soc., 44, 2480-2483.
8. Conant, J.B., Kahn, H.M., Fieser, L.F. and Kurtz, S.S. (1922) J. Am. Chem. Soc., 44, 1382-1389.
9. Bennet, R.M., Bagby, G.C. and Davis, J. (1981) Biochem. Biophys. Res. Comm., 101, 88-95.

Closing Remarks

© 1982 Elsevier Biomedical Press B.V.
Cytochrome P-450, Biochemistry, Biophysics
and Environmental Implications, E. Hietanen,
M. Laitinen and O. Hänninen editors

Klaus Ruckpaul
Department of Biocatalysis
Central Institute of Molecular Biology
Acad. Sci. of DDR
1115 BERLIN-BUCH
DDR

We have come to the end of 3 stimulating and awarding days in this beautiful landscape with lots of clean lakes and wonderful forests. Despite lack of time due to an exhausting sequence of lectures, posters and discussions we got a feeling of undisturbed nature surrounding our meeting.

The aims of this European series of P-450 meetings were (1) to enable people from different countries in the east and in the west working in the same field to exchange ideas, to discuss results in order to stimulate further scientific development. After or almost after the 4th conference we have to ask, if we have been successful in this point. Increasing numbers of participants starting with about 40 people in Promosten (Yugoslavia, 1975) we have now reached a number of more than 200 with about the same number of contributions. And what may be more important with a lot of new and especially young colleagues. In this respect the decision has proved right. Let us not be discouraged that not all countries are represented in a way that reflects their scientific activities. Just by this fact our responsibility should be challenged to work for improvements in forthcoming meetings.

The second aim of the first conference was to focus on biophysical and biochemical problems. In the meantime, however, we felt that we are all obliged to keep our professional vision wideangled for effective usefulness and therefore important topics as molecular biology and environmental implications were included. Also in this respect we have developed.

If I come to summarize briefly the most important results of the meeting I can do it only from a very personal view and try to select some data having in mind where came we from, where are we now and where shall we go.

It seems to me that all results presented in the last three days can be regarded as an attempt to better understand sturcture/function relationships at different levels of integration: at the molecular, the subcellular and cellular and at the level of the whole organism. Finally with the aim to get insight which way the system is regulated and controlled to draw conclusions via which mechanisms it may be influenced. Let me start with the structure at the molecular level. With the detection of the primary structure of P-450 CAM and the coding sequence of P-450 b from rats remarkable breakthrough has been achieved which together with the concomitant partially protein sequencing of rat and rabbit isozymes provides a basis for developmental analysis and experimental evidence for genetically determined isozymes. In this connection let me remind you the elegant N-terminal analysis of the LM4 fraction evidencing the polypeptide chain as homogeneous gene product. The work which has been done is enormous. The question how many isozymes do exist we put in Stockholm has not been answered at this meeting and I will not try to extrapolate any number. On the other hand we have learnt that not only inducer dependent fractions up to about 13 in the rat have to be considered but moreover sex dependent fractions too. Despite the increasing work to be done with each new fraction we must be aware that just the multiplicity of the molecular structure is of principle importance to be adapted to the function of the enzyme system. The sequence of the polypeptide chain, however, is not the only structural parameter of functional importance. The attack of the active center proceeds. Experimental evidence by different approaches for the existence of an oxygen as 6th axial heme iron ligand has been provided and this oxygen has been attributed to the phenolid hydroxyl group of a tyrosine. Moreover first detailed analysis of the thiol group has been evidenced from the comparison of the structure from different P-450's. That is indeed the first time that numerous indirect results indicating thiol as 5th ligand have been ascertained by the sequencing analysis and further supported by Raman Resonance Detection of a Fe-S bond in P-450 CAM. Selective chemical modification has been used to continue the analysis of the heme crevice in P-450 CAM and to identify such residues which are functionally involved in protein/protein interactions in the liver microsomal system. What functional implications can be derived from these molecular structural results. As it has already been shown yesterday by comparing the amino acid residues near the thiol group the structural basis of species dependent functional differences and those between various isozymes has become available and I would expect in the near future remarkable insight in the molecular basis of substrate and species specificities. I would also mention at this state of consideration the elegant work to clarify the way of oxygen activation by proposing the homolytic way of cleavage as most suitable. The experiments on the mechanism of action of the LM3 a fraction contribute to understand the multiplicity of action. One of the most remarkable progress has occurred in the field of molecular biology and genetics of P-450. P-450 is going to become one of the most intense studied models in molecular genetics. I agree with Dr. Gustafsson's estimation of this session that the results presented there have opened a new era in P-450 research. At the cellular level of integration the results may not only help to answer questions of the primary structure, the insertion mechanism and possible processing involved therein but also to understand the mechanism of induction and this way we may get access to the genetically determined inducibility. Thus the molecular genetics have relevant implications to basic research and to more applied questions as well. Reliable access to quantitate the individual inducibility and studies on endogenous substrates (steroids, prostaglandins) which are of essential biological importance, may be a future goal. Half times between 15 and 30 hours of P-450 isozymes may induce new considerations with respect to drug administration and together with the inducibility may provide an exact basis for a rationally based therapy.

The third level of integration I have to discuss with you was introduced by the organizers with environmental implications. We have learnt interesting data about P-450 of different fishes sensitively monitoring environmental pollutants and the hormonal control of steroid metabolism by P-450 in the whole organism. I would also stress on the results obtained from P-450 of higher plants and different yeast strains evidencing P-450 as a ubiquitous system in all living beings with highly relevant possibilities for application. These results together with the numerous contributions devoted to analyse the involvement of P-450 in the metabolism of different drugs and toxic compounds presented in the to day sessions clearly indicate that the basic research done so far and further to do in the P-450 field becomes more and more relevant for application to human welfare. Just this characterizes a scientific field which is dynamically developing and should encourage the efforts of all of us working therein.

Dear colleagues, let me come to the end. The scientific success of each meeting - and the 4th International Conference on Cytochrome P-450 was successful - results from an efficient interaction between hosts and guests. Let me say that all of you during the lectures, the poster presentations and in many discussion contributed to make this conference awarding and stimulating. But let me especially appreciate the efforts of our Finish friends to surround us with warm hospitality, taking the organizational burden to make us enjoy these three days here in Kuopio. Congratulations and many thanks to Dr. O. Hänninen, Dr. E. Hietanen and to Dr. M. Laitinen and the whole staff before and behind the stage.

Author Index

Abe, T. 649
Ade, P. 387
Adesnik, M. 143
Agosin, M. 661,797
Ahotupa, M. 157,161,705,
749,773
Akagawa, S. 193
Akhrem, A.A. 87,413
Altieri, M. 119,181
Amelizad, Z. 63
Andersson, B. 637
Andersson, L.A. 523,589
Andersson, T. 717
Andres, J. 441
Anzenbacher, P. 611
Archakov, A.I. 487,551,615
Aström, A. 395
Atanasov, B. 607
Atchison, M. 143
Autrup, H. 643
Azari, M.R. 369

Bachmanova, G.I. 551
Balny, C. 463
Banchelli Soldaini, M.G. 387
Bandiera, S. 441
Bässmann, H. 329
Bayney, R.M. 165,169
Bellomo, G. 769
Belvedere, G. 679,689
Bend, J.R. 233
Bengtsson, M. 653
Benveniste, I. 201
Berner, H. 555
Bernhardt, R. 513,581
Bespalov, I.A. 413
Betner, I. 701
Birgersson, C. 459
Black, S.D. 277
Blanck, J. 585
Boggaram, V. 631
Bonati, M. 689
Bonfils, C. 463,751
Bornheim, L.M. 307
Bösterling, B. 497,619
Boström, H. 333,429
Böttcher, J. 329
Bresnick, E. 127
Burka, L.T. 27

Campbell, M.A. 341,441
Castelli, M.G. 387
Champion, P.M. 547
Chashchin, V.L. 413
Chen, Y.-T. 119
Cheng, B. 695
Chiesara, E. 387
Chou, M.W. 39,71
Clementi, F. 387
Cohen, G.M. 725
Conner, J.W. 245
Coon, M.J. 277,481
Copp, L. 341

D'Arcy Doherty, M. 725
Dalet, C. 751
Dalet-Beluche, I. 751
Danielsson, H. 373
Dansette, P.M. 95
Das, M. 671
Dawson, J.H. 523,589
Depierre, J.W. 395
Dieter, H.H. 283,337
Dock, L. 67
Donohue, A.M. 149
Drangova, R. 47
Dubé, A.W. 173
Durst, F. 201
Dus, K.M. 377

Egger, H. 555
Eichinger, V. 597
Eisen, H.J. 91,567
Eklöw, L. 793
Ekström, G. 19
Elenkova, E. 607
Elovaara, E. 765
Embree, C.J. 99
Engström, K. 765
Etzold, G. 581

Fanelli, R. 387
Fasco, M.J. 291
Feuer, G. 47,471
Fiechter, A. 453
Fonne, R. 201
Förlin, L. 217,717
Forstmeyer, D. 95
Foureman, G.L. 233

Frahseck, J. 361
Franklin, M.R. 307
Frederick, C.B. 27
Friedrich, J. 513
Fu, P.P. 39,71
Fucci, L. 531
Fujii-Kuriyama, Y. 135
Funari, E. 387
Funkenstein, B. 631

Gabriac, B. 201
Gallagher, E. 153
Galletti, F. 689
Gelb, M.H. 573
Gibson, G.G. 437
Glaumann, H. 789
Göransson, M. 459
Goshal, A. 471
Govindwar, S. 675
Graf, H. 103
Grafstrom, R.C. 643
Gromova, O.A. 517
Guengerich, F.P. 27,291
Gunsalus, I.C. 547
Gustafsson, J.-A. 217,353, 789
Guzelian, P. 153

Haaparanta, T. 789
Hagbjörk, A.L. 19,75
Hahn, J.E. 589
Hajek, K.K. 713
Hallberg, E. 653,785
Halpert, J. 353,701,789
Hankinson, O. 567
Hannah, R.R. 567
Hänninen, O. 209,251,255,259, 263,801
Hansson, R. 429
Hansson, T. 217
Harada, N. 405
Hartzell, P. 729
Heikelä, A. 705
Heinonen, T. 733
Hietanen, E. 161,263,705,773 777,781
Hietaniemi, L. 477
Hildebrandt, A.G. 743
Hlavica, P. 563
Hodgson, K.O. 589
Honeck, H. 445
Hook, J.B. 721
Hynninen, P.H. 345

Ignesti, G. 387
Ikeda, T. 119,181
Ingelman-Sundberg, M. 19,75
Ioannides, C. 437
Ishimura, Y. 627
Izumi, S. 627

Jaccarini, A. 53
Jahn, F. 189,467
Jänig, G.-R. 513,581
Järvisalo, J. 733
Jeffery, E. 59
Jernström, B. 67
Jewell, S.A. 769
Johansson, I. 19
Johnson, E.F. 283,337
Jritano, L. 689
Jung, C. 607
Juvonen, R. 263

Kachole, M. 675
Kadlubar, F.F. 27
Kalles, I. 373
Kamataki, T. 421
Kaminsky, L.S. 291
Kanaeva, I.P. 551
Kappeli, O. 453
Kärenlampi, S.O. 91
Karge, E. 467
Kärki, N.T. 759
Karppinen, A. 759
Kato, R. 421
Kawajiri, F. 135
Kawano, S. 509
Kharasch, E.D. 623
Kimura, T. 695
King, D.J. 369
Kisel,M.A. 87
Kiselev, P.A. 87
Kling, L. 365
Klinger, W. 189,467
Ko, A. 381
Koivusaari, U. 209,251,259
Komatsu, N. 627
Koop, D.R. 269
Kramer, R.E. 631
Krueger, R.J. 559
Kumar, A. 143
Kurchenko, V.P. 79
Kuthan, H. 509

Laaksonen, M. 801
Laitinen, M. 263,705,781
Lambert, I. 341,441
Lang, M.A. 391
Larsson, R. 637
Lau, P.P. 321
Lech, J.J. 225
Leclaire, J. 95
Lee, R.F. 245

Legraverend, C. 567
Legrum, W. 361,365
Levin, W. 127
Levine, R. 531
Lidström, B. 373
Liebler, D.C. 27
Liimatainen, A. 255
Lindeke, B. 637
Lindström-Seppä, P. 209,251
Lippman, A. 143
Lock, E.A. 721
Lorr, N.A. 35
Lu, A.Y.H. 39,149,357,405
Lundell, K. 333,429
Luthman, H. 75
Lyakhovich, V.V. 501,517

Macdonald, T.L. 27
MacGeoch, C. 353
Mälkönen, P. 477
Mankowitz, L. 653,755
Mannering, G.J. 59
Mansuy, D. 95
Mäntyjärvi, R. 801
Mäntylä, E. 157,161,773
Marabini, A. 387
Martsev, S.P. 413
Masson, H. 437
Maurel, P. 463,751
McCarthy, J.L. 631
Meijer, J. 395
Mellström, B. 459
Metelitza, D.I. 3,79
Miller, R.E. 27
Mishin, V.M. 517
Mitani, F. 627
Miwa, G.T. 357,405
Mizukami, Y. 135
Moldéus, P. 67,637,793
Momoi, H. 193
Montelius, J. 349,653
Morgan, E.T. 269
Mukhtar, H. 671
Müller, H.-G. 445
Muller-Eberhard, U. 283,337
Muramatsu, M. 135
Murray, R.I. 377

Nakamura, M. 119,181
Narbonne, J.F. 63
Näslund, B. 701
Nebert, D.W. 91,109,119,177,
181,391,567
Negishi, M. 119,177,181,391
Netter, K.J. 365
Newman, S. 153
Nieminen, M. 263
Nikkilä, H. 345
Novak, R.F. 623,713

Ohlsson, P.-I. 601,805
Ohmachi, T. 649
Ohyama, T. 119,177
Okey, A.B. 173
Oliver, C. 531
Omiecinski, C.J. 127
Omura, T. 299
Orme-Johnson, N.R. 559
Oronesu, M. 387
Orrenius, S. 729,769,793
Owens, I.S. 567

Palmero, S. 387
Parkinson, A. 441
Parkki, M.G. 749
Pasanen, M. 317,449
Paul, K.G. 601,805
Pawar, S. 675
Pelkonen, O. 317,449
Peng, R. 35
Pershing, L.K. 307
Pesonen, M. 259
Phillips, I.R. 165,169
Pickett, C.B. 149
Pike, S.F. 165,169
Pirisino, R. 387
Popova, V.I. 517
Potter, D.W. 43
Prough, R.A. 27
Pyykkö, K. 709
Rabin, B.R. 165,169
Rafter, J. 217
Rahal, S. 99
Ramundo Orlando, A. 387
Rane, A. 785
Raphael, C. 143
Rauramaa, R. 781
Reed, D.J. 43
Reichhart, D. 201
Rein, H. 585,607
Reitze, H. 365
Reubi, I. 283,337
Riege, P. 445
Ristau, O. 513,585,607
Rivkin, E. 143
Robertson, L. 441
Roots, I. 743
Ruckpaul, K. 87,513,581,585
607,813
Ruf, H.H. 597
Ryan, D.E. 127
Rydström, J. 349,653,755,785

Saarni, H. 739

Sadano, H. 299
Safe, L. 441
Safe, S. 341,441
Sagami, I. 649
Salaün, J.-P. 201
Saldana, J.L. 463
Sauer, M. 453
Sawyer, T. 441
Scheler, W. 585
Schiller, F. 189
Schunck, W.-H. 445
Schüppel, R. 329
Schuster, I. 555
Schwab, G.E. 283,337
Schwarze, W. 581
Senciall, I.R. 99
Seth, P.K. 671
Sheets, J.J. 539
Shephard, E.A. 165,169
Shively, J.E. 127
Silano, V. 387
Simon, A. 201
Simpson, E.R. 631
Sípal, Z. 611
Skotzelyas, E.D. 615
Sligar, S.G. 573
Smettan, G. 87,513
Smith, M.L. 601
Smith, M.T. 725,729
Sogawa, K. 135
Sommer, M. 585
Soni, M. 675
Sono, M. 523
Soodaeva, S.K. 615
Stadtman, E.R. 531
Stallard, B.R. 547
Stiel, H. 581
Strauch, B. 611
Strobel, H.W. 321
Stupans, I. 391
Suwa, Y. 135

Tamburini, P.P. 437
Tarr, G.E. 277
Telakowski-Hopkins, C.A. 149
Terelius, Y. 19
Thomas, P.E. 127
Thor, H. 729,769
Timbrell, J.A. 725
Toikkanen, S. 749
Toscano Jr., W.A. 573
Trevor, A.J. 619
Trudell, J.R. 497,619
Tu, Y.Y. 35
Tukey, R.H. 119,177
Tursi, F. 679,689
Twardowski, J. 611

Ullrich, V. 103
Usanov, S.A. 79
Uvarov, V. Yu. 551

Vainio, H. 679,733,765,773,777
Vapaatalo, H. 709
Viarengo, A. 387
Vickery, L.E. 539
Vittozzi, L. 387
Vlasuk, G.P. 127
Vodicnik, M.J. 225
von Bahr, C. 459

Wagner, G.C. 547
Walsh, C. 311,381
Walsh, J.S.A. 405
Walz Jr., F.G. 127
Watanabe, K. 627,649
Watanabe, M. 649
Waterman, M.R. 631
Waxman, D.J. 311,381
Weiner, L.M. 501,517
Werringloer, J. 509
White Jr., R.A. 797
White, R.E. 18
Wikvall, K. 333,373,429
Wiseman, A. 369
Wislocki, P.G. 39
Wittenberger, M. 531
Woldman, Ya.Yu. 501
Wolf, C.R. 721

Yagi, M. 193
Yamakura, F. 695
Yang, C.S. 35
Yang, S.K. 39
Yokota, H. 185
Yuan, P.-M. 127
Yuasa, A. 185

Zacharias, P.S. 47,471
Zhukov, A.A. 615
Zirvi, K. 27
Zitting, A. 657
Zubairy, Y. 675